6565
2

THE PICTURE OF THE TAOIST GENII PRINTED ON THE COVER
of this book is part of a painted temple scroll, recent but traditional, given to
Mr Brian Harland in Szechuan province (1946). Concerning these four divinities,
of respectable rank in the Taoist bureaucracy, the following particulars have been
handed down. The title of the first of the four signifies 'Heavenly Prince', that
of the other three 'Mysterious Commander'.

At the top, on the left, is Liu *Thien Chün*, Comptroller-General of Crops and
Weather. Before his deification (so it was said) he was a rain-making magician
and weather forecaster named Liu Chün, born in the Chin dynasty about +340.
Among his attributes may be seen the sun and moon, and a measuring-rod or
carpenter's square. The two great luminaries imply the making of the calendar, so
important for a primarily agricultural society, the efforts, ever renewed, to reconcile
celestial periodicities. The carpenter's square is no ordinary tool, but the gnomon
for measuring the lengths of the sun's solstitial shadows. The Comptroller-General
also carries a bell because in ancient and medieval times there was thought to be
a close connection between calendrical calculations and the arithmetical acoustics
of bells and pitch-pipes.

At the top, on the right, is Wên *Yuan Shuai*, Intendant of the Spiritual Officials
of the Sacred Mountain, Thai Shan. He was taken to be an incarnation of one of
the Hour-Presidents (*Chia Shen*), i.e. tutelary deities of the twelve cyclical characters
(see p. 297). During his earthly pilgrimage his name was Huan Tzu-Yü and he was
a scholar and astronomer in the Later Han (b. +142). He is seen holding an
armillary ring.

Below, on the left, is Kou *Yuan Shuai*, Assistant Secretary of State in the Ministry
of Thunder. He is therefore a late emanation of a very ancient god, Lei Kung.
Before he became deified he was Hsin Hsing, a poor woodcutter, but no doubt an
incarnation of the spirit of the constellation Kou-Chhen (the Angular Arranger),
part of the group of stars which we know as Ursa Minor. He is equipped with
hammer and chisel.

Below, on the right, is Pi *Yuan Shuai*, Commander of the Lightning, with his
flashing sword, a deity with distinct alchemical and cosmological interests. According
to tradition, in his early life he was a countryman whose name was Thien Hua.
Together with the colleague on his right, he controlled the Spirits of the Five
Directions.

Such is the legendary folklore of common men canonised by popular acclamation.
An interesting scroll, of no great artistic merit, destined to decorate a temple wall,
to be looked upon by humble people, it symbolises something which this book has
to say. Chinese art and literature have been so profuse, Chinese mythological
imagery so fertile, that the West has often missed other aspects, perhaps more
important, of Chinese civilisation. Here the graduated scale of Liu Chün, at first
sight unexpected in this setting, reminds us of the ever-present theme of quantitative measurement in Chinese culture; there were rain-gauges already in the Sung
(+12th century) and sliding calipers in the Han (+1st). The armillary ring of
Huan Tzu-Yü bears witness that Naburiannu and Hipparchus, al-Naqqāsh and
Tycho, had worthy counterparts in China. The tools of Hsin Hsing symbolise that
great empirical tradition which informed the work of Chinese artisans and technicians
all through the ages.

SCIENCE AND CIVILISATION IN CHINA

The Chymists are a strange Class of Mortals, impelled by an incomprehensible Impulse to take their Pleasure amid Smoke and Vapour, Fume and Flame, Poisons and Poverty—yet among all these Evils, I seem to live so sweetly that may I die if I would change places with the Persian King!

Johann Beccher
Physica Subterranea, 1703

Quasi nimirum in Fatis esset, Sal hoc admirabile non minus in philosophia quam bello strepitus aderet, omniaque sonitu suo implere. (As if ordained by Fate, Nitre, that admirable salt, hath made as much noise in Philosophy as in War, all the world being filled with its thunder).

John Mayow
Tractatus Quinque Medico-Physici, 1674

For it is now certainly known that the great Kings of the uttermost East, have had the use of the canon many hundreds of years since, and even since their first civilitie and greatnesse, which was long before Alexander's time. But Alexander pierc'd not so far into the East.

Sir Walter Raleigh
History of the World, 1614

Dr John Bell of Antermony asked the Khang-Hsi Emperor's Tartar General of Artillery: 'How long the Chinese had known the use of gunpowder? He replied, above 2000 years, in fire-works, according to their records; but that it's application to the purposes of war, was only a late introduction. As the veracity and candour of this gentleman were well known, there was no room to question the truth of what he advanced on the subject.'

John Bell's   1 Jan. 1721
Travels from St. Petersburg in Russia to Diverse Parts of Asia, 1763

And though it be very true that man is but the Minister of Nature, and can but duely apply Agents to Patients (the rest of the Work being done by the applyed Bodies themselves), yet by his skill in making these Applications, he is able to perform such things as do not only give him a Power to master Creatures otherwise much stronger than himselfe; but may enable one man to do such wonders, as another man shall think he cannot sufficiently admire. As the poor Indians lookt upon the Spaniards as more than Men, because the Knowledg they had of the Properties of Nitre, Suplphur and Charcoal duely mixt, enabled them to Thunder and Lighten so fatally, when they pleas'd.

Robert Boyle
Some Considerations touching the Usefulnesse of Experimental Philosophy, propos'd in a Familiar Discourse to a Friend, by way of Invitation to the Study of it, 1663

中國科學技術史

李約瑟 著

冀朝鼎

SCIENCE AND CIVILISATION IN CHINA

BY

JOSEPH NEEDHAM, F.R.S., F.B.A.

SOMETIME MASTER OF GONVILLE AND CAIUS COLLEGE, CAMBRIDGE, DIRECTOR OF THE EAST ASIAN
HISTORY OF SCIENCE LIBRARY, CAMBRIDGE, HONORARY PROFESSOR OF ACADEMIA SINICA

With the collaboration of

HO PING-YÜ (HO PENG YOKE), Ph.D.

PROFESSOR OF CHINESE IN THE UNIVERSITY OF HONGKONG

LU GWEI-DJEN, Ph.D.

FELLOW OF ROBINSON COLLEGE, CAMBRIDGE,
ASSOCIATE DIRECTOR OF THE EAST ASIAN HISTORY OF SCIENCE LIBRARY

and

WANG LING, Ph.D.

EMERITUS PROFESSORIAL FELLOW, DEPARTMENT OF FAR EASTERN HISTORY, INSTITUTE OF
ADVANCED STUDIES, AUSTRALIAN NATIONAL UNIVERSITY, CANBERRA

VOLUME 5

CHEMISTRY AND CHEMICAL TECHNOLOGY

Part 7: MILITARY TECHNOLOGY;
THE GUNPOWDER EPIC

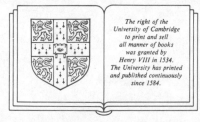

The right of the
University of Cambridge
to print and sell
all manner of books
was granted by
Henry VIII in 1534.
The University has printed
and published continuously
since 1584.

CAMBRIDGE UNIVERSITY PRESS

CAMBRIDGE

LONDON   NEW YORK   NEW ROCHELLE
MELBOURNE   SYDNEY

Published by the Press Syndicate of the University of Cambridge
The Pitt Building, Trumpington Street, Cambridge CB2 1RP
32 East 57th Street, New York, NY 10022, USA
10 Stamford Road, Oakleigh, Melbourne 3166, Australia

© Cambridge University Press 1986

First published 1986

Printed in Great Britain at
the University Press, Cambridge

British Library cataloguing in publication data
Needham, Joseph
Science and civilisation in China.
Vol. 5: Chemistry and chemical technology
Pt. 7: Military technology: the gunpowder epic.
1. Science—China—History   2. Technology—
China—History
I. Title   II. Ho, Ping-Yü   III. Lu, Gwei-Djen
IV. Wang, Ling
509'.51   Q127.C5

Library of Congress cataloguing in publication data
(Revised for volume 5, part 7)
Needham, Joseph, 1900–
Science and civilization in China.
Includes bibliographies.
Contents: v. 1. Introductory orientations. – v. 2.
History of scientific thought. – [etc.]'– v. 5. Chemistry
and chemical technology. – [etc.] – pt. 7. Military
technology, with the collaboration of Ho Ping-Yü,
Lu Gwei-Djen, and Wang Ling. – [etc.]
1. China – Civilization – Collected works.
2. Science – China – History – Collected works.
3. Technology – China – History – Collected works.
4. Science and civilization – Collected works.
I. Wang, Ling.   II. Title.
DS721.N39   509.51   54-4723

ISBN 0 521 30358 3

To
the memory of
FU SSU-NIEN

eminent scholar of history and philology,
then at Lichuang in Szechuan, and most friendly
welcomer to war-time China,
who led a discussion one evening while we were
there on the history of gunpowder in China;

and to
YÜ TA-WEI

physicist, then
Ping-kung-shu Shu-chang (Intendant-General of Arsenals)
1942–1946
whose 'field coffee' I used to drink with him in his
office, and with whom we had a happy reunion in 1984,

this volume is dedicated.

CONTENTS

LIST OF ILLUSTRATIONS

LIST OF TABLES

LIST OF ABBREVIATIONS

The following abbreviations are used in the text and footnotes. For abbreviations used for journals and similar publications in the bibliographies, see pp. 584 ff.

BN — Bibliothèque Nationale, Paris.

CC — Chia Tsu-Chang & Chia Tsu-Shan (*1*), *Chung-Kuo Chih Wu Thu Chien* (Illustrated Dictionary of Chinese Flora), 1958.

CCL — *Chê Chiang Lu* (Biographies of [Chinese] Engineers, Architects, Technologists and Master-Craftsmen).
1 to 6 See Chu Chhi-Chhien & Liang Chhi-Hsuing (*1* to *6*);
7 See Chu Chhi-Chhien, Liang Chhi-Hsuing & Liu Ju-Lin (*1*);
8, 9 See Chu Chhi-Chhien & Liu Tun-Chên (*1, 2*).

CCT — Chao Shih-Chên, *Chhê Chhung Tau* (Illustrated Account of Muskets, Field Artillery and Mobile Shields, etc.), Ming, *c.* +1585.

CHHS — Chhi Chi-Kuang, *Chi Hsiao Hsin Shu* (New Treatise on Military and Naval Efficiency), Ming, +1560, pr. +1562, often repr.

CHS — Pan Ku, (and Pan Chao) *Chhien Han Shu* (History of the Former Han Dynasty), H/Han, *c.* +100.

CHSK — Ting Fu-Pao (ed.), *Chhüan Han San-Kuo Chin Nan-Pei-Chhao Shih* (Complete Collection of Poetry from the Han, Three Kingdom, Chin and Northern and Southern Kingdoms), Peking, *c.* 1935. Index by Tshai Chin-Chung, Harvard–Yenching Institute, Paping, 1941 repr. Taipei, 1966.

CHTP — Chêng Jo-Tsêng, *Chhou Hai Thu Pien* (Illustrated Seaboard Strategy and Tactics), Ming, +1562, repr. +1572, +1594, +1624, etc.

CKKCSL — *Chung-Kuo Kho Chi Shih Liao* (Materials on the History of Science and Technology in China), a journal.

CSHK — Yen Kho-Chün (ed.), *Chhüan Shang-Ku San-Tai Chhin Han San-Kuo Liu Chhao Wên* (Complete Collection of prose literature (including fragments) from remote antiquity through the Chhin and Han Dynasties, the Three Kingdoms, and the Six Dynasties), 1836.

CLPT — Thang Shen-Wei *et al.* (ed.), *Chêng Lei Pên Jahao* (Reorganised Pharmacopoeia), Sung, ed. of +1249.

CTS   Liu Hsü, *Chiu Thang Shu* (Old History of the Than Dynasty), Wu Tai, +945.

CYMTYL   Attrib. Chêng Ssu-Yuan, *Chen Yuan Miao Tao Yao Lüeh* (Classified Essentials of the Mysterious Tao of the True Origin of Things), Ascr. Chin (+3rd) but probably mostly Thang (+8th and +9th).

DSB   *Dictionary of Scientific Biography* (16 vols.), ed. C.G. Gillespie et al., (Scribner, New York, 1970).

HCC   Hsü Tung, *Hu Chhien Ching* (Tiger Seal Manual, a Military Encyclopaedia), Sung, begun +962, finished +1004.

HCT   *Huo Chhi Thu* (Illustrated Account of Gunpowder Weapons and Firearms), running-head title of the Hsiang-yeng edition of the *Huo Lung Ching*, q.v.

HHPT   Su Ching *et al.* (ed.), *Hsin Hsiu Pên Tshao* (Newly Improved Pharmacopoeia), Thang, +659.

HHS   Fan Yeh & Ssuma Piao, *Hou Han Shu* (History of the Later Han Dynasty), +450.

HKPY   *Huo Kung Pei Yao* (Essential Knowledge for the Making of Gunpowder Weapons), alternative title of Pt. 1 of the *Huo Lung Ching*, q.v.

HKTC   Wei Yuen & Lai Tsê-Hsü, *Hai Kuo Thu Chih* (Illustrated Record of the Maritime [Occidental] Nations), Chhing, 1844, enlarged 1847, further enlarged, 1852, abridged edition, 1855.

HLC   Chiao Yü, *Huo Lung Ching* (The Fire-Drake (Artillery) Manual). Ming, +1412, but probably continuing information dating from the previous half-century, perhaps back to +1300. In three parts. The first part of the back is fancifully attributed to Chuko Wu-Hou (Chuko Liang, +3d cent.) & Liu Chi (+1311 to +1375). The latter appears as editor but was perhaps co-author. The second part of the book is attributed to Liu Chi, but Mao Hsi-Ping (+1632) was probably the writer. The third part is by Mao Yuan-I (fl. +1628) and has a preface by Chuko Kuang-Lung dated +1644.

HSCH/TCTC   Liu Shih-Chi: *Hsü Sung Chung-Hsing Pien Nien Tzu Chih Thung Chien* (Continuation of the 'Mirror of History for Aid in Government' for the Sung dynasty from its Restoration Onwards), i.e. The Southern Sung from +1126. Sing, *c.* +1250.

HTCTC/CP   Li Tao *Hsü Tzu Chih Thung Chien Chhang Phien* (Continuation of the 'Comprehensive Mirror (of History) for Aid in Government), dealing with events from +960 to +1126, i.e. the Northern Sung dynasty, Sung, +1183.

HTS Ouyang Hsiu & Sung Chhi, *Hsin Thang Shu* (New History of the Thang Dynasty), Sung, +1061.

HWHTK Wang Chhi (ed.) *Hsü Wên Hsien Thung Khao* (Continuation of the 'Comprehensive Study of (the History of) Civilisation'), Ming, +1586, pr. +1603.

LPSC (TC) Chhi Chi-Kuang, *Lien Ping Shih Chi Tsa Chi* (Miscellaneous Records concerning Military Training and Equipment), an appendix to *Lien Ping Shih Chi*, Ming, +1568, pr. +1571.

MCPT Shen Kua, *Mêng Chhi Pi Than* (Dream Pool Essays), Sung, +1089.

NKKZ *Nihon Kagaku Koten Zenshi* (Collection of Works concerning the History of Science and Technology in Japan), 12 vols., 1944; 10 vols. 1978.

PL Ho Ju-Piu, *Ping Lu* (Records of Military Art). Ming, +1606, pr. +1628, later eds. +1630, +1632.

PPT/NP Ko Hung, *Pao Phu Tzu (Nei Phien)*, (Book of the Preservation-of-Solidarity Master; Inner Chapters), Chin, c. +320.

PTKM Li Shih-Chen. *Pên Tshao Kang Mu* (The Great Pharmaco-poeia), Ming, +1569.

PTKMSI Chao Hsüeh-Min, *Pên Tshao Kang Mu Shih I* (Supplementary Amplifications for the 'Great Pharmacopoeia' of Li Shih-Chen), Chhing, begun c. +1760, first prepared +1765, prolegomena added +1780, last date in text 1803. First pr. 1871.

R Read, Bernard E. *et al.*, Indexes, translations and précis of certain chapters of the *Pên Tshao Kang Mu* of Li Shih-Chen. If the reference is to a plant see Read (1); if to a mammal, see Read (2); if to a bird see Read (3); if to a reptile see Read (4 or 5); if to a mollusc see Read (5); if to a fish see Read (6); if to an insect see Read (7).

RARDE Royal Armament Research and Development Establishment, Fort Halstead, Kent.

SCP Chao Shih-Chên, *Shen Chhi Phu* (Treatise on Extraordinary (lit. Magical) Weapons, i.e. Muskets), Ming, +1598.

SF Thao Tsung-I (ed.), *Shuo Fu* (Florilegium of (Unofficial) Literature), Yuan, c. +1368.

SKCS *Ssu Khu Chhüan Shu* (Complete Library of the Four Categories), Chhing, +1782; here the reference is to the *tshung-shu* collection printed as a selection from one of the seven imperially commissioned MSS.

SKCS/TMTY Chi Yün (ed.), *Ssu Khu Chhüan Shu Tsung Mu Thi Yao* (Analytical Catalogue of the *Complete Library of the Four Categories*), +1782; the great bibliography of the imperial MS collection ordered by the Chhien-Lung emperor of Chhing in +1772.

STTH   Wang Chhi, *San Tshai Thu Hui* (Universal Encyclopaedia), Ming, +1609.

TCKM   Chu Hsi *et al.* (ed.), *Thung Chien Kang Mu* ((Short View of the) *Comprehensive Mirror (of History), for Aid in Government*), classified into Headings and Subheadings); the *Tzu Chih Thung Chien* condensed, a general history of China, Sung, +1189; with later continuations.

TKKW   Sung Ying-Hsing, *Thien Kung Khai Wu* (The Exploitation of the Works of Nature), Ming, +1637.

TPKC   Li Fang (ed.) *Thai-Phing Kuang Chi* (Copious Records collected in the Thai-Phing reign-period), Sung, +978.

TPYC   Li Chhüan, *Thai Pai Yin Ching* (Manual of the White and Gloomy Planet (of War, Venus)), treatise on military and naval affairs, Thang, +759.

TPYL   Li Fang (ed.), *Thai-Phing Yü Lan* (the Thai-Phing reign-period Imperial Encyclopaedia), Sung, +983.

TSCC   Chhen Mêng-Lei et al. (ed.), *Thu Shu Chi Chhêng*; the Imperial Encyclopaedia of +1726). Index by Giles, L. (2). References to 1884 ed. given by chapter (chüan) and page. References to 1934 photolitho reproduction given by tshê (vol.) and page.

TT   Wieger, L. (6), *Taoïsme*, vol. 1, Bibliographie Générale of the works continued in the Taoist-Patrology, *Tao Tsang*).

TTSLT   *Thai Tsu Shih Lu Thu* (Veritable Records of the Great Ancestor (Nurhachi, d. +1626, retrospectively emperor of the Chhing), with illustrations). Ming, +1635, revised in Chhing, +1781.

WCTY   Tsêng Kung-Liang (ed.), *Wu Ching Tsung Yao* (The Most Important Affairs to the Military Classics—a military encyclopaedia). Sung, +1044.

WCTY/cc   Tsêng Kung-Liang (ed.), *Wu Ching Tsung Tao (Chhien Chi)*, military encyclopaedia, first section, Sung, +1044.

WHTK   Ma Tuan-Lin, *Wên Hsien Thung Khao* (Comprehensive Study of (the History of) Civilisation), Yuan, +1319.

WPC   Mao Yuan-I, *Wu Pei Chih* (Treatise on Armament Technology), Ming, +1628.

WPHLC   Attrib. Chiao Yü, *Wu Pei Huo Lung Ching* (The Fire-Drake Manual and Armament Technology), Ming, after +1628, but containing much material from earlier versions of the *Huo Lung Ching*.

YCLH   Chang Ying (ed.), *Yuan Chien Lei Han* (encyclopaedia), Chhing, +1710.

YH   Wang Ying-Lin, *Yü Hai* (Ocean of Jade, an encyclopaedia of
            quotations). Sung, +1267 but not pr. till Yuan, +1337/
            +1340, or perhaps +1351.

YHSF   Ma Kuo-Han (ed.), *Yü Han Shan Fang Chi I Shu* (Jade-Box
            Mountain Studio Collection of (reconstituted and some-
            times fragmentary) Lost Books), 1853.

AUTHOR'S NOTE

This volume has been forty-three years in the gestating. On 4th June 1943 Huang Hsing-Tsung[1a] and I landed at Lichuang[2] in Szechuan after a rather adventurous journey down the Min-chiang and the Yangtse River from Wu-thung-chhiao[3].[b] There, near the delightful little town, were the Chinese–German Thung-Chi[4] University, and also the evacuated National Institutes of History and Sociology of Academia Sinica. These were then headed by two very famous scholars, Fu Ssu-Nien[5] and Thao Mêng-Ho[6] respectively, whom I was honoured to meet. Also in the neighbourhood were the evacuated National Archaeological Museum directed by Li Chi[7], and the Institute for the History of Chinese Architecture under Liang Ssu-Chhêng[8]. One evening the talk turned to the history of gunpowder in China, and Fu Ssu-Nien himself copied out for us the earliest printed passages on its composition from the *Wu Ching Tsung Yao* of +1044, a book which we did not then possess.[c] It was at Lichuang also that I first met Wang Ling[9] (Wang Ching-Ning[10]), who was destined to be my initial collaborator in the writing of *Science and Civilisation in China*, from 1948 to 1957, in Cambridge. At that time, he was a young research worker in the History Institute of Academia Sinica, and made the history of gunpowder, in all its ramifications, a lifelong study. Later on, he pursued a distinguished career as research professor of the Institute of Advanced Studies at the Australian National University at Canberra.

Of the other two collaborators whose names are on the title-page of this volume, Ho Ping-Yü[11], now Professor of Chinese at Hongkong University, has the great merit of having written the first draft of it. Having grown up in Singapore, he became an eminent historian of science, and later professor at Kuala Lumpur and Brisbane successively, since when he has produced many excellent books of his own. Finally, Lu Gwei Djen[12] was one of the first who converted me to Chinese studies from 1937 onwards, at which time we planned the present series of volumes; and when, twenty years later, she returned from UNESCO in Paris to Cambridge, she succeeded Wang Ling as my chief collaborator. This she still is. For the present book we have together checked all the battle accounts and the entries in the military encyclopaedias.

With the production of this volume, it will be been that all three of the fun-

---

[a] My first colleague in the Sino-British Science Cooperation Office, and in recent years our collaborator in Botany and Nutritional Science.

[b] There are fuller accounts in Needham & Needham (1), pp. 40ff., 119, and Huang Hsing-Tsung (1), pp. 45 ff.

[c] See pp. 117–26 below.

[1] 黃興宗　　　[2] 李庄　　　[3] 五通橋　　　[4] 同濟　　　[5] 傅斯年
[6] 陶孟和　　　[7] 李濟　　　[8] 梁思成　　　[9] 王鈴　　　[10] 王靜寧
[11] 何丙郁　　　[12] 魯桂珍

damental inventions enumerated by Francis Bacon in +1620 have now been dealt with in detail. We quoted the *Novum Organon* fully in Vol. 1[a], but the passage is well worth reproducing in shortened form here.[b]

Discoveries are to be seen nowhere more conspicuously than in those three which were unknown to the ancients, and of which the origin, though recent, is obscure and inglorious; namely printing, gunpowder, and the magnet. For these three have changed the whole face and state of things throughout the world, the first in literature, the second in warfare, the third in navigation; whence have followed innumerable changes; insomuch that no empire, no sect, no star, seems to have exerted greater power and influence in human affairs, than these three mechanical discoveries.

So, looking back, we dealt with the magnetic compass first, in Vol. 4, pt. 1, then with paper and printing, by the care of our valued collaborator, Professor Chhien Tshun-Hsün[1], in Vol. 5, pt. 1, and now finally with gunpowder in Vol. 5, pt. 7. Francis Bacon died without knowing that every one of the discoveries which he singled out had been Chinese. And although we have not been able to identify the personal name of any individual as the *fons et origo* of the three discoveries, no doubt whatever can remain about the people in the midst of whom they first came into being.

The present volume is the middle one of three on military technology. It is appearing ahead of the others simply because it is now ready. The first (Vol. 5, pt. 6), after an introduction, will deal with (b) Chinese literature on the art of war, (c) basic concepts of the classical Chinese theory of war, (d) distinctive features of Chinese military thought, (e) projectile weapons, the bow and crossbow, (f) ballistic machinery—pre-gunpowder artillery, and (g) early poliorcetics—the siege and defence of cities. I owe a great deal of gratitude to my collaborators in these subjects, Wang Ching-Ning, Robin Yates, Krzysztof Gawlikowski and Edward McEwen.

The third part (Vol. 5, pt. 8) wil deal with (i) close-combat weapons, (j) chariot warfare, (k) cavalry techniques, including the invention of the stirrup and its spread, (1) armour and caparison, (m) camps and formations, (n) signalling and other forms of communication; and the whole will end with some comparisons and conclusions. Here my principal collaborators have been Wang Ching-Ning, Robin Yates, the late Lo Jung-Pang[2] and Albert Dien. Professor Robin Yates of Harvard is taking charge of the general editing of both these.

It is natural enough that the present volume should take its place among the military three because the finding of the gunpowder mixture in the middle of the +9th century was no doubt the greatest of all Chinese military inventions. The gunpowder rocket might indeed turn out, as we venture to say in this volume, to be the greatest single invention ever made by man, for if the sun cools or over-

---

[a] P. 19.
[b] Montagu ed. (Latin), vol. 9, pp. 381–2; Ellis & Spedding ed. (English), p. 300. It is in Bk. 1 of the original work, Aphorism 129.

[1] 錢存訓   [2] 羅榮邦

heats, and we have to go somewhere else, the rocket will be our only means of doing so, since it is the sole vehicle known to man capable of navigating in outer space. Not of course the gunpowder rocket, as the Chinese military engineers knew it in the middle of the +12th century, but the rocket vehicles of today and tomorrow, powered by liquid fuels, or more probably by sub-atomic nuclear reactions.

In the same way, the story we tell here is far more exciting than it could have been if warlike applications alone had been in question. Quite apart from the application of explosions in mining, quarrying, and the building of human lines of communication—all civil engineering tasks—gunpowder, the first chemical explosive known to man, had a vital role in the development of all heat engines. Mechanical engineers were therefore also involved. Not everyone realises that before the steam-engine came into its heyday, Christiaan Huygens and Denis Papin in the late +17th century tried to make successful gunpowder-engines; and although they could never get them to work, it put them in mind of simple water and condensible steam. Hence Thomas Newcomen's success in +1712.

We have also tried to tell the story of the internal-combustion engine, which followed upon his triumph, though long afterwards; and how by 1830 Luigi de Cristoforis suggested fuelling it with petrol. The oldest internal-combustion engine was of course the cannon, but from the engineering point of view its piston was not tethered, and the work it did was not useful work. With petrol and similar fuels the internal-combustion engine came into its own, permitting, among other things, the successful aviation of today. But petrol was nothing else than the old Greek Fire, first distilled from petroleum by Callinicus in +7th-century Byzantium. This was the greatest incendiary predecessor of gunpowder; and in fact the first use of the latter in warfare was as a slow-match in the ignition-chamber of a Chinese Greek-Fire projector. This event we date at +919. So the wheel had at last come full circle, and the only tragic aspect of the affair was the centuries of time it had taken for men to see the beneficent uses of a discovery, and the celerity with which its evil uses were found out and put into practice.

We end this volume with an excursus on the travel of the knowledge of gunpowder from east to west. Perhaps the most extraordinary fact is that all the stages, from the incendiary uses of the mixture right through to the metal-barrel hand-gun or bombard, with the projectile fully occluding the bore, were passed through in China, before Europeans knew of the mixture itself. Probably there were three comings. Roger Bacon by +1260 or so was able to study fire-crackers, doubtless brought west by some of his brother friars; and the Arabian military engineers in the Chinese service must have let Ḥasan al-Rammāḥ know about bombs and rockets by +1280. Then, within the following twenty years, came the cannon, quite possibly directly overland through Russia.

The preparation of this volume has been accompanied by many changes in our group. First I must refer to the much-lamented death of Peter Burbidge in

May 1985. He had been not only Executive Vice-Chairman of the East Asian History of Science Trust, but also from 1984 onwards the presiding genius, and benign protector, of all our volumes, the publication of which he guided as Production Director of the Cambridge University Press. At our weekly meetings we have missed him tremendously. But we are fortunate that Colin Ronan, our collaborator in the *Shorter Science and Civilisation in China* series, has taken over as Project Co-ordinator.

Next, this volume has been passing through the press alongside the erection of a new and permanent building for the East Asian History of Science Library, on the basis of funds most generously subscribed both in Hongkong and Singapore. We owe particular gratitude to Dr Mao Wên-Chi[1], Chairman of the East Asian History of Science Foundation Ltd. in Hongkong, with its members and benefactors; and to the outstandingly liberal beneficence of Tan Sri Tan Chin Tuan[2] of the Overseas Chinese Banking Corporation in Singapore.

Similarly, our East Asian history of Science Board, Inc. of New York, headed by Mr John Diebold, has concentrated rather on raising funds for the endowment and research necessitated by the *Science and Civilisation in China* project, and it is due to them that the National Science Foundation, the Luce Foundation and the Mellon Foundation, have contributed generously to this end. And here Japan has also joined in, for the National institute for Research Advancement (NIRA) of Tokyo has given a noble benefaction directed mainly for Vol. 7. Our deepest thanks are due to this organisation, directed by Dr Shimokobe Atsushi[3]. One cannot be too grateful for such help in the payment of necessary emoluments and research expenses for our far-flung collaborators.

As usual, we would like also to thank those who have been of special help to us in the preparation of this volume. Thus we are glad to number among our friends Mr Howard Blackmore, formerly Deputy Keeper of the Armouries at H.M. Tower of London, who gave us valuable criticism throughout; Dr Nigel Davies, who arranged for experimental trials of gunpowders containing different nitrate percentages, at the Royal Armament Research and Development Establishment at Fort Halstead in Kent; and Dr Graham Hollister-Short, who greatly helped us in our work on the old gunpowder triers or testers, precursors of the gunpowder-engines, as also with the history of blasting in mines and quarries. Similarly, Dr Nakaoka Tetsurō[4] gave us much help with the Japanese context of the *Mōko Shūrai Ekotoba* (p. 177), the only surviving picture of a +13th-century bursting bomb-shell. A special debt is owing to De Clayton Bredt of Brisbane, the discoverer of the +10th-century painting of a fire-lance, who read through the whole volume and offered numerous amendments.

Next we wish to record our indebtedness to all the staff of the East Asian History of Science Library. In particular we want to thank Mrs Liang Chung Lien-Chu[5] who has attended to all the cross-references, as well as checking the

---

[1] 毛文奇   [21] 陳振傳   [3] 下河邊淳   [4] 中岡哲郎   [5] 梁鍾連杼

proofs of Bibliographies A and B. When we have had occasion to seek linguistic
help, we have turned, as before, to Prof. D.M. Dunlop for Arabic, the late Dr
Charles Sheldon and Dr Ushiyama Teruyo[1] for Japanese, and Prof. Shackleton
Bailey for Sanskrit.

So now let us pull the lanyard and fire off this unpowder volume (to use an
appropriate analogy) upon the Republic of Learning, not indeed with the intention of doing any damage, but rather hoping that it may help those still looking
for enlightenment about the history of gunpowder-weapons and heat-engines.
War may or may not have been a decisive factor in human evolution and social
progress, but what cannot be denied is that the steam-engine and the internal-
combustion engine have been this, and all were children of the cannon. And that
in turn was one development of the fire-lance, while the other was the rocket, on
which all space travel depends. Gunpowder-engines and the steam-engine no
less than the rocket vehicle were thoughts springing from the European Scientific Revolution—but all the previous developments, through eight preceding
centuries, had been Chinese.

---

[1] 牛山燿代

30 MILITARY TECHNOLOGY (*cont.*)

(*f*) PROJECTILE WEAPONS, III, THE GUNPOWDER EPIC

(1) INTRODUCTORY SURVEY

The development of gunpowder was certainly one of the greatest achievements of the medieval Chinese world. One finds the beginning of it towards the end of the Thang, in the +9th century, when the first reference to the mixing of salt-petre (i.e. potassium nitrate), sulphur, and carbonaceous material, is found. This occurs in a Taoist book which strongly recommends alchemists not to mix these substances, especially with the addition of arsenic, because some of those who have done so have had the mixture deflagrate, singe their beards, and burn down the building in which they were working.

The beginnings of the gunpowder story take us back to those ancient practices of religion, liturgy and public health which involved the 'smoking out' of undesirable things in general. The burning of incense was only part of a much wider complex in Chinese custom, fumigation as such (*hsün*[1]).[a] That this procedure, carried on for hygienic and insecticidal reasons, was much older than the Han, appears at once from a *locus classicus* in the *Shih Ching*[2] (Book of Odes), where the annual purification of dwellings is referred to in an ancient song, datable to the −7th century or somewhat earlier.[b] It is perhaps the oldest mention of the universal later custom of 'changing the fire' (*kuan huo*[3], *huan huo*[4]), a 'new fire' ceremony annually carried out in every home.[c] The medical fumigation (*han*[5]) of houses, after sealing all the apertures, with *Catalpa* wood (*chhiu*[6]), is referred to in the *Kuan Tzu*[7] book not many centuries later;[d] and the *Chou Li*[8], archaising in character even if a Former Han compilation, has several descriptions of officials superintending fumigation with the insecticidal principles of the plants *Illicium*

---

[a] We had a good deal to say about this whole subject in Vol. 5, pt. 2, pp. 148 ff.

[b] Mao no. 154, tr. Legge (8), vol. 1, p. 230; Karlgren (14), p. 98; Waley (1), p. 166. We quoted the text in Vol. 5, pt. 2, p. 148 f.

[c] Cf. Bodde (12), p. 75; Fan Hsing-Chun (1), pp. 24–5. There are of course also references to fire ceremonies of various kinds in the oracle-bone writings, attesting their existence already in the Shang.

[d] Ch. 53, p. 11b, tr. Needham & Lu Gwei-Djen (1), p. 449. Various more or less fragrant composites (*chhiu*[9]), e.g. may-weed, cud-weed or chamomile (*Antennaria, Gnaphelium* or *Anthemis*) were burnt in the same way. The process was also used for the drying-out of new houses. Not all the smokes were balmy, however, for as Harper (2) has shown, the newly discovered Han almanac texts (*jih shu*[10]) prescribe, under restraints and punishments (*chieh*[11]) the burning of various types of faeces (*shih*[12]) to exorcise demons from houses. This is particularly interesting because of the practice later common of adding faeces to incendiary, and even explosive, gunpowder (pp. 124–5, 343–4 below). Might there not be some significance here in the fact that the main meaning of *shih*[12]) has always been 'arrow'?

[1] 燻   [2] 詩經   [3] 爟 火   [4] 換 火   [5] 熯
[6] 楸   [7] 管子   [8] 周禮   [9] 萩   [10] 日書
[11] 詰   [12] 矢

I

and *Chrysanthemum*.[a] From later literature we know that among Chinese scholars it was long the custom to fumigate their libraries to minimise the damage caused by bookworms, a great pest, especially in the centre and south.[b]

As an extension of techniques like these, we find that the uses of scalding steam in medical sterilisation were appreciated as early as the +10th century. In his *Ko Wu Tshu Than*[1] (Simple Discourses on the Investigation of Things) about +980, Lu Tsan-Ning[2] wrote:

When there is an epidemic of febrile disease, let the clothes of the sick persons be collected as soon as possible after the onset of the malady and thoroughly steamed; in this way the rest of the family will escape infection.

How general this practice was it would be hard to say, but it probably formed part of traditional hygienic usages from the Thang onwards.[c]

Not only in peace, moreover, but also in war, the ancient Chinese were great smoke-producers. Toxic smokes and smoke-screens generated by pumps and furnaces for siege warface occur in the military sections of the *Mo Tzu*[3] book (−4th century), especially as part of the techniques of sapping and mining;[d] for this purpose mustard and other dried vegetable material containing irritant volatile oils was used. There may not be sources much earlier than this, but there are certainly abundant sources later, for all through the centuries these strangely modern, if reprehensible, techniques were elaborated *ad infinitum*. For example, another device of the same kind, the toxic smoke-bombs (*huo chhiu*[4]) of the +15th century, recall the numerous detailed formulae given in the *Wu Ching Tsung Yao*[5] of +1044.[e] The sea-battles of the +12th century between the Sung and the Chin Tartars, as well as the civil wars and rebellions of the time, show many further examples of the use of toxic smokes containing lime and arsenic. Indeed, the earth-shaking invention of gunpowder itself, some time in the +9th century, was closely related to these, for it was at once seen to be connected with incendiary preparations, and its earliest formulae sometimes contained arsenic.

The whole story from beginning to end illustrates a cardinal feature of Chinese technology and science, the belief in action at a distance.[f] In the history of naval warfare, for instance, one can show that the projectile mentality dominated over ramming or boarding, with its close-contact combat.[g] Smokes, per-

---

[a] Ch. 9, *pp.* 5*b*, 6*b*, ch. 10, pp. 7*a*, 9*a*, tr. Biot (1), vol. 2, pp. 386 ff., discussed by Needham & Lu Gwei-Djen (1), pp. 436–7. Cf. Shih Shu-Chhing (2).

[b] For further information on this subject see Vol. 5, pt. 2, pp. 148–9.

[c] We shall return to the matter in Sect. 44 on medicine and hygiene; in the meantime there is much relevant information in Needham & Lu Gwei-Djen (1).

[d] See Vol. 4, pt. 2, pp. 137–8, Vol. 5, pt. 6 and Yates (3), pp. 424 ff.

[e] Cf. pp. 117 ff. below.

[f] Cf. Vol. 4, pt. 1, pp. 8, 12, 32–3, 60, 233 ff. If this had not been so, the polarity of the magnet would never have been discovered, for in China it was never thought odd that an earthly sublunary stone or metal needle should direct itself towards the pole-star on high.

[g] Cf. Vol. 4, pt. 3, pp. 682 ff., 697.

[1] 格物麤談   [2] 錄贊寧   [3] 墨子   [4] 火毬   [5] 武經總要

fumes, hallucinogens,[a] incendiaries, flames, and ultimately the use of the propellant force of gunpowder itself, form part of one consistent tendency discernible
throughout Chinese culture from the earliest times to the transmission of the
bombard, gun and cannon to the rest of the world about +1300. And indeed we
believe that the following sub-sections will demonstrate beyond doubt that the
entire development from the first discovery of the gunpowder formula to the
perfection of the metal-barrel gun emitting a projectile of dimensions closely
fitting the bore, took place in China before other peoples knew of the inventions
at all.[b]

Now in order that the reader may the more easily dominate the mass of detail
necessarily appearing in evidence as we go on, it may be desirable to explain a
chart (Fig. 1) which sets forth the whole course of events as we have found them.
This may correspond to another chart (Fig. 233 on p. 569 below) to be considered at the end of our enquiry, which illustrates the inter-cultural transmissions which took place.

But before going any further it must be emphasised that although naturally
this Section is placed in a Volume on military technology, the invention of
gunpowder had implications far transcending military history. The viewpoint of
the civil engineer is not to be ignored. His attitude towards explosives is very
different from that of the soldier, for he thinks of them as rock-blasting and
earth-moving facilities, means for carving out the formations for roads, waterways, railways, pipe-lines and all the multifarious veins and arteries of civilised
intercourse; nor could the achievements of modern mining and quarrying be
thinkable without the use of explosives. These things we shall take a look at later
on (p. 533) as we see them growing out of the very ancient technique of 'fire-
setting'. Other civil uses of gunpowder and the more sophisticated explosives
that derived from it can be found in religious, ceremonial and meteorological
rockets, whether exploratory or weather-modifying (p. 527).[c] But the mechanical
engineer is also in the picture. Later on (p. 544) we shall have something to
say about the efforts to make gunpowder-engines before steam-engines came
into their own, and indeed it was the former that led directly to the latter.
As everyone knows, the steam-engine had its day, and it was a great one, not
yet quite over; but when men's thoughts returned to internal combustion
a fuel was needed to explode obediently in the cylinder, and what was it?
Nothing other than the antecedent of gunpowder, namely the distilled petroleum that had consituted Greek Fire. And so these substances the effects of
which have been so terrible in warfare, turn out to be most intimately related
to the development of the heat-engine, on which all modern civilisation has
depended.[d]

[a] Cf. Vol. 5, pt. 2, pp. 150 ff.
[b] As we shall see (p. 51 below), Berthold Schwartz is a pure myth.
[c] We say nothing here of the sublime function of the rocket as the only space-vehicle known to man, but in
due course (pp. 506, 521 ff) we shall.
[d] Cf. e.g. Prigogine & Stengers (1), pp. 111 ff.

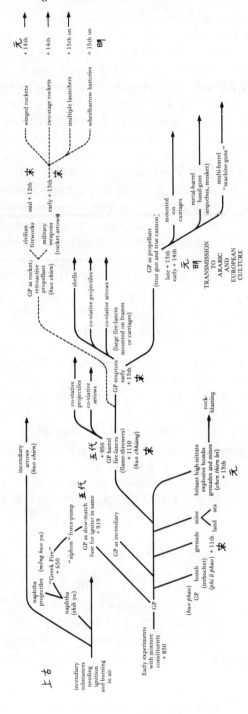

Fig. 1. Chart to illustrate the development of gunpowder technology.

Clearly this entire subject is concerned with power,[a] might and power placed in the hands of man as social evolution has gone on, power and might which form a couple of chapters only in the line of development which in the end has now given him mastery over the sub-atomic processes of suns, sources of inextinguishable energy, a mastery which has outstripped (it may be greatly feared) his ethical and moral maturity. Yet mastery over Nature remains the second grandest of ideals, as Robert Boyle wrote long ago, in 1664. He is well worth listening to.

And though it be very true [said he][b] that man is but the Minister of Nature, and can but duely apply Agents to Patients (the rest of the Work being done by the applyed Bodies themselves) yet by his skill in making those Applications, he is able to perform such things as do not only give him a Power to Master Creatures otherwise much stronger than himselfe; but may enable one man to do such wonders, as another man shall think he cannot sufficiently admire. As the poor Indians lookt upon the Spaniards as more than Men, because the knowledg they had of the Properties of Nitre, Sulphur and Charcoale duely mixt, enabled them to Thunder and Lighten so fatally, when they pleas'd.

And this Empire of Man, as a Naturalist, over the Creatures, may perchance be, to a Philosophical Soul preserved by reason untainted with Vulgar Opinions, of a much more satisfactory kind of Power or Soveraignty than that for which ambitious Mortals are wont so bloodily to contend. For oftentimes this Latter, being commonly but the Gift of Nature, or Present of Fortune, and but too often the Acquist of Crimes, does no more argue any true worth or noble superiority in the possessor of it, than it argues one Brasse Counter to be of a better Metal than its Fellows, in that it is chosen to stand in the Account for many Thousand Pounds more than any of them. Whereas the Dominion that Physiologie gives the Prosperous Studier of it (besides that it is wont to be innocently acquired, by being the Effect of his knowledge), is a Power that becomes Man as Man. And to an ingenious spirit, the Wonders he performes bring perchance a higher satisfaction, as they are Proofes of his Knowledge, than as they are Productions of his Power, or even bring Accessions to his Store.

Here at the outset it would not be inappropriate to say something, for the benefit of those less familiar with the Chinese literary tradition than others, on what we might call the 'philological network'. Chinese historical writing cannot just be dismissed as unreliable, for no civilisation has had a greater historical tradition than China, and the accounts of what really happened in all the ages have been the work of thousands of meticulous and painstaking scholars. All that historians can do, they did, and archaeological finds have proved them right again and again, sometimes spectacularly. No other civilisation produced a

[a] Everyone will remember the delicious but bitter satire of Jonathan Swift (2), in the 7th chapter of pt. 2 of his *Gulliver's Travels* (+1726); where the voyager tells how he explained the nature and effects of gunpowder weapons to the King of Brobdingnag, and how utterly shocked and horrified this prince was when he heard about them. Gulliver affects to despise him for that—but Swift himself never knew that the explosive properties of gunpowder had already led a couple of dozen years earlier to the development of the steam-engine, which in its turn would generate the internal-combustion engine, with all the inestimable benefits which they have brought to mankind.

[b] (8), pt. 1, p. 20.

body of work like the twenty-four dynastic histories (*erh shih ssu shih*[1]), and these were supplemented by a vast body of unofficial historical writings; besides which there were encyclopaedists with high scholarly values in all ages, as well as biographers and authors of memorabilia. Modern philology has had a great part to play in the evaluation of all this, for the authenticity of texts can be cross-checked in many ways—who quotes whom, and is quoted by whom, who was a contemporary of whom, and what do we know about their life and times. Occasional false attributions and anticipatory ascriptions of course there are, but a whole literature of historical criticism and elucidation is available in Chinese, whereby the texts of erroneous, composite or doubtful date (*wei shu*[2]) can be distinguished from the majority which have impeccable authenticity. As we noted at an earlier point,[a] the study of the history of science and technology in China is in fact aided by the very circumstance that these pursuits were not highly regarded by the Confucian *literati*, so it would not have occurred to anyone that credit could be gained by falsifying matters so as to ascribe a given discovery or invention to a date earlier than that at which it actually happened. The same circumstance prevented dealers from forging non-artistic objects such as scientific equipment or military weapons so as to give an erroneous appearance of antiquity. No one wanted to collect such things; there was no profit in it.[b] The Confucian bureaucrats always had a supercilious attitude towards the soldiers, whose commanders were invariably lower than the corresponding civilians in official rank. From the texts of the military compendia one gets the impression that they were in deadly earnest, lacking the allusions and literary graces which other books possessed.[c] Interpolations in them are very rare indeed.[d] All in all, we believe that what the Chinese historians and military writers say is almost always credible.[e] Such is our view of the reliability of what we shall be telling in the following sub-sections.

It is well to be clear from the beginning that broadly speaking the term 'fire-chemical' or 'fire-drug' (*huo yao*[3]) never means anything other than that mixture of saltpetre, sulphur and charcoal which we call gunpowder. To this there is, so far as we know, but a single exception—a recondite one—and that lies in the field of physiological alchemy, or the making of the 'inner elixir' (*nei tan*[4]), where

---

[a] Vol. 1, p. 77.

[b] Of course where Shang or Chou bronze vessels of artistic merit were concerned, the case is different. There was a great growth of antiquarianism from Sung times onwards (cf. Vol. 2, pp. 393 ff.), and forgeries certainly occurred.

[c] This does not mean that every device described in late books such as the *Wu Pei Chih* (see p. 34 below) was necessarily used at the time; they are often liable to describe, with antiquarian zeal, inventions of the past, even when sometimes long disused. This has to be allowed for when reconstructing Chinese military engineering history.

[d] One has to allow of course for legendary attributions to figures such as Yao, Shun, Huang Ti and even Chuko Liang (cf. p. 25 below); but these are easily recognised.

[e] Though the latter may exaggerate a bit now and then about the ranges of their weapons. It is usually fairly easy to correct for such things.

[1] 二十四史   [2] 僞書   [3] 火藥   [4] 內丹

the juices and fluids of the human body, wrought upon by divers techniques and exercises, were believed to generate an enchymoma or macrobiotic drug which would confer material immortality upon the adept.[a] Here, in order to make manifest the intimate relations of these entities with the Five Elements, it was necessary to coin special adjectives, and to translate *chin i*[1] as 'metallous juice' (not as potable gold), or *mu yao*[2] as 'lignic medicine'. Accordingly, encountering *huo yao* in *nei tan* texts of late date, it has to be translated as 'pyrial salve',[b] i.e. the salivary Yang descending, in contrast with the 'aquose salve', the seminal Yin, ascending;[c] essential components of the enchymoma to be formed at the centre of the body. But the lore of these two pro-enchymomas had an extremely limited readership, and we can be sure that very few Chinese scholars throughout history ever understood *huo yao* in any sense other than gunpowder.

Fig. 1 runs from left to right. Out of the remote depths of history come the incendiary substances, needing ignition, and burning, sometimes quite fiercely, in air. Attached to arrows (*huo chien*[5]), they cross the stage and must have lasted down well into the Sung time or even later. One of these incendiaries was naphtha, derived from natural petroleum seepages;[d] but a great step forward was made in +7th-century Byzantium, when Callinicus successfully distilled it to give low boiling-point fractions something like our petrol, which could be projected at the enemy by pumps which constituted flame-throwers.[e] We think we can identify naphtha under the name *shih yu*[6], and Greek Fire, as it was called, under that of *mêng huo yu*[7]. The 'siphon', or force-pump, was of particular importance because it was the site of the first use of gunpowder in war; this was the appearance of a slow-match impregnated with the material in the ignition chamber (*huo lou*[8]) of the machine—and the date was +919. That was a century which saw great commerce in these petrol fractions; they often came through from the Arab trade, but so much of the spirit was circulating among the rulers of the Five Dynasties period that the Chinese must surely have been distilling it themselves.[f]

Without doubt it was in the previous century, around +850, that the early alchemical experiments on the constituents of gunpowder, with its self-contained oxygen, reached their climax in the appearance of the mixture itself. We need not harp upon the irony that the Thang alchemists were essentially looking for elixirs of life and material immortality.[g] But it is only reasonable to recognise that once their elaboratories had jars containing (among many other things) all the constituents (more or less purified) of the deflagrative and

---

[a] Full details on this have been given in Vol. 5, pt. 5.
[b] Cf. p. 100 below.
[c] E.g. *Shih Chin Shih*[3], p. 14*b*, and all the writings of Fu Chin-Chhüan.[4]
[d] See Vol. 3, pp. 608 ff.   [e] See Vol. 4, pt. 2, pp. 144 ff.
[f] See Vol. 5, pt. 4, pp. 158 ff.   [g] See Vol. 5, pt. 2, pp. 77 ff.

[1] 金液   [2] 木藥   [3] 傅金銓   [4] 試金石   [5] 火箭
[6] 石油   [7] 猛火油   [8] 火樓

explosive substance on their shelves, and once the alchemists started mixing them in all possible combinations, gunpowder was sure to be found one day. If its first formulae did not appear in print until +1044, that was a full two hundred years before the first mention of the mixture in the Western world,[a] and even then no information was available there about the proportions necessary.

By about +1000 the practice was coming into use of putting gunpowder in simple bombs and grenades, especially those thrown or lobbed over from trebuchets[b] (huo phao[1]).[c] Here the progression was from bombs with weak casings (phi li phao[2] or 'thunderclap bombs') to those with strong ones (chen thien lei[3] or 'thunder-crash bombs'). This paralleled a slow but steady rise of the percentage of saltpetre (potassium nitrate) in the composition, so that by the +13th century brisant explosions became possible. In the meantime there was also a development of devices for mines, both on land and in the water. As long as the nitrate content remained low, there was a tendency to use gunpowder just as an incendiary better than those before available, but this did not outlast the +12th century.

So far all the containers had been in principle spherical, but the way to the true barrel gun—and to the piston of all engines too—lay through the cylindrical container. Biological analogies must always have been in men's minds (at least subconsciously); the cylindrical tubes through which excretion and emission occur.[d] But in China people had a natural cylinder ready to hand, the bamboo stem, once cleared of its septa, and any contents of the internode removed.[e] This transition occurred first in the middle of the +10th century as we know from a silk banner belonging to one of the Buddhist cave-temples at Tunhuang in Kansu (p. 222 below). The scene depicts the temptation of a Buddha by the hosts of Mara the Tempter, many of whose demons are in military uniforms and carry weapons, all aiming to distract him from his meditation. One of them, wearing a head-dress of three serpents, is directing a fire-lance (huo chhiang[4]) at the seated figure, holding it with both hands and watching the flames shoot out horizontally. This is the earliest representation we have of a weapon which had enormous repercussions between +950 and +1650; it played a very prominent part, for example, in the wars between the Sung and the Jurchen Chin Tartars from +1100 onwards. It was then for the first time described,

---

[a] By Roger Bacon, of course, cf. p. 47 below.

[b] Or mangonels, not the torsion type of pre-gunpowder artillery, but depending on the swape principle. Cf. Vol. 4, pt. 2, pp. 331 ff.

[c] Note that this term applied also to the projectile, hence much grief for the historians. We shall return to the problems of vocabulary and terminology in a moment.

[d] There are even examples of missiles, if anybody had known about it, or thought of it, in the animal world. One could mention the dart-sacs of gastropod molluscs, which emit calcareous pencil-like rods into the body of the partner during sexual intercourse (Shipley & McBride (1), 1st ed., p. 208, 4th ed., p. 295; Marshall & Hurst (1). fig. 32, pp. 128–9, 133); or the nematocysts of coelenterates, which send out poisonous lassoes (Lulham (1), p. 27). But it is doubtful whether any of these cases would have been known in the Middle Ages.

[e] Cf. Burkill (1), vol. 1, pp. 289 ff.

[1] 火砲   [2] 霹靂砲   [3] 震天雷   [4] 火槍

about +1130, in the *Shou Chhêng Lu*[1] of Chhen Kuei[2], relating the defence of a certain city north of Hankow.[a] Essentially the fire-lance was a tube filled with rocket composition, relatively low in nitrate, but not allowed to fly loose, held instead upon the end of a spear. An adequate supply of these five-minute flame-throwers, passed from hand to hand, must have been an effective discouragement to enemy troops from storming one's city wall.

The development of the fire-lance from the petrol flame-thrower pump must have been an easy and logical process. It turned that flame-projector into a portable hand-weapon for spouting fire, and since gunpowder, even though very low in saltpetre, had been used in the projector as a slow-match igniter, the new development was not far to seek. Also it was in a way a more effective method of using the incendiary properties of gunpowder, which must have been apparent even before the +10th century had begun. But the basic point was that the cylinder had been born. Most probably it originated with the natural gift of the bamboo tube, but as time went on all kinds of materials were employed for it, even paper (another Chinese invention), a substance which by appropriate treatment can be made so hard that it was actually used for armour.[b] What is important to note is that as the fire-lance period went on, through the +10th and +12th centuries, metal, both bronze and cast iron, perhaps also brass, was used to make the tube. This was one outstanding precursor aspect of the true metal-barrel gun or cannon, but the other was the addition of projectiles which issued forth along with the flames.

Here in this phase we have been obliged to coin two technical terms. The projectiles which were spurted forth in this way needed a special name, so we call them 'co-viative', distinguishing them thus from the true bullet or cannon-ball, which in order to use the maximum propellant force of the gunpowder charge, must fill the bore of the barrel. The fire-lance projectiles could be anything offensive, such as bits of scrap metal or broken porcelain, but they could also be arrows. None of them would have issued with great velocity, but they could have been effective enough against unarmoured attackers, especially if the arrows were poisoned, as the texts often say they were. Secondly, when the fire-lances grew large, they were mounted on specially designed frames or carriages, almost like field-guns, and these we call 'eruptors'. These in their turn emitted miscellaneous co-viative projectiles, including arrows and containers of poisonous smokes, containers which in some cases may have been explosive, and therefore merit the name of shells or proto-shells. We often have to utilise these ambiguous prefixes, for example a gunpowder which contains carbonaceous material rather than charcoal may be usefully called proto-gunpowder. Similarly, we cannot always be sure whether a projectile fitted the bore of a gun or not, in which case it is convenient to call the weapon a quasi-gun or a proto-gun.

[a] P. 222 below.   [b] Cf. Vol. 5, pt. 1, pp. 114–6; pt. 8 (*l*) 2, iii.

[1] 守城錄   [2] 陳規

Thus the fully developed firearm had three basic features: (1) its barrel was of metal; (2) the gunpowder used in it was rather high in nitrate; and (3) the projectile totally occluded the muzzle so that the powder charge could exert its full propellant effect. This device may be called the 'true' gun, hand-gun, or bombard, and if it appeared in late Sung or early Yuan times, about +1280, as we believe it did, its development had taken just about three and a half centuries since the first cylindrical barrels of the fire-lance flame-throwers. This was not bad going for the Middle Ages, and it is important to realise that none of these early tentative phases had existed in Islam or Europe at all. The bombard appears quite suddenly full-fledged in the famous illustration of Walter de Mila-mete's Bodleian MS. of +1327. Give or take a few decades, the bombard cannot have come to Europe much before +1310.

There, however, great sociological changes were about to happen—the Re-naissance, the Reformation, the growth of capitalism, and the scientific revolu-tion. Hence the speed of change in Europe began to outstrip the slow and steady rate of advance dictated by Chinese bureaucratic feudalism. The merchant-adventurers and the bourgeois entrepreneurs were to the fore once the +15th century had begun; the patricians of the mercantile city-States, the ironmasters, the mining proprietors and the factory builders, all these took charge as Euro-pean aristocratic military feudalism died. Hence the way in which the gunpow-der weapons first worked out by the Chinese began to come back to them in improved form. The serpentine lever, which applied the smouldering match to the touch-hole of guns (p. 459) may have been invented in China, and the Turks may have improved it into the matchlock musket; certain it is that this superior weapon reached China either direct through Central Asia by +1520, or at the latest via the Portuguese and Japanese by +1548. Similarly, the Portuguese breech-loading culverin[a] or small cannon came up from Malaya by +1510 or so, and its replaceable chambers were greatly appreciated by the Chinese gunners. And later the flintlock musket appeared, and later still the rifle. In the +17th century the Jesuits were 'drafted', so that John Adam Schall von Bell could be seen superintending the Western-style cannon foundry of the last Ming emperor in +1642–3, while Ferdinand Verbiest had to undertake the same duty for the Chhing court in +1675. Thus did the inventiveness of the Chinese re-verberate and recoil across the length of the Old World. Some eastern nations in modern times have been accused of being able only to copy and improve; but of no one was this more true in the +15th and +16th centuries than the Wester-ners. To be sure, with ballistics and dynamics they soon became 'airborne', but that was quite a time after the first knowledge of the first of all chemical explo-sives reached Europe.

It may seem surprising that until now nothing has been said about the rocket.

---

[a] The word is improper, but there is no generally available equivalent (cf. p. 367).

In this day, when men and vehicles have been landed on the moon, and when the exploration of outer space by means of rocket-propelled craft is opening before mankind, it is hardly necessary to expatiate upon what the Chinese engineers started when they first made rockets fly. After all, it was only necessary to attach the tube of the fire-lance to an arrow, with its orifice pointing in the opposite direction, and let it soar away free, in order to obtain the rocket effect. Exactly at what date this 'great reversal' happened has been a debatable question.[a] Twenty years ago, when our contribution to the *Legacy of China* was written,[b] we thought that rocket-arrows were developed by about +1000, in time for the *Wu Ching Tsung Yao*. That depended on one's interpretation of the 'gunpowder whip-arrows' (*huo yao pien chien*[1]) described therein, but we now believe that these were not rockets, nor yet the *huo chien*[2] either, which it also mentions and illustrates. All these were still incendiary arrows, designed to set on fire from a distance the enemy's camps or city buildings; but in later times this same phrase was universally used to mean rockets. Here was another example of terminological confusion, when the thing fundamentally changed, while the name did not.[c]

There would be a very good case for a linguistic analysis of such problems over the whole range of science and technology, and Hollister-Short (2) has made a valuable contribution to it. How, he inquires, is a technical vocabulary generated in order to denote some new machine or technique? Language has often failed to keep up with technical change. Already we have come across the difficulties of precise nomenclature with regard to water-raising machinery,[d] and vertical or horizontal wheels.[e] We had to define our terms. Hime long ago

---

[a] Probably the best estimate would be some time between +1150 and +1180, and earlier rather than later.

[b] Needham (47).

[c] A classical example of this is the word *phao*[3], which meant anciently both the trebuchet and the projectile which was hurled from it. When fire came into warfare, *huo phao*[4] (as we have just seen) was used still to mean the engine and the missile which it threw. But the term remained the same through all the following stages: (1) incendiary projectiles, (2) gunpowder used as incendiary in projectiles, (3) explosive projectiles in weak casings, (4) higher-nitrate explosive projectiles in strong casings, and finally (5) bombards and cannon, where gunpowder was used as propellant, and no trace remained either of the trebuchet or the explosive projectile. This gives some idea of the terminological toils which it has been necessary to unravel in the sub-sections which follow.

[d] Cf. Vol. 4, pt. 2, p. 330 ff.

[e] Vol. 4, pt. 2, p. 367. These were both examples of floating terminology among historians of engineering. But plenty of ambiguity can be found in Chinese writings too—as we have already pointed out (Vol. 4, pt. 2, pp. 267, 278 and passim) the word *chhê*[5] was used quite indifferently for 'vehicle' and 'machine'. A closer parallel to the point at issue here is the way in which the term *tho*[6] continued to mean the axial rudder after having for centuries meant something entirely different, namely the steering-oar (cf. Vol. 4, pt. 3, pp. 638 ff.). And to come nearer home, we may note that *hsiao*[7] was as much a witch-word in early chemistry in China as 'nitre' was in the West. We decided (Vol. 5, pt. 4, pp. 193–4) that the only way to be sure that an early writer meant 'saltpetre' when he said *hsiao shih*[8] ('solve-stone') is to see what he said about its properties. China was not short of good technical terms, but all through history there has been everywhere a great reluctance to coin new ones when they were needed.

[1] 火藥鞭箭　　[2] 火箭　　[3] 砲　　[4] 火砲　　[5] 車
[6] 柂　　[7] 消　　[8] 消石

encountered the same problem in relation to the subject of the present Section. He wrote:[a]

Take for example a word *W*, which has always been the name of a thing *M*. It is then applied to some new thing, *N*, which has been devised for the same use as *M* and answers the purpose better. *W* thus represents both *M* and *N* for an indefinite time, until *M* eventually drops into disuse, and *W* comes to mean *N*, and *N* only. The confusion necessarily arising from the equivocal meaning of *W* during this indefinite period is entirely due, of course, to (the failure of people) to coin new names for new things. If a new name had been given to *N* from the first, no difficulty would have ensued ... But as matters have fallen out, not only have we to determine whether *W means M* or *N* when it is used during the transition period, but we have to meet the arguments of those ... who insist that because *W* finally meant *N* it must have meant *N* at some bygone time when history and probability alike show that it meant *M*, and *M* only.

This is exactly the case with the fire-arrow and the rocket. We can recall a similar situation in China when the invention of the escapement for mechanical clocks was made, yet no one could think of a new name to distinguish such horological machines from clepsydras.[b] As for Hollister-Short, he took for his study the term *Stangenkunst* (rod-engine), which had two entirely different meanings, (1) a water-wheel placed above a mine-shaft, with rods descending from its cranks to actuate tiers of suction-pumps, and (2) the transmission of power across country from a water-wheel by means of horizontal rocking pantograph-like 'field-rods'.[c] It took all of two and a half centuries to clarify this. Fifty years ago I drew attention to the development of technical terms as a prime limiting factor in the history of science.[d]

So when then did the rocket really start on its prestigious career? It is clear now that the fire-lance long preceded it; the Tunhuang banner of about +950 settled that question. We have to search for rocket beginnings in a rather different direction, and a couple of centuries later. During the second half of the +12th century we find the appearance of two kinds of fireworks, the one called

[a] (2), p. 8. The words in brackets are a simplification introduced by us. Hime actually quoted Horace (*Ars Poetica*, ll. 48–53):

.... Si forte necesse est
Indiciis monstrare recentibus abdita rerum
Fingere cinctutis non exaudita Cethegis
Continget, dabiturque licentia sumpta pudenter....

I.e. 'If by any chance it's necessary to reveal hidden things by new indications, making up words which were never heard by the old Cethegi in their antique robes; this is permissible so long as it's not overdone, and these new and made-up words will have authority if they fall sparingly as drops from the fountains of the Greek tongue'. This must have inspired Linnaeus, who would not admit into his binomial nomenclature any word that did not come from Greek or Latin—or looked as if it did. On this cf. Vol. 6, pt. 1, p. 168. What inspired Hime to make his 'poignant cry' was the fact that so many European words had remained the same though the sense fundamentally changed. Thus 'artillerie' could mean in old times bows and arrows, while 'gonne' was used for the projectile of a ballista. He gave examples from Arabic, and even from Chinese, too. On the passage from Horace, see Brink (2), Vol. 2, pp. 57, 138.
[b] Vol. 4, pt. 2, p. 465.
[c] Vol. 4, pt. 2, p. 351.
[d] Needham (2), p. 215, (27).

'ground-rats' (*ti lao shu*[1]), and the other 'meteors' (*liu hsing*[2]). Probably the former was the older, just a tube, probably of bamboo,[a] filled with gunpowder and having a small orifice through which the gases could escape; then when lit, it shot about in all directions on the floor at firework displays. Alternatively, if attached to a stick, it flew off into the air, as at the night-time celebrations on the West Lake at Hangchow. That the two things were closely connected appears from late appellations such as 'flying rat' (*fei shu*[3]) and 'meteoric ground-rat' (*liu hsing ti lao shu*[4]). Ground-rats are contained in many specifications for bombs, where they are often equipped with hooks, and they must have been quite effective, especially when used against cavalry. As a firework they were certainly capable of frightening people, as we know from the story of a Sung empress who was 'not amused' by them (p. 135).

Such civilian uses would have reminded the soldiers of the recoil effect of fire-lances which they must always have had to withstand, whereupon someone in the last decades of the century, perhaps about +1180, tried a fire-lance fitted backwards on a pike or arrow, with the result that it whizzed away into the air towards a target. Thenceforward, rockets were very commonplace, both in peace and war, through the Southern Sung, the Yuan and Ming, indeed down to the late Chhing, when they appeared in action against the foreign invaders in the Opium Wars. Many developments of great interest occurred during this long period. First, there were several types of multiple rocket-arrow launchers, designed so that a single fuse would ignite and despatch more than fifty projectiles. Later on these were mounted on wheelbarrows, so that whole batteries could be trundled into action positions like regular artillery in modern times. But even more interesting were the rockets provided with wings, and carrying a bomb with a bird-like shape, early attempts to give some aerodynamic stability to the missile's flight, prefiguring the fins and wings of modern rocket vehicles. And just as the Chinese had invented the rocket itself, so it was natural that they should be the first to construct large two-stage rockets; propulsion motors ignited in successive stages, and releasing automatically towards the end of the trajectory a swarm of rocket-arrows to harass the enemy's troop concentrations. This was a cardinal invention, foreshadowing the Apollo space-craft, and the exploration of the extraterrestrial universe.

Like all the other stories, we shall tell this one in its place (p. 472), but a word may be said here of how the path led from the ground-rat to the space-rocket. We shall see how for a time it was the Indians who excelled in the use of rocket missiles, a circumstance which led to a great development of warhead rockets in the first half of the nineteenth century in Europe. But this was a phase which came and went, for high explosive and incendiary shells could be fired from

---

[a] But carton, or paper even, could have been used, as it certainly was later.

[1] 地老鼠　　　[2] 流星　　　[3] 飛鼠　　　[4] 流星地老鼠

more advanced artillery with much greater accuracy of aim; so that the rocket batteries of the West died out after about 1850, and little use was made of rockets during the First World War. Meanwhile, however, another fundamental step forward had been made, to join the cluster of inventions which had happened in China in the first place; this was the study and development of liquid fuels, rather than the deflagrative gunpowder with which it had all begun. And this development was not inspired by war, rather by the science fiction writers, some of whom had appreciated the crucial fact that the rocket is the only vehicle known to man which can overcome earth's gravity, leave earth's atmosphere, and voyage among the planets and the stars. Truly, 'meteoric' was no bad name that the Chinese of the +12th century had coined for their 'flying rats'.

We have now passed in review the whole procession of inventions, with all their implications so fateful for the human race, between the earliest experiments with the gunpowder mixture in the +9th century and the appearance of the multi-stage rocket in the +14th. This had occupied some five centuries or so, with the transmission to the Western world coming right at the end of the period. And so, as we view the wheelbarrow rocket-launcher batteries passing off behind the curtains on the right of the stage, we must feel bound to salute those ingenious men of the Chinese Middle Ages 'that were Authours of such great Benefits to the universal World'. For benefits there really were in store, and great ones, even though the warlike applications of gunpowder dominated for a very long time.

With this, our introduction may be ended; but before throwing open to the reader the vast museum of historical detail which justifies the statements that have been made, there are a very few concluding considerations we ought not to omit.

For example, there is a classical notion, a cliché perhaps, an *idée reçue*, a vulgarism, a false impression, which still circulates in the wide world—namely that though the Chinese discovered gunpowder, they never used it for military weapons, but only for fireworks.[a] This is often said with a patronising undertone, suggesting that the Chinese were just simple-minded; yet it has an aspect of admiration too, stemming from the Chinoiserie period of the eighteenth century, when European thinkers had the impression that China was ruled by a 'benevolent despotism' of sages. And indeed it was quite true that the military were always (at least theoretically) kept subservient in China to the civil officials.[b] Like scientists in England during the Second World War, the soldiers and their commanders were supposed to be 'on tap, but not on top'. No other

---

[a] Actually it may well be that the application of it to rock-blasting preceded the first warlike use of it by half a century or so (cf. p. 538).

[b] Of course there were periods of anarchy and warlordism from time to time; and especially in early periods men such as Tu Yü[1] (+222 to +284) were successful military commanders as well as great scholars and civil officials. But in spite of the tendency of *literati* historians to exalt their own estate, the generalisation holds good.

[1] 杜預

civilisation in the world succeeded as well as China in keeping the military under tight control for all of two millennia, in spite of massive and extended foreign invasions, as well as peasant rebellions ever renewed.[a] So the cliché could have been justified, but as we shall abundantly see, it never was.

Then the Chinese invention of the first chemical explosive known to man should not be regarded as a purely technological achievement. Gunpowder was not the invention of artisans, farmers or master-masons; it arose from the systematic, if obscure, investigations of Taoist alchemists. We say systematic most advisedly, for although in the +6th and +8th centuries they had no theories of modern type to work with, that does not mean that they worked with no theories at all. On the contrary, we have shown that the theoretical structure of medieval Chinese alchemy was both complex and sophisticated.[b] An elaborate doctrine of categories, foreshadowing the study of chemical affinity, had grown up by the Thang, reminiscent in some ways of the sympathies and antipathies of the Alexandrian proto-chemists; but more developed and less animitsic.[c] Thus it remains to be seen what elements in this thought-complex were dominant when the fateful mixture was for the first time made. To sum the matter up, its first compounding arose in the course of century-long systematic exploration of the chemical and pharmaceutical properties of a great variety of substances, inspired by the hope of attaining longevity or material immortality. The Taoists got something else, but in its devious ways also an immense benefit to humanity.

Robert Boyle had something to say on this subject in +1664.

Those great Transactions [he wrote][d] which make such a Noise in the World, and establish Monarchies or ruine Empires, reach not so many persons with their influence, as do the Theories of Physiology.

To manifest this Truth, we need but consider what changes in the Face of things have been made by two Discoveries, trivial enough, the one being but of the inclination of the Needle, touched by the Load-stone, to point toward the Pole; the other being but a casual Discovery of the supposed Antipathy between Salt Petre, and Brimstone.[e] For without the knowledge of the former, those vast Regions of America, and all the Treasures of Gold, Silver, and precious Stones, and much more precious Simples they send

[a] Cf. the words of Lu Chia and Shusun Thung to the first Han emperor, Liu Pang, quoted in Vol. 1, p. 103.

[b] The role of time was paramount in it (cf. Vol. 5, pt. 4, pp. 221 ff., 231 ff.), and the alchemists believed that they could accelerate and decelerate temporal processes at will (*ibid.* p. 244 and Fig. 1516). They also recognised what we should call a basic law, namely that the maximum state of a variable is inherently unstable. Thus Yin begins to go over to Yang as soon as its apogee is reached (cf. *ibid.* p. 226 and Fig. 1515). The alchemists also made much use in their apparatus of cosmic models (*ibid.* pp. 279 ff.). It is often said that with all the jars of purified substances on the shelves, the gunpowder constituents were probably mixed for the first time 'by chance'; but who can tell what train of thought the macrobiotic experimenter was following when he did it?

[c] Cf. Vol. 5, pt. 4, pp. 305 ff.

[d] (8), pt. 2, p. 5.

[e] Here he was quoting almost verbally from Francis Bacon's *Novum Organum*, published forty-four years earlier (cf. Vol. 1, p. 19). Bacon of course added printing to make the three inventions which had upset the whole world.

us,[a] would have probably continued undetected; And the latter giving an occasional rise to the invention of Gunpowder, hath quite altered the condition of Martial Affairs over the World, both by Sea and Land. And certainly, true Natural Philosophy is so far from being a barren Speculative Knowledg, that Physick, Husbandry, and very many Trades (as those of Tanners, Dyers, Brewers, Founders, &c.) are but Corollaries or Applications of some few Theorems of it.

Thirdly, in the gunpowder epic we have another case of the socially devastating discovery which China could somehow take in her stride, but which had revolutionary effects in Europe. For decades, indeed for centuries, from Shakespeare's time onwards, European historians have recognised in the first salvoes of the +14th-century bombards the death-knell of the castle, and hence of Western military aristocratic feudalism. It would be tedious to enlarge on this here. In one single year (+1449) the artillery train of the King of France, making a tour of the castles still held by the English in Normandy, battered them down, one after another, at the rate of five a month.[b] Nor were the effects of gunpowder confined to the land. They also had profound influence at sea, for in due time they gave the death-blow to the multi-oared war galley of the Mediterranean, which was unable to provide sufficient space for the numerous heavy guns carried on the full-rigged ships of the North Sea and the Atlantic.[c] Chinese influence on Europe even preceded gunpowder by a century or so, because the counter-weighted trebuchet,[d] an Arabic improvement on the projectile-hurling device most characteristic of China (the *phao*[1]), was also most dangerous for even the stoutest castle walls.

Here the contrast with China is particularly noteworthy. The basic characteristics of bureaucratic feudalism remained after five centuries of gunpowder weapons just about the same as they had been before the invention had developed. The birth of this form of chemical warfare had occurred before the end of the Thang, but it did not find wide military use before the Wu Tai and Sung, and its real proving-grounds were the wars between the Sung empire, the Chin Tartars, and the Mongols, from the +11th to the +13th centuries. There are plenty of examples of its use by the forces of agrarian rebellions, and it was employed at sea as well as on land, in the siege of cities no less than in the field. But since there was no heavily armoured knightly cavalry in China, nor any aristocratic or manorial feudal castles either, the new weapon simply supplemented those which had been in use before, and produced no perceptible effect upon the age-old civil and military bureaucratic apparatus, which each new foreign conqueror had to take over and use in his turn, if he could. If he

---

[a] An echo here of the Buddhist missionaries to the Hellenistic world, who came with 'healing herbs, and yet more healing doctrine' (Vol. 1, p. 177).

[b] Oman (1), vol. 2, pp. 226, 404.

[c] Cf. Guilmartin (1), pp. 39, 175. On this Gibson (2) is still well worth reading.

[d] Cf. Hollister-Short (1) for a look at all the machinery which developed from it in later times.

[1] 砲

could not, his dynasty would not last very long, and always the Confucian bureaucracy, with their more or less obedient military inferiors, were ready to sweep back and run the country as it had been run from the very beginning of the Empire.

Finally, the sting in the tail, which shows once again how unstable Western medieval society was in comparison with that of China, is the foot- (or boot-) stirrup (têng[1]). As we shall see (Vol. 5, pt. 8 below), after many discussions involving the nomadic Central Asian peoples, the conclusion now is that it was a Chinese invention, for tomb-figures of about +300 clearly show it,[a] and the first textual descriptions come from the following century (+477), about which time there are numerous representations, Korean as well as Chinese.[b] Foot-stirrups did not appear in the West (or Byzantium) till the +8th century, but their sociological influence there was quite extraordinary.[c] The foot-stirrup welded the horseman and the horse together, and applied animal-power to shock combat. Such riders, equipped with the spear or the heavy lance, and more and more enveloped in metal armour, came in fact to constitute the familiar feudal chivalry of nearly ten European medieval centuries; that same body of knights which the Mongolian archers overcame on the field of Liegnitz. There is no need to stress all that the equipment of the knights had meant for the institution of medieval military aristocratic feudalism. Thus one can conclude that just as Chinese gunpowder helped to shatter this form of society at the end of the period, so Chinese stirrups had originally helped to set it up. But the mandarinate went on its way century after century unperturbed, and even at this very day the ideal of government by a non-hereditary, non-acquisitive, non-aristocratic élite holds sway among the thousand million people of the Chinese culture-area.

The social effects of gunpowder have of course often been meditated. A great Victorian writer, H. T. Buckle,[d] saw its chief effect in 1857 as the professionalisation of warfare. Gunpowder technology was complicated and difficult to handle, therefore there inevitably arose a separate military profession, and ultimately standing armies; no longer was every man potentially a soldier. Hence there occurred a reduction in the proportion of the population entirely devoted to war, with the result that more people were shunted into peaceful arts, techniques and employment, hence also a 'diminution of the warlike spirit, by diminishing the number of persons for whom the practice of war was habitual'. Gunpowder technology was also expensive, more so than any individuals could afford, so only wealthy republics, or kings backed by merchants and endowed

---

[a] See Kao Chih-Hsi, Liu Lien-Yin et al. (1); Yang Hung (1), p. 101 and figs. 31, no. 1, 81, no. 5; and Needham (47), pp. 268 ff., pl. 20; and further in Vol. 5, pt. 8. The figures are mounted military bandsmen, from the Chhangsha tomb of a Chin general dating from +302, and it seems clear that their foot-stirrups were used primarily for mounting, because hanging at the front of the saddle on the left side only.

[b] Lynn White (7), p. 15.   [c] Ibid. pp. 28 ff.

[d] (1), vol. 1, pp. 185 ff. We are much indebted to Dr Elinor Shaffer for drawing our attention to his ideas.

[1] 鐙

with rich estates, could manufacture, own and operate musketry and artillery.[a] Hence the rise of what Buckle called the 'middle intellectual class', so that 'the European mind, instead of being, as heretofore, solely occupied with either war or theology, now struck out into a middle path, and created those great branches of knowledge to which modern civilisation owes its origin'.

As a description of one aspect of the rise of the bourgeoisie this was all well said, but Victorian optimism erred only in the belief that the situation would last. It might have been better to note what Robert Boyle had said in his *Usefulnesse of Experimental Natural Philosophy* (1664). Speaking of 'Engines so contriv'd, as to be capable of great Alterations from slight Causes', he wrote:[b]

> The faint motion of a mans little finger upon a small piece of Iron that were no part of an Engine, would produce no considerable Effect; but when a Musket is ready to be shot off, then such a Motion being applied to the Trigger by virtue of the contrivance of the Engin, the spring is immediately let loos, the Cock fals down, and knocks the Flint against the Steel, opens the Pan, strikes fire upon the Powder in it, which by the Touch-hole fires the Powder in the Barrel, and that with great noise throws out the ponderous Leaden bullet with violence enough to kill a Man at seven or eight hundred foot distance.

Thus a single touch could already mean life or death; and the touch would in time be open to everyone. It might have been wiser to foresee that science and technology would, as time went on, and by the very impetus of the industrial revolution itself, which Buckle so much admired,[c] immensely improve, and enormously cheapen, the production of these lethal weapons, not only on the mechanical side but also on the chemical, producing a vast variety of explosives which would come within the reach of almost every man, whether dubbed 'terrorist' or 'freedom-fighter'. History has passed through a complete cycle, and alas, once again, 'every man is potentially a soldier'. This is our plight today, and nothing but universal social and international justice will relieve it.

(2) THE HISTORICAL LITERATURE

(i) *Primary sources*

The fundamental authorities for the gunpowder epic are the Chinese military compendia. The earliest mention of gunpowder in this genre of writing, and of fire-weapons depending upon it, can be found in the *Wu Ching Tsung Yao*[1] (Col-

---

   [a] Bernal (1), p. 238.   [b] (8), pt. 2, p. 247.

   [c] It is true, as Nef (1) has shown in a classical work, that the industrial revolution itself was connected with peaceful development much more than with war, and that large-scale factory production was only partially stimulated by military demands, but the trend towards ever greater cheapness, efficiency and abundance of lethal weapons was surely implicit in modern science and technology from the first, if uncontrolled by enlightened world government.

   [1] 武經總要

lection of the Most Important Military Techniques), compiled under an order from the Sung emperor in +1040 and completed in the year +1044, under the editorship of Tsêng Kung-Liang[1] with the assistance of the Astronomer-Royal Yang Wei-Tê[2].[a] It is one of no less than 347 titles of military works listed in the bibliographical chapters of the *Sung Shih*[b], but apart from some fragments of a few other similar works incorporated in the *Yung Lo Ta Tien*[3], it is now, with the *Hu Chhien Ching*[4] (Tiger Seal Manual) written by Hsü Tung[5] in +1004, the only substantial Sung military writing extant.[c]

Of course, the *Wu Ching Tsung Yao* was not the first military treatise to speak about attack by fire. There is plenty about this in the ancient books on warfare, even though it may still be debatable how far the incendiary arrow had developed by the time of the *Sun Tzu Ping Fa*[d] and the *Mo Tzu*.[e] But by the Thang period fire-arrows (*huo chien*[6], *huo shih*[7]) had become a commonplace, as appears from Li Chhüan's book *Thai Pai Yin Ching* (Manual of the White (and Gloomy) Planet),[f] which we have already described (Vol. 5, pt. 6 above); the oldest of the military encyclopaedias still available. Dating from +759, it contains no word on gunpowder or anything remotely like it.

All the other important military compendia of the Sung are now lost. Among the works of the early Southern Sung were the *Yü Chhien Chün Chhi Chi Mo*[8] (Imperial Specifications and Models for Army Equipment) of unknown authorship, the *Wu Ching Shêng Lüeh*[9] (Essence of the Five Military Classics, for Imperial Consultation) by Wang Shu[10], the *Chung Hsi Pien Yung Ping*[11] (Military Practice on the Central and Western Fronts) by Fang Pao-Yuan[12], and two other works both by anonymous writers, i.e. the *Tsao Chia Fa*[13] (Treatise on Armour-Making) and the *Tsao Shen Pei Kung Fa*[14] (Treatise on the Making of the Strong Bow). The loss of the first of these books is particularly lamentable, as it would have filled the great gap between the *Wu Ching Tsung Yao* and the *Huo Lung Ching*. Another of the missing books mentioned in the *Sung Shih*, the *Phao Ching*[15] (Trebuchet Manual), would have been of much interest as it might have thrown light on the uses of gunpowder which led to the term 'fire trebuchet' (*huo phao*[16, 17]).

The original text of the *Wu Ching Tsung Yao* was preserved in the Imperial

---

[a] Who was responsible for the details on military prognostication in the Hou Chi; cf. Franke (24), p. 195.

[b] Ch. 207, pp. 3a–6a. Of course, many of those listed were pre-Sung.

[c] See also Arima (1), p. 28; and Wang Hsien-Chhen & Hsü Pao-Lin (1).

[d] The texts are notoriously difficult, and commentators have had a variety of opinions. For translations, see Griffith (1), p. 141; L. Giles (11), pp. 151–2; Machell-Cox (1), p. 50.

[e] Cf. Yates (1), pp. 152 ff.

[f] Descriptions of fire-arrows occur in ch. 4 (ch. 35), p. 2b (ch. 38), p. 8b, and ch. 5 (ch. 46), p. 2b. There is much also on arcuballistae and trebuchets, e.g. ch. 4 (ch. 35), pp. 1b ff.; and molten iron as a weapon in sieges is mentioned in ch. 4 (ch. 35), p. 4a.

---

[1] 曾公亮 [2] 楊惟德 [3] 永樂大典 [4] 虎鈐經 [5] 許洞
[6] 火箭 [7] 火矢 [8] 御前軍器集模 [9] 五經聖略
[10] 王洙 [11] 中西邊用兵 [12] 方寶元 [13] 造甲法 [14] 造神臂弓法
[15] 礮經 [16] 火砲 [17] 火炮

Library (Chhung Wên Yuan). A limited number of hand-written copies of the compendium could have been made, because we read in the *Sung Shih* that in the year +1069 the emperor gave copies of several military compendia, one of which was the *Wu Ching Tsung Yao*, to Wang Shao[1]. In the year +1126 the Sung capital fell and the Imperial Library lost all its books. Thus the original of the *Wu Ching Tsung Yao* disappeared, but a few copies still existed in different parts of China, though as it was a military text the book was not reproduced in large numbers for security reasons. But there was certainly an edition in +1231. During the Ming period the book was printed several times and in the +18th century it was included in the *Ssu Khu Chhüan Shu*. At present there are the following editions:

(*a*) A reprint of a Ming edition produced during the Hung-Chih and Chêng-Tê reign-periods (+1488 to +1521). A rare copy of this Ming edition, usually assigned to +1510, once belonged to the eminent archaeologist Chêng Chen-To[2] and from this it was reprinted at Shanghai in 1959. It is undoubtedly the most reliable version of the *Wu Ching Tsung Yao* still available to us, since it was made from blocks re-carved directly from tracings of the +1231 edition.[a]

(*b*) The Chia-Ching (+1522 to +1566) edition.

(*c*) The Wan-Li (+1573 to +1619) edition produced in Chhüan-chow[3].

(*d*) The Wan-Li (+1573 to +1619) edition produced in Chin-ling[4] by Thang Hsin-Yün[5] and preserved in the Tsun-ching-ko[6].

(*e*) Another possibly Wan-Li edition produced in Shan-hsi-fu[7] under the title *Wu Ching Yao Lan*[8] (Essential Readings in the most important Military Techniques).

(*f*) Chhing edition produced by the Fu-chhun-thang[9] of Chin-ling and preserved in the Tsun-ching-ko.

(*g*) Chhing edition produced by the Ching-chia-thang[10].

(*h*) The *SKCS* Wên-su-ko edition.

(*i*) The *SKCS* Wên-yuan-ko edition[b] reproduced at Shanghai, in 1934.

The book consists of two collections, Chhien Chi[11] and Hou Chi[12], the former being by far the more important, as it deals with all kinds of military equipment, weapons and machines, while the latter recounts stories of battles and combats, together with principles of strategy and tactics, drawn from history and tradition.

Besides all these editions, it is possible to find some curious partial printings of the *Wu Ching Tsung Yao*. In 1952 I purchased from a bookshop in the Liu-Li Chhang in Peking a copy of what seemed to be a very early edition of the book,

---

[a] This is certain, if only from the Sung tabu characters which appear in it. And it has the colophon that Chao Wei-Thing[13] wrote for it in +1231.

[b] A copy of this was presented to the East Asian History of Science Library, Cambridge, by the late Dr Kuo Mo-Jo and Dr Thao Mêng-Ho on behalf of Academia Sinica in 1955.

[1] 王韶   [2] 鄭振鐸   [3] 泉州   [4] 金陵   [5] 唐心雲
[6] 尊經閣   [7] 山西府   [8] 武經要覽   [9] 富春堂   [10] 靜嘉堂
[11] 前集   [12] 後集   [13] 趙魏挺

with a preface of +1439, and I presented it to the library of Academia Sinica. In this peculiar version the first ten chapters of *WCTY* were replaced by two other books, the *Hsing Chün Hsü Chih*[1] (What an Army Commander in the Field should Know) written by an unknown author about +1260 but with a preface by Li Chin[2], and the *Pai Chan Chhi Fa*[3] (Wonderful Methods for Victory in a Hundred Combats) of similar date and equally unknown authorship.[a] Then *WCTY* began suddenly in the middle of ch. 11, and omitted the second half of ch. 12, but apart from a few smaller gaps, the rest was apparently complete. Further investigation[b] showed that Li Chin's preface applied only to the *Hsing Chün Hsü Chih*, and therefore that the +1439 date could not apply to *WCTY*; though the connection between the two books was quite close, since the +1510 edition of the latter had both the preface and the book about army commanders suffixed to it.[c] Moreover, what there was of the *WCTY* text and illustrations turned out to be identical with those of the +1510 edition. Then on the backs of some of the pages there are fragments of much later works, notably one by Hui Tung[6] (+1697 to +1758). Therefore the whole thing must have been put together by some printer or book-dealer not earlier than his time, using miscellaneous old blocks and not caring too much whether they fitted together perfectly or not. So this version was a late jumble, and there was no edition of *WCTY* in +1439.

According to the *Ssu Khu Chhüan Shu Thi Yao*, the compilers of the *SKCS* knew of only one version of the *Wu Ching Tsung Tao*. Unfortunately it seems that the +18th-century editors tampered mildly with the work in a feeble attempt to up-date the +11th-century material, adding two illustrations of metal-barrel cannon. These are, of course, gross anachronisms, easily betrayed, moreover, by the fact that no description of the weapons was inserted at the time when the drawings were added. Arima noted that these pictures do not appear in the *Wu Ching Yao Lan*.[d] So in both the *SKCS* editions, there are illustrations of metal-barrel cannons, namely the 'mobile gun-carriage' (*hsing phao chhê*[7]) and the 'high-fronted cannon-cart' (*hsien chhê phao*[8]).[e] It seems to us that the reason for their insertion at this particular place was because of the slanting mobile bridge equipment, carriages and scaling-ladders shown near by, and that put the editors in mind of the frames or carriages of cannons also slanted for howitzer-style aiming.[f]

The 'squatting-tiger cannon' (*hu tun phao*[9]) arises here as a good example of the

---

[a] But it has a preface written by Li Tsan[4] in +1504. These two books are interesting because the term *huo thung*[5], 'fire-tube', appears several times in them, referring probably to fire-lances or eruptors, as we shall see (p. 230), but perhaps also to metal-barrel guns or bombards (p. 276).

[b] For part of which we are greatly indebted to the late Dr Fêng Chia-Shêng in Peking.

[c] Both these Sung books were generally appended to the Ming editions of *WCTY*, but neither has been reprinted in our own time, nor the Hou Chi of *WCTY* either.

[d] (1), pp. 60 ff.

[e] *WCTY/CC*, ch. 10, pp. 13a, 13b respectively. We reserve the illustrations for Figs. 77, 79.

[f] Goodrich & Fêng (1), pp. 116–7 recognised the anachronism of these cannon, but thought that the pictures of trebuchets were also late, which was not the case.

[1] 行軍須知　　[2] 李進　　[3] 百戰奇法　　[4] 李贊　　　[5] 火筒
[6] 惠棟　　　　[7] 行砲車　　[8] 軒車砲　　　[9] 虎蹲砲

way in which the thing changed fundamentally while the terminology did not. The Ming edition of the *Wu Ching Tsung Yao* contains a diagram of a trebuchet under that name.[a] On this Mao Yuan-I had the following to say in +1628:[b]

The Sung people used the turntable trebuchet (*hsüan fêng phao*[1]), the single-pole tre-buchet (*tan shao phao*[2]) and the squatting-tiger trebuchet (*hu tun phao*). They were all called 'fire trebuchets' (*huo phao*[3]) because they were used to project fire-weapons like the (fire)-ball (*huo chhiu*[4]), (fire)-falcon (*huo yao*[5]) and (fire)-lance (*huo chhiang*[6]).[c] They were the ancestors of the cannon (*phao chih tsu*[7]).

Thus the *hu tun phao* was at first a kind of trebuchet. Later on, perhaps about the middle of the +14th century, when Chiao Yü[8] wrote the *Huo Lung Ching*, where it appears,[d] the same name was given to another weapon, an early form of Chinese iron cannon, almost an eruptor, with many projectiles. In +1571 Chhi Chi-Kuang[9] described it again under the same name in his *Lien Ping Shih Chi, Tsa Chi*.[e]

The version closest to the original Sung book of +1044 is, according to Arima (1), the Ming edition entitled *Wu Ching Yao Lan*, the only copy of which is pre-served in the library of the Bōei Daigakkō[10] Military Academy in Japan. How-ever, the +1510 edition (copying that of +1231) was not known when Arima wrote his book.

A quite different genre of literature, which is nevertheless also of great impor-tance for the history of gunpowder weapons, is that which deals with what might be called practical poliorcetics; in other words, eye-witness accounts of some of the great sieges in Chinese history.[f] Here a few examples may suffice. From +1127 to +1132 Chhen Kuei[11] held the city of Tê-an (half-way between the Huai and the Yangtze Rivers) for the Sung against the Jurchen Tartars, and after-wards he wrote a book about it entitled *Shou Chhêng Lu*[12] (Guide to the Defence of Cities). Later on, a military officer named Thang Tao[13] went through all the records again, and wrote another book on the same siege with the title *Chien-Yen Tê-an Shou Yü Lu*[14] (Account of the Defence and Resistance of Tê-an City in the Chien-Yen reign-period).[g] Then in +1225 the two works were combined under Chhen Kuei's title, Thang Tao's text becoming chs. 3 and 4.[h] This was the book

---

[a] Ch. 12, p. 45*a*.
[b] *Wu Pei Chih*, ch. 122, p. 4*a*; tr. auct. He took the lists directly from *WCTY/CC*, ch. 12, p. 50*a* (Ming ed.).
[c] As we shall see, the term *huo chhiang* (fire-lance) was normally used for flame-throwers filled with low-nitrate gunpowder, but it also occurs in names of rockets (cf. e.g. *Wu Pei Chih*, ch. 128, pp. 16*b*, 17*a*), and here it must mean a projectile, presumably containing rocket composition and flaming at both open ends.
[d] In pt. 1, ch. 2, p. 3*a, b*.   [e] Ch. 5, pp. 19*a*–21*a*. Cf. p. 277 and Fig. 75.
[f] See the valuable discussion of H. Franke (24); and here, Vol. 5, pt. 6.
[g] Thang Tao's book probably embodied one with the same title which had been produced by Liu Hsün[15] in +1172.
[h] Cf. Balazs & Hervouet (1), p. 237. There is a special study of the whole work by Mikami Yoshio (21).

[1] 旋風砲   [2] 單梢砲   [3] 火砲   [4] 火毬   [5] 火鶴
[6] 火槍   [7] 砲之祖   [8] 焦玉   [9] 戚繼光   [10] 防衛大學校
[11] 陳規   [12] 守城錄   [13] 湯璹   [14] 建炎德安守禦錄
[15] 劉荀

which first gave a clear description of the *huo chhiang*[1] or fire-lances, five-minute flame-throwers filled with rocket composition (low-nitrate powder), though we now believe that this weapon had been invented at a much earlier date.[a]

Then, very nearly a century later, a second celebrated siege occurred at the same place. In his *Khai-Hsi Tê-an Shou Chhêng Lu*[2] (Account of the Defence of Tê-an in the Khai-Hsi reign-period, +1206 to +1207), Wang Chih-Yuan[3], son of the chief defender, Wang Yün-Chhu[4], gave the details of the action, in which the Jurchen Chin troops under Wanyen Khuang[5] had been unable to wrest the city from the Sung.[b] This was in the war which had been precipitated by the Sung side's premier Han Tho-Chou[6], a leader of the war party, and the opponent of the philosopher-politician Chu Hsi[7].

Next comes the *Hsiang-yang Shou Chhêng Lu*[8] (Account of the Defence of Hsiang-yang City) in the same campaign, and the same years, +1206 and +1207. This again held the city for the Sung against the Jurchen Chin,[c] and should not be confused with the still more famous siege of +1268 to +1273 when it eventually fell to the Yuan Mongols. And as in the case of Tê-an, the book was written by Chao Wan-Nien[9], the son of the general commanding the defence, Chao Shun[10].

Finally, mention may be made of the *Pao Yüeh Lu*[11] (Defence of the City of Shao-hsing), due to Hsü Mien-Chih[12], which described the gallant defence of this fortified place by Lü Chen[13] (Lü Kuo-Pao[14]) for the cause of Chang Shih-Chhêng[15] against the generally victorious troops of Chu Yuan-Chang[16] in +1358–9.[d] By this time gunpowder is very much in evidence, and there is much on the 'fire-tubes' (*huo thung*[17]) which by this time must have meant metal-barrel hand-guns and bombards.[e] All in all, this poliorcetic literature cannot be neglected in the study of the beginnings of gunpowder weapons and firearms.

We know little about writings on military matters published during the Mongol period. Sung Lien[19] and his colleagues did not include a bibliographical chapter when they compiled the official history of the Yuan Dynasty about +1367, nor did the *Ssu Khu Chhüan Shu* mention any work of this kind written during that period. But the *Pu Liao Chin Yuan I Wên Chih*[20] originated by Ni Tshan[21] and continued by Lu Wên-Chao[22] did list more than ten military books, among which

---

[a] Cf. pp. 222 ff. below.

[b] This is the only one of these books of which an integral translation has been published—by Korinna Hana (1).

[c] There is a short paper on it by Franke (25).

[d] See Franke (24), p. 188. There is a valuable unpublished translation of this book by H. Franke (23). The Ming siege army was commended by Hu Ta-Hai[18].

[e] After all, it was about seventy years after the first known Chinese example and the textual evidence associated with it.

[1] 火槍	[2] 開禧德安守城錄	[3] 王致遠	[4] 王允初
[5] 完顏匡	[6] 韓托冑	[7] 朱熹	[8] 襄陽守城錄	[9] 趙萬年
[10] 趙淳	[11] 保越錄	[12] 徐勉之	[13] 呂珍	[14] 呂國寶
[15] 張士誠	[16] 朱元璋	[17] 火筒	[18] 胡大海	[19] 宋濂
[20] 補遼金元藝文志	[21] 倪燦	[22] 盧文炤	

there is one with the title *Huo Lung Shen Chhi Thu Fa*[1] (Fire-Drake Illustrated Technology of Magically (Efficacious) Weapons).[a] It has long been lost, but if it was the predecessor of, or the model for, the work entitled *Huo Lung Ching*[2], which we shall have to discuss in detail on many following pages, it might well take the content of that back from +1412 by a whole century or even more, perhaps to the neighbourhood of +1270 or so. The paucity of military compendia during the reign of the Mongols might be accounted for either by their lack of interest in literary pursuits, or on the other hand a fear among the people of publishing anything that might arouse suspicion among the Mongols that preparations for a rebellion were going on. It is also quite possible, even likely, that new weapons were being designed in secret towards the later part of the Yuan Dynasty. Otherwise it is difficult to see why so many new fire-weapons suddenly emerged in early Ming.

The next series of Chinese military compendia came indeed from that dynasty. The historians of the *Ming Shih* listed fifty-eight titles in the sub-section on military writings in the bibliographical chapters. However, their knowledge of military books in the period they were writing about could not have been very complete, because they omitted most of the titles on the subject given by Chiao Hsü[3] in the preface of his *Huo Kung Chieh Yao*[4] in +1643 (p. 310), in spite of having mentioned the same work in the Ming official history themselves.

Chiao Hsü mentions three military books belonging to the early Ming period, namely the *Huo Lung Ching*[5] (Fire-Drake Manual), the *Chih Shêng Lu*[6] (Records of the Rules for Victory), and the *Wu Ti Chen Chhüan*[7] (Reliable Explanations of Invincibility). But the only military work of the early Ming still available to us is the first of these, the 'Fire-Dragon Manual'. This book is especially important because it comes from the +14th century, while all the other Ming military texts still extant belong to the +16th century.

Many books and articles have been written on the development of gunpowder and firearms, but with the exception of Fêng Chia-Shêng and Arima Seihō, no one seems to have referred to this interesting mid-14th-century book. It seems to have been practically unknown to all Western writers on the subject of fire-weapons or gunpowder. The version used by Arima (*1*) bears the title *Wu Pei Huo Lung Ching*[8]. There are several other different versions of the 'Fire-Drake Manual', but all are rare; for example, a modern catalogue of Chinese military books lists only one of them.[b] Since no one has yet made a comparison of the texts, it is necessary to go into this question in some detail.

---

[a] The title is strange, because the expression *shen chhi* came to be applied specifically to metal-barrel guns and light cannon towards the end of the Ming, and it does not usually occur so early. But as we shall see (p. 346), weapons using the propellant force of high-nitrate gunpowder did originate before the end of the Sung, so the term may have dropped out of use and been revived much later. Or some MSS may have been re-titled at a subsequent date.

[b] See Lu Ta-Chieh (*1*), p. 3.

[1] 火龍神器圖法 [2] 火龍經 [3] 焦勗 [4] 火攻絜要
[5] 火龍經 [6] 制勝錄 [7] 無敵眞銓 [8] 武備火龍經

In the course of visits to China during the past thirty-five years I succeeded in obtaining four different texts with a more or less similar title, *Huo Lung Ching*. These are as follows:

(a) *Huo Lung Ching* (Fire-Drake Manual), printed from blocks preserved in Hsiang-yang, bearing the words '*Hsiang-yang-fu tshang pan*'[1]. It carries the running title *Huo Chhi Thu*[2] (Illustrations of Fire-arms) and by this it is often quoted; it contains no preface and does not give the year of publication. It is attributed anachronistically to the +3rd-century Captain-General of Shu, Chuko Liang[3],[a] and edited by two early Ming personalities, Liu Chi[4] and Chiao Yü[5], then re-edited by Li Thien-Chên[6] of Chhien-chiang[7]. The text includes quotations from Liu Chi[b] and Chiao Yü.[c]

(b) *Huo Lung Ching Chhüan Chi*[8] (Fire-drake Manual in One Complete Volume), the Nanyang version, bearing the words '*Nanyang shih-shih tshang pën*'[9]. It contains a preface by Chiao Yü dated +1412, but gives no year of publication; otherwise its text is more or less similar to that of the Hsiang-yang-fu version. The anachronistic attribution to Chuko Liang (Chuko Wu-Hou[10]) is also prominent. The Tōyō Bunko has a copy of this book under the simple title *Huo Lung Ching*.

(c) *Huo Lung Ching Erh Chi*[11] (Fire-Drake Manual, Second Part), compiled by Mao Hsi-Ping[12] and carrying a preface by him written in the year

---

[a] Often known by his other name, Chuko Khung-Ming.[16] His association with gunpowder weapons was a widespread folk tradition, and it misled all the early Western sinologists into believing that these were Han in origin. To say nothing of Amiot, Cibot, Gaubil and de Mailla, one may cite Grosier (1), vol. 7, pp. 176 ff., Castellano & Campbell-Thompson tr., pp. 105 ff.; and Williams (1), vol. 2, pp. 89 ff.

[b] The appearance of Liu Chi (+1311 to +1375) here is of great interest, for he was a striking personality, of remarkable qualities both civil and military. In philosophy he was a sceptical naturalist, interested in all kinds of science and proto-science—astronomy, the calendar, magnetism and geomancy—and a friend of the eminent mathematician and alchemist Chao Yu-Chhin[13] (cf. Vol. 5, pt. 3, p. 206). But he was also concerned with administration, and for long an adviser to the first Ming emperor. In war he commanded at battles both on land and afloat, having in one instance (+1363) his flagship destroyed by a 'flying shot' (*fei phao*[14]) just after he had transferred to another vessel (*Ming Shih*, ch. 128, p. 6a, Forke (9), p. 307). Thunder was simply, Liu Chi said in one place, 'like fire shot from a *phao* (*yu huo chih chhu phao*[15])', (Chung Thai (1), vol. 2, p. 79). These may just have been references to trebuchets and explosive bombs thrown from them, but by this time, the mid +14th century, it is really much more likely that metal-barrel cannon were meant.

Already in Vol. 1, p. 142, we surmised that gunpowder firearms played a particularly important part in the triumph of Chu Yuan-Chang and the founding of the Ming dynasty; the present Section not only confirms what we then wrote, but goes a long way beyond it. Unfortunately, Liu Chi's biography (*Ming Shih*, ch. 128, pp. 1a ff.) is on the whole purely political, with only incidental references to his scientific and technological interests. Among these, gunnery must certainly have been one. The best authority on him was Chung Thai (1), whose book was much used for the account in Forke (9), pp. 306 ff.

Liu Chi was the sort of man who could successfully conjure a change in the wind just when the commander-in-chief needed it. This desideratum was not available to all Shakespearean armies, although in the play of Shaw success attends the prayers of St Joan. Cf. Dreyer (2), pp. 228, 359; Chhen Ho-Lin (1).

[c] This text was afterwards reprinted, sometimes in condensed form, as by the Wên Hui Thang[17] towards the end of the nineteenth century.

[1] 襄陽府藏版	[2] 火器圖	[3] 諸葛亮	[4] 劉基	[5] 焦玉
[6] 李天楨	[7] 潛江	[8] 火龍經全集	[9] 南陽石室藏本	
[10] 諸葛武侯	[11] 火龍經二集	[12] 毛希秉	[13] 趙友欽	[14] 飛礮
[15] 猶火之出礮	[16] 諸葛孔明	[17] 文滙堂		

+1632.[a] Its text differs widely from the First Part in the Hsiangyang-fu[1] and the Nanyang[2] versions. It talks about the bird-beak musket, *niao chhung*[3], and the *fo-lang chi*[4] breech-loading cannon,[b] which do not appear at all in the first two texts.[c] Its sections on the 'making of fire-weapons' and the 'testing of fire-weapons' are somewhat similar to the corresponding sections in the *Wu Pei Chih*.[d]

The Tōyō Bunko possesses a copy of this book under the title *Huo Lung Ching*.

(d) *Huo Lung Ching San Chi*[7] (Fire-Drake Manual, Third Part), another Nanyang publication bearing the words '*Nanyang Lung-chung chen tshang*[8]'. It was compiled by one Chuko Kuang-Jung[9]. It gives no year of publication, but it cannot have been written before the early +17th century since it quotes Mao Yuan-I, the author of the *Wu Pei Chih*. Again its text differs widely from the Hsiang-yang-fu and Nanyang versions of Pt. 1. A copy of this book also is in the Tōyō Bunko.

(e) *Huo Kung Pei Yao*[10] (Essential Knowledge for the Making of Gunpowder Weapons), reprinted in the year 1884 and bearing the words '*Tun Huai Shu Wu Chhung chien*'[11] showing that it derived from earlier blocks. It carries the preface by Chhiao Yü, and its text is similar to those in the Hsiang-yang-fu and the Nanyang versions of Pt. 1.

(f) Lastly comes the version of the 'Fire-Drake Manual' used by Arima (1) and entitled *Wu Pei Huo Lung Ching*[12]. It was produced in 1857 from an earlier first impression, bearing the words '*Pao Phu Shan Fang hsin hsien*'[13]; and it carries the preface by Chiao Yü. It appears that this book is available only in Japan, in the Bōei Daigakkō[14] Military Academy.[e]

Hence there are at least three different portions of Chiao Yü's 'Fire-Drake Manual'. The work should indeed be considered a main nucleus with two supplements, summarising the development of successive gunpowder weapons after about +1280. Chiao Yü had been, as we shall see, a leading artillery officer in the army of Chu Yuan-Chang which finally conquered China for the Ming in +1367. Arima noticed that the *Wu Pei Huo Lung Ching* contains later additions,

---

[a] Mao Hsi-Ping is quoted in the *Wu Pei Chih*, ch. 117, p. 11 *a*, *b*.

[b] Cf. Reid (1), pp. 12–13.

[c] But there is a mention of *niao chhiang*[5], i.e. bird-(beak fire-)lances, in Pt. 1, ch. 1, p. 11*b*, in connection with poisonous smoke attacks (*wu li wu*[6]). Either this was a later interpolation, or the 'bird-beak' epithet applied to a fire-lance before it applied to a true gun.

[d] Ch. 1, p. 24*b*, p. 26*b* and p. 27*a* resemble *Wu Pei Chih*, ch. 119, p. 4*b* to p. 6*a*.

[e] One of us (H. P. Y.) obtained a photocopy of this text through the courtesy of this institution. For a preliminary report see Ho Ping-Yü & Wang Ling (1). All the others are in the East Asian History of Science Library at Cambridge, and the University Library has a copy of (*e*). They have been indispensable sources for the account which here follows.

[1] 襄陽府   [2] 南陽   [3] 鳥銃   [4] 佛狼機   [5] 鳥鎗
[6] 五里霧   [7] 火龍經三集   [8] 南陽隆中珍藏   [9] 諸葛光榮
[10] 火攻備要   [11] 敦懷書屋重鐫   [12] 武備火龍經   [13] 抱樸山房新鐫
[14] 防衛大學校

for example, not only the Portuguese *fo-lang chi* breech-loading cannon, known and used in China from about +1510; but also the Japanese arquebus, known to the Chinese as the 'bird-beak gun' or *niao chhung* musket, which was not introduced to China via Japan until +1548. Hence this part of the 'Fire-Drake Manual' could not have been compiled before the middle of the +16th century. It was precisely because of the inclusion of these muskets and cannon in the text which he happened to come across that Fêng Chia-Shêng did not at first think much of the *Huo Lung Ching*.[a] Indeed the *Wu Pei Huo Lung Ching* must be later than +1628 because it mentions the *Wu Pei Chih* in several places. However, Arima rightly believed that the original text must have come from about the middle of the +14th century, especially the preface by Chiao Yü, which includes, as we shall see, a reference to events of +1355 in which he himself participated. Obviously there were many later additions to the text.

Chiao Yü himself manufactured firearms for the first Ming emperor during the middle of the +14th century, and he was eventually put in charge of the Shen Chi Ying[1] armoury, where all the guns and artillery were deposited and kept secret. Although the Ming dynastic history does not contain his biography, Chiao Yü is mentioned in Chao Shih-Chên's[2] *Shen Chhi Phu*[3] (+1598) and Chiao Hsü's[4] *Tsê Kho Lu*[5] (otherwise known as *Huo Kung Chhieh Yao*[6]) of the year +1643.[b] His name is also referred to by Ho Ju-Pin[7] in the *Ping Lu*[8] in +1606. Arima (*1*) concluded that much of the text of the 'Fire-Drake Manual' must have been written by Chiao Yü in the middle of the +14th century. This is of great importance when one remembers the key date of +1327 for the first picture of a bombard in Europe.

In the Preface Chiao Yü says that there were no firearms during Han times, but Chuko Liang[11] (in the +3rd century) met an extraordinary person who revealed to him the secrets of attacking with fire. Chiao himself met an adept named Chih-Chih Tao-Jen[12], who told him to support Chu Yuan-Chang[13], and gave him a book on fire-weapons and their uses. Chiao Yü presented to Chu Yuan-Chang several fire-weapons which he had cast according to his teacher's instructions. Chu ordered Hsü Ta[14] to prove them, and himself watched the tests, which pleased him much. After the conquest of the Mongols standard gunpowder factories were established in the capital, and arsenals were made to keep the 'magical weapons'. Thus gunpowder weapons were an important factor in the rise to imperial power of Chu Yuan-Chang.

---

[a] Private correspondence with Dr Fêng. See also pp. 440ff. below.
[b] In the biographical section on meritorious officials in the *Ming Shih*, there is a man named Chiao Cha[9]. He is the only person to be found bearing the title Tung-Ning Po[10] (Count of Tung-ning), an appellation which Chiao Yü also bore, so it must have been the same family, and perhaps Chiao Cha was the gunner's father or grandfather. On Chiao Yü's life and writings there is an interesting study by Chhêng Tung (*1*).

[1] 神機營	[2] 趙士禎	[3] 神器譜	[4] 焦勖	[5] 則克錄
[6] 火攻挈要	[7] 何汝賓	[8] 兵錄	[9] 焦札	[10] 東寧伯
[11] 諸葛亮	[12] 止止道人	[13] 朱元璋	[14] 徐達	

We translate the Preface in full as follows:[a]

In the days of old when the Yellow Emperor fought the battle at Cho-lu[1], he had Fêng Hou[2] as his teacher; when Yü[3] the Great waged war on the San Miao[4] (tribes) he had Po I[5] as his teacher; at the battle of Ming-thiao[6] Chhêng Thang[7] had a teacher in I Yin[8], and during his invasion of Mu-yeh[9] (King) Wu Wang[10] had a teacher in Lü Wang[11]. Such was the beginning of military tactics. When both sides have equal strength one side can win if it has superior virtue: in the case of equal virtue the righteous (i[12]) side will win.[b] (The ancient victors) resonated with the mandate of Heaven above, and abided by the will of the people below. Then when it came to the Spring-and-Autumn period there were struggles among the Five Hegemons, and during the time of the Warring States the Seven Powers waged war among themselves, endangering the lives of the people—there was hardly a single day of peace. Yet we learn no details concerning the deployment of fire in battle.[c] Then, with Chang Liang[13] as his teacher, the (Han emperor) Kao Tsu[14] fought at the battle of Ssu-shang[15], brought about the doom of Hsiang (Yü[16]) and founded the empire (of Han). (The emperor) Kuang Wu (-Ti)[17] began his campaign at Khun-yang[18] with Têng Yü[19] as his teacher, and suppressed (Wang) Mang[20] to restore the dynasty of Han. But again, nothing concerning fire-weapons (in those days) has been heard of.

When it came to the time of the Three Kingdoms we saw the rise of many tactician-advisers and great soldiers. Tshao Tshao[21] with villainous might controlled the central part of the empire, while Sun Chhüan[22], inheriting from his father and elder brother, firmly occupied the eastern part of the empire around the Yangtze River. No one else could match their power. At that time, when the 'Crouching Dragon' (i.e. Chuko Liang[23]) was farming in Nanyang, without any desire to seek fame, he met an extraordinary man who secretly taught him the use of fire in warfare and the tactics of battle formations. Then, touched by the sincerity of the First Ruler (of the Shu Han[24] Kingdom, i.e. Liu Pei[25]), who thrice visited him, he exerted every ounce of his strength to serve him. He set the military farms ablaze in Po-wang[26]; he deployed his troops at Chhih-pi[27]; and he burnt (the soldiers of Mêng Huo[28] by setting fire to) the rattan armour (worn by them). He attacked Shang-fang[29] and led an expedition beyond the Chhi-shan[30] mountains. All this resulted in a partition of the Empire into three Kingdoms.

(Chuko Liang) won every battle that he fought. His tactics baffled his enemies more and more, frightening Tshao Tshao out of his wits, and Sun Chhüan too. Incendiary techniques in warfare reached perfection in the hands of Khung-Ming[31] (Chuko Liang). As for his mine-setting[d] in (the Battle of) Hu-lu-ku[32] valley, both Ssuma (I[33] and Ssuma

[a] From the *Huo Kung Pei Yao* version of the 'Fire-Drake Manual'; tr.auct.

[b] The word 'virtue' here can also be interpreted as 'element', in which case the Law of Mutual Conquest comes into play. See Vol. 2, p. 256.

[c] This suggests that Chiao Yü had little literary learning, or he would hardly have ignored the *Sun Tzu Ping Fa* here. Yet later he indirectly quotes the *Shih Chi* about Chao Shê and Chao Kua.

[d] This is another reference to the unacceptable tradition that Chuko Liang knew of gunpowder in the +3rd century, and used it to make land-mines.

[1] 涿鹿	[2] 風后	[3] 禹	[4] 三苗	[5] 伯益
[6] 鳴條	[7] 成湯	[8] 伊尹	[9] 牧野	[10] 武王
[11] 呂望	[12] 義	[13] 張良	[14] 高祖	[15] 泗上
[16] 項羽	[17] 光武帝	[18] 昆陽	[19] 鄧禹	[20] 王莽
[21] 曹操	[22] 孫權	[23] 諸葛亮	[24] 蜀漢	[25] 劉備
[26] 博望	[27] 赤壁	[28] 孟獲	[29] 上方	[30] 祈山
[31] 孔明	[32] 葫蘆谷	[33] 司馬懿		

Chao[1]), father and son, would have been burnt to ashes if it had not been for an (unexpected) sudden downpour of heavy rain. (It was mainly due to his efforts that) people were prevented from forgetting the Han (Dynasty) completely. If it had not been for the will of Heaven that the empire should be divided into three (kingdoms), he could easily have marched his army right through, and brought about a re-unification of the Empire. At that time, if it had not been for the fire-weapons of Khung-Ming, even though the Shu Kingdom had the famous Five Tiger Generals, the Wu and Wei Kingdoms, each with their own strengths, might not necessarily have feared the Shu Kingdom as a veritable tiger.[a] Hence to be invincible nothing excels the expertise of using fire-weapons.

As for fire-weapons, there are those used only for combat, those that are set buried in the ground, those used only for attack, those used for defence, those used only on land, those used on water, and finally those used on city-walls. For charging and annihilating the enemy the fire must be intense and the weapons far-reaching. For sniping at enemy camps, and producing chaos among the enemy, the fire must be far-reaching and the weapons sharp. For guarding a city-wall and holding a fort, the fire must be strong and the weapons heavy. Those that fly overhead are called 'heavenly thunder' (*thien lei*[7]) (i.e. projectiles from bombards, or grenades and bombs hurled by trebuchets); those that are buried in the ground are called 'earthly thunder' (*ti lei*[8]) (i.e. mines); those that are set off in water are called "water thunder" (*shui lei*[9]); and finally those carried as weapons by the soldiers themselves are called 'human thunder' (*jen lei*[10]) (i.e. hand-guns and arquebuses). How fierce these weapons are depends on the nature of the fire, while the intensity and direction of the fire depend on the wind. When used openly they should be set off just at the right moment, and when they are used secretly they should be set to explode at a precisely predetermined time. The very existence or destruction of the Empire, and the lives of the whole armed forces depend on the exact timing of these weapons. This is what fire-weapons are all about.

From my early days onwards I read the Confucian classics, and studied books on military affairs. I roamed about the whole country, hoping to meet someone who had acquired the Tao. One day, when I was travelling in the Thien-thai[11] mountains, I came across a Taoist wearing a yellow cap and a black robe, with blue-green eyes and a grey beard, humming and dancing under a pine-tree. I approached and bowed to him. With his gown fluttering in the wind, he gave me the impression of being truly one of the holy immortals. Clearing a space on a great rock, I sat together with him, and tried to find out what he knew. (I discovered that) in the arts (he took) Confucius and Mencius as his teachers, but in military affairs (he had) inherited (the skill of) Sun Wu[12]; above, he had exhausted the knowledge of the stars and asterisms, below, he could distinguish between all the different mountains and streams. I paid homage and kowtowed to him asking him to be my teacher. Later, we travelled the four quarters together, for three years. He styled himself Chih-Chih Tao-Jen[13] (the 'Knowing-when-to-stop Taoist')[b] and never spoke about his personal name or surname. One day we visited the Shêng Chen Yuan Hua Tung Thien[14] cave[c] in the Wu-i[15] mountains, and he looked at me, saying: 'When I

---

[a] The 'Five Tiger Generals of the Shu Kingdom' were Kuan Yü[2], Chang Fei[3], Chao Yün[4], Ma Chhao[5] and Huang Chung[6].
[b] Cf. Vol. 2, p. 566. We assume that Chih-Chih[16] was intended.
[c] 'Rising to the truths of universal change.'

[1] 司馬昭   [2] 關羽   [3] 張飛   [4] 趙雲   [5] 馬超
[6] 黃忠   [7] 天雷   [8] 地雷   [9] 水雷   [10] 人雷
[11] 天台   [12] 孫武   [13] 止止道人   [14] 昇真元化洞天
[15] 武夷   [16] 知止

was only twelve I passed the junior examination, but later I acquired the mysterious
Tao (of the immortals). Already many years ago I lost interest in all worldly fame (and
the life of an official). However, my own teacher in great confidence gave me a book, the
use of which will help a man to express by action his loyalty to the emperor, and render
service to his country, to administer the realm and bring peace to the people, and to
establish himself and put the Tao into practice. I cannot bear to keep this secret, and now
I wish to impart it to you. At this present time (according to all divination) both heaven
and earth are blocked, and the reigning emperor's mind is tired and muddled. But in a
few years a new emperor will arise in the Huai Valley. Go and help him to accomplish
the meritorious task (of founding a new dynasty), and do not disappoint me.' I saluted
him over and over again, and on looking at the book I found that it was devoted to the
deployment of fire-weapons in warfare. Three days later we came out of the mountains,
and having said good-bye, I walked away a distance of less than a hundred paces, then
when I glanced back towards him, I saw only cloud and mist among the trees of the
forest. I did not know where he had gone.

During an *i-wei* year, the 15th of the Chih-Chêng reign-period (+1355) the sage-
founder of our (dynasty), emperor Kao Huang Ti[1],[a] took command in Ho-chou[2]. Crossing the Yangtze he captured Tshai-shih[3] and Thai-phing[4].[b] At that time Han Lin-Erh[5]
and Han Shan-Thung[6] were occupying Hao-chou[7], Pien(-chou)[8] and Liang(-chou)[9];
Chhen Yu-Liang[10] was controlling Hu-kuang[14];[c] and Liu I[15] was in charge of Liaoyang[16]. Meanwhile, Chang Shih-Chhêng[17][d] occupied Western Chê;[e] Mao Kuei[18] held
the left side of (Thai-)Shan (mountain), Fang Kuo-Chen[19] had Eastern Chê, Wang
Ming[20] governed Szechuan, Chhen Yu-Ting[21] Fukien and Li Ssu-Chhi[22] Kuangtung.[f]
Bandits such as these came out like bees (from their hives) assuming false titles of kings
and rulers to sub-divide (the empire).[g]

Accordingly, following the methods (taught by) my teacher, I cast several types of
fire-weapons (*huo chhi*[23])[h] and presented them (to the founding emperor of our Ming
dynasty). The emperor ordered Hsü Ta[24] (+1372 to +1355), a leading general, to test
them.[i] They were found to behave like flying dragons, able to penetrate several layers

---

[a] The name Kao was not assumed until +1398, but Thai Tsu is meant.

[b] Chu Yuan-Chang[11] joined the service of Kuo Tzu-Hsing[12] in +1352. In +1355 he was given charge of the
army after the capture of Ho-chou. The same year Kuo Tzu-Hsing died, and Han Lin-Erh, a son of Han
Shan-Thung, stepped in and was made king of Sung at Hao-chou. Dissatisfied with the development of events,
Chu Yuan-Chang led his followers across the Yangtze and captured first Tshai-Shih and then Thai-phing. He
established himself at the latter place and made himself commanding general (Yuan Shuai[13]) of the revolutionary forces, still under the flag of Sung. See *Ming Shih*, ch. 1, pp. 2 a ff. and ch. 122, all on Kuo Tzu-Hsing
and Han Lin-Erh.

[c] One of the twelve administrative provinces during the time of the Yuan dynasty, comprising the greater
parts of modern Hunan, Kuangsi and Kueichow provinces.

[d] We shall meet this would-be ruler again in connection with the casting of iron cannon which are still
preserved (p. 295 below).

[e] Modern Chekiang and parts of Kiangsu, Anhwei, Kiangsi and Fukien provinces.

[f] See Goodrich & Fang Chao-Ying (1), vol. 1, pp. 485–588 for Han Lin-Erh, pp. 185–8 for Chhen Yu-
Liang, pp. 99–103 for Chang Shih-Chheng, and pp. 433–5 for Fang Kuo-Chen.

[g] For some idea of the affairs of these contending warlords, see Dreyer (2), pp. 203 ff.; Dardess (1).

[h] Instead of *huo chhi*, the *Wu Pei Huo Lung Ching* says *huo lung chhiang*[25] (fire-dragon lances).

[i] See Goodrich & Fang Chao-Ying (1), vol. 1, pp. 602–8, for a biography of Hsü Ta.

[1] 高皇帝	[2] 和州	[3] 采石	[4] 太平	[5] 韓林兒
[6] 韓山童	[7] 亳州	[8] 汴州	[9] 梁州	[10] 陳友諒
[11] 朱元璋	[12] 郭子興	[13] 元帥	[14] 湖廣	[15] 劉益
[16] 遼陽	[17] 張士誠	[18] 毛貴	[19] 方國珍	[20] 王明
[21] 陳友定	[22] 李思齊	[23] 火器	[24] 徐達	[25] 火龍鎗

of armour. The emperor Thai Tsu was delighted at the result, and said, 'With these types of fire-weapons I shall be able to conquer the whole empire as easily as turning the palms of one's hands upside down. When we have accomplished this, I shall bestow upon you high honour as a Founding Officer of the Empire.'

From then on in one expedition we captured Ching(-chou) and Hsiang(-chou), and in another we took (the administrative provinces of) Chiang and Chê, while in a third Fukien and all its surrounding waters surrendered.[a] In a fourth campaign we stormed the whole of Chhi (i.e. Shantung).[b] We also annihilated (Chhen) Yü-Liang and took the whole region of Chhin, Chin, Yen and Chao.[c] The Mongolian barbarians fled to the north and our capital was established at Chin-ling (Nanking). (Thus Thai-Tsu) re-unified the whole empire, and began reigning over a new dynasty that will last for thousands of years. In the capital he set up a Gunpowder Department (*huo yao chü*[1]) for the manufacture of the explosive, and an Armoury (*nei khu*[2]) for storing the magically effective weapons (*shen chhi*[3]). Such was the attention our first sage-emperor paid to military matters.

The types of fire-weapons (made for the emperor), however, did not fully represent all the secrets passed on to me by the holy immortal. The sacred accomplishments and military exploits (of our first emperor) should ensure peace in the Empire for ten thousand generations. Yet in order to safeguard it one must not forget in time of peace about protection against dangers. Lest these fire-weapon techniques might be lost during a long period (of peace), I have endeavoured to illustrate them in diagrams, and de-scribe them accordingly in writing, for the benefit of soldiers and tacticians who will serve our country as loyal subjects ready to die for its cause. They will be able to appreci-ate the immeasurable (applications) of the secrets handed down by the holy immortal (and the rare opportunity that I, their predecessor, enjoyed). It is not easy to meet an emperor and to become Master of his Ordnance. I fervently hope that none will be like Chao Khua[4], who only read his father's writings (but failed badly when he tried to put them into practice).[d] (May the reader) bear this in mind.

Preface (written during the) 10th year of the Yung-Lo reign-period (+1412) by the Count of Tung-ning[5], Chiao Yü[6].

One or two interesting problems arise out of the versions of the 'Fire-Drake Manual' that Arima (*1*) did not see. In the preface of the *Wu Pei Huo Lung Ching* it is stated that Chiao Yü himself made for the first Ming emperor a *huo lung*

[a] Ching-chou and Hsiang-chou referred to the territory formerly occupied by the state of Chhu in the Spring-and-Autumn Period, including the modern provinces of Hupei and Hunan and portions of Honan, Anhwei and other provinces. The occupation of these regions by the Ming army occurred in the years +1364 and +1365 under the able commanders Hsü Ta and Chhang Yü-Chhun. See *Ming Shih*, ch. 1, p. 11*a* to p. 12*b*. Fukien surrendered to the forces of Chu Yuan-Chang in +1361. See *Ming Shih*, ch. 1, p. 7*a*.

[b] The *Ming Shih*, ch. 2, p. 1*b*, says: 'On a *kuei-mao* day in the second month of the first year of the Hung-Wu reign-period (20 February +1368) Chhang Yü-Chhun captured Tung-chhang and thus conquered Shantung.'

[c] i.e. roughly the whole of Shensi, Shansi, Honan, and Hopei provinces. Chhen Yu-Liang was killed by a stray arrow on 3 October +1363 during a battle against the Ming fleet. See *Ming Shih*, ch. 123, p. 4*b*; Dreyer (*1*), p. 238.

[d] Chao Khua was a military commander of the State of Chao in the −3rd century. In his youth he talked so much about military matters that he worried his father Chao Shê[7], who had previously led an army of Chao State to rescue the State of Han which was under attack from the Chhin army. However, when it later came to Chao Khua's turn to face the Chhin invaders, he was slain and his whole army annihilated by Pai Chhi[8]. See *Shih Chi*, ch. 81, pp. 6*b* ff.; tr. Kierman (*1*), pp. 31 ff.

[1] 火藥局　　　[2] 內庫　　　[3] 神器　　　[4] 趙括　　　[5] 東寧伯
[6] 焦玉　　　[7] 趙奢　　　[8] 白起

chhiang[1] (fire-drake spear or lance), which Arima believed was the earliest form
of the musket or arquebus. But this weapon is not mentioned anywhere in the
text of the *Wu Pei Huo Lung Ching*. Arima thought that it had been purposely left
out because of the highly secret nature of the weapon, one which writers of later
military compendia like the *Shen Chhi Phu* and *Tsê Kho Lu* of course knew well.
But the term *huo lung chhiang* does not appear in the preface in any of the other
versions. The preface in the Nanyang edition and that in the *Huo Kung Pei Yao*
both use the term *huo chhi*[2] (fire-weapons). So this evidence by itself will not
establish the existence of the arquebus or musket in +1355, even though it may
perhaps have been known to Chiao Yü.[a]

There is room for some speculation about the oldest version of the *Huo Lung
Ching*. A few pages earlier we mentioned the *Huo Lung Shen Chhi Thu Fa*[3] (Fire-
Drake Illustrated Technology of Magically (Efficacious) Weapons) which Lu
Wên-Chao listed in his completion of the Yuan bibliography,[b] and if this went
back to the beginning of the dynasty it could mean about +1280. Another work,
the *Huo Lung Wan Shêng Shen Yao Thu*[4] (Illustrated Fire-Drake Technology for a
Myriad Victories using the Magically (Efficacious) Gunpowder)[c] is known only
by title from Chhien Tsêng's[5] *Tu Shu Min Chhiu Chi*[6] catalogue of Sung and Yuan
editions finished in +1684, but it must have belonged to the same family of texts.
Most interesting is the *Huo Lung Shen Chhi Yao Fa Pien*[7] (Fire-Drake Book of
Magically (Efficacious) Weapons, with the Method of Making Gunpowder),
which exists as an anonymous MS. in the library of the History of Science Institute of Academia Sinica at Peking. This has illustrations similar to those in the
printed *Huo Lung Ching* editions, but more delicate and precise.[d] Its relationship
to these has not yet been elucidated, but perhaps further study will establish it as
a Yuan work or an early copy of one.

Under the Chhing, the *Huo Lung Ching*, like the *Wu Pei Chih*, was of course for
centuries a prohibited book. It is full of expressions such as 'northern barbarians' (*pei i lo*[8]), which would have made it impossible to reprint under the
Manchu rule. Only when the Ming period had receded so far into the past that it
had no contemporary relevance at all was it possible to give the book historical
study and bring out new editions of it.

But perhaps the most interesting feature about it was that it was distinctly
earlier than any component of the *Büchsenmeisterei*[e] literature of Europe. From
the Master-gunners of the West hardly anything has come down to us before

---

[a] If so, it must have been in one of its most primitive and archaic forms, hardly distinguishable from the
hand-guns of the beginning of the century.
[b] Cf. Lu Ta-Chieh (*1*), p. 108.
[c] Lu Ta-Chieh (*1*), p. 169.
[d] We had the opportunity of examining it briefly in September 1981.
[e] The word 'Büchse' probably came from pyx or pixis, a tubular container (cf. Partington (5), p. 116).

[1] 火龍鎗   [2] 火器   [3] 火龍神器圖法
[4] 火龍萬勝神藥圖   [5] 錢曾   [6] 讀書敏求記
[7] 火龍神器藥法編   [8] 北夷虜

+1400, though it is reasonable to place Chiao Yü's composition between +1360 and +1375, even though it was not first printed till +1412. The nearest approach to this is an untitled manuscript of about +1395.[a] Then comes the well-known *Bellifortis* of Konrad Kyeser, dating from +1405.[b] After that the European artillerists wrote and limned copiously; the book of the Anonymous Hussite engineer and that of Giovanni da Fontana both came about +1430.[c] Many interesting works, still in manuscript, followed—a *Streydtbuch* in +1435,[d] a *Feuerwerkbuch* in +1437,[e] the *Kunst aus Büchsen* ... in +1471,[f] and a similar treatise in +1496.[g] Meanwhile there was the *Mittelalterliche Hausbuch*, often previously referred to, in +1480.[h] We need not pursue this literature further here, but the fact is that those who towards the end of the +14th century in Europe began to write down what they knew about bombards, hand-guns, ribaudequins and gunpowder, had all been preceded by Chu Yuan-Chang's Master of Ordnance from +1355 onwards.

After the *Huo Lung Ching* there came about a dozen books on military affairs which touch on firearms to a varying extent, from the first half of the +16th to the first half of the +17th century. One of the earliest among them is the *Wu Pien*[1] (Military Compendium) by Thang Shun-Chih[2] (+1507 to +1560).[i] It was included in the *Ssu Khu Chhüan Shu* collection, but the author is criticised by its bibliographers as being too bookish, for he met with disastrous results when he led an army against the Japanese *wo-khou*[3] pirates who raided the Chinese coast during his time.[j] He had to be rescued by Hu Tsung-Hsien in +1559. We do not know the exact date of the *Wu Pien*, though we can infer that it must be some time between +1548 and +1558, since Thang Shun-Chih mentions the 'bird-beaked gun' which was first introduced to China from Japan in the former year.

About +1561 Chêng Jo-Tsêng[4] wrote the *Chhou Hai Thu Pien*[5] (Illustrated Seaboard Strategy and Tactics). There are five different editions of this book, one of which was published by the grandsons of Hu Tsung-Hsien[6], who dropped Chêng's name.[k] Then came the *Chiang-nan Ching Lüeh*[7] (Military Strategies in Chiang-nan), again written by Chêng Jo-Tsêng in +1566. The next writer of

[a] Partington (5), p. 144. Munich, Cod. Germ. 600. The drawings are very crude.

[b] Cf. Vol. 4, pt. 2, p. 113 and *s.v.* Many of the manuscripts mentioned in this paragraph were the subjects of three classical papers by Berthelot (4, 5, 6). On Kyeser and gunpowder see Partington (5), pp. 146 ff.

[c] Cf. Vol. 4, pt. 2, p. 82 and passim for the Hussite. Partington (5) describes both works, the former on p. 144, the latter on pp. 160 ff.

[d] *Streydtbuch von Pixen, Kriegsrüstung, Sturmzeuch und Feuerwerckh*, the writer of which is not known. Partington (5), p. 159.

[e] Partington (5), p. 152, Hassenstein (1).

[f] *Kunst aus Büchsen zu Schiessen*; this was by Martin Mercz. Cf. Partington (5), p. 159.

[g] *Buch der Stryt und Büchsse*; this was by Philip Monch. Cf. Partington, (5), p. 160.

[h] Cf. Vol. 4, pt. 2, p. 216 and *s.v.*

[i] Thang Shun-Chih was also the author of a book on geometry, the *Kou Ku Têng Liu Lun*[8] (Six Discourses on the Base and Vertical Side of Right-Angled Triangles). See *Ming Shih*, ch. 96, p. 23a.

[j] *SKCS/TMTY*, ch. 99, p. 42b.

[k] Hence Hu Tsung-Hsien has sometimes been mistaken as the author. See W. Franke (4), p. 224. On his campaigns against the allied Chinese-Japanese marauders in +1556 see the interesting study of Hucker (5).

¹ 武編   ² 唐順之   ³ 倭寇   ⁴ 鄭若曾   ⁵ 籌海圖編
⁶ 胡宗憲   ⁷ 江南經略   ⁸ 勾股等六論

importance was the famous Ming general Chhi Chi-Kuang[1] (+1528 to +1587),[a] who wrote the *Lien Ping Shih Chi*[2] (Treatise on Military Training) in +1571,[b] and the *Chi Hsiao Hsin Shu*[3] (New Treatise on Military and Naval Efficiency), *c.* +1575. These two are valuable books for the study of Chinese firearms,[c] but they have received less attention than the *Wu Pei Chih*[5] (Treatise on Armament Technology), though they preceded the latter by half a century.[d] Another book of Chhi Chi-Kuang's was the *Wu Pei Hsin Shu*[6] (New Book on Armament Technology). All these three military works of Chhi were included in the *Ssu Khu Chhüan Shu* collection. Similarly incorporated was the *Chen Chi*[7] (Records of Battle Arrays), written in +1591 by Ho Liang-Chhen[8]; it gives a long list of firearms.

Before the end of the century came two further military works, namely the *Shen Chhi Phu*[9] (Treatise on Magically (Efficacious) Weapons), i.e. firearms, written by Chao Shih-Chên[10] in +1598;[e] and the *Têng Than Pi Chiu*[11] (Knowledge Necessary for Army Commanders), written by Wang Ming-Hao[12] in +1599.[f] Wang was also the editor of another military book, the *Ping Fa Pai Chan Ching*[13] (Manual of Military Strategy for a Hundred Battles). A multitude of fire-weapons and firearms, too, are illustrated and described in the *Ping Lu*[23] (Records of Military Art) that Ho Ju-Pin[14] wrote in +1606.[g] This work describes the theory of the gunpowder formula in great detail, treating it like a medical prescription by regarding saltpetre and sulphur as the 'sovereign' (*chün*[15]) components, carbon as 'minister' (*chhen*[16]), and other substances added to the mixture as the 'adjutants' (*tso*[17]).[h] In +1607 came the *Chiu Ming Shu*[18] (Book on Saving the Situation) written by Lü Khun[19].[i]

Probably the most comprehensive Chinese military compendium ever written was the *Wu Pei Chih* (Treatise on Armament Technology) in 240 chapters, completed by Mao Yuan-I[24] in +1628.[j] He also wrote an interesting *Huo Yao Fu*[25]

---

[a] Perhaps the best biography of him in a Western language, absorbing reading, is that of Huang Jen-Yü (5), pp. 159 ff.

[b] It had a second part, almost as long, the *Lien Ping Shih Chi, Tsa Chi*[4].

[c] Partial translations are contained in the special study of Chhi Chi-Kuang by Werhahn-Mees (1).

[d] For example, the section in the *Wu Pei Chih* on the *niao tsui chhung* musket in ch. 124 is but a repetition of the *Chi Hsiso Hsin Shu*.

[e] It also had a supplement, the *Shen Chhi Phu Huo Wên*[20].   [f] See W. Franke (4), p. 208.

[g] Three chapters of the work (11 to 13) are devoted to this subject.

[h] Cf. our account of the classification of drugs in the *Shen Nung Pên Tshao Ching* in Sect. 38 (Vol. 6, pt. 1). On the theory of gunpowder explosives, see p. 163 below.

[i] It was divided into two parts, the *Hsiang Ping Chiu Ming Shu*[21] on the raising of militia, and the *Shou Chhêng Chiu Ming Shu*[22] on the defence of cities.

[j] Various works derivative from this came out later. For example, Fu Yü[26], about +1660, produced a *Wu Pei Chih Lüeh*[27] (Classified Material from the Treatise on Armament Technology). And the Cambridge University Library has a MS. of 1843 entitled *Wu Pei Chih Shêng Chih*[28] (The Best Designs in Armament Technology). W. Franke (4), p. 209, thinks that this was then printed.

[1] 戚繼光	[2] 練兵實紀	[3] 紀効新書	[4] 練兵實紀雜集	
[5] 武備志	[6] 武備新書	[7] 陣紀	[8] 何良臣	[9] 神器譜
[10] 趙士禎	[11] 登壇必究	[12] 王鳴鶴	[13] 兵法百戰經	[14] 何汝賓
[15] 君	[16] 臣	[17] 佐	[18] 救命書	[19] 呂坤
[20] 神器譜或問	[21] 鄉兵救命書	[22] 守城救命書	[23] 兵錄	[24] 茅元儀
[25] 火藥賦	[26] 傅禹	[27] 武備志略	[28] 武備制勝志	

(Rhapsodical Ode on Gunpowder) about the same time.[a] Almost simultaneously, another military encyclopaedia with a rather unusual title, the *Phing Phi Pai Chin Fang*[1] (The Washerman's Precious Salve; Appropriate Techniques of Successful Warfare), edited by Hui Lu[2], also appeared.[b] It reproduces many illustrations from the older books, adding also new ones, on the trebuchet principle[c] and on the telescope,[d] but it is not particularly good on firearms. Yet another late Ming military book was the *Chin Thang Chieh Chu Shih-Erh Chhou*[3] (Twelve Suggestions for Impregnable Defence),[e] by Li Phan[4].

Towards the end of the Ming Dynasty Chiao Hsü[5], with the help of the Jesuit Adam Schall von Bell, wrote the *Huo Kung Chhieh Yao*[6] (Essentials of Gunnery) in the year +1643.[f] From the 1841 reprint onwards, this work has also borne another name, the *Tsê Kho Lu*[7] (Book of Instantaneous Victory). Chiao Hsü mentions the 'Fire-Drake Manual', and his name leads one to speculate on a possible relationship between him and Chiao Yü[8]; he could have been a descendant working in the Imperial Arsenal, but there is, as yet, no positive evidence for this. Jesuit intermediation was not the only channel which brought the knowledge of Western firearms to China. Contact with the Portuguese led to at least one book, the *Hsi-Yang Huo Kung Thu Shuo* (cf. p. 393 below); and there were also the Vietnamese, the Japanese and the Turks, as will appear in due course (pp. 310, 429 440).

We have only mentioned those military compendia of the Ming which describe firearms and are still available to us. There were other military writings now lost or extremely rare, as for example, the *Huo Chhi Thu*[10] (Illustrated Account of Gunpowder Weapons and Firearms)[g] written by Ku Pin[11]. The *Wu Pei Chih* and some of the others mentioned were naturally prohibited books in the Chhing period. In general military works were regarded as 'classified' items during the early part of that time, a fact which explains the difficulty of gaining access to them now.

Besides all the above, many monographs on guns and cannon were produced during this century. Hu Tsung-Hsien[12], the commander-in-chief in the Southeast from +1556 to +1562,[h] himself wrote two, the *Wu Lüeh Shen Chi Huo Yao*[13]

---

[a] It can be found in *TSCC, Jung chêng tien*, ch. 96.

[b] The title was taken from a story in *Chuang Tzu*, ch. 1 (tr. Legge (5), vol. 1, p. 173; Fêng Yu-Lan (5), p. 39). A man of Sung State invented a salve for chapped hands, and it was used in his family for several generations as they were professional washers of silk. A stranger bought the formula for a hundred pieces of gold (hence the title), went down to Wu State, and being made Admiral there, employed it for his sailors so that they won a great victory over the fleet of Yüeh. One application brought little gain; the other gained great reward and a noble title. The work seems to be rather rare; it was not mentioned in *SKCS/TMTY*.

[c] Ch. 4, p. 35b.

[d] Ch. 12, p. 24b. This dates it after +1626 (see Vol. 4, pt. 1, p. 117).

[e] The first two words of the title recall the phrase *chin chhêng thang chhih*[9], 'adamantine walls and scalding moats', hence impregnable.

[f] On this collaboration we have already written something (Vol. 5, pt. 3, pp. 240–1).

[g] See *SKCS/TMTY*, ch. 100, p. 48b.   [h] Cf. the account of Hucker (5).

[1] 洴澼百金方   [2] 惠麓   [3] 金湯借箸十二籌   [4] 李盤
[5] 焦勗   [6] 火攻挈要   [7] 則克錄   [8] 焦玉   [9] 金城湯池
[10] 火器圖   [11] 顧斌   [12] 胡宗憲   [13] 武略神機火藥

and the *Wu Lüeh Huo Chhi Thu Shuo*[1], both on muskets and gunpowder compositions and their use in various tactical situations. These were embodied in a very
rare collection edited by Phan Khang[2] and entitled *Wu Pei Chhüan Shu*[3], which
contains a number of other interesting books as well.[a] Another book with a
closely similar title but of rather earlier date, the *Wu Lüeh Shen Chi*[4] by Hu Hsien-
Chung[5] also dealt with the tactical employment of musketeers. Then there was
Huang Ying-Chia's[6] *Huo Chhi Thu Shuo*[7], closely similar to the later forms of the
'Fire-Drake Manual'; and two works the names of the authors of which have not
come down to us, a *Huo Kung Chen Fa*[8] on the tactical use of guns and artillery,
and a *Huo Yao Miao Phin*[9] on gunpowder compositions and what they were good
for. Thus all in all the late Ming was a very prolific period for works on gunpowder weapons.[b] They help to explain the high figures for military writings of all
kinds during this dynasty which we noted at an earlier stage.[c]

Firearms are also described in some general technological writings, and other
encyclopaedic works. For example, the *San Tshai Thu Hui*[12], written by Wang
Chhi[13] in +1609, illustrates the 'bird-beaked gun' matchlock musket. A small
section on firearms is contained in Sung Ying-Hsing's[14] *Thien Kung Khai Wu*[15]
(The Exploitation of the Works of Nature) of +1637.[d] Fang I-Chih's[16] *Wu Li
Hsiao Shih*[17] (Small Encyclopaedia of the Principles of Things), finished by
+1643, also has something to say about fire-weapons. However, none of these
gives as much information on firearms as is found in the specialist military
treatises.

During and after the Ming dynasty, Chinese military compendia seem to have
made their appearances mainly during times of particular need or emergency.
Early in the Ming the *Huo Lung Ching* described some of the weapons used to
overthrow the Mongols. During the mid +16th century a host of military works
appeared soon after the introduction of the Portuguese breech-loading cannon
(*fo-lang chi*[18]), and the arquebus, the so-called 'bird-gun' or 'bird-beaked gun'.
Such were the *Wu Pien*, the *Chhou Hai Thu Pien*, the *Chiang-Nan Ching Lüeh*, the
Lien Ping Shih Chi and the *Chi Hsiao Hsin Shu*. That was also the time when the
Chinese coasts were plagued by the Japanese *wo-khou*[19] pirates, often led by
Chinese renegades.[e] In the 1590s the Chinese were engaged in helping Korea
against Japanese invasions led by Toyotomi Hideyoshi (+1536 to +1598). That

---

[a] Such as the *Hai Fang Tsung Lun*[10] on coastal defence, and the *Wu Shih Thao Lüeh*[11] on military examinations.
[b] For the elucidation of these byways of the gunnery literature we have to thank our friends Dr Phan
Chi-Hsing of the History of Science Institute of Academia Sinica in Peking and Dr Miyashita Saburō of the
Takeda Science Foundation at Osaka.
[c] Cf. Vol. 5, pt. 6 above.
[d] This we shall have occasion to discuss later, on pp. 187, 437 below.
[e] On this subject see the book of So Kwan-Wai (1).

[1] 武略火器圖說	[2] 潘康	[3] 武備全書	[4] 武略神機
[5] 胡獻忠	[6] 黃應甲	[7] 火器圖說	[8] 火攻陣法	[9] 火藥妙品
[10] 海防總論	[11] 武試韜略	[12] 三才圖會	[13] 王圻	[14] 宋應星
[15] 天工開物	[16] 方以智	[17] 物理小識	[18] 佛郎機	[19] 倭寇

was the time when Ho Liang-Chhen wrote the *Chen Chi*, and when Chao Shih-Chên submitted blueprints for the making of more powerful muskets in his *Shen Chhi Phu*. That was also the time when the *Têng Than Pi Chiu* made its appearance.

By the beginning of the +17th century the Ming military power had waned and it never recovered. Some scholars, seeing the urgency of rearmament and the need for acquiring military knowledge, hoped to restore the dynasty by compiling military works. We have a new crop of publications in late Ming including the *Ping Lu*, the *Chiu Ming Shu*, the *Phing Phi Pai Chin Fang*, the *Chin Thang Chieh Chu Shih-Erh Chhou*, the *Wu Pei Chih*, and the *Huo Kung Chhieh Yao*. The first six included accounts of Western firearms which became known through Japanese contacts, while the last incorporated knowledge of Western guns and cannon introduced directly to China by the Jesuits. Our list is of course incomplete. We know of other military books now quite lost, for example the *Hsi-Yang Huo Kung Thu Shuo*[1] (Illustrated Treatise on European Gunnery) written by Chang Tao[2] and Sun Hsüeh-Shih[3] before the year +1625.[a]

By and large the Chhing period was one of peace, so that military compendia were produced less frequently[b], but before the echoes of the Manchu conquest had completely died away Lü Phan[4] and Lu Chhêng-Ên[5] produced in +1675 their *Ping Chhien*[6] (Key to Martial Art).[c] By this time artillery had entered the modern world (cf. Figs. 145 and 147 below). These authors gave a good deal of their attention to naval affairs, incorporating in their book several important rutters, and registers of compass-bearings.[d] Moreover, the *Man-Chou Shih Lu*[7] (Veritable Records of the Manchu Dynasty) contains a number of valuable illustrations showing the use and disposition of artillery pieces in the field, some of which we reproduce below (Figs. 152, 155). From the study of Chhen Wên-Shih (*1*) we know that all the technical handicrafts and industries were poorly developed among the Manchus before the time of Nurhachi, but Chinese and Korean craftsmen were attracted to give their aid, and the first Manchu cannon was cast in +1631. Thereafter military needs long dominated. To trace the development of military writing in Japan would take us too far from our present theme, but this may perhaps be the place to mention the *Honchō Gunkikō*[8] (Investigation of the Military Weapons and Machines of the Present Dynasty), a famous work by Arai Hakuseki[9] begun about +1705 and printed in +1737. It is

---

[a] See Bernard-Maitre (7), p. 446, and Pelliot (55), p. 192. This work had no connection with the Jesuits, and probably emanated from friends of the Portuguese gunners who were sent up from Macao to help the Ming.

[b] Some that we should like to see have been lost, for example the *Huo Chhi Lüeh Shuo*[10] (Classified Explanations of Firearms) by Wang Ta-Chhüan[11], and the *Huo Chhi Chen Chüeh Chieh Chêng*[12] (Analytical Explanations of Firearms and Instructions for Using Them) by Shen Shan-Chêng.[13] We do not know their exact dates.

[c] Preface of +1669.

[d] Two of these have been reprinted and edited by Hsiang Ta (5). Cf. Vol. 4, pt. 3, pp. 581 ff.

[1] 西洋火攻圖說   [2] 張燾   [3] 孫學詩   [4] 呂磻
[5] 盧承恩   [6] 兵鈐   [7] 滿洲實錄   [8] 本朝軍器考   [9] 新井白石
[10] 火器略說   [11] 王達權   [12] 火器眞訣解證   [13] 沈善蒸

very detailed on slat-armour and close-combat weapons, but it has almost nothing at all on gunpowder and firearms.[a]

The publication of military compendia during the Chhing dynasty was also correlated with national emergencies. For example, the *Hai Kuo Thu Chih*[1] (Illustrated Record of the Maritime Nations), written by Wei Yuan[2] and incorporating articles on Western firearms and gunboats, appeared in 1841, immediately after China's defeat in the Opium War. Lin Tsê-Hsü[3] lent a hand to get it published, and indirectly contributed much to it.[b] This was the time of Li Shan-Lan's[4] tractate *Huo Chhi Chen Chüeh*[5] (Instructions on Artillery); he was an outstanding mathematician and technologist, later one of the group at Anking, predecessor of the Kiangnan Arsenal.[c] It is not generally known that in the *Chiang-nan Chih Tsao Chü Chi*[6] (Records of the Kiangnan Arsenal) of 1905 Wei Yün-Kung gave brief histories of gunpowder weapons in China, including some which the Arsenal may never have had occasion to make, such as the fire-lance and the war-rocket.

But the *Ko Chih Ching Yuan*[7] (Mirror of Scientific and Technological Origins), written by Chhen Yuan-Lung[8] in +1735, had already had much to say on fire-weapons.[d] It quotes from books like the *Wu Yuan*, the *Pai Phien*, the *Shih Wu Chi Yuan* and the *Chiu Thang Shu* on the *phao* trebuchets, mentioning battles where mines were used, incendiary arrows, rockets, early Chinese bombs, guns, cannons, breech-loaders, the 'bird-beaked' musket and multiple-barrel guns. Of course, books such as this are only late secondary sources.

In looking back over the whole of this literature, several interesting thoughts present themselves. It is quite remarkable that after the first appearance of gunpowder in war it took about a century and a half before anything concerning it was put down on paper; the literate scholars so long failed to take notice of what the technicians were doing. Gunpowder appears first as a low-nitrate composition used for slow-match ignition in pumped naphtha flame-throwers (p. 81 below). This is datable to +919.[e] Similarly, an account of a siege in +904 tells of 'flying fire launched from machines' (*fa chi fei huo*[9]), i.e. low-nitrate gunpowder used in the form of incendiary projectiles or 'bombs' hurled from trebuchets (p. 85 below).[f] Yet nothing much got into print about these things until the *Wu Ching Tsung Yao* of +1044. And then what a long gap there was until the writing

---

[a] This may well be connected with the Japanese aversion to firearms, which we shall consider later (p. 467 ff.). Cf. Perrin (1). There is a biography of Arai Hakuseki by Ackroyd (1).

[b] See Hummel (2), p. 851.

[c] Cf. Vol. 3, p. 106.

[d] Ch. 42, pp. 27 a ff.

[e] By deduction from the clear description of +1044 (*WCTY*). See p. 82 below.

[f] Following the interpretation given in +1004 (*HCC*). The projectiles could also have been gunpowder arrows (not rockets) shot from arcuballistae, or just possibly even naphtha flame-throwers were meant. Cf. Fêng Chia-Shêng (1), p. 46, (6), p. 73. See p. 85 below.

¹ 海國圖志   ² 魏源   ³ 林則徐   ⁴ 李善蘭   ⁵ 火器眞訣
⁶ 江南製造局記   ⁷ 格致鏡原   ⁸ 陳元龍   ⁹ 發機飛火

of the *Huo Lung Ching* three centuries later! Today we can only fill it by the judicious use of the narratives of the historians, both official and unofficial, as indeed the rest of this Section will show.[a]

(ii) *Arabic and Western sources*

One of the earliest European texts mentioning gunpowder is the famous *Liber Ignium ad Comburendos Hostes* (Book of Fires for the Burning of Enemies) attributed to Marcus Graecus (Mark the Greek, or Byzantine).[b] We have come across it before[c] in connection with a recipe for the distillation of strong alcohol, which occurs as one of the latest of its components, belonging to *c.* +1280. The gunpowder formulae also belong to this last stratum, perhaps as late as +1300, in contrast to the earliest entries which may well go back to the +8th century. Several Latin versions of the manuscript, only about six pages long, exist,[d] but none of them bears a Greek title, and there is no evidence that the author or compiler of the work was a Byzantine. Marcus Graecus was certainly not, as some have supposed, the Marcus mentioned by Galen (d. +201), nor the Graecus referred to by Mesue (d. 1015),[e] nor yet the +12th-century Mark of Toledo, who translated the Holy Koran into Latin, nor yet again the King Marqouch mentioned in late Arabic alchemical texts.[f] Perhaps he was not a real person at all—*nomen et praeterea nihil*, just a name for a collection. The *Liber Ignium* is more probably of Arabic origin, perhaps translated and put together gradually by Jewish scholars in Spain, for it mentions certain climatic conditions not found in Europe, and leaves a number of Arabic and Spanish words untranslated. Moreover it contains[g] several specifically Arabic +12th-century recipes for 'automatic fire'[h] and 'oil of bricks'.[i]

The *Liber Ignium* certainly belongs to the group of collections of 'secrets', lacking all classification or order. Of its 35 recipes, 14 are concerned with war, 11 for

[a] Judicious we say, because the pitfalls are innumerable, largely owing to the continuing lack of precise technical terminology. Cf. p. 22 above.

[b] All the modern work on it has been summarised and sifted by Partington (5), pp. 42 ff.; this largely supersedes previous accounts. But Sarton (1), vol. 2, p. 1037 and Thorndike (1), vol. 2, pp. 252, 738, 785 ff. are still worth consulting.

[c] Vol. 5, pt. 4, p. 123.

[d] There is a printed version of the text by de la Porte du Theil (1804), and others by Berthelot (10), pp. 89 ff. and Hoefer (1), vol. 1, pp. 491 ff., 2nd ed., pp. 517 ff. with translations. But none are wholly satisfactory, and there is no modern critical edition.

[e] Probably Masawayah al-Mardīnī of Baghdad.

[f] Berthelot (14); Berthelot & Houdas (1), pp. 15, 16, 124. But this personage was associated with sal ammoniac, another of the new things that came to the Arabs (like saltpetre) from China; cf. Vol. 5, pt. 4, p. 432.

[g] Partington (5), pp. 47, 50, 55–6, 156, 198.

[h] This was brought about by mixtures of quicklime with combustibles such as petroleum, sulphur and other things, which took fire when wetted in any way. Cf. p. 67 below, and Partington (5), pp. 53 ff.; Cahen (1), p. 147.

[i] This was half-recognised hydrochloric acid, going far back into the Middle Ages; see Vol. 5, pt. 3, pp. 237–8, pt. 4, p. 198. It got the name of *oleum benedictum* and Roger Bacon spoke of it as 'blessed' oil (*De Erroribus Medicorum*, §16; Welborn (1), p. 32)—not unreasonably in view of its use in pharmacy. Cf. Cahen (1), p. 146.

lamps and lights, 6 for preventing or curing burns, and 4 for preparing chemicals, especially saltpetre. Of the 14 military entries, 10 are for various incendiary mixtures, 3 of them containing quicklime; these recipes belong mainly to the early and middle strata, and it is directed that some of the mixtures should be shot off, after ignition, with javelins or arrows. The remaining 4 recipes all contain saltpetre.

It is noted in the *Liber Ignium* (§14) that

saltpetre is a mineral of the earth, and is found as an efflorescence on stones. This earth is dissolved in boiling water, then purified and passed through a filter. It is boiled for a day and a night and solidified, so that transparent plates of the salt are found at the bottom of the vessel.

The book contains two compositions of 'fire flying in the air' (*ignis volatilis*), which Berthelot (10, 14) interpreted as rockets. The first (§12) gives one part of colophonium resin, one of native sulphur, and 6 (?) parts of saltpetre. The second (§13) gives 1 lb of native sulphur, 2 lbs of linden or willow charcoal, and 6 lbs of saltpetre. There are also two compositions for *ignis volantis in aere* (§§32 and 33), one having equal parts of saltpetre, sulphur and linseed oil; the other 9 parts of saltpetre to one of sulphur and three of charcoal. If we take the carbonaceous materials as equivalent to charcoal,[a] this means, when tabulated:

§	% N	S	C[b]
12	75	12·5	12·5
13	66·5	11	22·5
32	33	33	33
33	69	8	23

The text of the first of these being of doubtful interpretation, we clearly have to do with low-nitrate gunpowders which would deflagrate and not explode, or if so but weakly.[c]

The descriptions certainly sound like primitive rockets, though these seem to have been primarily incendiary or intended to terrify, since there is no mention of a propelled arrow or its warhead; but as Hime[d] and Partington[e] pointed out, the composition may well have been used in fire-lances such as were known to the Arabs, having appeared three centuries earlier still in China.[f] The trouble with the expression 'flying fire', or 'fire flying in the air', here, is exactly the same as that met with in Chinese historical writings (cf. p. 22), where *fei huo*

---

[a] One could apply some kind of correction for this, which would make the percentage of saltpetre appear higher, but the mixtures would still deflagrate, not explode.

[b] We use this generally accepted convention henceforward to denote the percentages of saltpetre, sulphur and charcoal (or carbonaceous material) in that order; cf. Partington (5), p. 202.

[c] Two of them might be suitable for rockets, fire-lances or Roman candles.

[d] (2), p. 85.

[e] (5), p. 61.

[f] Reinaud & Favé (2), p. 316, suggested indeed a Chinese origin for the 'feu volant', saying that it had come with the Mongol invasions about +1250.

chhiang[1] might be either 'flying fire-spears' or (as is much more probable) 'flying-fire spears'. The historians cannot be blamed if, as is possible, they had never seen rockets, and knew only fire-lances. Lastly there is the thunder mentioned in the *Liber Ignium*, but that need not imply detonation, since deflagration in a confined space could produce a similar effect. Thus, to sum up, we have in this important, though unsystematic work, clear evidence of saltpetre and low-nitrate gunpowder, no more, and certainly no use of gunpowder as propellant in a gun. Prior to its probable date, +1280, only one other European reference to gunpowder is known, and presently (p. 47) we shall take a look at it.

From the Arabic world the literary monument most corresponding to the *Liber Ignium* is the *Kitāb al-Furūsīya wa'l-Munāṣab al-Ḥarbīya* (Treatise on Horsemanship and Stratagems of War),[a] written also about +1280 by Ḥasan al-Rammāḥ (the lancer) Najm al-Dīn al-Aḥdab (the hunchback), probably a Syrian.[b] Giving many formulae for gunpowder, it resembles Graecus in concentrating on incendiary compositions and deflagrating powders suitable for fire-lances and rockets, but it differs from him (or them) in giving many more overt signs of Chinese ancestry.[c] Although al-Rammāḥ does not call saltpetre *thalj al-Ṣīn* (Chinese snow)[d], only *bārūd*, he draws a great deal from Chinese practice, partly for recreational pyrotechny;[e] and he has a variety of Chinese habits such as incorporating arsenic sulphide, lacquer and camphor in his compositions, or using expendable birds to carry incendiaries.

If we examine some of the gunpowder formulae in the book of al-Rammāḥ, we find a tendency to have the saltpetre content rather higher than anything certain in Marcus Graecus.

	% N	S	C
flying fire	71	7	21
rocket	70	11	18
firework ('flower of China')	69	13	18[f]
'Chinese arrows'[g]	72	8.7	18.8

Thus, although not so high as the theoretical value of 75%, these powders were distinctly more lively than the slower blasting or rocket levels of 60 to 68%. If

[a] This was but one of many *furūsīya* treatises on military arts, the bibliography of which has been studied by Ritter (4).

[b] The two known MSS, both in Paris, have been studied by Quatremère (2), Hime (1) and others, but the most judicious assessment is that of Partington (5), pp. 200 ff. See also Sarton (1), vol. 2, p. 1039.

[c] As Partington (5), p. 202 well noticed. This was recognised even by Mercier (1), p. 117.

[d] Cf. Vol. 5, pt. 4, p. 432.

[e] Among the fireworks there are 'wheels of China', 'flowers of China', white and green lotuses, coloured smokes (as in *WPC*, ch. 120, pp. 5b ff.), etc.

[f] Plus 10 parts of 'Chinese iron'. If this was powdered cast iron, or iron filings, it was to give a white flame (cf. Audot (1); Brock (1), p. 23'; Davis (17), p. 67). But it could have been what the Arabic alchemists called 'Chinese iron', i.e. *hadād al-Ṣīnī* or *kharṣīnī* (cf. Vol. 5, pt. 4, p. 429), in which case it would have been cupro-nickel. And copper gives a blue or purple flame (Davis (17), pp. 65, 67).

[g] Certainly rockets, the *sahm al-Khiṭāi* of Vol. 5, pt. 4, p. 432. Cf. Reinaud & Favé (2), pp. 314 ff.; Partington (5), p. 203; Zaky (4).

[1] 飛火鎗

al-Rammāḥ never talks about bombs or detonations, accidental explosions were certainly to be feared in the Arabic arsenals of the time, since the exact proportions of saltpetre would naturally sometimes be inadvertently exceeded.[a] Again here gunpowder is an incendiary used on arrows or thrown in pots from trebuchets, and as flame-thrower filled into fire-lances; it is propellant only for rockets (much more certain than in Marcus Graecus), and not behind projectiles in guns or cannon of any kind. But it is very significant that the characteristically Chinese co-viative projectiles appear, thrown out from fire-lances as balls, 'chickpeas', of burning material.[b] In the book of al-Rammāḥ many important descriptions of things are given too, such as fuses (ikrīkh) and incendiary 'bombs' or naphtha pots (qidr). By itself alone the Kitāb al-Furūsīya ... would serve as striking evidence for the westward passage of gunpowder and all military pyrotechnics from the Chinese culture-area.

Finally, al-Rammāḥ it was who gave the first description of the purification of saltpetre (bārūd) by a Muslim writer.[c] The solution of the mixture containing potassium nitrate was treated with wood ashes to precipitate the deliquescent calcium and magnesium salts, then decanted or filtered and allowed to crystallise. So this knowledge was shared with the author of the Liber Ignium.

Ḥasan al-Rammāḥ's book was not of course the earliest Arabic treatise on military incendiaries (nufūṭ). A work by Murḍā ibn 'Ali ibn Murḍā al-Tarsūsī was composed for the famous commander Saladin (Ṣalāḥ al-Dīn) about +1185. It had a memorable title: Tabṣirat arbāb al-albāb fī Kaqfyyat al-Najāh fi'l-Ḥurūb wa-nashr a'lām al-i'lām fi'l-'udad wa-'l ālāt al-mu'īna 'alā liqā' al-a'dā' (Information for the Intelligent on how to Escape Injury in Combat; and the Unfurling of the Banners of Instruction on Equipment and Engines which assist in Encounters with Enemies).[d] Here the important point is that neither saltpetre nor mixtures containing it are mentioned—which is not surprising since the first mention of potassium nitrate among the Arabs comes with Ibn al-Bayṭar about +1240.[e] The book therefore parallels the Thai Pai Yin Ching[1], and the fact that that was written four centuries earlier simply points up the décalage between China and Western Asia or Europe. There is naturally much about naft (naphtha), though it is not called Greek Fire, and 'automatic fire' is prominent too.[f] Other Arabic

---

[a] This was also the opinion of Bonaparte & Favé (1), vol. 3, p. 33.

[b] This was later widely adopted in Western pyrotechny, cf. Brock (1), pp. 191 ff. Babington in +1635 spoke of 'trunckes of fire which shall cast forth divers fire balls', and this, it has been thought, is one of the earliest references to Roman candles. Cf. Bate (1), writing in the previous year.

It is generally said that the expression itself, 'Roman candle', first appears in Marryat's Peter Simple (1842), but this cannot be right because it was prominent at a display in +1769 (Brock (1), p. 192). The name originated, according to him, as a reference to a traditional pre-Lent carnival at Rome, where each merrymaker sought to extinguish the candle of his neighbour while keeping his own alight. But there is no continuity of sense, and one wonders whether, if the usage could be traced further back, the adjective 'Roman' would not be found to mean East Roman, i.e. Byzantine—like Greek Fire itself. Were fire-lances not known to the Byzantine forces between +1250 and +1450?

[c] Tr. Partington (5), p. 201.

[d] We are grateful to Dr J. F. C. Hopkins for the englishing of this. The work has been printed and in part translated by Cahen (1). A useful summary is in Partington (5), pp. 197–8.

[e] Cf. Vol. 5, pt. 4, p. 194.   [f] Cf. p. 39 above and p. 67 below.

[1] 太白陰經

Fig. 2. An illustration from the Arabic Rzevuski MS at Leningrad, showing what must be a rocket-arrow on the left, and a *midfa'* on the right, with perhaps in firework in the middle.

MSS before +1225 give directions for the distillation of low boiling-point fractions of natural petroleum and naphtha,[a] i.e. the essential process by which Greek Fire was made.[b] Hime was undoubtedly right in saying that there was no evidence for any gunpowder weapons throughout the Crusades (+1097 to +1291),[c] and none has appeared since his time.

Arabic works in manuscript approximately contemporary with that of Ḥasan al-Rammāḥ and paralleling it exist also,[d] but a considerable step forward is represented by the Rzevuski MS at Leningrad, which was copied for a Mamlūk Sultan of Egypt, and dates from about +1350.[e] Perhaps the writer's name was Shams al-Dīn Muḥammad, but this is uncertain; the work has a jejune title something like 'Collection of Notes on Various Branches of the (Military) Art'. Now enters the mysterious term *midfa'*, which certainly denoted a tube of some kind, though whether it was originally something analogous to the siphon of the Byzantines and the flame-thrower pump of the Chinese (p. 82), i.e. the apparatus called *al-zarrāq*,[f] or else a low-nitrate gunpowder fire-lance throwing coviative projectiles,[g] or else again a true hand-gun or bombard using high-nitrate gunpowder to propel a projectile of equal diameter to the bore; and also whether it was generally made of wood, bronze or iron, are questions which have never been decisively answered. Most probably it was all of these in chronological succession. But the composition given in the Rzevuski MS, (N, (S) C % 74; 11: 15) was quite fast enough for a true gun. We also hear of the projectiles shot out, *bundūq* (originally meaning a hazel-nut); and one of these, seated in the splayed mouth of a hand-held tube on the end of a stock (Fig. 2) does really indicate a

---

[a] Partington (5), p. 199.

[b] We shall return to this (p. 76), but Partington (5), pp. 30–2, proved that it was distilled petrol thickened with sulphur, resins, etc.

[c] (1), pp. 64 ff.   [d] Partington (5), p. 207.

[e] Or possibly a century later. See Reinaud & Favé (2), pp. 309 ff., Reinaud (1), p. 203. Other references in Partington (5), pp. 204 ff., 231.

[f] Cf. Wiedemann (7), p. 38, repr. in (23), vol. 1. p. 210. The expressions *bab al-midfa* (the mouth of the projector-pipe) and *bab al-mustaq* (the mouth of the Chinese flute-pipe) occur in the *Mafātiḥ al-'Ulum* (Key of the Sciences) written by Abū 'Abdallāh al-Khwārizmī al-Kātib in +976. These occur, it is said, in naphtha-throwers and ejectors (*al-naffātat wa'l zarrāqāt*).

[g] Arrows are mentioned among these (cf. p. 270 below).

Fig. 3. Pottery naphtha container (or incendiary 'bomb'). From Mercier(1),
with cross-section on opposite page. Arabic origin.

hand-gun. If the date is right, there is not much point in denying to the Arabs
the knowledge of true guns and bombards at this time.[a] Even if Quatremère (2)
was correct in saying that *midfaʿ* did not mean a cannon until +1383, it remains
that the technique of early artillery was spreading in the Middle East and the
Maghrib in the +14th century just as it was in Europe, having started in China
in the +13th, as we shall see (p. 294). Of course the older methods still continued, for the Rzevuski MS speaks of incendiary 'bombs' thrown from trebuchets or arcuballistae,[b] and also 'Chinese arrows', i.e. rockets.

The Mamlūk dynasty, centered on Cairo, lasted from +1250 to +1517, and
the process of chemicalisation of warfare during its sway has been studied in an
interesting book by Ayalon (1). The earliest mentions of *midfaʿ* that he could find
dated from +1342 and +1352, but he could not prove that they were true hand-
guns or bombards. However, the decisive eye-witness description by the ency-
clopaedist Shihāb al-Dīn Abū al-ʿAbbās al-Qalqashandī[c] of a metal-barrel cannon shooting an iron ball at Alexandria must lie between +1365 and +1376. By

---

[a] After all, Lavin (1) shows that cannon (*truenos*) were used by the Moors besieged in Algeciras in +1343.
[b] Many of these pottery containers, notably from the siege of Fustat in +1168, are illustrated in the plates of
Mercier (1), whose book was vitiated however by a dating of Marcus Graecus a couple of centuries too early,
and a belief that saltpetre was known and used in the West several centuries before it actually was. Cf. Figs. 3
and 4. On the naphtha 'bombs' see also Lenz (1) and Gohlke (3).
[c] Mieli (1), p. 276.

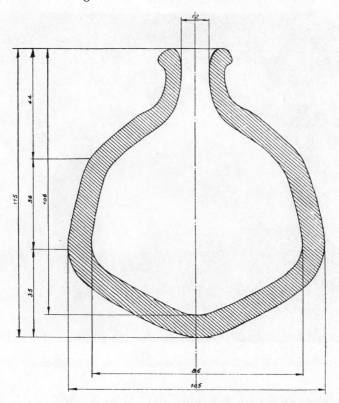

the end of the century the use of true artillery was becoming widespread just as in Europe.[a] One of the historian's troubles in this era is that as incendiary substances changed to deflagrating low-nitrate gunpowder, and as that in turn evolved into explosive and propellant high-nitrate gunpowder, the name did not change. So *naft* came to mean gunpowder, and they spoke of *midfa' al-naft*. Even when they gave that up, they transferred the term *bārūd* from saltpetre to gunpowder itself,[b] which made matters no better. True, the fact that in the beginning gunpowder was itself used as an incendiary substance renders the continuation natural enough, but can hardly condone it. Although the same unfortunate failure to develop new technical terms for new things[c] also bedevils the situation in China (cf. p. 11 above), this particular trouble does not arise there, and *huo yao*[1] (the fire chemical) invariably indicates the existence of the gunpowder mixture.[d]

[a] Fire-lances were certainly used at its beginning in the battles between Mamlūks and Mongols in +1299 and +1303, possibly hand-guns also; cf. Hassan (1) Ayalon (3). It is much to be wished that the relevant Arabic MSS may soon be translated and published.

[b] On this process see the valuable encyclopaedia article by Colin, Ayalon *et al.* (1).

[c] I came across this first many long years ago in the history of biology and embryology; cf. Needham (2), p. 215 (27).

[d] In these contexts, that is to say; for occasional mystical uses of the term can be found. See Vol. 5, pt. 5, pp. 240, 248; and pp. 6–7 above.

[1] 火藥

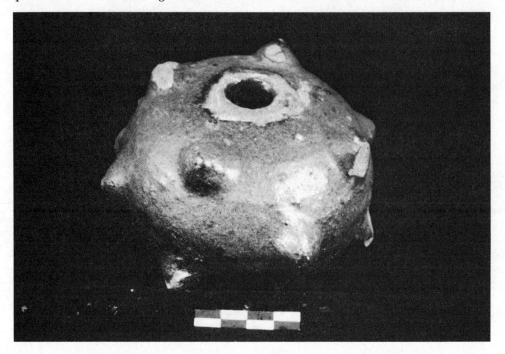

Fig. 4. A pottery naphtha or Greek Fire container from China (photo. Nat. Historical Museum, Peking)

Certain passages from Ibn Khaldūn and al-Qalqashandī are worth giving in full. The former does not occur in the famous *Muqaddimah*[a] but in his 'History of the Berbers and the North African Kingdoms'. Written about a century later (*c.* +1382), it relates to a year-long siege of the city of Sijilmāsa[b] in +1274.

There were [he says] trebuchets (*madjanikh*) and ballistas (*harradāt*) but also a 'naphtha engine' (*hindām al-nafṭ*), which hurled a kind of 'iron gravel' (*ḥasa al-ḥadīd*). This 'grape-shot' is thrown out from the tube or chamber (*khâzna*) of the thing by means of an inflammable powder (*bārūd*), the extraordinary effects of which rival the powers of the Creator himself.[c]

Reinaud & Favé[d] knew this passage, but did not believe it, thinking that Ibn Khaldūn was absent-mindedly attributing to the siege of +1274 a technique which he himself had seen by +1384. However, the account does not seem impossible to us if we interpret it as a description of a fire-lance or eruptor with co-viative projectiles (cf. pp. 220, 263) rather than a true gun or cannon.[e] This

[a] Tr. Rosenthal (1); Monteil (1).
[b] This was a town east of the Moroccan Atlas mountains, almost due south of Fez, and south-west of Tlemcen. The attacking force was that of the Marīnid Sultan Yaʿkūb.
[c] Tr. de Slane (3), vol. 4, pp. 69–70. Other accounts, even by Ibn Khaldūn himself (e.g. vol. 3, p. 356) make no mention of the engine. Nor does it appear in the previous siege of +1261 (vol. 4, p. 68). Another translation is in Ayalon (1), p. 21. Attention was again drawn to this passage by Hassan (1).
[d] (1), pp. 73 ff.   [e] Partington (5), pp. 191, 196 concurs.

would consort well enough with what Ḥasan al-Rammāḥ tells us (p. 41 above), as also with the passage of such gunpowder techniques from China previously. But gunpowder has no name of its own; saltpetre and naphtha both do duty for it.

Very different in material content, if not in phaseology, is the passage in al-Qalqashandī's geography and description of Egypt and its government written a decade or so before his death in +1418. He says:[a]

And (one kind of) these (siege instruments) is *makāhil al-bārūd*, and these are *al-madāfiʿ* from which one shoots by means of *naft*. In part they shoot big arrows, which almost pierce a stone,[b] and in part they shoot balls of iron weighing from 10 to over 100 Egyptian *ratls*. And (another kind of) these (instruments) is *qawārīr al-naft*, and these are *qudūr* and the like (pots), into which the *naft* is put, and then they are (lit and) thrown at fortresses with the purpose of burning with fire.

Here there were clearly true bombards as well as incendiary projectiles of 'petrol', but nothing could better show the confusion of the Arabic terminology, in which neither saltpetre nor naphtha was distinguished from gunpowder.

Indeed, current standard Arabic still to this day speaks of gunpowder as *bārūd*, and saltpetre is now called *milḥ al-bārūd* (the gunpowder salt).[c] Occasionally, however, there was used another term, *dawāʿ*, the 'drug',[d] normally applied to any prescription or therapeutic preparation, or even to wine;[e] but here it may have some historical significance since it evokes so closely the Chinese term *huo yao*, 'fire-drug'[f] or 'fire-chemical'.

We must now return to +13th-century Europe and take up the celebrated references to gunpowder in Roger Bacon and Albertus Magnus. One may recall that the Franciscan 'doctor mirabilis' was born probably in +1219[g] and died about +1292.[h] Among his earlier works was that commentary on the translation of Pseudo-Aristotle, *Secretum Secretorum*, which we came across in the context of elixir chemistry;[i] this belongs to the time between +1243 and +1257, at which

---

[a] Ayalon (1), pp. 21–2 (3). We could not find the passage in the translation of Wüstenfeld (1), so perhaps it only occurs in certain MSS.

[b] Note the similarity with Walter de Milamete's bombards (p. 287, Fig. 82).

[c] For this information we are indebted to Prof. Douglas Dunlop. The transition from *al-naft* to *bārūd* for what was essentially a new thing, different from either, was already pointed out by Casiri (1) in 1770, vol. 2.

[d] Cf. Partington (5), pp. 204–5, 313, following Reinaud & Favé (2), p. 310, from the Rzevuski MS. (c. +1350), cf. p. 43 above.

[e] Frankel (1), p. 163.   [f] Cf. the case of 'kraut'; p. 108.

[g] Some say +1214, and there is evidence for both dates.

[h] On Roger Bacon's life see DSB, vol. 1, pp. 377 ff.; Sarton (1), vol. 2, pp. 952 ff.; and Thorndike (1), vol. 2, pp. 616 ff. On his alchemical interests see Multhauf (1), p. 188; Welborn (1). Partington (5), pp. 64 ff. gives an elaborate discussion of Bacon and Albertus from the point of view of the history of explosives, but we cannot follow him in every respect. We recall, however, a fascinating discussion with him on the subject in Jan. 1959.

[i] Vol. 5, pt. 4, p. 494. This work, the title of which could be expanded as 'The Secret of all Secrets, which Aristotle expounded to Alexander the Great', seems to have been of Arabic origin, about +800. This *Kitāb Sirr al-Asrar* was translated into Latin by Philip of Tripoli (or, of Salerno) about +1200. It exists in a host of manuscripts, in several languages besides Latin; and discusses personal conduct, royal policy, medicine, astrology, and all kinds of real or supposed strange natural phenomena. See Thorndike (1), vol. 2, pp. 267 ff., 310, 633.

lacks all manuscript authority, and appears only in the earliest printing (+1542), where it may be a corruption of some Greek quotation.[a] It does not even belong to one of the probably authentic chapters. Still, in ch. 6 we find an approximate repetition of the passages about the gunpowder crackers just given, and prophecy of 'greater horrors' to come.[b]

As for the involvement of the great Dominican, Albertus Magnus, a scholar 'who had the honour of being beatified both by the Church and by science',[c] it turns out to be, like the Baconian cryptogram, a non-starter. Albert of Bollstadt was born in +1193[d] and died in +1280, but the *De Mirabilibus Mundi* (On the Wonders of the World) is of highly doubtful authenticity, and may not be +13th-century at all.[e] Perhaps it was written by Arnold of Liège about +1300, possibly by Albert of Saxony as late as +1350. In any case, what it says about gunpowder and 'flying fire' is verbally identical with §13 of Marcus Graecus,[f] so it adds nothing to our picture of the earliest European knowledge of the explosive mixture.

There are two outstanding parallels between Roger Bacon's knowledge of gunpowder and what happened in other fields. First, mechanical clocks. We know that almost the first reference to a time-keeper of this kind comes from Dante in +1319,[g] but also that Western men were hard at work about +1271 trying to arrest the motion of a wheel[h] so as to make it keep time with the apparent diurnal motion of the heavens.[i] So also the first bombard got into an illumination in +1327,[j] while Bacon knew approximately the gunpowder formula in +1268.[k] Secondly, just as he was the first person in the West to acquire this knowledge and to write about it, so he was also the first Westerner to talk like a Taoist, saying that if only we knew more about chemistry human life could be immeasurably prolonged.[l] Chinese influences in all these parallels are unmistakable. And the paradox was only repeating itself 120° of longitude West, that those who sought elixirs of longevity and immortality should find an explosive mixture as well—all knowledge and all skill inevitably fraught with danger if mankind should not be conscious of the ethics of the employment of such mastery.

Gunpowder formulae are of course given in all those early European treatises of military technology at which we have already had a look (p. 40). The oldest

---

[a] See the special studies of Steele (4); Sarton (1), vol. 2, pp. 958, 1038 and Thorndike (1), vol. 2, p. 688. Some of the passages to which Hime drew attention were not about charcoal at all, still less about saltpetre.

[b] Text and translation in Partington (5), pp. 75–6.

[c] Needham (2), p. 73.

[d] Some say +1206, but which is right we may never know. For his life, cf. DSB, vol. 1, p. 99.

[e] See Thorndike (1), vol. 2, pp. 720 ff., 724, 737.

[f] Hoefer (1), vol. 1, p. 390 was perhaps the first to notice this.

[g] Cf. Vol. 4, pt. 2, p. 445.

[h] Robertus Anglicus' commentary on Sacrobosco's *Sphaera*, cf. Needham, Wang & Price (1), p. 196.

[i] This had first been done, as we know, by I-Hsing and Liang Ling-Tsan about +725.

[j] Cf. Fig. 82.

[k] As Tsêng Kung-Liang had known it about +1040.

[l] See Vol. 5, pt. 4, pp. 492 ff.

illustrated MS.[a], which may be dated about +1395, contains a recipe for gun-powder[b], a method for the making, purifying and testing of saltpetre, together with crude coloured illustrations of guns. Konrad Kyeser's *Bellifortis*, written between +1400 and +1405, mentions rockets, guns and a number of peculiar formulae for gunpowder.[c] München Codex 197 is a composite work, the note-book of a military engineer writing in German, the Anonymous Hussite, and that of an Italian, probably Marianus Jacobus Taccola, writing in Latin; it con-tains dates such as +1427, +1438 and +1441. It gives gunpowder formulae and describes guns with accompanying illustrations.[d] A curious feature, very Chinese (cf. pp. 114, 361), is the addition of arsenic sulphides to the powder; this dates from fire-lance days but probably had the effect of making it more brisant, hence it could have been useful in bombs and grenades.[e] The +15th-century Paris MS.[f], supposedly before +1453, *De Re Militari*, perhaps by Paolo Santini, shows a gun on a carriage with a shield at the front, mortars shooting incendiary 'bombs' almost vertically to nearby targets, a bombard with a tail (cerbotane or tiller), and a mounted man holding a small gun with a burning match. But we need not pursue the European literature further here.

(iii) *Speculations and research contributions*

Roger Bacon and his friends could have had only the faintest idea about the location of those 'diverse parts of the world' which had produced the crackers, and where the invention of gunpowder had been made, though the travelling friars of the +13th century had been well aware of the existence of Cathay. But by the +15th, after an abundant use of firearms during the preceding century, Europeans were less oecumenically minded, and it seemed inconceivable to them that the invention could have occurred in any other continent—hence the legend of Berthold Schwartz, variously believed to have been an alchemical monk or friar, usually thought German, but in the earliest sources (perhaps significantly) a Byzantine Greek.[g]

Probably the earliest appearance of this personage is found in a MS. of about +1410, described by Köhler,[h] where he is a 'Meister von Kriechenland', Niger Berchtoldus. A more circumstantial relation was apparently given by Felix Hemmerlin, writing about +1450, in his *De Nobilitate et Rusticate Dialogus*,[i] but

---

[a] München, Cod. Germ. 600.
[b] Basically this is N: S: C % 71·4: 14·3: 14·3, but it had sal ammoniac and camphor as well.
[c] See Partington (5), pp. 146 ff.; Berthelot (5, 6).
[d] This was fully described by Berthelot (4); cf. Partington (5), pp. 144–5.
[e] Berthelot (14), p. 692, quoted an entry of +1342 in the *Account-Book of the Bonis Brothers*, vol. 2, p. 127, on the addition of arsenic to gunpowder. It was thought to increase the range.
[f] BN Latin 7239; described in Berthelot (4) and Partington (5), pp. 145–6.
[g] See Hansjakob (1); Feldhaus (1), pp. 78 ff. (30–32). The widespread literature has been assembled in the critique of Partington (5), pp. 91 ff.
[h] (1), vol. 3, pt. 1, p. 244. The MS. is in the Ambraser Coll. no. 67 (148); it was printed towards the end of the +15th century (Berlin Incun. 10117a).
[i] Not in the +1490 printing, however, only in the later editions of +1495 and +1497.

the details need not detain us. After this, the story was repeated by innumerable writers, including Polydore Vergil[a] from +1500 onwards, and Guido Panciroli[b] from +1600.[c] Panciroli was puzzled, like many others, by the long time which the invention had taken, if the story of its much earlier appearance in the East was true, to find dissemination among the European peoples.

Some Writers of the Indian History[d] tell us [he said], that Guns as well as Printing were found out by the Chinese many Ages ago. They say also that they were in Use among the Moors long before they were known in Germany: But how is it possible or credible, that an Instrument so necessary for the besieged to repel the Attacks of their Enemies, should lie dormant so long? Whereas, as soon as ever the Use of Guns was known to the Venetians, and Printing to the Romans, it was presently communicated to other People, so that now nothing is more common throughout the World.

Already in +1572[e] Sebastian Munster had posthumously popularised the account of Schwartz,[f] having obtained it, so he said, from his friend Achilles Gasser.[g]

For nearly five centuries the Berthold Schwartz story battled with the growing conviction that gunpowder, unknown to the ancients, had originated in the East. The only surprising thing is that it lasted as long, a tribute to the Europocentrism of Westerners. William Camden the antiquary, writing in +1605, was sceptical about the Eastern origin.[h]

If ever the witte of man [he said] went beyond belief itselfe it was in the invention of artillarie or Engines of warre...

Some have sayled a long course as farre as China, the farthest part of the world, to fetch the invention of guns from thence, but we know the Spanish proverb 'long waies, long lies'. One writeth, I know not upon whose credit, that Roger Bacon, commonly called Friar Bacon, knew how to make an engine which with Saltpeter and Brimstone, should prove notable for Batterie, but he, tendering the safety of mankind, would not discover it. The best approved authors agree that guns were invented in Germanie, by Berthold Swarte, a Monke skilful in Gebers Cookery or Alchimy, who tempering Brimstone and Saltpeter in a mortar, perceived the force by casting up the stone which covered it, when a sparke fell upon it....

---

[a] (1), bk. 2, ch. 11.   [b] (1), bk. 2, ch. 18, Eng. tr., p. 383.
[c] Cf. Vol. 4, pt. 2, p. 7 and Fig. 352 opposite it.
[d] The Further Indies included, of course.
[e] This was also the year of publication of Luis de Camoens' *Lusiados*. Some have thought that in that great poem he recognised the Chinese priority in firearms because of the lines (Canto 10, stanza 129)

> Here ere the cannons' range in Europe roared
> The bombard's thunder on the foe was poured.

But this seems not to be in the original, and cannot be found in the best translations, such as those of Fanshawe (1), p. 300 or Atkinson (1), pp. 243–4. It only occurs in Mickle (1), p. 342, who had a reputation for making unauthorised insertions and deletions, in this case accompanying it by a long and vitriolic footnote on Chinese culture and technology.
[f] *Cosmographia Universalis*, bk. 3, ch. 174. It is not in the earlier editions.
[g] Cf. Gasser (1), pt. 2, p. 108.
[h] Edition of +1614, pp. 238 ff. The passage was modernised and abridged in Ffoulkes (2), p. 92.

Camden was sure that firearms had been used at the siege of Calais in +1347, and here he was quite right.[a]

One can see the balance trembling in the writings of the Jesuit, Athanasius Kircher. In his *Mundus Subterraneus* of +1665 he reported (with some picturesque details, says Partington) that Berthold Schwartz had been an alchemical Benedictine of Goslar, who invented gunpowder in +1354 (!), and later made it known in Italy, where it was used in the wars between Venice and Genoa.[b] But two years later, in his *China Illustrata*, he spoke quite differently.[c]

Besides, many inventions were seen in China before we ever got them in Europe, and three in particular may be mentioned. First, printing, and what it is, I shall explain...

Another was the invention of gunpowder, which it is not possible to deny took place long before our times in China. The Fathers of our Society have testified that they have seen in many provinces great cannon, especially at Nanking, which were cast long ago, time out of memory; although the pyrabolical art was not brought to such a height as we Europeans have now attained. One thing, however, is sure, namely that the Chinese have been outstanding in the art of the foundry for the casting of guns and cannon, which appears partly in the making of statues of cast (iron and bronze), and partly in the casting of massy cannon, which occur only occasionally in other nations.

Then all through the eighteenth century the missionary writers, some in the West (such as de Faria y Sousa in +1731)[d], others in China (such as Gaubil in +1739, de Mailla in +1777 and Amiot in +1782) insisted that gunpowder and firearms had originated there and nowhere else. By the end of the seventeenth century this conviction was already strong enough to induce the military commander Louis de Gaya to aver that Berthold Schwartz had got them from the 'Tartars' when travelling in the East. Thus in +1678 he wrote:[e]

We have had the invention of Gun-powder from China, by means of the communication that a Monk named Bertholdus had with the Tartars in his Travels in Muscovy, about the year +1380. And therefore the Portugese were never so much surprized as when upon their accosting these unknown Countreys, they saw a great many Ships equipped and ranked in Bataillia ... but their surprize augmented when they heard the Guns fire; when they expected no such thing. So that it is not true that the Monk was the first inventor of Gun-powder; he was no more than the publisher of a Secret which he learnt from the Tartars, and which he had better kept to himself, without trying an experiment of it, that cost him so dear, and which buried him in the Furnace which he himself contrived.

[a] Partington (5), p. 109; Brackenbury (1), p. 303.
[b] (5), vol. 2, p. 467. This was the conventional wisdom followed by Thomas Sprat in his *History of the Royal Society* (+1722), pp. 260 ff., 277 ff., esp. p. 267.
[c] (1), p. 222, eng. auct.
[d] (1), p. 91. He also recognised how devoted the Chinese had been to alchemy and longevity elixirs (p. 26).
[e] (1), Harford tr., p. 53.

To make the Monk thus a mere transmitter was exactly what some of the scholars of China and Japan themselves propounded after they had come to hear of him, as we shall shortly see.[a]

Besides, the idea was not a new one. It had already been suggested, as far back as +1585, by Juan de Mendoza, in his famous *Historia de la Cosas mas Notables, Ritos y Costumbres del gran Reyno de la China*. One section of this was entitled: 'How that with them they have had the use of Artillery long before us in these parts of Europe'. In Robert Parke's translation:[b]

Amongst many things worthie to bee considered, which have beene and shal be declared in this historie, and amongst many other which of purpose I omit, because I would not be tedious with the reader, no one thing did cause so much admiracion unto the Portugals, when that they did first traficke in Canton, neither unto our Spaniards, who long time after went unto the Philippinas, as to find in this kingdome artillerie. And wee finde by good account taken out of their histories, that they had the use thereof long time before us in Europe.

It is said that the first beginning was in the yeare 1330, by the industrie of an Almane, yet howe he was called there is no historie that dooth make mention; but the Chinos saie, and it is evidently seene, that this Almaine dooth not deserve the name of the first inventer, but of the discoverer,[c] for that they were the first inventors, and from them hath the use thereof been transported unto other kingdomes, where it is now used...

Meanwhile Isaac Vossius, who did so much in the seventeenth century to give credit where credit was due, included in his *Variarum Observationum Liber* of +1685 two pieces about gunpowder in relation to China. The first was part of his *De Artibus et Scientiis Sinarum*, and the second followed immediately upon it—*De Origine et Progressu Pulveris Bellici apud Europaeos*.[d]

Vossius began by saying that 'the powder of nitre, with cannon great and small', usually considered an invention of Christians, had in fact been very well known to the Chinese sixteen centuries before his time; while guns of exquisite workmanship dated there at least eight centuries back. The first of these statements was a wild exaggeration, the second rather a good guess. The Siamese, as Tabernarius[e] had rightly affirmed, got their gunpowder originally from China, and better made too, than that of Christendom. The Europeans, though in most matters of military art long superior to the Chinese, must yield in part to them where the warlike uses of gunpowder were concerned. As for recreational fire-

---

[a] There were occasional writers in China who believed that gunpowder had been discovered somewhere else. For example, one of de Gaya's contemporaries, Fang I-Chih[1], in his *Wu Li Hsiao Shih*[2] (Small Encyclopaedia of the Principles of Things) of +1664 (ch. 8, p. 26a, b) said that gunpowder came from the outer barbarians (*wai i*[3]). But it must have come before the Thang, because he believed that the fireworks of that time had used it.

[b] (1), vol. 1, ch. 15, p. 99 (+1588 ed.), pp. 128–9 (Hakluyt Soc. ed.).

[c] I.e. the one who made it generally known.

[d] They are chs. 14 and 15 respectively, pp. 83–4 and pp. 86 ff.

[e] This was J. B. Tavernier, whose travels in Turkey, Persia, India and Siam were published (1) in +1676.

[1] 方以智   [2] 物理小識   [3] 外夷

works using the same powder, in these the Chinese truly excelled, what with flames of all colours, forms and figures of any kind they pleased, nay, whole pictures picked out with light in the empty air: 'The Europeans in all their wars have not lavished more gunpowder than the Chinese have in these joyful spectacles.' And then he goes on about the Great Wall, and how the dregs of society were sent to guard it, the Chinese being greatly given to peaceful literary pursuits; until during the past forty years the wars caused by the invading Manchu Tartars had led to great misery. In any case, he concluded, 'everything we have in the arts and sciences we owe either to the Greeks or to the Chinese'.

In the second of his essays Isaac Vossius was really rather sagacious. He was sure that gunpowder had not been known in the West above four hundred years past (which would make it +1285), and he did not believe that either Roger Bacon or any other nameable person had been responsible for it. Fire itself, of course, had been used in war much earlier; which was easy to show from the 'automatic fire' of Julius Africanus,[a] the naphtha of the Persian and Gothic wars, and the Greek Fire invented by Callinicus about +685 or a little before. It was still incendiaries that frightened the host of St Louis in the Crusades, as told by Joinville.[b] 'Of flame and thunder there is often mention, but of stone or metal balls, or explosions, all is silence.' So who among the Christians had first begun to use gunpowder bombards with projectiles of iron, lead or stone, we really did not know. The description of one of these guns of 20 in. calibre in Froissart[c] remained among the earliest references. The danger of bursting such pieces had led to the ribaudequins or multiple-barrel cannon, where the charges could be less. Who had first introduced corned powder[d] Vossius also did not know, but Tabernarius (Tavernier) had, he said, described the gunpowder compressed into little rods characteristic of Tonking and Siam, which was very good. On account of their love of peaceful humanism, the Chinese had neglected to adopt such improvements, and so when they were necessitated to oppose the Manchu Tartars, they prevailed upon Christian Masters to cast cannon for them.[e]

[a] Ca. +225. See Partington (5), pp. 5 ff.
[b] (1), pp. 235–6; cf. de Wailly (1); J. Evans (1). Joinville lived from +1224 to +1319, and wrote his *History of St Louis* in +1309, remembering the events of the Sixth Crusade (+1248 to +1254) of which he had been a member in the train of the sainted king. See Sarton (1), vol. 4, p. 928.
[c] It applied to a cannon used at the siege of Oudenarde in +1382 by the troops of Ghent (cf. Partington (5), p. 103), so it was not in fact particularly early. Brackenbury (1), p. 15 deduced the dimension from Froissart's expression 'cinquante trois pouces de bec'. The chronicler did give earlier examples, however, notably one at Quesnoy in +1340, which shot crossbow bolts like Walter de Milamete's, (1), bk. 1, ch. 111 in vol. 1, pp. 310–11; Brackenbury (1), p. 294. Froissart is also, of course, one of the main sources for the English use of artillery (three small bombards) at the Battle of Creçy, +1346 (de Lettenhove (1), vol. 1, pt. 2, p. 153; vol. 5, p. 46; Partington (5), p. 106; Brackenbury (1), pp. 297 ff.). Multiple-barrel gun-carriages (ribaudequins) are mentioned by Froissart too (Partington (5), pp. 116, 138; Brackenbury (1), p. 14). He was almost exactly contemporary (+1337 to +1410) with Timur Lang.
[d] Granular gunpowder may have been invented at Nürnberg about +1450 (Räthgen (1), pp. 77, 109 ff.); in any case, it was widely used by +1550 (Partington (5), p. 174). So we still do not know who started it.
[e] This was an obvious reference to the Jesuits Adam Schall von Bell and Ferdinand Verbiest. We have already said something about their work as gun-founders in China (Vol. 5, pt. 3, pp. 240–1) and shall return to it later (p. 395 below).

'But there are none of those cunning missiles which subvert cities, overthrow walls, and repel enemies from ramparts, used by the Europeans, which were not made long before by the Chinese, even though they preferred the arts of peace in which they were supreme.' So gunpowder had come to Europe from East Asia, said Isaac Vossius, but exactly how, he could not tell. Nor can we; though we can probably make a few better guesses, and we do now know the decades during which it must have happened.[a]

We may perhaps end this digression on Black Berthold (if such it is), today chiefly of antiquarian interest, though not without instructiveness on the mental compulsions of Europeans in former days, by quoting Hermann Boerhaave's *Elementa Chemiae* of +1732.[b]

It were indeed to be wish'd [he wrote] that our art had been less ingenious in contriving means destructive to mankind; we mean those instruments of war, which were unknown to the ancients, and have made such havoc among the moderns. But as men have always been bent on seeking each other's destruction by continual wars; and as force, when brought against us, can only be repelled by force; the chief support of war, must, after money, be now sought in chemistry.

Roger Bacon, as early as the twelfth century [*sic*], had found out gunpowder, wherewith he imitated thunder and lightning; but that age was so happy as not to apply so extraordinary a discovery to the destruction of mankind. But two ages afterwards, Barthol. Schwartz,[c] a *German* monk and chemist, happening by some accident to discover a prodigious power of expanding in some of this powder which he had made for medicinal uses,[d] he apply'd it first in an iron barrel, and soon after to the military art, and taught it to the *Venetians*. The effect is, that the art of war has since that time turned entirely on this one chemical invention; so that the feeble boy may now kill the stoutest hero: Nor is there anything, how vast and solid soever, can withstand it. By a thorough acquaintance with the power of this powder, that intelligent *Dutch* General *Cohorn*[e] quite alter'd the whole art of fighting; making such changes in the manner of fortification, that places formerly held impregnable, now want defenders. In effect, the power of gunpowder is still more to be fear'd.

I tremble to mention the stupendous force of another powder, prepar'd of sulphur, nitre, and burnt lees of wine;[f] to say nothing of the well-known power of *aurum fulminans*.[g] Some person taking a quantity of fragrant oil, chemically procured from spices, and mixing it with a liquor procured from salt-petre, discover'd a thing far more powerful than gun-powder itself; and which spontaneously kindles and burns with great fierce-

---

[a] Cf. pp. 568 ff. below.

[b] (1), vol. 1, pp. 99 ff., in the English translation of Peter Shaw, +1753, vol. 1, pp. 189 ff.

[c] Here came Shaw's footnote, given in full below.

[d] How exactly this corresponded with the true course of events in China many centuries earlier, cf. p. 117 below.

[e] B. van Cohorn, military engineer, +1641 to +1704.

[f] 'Fulminating powder', consisting of sulphur with potassium nitrate and carbonate. See Ure (1), first ed., 1821.

[g] A complex compound approximating to aurous ammonium hydroxide, discovered by Oswald Croll before +1609 (Partington (7), vol. 2, p. 176).

ness, without any application of fire.[a] I shall but just mention a fatal event which lately happen'd in Germany, from an experiment made with balsam of sulphur terebinthinated, and confined in a close chemical vessel, and thus exploded by fire: God grant that mortal men may not be so ingenious at their own cost, as to pervert a profitable science any longer to such horrible uses. For this reason I forbear to mention several other matters far more horrible and destructive, than any of those above rehearsed.

Whatever would Boerhaave have said, one wonders, about nuclear weapons? He was a high-minded and far-seeing chemist, but the part about Schwartz, which mainly interests us here, received a debunking footnote from Boerhaave's translator, Peter Shaw, in +1753, a footnote which we cannot omit, since it shows how careful history as well as sympathetic ethnology was proving the Schwartz story legendary. He simply said:

What evidently shows the ordinary account of its invention false is, that *Schwartz* is held to have first taught it to the *Venetians* in the year 1380; and that they first used it in the war against the *Genoese*, in a place antiently called *Fossa Caudeana*, now *Chioggia*.[b] For we find mention of fire arms much earlier: *Peter Messius* in his *variae lectiones*, relates that *Alphonsus XI*, king of Castile, used mortars against the *Moors*, in a siege of 1348;[c] and *Don Pedro*, bishop of *Leon*, in his chronicle, mentions the same to have been used above four hundred years ago[d] by the people of *Tunis*, in a sea-fight against the *Moorish* king of *Sevil*. *Du Cange* adds that there is mention made of this powder in the registers of the chambers of accounts in *France*, as early as the year 1338.[e]

Thus no one could fix his exact date, or find evidence of his existence.[f]

To sum it all up, Partington concluded[g] that 'Black Berthold is a legendary figure like Robin Hood (or perhaps better, Friar Tuck); he was invented solely for the purpose of providing a German origin for gunpowder and cannon'.[h] If we widened this to European in general we would not go far wrong.

[a] This and the ensuing experiment were examples of the explosive oxidation of organic substances by concentrated nitric acid; cf. Mellor (1), p. 512. What Boerhaave 'forbore to mention' is not altogether obvious, for mercuric fulminate was not found and studied till the early years of the following century. Cf. Partington (10), p. 400.

[b] Cf. Brackenbury (1), p. 29.

[c] Actually, the bombards were used by the Moors against the Spaniards at the siege of Algeciras, and the date was +1343; cf. Lavin (1).

[d] This would make it about +1350.

[e] Perhaps he got some of this from Gram (1), who had made the point earlier in Denmark.

[f] This was also the conclusion of Gohlke (1) in 1911. A vivid picture of these perplexities occurs (as Dr Michael Moriarty has reminded us) in Laurence Sterne's *Tristram Shandy*, bk. 8, p. 517 (+1765) where Uncle Toby is discussing the origins of gunpowder with Corporal Trim. He knows that Bacon long preceded the supposed date of Schwartz, and also gives several examples of the use of gunpowder weapons in war during that time, adding: '"And the Chinese embarrass us, and all accounts of it, still more, by boasting of the invention some hundreds of years even before him." "They are a pack of liars, I believe" cried Trim.' And Uncle Toby goes on to say that he thinks they must somehow be deceived, because of the backward state of fortifications among them. This of course took no account of the role of distinctively modern science and mathematics in such designs in the Western world alone.

[g] (5), p. 96.

[h] He lives on to this day, however, in the works of such writers as Laffin (1), p. 15; or is dismissed along with all the Chinese evidence in a common uncritical condemnation, as in Lindsay (1), p. 14.

One of the ironies of the situation was that Schwartz got transmigrated into East Asian literature to mystify the scholars of that part of the world. In 1832 J. N. Calten, a Dutch gunnery officer, wrote a book entitled *Leiddraad bij het Onderrigt in de Zee-artillerie* ... in which he said that gunpowder had been discovered accidentally in an alchemical laboratory by Schwartz in +1320, who later invented cannon to make use of it. This book was translated into Japanese by the chemist Udagawa Yōan[1],[a] and his colleagues, who incorporated it into the *Kaijo Hojitsu Zensho*[2] (Complete Treatise on Naval Artillery) about 1847.

But Schwartz was also figuring in Chinese dress. In the course of his campaign to demonstrate that all the arts and sciences had originated in China, Wang Jen-Chün[3] in his *Ko Chih Ku Wei*[4] of 1895 chose several interesting quotations on gunpowder.[b] His best came from the *Ying Huan Chih Lüeh*[5] (Geography of the Vast Sphere), a clearly written treatise by Hsü Chi-Yü[6],[c] which in 1848 devoted much space to Europe, with mention of machinery, steam-engines and industry in the various separate countries. On gunpowder Hsü wrote:[d]

The technique of artillery (*huo phao*[9]) was invented in China, and European people did not know of it. At the end of the Yuan period[e] an Jih-erh-man[10] person (i.e. Aleman, German) named Su-Erh-Ti-Ssu[11] (i.e. Schwartz) started to imitate the art, but hardly attained the right way of managing it. During the Hung-Wu reign-period[f] of Ming, Ti-Mu-Erh Wang[12](i.e. Timur Lang or Tamerlane) of Sa-Ma-Erh-Han[13] (i.e. Samarqand) was very powerful in the Western countries; some Europeans enlisted in his army, and afterwards returned home taking gunpowder and cannon (*huo yao phao*[14]) with them. They got to know the whole technique of it, changing and improving its methods, so that they developed the 'bird-gun' (*niao chhiang*[15]) musket, and used it in a multitude of battles to gain innumerable victories. And they built great ships to sail all the seas, so that they appropriated vast territories such as Siberia and Malaya, including the Indies and all the islands of the South Seas. Their victories criss-crossed the four quarters, and now they own more than ten (Eastern) countries.

All this was ingenious, but Tamerlane will not work, because the first of the Timurid emperors, Amir Taimur Sāhib Qirān, was not born till +1336 and did not set out on his conquests till +1370, by which time gunpowder weapons had

---

[a] +1798 to 1846; cf. Vol. 5, pt. 3, p. 255.

[b] Ch. 2, pp. 27*b*, 28*a*. He also quoted from the narrative of a journey in Russia in 1887, the *O Yu Hui Pien*[7] by Miu Yu-Sun[8] (ch. 12, p. 3*a, b*). But this was very confused. Miu did not want to believe that the saints and sages of China could have invented anything so poisonous to mankind as gunpowder, so he supposed that it had come from the Arabs; but here he fell into the common mistake of interpreting the counterweighted trebuchets used at the siege of Hsiang-yang (+1267 to +1273) as cannon. Then he brought in Timur Lang, suggesting, like Hsü Chi-Yü, that Russian soldiers serving under that conqueror had taken it back to the West. Nothing of this stands up today.

[c] +1795 to 1872; Hummel (2), p. 309.

[d] Ch. 4, p. 3*b*; cf. p. 8*b*.   [e] I.e. about +1350.   [f] +1368 to +1398.

[1] 宇田川榕庵	[2] 海上砲術全書	[3] 王仁俊	[4] 格致古微
[5] 瀛環志略	[6] 徐繼畬	[7] 俄遊彙編	[8] 繆祐孫	[9] 火炮
[10] 日耳曼	[11] 蘇爾的斯	[12] 帖木兒王	[13] 撒馬兒罕	[14] 火藥砲
[15] 鳥鎗			

been known and used in Europe for more than forty years.[a] Still, this voice from East Asia was echoing the conviction of Louis de Gaya long before that Schwartz had been a transmitter, not an originator. It was not given to Hsü Chi-Yü to know that in fact he had no real existence at all, but since this is indubitably the case we shall now dismiss him into the realm of legend, and speak no more of him in our history.

Meanwhile, all kinds of unacceptable accounts of the origins of gunpowder and firearms were circulating in China. For example, both Gaubil (12) and Amiot (2),[b] Jesuits working there in the +18th century, adopted the persistent legend that gunpowder had been known in the +3rd[c] and had been used by the Captain-General of Shu, Chuko Liang[1], for constructing land mines (ti lei[2]). Then in the +15th century the Ming book Wu Yuan[3] (Origins of Things) by Lo Chhi[4] averred that guns (chhung[5]) were first made by Lü Wang[6] (in the −11th century),[d] and sticks of fire-crackers (pao chang[7]) invented by Ma Chün[8] of the Wei Kingdom (in the +3rd century).[e] Lo Chhi also said that emperor Yang Ti[9] of the Sui dynasty (in the +6th century) had used gunpowder for fireworks and miscellaneous amusements;[d] while Liu An[10] (Huai Nan Tzu), the naturalist-prince of the −2nd century, first prepared saltpetre (yen hsiao[11]).[f] These sayings were all reproduced and elaborated by Tung Ssu-Chang[12] in his Kuang Po Wu Chih[13] (Enlargements of the 'Records of the Investigation of Things') of +1607.[g] Fêng Chia-Shêng rightly dismissed such claims as legendary, putting them in the same category as those of Europe which attributed the invention of gunpowder to Marcus Graecus, Albertus Magnus or Berthold Schwartz, if not Roger Bacon.[h] He agreed with Hallam's idea that gunpowder was discovered accidentally by several people rather than invented by any individual.[i] As for the legend that guns were introduced by Lü Wang, it was obviously self-

---

[a] Amir Taimur (Tamerlane) died in +1405 at his capital, Samarqand, having conquered Kandahar, all Persia, Baghdad, Delhi and Cairo. He was an enemy of the Ottoman Turks under Bajazet, whom he defeated in +1402, and consequently friendly with the Byzantine emperors, especially Manuel Palaeologus. It was one of his descendants, Babar, who founded the Mughal (Mogul) empire centred on Delhi. His extraordinary career stimulated two English plays, one by Christopher Marlowe in +1590 and another by Nicholas Rowe in +1702.

[b] Suppl., p. 336. Cf. Hime (1), pp. 86 ff.

[c] Already we have seen appearances of this, pp. 25, 28 above.

[d] P. 30.

[e] We translated a long passage on the life of this remarkable engineer in Vol. 4, pt. 2, pp. 39 ff. Bamboo crackers he would certainly have known, but crackers containing gunpowder, no. Attempts have been made in recent times to substantiate Ma Chün's connection with gunpowder, as by Wang Yü (1), but it cannot be done.

[f] P. 32. Whatever fireworks Sui Yang Ti had (cf. p. 136 below) they did not contain gunpowder; but a knowledge of saltpetre on the part of Liu An is not impossible, as we shall see (p. 96 below).

[g] Ch. 33, p. 51b, ch. 39, p. 33b. By way of commentary Tung added many excerpts from later literature. Cf. Wylie (1), p. 150.

[h] (1), pp. 30–1.

[i] (1), vol. 1, p. 479.

[1] 諸葛亮	[2] 地雷	[3] 物原	[4] 羅頎	[5] 銃
[6] 呂望	[7] 爆仗	[8] 馬鈞	[9] 煬帝	[10] 劉安
[11] 焰焇	[12] 董斯張	[13] 廣博物志		

contradictory, since in the next breath saltpetre was attributed to Liu An some eight centuries later.[a]

By +1780 there had come the first serious sinological discussion of the history of fire-weapons in China; it was in the 'Supplément' which de Visdelou & Galand added to the famous *Bibliothèque Orientale* of Barthélemy d'Herbelot.[b] They knew of the naval battle of Thang-tao[1] island between the J/Chin and Sung fleets in +1161,[c] and thought that the '*pao-à-feu*' (*huo phao*[2]) might have been cannon, especially firing red-hot shot, though they also recognised *huo chien*[3] rightly as incendiary arrows. They acutely remarked on the failure of terminology to adapt, pointing out that *tormentum* in Latin was just like *phao*[2], the thing fundamentally changing (trebuchets to cannon) while the old name continued in use. They knew that there was nothing on gunpowder to be found in Thang sources, but they also knew of the novel fire-weapons (whatever they were) invented by Fêng Chi-Shêng[4] in +970 (cf. p. 148 below), by Thang Fu[5] in +1000 (p. 149 below), and by Shih Phu[6] in +1002 (p. 149 below); realising that gunpowder was involved, but not being able to say whether as incendiary, explosive or propellant. It was clear to them, however, that the *chen thien lei*[7] (heaven-shaking thunderer) used by the J/Chin army when defending Khaifêng against the Mongols in +1232, was an explosive bomb or mine, though here also they did not feel they could exclude cannon.[d] By this time they were getting very near the bone. They also knew about the *thu huo chhiang*[12] (fire-spurting lances) invented and introduced in +1259 at Shou-chhun[13e] in Anhui,[f] and being well aware that some kind of tube was involved, they believed that these might have been true cannon. Again, they were not far wrong, though today we would call them eruptors or fire-lances with co-viative projectiles. Finally, they quoted from the *Ming Shih*[g] the reply of an emperor in response to a courtier who said that firearms had led to cowardice: 'No, the use of firearms has always been

[a] The legends could be pursued in collections such as the Yuan *Shih Lin Kuang Chi* and the Chhing *Ko Chih Ching Yuan*, if anyone were sufficiently interested.

[b] (1), Suppl. p. 117, 'De l'Invention des Canons en Chine'.

[c] This is described in the biographies of the two Sung commanders Li Pao[8] (*Sung Shih*, ch. 370, p. 4b; *WHTK*, ch. 158, pp. 1381·3, 1382·1) and Wei Shêng[9] (*Sung Shih*, ch. 368, pp. 11 b ff., 15b). The former mentions only fire-arrows (*huo chien*[3]), but the latter speaks of *huo shih phao*[10], which must mean trebuchets casting incendiaries and stones. *Huo phao*[2] at this battle are also mentioned in the biography of the J/Chin admiral, Chêng Chia[11] (*Chin Shih*, ch. 65, p. 16b), who jumped into the sea and was drowned when all his fleet was set ablaze. Cf. Fêng Chia-Shêng (1), p. 59; Lu Mou-Tê (1), pp. 30–1; Wang Ling (1), pp. 166, 169. There may well have been gunpowder in these bombs, but at that time it would probably have been low-nitrate incendiary rather than high-nitrate explosive.

[d] The description of these affairs is in *Chin Shih*, ch. 113, p. 19a; cf. Fêng Chia-Shêng (1), p. 80; Lu Mou-Tê (1), p. 32.

[e] Mod. Shou-hsien.

[f] *Sung Shih*, ch. 197, p. 15b; cf. Fêng Chia-Shêng (1), p. 71; Wang Ling (1), p. 172.

[g] They gave the reference as ch. 72, p. 51, but we have not been able to locate the passage.

[1] 唐島	[2] 火砲	[3] 火箭	[4] 馮繼昇	[5] 唐福
[6] 史普	[7] 震天雷	[8] 李寶	[9] 魏勝	[10] 火石砲
[11] 鄭家	[12] 突火槍	[13] 壽春		

one of the prerogatives that China has had over all other nations!' Such was the first serious sinological approach to the history of gunpowder weapons in China.

It was just about this time (+1774) that the witty but iconoclastic Cornelius de Pauw[a] came into collision with the witty and much better informed Chinese Jesuit Aloysius Ko (Kao Lei-Ssu[1])[b] who replied in +1777. Finding nothing about gunpowder in the *Sun Tzu Ping Fa*, and taking a poor view of the matchlock muskets still used in China, de Pauw wrote off all the Chinese gunpowder evidence, including the events of +1232 (p. 171 below), but the Jesuit knew also about those of +970, +1002 and many others as well, successfully defending the authenticity of Chinese historiography.

The nineteenth century saw a great intensification in the history of gunpowder weapons and artillery, but the pitfalls were many, and many historians fell into them. Thus Reinaud & Favé (2) in 1849 were convinced (quite rightly) from the descriptions that the 'heaven-shaking thunder' (*chen thien lei*[2]), used from +1231 onwards, was some kind of explosive. Mayers (6) in 1870 thought, on the other hand, that gunpowder went to China either from India or Central Asia in the +5th or +6th century, but that the Chinese were the last to realise its full implications, and only during the first quarter of the +15th century did they make use of its propellant power.[c] H. A. Giles fell into the misunderstanding that firearms were first used by the Chinese when the Ming general Chang Fu[3] defeated the Annamese in +1407;[d] while Geil, conceding the invention of gunpowder to China, maintained that cannon were cast only under foreign influence.[e] On the other hand, Greener (1) was prepared to credit China with a far too early knowledge of the properties of saltpetre, saying that 'the Chinese and Hindus contemporary with Moses are thought to have known even the more recondite properties of the compound'. Then at the beginning of this century (1902) Schlegel (12) well argued the case for the origin of gunpowder in China, but interpreted the term *chen thien lei*[2] wrongly as referring to cannon. His conclusion that 'the Chinese ... knew and employed fire-arms, cannon and guns, as early as the 13th...century', turned out however to be quite justified.

There were fierce controversies too. Some of these arose over the nature of Greek Fire;[f] others concerned the interpretation of the earliest evidence for guns and cannon in Europe.[g] On gunpowder history in India, Oppert (1) was duly

---

[a] (1), vol. 1, pp. 441 ff. [b] (1), p. 491.
[c] In the previous year an anonymous article in *Harper's Magazine* (Anon. 196) had got it somewhat more right than this, though still accepting that gunpowder was known in the San Kuo and the Sui. We shall mention W. F. Mayers again on p. 172 below, in connection with his recognition of the fire-lance, a weapon which so many other scholars did not understand. Dr Clayton Bredt tells us that Mayers' files and papers still exist, and are preserved along with other material from the old British Legation in Peking at the Public Record Office at Kew.
[d] (1), p. 21. See further, p. 240 below. [e] (3), p. 82. Cf. p. 394 below.
[f] E.g. Lalanne (1, 2) and Quatremère (2) against Reinaud & Favé (1, 2, 3) in the forties of the last century; here the question largely was whether it had contained saltpetre or not.
[g] E.g. Lacabane (1); Bonaparte & Favé (1), also in the forties.

[1] 高類思 [2] 震天雷 [3] 張輔

exploded by Hopkins (2).[a] Then came in the German writers, remarkable military historians,[b] but liable to get into trouble by claiming too much for Teutonic abilities. On the +15th-century European fire-book writers, Berthelot (4–7) was better, and on the +14th-century bombards Brackenbury (1) and Clephan (1–5) produced histories still useful today. In 1895 von Romocki (1) made a gallant effort to identify the origin of gunpowder weapons in Asia, with results satisfactory as far as they went, but he was impeded by little access to the original texts, and dependent on the work of the Jesuits and the earlier sinologists, not always reliable guides. Still, he did correctly interpret the *thu huo chhiang*[1] of +1259 as a gunpowder flame-thrower with what we should call co-viative projectiles,[c] though of course he knew nothing of the actual dated true Chinese hand-guns and cannon going back as far as +1290.[d]

In the early years of the present century much uncertainty continued. For example, Gohlke (1) believed that gunpowder originated in China, but that the Chinese did not arrive at making metal gun-barrels, nor did the Arabs, though he could not be quite sure what the *midfa'* was. According to him, firearms appeared almost simultaneously in several European countries, and it was not possible to determine the place, nor the person who first invented them. Next Pelliot and Chavannes were able to prove that the Chinese *huo phao* of the +12th century was a kind of bomb and not a cannon.[e] In 1915 there appeared a well-known monograph on the history of artillery by Colonel Henry Hime (2), who believed that 'in all probability gunpowder was not invented, but discovered accidentally, by (Roger) Bacon'.[f] At the same time he refused to accept the evidence brought forward by the +18th-century Jesuits on the origin of gunpowder, saying that the invention of gunpowder was probably carried from the West to China, by land or by sea, at the end of the +14th century or the beginning of the +15th and 'was falsely adopted as an old national discovery before the arrival of the Portuguese and the Jesuits in the +16th'. This was quite courageous of Hime, seeing that he had no access whatever to any of the original Chinese sources. One might say that until the end of the Second World War the theory of a European origin of gunpowder continued to hold its ground. In 1925, for example, Rathgen could write about the exclusively European origin of Indian gunpowder weapons.[g]

Forty years ago, however, decisive advances began to take place. One can see that the history of fire-weapons and gunpowder during the previous two centuries had been a welter of mistakes and misunderstandings, mistranslations,

---

[a] Cf. Partington (5), pp. 211 ff. That, however, did not prevent Oppert's mistakes being repeated by later writers such as Greener (1), p. 14.

[b] Jähns (1, 2, 3); Boeheim (1); Delbrück (1); Rathgen (1–4).

[c] Cf. p. 227 below.   [d] Cf. p. 290 below.

[e] Pelliot (59), p. 408; Chavannes (22), pp. 199, 200.

[f] Cf. p. 49 above.   [g] (1), p. 564, (5).

[1] 突火槍

legendary traditions, allegations unsupported by sources, false attributions and cultural prejudices. In the fifties and sixties this log-jam came under fire from two batteries of exceptionally heavy artillery, as it were: the writings of Fêng Chia-Shêng (*1–8*) from 1947 onwards, and Partington (*5*) in 1960. Fêng[a] and Partington[b] swept it all away, or rather amassed it in heaps and critically sifted it,[c] rejecting the nonsense and formulating some reasonably sure conclusions. Of course, some of these are today not beyond criticism, and there was much which Fêng and Partington never knew—indeed a great deal still remains to be found out. For instance, if only we knew the exact composition and physical character of the gunpowder in each of the many and various fire-weapons used in China from +900 onwards we would be much better off; as it is, we can only guess.

These heavy batteries were heralded and supported by lighter, but still extremely effective, field-guns. The new approach was pioneered by Wang Ling (*1*) and Goodrich & Fêng Chia-Shêng (*1*).[d] Abundant evidence from Chinese historical sources and the descriptions of gunpowder and firearms in the Chinese military compendia came to light. For example, Davis & Ware (*1*) studied some of the many firearms described in the *Wu Pei Chih*.[e] All of them came to the conclusion that gunpowder originated in China, a conclusion that Partington cautiously accepted, elucidating the part played by the Arabs in the transmission of the knowledge of gunpowder to Europe.[f] In Japan Arima Seihō (*1*) produced an interesting book on the origin and diffusion of cannon, in which he expressed the same view regarding the origin of gunpowder in China and drew further evidence from actual surviving examples of old Chinese cannon. Such guns, dated +1332, +1351 and +1372 were also cited by Wang Jung (*1*) to testify to the existence of bronze cannon in +14th-century China. Indeed most of the best work since Partington's book has appeared in Chinese[g] and Japanese. In 1968 a Japanese explosives chemist, Nambō Heizō (*1*) wrote an important monograph on the development of fire-weapons, gunpowder and firearms in East Asia, and their transmission to Europe, partly through the Arabs.[h]

[a] I had the honour of being personally acquainted with this inspiring scholar both in New York and in Peking; he was always most amiable in answering our many queries.

[b] Partington had been an engineer officer and a staff member of the Ministry of Munitions in both world wars, so besides being (like Berthelot) outstanding both as a chemist and a historian of chemistry, he knew about blowing things up in actual practice. He was not, as I was, an ignorant and inexperienced Adviser to the Ping Kung Shu during the second world war. Then in July 1956, Wang Ling and I enjoyed a conference with him of several days' duration, in which (with his coming book in mind) we went over all the evidence about China and gunpowder which we then had, and learnt a great deal from him as to how the history of the subject should be written.

[c] Two of Fêng's papers, (*3*) and (*8*), were devoted to critiques of earlier Western histories of fire-weapons and gunpowder.

[d] Wang Ling (Wang Ching-Ning) was already engaged in this when I first met him in 1944 at the History Institute of Academia Sinica, at that time evacuated to Lichuang in Szechuan. He had, I think, been stimulated to take up the subject by that eminent scholar Fu Ssu-Nien. Fêng Chia-Shêng worked with Carrington Goodrich before returning permanently to China.

[e] Similarly Davis & Chao Yün-Tshung (*9*) made a great contribution to the history of gunpowder fireworks in China.

[f] Partington's book has been much appreciated by later writers, e.g. J. E. Smith (*1*).

[g] One thinks of Chou Chia-Hua (*1*); Liu Hsien-Chou (*12*); Wei Kuo-Chung (*1*) and Wei Chü-Hsien (*7*).

[h] The English translation of this (*1*), however, contains many errors, and must be used with caution.

The progress of enlightenment can be traced in the comprehensive and synthetic study of Sarton (1). When he published his second volume in 1931 he thought that gunpowder had been found out in Western Europe or Syria towards the end of the +13th century; Chinese origins were not excluded, but unproven. The first guns did not come until the second half of the +14th. Sarton realised that the machines of the Hsiang-yang siege were trebuchets, but did not recognise them as counterweighted.[a] Then when he finished his third volume in 1947 he knew about Walter de Milamete,[b] and he was able to draw upon Wang Ling (1) and Goodrich & Fêng (1), so he knew of the Chinese cannons of +1356 and +1377.[c] Although he did not admit China's priority in so many words, and was evidently loth to give up the legend of Black Berthold,[d] his accounts clearly show that he moved a long way towards the standpoint which we now adopt.

It is interesting to read the following judicious comment from two Russian scholars, Vilinbakhov & Kholmovskaia (1) concerning Western writings.

Although much of this work made great contributions to the study of the history of gunpowder and firearms, it was characterised by very slight knowledge of Oriental sources, especially those in Chinese.... The statement by Western scholars that gunpowder weapons were known in China only after they had been introduced thither by Europeans, does not correspond at all with what actually occurred. The fire-weapons of mediaeval China pursued an independent course of development, the logical culmination of which was the invention of metal-barrel weapons making use of the propellant force of gunpowder.[e]

With this we entirely agree.

The conclusions to which we come in our gunpowder epic are generally similar to those arrived at by Fêng Chia-Shêng, Partington, Wang Ling, Goodrich, Arima Seihō, Nambō Heizō, and Okada Noboru.[f] We have, however, incorporated a study of the Chinese military compendia on a scale that has not been attempted before; and the results of recent archaeological findings in China have also been included.

It now remains only to direct the reader's attention to the most useful books on the nature and properties of gunpowder itself. Here our standby has been the work on the chemistry of powder and explosives by Tenney Davis (17), finalised in 1956. Since in modern times gunpowder has taken a back seat, as it were, to the nitrate and other organic compounds which give true molecular detonations with a supersonic rate of burning,[g] the most interesting modern books, such as that of Urbański (1),[h] which deal only with these, are not very useful in the

---

[a] (1), vol. 2, pp. 29, 766, 1034, 1036 ff.
[b] *Ibid.* vol. 3, pp. 722 ff.   [c] *Ibid.* vol. 3, pp. 1548 ff.   [d] *Ibid.* vol. 3, p. 1581.
[e] Their own paper, however, was not at all beyond criticism, containing as it does several mistakes and misunderstandings.
[f] Some of the older scholars, moreover, were uncommonly perspicacious, notably Laufer (47). A brief recent account in Chinese, Anon. (*214*), pp. 37 ff., is also to be recommended.
[g] To say nothing of nuclear explosions.
[h] Or Fordham (1). Cf. particularly Bowden & Yoffe (1, 2).

present context. On the other hand, it may not be desirable to go back too far, though there are books of value dating from before the first world war.[a] The two volumes of Marshall (1) in 1917, and the three of Faber (1) a couple of years later, we have found quite helpful; while those belonging to the second world war period, such as the very practical book of Reilly (1) which includes accounts of slow and quick match, and Weingart (1) on military pyrotechnics in general, may also be mentioned. This was the time when the historian of alchemy, John Read (3) gave an instructive popular exposition of the subject. As for civilian pyrotechny, we have used Brock (1, 2).

Lastly, in the following pages we shall be giving many accounts of battles in China in which gunpowder weapons were used between about +900 and +1600. It is therefore desirable to have at hand a comprehensive history of the campaigns of East Asian warfare so that one may gain some idea of the strategic background of these engagements. Fortunately we now have the valuable compilation of Chhen Thing-Yüan & Li Chen (1) in sixteen volumes, abundantly illustrated with maps and plans.[b]

(3) ANCESTRY (I): INCENDIARY WARFARE

In traditional Japan, fire, together with earthquakes, thunderbolts and paternal power, were regarded as the four most fearful things in life.[c] The awe-inspiring and destructive force of fire led to the deployment of incendiaries in warfare among all ancient people; and incendiaries of various kinds were assuredly the predecessors of gunpowder. Assyrian bas-reliefs dating back to the −9th century depict torches, lighted tow, burning pitch and fire-pots thrown at the siege engines of troops attacking a city.[d] In −480 the Persians used arrows tipped with burning tow to capture Athens,[e] and the first recorded use of incendiary arrows by the Greeks was in −429 at the siege of Plataea during the Peloponnesian war.[f]

Technologically speaking, the Greeks seem to have advanced more quickly than any other ancient people in the warlike employment of incendiary substances.[g] In −424, according to Thucydides, the Boeotians besieging Delium made use of a long iron tube, moved on wheels and carrying a vessel containing

[a] For example Bockmann (1) in 1880; Kedesdy (1) in 1909.

[b] Accounts of a somewhat similar kind can be found in the older literature, such as that of Hu Lin-I (1), but this is by far the most modern and complete.

[c] A striking passage on the subject was written by Shiba Kōkan[1] late in the +18th century, and translated by Waley (28), pp. 123–4.

[d] Cf. Barnett & Faulkner (1), pl. cxviii; from Layard's drawings. A similar depiction comes from the siege of Lachish by Sennacherib (r. −704 to −651); cf. Yadin (1), pp. 431, 434–5 (in colour).

[e] Herodotus, *History*, VIII, 52.

[f] Thucydides, *History*, II, 75.

[g] Cf. Finó (1).

[1] 司馬江漢

burning charcoal, sulphur and pitch, behind which was a large bellows which blew the flame forwards.[a] This recalls the bellows described in China by the −4th-century Mohist military writers which blew toxic or irritating smokes into the enemy's sapping tunnels.[b] In the early +2nd century Apollodorus described a similar apparatus using powdered charcoal and intended as a kind of fire-setting device against fortifications with stone walls.[c] A description of a similar apparatus was given by Heron of Byzantium as late as the +10th century.[d] About −360, Aeneas the Tactician gave the composition of war fire as a mixture of pitch, sulphur, pine-shavings and incense or resin filled into pots for throwing on the wooden decks of enemy ships or at wooden fortifications. Hooks on the containers helped them to stick fast.[e]

In those ancient ages the use of expendable animals also figured tactically from time to time. We have an example of this in a text of early Jewish history, dating from about −580. It concerns the wars with the people of Philistia.[f]

So Samson went and caught three hundred foxes, and setting them tail to tail, took torches and bound them to each pair of tails. And when he had lit the torches he loosed the foxes and let them go free, so that they entered into the standing corn of the Philistines, and burnt up both the shocks and the grain, and not only that but vineyards and olive groves too.

This is particularly interesting because, as will later be seen, expendable animals appear in all the medieval Chinese military compendia,[g] continuing on as a means of delivery of gunpowder as incendiary and later as explosive.[h] Indeed the winged rockets of China almost certainly derived their inspiration from the wings of birds made to carry incendiaries or explosive weapons.[i]

Fire-arrows were naturally part of the equipment of Roman armies. They were mentioned by Vergil (−70 to +19)[j] and Livy (−59 to +17).[k] There were also the *malleoli* or 'little hammers', a type of fire-arrow that could only be extinguished by sand but not water, mentioned by Ammianus Marcellinus about +390.[l] The inflammable material attached to the arrow consisted of sulphur, resin, bitumen and tow soaked in oil, according to Vegetius, writing about the same date.[m] After the invention of non-torsion catapults (*arcuballista* and *gastraphetes*) under Dionysius of Syracuse in −399, and of torsion catapults by Polyidus of Thessaly under Philip II about fifty years later,[n] this artillery was

---

[a] Thucydides, IV, 100 ff. Cf. Garlan (1), p. 141.
[b] See Vol. 4, pt. 2, pp. 137–8, and Yates (3), pp. 424 ff.
[c] *Poliorcetikon*, in R. Schneider (4). On fire-setting see p. 533 below.
[d] *Poliorcetika*, in Wescher (1), pp. 219, 244.
[e] *Poliorcetikon*, XXXIII, IV ff. On this and other ancient references see Hime (1), pp. 25 ff.
[f] Judges, 15, 4.   [g] P. 210 below.   [h] P. 213 below.
[i] P. 502 below.   [j] *Aeneid*, IX, 705.   [k] *History*, XXI, 8.
[l] *History*, XXIII, IV, 14–15.
[m] *Rei Militaris Instituta*, IV, 1–8, 18. Partington (5), p. 2, supposed that the oil was mineral petroleum, but no doubt vegetable oils would also do.
[n] Cf. Marsden (1), pp. 48 ff., 57, 60.

often employed, when need arose, to project pots containing incendiary material. Fire-ships and resinous torches had been used at the siege of Syracuse in −413;[a] and the Phoenicians also used fire-ships to burn the works on the mole made by the Macedonians at the siege of Tyre in −332.[b] After −323, the year of the death of Alexander the Great, the use of incendiary missiles became common practice among all troops of the Mediterranean cultures. In −304 fire-ships and resinous torches were again employed in the siege of Rhodes.[c] Burning spears (*ardentes hastae*) hurled by catapult artillery were described by Tacitus (*c*. +60 to +120).[d] And so it went on, down to the conclusion of the Gothic wars.[e]

'Automatic fire' (*pyr automaton*, πῦρ ᾿αυτόματον) was also used in antiquity, but how much military value it had is doubtful, for it depended on the spontaneous inflammation of quicklime mixed with combustibles such as sulphur and petroleum when wetted.[f] The heat evolved is enough to light the incendiary mixture. The term itself was first used by Athenaeus of Neukratis about +200.[g] According to Viellefond's edition of the *Kestoi* of Julius Africanus (*c*.+225), as interpreted by Partington,[h] it consisted of equal parts of native sulphur, rock salt, incense, thunderbolt stone or pyrites, all ground in a black mortar in the midday sun and mixed with equal parts of black sycamore resin and liquid Zakynthos asphalt to make a greasy paste. Some quicklime was then added and the mass stirred carefully at noon, the body being protected as the composition was liable to take fire quickly. It had to be kept in bronze boxes tightly covered until it was needed. It was to be smeared on the 'engines' (*hopla*, ὅπλα) of the enemy and when the morning dew wetted it, all would be burnt. Automatic fire recipes also appear in the *Liber Ignium*[i] and in *De Mirabilibus Mundi*,[j] +13th-century works already discussed (p. 40 above). One may conclude that mixtures of quicklime with combustible materials, if stowed away secretly in unexpected places, might produce some mysterious conflagrations, but the technique can never have been of much use either on land or sea; in the latter case (provided means were used to prevent the material from sinking) the combustion would have been mild, quiet and harmless, apart from some element of surprise.[k]

Fire-weapons were also used in the −1st millennium in India. The *Mahābhārata* epic often mentions the use of inflammable materials such as resin or tow in

[a] Thucydides, *History*, VII, 53.

[b] Arrian, *Exped. Alexander*, II, 19.

[c] Diodorus Siculus, XX, 86.

[d] *History*, IV, 23. Partington (5) calls these 'fire-lances', but in view of what is to come, the term would be very misleading here.

[e] Ammianus Marcellinus (*c*. +390), XXIII, iv, 14, 15.

[f] Many have tried to repeat this but not everyone has been able to do it. Marshall (1), vol. 1, pp. 12–13 could not make it work, but Partington's friend Richardson (1) fully succeeded.

[g] Talking about the tricks of one Xenophon the Wonder-worker, a conjuror.

[h] (5), p. 8.

[i] In § 9, *calx non extincta*, Partington (5), p. 47.

[j] Partington (5), p. 85, text and translation.

[k] Cf. Zenghelis (1). We shall hear later on (p. 165) of a famous Chinese naval-battle of +1161 at which quicklime was used in bombs of some kind, as also other examples of the same, but this seems to have been because of its irritant properties when dispersed in smoke rather than as an igniter of incendiary substances.

battles.[a] There are many recipes for incendiary mixtures, toxic smokes and similar devices in the *Arthaśāstra*,[b] including showers of firebrands, and fire-pots hurled from catapults of some kind. The troops of Alexander the Great encountered fire-weapons in India in −326. The Oxydraces, a people of the Punjab, were particularly renowned for this.[c] When Apollonius asked why Alexander the Great had refrained from attacking them, he was told that

these truly wise men dwell between the rivers of Ganges and Hyphasis. Their country Alexander never entered, deterred not by fear of the inhabitants but, as I suppose, by religious motives, for had he passed the Hyphasis he might doubtless have made himself master of all the country round—but their cities he never could have taken, though he had a thousand men as brave as Achilles, or three thousand like Ajax; for they come not out into the field to fight those who attack them, but rather these holy men, beloved of the gods, overthrow their enemies with tempest and thunderbolts shot from their walls. It is said that the Egyptians Hercules and Bacchus, when they invaded India, attacked this people also, and having prepared warlike engines attempted to conquer them; they in the meantime made no show of resistance, appearing perfectly quiet and secure, but upon the enemy's near approach they repulsed them with storms of lightning and flaming thunderbolts hurled upon their armour from above.[d]

This was a remarkable description of incendiary warfare. The element of 'thundering', which occurs not only here in the words of Philostratus (d. +244), but in the many accounts of Crusade battles a thousand years later, has deceived many into supposing that true explosions or detonations of gunpowder were meant;[e] but in fact the forced draught during the rapid aerial trajectory of large containers of combustibles is enough to produce the effect.

Much confusion also has been caused by Sanskrit terms such as *agni astra*, which undoubtedly meant 'fire-arrow' in the classics, but was later given the meaning of 'cannon'.[f] The word *śataghni* 'killer of hundreds', also appears in the Sanskrit classics, and led some scholars into believing that gunpowder was known and used in India before the end of the −1st millennium, a conclusion which cannot be sustained.[g] Again, it has been said that at the battle of Biyanagar in +1368 the Hindus used *'araba* against the Muslims. The modern meaning

[a] McLagan (1); Winter (1). The work contains material from −200 to +200, with later additions.

[b] Attributed to Kautilya, *c.* −300, but as we have it now containing further material as late as the +5th century. See the translation of Shamasastry (1), pp. 57, 92, 154, 424, 451, 458, 468; and Partington (5), p. 210.

[c] On the Oxydraces see the *Anabasis Alexandri* of Arrian (+96 to +180), v, 22 and vi, 4, 11, 14, but he makes no mention of incendiary weapons (tr. Brunt, 1).

[d] *Life of Apollonius of Tyana*, ii, 33. Accounts of this got into Japanese, cf. Arisaka Shōzō (1), vol. 2, pp. 113–14 and Arima Seihō (1), p. 3.

[e] It certainly deceived Francis Bacon, who in his essay on 'Vicissitude of Things' (+1625) wrote: '... we see that even weapons have returns and vicissitudes; for certain it is, that ordnance was known in the city of the Oxidrakes in India; and was that which the Macedonians called thunder and lightning and magic. And it is well known that the use of ordnance hath been in China above two thousand years' (Essay 58, Montagu ed., vol. 1, p. 192; Spedding & Ellis ed., vol. 6, p. 516).

[f] Here is yet another example of the tendency of things to change radically while the words denoting them remain the same.

[g] Details in Partington (5), pp. 211 ff. The authors and poets of ancient India had a particular fondness for fabulous weapons.

of this word is certainly 'gun-carriage', but originally it meant simply a cart as such. Hime saw that the historian Firishta (d. *c.*+1611) fell into this trap by interpreting the passage to imply field-artillery unjustifiably, and other historians did the same.[a]

In an abortive attack on the fortress of Rantambhor in +1290 the sultan Jalā al-Dīn ordered *maghrībīhā* machines (i.e. trebuchets)[b] to be erected, but later the besieged forces constructed their own. When the fort was successfully besieged in +1300 the Hindus inside

collected fire in each bastion; and every day the fire of those infernal (machines) fell on the light of the Muslims. As there was no means of extinguishing it they filled bags with earth and prepared entrenchments.... Later the royal army made vigorous attacks, rushing like salamanders through the flames that surrounded them....[c]

During the siege of Bhatnīr in +1398 the Hindus 'cast down arrows and stones, and (incendiary) fire-works' upon the heads of the assailants.[d] The elephants in the army of the sultan Mahmud, which Timur defeated at Delhi in +1399, carried throwers of grenades (*ra'd-andāzān*), fireworks (*ātish bāzī*) and launchers of rockets (*taksh-andāzān*).[e] By this time of course explosive gunpowder bombs would have been only too available, and rockets as well, but the second weapon mentioned looks like the old incendiary fire-pots.

In China fire as an arm of war has been recognised at least since the classical −4th-century military handbook, the *Sun Tzu Ping Fa*[1], where ch. 12 is entirely devoted to it.[f] Apart from incendiary methods to set alight the enemy's weapon-stores or provisions, the most interesting reference is to 'dropping fire' (*chui huo*[2,3]); a phrase which has caused a lot of trouble to commentators through the ages,[g] but which is most plausibly interpreted, as since the Thang it has been, to mean fire-arrows shot into the enemy's camp.[h] The use of fire in

---

[a] Hime (1), p. 80. Partington (5), p. 216 shows that copyists and eighteenth-century translators of Firishta were very uncritical in their use of words connected with guns and artillery.

[b] I.e. machines of Western origin. The Maghrīb included all the western regions of the Arabic culture-area in North Africa and Spain. But, as we saw above (Vol. 5 pt. 6.), the swape-principle embodied in the trebuchet and mangonel was much older in China than in Europe. That, however, the Muslims in India did not know.

[c] From the account of Amir Khusrū (d. +1325), in Elliott (1), vol. 3, p. 75, vol. 6, p. 465. Cf. Partington (5), p. 218, and Hime (1), p. 83.

[d] Timur's autobiography, the *Malfūzāt-i Tīmūrī*, in Elliott (1), vol. 3, p. 424; cf. Partington, *loc. cit.*

[e] The same work, in Elliott (1), vol. 3, pp. 430 ff., 439; cf. Partington, *loc. cit.*

[f] See the translations of Giles (11), pp. 150 ff. and Griffith (1), pp. 141 ff, as well as the transcription into modern Chinese done by Kuo Hua-Jo (1). A number of variant versions of the book and parallel texts have been found in recent years (see Anon. (210) pp. 86 ff.). Most of them have something about attack by fire.

[g] The texts all write *tui*[2] (division or battalion) but *chui*[3] was assumed since the characters are often taken as interchangeable. Unfortunately the newly discovered Early Han text has a lacuna at this point, but *WCTY/CC*, ch. 11, p. 19a, b opts for the first form and meaning. The character could also be *sui*[4], which would suggest an underground mine passage, unlikely here.

[h] Amiot, the eighteenth-century Jesuit, misled by tradition (cf. p. 59), attributed incendiary 'bombs' filled with weak gunpowder 'having the effect of Greek Fire', to Master Sun; ((2), p. 146, Suppl., p. 337, cf. pl. 16, fig. 77, explanation, p. 361). Here he was unquestionably wrong.

[1] 孫子兵法   [2] 隊火   [3] 墜火   [4] 隧

battle is also mentioned in a Chhin and Han military handbook, the *Liu Thao*[1] (Six Quivers), which has the semi-legendary Chiang Shang[2] as its putative author.[a] Two famous early battles deploying incendiaries are often retold in Chinese history. The first is the ingenious use of fire and expendable buffaloes by Thien Tan[3] in −279 when he defended the last stronghold of Chhi[4] State and repelled the superior invading force from the State of Yen[5].[b] After winning this decisive battle Thien Tan recovered more than seventy Chhi cities which had previously fallen into enemy hands.[c] The other is the complete destruction of Tshao Tshao's[6] Wei fleet by fire at the Battle of the Red Cliff in +208 by the forces of Shu[d] and Wu under the combined command of Chuko Liang[7] and Chou Yü[8].[e] Fire-ships (Fig. 5) were indeed very important in Chinese naval engagements through the centuries, for example the Po-yang Lake battles of +1363 in which Chu Yuan-Chang and his admirals overcame all their adversaries.[f] Incendiary arrows using burning tow are described in all the military handbooks, such as the Thang *Thai Pai Yin Ching*[9] written by Li Chhüan[10] in +759,[g] and the Sung *Hu Chhien Ching*[11] of +1004 by Hsü Tung[12].[h]

Incendiary weapons in the form of projectiles hurled towards the enemy lines, or let down from city-walls over besiegers, are described in the *Wu Ching Tsung Yao* of +1044. For example, it speaks of two of the second sort as follows:[i]

On the right is a drawing of the 'swallow-tail incendiary' (lit. torch, *yen wei chü*[16]). Straw is fastened together in a divided shape like the two parts of a swallow's tail, and soaked in oil and fat. After ignition it is let down on the enemy approaching the city walls so that it destroys their wooden structures (scantlets, etc.) by fire.

---

[a] Cf. *WPC*, ch. 5, p. 25a, b. Alternatively, Lü Wang.[15]

[b] A number of texts on incendiaries carried by expendable animals were collected in Pfizmaier (98), p. 6.

[c] *WPC*, ch. 29, pp. 7b, 8a; Giles (11), p. 91, translating a *Sun Tzu* commentary.

[d] Amiot also ascribed the use of explosive land-mines to Khung-Ming[13] (i.e. Chuko Liang), saying that he had set off 'earth-thunder' (*ti lei*[14]) about +200; (2), Suppl., pp. 331–2, 336. Indeed he was better at this than any other general of his time: 'On sait d'ailleurs, à ne pas en douter, que dans leur manière de combattre par le feu, ils employoient le salpêtre, le soufre et le charbon, qu'ils méloient ensemble en certaine proportion; d'où il résulte qu'ils savoient faire le poudre à tirer, bien des siècles avant même qu'on se doutat en Europe que cette invention existoit.' Amiot was justly criticised by Hime (1), p. 90, for not appreciating the difference between an explosive and an incendiary. But on the main issue he was quite right, albeit for the wrong reasons. Cf. Partington (5), pp. 238–9, 251–2. Amiot reproduced many drawings in copperplate form of fire-lances, bombards, mines, etc., from Chinese books that we know well and use in this Section (cf. pl. 15, figs. 67–71, pl. 16, figs. 72–80, pl. 29, fig. 136, this last the wheelbarrow rocket-launchers).

[e] Cf. Wieger (1), vol. 1, p. 827; *WPC*, ch. 26, pp. 21b, 22a.

[f] See Dreyer (2).

[g] E.g. ch. 35 (ch. 4), p. 2b, ch. 38 (ch. 4), p. 8b. The first of these accounts describes how arrows were first sent over having gourds of oil attached which on breaking spread it about all over the houses, towers and wooden structures of the enemy; then later volleys of burning arrows ignited it all. The second says that *arcuballistae* shooting with a range of 300 paces should be used.

[h] E.g. ch. 54 (ch. 6), p. 5a, ch. 66 (ch. 6), p. 14a.

[i] *WCTY/CC*, ch. 12, pp. 60a, 61a. For the *yen wei chü* and *fei chü* see also *WPC*, ch. 130, 23a, b, 24a,b.

---

[1] 六韜	[2] 姜尚	[3] 田單	[4] 齊	[5] 燕
[6] 曹操	[7] 諸葛亮	[8] 周瑜	[9] 太白陰經	[10] 李筌
[11] 虎鈐經	[12] 許洞	[13] 孔明	[14] 地雷	[15] 呂望
[16] 燕尾炬				

Fig. 5. Fire-ships, from *WCTY*, ch.11, p. 26*a*.

The 'flying incendiary' (*fei chü*[1]) is shaped like the swallow-tail incendiary, and let down on an iron chain from a swape lever set up on the city wall. These will burn enemy troops even when attacking in great numbers.

The significance of the swallow-tail shape is not evident unless one realises that battering-rams and other offensive machines were brought up under cover of temporary wooden structures with wheels and ridged roofs; the incendiary device would rest astride these and set them on fire (Fig. 6). Another page describes a projectile.[a]

[a] *WCTY/CC*, ch. 12, pp. 64*a*, 65*a*.
[1] 飛炬

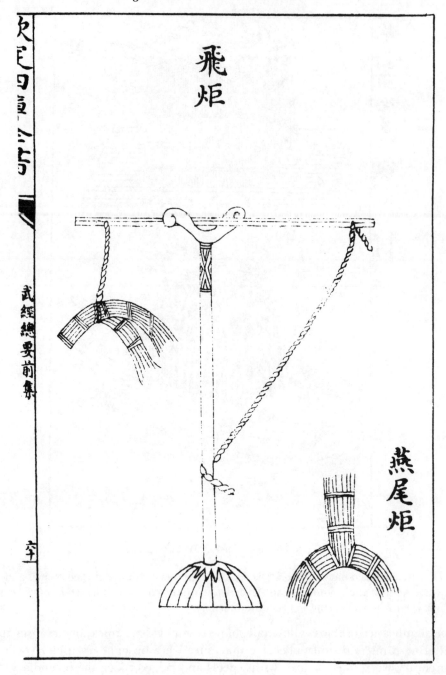

Fig. 6. The 'swallow-tail' incendiary device, for letting down on the roofs of siege machinery housings from the walls of besieged cities. *WCTY*, ch. 11, p. 60*a*.

The 'igniter ball' (*yin huo chhiu*[1]) is made of paper round like a ball, inside which is put in between three and five pounds of powdered bricks. Melt yellow wax and let it stand until clear, then add powdered charcoal and make it into a paste permeating the ball; bind it up with hempen string. When you want to find the range of anything, shoot off this fire-ball first, then other incendiary balls can follow.

Such a blazing projectile would certainly have set the enemy's huts or trebuchets on fire, as well as giving an idea of the distance at which your own trebuchets would have to aim (Fig. 17). But in the *Wu Ching Tsung Yao* there are not so many of these specifications, since most of the incendiary projectiles by this time contained low-nitrate gunpowder, as we shall see in the appropriate place (p. 149 below).

(4) Naptha, Greek Fire And Petrol Flame-Throwers

Among all the combustible substances which would be used in war, naturally occurring mineral oils came to take more and more importance. The knowledge of petroleum and its congeners goes back in all nations to high antiquity.[a] Already we have discussed it in relation to China in more than one place,[b] here we have to concentrate on its use in war. Seepages of natural oil were made use of both in east and west for many purposes, varying according to its composition, whether heavy oil, sulphurous or waxy, or the lighter, lower boiling-point, fractions that got the name of naphtha.

A Greek physician at the Persian court, Ktesias of Cnidus, writing in the neighbourhood of −398, reported a story about an oil derived from a gigantic worm (*scolex, σκώληξ*) living in the Indus River, an oil which was capable of setting everything on fire.[c] The tale was repeated by Aelian (d. +140)[d] and Philostratus (d. +244).[e] The latter said that the white worm was found in the Hyphasis River in the Punjab, and that the oil made by melting it down could be kept only in glass vessels; when once set on fire it could not be extinguished by any ordinary means. Naturally occurring naphtha was most probably the basis of the legend.[f]

Persian naphtha[g], which the Greeks called 'oil of Media', was well known in the time of Alexander the Great when he captured Babylon in −324. Pliny wrote about an 'inflammable mud' called *maltha* found at Samosata on the Euphrates.[h] Petroleum was described at length by Vitruvius,[i] and 'white naphtha' was prob-

---

[a] Here the studies of Forbes (20, 21) are important guides.
[b] Vol. 3, pp. 608 ff., Vol. 4, pt. 1, pp. 66–7; Vol. 5, pt. 4, p. 158.
[c] See McCrindle (2) and Partington (5), pp. 209, 231.
[d] *De Nat. Animalium*, v, 3.   [e] *Life of Apollonius of Tyana*, III, 1.
[f] Here the view of Partington, *loc. cit.*, has general assent.
[g] The word itself is of Iranian origin. The great oilfields of Batum and Baku must have been the origin of the material and the stories.
[h] *Nat. Hist.* II, 108–9.   [i] *De Architectura*, VIII, 3.

[1] 引火毬

ably petroleum purified by filtration through fuller's earth. These substances were all used as incendiaries in warfare, as for example against Maximinus when he captured the town of Aquileia.[a] They found employment more and more, as by Genseric, king of the Vandals, to destroy a Roman fleet in +468,[b] and in +551 when Petra in Colchis was being defended by the Persians.[c] By this time the composition was getting more complicated, sulphur, resins, bitumen and tow being mixed with the incendiary oil; as we know from the recipe that Vegetius gives[d] for fire-arrows about +385.

From the beginning of the Arab conquests their armies acquired particular skill in the use of naphtha as a war-weapon. Special corps of *naffāṭūn* in fireproof suits were formed to handle it. Already in +712 at the siege of Alor in India the Muslims used *ātish bāzī*, or incendiary projectiles developed on the basis of what they had seen in use by the Byzantines and Persians. They threw *huqqahā-i ātish bāzī*, probably naphtha pots,[e] at the howdahs on the elephants, making them rush away in panic.[f] In +904, at the siege of Salonika, they used earthenware grenades filled with pitch, oil, quicklime and other materials.[g] When Jerusalem was attacked in +1099 the Saracens hurled flaming balls of pitch, wax, sulphur and tow against the machines of the Crusaders.[h] When the Turks were besieged in turn in Nicaea they took similar defensive action.[i] At the siege of Assur, all in the same year, the Turks set ablaze a tower using iron stakes wrapped in tow soaked in oil, pitch, and other combustible substances, and it was said that the fire could not be extinguished with water.[j] During the Second Crusade (+1147 to +1149) the Arabs again used naphtha. In +1168 Shawar employed 20,000 barrels of petroleum to burn down the city of Fustāt (Cairo) to prevent its recapture by the Franks.[k] At the time of the Third Crusade (+1190, +1191) during the siege of Acre, 'boiled naphtha'[l] and other incendiaries contained in copper pots (*marmites*) were thrown at the attack towers of the Christians, successfully destroying them by fire.[m] Thundering tubs of incendiaries thrown from trebuchets were used in all the battles of the Seventh Crusade (+1249), when St Louis of France and the Sieur de Joinville were there to record it.[n] Such was the

[a] Herodianus (d. +240), *History*, VIII, 4.

[b] Lebeau (1), vol. 7, p. 16.

[c] Agathias (c. +570), *History*, III, 5; Lebeau (1), vol. 9, p. 211.

[d] *Rei Militaris Instituta*, IV, 1–8, 18. Many other references have been collected by Partington (5), pp. 3 ff.

[e] Mercier (1) has given us the most detailed survey of surviving specimens of these ceramic naphtha containers. Cf. Fig. 3.

[f] Elliott (1), vol. 1, p. 170, vol. 6, p. 462; Partington (5), pp. 189, 215. On the other hand, Shāh Rukh's ambassador to India in +1441, Abd al-Razzāq, reported on the naphtha-throwers mounted on the backs of elephants.

[g] According to Joannes Kameniata, *De Excidis Thessalonicensi*; cf. Partington (5), pp. 14, 37.

[h] Raymund de Agiles, in Bongars (1), p. 178; cf. Partington (5), pp. 22–3.

[i] William of Tyre, in Bongars (1), pp. 670–1.

[j] Albert of Aachen, in Bongars (1), pp. 193, 294–5.

[k] Mercier (1), p. 73.

[l] This sounds uncommonly like distilled petroleum, the essential secret of Greek Fire (p. 76 below).

[m] Bahā'al-Dīn, in Reinaud, Quatremère *et al.* (1), *Orientaux*, vol. 3, p. 155.

[n] Partington (5), pp. 25–6.

character of incendiary warfare down to the very century when the knowledge of gunpowder was making its way to the Arab and European cultures.[a]

Petroleum is called in Chinese *shih yu*[1], presumably as an abbreviation of the old term *shih nao yu*[2] (mineral-brain oil).[b] Natural oil seepages were already being used in China in late Chou times (−5th century onwards). Thang Mêng[3] described one about +190 in the district of Yen-shou[4], calling it *shih chhi*[5] (stone lacquer), because it was dark to begin with and gradually got darker.[c] A similar account was given in one of the commentaries on the (*Hou*) *Han Shu*, which says that

south of Yen-shou among mountain rocks there oozes out a liquid looking like uncoagulated fat. When burnt it generates an intense brightness, but it cannot be consumed as food (or used for frying). The local people call it 'mineral lacquer'.[d]

Not long afterwards Chang Hua recorded an event of about +270 when stores of oil in an arsenal caught fire, suggesting that petroleum was included in Chin army supplies.[e] In Thang times natural petroleum was still a wonder. Tuan Chhêng-Shih[6] had an entry for it in his *Yu-Yang Tsa Tsu*[7] finished about +860:[f]

Mineral lacquer is found in Kao-nu *hsien*; (they call it) 'rock-fat liquid' (*shih chih shui*[8]). It floats on the surface of the water like lacquer (i.e. dark in colour). People use it for greasing their cart axles, and when burnt in lamps it gives a bright flame.

Petroleum was produced at many places in China. Li Shih-Chen wrote:[g]

Mineral oil (*shih yu*[1]) is not found only in one location. In Shensi province it comes from Su-chou[9], Fu-chou[10], Yen-chou[11] and Yen-chhang[12], while many places in Yunnan and Burma produce it, as well as Nan-hsiung[13] in Kuangsi. It flows out from the rocks and mixes with the spring water, gushing and gurgling. It is oily like the juices of cooked meat. The local inhabitants sop it up with straw and put it in earthenware pots. It is black in colour, rather resembling fine lacquer, and has an odour of realgar and sulphur. Many local inhabitants use it for burning in lamps, which shine very brightly. When water is added, the flame only becomes more intense. This oil is inedible, but it gives a thick smoke. When Shen Tshun-Chung[16] (i.e. Shen Kua) was an official in the west he

---

[a] It is not always possible to be sure, from the numerous descriptions, when petroleum itself was being used, and when the effects were due to preparations of Greek Fire type.

[b] *PTKM*, ch. 9 (p. 94) ff. It first got into the pharmaceutical natural histories in the *Chia-Yu Pên Tshao*[14] of +1057. The term *shih nao yu* should not be confused with *shih nao*[15], the modern name for paraffin wax.

[c] We gave the translation in Vol. 3, p. 609.

[d] *Pên Tshao Kang Mu Shih I*, ch. 2, p. 61 *b*, where four pages of information are collected.

[e] *Po Wu Chih*, ch. 4, p. 3 *a*; we already gave the translation in Vol. 4, pt. 1, p. 66.

[f] Ch. 10, p. 2 *b*, tr. auct.

[g] *PTKM*, ch. 9 (p. 94), tr. auct. Cf. the new treatise edited by Shen Li-Sheng (*1*).

---

[1] 石油	[2] 石腦油	[3] 唐蒙	[4] 延壽	[5] 石漆
[6] 段成式	[7] 酉陽雜俎	[8] 石脂水	[9] 肅州	[10] 鄜州
[11] 延州	[12] 延長	[13] 南雄	[14] 嘉祐本草	[15] 石腦
[16] 沈存中				

collected the soot to make ink with; the product was black and lustrous like lacquer, and superior to that made from pinewood lamp-black[a].

Petroleum was discovered in different parts of China at different times in history. Li Shih-Chen quotes an example from the +16th century when oil was found in Chia-chou[1] (in modern Szechuan). He says:[b]

During the last year of the Chêng-Tê reign-period of the present (Ming) dynasty (i.e. +1521) oil was accidentally found during the process of digging salt-wells. When used for illumination at night it gave twice the brightness (of ordinary lamps). When water was sprinkled over it the flame became more intense than before, and it could only be extinguished by stifling it with ashes. It gave off an odour of realgar and sulphur, so that the locals called it *hsiung-huang yu*[2] and also *liu-huang yu*[3]. Several more wells have recently been opened and they are all managed by the government. This is also *shih yu* (petroleum) only it comes from wells.

Chinese scholars also noted the occurrence of petroleum in other countries. For example, quoting from a late Buddhist tractate, the *Chhiu Shêng Khu Hai*[4], Chao Hsüeh-Min says:[c]

Burma (Mien-Tien) also produces *shih yu*, which is the same as *shih nao yu*. It flows out from crevices in the rocks, and has an unbearably pungent smell. It is black in colour. It can be used to apply to sores, and is good for treating boils.

This was not surprising, in view of the great oilfields worked in Burma in modern times. And many other similar statements could be quoted.

There can be little doubt that naturally occurring petroleum obtained from seepages or wells was used in China through the centuries for military incendiary purposes.[d] But a different chapter opens when the phrase *mêng huo yu*[5] (fierce fire oil) makes its appearance, for while native petroleum, *shih yu*, had been known for so long, the new appellation is found only from the beginning of the +10th century onwards. We think that wherever it occurs it means preparations like Greek Fire.

What then was the difference between natural mineral oil, petroleum, as such, and the artificial inflammable gasoline that was called Greek Fire? The answer can today be given in a few words, because Partington demonstrated (in so far as it is ever possible to prove anything in the history of chemistry) that Greek Fire was essentially petroleum *distilled*.[e] This liquid rectified petroleum would have

---

[a] On this subject we gave a translation of the whole of Shen Kua's own account in a passage from *MCPT*, ch. 24, para. 2; see Vol. 3, p. 609.

[b] *PTKM*, ch. 9, *loc. cit.*, tr. auct.

[c] *Pên Tshao Kang Mu Shih I*, ch. 2, p. 62 *b*, tr. auct.

[d] There is no mention of it, however, in the Mohist military chapters. But Dr Phan Chi-Hsing informs us (priv. comm.) that the Lüshun Museum contains at least one hollow ceramic container similar to those described by Mercier (p. 44 above); it was excavated at Ta-lien about thirty years ago. There is a hole for filling and for the fuse. Cf. Fig. 4.

[e] (5), pp. 10 ff., 28 ff. Cf. Marshall (1), vol. 1, pp. 12–13. Lebeau (1) in 1827 (vol. 9, p. 211; vol. 11, p. 420) was perhaps one of the first to suggest that distillation was the key to the matter.

¹ 嘉州   ² 雄黄油   ³ 硫黄油   ⁴ 救生苦海   ⁵ 猛火油

been not unlike the volatile petrol which everyone is familiar with today, and consisted of the low boiling-point fractions containing relatively short-chain hydrocarbons which come over when petroleum is distilled. Undoubtedly many of the later accounts of naphtha 'grenades' had to do with its use in such break-able bottles. But we know that the Byzantines (who first invented it) used it in 'siphons' ($\sigma i\phi\omega\nu$),[a] i.e. projector-pumps or flame-throwers. As Partington reflected, petrol alone would float, still fiercely burning, around enemy hulls,[b] but it would dissipate rather quickly, and carry only a short distance; for these reasons (as the texts show) it was thickened with resinous substances dissolved in it,[c] and perhaps sulphur also.[d]

The significance of the distillation of petroleum in the Greek world is very considerable. At an earlier stage we described the four classical still types (the Chinese, Mongolian, Gandhāran and Hellenistic)[e], and we know now that all of them were about equally effective from the physico-chemical point of view.[f] The distillation of oils is not at all prominent in the Alexandrian-Byzantine *Corpus Alchemicorum Graecorum*[g], perhaps not even detectable, but there was no reason whatever why some daring experimenter should not have tried it by the middle of the +7th century.[h] Indeed, like gunpowder itself, it was almost bound to come.[i]

Greek Fire is one of those inventions which can be dated rather exactly. Theophanes, who finished his *Chronographia* in +815, described how the Arabs continually attacked Byzantium from +671 to +678. But they finally gave up, a major factor in their defeat being the chemical process introduced a few years earlier[j] by an architect-engineer named Callinicus who came from Heliopolis.[k]

[a] This was the word that had been used for the double-acting force-pump for liquids invented by Ctesibius in the −2nd century and improved by Heron of Alexandria. Cf. Vol. 4, pt. 2, pp. 141, 144, as also Vitruvius, *De Archit.* x, vii, and Neuburger (1), p. 299; Usher (1), 1st ed., p. 86, 2nd ed., p. 135. We generally call siphons as understood today examples of the 'true siphon'.

[b] Hence the name 'sea-fire' (*thalassion pyr*, θαλάσσιον πῦρ) in Theophanes.

[c] Closely similar reasoning led in contemporary times to the invention of 'napalm' (the word deriving from naphthenate + palmitate). This is essentially petrol or gasoline thickened to a jellylike consistency by the incorporation of a mixture of aluminium soaps. Its extremely controversial use in incendiary anti-personnel bombs need not be enlarged upon here.

[d] One of the greatest occasions of controversy has been whether or not Greek Fire contained saltpetre, as many, e.g. Lalanne (1, 2); Reinaud & Favé (1, 3); Berthelot (9), (10), p. 98, (13, 14); Mercier (1); Oman (1), p. 546; Brock (1), pp. 232–3; Forbes (21), have thought that it did. But the history of saltpetre makes this quite impossible. Von Romocki (1), vol. 1, p. 7, stood out against the idea even when it was most prevalent—but unfortunately he himself fell for quicklime.

[e] See Vol. 5, pt. 4, pp. 80 ff.

[f] Butler & Needham (1).

[g] Berthelot & Ruelle (1).

[h] On the distillation of essential oils, turpentine, pitch, etc. in Roman times, see Partington (5), pp. 30–1.

[i] Among further sources of information on Greek Fire (unenlightened by Partington's insight) we may mention Oman (1), vol. 2, pp. 46 ff.; Forbes (4a), pp. 28 ff. (4b), pp. 95 ff. (unreliable on China); Diels (1), pp. 108 ff.; von Lippmann (22), pp. 131–2; Hime (1), pp. 27 ff. In 1904 Hime (2) had been an adherent of quicklime, but abandoned it in favour of calcium phosphide, an even more implausible idea.

[j] The exact date is not clear, but it would have been in the neighbourhood of +675. Also it seems that the invention was perfected by Callinicus after his arrival in Byzantium.

[k] Whether that in Syria or in Egypt is uncertain. But in either case he would have been well in the Hellenis-tic proto-chemical tradition, described in Vol. 5, pts. 2 and 4.

The defending ships of the Romaioi (as the Byzantine Greeks spoke of themselves) were now all 'siphon-bearing' (*siphōnophoroi*, σιφωνοφόροι) and they sys-
tematically set the enemy craft on fire, as well as burning those aboard them.
Further information on these petrol flame-throwers[a] comes from many sources,
for example the *Tactica* of the Emperor Leo, written in the +8th or +9th century.[b] He tells us of the iron shields protecting the men working the bronze
flame-thrower pumps, and of the rumbling thunderous noise made by the blazing jets,[c] a notice which indicates that the apparatus must sometimes have been
of considerable size, though others were hand-held.[d] One account says that the
pumps were worked by compressed air, which could mean that the petrol was
forced out of the tanks by some sort of piston-bellows.[e] Another implies that
flexible pipes formed part of the apparatus.[f] It could be directed to left or right at
the will of the operator, or even at a howitzer trajectory to descend on enemy
ships from above.[g] The mouths of the tubes[h] were often given the shapes of
animal heads.[i]

A graphic account of the use of Greek Fire in a sea-fight between the Byzantines and the Pisans in +1103, based on Anna Comnena's book, is given by
Oman in his second volume, and it is worth reproducing here because it gives
us some idea of what the Chinese flame-throwers shortly to be discussed (p. 82)
were like in practice. By her time these had been standard army equipment in
China for a couple of centuries.

---

[a] This leads one to propose a question whether all the medieval tales of fire-breathing dragons may not
derive from the Byzantine petrol flame-thrower? For example, the Anglo-Saxon epic poem *Beowulf* has a rather
graphic description of flame as a weapon during the last combat of this Swedish hero with a fire-spouting drake
or 'wild worm'; cf. Morris & Wyatt tr., pp. 137 ff., esp. p. 152; Ebbutt (1). Although the oldest MS. of the poem
is of the late +10th century, and the historical characters referred to belong to the +6th, the composition itself
must be of the early +8th; cf. Klaeber (1), p. cxiii. As is well known, the Scandinavians were long in close
touch with Micklegard (Byzantium), and would at least have heard of Greek Fire 'siphons'.
    It is true that the ancient Greek giant-monster Typhoeus, who fought against Zeus, was said to send forth
fire from eyes and mouth (Homer, *Il.* II, 752; Hesiod, *Theog.* 306, 820; Pindar, *Pyth.* I, 15; Aeschylus, *Prom.* 355);
but Roscher's *Lexicon* takes him to have been a personification of volcanic flame. Significantly, he was a son of
Gaea and Tartarus. At any rate Callinicus may have been responsible for a considerable amplification of the
fire-breathing dragon motif. Thanks are due to Prof. Charles Brink for discussing this question with us.
    [b] Depending on his identification, whether Leo III the Isaurian (r. +717 to +741) or Leo VI the Armenian
(r. +886 to +911).
    [c] Cf. p. 68 above. It does not imply detonation.
    [d] *Tactica*, XIX, 6, 51–7.
    [e] Joannes Kameniata, *De Excidio Thessalonicense*, in *Corpus Script. Hist. Byzant.*, pp. 534, 536, speaking of the
siege of Salonika in +904.
    [f] Constantine VII (r. +912 to +945), *Tactica*, in his *Opera*, VI, 1348. Cf. Sarton (1), vol. 1. p. 656; Previté-
Orton (1), vol. 1, p. 257.
    [g] Leo says this, as does Anna Comnena (b. +1083), daughter of Alexios I Komnenos, in her biography of
her father (*Alexias*, XI, 10). Cf. Rose (1).
    [h] Recently a tapered bronze pipe, possibly part of a 'siphon' pump, has been found in the underwater
excavation of a +7th-century Byzantine ship, the 'globe wreck', west of Bodrum and north of Cos. See Frost
(1), pp. 166–7, 173.
    [i] Apart from the military 'Flammenwerfer' (used mainly by the Germans in the first world war, and by the
Americans in the second), the chief lineal descendant of the device is the humble blow-torch or blow-lamp,
which emits the flames of methyl alcohol under pumped air-pressure, for burning off old paint and suchlike
uses.

She says that her father [Alexios], knowing that the enemy were skilled and courageous warriors, resolved to rely on the use of the device of fire against them. He had fixed to the bows of each of his galleys a tube ending in the head of a lion or other beast wrought in brass or iron, 'so that the animals might seem to vomit flames'. The fleet came up with the Pisans between Rhodes and Patara, but as its vessels were pursuing them with too great zeal it could not attack as a single body. The first to reach the enemy was the Byzantine admiral Landulph, who shot off his fire too hastily, missed his mark and accomplished nothing. But Count Eleemon, who was the next to close, had better fortune; he rammed the stern of a Pisan vessel, so that the bows of his ship got stuck in its steering-oar tackle. Then, shooting forth the fire, he set it ablaze, after which he pushed off and successfully discharged his tube into three other vessels, all of which were soon in flames. The Pisans then fled in disorder, 'having had no previous knowledge of the device, and wondering that fire, which usually burns upwards, could be directed downwards or to either hand, at the will of the engineer who discharged it'. That the Greek Fire was a liquid, and not merely an inflammable substance attached to ordinary missiles, after the manner of fire-arrows, is quite clear from the fact that Leo proposes to cast it on the enemy in fragile earthen vessels which may break and allow the material to run about—as also from the name *pyr enygron* (πῦρ ἔνυγρον) or 'liquid fire' which Anna uses for it.[a]

Extremely few illustrations of the Byzantine flame-thrower apparatus (the *siphōn* or *strepta*) have survived,[b] and no accounts whatever of their construction or manner of operation. Perhaps this is because they were for so long a time classified as 'restricted information' by the Byzantine War Office,[c] and then, in the +11th or +12th century, when they could have been described in Arabic,[d] the era of gunpowder, even if weak in nitrate, was already on the horizon. So it is a precious circumstance that we do have a complete account of such a pump—in the *Wu Ching Tsung Yao*, as we shall shortly see (p. 82).

Greek Fire was used again in naval fight with great success to repel a Russian attack on the city in +941,[e] and again in +1103 against the Pisans near Rhodes,[f] and on many other occasions. After the Third Crusade (+1192) the Venetians in Byzantium learnt the secret of distilled, low boiling-point, petro-

---

[a] (1), vol. 2, p. 47, mod. auct.

[b] Among the best known is a picture in an +11th-century MS. (Vatican Cod. 1605), showing a soldier with a flame-thrower on top of a wooden structure outside a castle. The weapon looks rather heavy, but it is certainly hand-held. See Feldhaus (1), col. 303, (2), p. 232, fig. 264; Zenghelis (1); Wescher (1), p. 262; Cheronis (1).

Another picture, of a boat with three rowers, and two men at the bow doing something to a wide-mouthed tube from which spout diffuse flames enveloping another boat, is also known. It has been published by Mercier (1), pl. opp. p. 28 and Previté-Orton (1), vol. 1, p. 214, fig. 37. It comes from a +14th-century MS. of John Skylitzes in the Bib. Nat. Madrid, MS 5–3, N 2. On this late +11th-century work see de Hoffmeyer (4).

A third is mentioned by Byron, (1), p. 280; a hand-held pipe about 5 ft long with flames issuing from a funnel-shaped mouth, in a late MS. (c. +1460) in the Bibliothèque de l'Arsenal in Paris.

[c] Cf. Partington (5), pp. 20–1.

[d] Perhaps they were, but the writings have not come down to us. Cf. pp. 41 ff. above.

[e] Luitprand, *Historia ejusque Legatio ad Nicephorum Phocam*, v, 6; cf. von Romocki (1), vol. 1, p. 15.

[f] As we have just seen. Anna Comnena, *Alexias*, loc. cit.

leum fractions,[a] and by then it was passing over to the Arabs too.[b] Or rather, it was getting widely known there, for already by about +900 there had been directions for the distillation of *naft* in al-Razī's work *Kitāb Sirr al-Asrār* (Book of the Secret of Secrets).[c] But by +1200 references are numerous, for example in the writings of the pharmacist Ibn Muḥammad al-Shaizārī al-Nabarāwī (d. +1193),[d] again in the work of the agriculturist Ibn al-'Awwām[e] about +1230, then twenty years or so later in the mineralogy of Zakarīya ibn Maḥmūd al-Qazwīnī,[f] and the pharmacy of Ibn al-Baiṭhar,[g] finally in the cosmography of Shāms al-Dīn al-Dimashqī (d. +1327).[h] This looks as if it was kept rather dark in the +10th and +11th centuries before becoming widely known in the +13th, just as it was about to be replaced by the perhaps more dependable and controllable weapon of gunpowder, first incendiary, then explosive, finally propellant.

When Richard I of England was sailing from Cyprus to Acre in +1191, he captured a Saracen transport ship laden with all kinds of armaments, including an abundance of Greek Fire petrol in bottles, which a witness had seen put aboard at Beirut.[i] Later the historian 'Abd al-Laṭīf al-Baghdādī described a great parade held in Baghdad in +1228 on the occasion of the reception of a Mongolian ambassador; there were 'soldiers with glass flasks of *naft*, who filled the whole plain with fire'.[j] The *naffāṭūn* troops certainly had now something else in their armoury than ordinary unprocessed mineral oil, and thence the line ran straight and quick not only to the *Liber Ignium* (p. 39 above) but also to the book of Ḥasan al-Rammāḥ (p. 41 above), towards the end of the +13th century.

If we are right in our identification of *mêng huo yu*,[1] Greek Fire came into China by about +900, just the time when various Byzantine emperors were writing about it in their military treatises (p. 78). How it came we shall consider presently. So far as we know, the very first mention of the 'fierce fire oil' occurs in connection with a gift from a southern Chinese State to a prince of the Chhi-tan Liao Tartars up in the north; and the date was +917. In his *Shih Kuo Chhun Chhiu*[2] (Spring and Autumn Annals of the Ten Independent States between Thang and Sung), Wu Jen-Chhen[3] says:[k]

---

[a] Yet as late as the early nineteenth century in England there was great uncertainty about the effects of heating and distilling on all kinds of oils, animal and vegetal no less than mineral. This clearly appears from a legal case analysed in fascinating detail by Fullmer (1).

The 'cracking' of the long hydrocarbon-chain oils of petroleum into shorter chain-molecules is of course something else again. It was a characteristic discovery or invention of the modern petrochemical industry, not put into practice industrially until about 1913 (Sherwood Taylor (4), pp. 270–1, 420). It involves high temperatures and high pressures, as well as the use of metallic catalysts, a field pioneered by Russian chemists such as V. N. Ipatiev (1867 to 1952).

[b] Cf. Partington (5), p. 197, on the Saladin military compendium of +1193 (p. 42 above).

[c] Ruska (14), p. 221. [d] Wiedemann (28); Wiedemann & Grohmann (1).

[e] Wiedemann (23); *passim.* [f] Ruska (24). [g] Leclerc (1).

[h] Mehren (1). All these references were assembled by Forbes (20).

[i] *Ricardi Regis Itinerarium Hierosolymorum*, Gale ed., +1687, vol. 2, p. 329. Cf. Partington (5), pp. 25, 39.

[j] Von Somogyi (1), p. 119.

[k] Ch. 2, p. 16a. The passage was quoted by Fêng Chia-Shêng (2), p. 17. Tr. auct.

---

[1] 猛火油   [2] 十國春秋   [3] 吳任臣

In that year the king of Wu-Yüeh sent an envoy with fierce fire oil to the Chhi-tan. He said that when they attacked cities they should use this oil to start fires, which would burn the buildings and the watch-towers. If the enemy poured water on it, it would burn all the more fiercely. The ruler of the Chhi-tan was delighted.

The fullest version of this occurs, naturally, in the *Liao Shih*[1] (History of the Liao Dynasty),[a] but it is preferable to translate the somewhat less diffuse text edited for a later historical work.[b] In this we read a rather amusing story:

The ruler of Wu State (Li Pien[2])[c] sent to A-Pao-Chi[3], ruler of the Chhi-tan (Liao), a quantity of furious fiery oil (*mêng huo yu*) which on being set alight and coming in contact with water blazed all the more fiercely. It could be used in attacking cities. Thai Tsu (A-Pao-Chi) was delighted, and at once got ready a cavalry force thirty thousand strong with the intention of attacking Yu-chou[4].[d] But his queen, Shu-Lü[5] laughed and said: 'Whoever heard of attacking a country with oil? Would it not be better to take three thousand horse and lie in wait on the borders, laying waste the country, so that the city will be starved out? By that means they will be brought to straits infallibly, even though it takes a few years. So why all this haste? Take care lest you be worsted, so that the Chinese mock at us, and our own people fall away.' Therefore he went no further in his design.

Here we can see how the nomadic traditions of cavalry strategy found it hard to absorb the new-fangled siege weapon.

So far nothing has been said about any siphon-like projector-pump. But sure enough it appears only a few years later, in fact in +919. The *Wu Yüeh Pei Shih*[7] (Materials for the History of the Wu-Yüeh State in the Five Dynasties Period), written by Lin Yü[8] only a few decades later, gives us an extremely interesting passage. The Wên-Mu King (Wên-Mu Wang[9]) was in command at an important naval battle when with more than five hundred dragon-like battleships he attacked the men of Huai[e] at a place called Lang-shan Chiang[10] (Wolf Mountain River). This was Chhien Yuan-Kuan[11], the seventh son and later (+932) the successor of the Wu-Su king (Wu-Su Wang[12]), Chhien Chhiu.[13] They won a great victory over the other side's forces 'because fire oil (*huo yu*[14]) was used to

<hr />

[a] Ch. 71, pp. 2*b*, 3*a*, tr. Wittfogel & Fêng Chia-Shêng (1), pp. 564–5. A closely similar text is to be found in *Chhi-Tan Kuo Chih*, ch. 13, p. 1*b*.

[b] *TCKM*, ch. 54, p. 85*b*, in the commentary to the history of Liang State, but referring to two years later. The passage, verbally identical, comes again in *Wu Pei Chih*, ch. 43, p. 10*a*, *b*. Tr. auct., adjuv. Mayers (6), p. 86.

[c] There is a historical mix-up here, for Li Pien was not the ruler of any State until he set up the Southern Thang (Nan Thang) dynasty on the ruins of Wu (= Huainan) in +937. Of course he might have acted as an official of Wu before A-Pao-Chi died, in +926. But it is more likely that the person really meant was Chhien Chhiu[6] (or Liu), the ruler of another small contemporary State, Wu-Yüeh, with its capital further south, at Hangchow. And indeed it is he who appears in the next passage.

[d] Mod. Peking.

[e] Presumably the forces of Huainan (or Wu) State, with its capital at Yangchow. Wu-Yüeh was centred on Hangchow.

[1] 遼史　　　　　[2] 李昇　　　　　[3] 阿保機　　　　[4] 幽州　　　　　[5] 述律
[6] 錢鏐　　　　　[7] 吳越備史　　　[8] 林禹　　　　　[9] 文穆王　　　　[10] 狼山江
[11] 錢元瓘　　　　[12] 武肅王　　　　[13] 錢鏐　　　　　[14] 火油

burn them up'. Then the author's commentary goes on:[a]

What is 'fire oil'? It comes from Arabia (Ta-Shih Kuo[1]) in the southern seas. It is spouted forth from iron tubes, and when meeting with water or wet things it gives forth flame and smoke even more abundantly. Wu-Su Wang used to decorate the mouths of the tubes with silver, so that if (the tank and tube) fell into the hands of the enemy, they would scrape off the silver and reject the rest of the apparatus. So the fire oil itself would not get into their hands (and could be recovered later).

And the text goes on to say that in this battle more than seven thousand men were captured and over four hundred naval ships destroyed in the conflagration.

The reason why this passage is of such signal importance is that it probably implies the first use of gunpowder in warfare in China. For just over a century later we come upon the only surviving description of a Greek Fire flame-thrower pump, in the *Wu Ching Tsung Yao* of +1044, and there gunpowder (*huo yao*[2]) makes its first appearance on the stage, used as a slow match to ignite the petrol when pumped forth. This description of the flame-thrower, which constituted a double-acting double-piston single-cylinder force-pump for a liquid, has already been given earlier, in Section 27;[b] but the translation of the passage in Tsêng Kung-Liang's text is so essential for our argument that we must repeat it here (Fig. 7).[c]

On the right is the petrol flame-thrower (lit. fierce fire oil shooter, *fang mêng huo yü*[3]). The tank is made of brass (*shu thung*[4]),[d] and supported on four legs. From its upper surface arise four (vertical) tubes attached to a horizontal cylinder (*chü thung*[5]) above; they are all connected with the tank. The head and the tail of the cylinder are large, (the middle) is of narrow (diameter). In the tail end there is a small opening as big as a millet grain.[e] The head end has (two) round openings $1\frac{1}{2}$ in. in diameter. At the side of the tank there is a hole with a (little) tube which is used for filling, and this is fitted with a cover. Inside the cylinder there is a (piston-) rod packed with silk floss (*tsa ssu chang*[6]), the head of which is wound round with hemp waste about $\frac{1}{2}$ in. thick. Before and behind, the two communicating tubes[f] are (alternately) occluded (lit. controlled, *shu*[7]), and (the mechanism) thus determined. The tail has a horizontal handle (the pump handle), in front of which there is a round cover. When (the handle is pushed) in (the pistons) close the mouths of the tubes (in turn).[g]

[a] Ch. 2, p. 4*a*, *b*, tr. auct., adjuv. Wang Ling (*1*), p. 167. An abbreviated quotation was given by Fêng Chia-Shêng (*2*), p. 17.

[b] Vol. 4, pt. 2, p. 145. We refrain, however, from repeating the explanations of the mechanism of the pump, for which the reader is referred to pp. 147 ff. there.

[c] (Chhien chi) ch. 12, pp. 66*a*ff.; tr. auct. This flame-thrower was, I think, first introduced to Western scholars in the paper of Wang Ling (*1*), pp. 166 ff. I like to remember that the text describing it was copied out for us more than forty years ago by a great scholar, the late Dr Fu Ssu-Nien, long before we possessed a copy.

[d] This interpretation is fixed by *TKKW*, ch. 8, p. 4*a*, ch. 14, p. 7*b*, etc. and other late Ming sources; cf. Chang Hung-Chao (*3*), p. 22. The practical use of brass at this date may be noted.

[e] If this is not the hole in the back wall through which the pump-rod passed (and for this purpose it seems rather too small), we cannot explain it.

[f] Reading *thung*[8] for *thung*[9].   [g] Like slide-valves.

| [1] 大食國 | [2] 火藥 | [3] 放猛火油 | [4] 熟銅 | [5] 巨筒 |
| [6] 拶絲杖 | [7] 束 | [8] 筒 | [9] 銅 | |

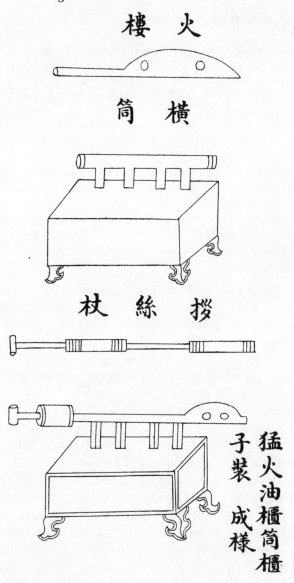

Fig. 7. Greek Fire (*mêng huo yu*) flame-thrower, with tank for the petrol-like liquid and double-acting pump with two pistons so that it works continuously. From *WCTY*.

Before use the tank is filled with rather more than three catties of the oil with a spoon through a filter (*sha lo*[1]); at the same time gunpowder (composition) (*huo yao*[2]) is placed in the ignition-chamber (*huo lou*[3]) at the head. When the fire is to be started one applies a heated branding-iron (*lao chui*[4]) (to the ignition-chamber), and the piston-rod (*tsa chang*[5]) is forced fully into the cylinder—then the man at the back is ordered to draw the piston-rod fully backwards and work it (back and forth) as vigorously as possible.

---

[1] 沙羅   [2] 火藥   [3] 火樓   [4] 烙錐   [5] 拶杖

Whereupon the oil (the petrol) comes out through the ignition-chamber and is shot forth as blazing flame.

When filling, use the bowl, the spoon and the filter; for igniting there is the branding-iron; for maintaining (or renewing) the fire there is the container (*kuan*[1]).[a] The branding-iron is made sharp like an awl so that it may be used to unblock the tubes if they get stopped up. There are tongs with which to pick up the glowing fire, and there is a soldering-iron for stopping-up leaks.

[*Comm.* If the tank or the tubes get cracked and leak they may be mended by using green wax. Altogether there are 12 items of equipment, all of brass except the tongs, the branding-iron and the soldering-iron.]

Another method is to fix a brass gourd-shaped container inside a large tube; below it has two feet, and inside there are two small feet communicating with them.

[*Comm.* all made of brass],

and there is also the piston (*tsa ssu chang*[2]).[b] The method of shooting is as described above.

If the enemy comes to attack a city, these weapons are placed on the great ramparts, or else in outworks, so that large numbers of assailants cannot get through.

And he goes on to say that in the defence of cities rolls of blazing straw should first be thrown down from the walls on to the assault bridges. The burning petrol does immense damage to enemy personnel, and water will not put it out. In naval fights it will burn and destroy floating bridges as well as wooden battle-ships. Also if directed upwards, matting fragments and chaff and any dry vegetable material should be thrown up first into enemy towns or camps; this will quickly catch fire and give rise to conflagrations.

As we pointed out before,[c] piston-pumps for liquids were not a characteristic constituent of Chinese engineering traditions, though the piston syringe was known and used in Han times, and bamboo had always been available for cy-linders. Moreover the double-acting piston-bellows goes back as far as the −4th century in China. One might for a moment, therefore, be tempted to think that this flame-thrower pump was a direct technical loan from Byzantium through the Arabs. But its design was too original, and if the 'siphon' pump gave forth a continuous jet, as most probably it did, that was assuredly accomplished rather by a combination of two cylinders in a Ctesibian force-pump system of true Graeco-Roman style. Even more original was the presence of a gunpowder slow-match in the ignition chamber of the petrol flame-thrower, identified all but infallibly by the term *huo yao*. Coarse twine impregnated with saltpetre and slow-ly burning will of course also do, and it touched off many a round during the first three hundred years of gunpowder weapons in the West, but very low-nitrate

---

[a] This must have been a jar in which glowing charcoal was kept, or else perhaps the glowing composition. It is seen at the top of Fig. 8, taken from the *San Tshai Thu Hui* encyclopaedia (+1609). That details about the Greek Fire petrol flame-thrower were still being given as late as this is remarkable in itself.

[b] This must describe some other design for a double-acting force-pump, but the account is too brief to allow of any visualisation or reconstruction.

[c] Vol. 4, pt. 2, pp. 143–4.

[1] 罐   [2] 捘絲杖

gunpowder would work in the same way, and that presumably is what we are dealing with here. And if this may be considered established for +1044 there is good reason to suppose that gunpowder slow-match was also used in the projector-pump of +919. After all, the earliest evidence of the composition goes back to about +850 (p. 112), so the historical sequence is quite reasonable.

There is one passage referring to events earlier than +919 which may be significant for the first use of gunpowder in war, but it is somewhat ambiguous. In his *Chiu Kuo Chih*[1] (Historical Memoir on the Nine States, in the Wu Tai period) Lu Chen[2] about +1064 wrote a series of biographies of the notable men who had served those warring principalities. Chêng Fan[3] was a general of Wu, and Lu Chen relates that:

at the beginning of the Thien-Yu reign-period (+904) ... in the course of the attack on Yü-chang (mod. Nan-chhang) Chêng Fan's men let off flying fire machines (*fa chi fei huo*[4]), which burnt the Lung-Sha Gate; then leading a troop of brave soldiers he entered the city, but was himself badly burnt by the flames. For this he afterwards received promotion.[a]

The difficulty here is to interpret the description. The phrase 'flying fire', which must have come so naturally to the pens of chroniclers, is as troublesome here as anything that Joinville had to say about Crusader battles. Hsü Tung[5], who started writing his military treatise *Hu Chhien Ching*[6] (Tiger Seal Manual) in +977 and finished it by +1004,[b] has a brief note on *fei huo*[4] in which he says that 'Flying fire is of the nature of trebuchet "bombs" (*huo phao*[7]) and incendiary arrows (*huo chien*[8])'.[c] This has led several writers to suppose that what Chêng Fan used were incendiary projectiles fired from *arcuballistae* or trebuchets, and these could of course by this date have contained gunpowder, even if probably low in saltpetre.[d] But the possibility also seems open that what he and his men employed were Greek Fire (distilled petroleum) flame-throwers,[e] and if the Wu-Yüeh people in +919 were using the gunpowder slow-match in the ignition chambers of their pumps, then Chêng Fan might well have done so too. But the incident remains ambiguous, though gunpowder in one form or another was probably involved.[f]

That the flame-thrower was fully in use in the first years of the +11th century appears from a story in which certain officials were laughed at for being more expert with it than with their writing-brushes. The *Chhing Hsiang Tsa Chi*[9] mentions two Sung officials Chang Tshun[10] and Jen Ping[11] who gained promotion

[a] Ch. 2, p. 13*a* (p. 29), tr. auct.
[b] It seems to have been based on drafts by earlier writers going back to +962 (Fêng Chia-Shêng (*1*), p. 46).
[c] Ch. 6 (ch. 53), p. 4*b* (p. 44), tr. auct.
[d] Fêng Chia-Shêng (*1*), *loc. cit.*; Tshao Yuan-Yü (*4*), p. 196.
[e] Though the date is rather early for Chinese-distilled petrol (cf. p. 76 above).
[f] We shall return to the incident, p. 148 below.

[1] 九國志　　　[2] 路振　　　[3] 鄭璠　　　[4] 發機飛火　　　[5] 許洞
[6] 虎鈐經　　　[7] 火炮　　　[8] 火箭　　　[9] 青箱雜記　　　[10] 張存
[11] 任幷

because of their expertise in using weapons such as the flame-thrower. Wu Chhu-Hou[1] says:[a]

In the Ching-Tê reign-period (+1004 to +1007) all the Ho-shuo scholars with Chü-jen (de grees) obtained official positions because of their services in defending cities. After Fan Chao[2] became the Optimus Scholar (Chuang-yuan[3]) in the State examinations (his friends) Chang Tshun and Jen Ping both received promotion though they had very much neglected their studies. Whereupon a certain (writer who styled himself) Anonymous-Master (Wu Ming Tzu[4]) composed an ironical poem about them, which included the lines: 'Chang Tshun knows only how to shoot with the whirlwind trebuchet (hsüan fêng phao[5]),[b] and all that Jen Ping can do is to set off the fierce fire oil flame-thrower (mêng huo yu[6]).' But afterwards (Chang) Tsun rose as high as Secretary of State (Shang-shu[7]) and (Jen) Ping became Inspector of Military Colonies and eventually Governor of Yaochou, in which office he died.

From this it is clear that the petrol flame-throwers were very familiar military equipment about +1000, even though the technicians who used them tended to be despised by the Confucian scholars.

At this point let us think about the route by which Greek Fire (distilled pet-roleum) came into China, and how long it was before it became indigenous there. The impression grows that South-east Asia was the way-station, and that the fierce fire oil travelled along with Arabic merchants by the sea route. From an important passage in the Wu Tai Shih Chi[8] (Hsin Wu Tai Shih[9]) we learn of a presentation of it by the King of Champa in +958 to the imperial court of the Later Chou dynasty (+951 to +960) which had its capital at Khaifêng in the north. The text reads:[c]

Champa (Chan-Chhêng[10]) lies by the south-eastern sea.... In the 5th year of the Hsien-Tê reign-period the king of that country, Yin Tê-Man[11] (Sri Indravarman III) sent an envoy named Phu-Ho-San[12] (Abū'l Hassan) with a tribute gift of 84 bottles of fierce fire oil (mêng huo yu) and 15 bottles of rose-water. The letter of presentation was written on many large (palm-) leaves, and enclosed in a box of fragrant wood. (It was said that) the

---

[a] Ch. 8, p. 6a; tr. auct.
[b] I.e. the fixed single-pole trebuchet rotatable so as to face all directions; cf. WCTY/CC, ch. 12, p. 50a.
[c] Ch. 74, p. 17a, tr. auct. The passage is quoted in Fêng Chia-Shêng (2), p. 17. It dates from about +1070, but there is a more nearly contemporary one in Chang Pi's[13] Chuang Lou Chi[14] (Records of the Ornamental Pavilion) written about +960, though it only speaks of the rose-water (in Thang Tai Tshung Shu, chi 7, ch. 81, p. 34a). The Thai-Phing Huan Yu Chi of about +980 also has the story, however, adding that the liquids were contained in glass bottles, and that the people of Champa were accustomed to use the Greek Fire in sea-fights (ch. 179, p. 16a, b). Its words were quoted in Tshê Fu Yuan Kuei (+1013), ch. 972, pp. 22a, b; and again in the Champa monograph in Sung Shih, ch. 489, p. 3b (though not in Sung Hui Yao); as also in WHTK, ch. 332, p. 18a (p. 2608·1), whence d'Hervey de St Denys (1), vol. 2, p. 545. Afterwards the event was well known and noted in many places, e.g. Tung Hsi Yang Khao, ch. 2, p. 6b. We had occasion to touch upon it in connection with distillation at an earlier point (Vol. 5, pt. 4, p. 158). On the rose-water see also Schafer (13), p. 173, (16), p. 75.

---

¹ 吳處厚   ² 范昭   ³ 狀元   ⁴ 無名子   ⁵ 旋風砲
⁶ 猛火油   ⁷ 尙書   ⁸ 五代史記   ⁹ 新五代史   ¹⁰ 占城
¹¹ 因德漫   ¹² 莆訶散   ¹³ 張泌   ¹⁴ 妝樓記

fierce fire oil could be used for sprinkling over things,[a] and when in contact with water it would burst into flame. The rose-water was said to come from the western regions, and if it was sprinkled over clothes the perfume would still remain even when they became old and worn-out.

Thus here again the petrol came up from the south.

One can follow this association of Greek Fire petrol, or 'naphtha' fighting, with south-east Asia all through the centuries. During the last few years of the +13th century, Chou Mi[1] was writing his *Kuei Hsin Tsa Chih*[2] (Miscellaneous Information from Kuei-Hsin Street, in Hangchow) with its various parts and supplements. This contained a graphic passage on sea-fights in the south seas, but as it has dropped out from nearly all the editions of this work we quote it from the *Yü Chih Thang Than Wei*[3] (Thickets of Talk from the Jade-Mushroom Hall), a Ming collection or anthology gathered together by Hsü Ying-Chhiu[4]. The text runs:[b]

Most of the countries of the south seas have what is called 'mud oil' (*ni yu*[5]). Nowadays their people who sail in shallow-draught vessels (*chhien fan chhuan*[6]) all keep it, and when they encounter another ship they fight with it, if they think they are the stronger of the two. This is called *ping chhuan*[7], a 'ship-collision'. When this happens, four men hoist up the mud-oil into the crow's-nest.[c] Little bottles are filled with it, and a roll of betel-nut husk (*pin-lang phi*[8])[d] is used as a stopper. When this is lit it acts like a fuse. Then the bottles are thrown down from on high, and when the mud-oil (bottles) hit the deck (of the other ship) they (break and) burst into flames which spread everywhere and continue to burn. If water is thrown on it, it blazes all the more fiercely, and nothing but dried earth and stove ashes will put it out.

Nowadays our official naval ships do not like to approach these shallow-draught barbarian vessels, because of this fearsome weapon.

This would have been written in the neighbourhood of +1298. The expression 'mud oil' for distilled petroleum is at first sight somewhat mysterious, but there are several possible explanations. The most obvious one would refer back to the appearance of petroleum at the natural seepages themselves, but another suggestion would see in the term a reference to the thick oil or sludge remaining in the retort after the distillation of the low boiling-point fractions. There must have been some tradition here other than that which gave rise to the commoner term 'fierce fire oil' (*mêng huo yu*[9]). At any rate the description shows that people were talking about the same thing. What is to be noticed here, however, is that there is no flame-thrower or projector pump, just the throwing of naphtha

---

[a] To remove stains? Or was it with the aim of setting them alight? Possibly the words were a confusion with those used of the rose-water.

[b] Ch. 27, p. 13*a*, tr. auct. We record the kindness of Dr Werner Eichhorn in bringing this passage to our attention twenty-five years ago.

[c] The peck-shaped box so characteristic of Chinese masts and flagstaffs.

[d] *Areca catechu*, the betel-nut palm indigenous to Malaysia. See Burkill (1), vol. 1, pp. 222 ff.

[1] 周密　　　　[2] 癸辛雜識　　　[3] 玉芝堂談薈　　[4] 徐應秋　　　　[5] 泥油

[6] 淺番舡　　　　[7] 併船　　　　　[8] 檳榔皮　　　　[9] 猛火油

grenades, as among the Arabic troops in the contemporary time of Ḥasan al-Rammāḥ.

Just about the same time that Hsü Ying-Chhiu was assembling his collection, the navigator and geographer Chang Hsieh[1] was writing about the *mêng huo yu* still in use for sea-warfare in south-east Asia.[a] In his *Tung Hsi Yang Khao*[2] (Studies on the Oceans East and West) of +1618, he wrote:[b]

San-Fo-Chhi[3] (Palembang in Sumatra) ...[c] is situated in the south-eastern seas.... Originally (the people) belonged to a special tribe of southern barbarians intermediate between those of Cambodia and Java ... Later it was defeated by the Javanese, and its name was changed to Chiu-Kang[4], which is still in use now.... It produces the furious fiery oil (*mêng huo yu*[5]), which according to the *Hua I Khao*[6] is a kind of tree secretion (*shu chin*[7]), and is also called mud oil (*ni yu*[8]).[d] It much resembles camphor, and can corrode human flesh. When ignited and thrown on water, its light and flame become all the more intense. The barbarians use it as a fire-weapon and produce great conflagrations in which sails, bulwarks, upperworks and oars all catch fire and cannot withstand it. Fishes and tortoises coming in contact with it cannot escape from being scorched.

Late in the following century, Chao Hsüeh-Min[9], who quoted the passage in abbreviated form[e] thought (wrongly) that this oil was a reference to natural petroleum (*shih yu*[10]). 'But from one of its names', he went on, 'it is obvious that "mud oil" cannot be any sort of vegetable exudate. In the (*Tung Hsi*) *Yang Khao* (Chang Hsieh) made a mistake about this.' Here again there was no talk of projector pumps, so it was probably a matter of breakable bottles with fuses once more.

These texts have surely demonstrated that the use of Greek Fire petrol or naphtha in war went on till quite a late date in south-east Asia, and that the distilled petroleum reached China through that region in the first place rather than over the land route. But now, before drawing all the threads together in a coherent picture we must fill it out by one or two more accounts of the techniques in use in China itself. For example, going back to the +10th century, petrol flame-throwers were prominent on both sides in the suppression of the

---

[a] Cf. Vol. 4, pt. 3, pp. 582 ff. and *passim*.
[b] Ch. 3, pp. 13*b*, 17*a*, tr. auct.   [c] Cf. Gerini (1), *passim*.
[d] Here he must have been referring to the *Hua I Hua Mu Niao Shou Chen Wan Khao*[11] (Useful Examination of the Flowers, Trees, Birds and Beasts found among the Chinese and Neighbouring Peoples), written by Shen Mou-Kuan[12] in +1581. Shen's ideas of precision in the natural sciences were none too exacting, as Wylie (1), p. 135 noted.
[e] *Pên Tshao Kang Mu Shih I*, ch. 2, p. 62*b*. In the same entry he quotes eight other sources, all about natural petroleum, including Chu Pên-Chung's[13] *Ko Wu Hsü Chih*[14] (What One should Know about Natural Phenomena), a late Chhing book. This repeats the stories about the fishes and the ashes, saying that these mineral oils can be kept only in glass vessels, and that they burn when floating on water so that water will not put them out.

[1] 張燮	[2] 東西洋考	[3] 三佛齊	[4] 舊港	[5] 猛火油
[6] 華夷考	[7] 樹津	[8] 泥油	[9] 趙學敏	[10] 石油
[11] 華夷花木鳥獸珍玩考		[12] 愼懋官	[13] 朱本中	[14] 格物須知

Southern Thang dynasty (+937 to +976) by the great Sung, established in +960. We have descriptions of the naval battle on the Yangtze in +975 near Nanking (Chinling), its capital,[a] which sealed its fate. In his *Tiao Chi Li Than*[1] (Talks at Fisherman's Rock) Shih Hsü-Pai[2] wrote:[b]

Chu Ling-Pin[3] (admiral of Nan Thang) was attacked by the Sung emperor's forces in strength. Chu was in command of a large warship more than ten decks high, with flags flying and drums beating. The imperial ships were smaller but they came down the river attacking fiercely, and the arrows flew so fast that the (Nan Thang) ships were like porcupines. Chu Ling-Pin hardly knew what to do. So he quickly projected petrol from flame-throwers (*fa chi huo yu*[4]) to destroy the enemy. The Sung forces could not have withstood this, but all of a sudden a north wind sprang up and swept the smoke and flames over the sky towards his own ships and men. As many as 150,000 soldiers and sailors were caught in this and overwhelmed, whereupon (Chu) Ling-Pin, being overcome with grief, flung himself into the flames and died.

The sailors of Byzantium would have felt very much at home in this battle. Then from another 'History of the Southern Thang Dynasty', that of Ma Ling[7], we hear more about another admiral, Tshao Pin[8], in this case on the imperialist side.[c]

In the 8th year of the Khai-Pao reign-period (+975), Tshao Pin[d] came down upon Chinling. He had large ships furnished with (bundles of) reeds saturated with thick oil, with the intention of taking advantage of the wind to start conflagrations; these were called 'rock-oil' devices' (*shih yu chi*[9]). But in urgent situations, then they used the machines to shoot the fire-oil forwards to resist the enemy (*huo yu chi chhien chü*[10]).

This was a clear mention of Greek Fire flame-throwers. Finally, rather more than a century later, Li Kang[11] brought them into action when trying to prevent the crossing of the Yellow River by the Chin Tartars before the siege of Khaifêng in +1126.[e]

Here we need not follow the petrol flame-throwers very far beyond the gunpowder era, but one reference imposes itself. When the Mongol ruler, Hulagu Khan, was setting forth in +1253 for the conquest of Persia, 'couriers were sent to Cathay to bring from thence a thousand men skilled in war machines (trebuchets), naphtha throwing (or projecting), and crossbow (or *arcuballista*) shoot-

---

[a] Grousset (1), vol. 1, pp. 367–8.
[b] Pp. 30b, 31a. We translate conflating with two parallel passages in Lu Yu's[5] *Nan Thang Shu*[6], ch. 5, pp. 3a, b; ch. 8, p. 4b.
[c] Ma Ling's *Nan Thang Shu*, ch. 17 (p. 117), tr. auct. The passage is quoted in Fêng Chia-Shêng (2), p. 17.
[d] Tshao Pin's biography is in *Sung Shih*, ch. 258, p. 1a, but it does not describe his use of incendiary or fire weapons.
[e] *Ching-Khang Chhuan Hsin Lu*[12] (Record of Events in the Ching-Khang reign-period), ch. 1 (p. 6).

[1] 釣磯立談   [2] 史虛白   [3] 朱令贇   [4] 發急火油   [5] 陸游
[6] 南唐書   [7] 馬令   [8] 曹彬   [9] 石油機   [10] 火油機前拒
[11] 李綱   [12] 靖康傳信錄

ing, with their families …[a] And as late at +1609 the *San Tshai Thu Hui* encyclopaedia gave a full account of the flame-thrower (Fig. 8), with illustrations.[b]

Perhaps the most curious story about the petrol occurs in the *Tso Mêng Lu*[1] written by Khang Yü-Chih[2] about +1137, a book which, as its title indicates, 'Dreaming of the Good Old Days', was written in the south after the victory of the Tartars, and recalls life in the former capital of the Northern Sung.[c] Khang had some rather wild ideas about the origin of distilled petroleum, but he remembered the way it was stored in the arsenals of the north-west in Northern Sung times,[d] and vividly sketched manoeuvres with the use of it against enemy encampments. What he said was as follows:[e]

Near the Wall Defence Arsenal (Fang Chhêng Khu[5]) at the northwest frontier Wall (the military engineers) used to dig out earth and make a large reservoir more than ten feet square in order to store 'fierce fire oil' (*mêng huo yu*[6]). In less than a month the surrounding earth would turn orange in colour, so further reservoirs were dug and the oil transferred to them; if this had not been done fire would have broken out and set light to the pillars supporting the roof (of the shed over the reservoir).[f]

I have heard that this fierce fire oil comes from a region several thousand li east of Korea.[g] When the sun begins to shine with all the strength of full summer, it makes the stones so hot that this oil oozes out from them. If it comes into contact with anything else it bursts into flame. It should only be stored in real glass vessels.

West of Chung-shan Fu[7h] there was a large body of water called Ta-po Chhih[8] (Big Wave Pond), so large that the local people called it a 'lakelet' (*hai tzu*[9]). I myself still remember the district commanders coming to it to study (and practise with their troops) water-combat, and to test the fierce fire oil. The opposite bank of the lake represented the fortified camp of the enemy. Those who were in charge of the oil sprayed it about, and as it was ignited it broke into a sheet of flame, so that the (fictive wooden) fortifications of the enemy were all in a short time completely destroyed. What is more, the oil had a secondary effect on the water, for all the water-plants were killed, and the fishes and turtles died.

[a] Quatremère (1), p. 133, translating Rashīd al-Dīn's *Jāmi 'al-Tawārīkh*. A parallel passage occurs in the *Ta'rīkh-i Jahān-Gusha* (History of the World Conqueror, Chingiz Khan) by 'Alā' al-Dīn al-Juwaynī (d. +1283); tr. Boyle (1), vol. 2, p. 608.

[b] Chhi yung sect., ch. 7, pp. 18a–21b. Greek Fire petrol also appears in the *Wu Pei Chih* of +1628, ch. 122, p. 21b, but in connection with an eruptor (cf. p. 267 below).

[c] There is a good account of the book in *SKCS/TMTY*, ch. 143, p. 72b. The editors knew the passage in the *Liao Shih* (p. 81 above), but doubted whether anyone knew much about Greek Fire petrol before the capture of Li Lun[3] (or Yin Li-Lun[4]). We have not been able so far to unearth any information about this character, who might be of considerable interest for this history.

[d] One would like to know what measures were taken to reduce evaporation losses.

[e] In *SF*, ch. 21, pp. 23b, 24a (ch. 34, p. 11b), also reproduced in *Kuang Pai Chhuan Hsüeh Hai*, vol. 2 (pp. 1052–3), and in *Ku Chin Shuo Hai*, whence it was kindly brought to our attention by Dr Werner Eichhorn in 1956. It is also quoted in Fêng Chia-Shêng (2), p. 18.

[f] We have not been able to think of an explanation for these phenomena, but unless the reservoirs were bricked the surrounding earth would certainly have become wastefully saturated.

[g] Curiously, this was diametrically opposite to the real point of origin.

[h] The present Chen-ting[10].

[1] 昨夢錄	[2] 康譽之	[3] 李倫	[4] 尹李倫	[5] 防城庫
[6] 猛火油	[7] 中山府	[8] 大波池	[9] 海子	[10] 眞定

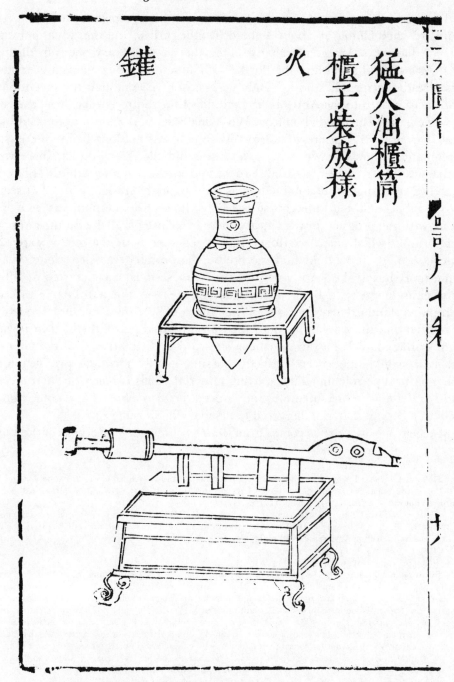

Fig. 8. Greek Fire projector. From *STTH*, Chhi yung sect., ch. 7, p. 18*b* (+1609).

To recapitulate, it seems quite clear that Greek Fire petrol (distilled petroleum) reached China by about +900 or rather earlier, and that it came by the
intermediation of Arabic merchants along the sea-route, and then up through
East Asia from the south to the north.[a] Of Chinese–Arab contacts a great deal
has already been said,[b] and it is only necessary to assume that the invention of
Callinicus passed to the Arabs by the middle of the +9th century. We have seen
how the petrol was handed on from Wu-Yüeh State in the south to the Chhi-tan
Liao Tartars in the far north in +917 (p. 80). It was not long before there were
diplomatic contacts between the Liao State and the Arabs direct; there were
embassies in +1019, +1021 and +1027, and in the same decade the Liao despatched the daughter of a nobleman as spouse for an Arabian prince.[c] It seems
most probable that in due course the know-how of distillation was conveyed
along with the product, in which case there is a close parallel with the gunpowder formula which reached Roger Bacon together with the gift of explosive
crackers (p. 48 above). At first, no doubt, the product travelled alone, but at
some time during the +10th century the Chinese must have started distilling
petroleum themselves.[d] By +1000, certainly by +1040, the petrol flame-thrower
pump was standard army equipment in China (p. 82), and it is quite impossible
to believe that these weapons depended on imported petrol distilled in Byzantium or Baghdad. It has already been shown that distillation with the Chinese
still-type would have been perfectly possible—with adequate precautions—
using natural petroleum as the starting material.[e] This is borne out by the *Sung
Hui Yao Chi Kao*[1], which mentions, not indeed the 'fierce fire oil' by name, but oil
itself as a supply material delivered to arsenals for working up.[f]

Much stronger evidence comes from the *Tung Ching Chi*[7] (Records of the East-

---

[a] This was the view of Mayers long ago, (6), p. 87, and we are much inclined to think that he was right.

[b] Vol. 1, pp. 214 ff.; Vol. 4, pt. 3, pp. 486 ff. After all, the Arab settlement at Canton started in the +8th
century if not in the +7th. Cf. too Schafer (16), p. 75.

[c] Wittfogel & Fêng Chia-Shêng (1), pp. 51, 357 and *passim*; Minorsky (4), pp. 5, 21, 76 ff. tr. and comm.
al-Marwazī, viii, 22–25. The embassy of +1026/7 was from the Chhi-tan to Mahmūd of Ghazni; it was well
enough received but the prince sent a cold reply.

[d] Cf. Vol. 5, pt. 4, pp. 129, 158 ff., 206.

[e] Vol. 5, pt. 4, pp. 68 ff. Cf. Butler & Needham (1).

[f] Tshao Yuan-Yü (4), p. 199. His reference was probably *Sung Hui Yao*, tsê 187, fang yü, ch. 3, p. 52 b, under
the heading Tung Hsi Tso Fang[2] (Eastern and Western Arms Factories), included strangely enough in the
Section on Geography. Here are mentioned many manufactures, e.g. armour for men and horses, spears,
swords, bows and crossbows, trebuchet artillery, drums and flags, with all kinds of works using leather, horn,
rattan, lacquer and glue. The *Sung Hui Yao* is of course a mine of information, but more for the bureaucratic
apparatus than for technical detail, since it consists so largely of imperial rescripts and decrees. We have found
only one mention of gunpowder (and even then not openly by name), a short piece on the inventions of Thang
Fu[3] (cf. p. 149 below), in tsê 185, ping, ch. 26, p. 37 a. The arsenals (*ta chün khu*[4]) come in tsê 146, shih huo, ch.
52, pp. 8 a ff., 25 a, 27 a, 30 a and 32 a; and the office controlling them (*chün chhi so*[5]) in tsê 68, chih fang, ch. 14,
pp. 1 a ff, and tsê 69, chih fang, ch. 16, pp. 4 a ff. There is also similar information (*chün chih*[6]) in tsê 171, hsing fa,
ch. 7, pp. 1 a ff. From the quotation immediately following above one can see good reasons why the *Sung Hui Yao*
should have been so reticent on Greek Fire and gunpowder.

[1] 宋會要輯稿   [2] 東西作坊   [3] 唐福   [4] 大軍庫   [5] 軍器所
[6] 軍制   [7] 東京記

ern Capital), written by Sung Min-Chhiu[1] some time before his death in +1079. In that he wrote:[a]

Apart from the Offices of the Eight Workshops (Pa Tso Ssu[2]), there are also other (government factories), notably the Workshops for General Siege Train Material (Kuang Pi Kung Chhêng Tso[3]), and now both of these, Eastern and Western, are under the authority of the Arsenals Administration (Kuang Pi Li Chün Chhi Chien[4]). Their work comprises ten departments, item, the gunpowder factory (*huo yao tso*[5]), item, the pitch, resin and charcoal department (*li chhing tso*[6]),[b] item, the fierce fire oil laboratory (*mêng huo yu tso*[7]), item, the metal shop (*chin tso*[8]),[c] item, the incendiaries plant (*huo tso*[9]), item, the joinery for timber large and small (*ta hsiao mu tso*[10]), item, the foundry for stoves large and small (*ta hsiao lu tso*[11]),[c] item, the leather yard (*phi tso*[12]), item, the hemp ropewalk (*ma tso*[13]), item, the brick and pottery kilns (*yao tzu tso*[14]).

All these have their regular specifications and procedures, so that those in charge can learn them by heart. But it is absolutely forbidden to divulge the text to outsiders.

This is certainly a vital text both for the petroleum distilleries and the gunpowder mills of the Northern Sung in the +11th century. It is, moreover, confirmed by several passages in the *Sung Shih* itself, which verifies in so many words the precautions about security when speaking of the Arsenals Administration (Chün Chhi Chien[26]).[d] The eight government arms factories (Tung Hsi Pa Tso Ssu[27]) were operated by a War Office department named Chiang Tso Yuan[28],[e] and the

[a] As quoted by Wang Tê-Chhen[15] in his *Chu Shih*[16] (Conversations on Historical Subjects), prefaced in +1115, ch. 1, pp. 4*b*, 5*a*, tr. auct. The passage is reproduced in Fêng Chia-Shêng (*1*), p. 53.

[b] This is an emendation by Fêng Chia-Shêng (*6*), p. 18, for the original text has *chhing yao*[17], a redundancy with the tenth workshop.

[c] Although these workshops may have had much else to do also, the functions of both indicate the preparation of equipment for making molten iron (*thieh chih*[18]) for assaulting the enemy and his wooden structures. This procedure is attested for many centuries in Chinese military history. In the beginning it was probably mainly a matter of pouring vessels of the hot metal on besiegers from above (as was proverbial in Europe), but already in the *Thai Pai Ying Ching* of +759 (ch. 4, p. 4*a*) 'bombs' (*phing*[19]) filled with it are hurled (*phao*[20]) from trebuchets. The *Wu Ching Tsung Yao* of +1044 (Chhien chi, ch. 12, pp. 59*a*, 62*a* ff.) gives more detailed specifications of the refractory clay containers (*kuan*[21]), projectiles breaking on impact; and a similar account is in the *Wu Pei Chih* of +1628 (ch. 132, pp. 14*b* ff.). Both these describe and illustrate the 'mobile furnace' (*hsing lu*[22]) for filling them, which could be drawn back and forth along city-walls. In considering all this it should be remembered that the melting-point of cast iron is 1130° C., some 400° lower than that of pure iron, but as was pointed out in Needham (*32*), pp. 14, 19, early Chinese cast iron (which long preceded cast iron anywhere else in the world) probably contained up to 3 % of phosphorus, and this reduces the melting-point still further, to about 950°. Of course, as an incendiary and offensive weapon this would be quite bad enough. That it was habitually used in medieval Chinese warfare appears from many texts, for example the Mongol general Bayan (Po Yen[23]) besieging Ying-chhêng[24] (mod. Wu-chhang in Hupei) about +1280 with trebuchet-launched molten-metal 'bombs' (*chin chih phao*[25]). Cf. Lu Mou-Tê (*1*), p. 32; Fêng Chia-Shêng (*1*), p. 47.

[d] *Sung Shih*, ch. 165, p. 23*b*: 'The people in the Arsenals Administration have designs of military equipment handed down, and no person below a certain rank was ever allowed to see them or copy down texts. Thus the secrets were not divulged.'

[e] *Ibid.* ch. 165, p. 21*b*; ch. 189, p. 19*b*.

[1] 宋敏求　　　[2] 八作司　　　[3] 廣備攻城作　　[4] 廣備隸軍器監
[5] 火藥作　　　[6] 瀝青作　　　[7] 猛火油作　　　　金作　　　　[9] 火作
[10] 大小木作　[11] 大小爐作　[12] 皮作　　　　[13] 麻作　　　[14] 窰子作
[15] 王得臣　　[16] 塵史　　　[17] 青窰　　　[18] 鐵汁　　　[19] 瓶
[20] 拋　　　　[21] 罐　　　　[22] 行爐　　　[23] 伯顏　　　[24] 郢城
[25] 金汁礮　　[26] 軍器監　　[27] 東西八作司　[28] 將作院

numbers of artisans and craftsmen who worked in them are discussed in yet another place.[a] So the Chinese were doing their own distilling.

The reason why we have considered Greek Fire so carefully is not just that as an incendiary weapon it was a predecessor of gunpowder. Its focal importance in the whole evolutionary story is that in the Chinese petrol projection-pumps its ignition took place by means of a gunpowder slow-match. The 'siphon' provided the occasion for the first use of the mixture; this is firmly established for the neighbourhood of +1040, and by implication to earlier dates such as +1004 back to +919.[b] Such was the first appearance of gunpowder on the practical stage of war, after its invention and elaboration by the Taoist alchemists of the +9th century, perhaps about +850, in their search for life-elixirs. And just as their work took place in China alone, so also exclusively Chinese was the practical use of the explosive mixture, even though as yet only in glowing and smouldering low-nitrate form.

And then, a whole millennium later, gunpowder and Greek Fire met together once again in a strange reincarnatory partnership—but what that was will be told in its place. As for that partnership which we have here been examining between the Byzantines and the Chinese—if one might figuratively call it so— the saying of the Persian Sharaf al-Zamān Ṭāhir al-Marwazī[c] about +1115 was exceedingly apposite:

The people of China are the most skilful of men in handicrafts. No other nation approaches them in this. The people of Rūm (the Eastern Roman Empire) are highly proficient (in technology) too, but they do not reach the standards of the Chinese. The latter say that all men are blind in craftsmanship, except the men of Rūm, who however are one-eyed, that is to say, they know only half the business.[d]

(5) ANCESTRY (II); THE RECOGNITION AND PURIFICATION OF SALTPETRE

The pre-requirement of the discovery of gunpowder was the recognition and purification of saltpetre (potassium nitrate); and before saltpetre could be used in any gunpowder composition, techniques had to be developed to extract it from its sources. It happened that medieval and pre-medieval chemical knowledge in China was in many ways more advanced than it was in Europe; and thus it is possible to trace the first uses of saltpetre from there eastwards leading

---

[a] *Ibid.* ch. 197, p. 13 *a*.
[b] Or even +904; p. 85 above.
[c] Born probably about +1046, d. after +1120.
[d] Minorsky (4), pp. 14, 65, tr. and comm. VIII, 4 of the *Tabā'i' al-Hayawān* (Natural Properties of Animals, Men and Places). This saying was more or less a *locus communis*; cf. Vol. 4, pt. 2, p. 602.

back to their origin in East Asia. In fact the early history of saltpetre is much more fully documented in China than in any other civilisation.

Broadly speaking, the word 'saltpetre' has been used to refer to potassium nitrate or sodium nitrate, or even calcium nitrate.[a] But its primary significance was potassium nitrate, the most singular ingredient of the gunpowder mixture. In nature potassium nitrate (prismatic saltpetre) is usually found mixed with sodium and magnesium salts. It needs the right climatic conditions to form, namely a sufficiently high temperature and a suitable humidity in which organic matter, especially excreta, can decompose; these conditions existed in Arabia, India and China, but not so much in Europe. For example, before the discovery of the nitre source in Chile, the Ganges valley was the greatest source of supply of saltpetre to England in the +18th and early 19th centuries.[b] The nitre from Chile, sodium nitrate (also known as cubical saltpetre), is far inferior to the potassium salt as an ingredient for gunpowder because it readily absorbs moisture, but it can be converted to potassium nitrate by treatment with strong solutions of potassium chloride. As for calcium nitrate, it is commonly found forming in the earth and on walls, together with many impurities, but again it can be converted to potassium nitrate by ionic exchange with potassium carbonate. These are, of course, modern techniques; in the early pre-gunpowder centuries, the problem was to distinguish it from other salts such as sodium or magnesium sulphate.

At an earlier stage we gave a rather thorough account of the recognition and isolation of saltpetre in China,[c] so we must not repeat it here. In the previous paragraph the word 'nitre' was used, coming in naturally enough, but historians of chemistry know well what multifarious meanings were attached at different times to that term, originating as it did through Greek and Latin from ancient Egyptian *ntry* or *natron*. This was sodium carbonate, mixed with the sulphate and common salt, together with a little bicarbonate, occurring naturally in desert regions; and it was known in China too, as *chien* or *shih chien*[1]. 'Nitre' could mean almost anything, generally soda, more rarely potash, and often mixed with the chloride. Its parallel term in Chinese was *hsiao*[2] or *hsiao*[3],[d] but instead of translating this in ancient and medieval texts as nitre it is better to use a word of equivalent vagueness, such as 'solve'. This is suggested not only by the etymology of congeneric characters and usages[e] but by the facts that potassium nitrate acts

[a] See Mellor (1), pp. 503 ff.
[b] Cf. Ray (1), 2nd ed. p. 229; Multhauf (9).
[c] Vol. 5, pt. 4, pp. 179 ff.
[d] Although several scholars have believed that they could establish a consistent semantic difference between the use of the water radical and the stone radical in *hsiao*, this has not been our experience. We fear that these were used indiscriminately both by the alchemists and the pharmacists in Chinese texts.
[e] Cf. Vol. 5, pt. 4, p. 5.

[1] 石鹼   [2] 消   [3] 硝

strongly as a flux in smelting, and participated in the procedures for bringing insoluble mineral substances into solution.[a]

The most characteristic name for saltpetre in Chinese texts was *hsiao shih*[1] (solve-stone), but to be sure of the identification of what they were talking about it is always necessary to see what they said about its properties. Hence the significance of other names such as *yen hsiao*[2] (blaze-solve), *huo hsiao*[3] (fire-solve), *khu hsiao*[4] (bitter-solve), *shêng hsiao*[5] (natural solve),[b] and *ti shuang*[6] (ground frost).[c] In China potassium nitrate was never confounded with sodium carbonate, as in the West, but rather with sodium sulphate and magnesium sulphate. The first of these generally went by the name of *phu hsiao*[7] (crude-solve), the second was *mang hsiao*[8] (prickle-solve)[d]—Glauber's Salt and Epsom Salt respectively In Thang and post-Thang texts, however, this last term came often to be applied to saltpetre;[e] and indeed it must be emphasised in general that none of the names can be taken as entirely reliable by themselves; always the description of the product must be examined in order to be sure what it was that the alchemical adepts or the pharmaceutical natural history writers had in hand.

Already the name *hsiao shih*[9] appears in the *Chi Ni Tzu*[10] book of the −4th century in its list of drugs and chemicals, but without details of its properties.[f] Then in the *Shen Nung Pên Tshao Ching*[11], the first of the pharmacopoeias, a couple of centuries later, it occurs again, but mainly in a medical and macrobiotic context.[g] Next comes the *Lieh Hsien Chuan*[12] (Lives of Famous Hsien), datable about the +2nd century, where a famous adept achieved immense longevity by

[a] Vol. 5, pt. 4, pp. 167 ff. See now also Butler, Glidewell & Needham (1). Li Shih-Chen, in *PTKM*, ch. 11, p. 28b quotes the *Pao Phu Tzu* book (c. +320) as saying that saltpetre can 'dissolve and soften the five metals, and bring the seventy-two minerals into aqueous solution'. Though saltpetre is mentioned there several times (e.g. ch. 11, p. 8b, ch. 16, pp. 7b, 9a), this passage seems not to be there now, but it certainly referred to the liquefaction of silicate gangues by a flux, and to the solution of inorganic substances normally very insoluble, like cinnabar.
   In +16th-century Europe a significant parallelism with gunpowder was constituted by the 'black' or 'defla-grating' flux, so often described by the metallurgist Lazarus Ercker (1) in +1574. Here saltpetre was mixed with 'argol' (potassium tartrate) and charcoal, giving, upon ignition, potassium carbonate, carbon and nitric oxide. This flux was used for smelting and purifying silver (Sisco & Smith (2), pp. 44, 81), gold (p. 110), copper (pp. 207, 215), bismuth (p. 275) and tin (p. 280). We should like to thank Prof. Cyril Stanley Smith for bringing this to our remembrance.
   [b] Chang Hung-Chao (1), p. 241, recognised all these names as synonyms.
   [c] As in all countries, saltpetre was at first scraped from the surface of the ground as a white crust or powdery efflorescence; cf. Kovda (1), pp. 121–2. As it happens, China possesses rather substantial areas of saltpetre solonchak soils in Honan province yielding more than 30,000 lbs. of saltpetre per acre p.a.; cf. K. C. Hou (1) and Yoneda (1), together with Wei Chou-Yuan (1), pp. 468–9 and Torgashev (1), pp. 380 ff. It is now more than thirty-five years since I first heard about these deposits (which produce sodium nitrate too) from Dr Wu Ching-Lieh of the 23rd Arsenal at Lu-hsien.
   [d] Presumably because of the shape of the crystals.
   [e] Cf. what Su Ching (Su Kung) says about *mang hsiao* and *hsiao shih* (+659) in *PTKM*, ch. 11, p. 26b. Li Shih-Chen abbreviated the passage, as one can see from *HHPT*, ch. 3 (vol. 1, pp. 21–2).
   [f] Ch. 3, p. 3a, in *YHSF*, ch. 69, p. 36a. Cf. Vol. 2, pp. 275, 554; Vol. 5, pt. 3, p. 14.
   [g] Mori ed., ch. 1 (p. 24). Cf. Vol. 6, Sect. 38.

[1] 消石   [2] 焰消   [3] 火消   [4] 苦消   [5] 生消
[6] 地霜   [7] 朴消   [8] 芒消   [9] 消石   [10] 計倪子
[11] 神農本草經   [12] 列仙傳

ingesting solve-stone as an elixir ingredient.[a] Reference to the extraction of saltpetre is already made in a list of seasonal prohibitions occurring in the *Hou Han
Shu* (History of the Later Han Dynasty).[b]

> From the day of the summer solstice onwards strong fires are forbidden, as well as the
> smelting of metals with charcoal. The purification of saltpetre (*hsiao shih*) has to cease
> altogether, until the beginning of autumn ....

This would have referred to the +1st and +2nd centuries, and shows that the practice must have been quite widespread, for otherwise the government would
hardly have issued an order interdicting it at this time of the year. But the
turning-point is reached in +492, when Thao Hung-Ching[1] described in his *Pên
Tshao Ching Chi Chu*[2] (Collected Commentaries on the Classical Pharmacopoeia
of the Heavenly Husbandman) the purple potassium flame test given by the
solve-stone,[c] together with its strong deflagration on charcoal. Since this and the
closely related work of +510, *Ming I Pieh Lu*[3] (Additional Records of Famous
Physicians on Materia Medica),[d] recorded so much +3rd-century knowledge, it
is likely that the flame test and other criteria such as the flux effect went back to
that time; in any case it is surely by far the oldest reference to the potassium
flame in any civilisation.[e]

After that there are many references to the purple potassium flame, the
deflagration, and the properties of flux and solubilisation. The *Hsin Hsiu Pên
Tshao*[4] (Newly Improved Pharmacopoeia) of +695 repeats them,[f] and soon afterwards, in +664, we have the Thang alchemical text about the wandering
monks.[g] Here in the *Chin Shih Pu Wu Chiu Shu Chüeh*[5] (Explanation of the Inventory of Metals and Minerals according to the Numbers Five and Nine) we read
about the Śaka or Sogdian monk Chih Fa-Lin[6], who recognised the presence of
saltpetre in northern Shansi and knew of its properties. Probably belonging to
the same century is another alchemical book, the *Huang Ti Chiu Ting Shen Tan
Ching Chüeh*[7] (Yellow Emperor's Canon of the Nine-Vessel Spiritual Elixir, with
Explanations) which has an important passage about saltpetre.[h] And then, not
later than about +850, there is a further text in the *Chen Yuan Miao Tao Yao Lüeh*[8]

---

[a] Kaltenmark (1), p. 171, translating ch. 2, p. 11a.
[b] Ch. 15, p. 5a, tr. auct. The passage was first noted by Wang Ling (1).
[c] *PTKM*, ch. 11, p. 25b (p. 54); *CLPT*, ch. 3 (p. 85·2). Cf. Chang Hung-Chao (1), p. 243, and Fenton (1),
p. 23.
[d] This is the oldest work in which the name *mang hsiao* (prickle-solve) appears. Li Shih-Chen (*PTKM*,
ch. 11, p. 29a) quotes it as saying that saltpetre can change (i.e. dissolve) the seventy-two minerals.
[e] The first European mentions seem to be of Renaissance time, or at least not earlier than Latin Geber.
[f] Ch. 3, pp. 10b, 11a, b.
[g] *TT* 900, pp. 5b, 6a, translated in full in Vol. 5, pt. 3, p. 139.
[h] *TT* 878, ch. 8, p. 12a, translated by us in Vol. 5, pt. 4, pp. 186–7.

[1] 陶弘景   [2] 本草經集注   [3] 名醫別錄   [4] 新修本草
[5] 金石簿五九數訣   [6] 支法林   [7] 黃帝九鼎神丹經訣
[8] 眞元妙道要略

(Classified Essentials of the Mysterious Tao of the True Origin of Things),[a] the same Thang book which has the oldest reference to a gunpowder mixture, words of great significance which will naturally be given below (p. 111).[b] By +1150, when gunpowder had already been in general use for some two and a half centuries, Yao Khuan[1] was writing his extended account of 'solve' or 'nitre' in the *Hsi Chhi Tshung Hua*[2] (Collected Remarks from the Western Pool), and in it he gave the first extant reference to artificial 'nitre-beds' or 'saltpetre-plantations'.[c] They must in fact have been earlier than this because Yao Khuan quotes the *Fu Hung Thu*[3] (Illustrated Manual on the Subduing of Mercury) by an adept named Shêng Hsüan Tzu[4] concerning them, but he and his book are exceptionally difficult to date, though probably of some time in the Thang or Wu Tai period.

There is perhaps no need to multiply examples, but it is noteworthy that all the alchemical books of the Thang period mention solve-stone (saltpetre) as a matter of course. For instance, the *Thai-Chhing Ching Thien-Shih Khou Chüeh*[5] (Oral Instructions from the Heavenly Masters on the Thai-Chhing Scriptures) of unknown authorship, describes a procedure for turning lead and tin into 'mercury' by the aid of saltpetre.[d] Solve-stone (*hsiao shih*[6]) appears time and again in the *Lung Hu Huan Tan Chüeh*[7] (Explanation of the Dragon-and-Tiger Cyclically Transformed Elixir)[e], by an alchemist otherwise unknown, the Golden-Tombs Master, Chin Ling Tzu[8], who was probably writing just after the Thang or in early Sung. He describes a procedure for 'subduing' saltpetre (*fu hsiao shih fa*[9]).

Take an ounce of saltpetre and comminute it to a fine powder, then place it in a porcelain container and compress it until the surface is flat. Cover it with two-tenths of an ounce of common salt and again flatten the surface. Place a porcelain cover over the top but do not seal it. Heat first with a gentle fire and then with a strong one until the salt is gone. (The saltpetre) is thus subdued.[f]

No chemical reaction would have taken place here, and the two substances probably just formed a melt. In the same book, solve-stone (*hsiao shih*[6]) and crude-solve (*phu hsiao*[10]) occur together at least three times in different chemical procedures, showing that the two were certainly differentiated by Chin Ling Tzu. And again, as we have previously seen, saltpetre occurs as a constituent of a hunger-prevention formula in a Thang manuscript.[g]

This same book, the *Lung Hu Huan Tan Chüeh*, if really of the +9th century or

---

[a] *TT* 917, p. 9*b*, translated in Vol. 5, pt. 4, p. 187. Cf. Fêng Chia-Shêng (*4*), p. 36.
[b] Cf. Vol. 5, pt. 3, p. 78. This text is also discussed by Yoshida Mitsukuni (7), p. 259.
[c] Ch. 2, pp. 36*a*ff., translated fully in Vol. 5, pt. 4, pp. 188ff. For an experimental study of nitre-beds see Williams (1).
[d] *TT* 876, p. 2*a*, *b*.   [e] *TT* 902.
[f] Cf. Vol. 5, pt. 4, p. 5. The salt of course did not 'go', but the diluted saltpetre would probably no longer ignite. Cf. also p. 115 below.
[g] Vol. 5, pt. 4, p. 146.

[1] 姚寬   [2] 西溪叢話   [3] 伏汞圖   [4] 昇玄子
[5] 太清經天師口訣   [6] 消石   [7] 龍虎還丹訣   [8] 金陵子
[9] 伏硝石法   [10] 朴消

before, may contain the earliest appearance of the phrase *huo yao*,[a] afterwards so universal as the appellation for gunpowder. But it occurs as a sub-title, i.e. a 'method for subduing chemicals by fire' (*fu huo yao fa*[1]). The experiment required white alum, potassium nitrate, sodium and magnesium sulphates, and mercury, so far as we can tell, producing in the end a purple sublimate. In general, Chu Shêng suggests that the procedures of 'subduing by fire' (*fu huo*[2])[b] were worked out by the alchemists of the Chin and Liu Chhao periods in their search for elixirs, precisely for the purpose of avoiding those deflagrations and proto-explosions which the military afterwards found so useful. For example, if the crude saltpetre contained much carbonaceous material, as it may well have done, heating it would produce potassium carbonate, something reasonably innocuous. Many mishaps were probably wrapped in silence, and the first actual record of such 'calamities' is in the *Chen Yuan Miao Tao Yao Lüeh*, which we shall speak of below (p. 111).

A rather clear statement was that given by Ma Chih[3] in the *Khai-Pao Pên Tshao*[4] (Pharmacopoeia of the Khai-Pao Reign-Period) in +973. He wrote:[c]

It was because saltpetre can dissolve (*hsiao*[5]) and liquefy (lit. change, *hua*[6]) minerals that it was given the name of solve-stone (*hsiao shih*[7]). When it is first boiled and refined it crystallises in small prickly (*mang*[8]) shapes, and in appearance resembles crude-solve (*phu hsiao*[9]);[d] therefore it has the synonym of prickle-solve (*mang hsiao*[10])....

Solve-stone (*hsiao shih*[7]) is in fact a 'ground frost' (*ti shuang*[11]), an efflorescence of the soil. It occurs among mountains and marshes, and in winter months it looks like frost on the ground. People sweep it up, collect it, and dissolve it in water, after which they boil to evaporate it, and so it is prepared. (The crystals) look like the pins of a hair-ornament. Good ones can be as much as five *fên* (about half an inch) in length. Thao Hung-Ching said all sorts of things (about these salts) because he did not know the facts ...

Actually solve-stone is produced among the rocks and cliffs in the mountains west of Mou-chou[12] in Szechuan. The size of the pieces (after purification) varies, but its colour is bluish-white, and it can be collected at all seasons.

It should be borne in mind that this was written some three hundred years before the Arabs and the Franks knew anything about saltpetre at all. For Ma Chih, solve-stone was unquestionably a substance different from crude-solve and prickle-solve, though the latter term could be applied to it as a synonym. In a moment we shall give some more detailed accounts of the manner of its purification.

One of the most interesting accounts of the whole subject, indeed remarkable

---

[a] As has been pointed out by Chu Shêng (*1*). In this paper he attributes to me doubts about the origin of gunpowder in China; but this is due to a mistranslation of Needham (86). Cf. Chu Shêng & Ho Tuan-Shêng (*1*).

[b] Cf. Vol. 5, pt. 3, p. 159; pt. 4, pp. 5, 250, 256, 262.

[c] Quoted in *PTKM*, ch. 11, p. 25*a*, *b*, tr. auct.

[d] Sodium sulphate?

[1] 伏火藥法	[2] 伏火	[3] 馬志	[4] 開寶本草	[5] 消
[6] 化	[7] 消石	[8] 芒	[9] 朴消	[10] 芒消
[11] 地霜	[12] 茂州			

for its insight, was that of Li Shih-Chen[1] in his *Pên Tshao Kang Mu*[2], published in +1596. His words were these:[a]

Ever since Chin and Thang times,[b] the different (substances) the names of which contain the term 'solve' (*hsiao*) have been the subjects of guesswork by most writers (on pharmaceutical natural history), who simply named them at random, with little justification. Only Ma Chih in the *Khai-Pao Pên Tshao* recognised that solve-stone was refined from 'ground frost', while prickle-solve and horse-tooth solve (*ma ya hsiao*[3]) were for the most part refined from crude-solve. His statements ought to have cleared up all the doubts and hesitations of these people. It was because solve-stone (*hsiao shih*[4]) was often given the synonym of prickle-solve (*mang hsiao*[5]), and because crude-solve (*phu hsiao*[6]) had the synonym of crude solve-stone (*hsiao shih phu*[7]), that the pundits got their names mixed up, and could not decide how to express the situation.

What they did not know was that the solves (*hsiao*) can be divided into two classes,[c] one aquose (*shui*[8]) and one pyrial (*huo*[9]).[d] Although their outward appearances are similar they differ completely in their nature (*hsing*[10]) and their *chhi*. Only the two items under crude-solve (*phu hsiao*) and solve-stone (*hsiao shih*) as set out in the *Pên Ching*[e] are correct. The rest, like prickle-solve (*mang hsiao*) in the (*Ming I*) *Pieh Lu*,[f] or horse-tooth-solve (*ma ya hsiao*) in the *Chia-Yu Pên Tshao*,[g] or natural solve (*shêng hsiao*) in the *Khai-Pao Pên Tshao*, were the outcome of unnecessary distinctions. I have therefore put them back where they ought to belong.

Now the crude-solve (*phu hsiao*) of the *Pên Ching* is an aquose solve (*shui hsiao*[11]). It is of two types. After the solution is evaporated by boiling, the (crystalline) produce appearing like prickles is called *mang hsiao*, and that appearing like horse teeth is called *ma ya hsiao*. Crude-solve (*phu hsiao*) is the solid which finally settles at the bottom (of the vessel); it has a salty sapidity and is (pharmaceutically) cooling (*han*[12])[h]

But the solve-stone (*hsiao shih*) of the *Pên Ching* is a pyrial solve (*huo hsiao*[13]). This also is of two types. After the solution has been evaporated by boiling, the crystals which appear like prickles are also called *mang hsiao*, and those which look like horse teeth are called (*ma*) *ya hsiao*. They are also named natural solve (*shêng hsiao*). What settles as a

---

[a] Ch. 11, p. 27b, tr. auct.
[b] I.e. from the +4th century onwards.
[c] Yin and Yang of course, as is evident from what follows.
[d] The obvious translation here would be 'watery' and 'fiery' respectively, but that would not do justice to the relationship of these in Li Shih-Chen's mind with the elements Water and Fire respectively. So what is needed is a set of specific adjectives, either invented, or adopted from obsolete words in old dictionaries, to signify connections with the Five Elements (cf. Vol. 2, pp. 242 ff., 253 ff.). In Vol. 5, pt. 5, on physiological alchemy, we had particular need of two of these, those for Metal and Wood (cf. pp. 56, 60 and *passim*). Accordingly we make use of the following series:

Metal	(M)	metallous
Wood	(W)	lignic
Water	(w)	aquose, aquescent
Fire	(F)	pyrial
Earth	(E)	terrene

[e] The usual abbreviation for the +2nd-century *Shen Nung Pên Tshao Ching*.
[f] Finished c. +510.
[g] Finished in +1060.
[h] A Yin property.

[1] 李世珍   [2] 本草綱目   [3] 馬牙消   [4] 消石   [5] 芒消
[6] 朴消   [7] 消石朴   [8] 水   [9] 火   [10] 性
[11] 水消   [12] 寒   [13] 火消

solid on the bottom is the solve-stone (*hsiao shih*). Its sapidity is acrid and bitter, and it is very heating (*ta wen*[1])[a] medically.

Both the solve classes give rise (during processing) to *mang hsiao* and to *ya hsiao*. For this reason the old prescriptions (of the Chin and Thang periods) took the salts as interchangeable, but since the Thang and Sung the prickle-solve (*mang hsiao*) and tooth-solve (*ya hsiao*) have been of the aquose class (*shui hsiao*).[b]

Thus it would seem that Li Shih-Chen distinguished rather clearly, and very justifiably, between the 'watery', aquose, sulphates, and the 'fiery', pyrial, nitrates. What he was really saying was that the traditional Chinese terminology of solves (*hsiao*) had often depended as much on crystal form (none too accurately observed) as on other properties. He was also clear that the same salt could crystallise in more than one form;[c] and by his time it was becoming evident that similarity of crystal shape could easily be very confusing.[d] It was because of the needle-like and other crystal forms that the name *mang hsiao* had got transferred both to saltpetre and to different sulphates at various times. Only the chemical properties themselves—and Li Shih-Chen knew them well—could really distinguish the salts. After all, he was writing in the century of Paracelsus and Agricola, no more able to attain modern scientific knowledge than they could, yet in distinguishing so well between the sulphates and the nitrates he reminds one of Paracelsus preparing his series of coloured metallic chlorides.[e] Both were early steps in the recognition and preparation of distinct chemical classes of salts.

One might not suppose that so great a naturalist as Li Shih-Chen would have anything to say about the military, but he has. A couple of pages later,[f] he is ruminating about medicinal properties, and says that since *phu hsiao* is a Yin or aquose substance, cold and salty, it tends to go downwards, and so can soothe and clean the intestinal tract, expelling malign pyrial *chhi* (*hsieh huo chhi*[2]) from the three coctive regions (*san chiao*[3]). Conversely, *hsiao shih* is a yang or pyrial substance, hot and sulphurous, so it tends to go upwards and can cure stasis of Fire in the three coctive regions as well as dispersing all kinds of accumulations. Then we read:

Now the military technicians (*ping chia*[4]) when making fire-weapons (lit. beacons, cannons and suchlike machinery, *fêng huo thung chi têng wu*[5]) use compositions containing

---

[a] A Yang property.
[b] I.e. Glauber's and Epsom salts.
[c] Cf. Mellor (1), p. 504, on the dimorphous character of potassium nitrate.
[d] Chang Hung-Chao (1), pp. 241 ff., reported in 1927 the analyses of seven contemporary traditional samples of *mang hsiao* from different localities, finding varying proportions of Na, Mg, Ca and K sulphates. The first named was always preponderant, the second did not exceed 7%, the third 1% and the fourth 5%. Up to 5% of common salt could also be present. Yet a Thang specimen (+756) turned out to be almost pure magnesium sulphate (Vol. 5, pt. 4, p. 181).
[e] See Sherlock (1); Pagel (10), p. 274. Cf. Vol. 5, pt. 4, p. 322.
[f] *PTKM*, ch. 11, p. 29*b*, tr. auct.

[1] 大溫　　　[2] 邪火氣　　　[3] 三焦　　　[4] 兵家　　　[5] 烽火銃機等物

saltpetre; so they fly up high, as if straight to the clouds and the Milky Way. For it is their very nature, as we know, to go upwards....

Here was a prescience, one might feel, of that escape from Planet Earth which Chinese rockets in their mature development would permit.

The passage puts us in mind of two things—first al-Juwaynī saying, about +1260, that the

trebuchet artillerymen of Cathay could with a stone missile convert the eye of a needle into a passage for a camel, and could fasten the woodwork of their trebuchets so firmly together with sinews and glue that when they aimed from the nadir to the zenith the missile did not return.[a]

But secondly, and more seriously, Li Shih-Chen's Aristotelian conviction of up-ward tendency reminds us of the Paracelsian theories of 'aerial nitre', that 'vola-tile saltpetre' somewhere up above us. In +16th- and +17th-century Europe this played a considerable part in physiological as well as meteorological spe-culation; for on the one hand it was appealed to as the element in the air essen-tial for respiration and muscular motion, while on the other it was thought to be the cause of thunderstorms and lightning. After all, for Paracelsians such as Joseph Duchesne and Robert Fludd, the atmosphere was the medium through which the heavenly and starry influences had to pass to reach us, so it was not hard to suppose that 'sophic fire' or 'vital nitre' was generated by them there. The gunpowder theory of thunder and lightning lasted on well into the +17th century, and the vital nitrous element led directly to John Mayow's classical work on the nature of respiration (+1668). Once again the quasi-mystical spe-culations of the Paracelsian tradition helped to generate modern science;[b] and once again the thoughts of Chinese naturalists of +1590 strangely recall the ideas of their contemporaries in Europe, isolated from each other though they were.

Lastly, what detailed accounts of the practical preparation of saltpetre for the gunners can we find in Chinese literature? We translate two, both from the neighbourhood of +1630, and first the indispensable passage from the Diderot of China in his technological book, *Thien Kung Khai Wu*[1] (Exploitation of the Works of Nature), about saltpetre. Sung Ying-Hsing[2] wrote:[c]

Saltpetre (solve-stone, *hsiao shih*[3]) is found both in China and in the lands of neighbour-ing peoples, all have it. In China it is chiefly produced in the north and west. Merchants who sell (saltpetre) in the southern and eastern (parts of the country) without first paying for the official certificate are punished for illegal trading. Natural saltpetre has

---

[a] *History of the World Conqueror*, pt. 3, ch. 6, tr. Boyle (1), vol. 2, p. 608.
[b] On this whole subject see Debus (9, 10), (18), pp. 32, 115 ff., 134; as also Guerlac (1, 2); and Partington (20).
[c] *TKKW* ch. 15, pp. 6a, b, 7a (Ming ed. ch. 3, pp. 32a, b, 33a), tr. auct. adjuv. Sun & Sun (1), pp. 269, 271.

[1] 天工開物   [2] 宋應星   [3] 消石

the same origin as common salt. Subterranean moisture streams up to the surface, and then in places near water (e.g. the sea), and where the earth is thin, it forms common salt, while in places near the mountains, and where the earth is thick, it forms saltpetre. Because it dissolves (*hsiao*[1]) immediately in water it is called solve-stone (*hsiao* (*shih*)). In places north of the Yangtze and the Huai river, after the mid-Autumn fortnightly period, (people) just have to be at home and sweep the (earthern) floors on alternate days to collect a little for purifying. Saltpetre is most abundant in three places. That produced in Szechuan is called *chhuan hsiao*[2]; that which comes from Shansi is commonly called *yen hsiao*[3]; and that found in Shantung is commonly called *thu hsiao*[4].

After collecting saltpetre by scraping or sweeping the ground ([*Comm.*] as also from walls) it is immersed in a tub of water for a night, and impurities floating on the surface are skimmed off. The solution is then put into a pan (*fu*[5]). After boiling until the solution is sufficiently concentrated, it is transferred to a container, and overnight the saltpetre crystallises out. The prickly crystals floating on the surface are called *mang hsiao*[6] and the longer crystals are *ma ya hsiao*[7]. ([*Comm.*] The amount of these varies with the places where the raw material has been collected.) The coarse (powder or crystals) left at the bottom as a residue is called *phu hsiao*[8].

For purification the remaining solution is again boiled, together with a few pieces of turnip, until the water has evaporated further. This is then poured into a basin and left overnight so that a mass of snow-white (crystals) is formed, and that is called *phên hsiao*[9]. For making gunpowder this *ya hsiao* and *phên hsiao* have a similar effect. When saltpetre is used for making gunpowder, if in small quantity it has to be dried on new tiles, and if in large quantity it should be dried in earthenware vessels. As soon as any moisture has all gone, the saltpetre is ground to a powder, but one should never use an iron pestle in a stone mortar, because any spark accidentally produced could cause an irretrievable catastrophe. One should measure out the amount of saltpetre to be used in a particular gunpowder formula, and then grind it together with (the right amount of) sulphur. Charcoal is only added later. After saltpetre has been dried, it may become moist again if left over a period of time. Hence when used in large cannons it is usually carried separately, and the gunpowder prepared and mixed on the spot.

Here we are back again in the terminological morass of the different 'solves', just like 'nitres'. If Sung Ying-Hsing thought that sulphates, chlorides or other salts would do instead of the nitrate, he was far astray, and may have been confused by his informants; but surely he was not so artless. He may not have been entirely wrong in what he said, for potassium nitrate does crystallise in two different forms, and in his time these may well have had different names. The salt is dimorphous, giving both rhombic crystalline plates and needle-like rhombohedral (trigonal) crystals isomorphous with those of sodium nitrate. So crystals or crystalline precipitates that looked different may all have been nitrate.

Of course the definitive identification and isolation of inorganic salts had to await the coming of modern chemistry, but Debus has shown[a] how great an

---

[a] (13), (18), pp. 137, 158 ff.

[1] 消　　　[2] 川消　　　[3] 鹽消　　　[4] 土消　　　[5] 釜
[6] 芒消　　　[7] 馬芽消　　　[8] 朴消　　　[9] 盆消

interest there was in late medieval Europe in the nature and composition of spa and mineral waters before the time of Robert Boyle. Here, besides such great names as Paracelsus and Agricola, advances were made by Edward Jorden (+1569 to 1632) and Gabriel Plattes (*fl.* +1639). Jorden was interested in salt-petre, and said that only when pure will it 'shoot forth needles.' Indeed, Debus can say[a] that all modern chemical analysis developed from dry metallurgical assays and wet analyses of mineral waters.

Another element of much interest in Sung Ying-Hsing's account, and one which we shall see even more prominently in the following text, is the use of colloidal organic material for clarifying the solution of the salt before crystallisation.[b] This industrial 'de-gunking' (as modern scientific colloquial might say) is a subject of far-reaching interest, on the history and theory of which singularly little seems to have been written. In the preparation of salts, sugar, etc., the problem of removing colloidal suspensions of organic origin, as well as turbidity due to unwanted inorganic substances, was empirically solved by the addition of other organic colloids; generally these formed a scum which could be scooped off from the surface of the solution.[c]

The Chinese saltpeterers used the soluble constituents of turnip slices, and also, as we shall see, glue,[d] but a great variety of other substances found employment in the Chinese table-salt industry, as we note at length in Sect. 37. For example, the briners there used ground soap-bean pods[e] and millet chaff,[f] but also hen's eggs, bodhi-seeds[g] and ground whole soya-bean suspensions.[h] European salt-boilers in their turn used the blood of bulls, calves and bucks, together with ale or beer (in moderation), as Agricola tells us in his *De Re Metallica*, finished in +1550.[i] This was still going on in the nineteenth century, but Dutch briners used sweet whey and many English ones egg-white.[j] Bull's blood was used too in sugar-refining,[k] while in culinary techniques egg-white reigned supreme, as for the making of clear consommé and aspic from thick broths and meat-extracts.[l] Since all these methods belong to a time long before modern industrial chemistry, it would take a special research to elucidate their origin, but they must surely go back well into the Middle Ages, both in China and in Europe.

As for the explanation of the effect, we doubt if the simple coagulation of

---

[a] (18), p. 28.

[b] Turnip tissue is also referred to by Li Shih-Chen in the preparation and purification of *phu hsiao*, *PTKM*, ch. 11, p. 18*b*. Probably *Brassica rapa*.

[c] This was a convenient property as it enabled crystallisation to start on the bottom and at the sides of the vessel.

[d] Glue and 'radishes' were mentioned by d'Incarville (1) in +1763 in his account of the making of pure saltpetre in China.

[e] *Gleditsia sinensis*; cf. p. 115 below.

[f] As Sung Ying-Hsing well knew; *TKKW*, ch. 5, p. 2*b* on edible salt, tr. Sun & Sun (1), p. 115.

[g] *Sapindus mukurossi*.   [h] *Glycine soja*; cf. Vol. 6, pt. 2, pp. 512ff.

[i] Bk. 12, tr. Hoover & Hoover (1), p. 552. In +1669 W. Jackson (1) recounted the same.

[j] Clow & Clow (1), pp. 56–7.   [k] *Ibid.* pp. 521, 526–7.

[l] Thudichum (1), pp. 155, 266. Also, of course, for clarifying wine.

proteins by heat, and the mechanical entanglement of the substances causing the inorganic or organic turbidity, will suffice to account for it. It seems more likely that the mutual precipitation of oppositely charged colloids[a] plays a large part. The recognition of the electric charges on colloid particles was one of the foundation-stones of colloid chemistry,[b] and in this clearing or clarification of salt solutions we must have one of the earliest empirical applications of it.

Let us look now at the last of our accounts of saltpetre making and testing; it comes from the military compendium *Phing Phi Pai Chin Fang*[1] (The Washerman's Precious Salve; Appropriate Techniques of Successful Warfare), edited by Hui Lu[2] at some time not long after +1626. He says:[c]

Take half a pan (*kuo*[4]) of crude saltpetre (in water) and boil it until the salt dissolves (completely). Then take one large red turnip, cut it up into four or five slices and put them into the boiling liquid. Remove the turnip (slices) when they have become cooked (and turn soft). Then mix the white of three eggs with two or three bowlfuls of water, and pour it into the pan while stirring with an iron spoon. Remove all the solid material (*cha tzu*[5]) floating on the surface. Then take about two ounces of the best clear liquefied glue, and pour it into the pan. After bringing to the boil again 3 to 5 times the contents of the pan are poured into a porcelain basin and then covered. The (precipitated) solids should not be allowed to flow out together with the water. The basin must not be moved or else the *chhi* may leak away. It is put into a cool place overnight.

If the needle-like crystals (*chhiang*[6], lit. spears)[d] which form look extremely fine and lustrous, (the saltpetre) is fit for use. If the crystals are not fine, or if they still have a salty taste, the chemical is not ready for making (gun)powder; and the above process should be repeated to refine (the saltpetre) once more.

Next Hui Lu mentions three methods of testing the nitrate. He says:[e]

There are only three ways to test solve-stone (saltpetre). The needle-like crystals should be extremely fine, the colour should be very lustrous and the taste should be insipid. If the product is white and without lustre, the impurities have not all been removed. If one tastes the crystals with the tip of the tongue and still finds them salty and tart, that is a sign that the salt has not all been removed. These two factors very often give rise to confusion and much harm can be caused as a result. However, saltpetre manufacturers often think in terms of profit and consequently it is very difficult to obtain purified saltpetre. But we can test the saltpetre by asking the manufacturer himself to hold the saltpetre (which he claims to be pure and genuine) on the palm of his hand and ignite it.

---

[a] Perhaps more correctly, of oppositely charged hydrophilic suspensoids. See Findlay (1), p. 282; Bull (1), pp. 224 ff., 327; Alexander & Johnson (1).

[b] Cf. Hardy (1, 2, 3).

[c] Ch. 4, p. 4a, b. The passage is almost identical with an earlier one in the *Ping Lu*[3] of +1606 (ch. 11, p. 5b). On the title see p. 35 above.

[d] It is interesting to note the term used here to denote the needle-like crystals of saltpetre. Obviously there was no technical term for these, and Hui Lu borrowed the word *chhiang* for this purpose. We have not found any dictionary which gives it the meaning it has here. In the *Ping Lu* the word *chhiang*[7] is written with the 'metal' radical.

[e] Ch. 4, p. 5a; tr. auct.

¹ 洴澼百金方　　² 惠麓　　　　³ 兵錄　　　　⁴ 鍋　　　　⁵ 渣滓
⁶ 槍　　　　　　⁷ 鎗

One should only (buy and) keep the saltpetre when it is found to burn on the palm without the hand becoming hot. Who would want to risk physical injury by thinking only about profit? This is the (third) method (of testing saltpetre).

This passage has a distinctly more professional air than that of Sung Ying-Hsing, as might perhaps be expected. On the glue as well as the turnip extractives we have already remarked. The test of the hand, however, is something new, and must have been very widespread, since we meet with it again in +17th-century Syria,[a] but China may well have been its home, like saltpetre itself. And now we shall give no more quotations in this sub-section, concluding it with a few words on India in comparison with the Arabic culture-area.

At what time saltpetre was first recognised, isolated and crystallised among the people of India remains obscure. At a guess, in view of their proximity to China, it can hardly have been later than the beginning of the +13th century, when the Arabs first understood the matter. On the other hand, the transmission of the knowledge from the Portuguese at the end of the +15th[b] seems much too late. The Chinese text of +664 about the wandering Śaka monk,[c] just referred to, indicates clearly that saltpetre was well known and produced in quantity by that time in Wu-Chhang[1],[d] i.e. Udyāna, a region of the high Indus valley near Gandhāra and Tokharestan,[e] and by this time a Kushan or Śaka principality.[f] But that is not really evidence for India.

Many years ago, Berthelot[g], translating a Latin MS. entitled *Liber Secretorum Bubacaris*,[h] probably of the +13th century, conjectured that the 'Indian salt' named in a list of salts was saltpetre.[i] Berthelot rightly identified the author as the great Abū Bakr ibn Zakariyā al-Rāzī (hence the title), and it was in fact more or less a translation of his systematic chemical treatise *Kitāb Sirr al-Asrār* (Book of the Secret of Secrets), written about +910.[j] When we look at the direct translation of this from the Arabic by Ruska, we find that in fact two salts are mentioned—'Chinese salt' as well as 'Indian salt'.[k] If one remembers the date, either, or indeed both, of these, could have been saltpetre; but the descriptions

---

[a] Description in Rafeq (1), in Parry & Yapp (1), p. 299. Apparently it was part of Syrian folklore.

[b] So Arima Seihō (1), p. 4; Hime (1), p. 74.

[c] Translation in Vol. 5, pt. 3, p. 139.

[d] Or Wu-Chha.[2]

[e] It lies south of the Hindu Kush mountains, on the border of modern Afghanistan and Pakistan, north of the road between Kabul and Peshawar.

[f] Cf. Vol. 1, p. 173.   [g] (10), pp. 306 ff.

[h] Paris Bib. Nat. MS. 6514, fols. 101–12, cf. 7156, fol. 114. Surprisingly, there seems to be no mention of this book anywhere in either Sarton, Thorndike or Ferguson.

[i] P. 308. Cf. Berthelot & Duval (1), p. 146. These authors, translating a Syriac version of al-Rāzī's book, included saltpetre among the boraxes (pp. 145, 154, 164, 198), but without justification from the original Arabic, as one can see from Ruska (14), pp. 84, 89. Similarly, Berthelot & Houdas (1), p. 155, found the word *bārūd* in a Jābirian text, and naturally translated it as saltpetre, but it must have been a later interpolation because the term never occurs in the Jābirian Corpus, judging from the exhaustive studies of Kraus (2, 3). On the Corpus see Vol. 5, pt. 4, pp. 391 ff.

[j] See Vol. 5, pt. 4, p. 398.

[k] (14), pp. 84, 90.

¹ 烏萇   ² 烏茶

are not promising, for the Indian salt is 'black and friable, with very little glitter', while of the Chinese salt al-Rāzī said that 'all we know about it is that it is white and hard, and has a smell like that of boiled eggs'.[a] On the whole, then, this suggestion of Berthelot's leads to a dead end.

And so does all the other evidence examined by Partington, who carefully weighed the legendary attributions, the undatable books, and the earliest Indian technical terms of problematical meaning.[b] There is no word in classical Sanskrit for saltpetre, *shoraka* being derived from the Persian *shurāj*. By +1526, the beginning of the Mughal (Mogul) Empire, there were plenty of guns and cannon in India, and therefore saltpetre for the gunpowder too, but that is much too late for our purpose; and before that time there seems to be little positive evidence for firearms other than Greek Fire and incendiaries. The crucial period, where future research will have to concentrate, lies between +1200 and +1400. In the meantime the obscurity remains.

All this indicates that between about +200 and +1200 the Chinese alchemists and pharmacists were slowly and painfully working out methods of isolating and purifying inorganic salts of many kinds, particular progress being made after the time of Thao Hung-Ching in +500, so that recognisable fairly pure saltpetre was available for the first gunpowder mixtures in the middle of the +9th century. By contrast the oldest Arabic mention of saltpetre occurs in the *Kitāb al-Jāmiʿ fī al-Adwiya al-Mufrada* (Book of the Assembly of Medical Simples) completed by Ibn al-Baithār about +1240,[c] after which many other mentions soon follow. There is, however, some reason to place the first knowledge of saltpetre among the Arabs in the first decades of that century,[d] while the earliest account of its use in war, especially for making gunpowder, belongs to the last decades of the

---

[a] This must refer to the presence of traces of hydrogen sulphide, H_2S. As we see in Sect. 37 on the salt industry, there were several kinds of common salt in China with names betraying this, e.g. *hei yen*[1] (black salt) and *chhou yen*[2] (stinking salt). In some processes the salt became contaminated with sulphides formed by bacterial action on the Ca and Fe sulphates precipitated first during the evaporation.

[b] (5), pp. 211 ff.

[c] Cf. Vol. 5, pt. 4, p. 194. Already by +1220 Bokhara and Samarqand had fallen to the Mongols, who were already established in Turkestan, so that many possibilities for the transmission of knowledge overland were open, as well as by the intermediation of the Arabic merchants along the sea route. Indeed there may be a reference to saltpetre in the book of ʿAbd al-Rahīm al-Jaubarī entitled *Kitāb al-Mukhtār fī Kashf al-Asrār* ... (The Revelation of Secrets), which was finished in +1225 (cf. Mieli (1), p. 156). He mentions *barūd al-thalji* (snowy saltpetre, if that was what he meant) and says that it was used by the magi in a fire-protective composition. *Barūd* was certainly the old name for saltpetre among the Arabs, though it came to mean gunpowder later on (cf. p. 45). Cf. Partington (5), pp. 190–1 and *passim*. This looks as if al-Jaubarī actually knew very little about the stuff, though with a vague impression that it had something to do with fire. For this reference we are indebted to Dr Ahmad Hassan of Aleppo.

[d] That saltpetre in its natural state mixed with other salts had been available in the Near East long before has been pointed out by Dr M. R. Bloch of the Negev Institute for Arid Zone Research in a private communication. In one of the caves beneath a Byzantine stable in Avdat he recovered a mixture containing 70 % of potassium nitrate, the rest being mostly sodium chloride. He thinks that this probably originated from Dead Sea sylvinite (a mixture of KCl and NaCl) acted upon by the urine which seeped through from the stable above. A similar situation seems to have occurred at Beth Govrin, between Jerusalem and Gad, though the archaeologists in 1965 regarded the very substantial amounts of saltpetre as a geological deposit. But although the salt may have been used for various purposes in ancient times, it was certainly not recognised and named until the +13th century, when knowledge transmitted from China was surely responsible.

[1] 黑鹽　　[2] 臭鹽

same; this is the book of Ḥasan al-Rammāḥ which we have already described (p. 41 above). The passage of knowledge of saltpetre to the world of the Franks and Latins must have taken place quite soon after its first appearance among the Arabs, for as we also saw, Roger Bacon about +1260 knew of it, and the *Liber Ignium* followed on before the end of the century.[a] The Westerners may not have called it 'Chinese snow', but the Arabs certainly did (*thalj al-Ṣīn*), and with great justice, for clearly it was recognised and prepared there long before anywhere else. This reason alone goes far to explain why China was the original home of all chemical explosives, starting with low-nitrate gunpowder.[b]

(6) GUNPOWDER COMPOSITIONS AND THEIR PROPERTIES

The word 'gunpowder', widely defined, should include all mixtures of saltpetre, sulphur and carbonaceous material; but any composition not containing charcoal, as for example those which incorporated honey, may be termed 'proto-gunpowder'. Our word gunpowder arises from the fact that Europe knew it only for cannon or hand-guns. In China, however, prototype mixtures were known to alchemists, physicians, and perhaps fireworks technicians, for their deflagrative properties, some time before they began to be used as weapons. Hence the Chinese name for gunpowder, *huo yao*[1], literally 'fire-chemical' or 'fire-drug'.[c] One also has to note that although a couple of centuries of the earlier stage of proto-gunpowder occurred in China, it never appeared in Europe at all—this in itself is an argument of some weight for diffusion from Asia.

All the conditions necessary for the first discovery of gunpowder were present in China by Han times. Saltpetre, as we have seen, was then already known, and fully recognised by +500. Sulphur too appears in the +2nd-century *Shen Nung Pên Tshao Ching*[2] (Pharmacopoeia of the Heavenly Husbandman),[d] and in the natural history of Wu Phu[3], the *Wu Shih Pên Tshao*[4] of about +235.[e] Charcoal was a substance commonly used in China from high antiquity. Alchemists were there also, busy from Chhin times onwards in the search for life-elixirs, which

---

[a] Cf. pp. 48, 39, above.

[b] As J. F. Davis well saw, (1), vol. 3, pp. 8 ff.

[c] We must not anticipate arguments the proper place of which is with transmissions (p. 568 below), but here it is impossible to overlook the fact that the earliest names for gunpowder in Europe, names which lasted long in the Germanic languages, denoted a plant-drug, i.e. (Germ.) *kraut*, (Dan.) *krud*, (Flem.) *kruyt*, etc. This was noted by Partington (5), pp. 95 ff., but he offered no comment on it. That the earliest European gunners (+1325 onwards) should have used a word meaning plant or plant-drug seems a coincidence passing strange unless it was a direct translation of *yao*. This might be considered a reason for thinking that some transmission came overland rather than by way of the Arabs. Cf. Nielsen (1), p. 208; Falk & Torp (1), pp. 583, 585.

[d] Ch. 2 (p. 57).

[e] *PTKM*, ch. 11 (p. 62). On sulphur-production methods in mediaeval China see Chang Yün-Ming (1).

[1] 火藥   [2] 神農本草經   [3] 吳普   [4] 吳氏本草

naturally involved the putting together of chemical products in all combinations and permutations. The only question is, when exactly these three substances were first mixed, and the incendiary or explosive property of the mixture realised. Since substances used in early time could not have been very pure, especially in the case of saltpetre, and also to some extent sulphur, one would rather expect the first Chinese mixtures to have been incendiary rather than explosive.

But it is time to define our terms more clearly.[a] We may reasonably draw up a scheme of combustible substances on the following criteria, depending on the character of the combustion.

(1) *Slow burning*. The old incendiaries: oils, pitch, sulphur, etc., used doubtless on the earliest incendiary arrows, as well as by other methods of delivery. See pp. 75 ff. above, on *shih yu*[1] and the like.

(2) *Quick burning*. Distilled petroleum or naphtha (Greek Fire, *mêng huo yu*[2]), either hurled in breakable pots with fuses or projected from mechanical flame-throwers. Still basically incendiary, though more effective against personnel.

(3) *Deflagration*. Low-nitrate powders, containing (*a*) carbonaceous material, or (*b*) charcoal as such.[b] To deflagrate is to burn with a sudden and sparkling combustion, producing a 'whoosh' like a rocket; and indeed as the nitrate proportion is increased these mixtures become suitable for rockets, as also for 'Roman candles', fire-lances (*huo chhiang*[3]) or 'eruptors', as we shall call them when of large dimensions. They could project incendiary balls, poisoned smoke-balls, and pieces of broken pottery and metal; though again essentially incendiary, they were as flame-throwers still more offensive against enemy troops, though not very prolonged in action. But here enters in the beginning of gunpowder's propellant property, since it carried the rocket *huo chien*[4] retroactively to its destination.[c]

(4) *Explosion*. This occurs with mixtures having higher proportions of nitrate, best with sulphur and charcoal alone as combustibles, but also sometimes in the presence of other substances such as arsenic. This may be termed 'weak explosion', giving the 'explosive puff', but if the firing is done in a closed space a considerable amount of noise can be produced, in fact a 'bang', and thin-walled containers of cast iron or other metal ('bombs', *huo phao*[5]) can be broken by the explosion.

---

[a] What follows is based upon a conference which two of us (J. N. and W. L.) had with the late Professor J. R. Partington on the 18 and 19 July 1956. A somewhat more condensed version was published in Partington (5), p. 266.

[b] This is the point at which proto-gunpowder (in our terminology) turns into gunpowder.

[c] A typical composition night be (in percentages of saltpetre, sulphur and carbon) 60:10:30 (Malina). Blasting powders also belong in this region, a famous French formula having saltpetre as low as 40:30:30 (Davis (17), p. 48).

Throughout this volume we give percentages in the following order: N (for saltpetre), S for sulphur, and C for carbon. This is the normal usage of explosives chemists (Partington (5), p. 324). It differs from that of the organic chemists, who use the order C:N:S for the ratio of the elements in organic compounds.

[1] 石油　　　[2] 猛火油　　　[3] 火槍　　　[4] 火箭　　　[5] 火砲

(5) *Detonation.*[a] When the nitrate content reaches the level of 'modern gunpowder', i.e. a suitably prepared mixture of saltpetre, sulphur and charcoal[b] in the proportions 75 : 15 ; 10, a 'brisant' explosion results upon firing. Metal containers burst with a loud noise, tearing and scattering, but leaving débris, and holes are blown in earth or masonry. The gunpowder is now a full propellant for projectiles launched from metal-barrel cannon or guns with walls of adequate strength (*huo thung*[1], *huo chhung*[2]). It is much too 'fast' for use in rockets.[c]

The proportions of the components just given are those of 'service gunpowder', which has great propulsive force, but the 'theoretical' percentage figures are considered to be 75 : 13 : 12.[d] This mixture, gradually attained through some ten centuries from a probable starting-point of equal quantities of the three constituents,[e] constituted the first chemical explosive known to man. An explosion may be defined as a loud noise accompanied by the sudden going away of things from the places where they were before. An explosive substance of this classical type[f] is something capable of giving rise to a sudden release of its own energy and a vast increase of its volume; in black powder the nitrate, which constitutes built-in oxidising power for the combustion, produces suddenly 3000 times its bulk of gas, giving off white smoke and including nitrogen, the oxides of carbon, and many salts of potassium in particulate form. The temperature reached in

[a] Here we follow the formulation of Partington, but we are aware that it does not quite represent the usage of contemporary explosives chemists.

Burning, deflagration and explosion are terms used to describe (in order of increasing violence) the release of energy from a reacting system by the same type of mechanism which involves a hot reaction zone moving through the reactive material. This may be solid, liquid or gaseous, a single compound or a mixture. The reaction propagates because the unreacted material near the reaction zone is heated to above its decomposition temperature by heat conducted from the hot reaction zone. The propagation rates (rates of advance of the reaction zone) are typically less than 1 mm/sec. for burning, and about 1000 m/sec. or so for explosion. However, the three types should be considered as overlapping regions of one single range.

Detonation, on the other hand, is a specific term applied to a process that propagates too quickly for heat conduction to cause decomposition ahead of the reaction zone. The necessary decomposition of the unreacted material is brought about by immensely rapid compression, by a shock wave travelling through it at the speed of sound. This speed, and hence the velocity of detonation, can be calculated and does not vary much for a given material. Detonation velocities are observed in the range (say) 7000–10,000 m/sec. The result of a detonation is markedly different from that of an explosion; it is an obviously more violent event, and produces much smaller fragments from a metal container with much more extreme fluctuations of air pressure at a distance.

Thus in modern parlance gunpowder can never do more than explode, but silver fulminate, and organic 'high explosives', detonate. On these latter see Urbański (1) and McGrath (1).

We are indebted to Dr Nigel Davies of Fort Halstead for the information in this note.

[b] Some importance attaches to the physical state of the constituents. On 'corning' see p. 349 below.

[c] Partington was rather conservative, suggesting that 'true gunpowder' should be taken as something between (4) and (5) rather than between (3) and (4); but this viewpoint is not adopted here, since rocket-compositions can hardly be denied the name of gunpowder, and indeed commonly bear it.

[d] See immediately below for the reason of our inverted commas. The best discussions are those of Mellor (2), vol. 2, pp. 820, 825 ff.; Davis (17), pp. 39, 43; Marshall (1), vol. 1, pp. 73 ff.; Ellern & Lancaster (1). We draw on these expositions in the following lines. We are indebted to Dr Peter Gray for an initiation into the subject in June 1953.

[e] As we shall see, pp. 120 ff. below.

[f] Not all the high explosives now known produce gas, but all do generate heat, so that the sudden heating of the surrounding air has a similar effect.

[1] 火筒   [2] 火銃

the explosion is of the order of 3880°C. The intrinsic energy may of course be liberated without explosion, since black powder will burn when uncompressed or unconfined; in fire-crackers for instance the rupture of the container because of the evolved gases makes the noise that shocked Roger Bacon so much,[a] not strictly the explosion of the powder.[b] Indeed, in comparison with high explosives, some of which are not combustible at all, gunpowder always essentially burns, but it can do this at so fast a speed that it can generate veritable explosions.

In gunpowder it is the sulphur which lowers the ignition temperature to 250°C. and on combustion raises the temperature to the fusion point of saltpetre (335°C.); sulphur also helps to increase the speed of combustion. The more the saltpetre the quicker the ignition and combustion are, and later on (p. 342) we shall document a continuous increase in the nitrate proportion which took place between the first warlike use of gunpowder in China in the +10th century, when the saltpetre hardly exceeded 50%, up to the 'theoretical' figure of 75%.[c] Thus the development was consistently from the slower and less vigorous effects to the maximum explosive power. As for the charcoal, its physical state, grain size, degree of aggregation and surface area, all turned out to be of much importance. Often quoted is the aphorism of John Bate in +1634: 'The Saltpeter is the Soule, the Sulphur the Life, and the Coales the Body of it....' But still in his time the optimal nitrate content was only just being approached in Europe.[d]

(7) PROTO-GUNPOWDER AND GUNPOWDER

(i) *The earliest alchemical tentatives and experiments*

The legend of Berthold Schwartz was in China no legend. There really were at least six centuries of alchemical experimentation before the first gunpowder explosion, and here we must take a look at some of the records which have come down to us from those times. The climax of this alchemical prelude is found in one of the books in the Taoist Patrology (*Tao Tsang*) entitled *Chen Yuan Miao Tao Yao Lüeh*[1] (Classified Essentials of the Mysterious Tao of the True Origin of Things).[e] This work details thirty-five elixir formulae or procedures which the writer considered wrong or dangerous, though some of them were popular in his time. At least three concern saltpetre, treated with quartz or blue-green rock

---

[a] P. 48 above.

[b] In Chinese fire-crackers the nitrate of the mixture is usually low, e.g. 66·6:16·6:16·8; Davis (17), pp. 111 ff. Of their ubiquity in Chinese culture we need say little; cf. Brewer (1), pp. 369–70.

[c] Many attempts have been made to represent the explosion of gunpowder in a single, if complex, chemical formula, but we need not enlarge on this here; perhaps the best known is that of H. Debus in 1882. Several alternatives are possible, which is why there cannot be any definitive theoretical set of proportions, only a certain range.

[d] Comparison is made on p. 358 below between Chinese and Western theories of gunpowder explosion.

[e] *TT* 917. We gave a fuller account of it in Vol. 5, pt. 3, pp. 78–9 but did not tranlate the key passenge.

[1] 眞元妙道要略

salt,[a] and then the text goes on to say:[b]

Some have heated together sulphur, realgar,[c] and saltpetre with honey;[d] smoke (and flames) result, so that their hands and faces have been burnt, and even the whole house (where they were working) burned down.

Evidently this only brings Taoism into discredit, and Taoist alchemists are thus warned clearly not to do it.[e]

These words are of focal importance for our history, and their approximate date is therefore of much importance. The book is attributed to Chêng Yin[1] (Chêng Ssu-Yuan[2]) who lived between +220 and +300, the putative teacher of the prince of alchemists, Ko Hung[3],[f] but very little of it can seriously be attributed to him. One modern scholar has placed it about the middle of the Wu Tai period,[g] which would mean the neighbourhood of +930, but in view of military descriptions which we give elsewhere (pp. 80, 81, 85) this must be too late. Perhaps therefore the best date for our passage would be c. +850, and this we shall retain.

It so happens that we have a circumstantial account of an alchemical disaster, quasi-fictional though it may be, just about contemporaneous with the foregoing fundamental deflagration statement, or even a little earlier. Li Fu-Yen[5] was a scholar from Kansu who was living and writing in +831, and in his *Hsüan Kuai Hsü Lu*[6] (Continuation of the Record of Things Dark and Strange) he told the story of a young man named Tu Tzu-Chhun[7], a story which was taken up and reprinted in the *Thai-Phing Kuang Chi*[8] collection.[h] Tu was rescued from poverty by a strange old alchemist, in return for which he was called upon to help him in his elixir experiments. While the reactions in the stove were going on, he had to take certain drugs and meditate in front of a blank wall, but terrifying nightmares supervened, including the apparent death of his own son, and although he

---

[a] The book also gives a test for saltpetre.

[b] P. 3a, tr. auct. Fêng Chia-Shêng was the first to note the passage, (1), p. 42, (5), p. 38. That was in 1947, but our collaborator Tshao Thien-Chhin discovered it independently while working in Cambridge in 1950.

[c] Arsenic disulphide.

[d] The drier the honey was allowed to become the better it would be as a source of carbon in this experiment.

[e] Perhaps not surprisingly, identical prohibitions occur in Europe, but in the late +13th century, not the +9th. As Berthelot (14), p. 694 pointed out, some versions of the Marcus Graecus text have the warning: 'Caveas ne flamma tangat domum vel tectum' (Beware lest the flames set the house and roof on fire!). And there is another statement: 'Haec autem sub tecto fieri prohibentum quoniam periculum immineret' (It is forbidden to make this (mixture) under a roof, because of the danger). Such parallels are very noteworthy; cf. Partington (5), p. 45.

[f] Cf. Vol. 5, pt. 3, pp. 76–7.

[g] Ong Thung-Wên (1). Besides Thang historical references, he discovered quotations from Yen Lo Tzu[4] (the Smoky Vine Master), whose *floruit* was Wu Tai, +936 to +943. But these could have been later interpolations. Apparently also there was a Chêng Ssu-Yuan in the +10th century. But he was not necessarily the real author. The doubts of Ong are shared to some extent by Kuo Chêng-I (1) and Wang Khuei-Kho & Chu Shêng (1).

[h] *TPKC*, ch. 16, pp. 1b ff. (vol. 1, pp. 132–3). We gave a fuller account of the proceedings in Vol. 5, pt. 4, p. 420, but we did not enlarge on the final explosion. It was Fêng Chia-Shêng (1), p. 43, who first saw the significance of it.

| [1] 鄭隱 | [2] 鄭思遠 | [3] 葛洪 | [4] 煙蘿子 | [5] 李復言 |
| [6] 玄怪續錄 | [7] 杜子春 | [8] 太平廣記 | | |

had been strictly warned not to make any sound.

as he awoke from his terrible visions near daybreak he cried out 'ai! ai!', and then saw purple flames already enveloping the house. Fierce flame burst forth from the chemical furnace (*yao lu*[1]), and set light to all the rooms around the courtyard. The foreign Taoist jumped into a water-butt and disappeared. Previously he had said that whatever the emotion and temptations the young man should say absolutely nothing, but in the end he had not been able to contain himself.

So again this seems like an account of some deflagrative composition. Perhaps the Taoist had been mixing saltpetre, sulphur and some source of carbon.[a]

When could we find the earliest account of saltpetre and sulphur together in an alchemical process? The answer could be in the neighbourhood of +300, for the *Pao Phu Tzu*[2] book of Ko Hung, already mentioned, has this in one of his aurifactive procedures. It is worth while giving the recipe in full, since we have not before done so. The text reads:[b]

Child's-play Method for making alchemical Gold (Hsiao erh tso huang chin fa[9]).
Prepare one iron cylinder 12 in. in diameter and 12 in. deep, and another smaller one 6 in. in diameter and highly polished. Grind and sieve 1 lb. each of (dry) red bole clay, saltpetre (*hsiao shih*[10]), mica, red haematite and calcareous spar, and mix them with $\frac{1}{2}$ lb. of sulphur and 4 oz. of laminar malachite. Make the powder into a paste with vinegar. Then coat the interior of the small cylinder with it to a thickness of $\frac{2}{10}$th of an inch.
Take 1 lb. of mercury and $\frac{1}{2}$ lb. each of cinnabar and lead amalgam.... Mix these thoroughly together with the mercury until (globules) are no longer visible, then place the material in the smaller cylinder and cover over with mica, closing it with an iron lid. Place the smaller cylinder within the larger one, set them both on a stove, and pour in enough molten lead to cover the smaller container and reach within $\frac{1}{2}$ in. of the brim of the larger one. Then heat over a raging fire for three days and three nights. What forms is called 'purple powder' (*tzu fên*[11]).
Seven inch-square spatulae of this, used for projection, will immediately turn 10 lb. of molten lead into gold; but the lead must have been held in the liquid state for 20 days beforehand in an iron vessel, and transferred to a copper vessel for the projection. Again, three inch-square spatulae of the same purple powder will at once turn heated mercury in an iron vessel to silver.

---

[a] There are plenty of accounts of alchemical explosions in the literature, but of course one can never be sure that mixtures of the gunpowder type were concerned, though very probably they sometimes were. For example, in a book entitled *Kuei Tung*[3] (The Control of Spirits), written by a Mr Shen[4] probably about +1185 but not printed till +1218, there is a discussion between a doughty soldier Wei Tzu-Tung[5] and his friend Tuan Kung-Chuang[6] about ghosts and apparitions, weird animals, and Taoist alchemical laboratories in caves, such as those of Thai-pai Shan[7], where they were talking. The story was that a certain Thang alchemist had experienced an explosion in which the furnace was blown asunder by the blast (*yao ting pao lieh*[8]). If the tradition was trustworthy it could well have been someone playing about with saltpetre, sulphur and sources of carbon.
[b] Ch. 16, p. 9a, b, tr. auct. adjuv. Ware (5), pp. 274–5. Cf. elixir no. 54 in Vol. 5, pt. 3, p. 95. Wang Ling was the first to draw attention to this passage, (1), p. 161.

[1] 藥爐　　　[2] 抱朴子　　　[3] 鬼董　　　[4] 沈氏　　　[5] 韋自東
[6] 段公莊　　　[7] 太白山　　　[8] 藥鼎爆烈　　　[9] 小兒作黃金法
[10] 消石　　　[11] 紫粉

What was really happening here is anyone's guess, and only a repetition under laboratory conditions could decide. Of the $7\frac{3}{4}$ lb. of reactants, only $1\frac{1}{2}$ lb. were nitrate[a] and sulphur, and although some carbon would have been present in the form of the carbonates of calcium and copper, it is unlikely that any fireworks would have resulted. Too many other elements were present—aluminium, silica, and iron, with probably small amounts of chromium and manganese. The lead would have held the temperature between 325° and 1500° C., while the heated mercury would have been below 360°. What the purple powder was remains uncertain,[b] but one thing *is* certain—sulphur and saltpetre were both ingredients in an alchemist's formula at the beginning of the +4th century. If this was going to be pursued, the spagyrical masters would infallibly hit one day, 'accidentally' it would be said, upon the inflammable nature of proto-gunpowder.

But more still, we can find in the *Pao Phu Tzu* book a combination of the three essential constituents, nitrate, sulphur and carbonaceous material. This occurs in a process for getting elementary metallic arsenic, the passage on which,[c] misunderstood by all translators so far,[d] has now been elucidated by Wang Khuei-Kho & Chu Shêng (*1*). After saying that realgar (*hsiung huang*[1], arsenic disulphide) can at need be consumed in hot water or wine, Ko Hung goes on to direct three treatments, with recrystallised saltpetre (*hsiao shih*[2]), with (dried and powdered) large intestine of the pig (*chu tung*[3]) heated in a red clay stove, and finally with pine resin (*sung chih*[4]). Then come the words: 'if you transmute with it these three things, (arsenical vapours) will arise like wisps of cloth, and (arsenic) sublimes as white as ice'.[e] Two separate points arise here also, first the question of the earliest preparation of pure metallic arsenic,[f] but second, more important for us here, the fact that the three gunpowder components are present all at once. No mention of a deflagration, or the possibility of one, is made by Ko Hung, but perhaps the presence of that danger was why he may have done it in three steps. Saltpetre oxidises the sulphide to arsenious oxide, 'white like ice',

---

[a] Saltpetre, as we have said (p. 96), appears again in three other places in *Pao Phu Tzu*, (ch. 11, p. 8b, ch. 16, pp. 7b, 8b), but mainly in connection with vinegar in solubilisation techniques (cf. Vol. 5, pt. 4, pp. 167 ff.), not with sulphur. Cf. Kuo Ching-I (*3*).

[b] Ware (*5*) thought litharge (PbO); Hsüeh Yü (*1*) thought cinnabar (HgS).

[c] Ch. 11, p. 9a, b.

[d] Ware (*5*), p. 188; Davis & Chhen Kuo-Fu (*1*), p. 316. Feifel (*3*), p. 17, got nearest, if not very near, saying 'knead it with these three things.'

[e] *I san wu lien chih, yin chih jo pu, pai jo ping*[5].

[f] This is generally ascribed to one of the alchemical writings, probably of the +14th century, attributed to Albertus Magnus (Mellor (*1*), p. 605; Multhauf (*5*), p. 189; Jagnaux (*1*), vol. 1, p. 656; Sarton (*1*), vol. 2, p. 937). But Sivin (*1*), pp. 180 ff. has shown experimentally that it was already accomplished in the mid +7th-century *Thai-Chhing Tan Ching Yao Chüeh*, probably written by Sun Ssu-Mo (cf. Vol. 5, pt. 3, pp. 132–3). Wang and his collaborators have repeated and confirmed this in the laboratory. Now it seems that we can trace its preparation back to Ko Hung about the beginning of the +4th century.

---

[1] 雄黃   [2] 硝石   [3] 豬胴   [4] 松指
[5] 以三物煉之引之如布白如冰

and then the carbonaceous materials reduce this to volatile elementary arsenic.[a] Perhaps it might not be too far-fetched to trace back the persistent inclusion of arsenic in later Chinese gunpowder compositions to ancient experiments of this kind.

We must next come down to a period between Ko Hung and the Taoist who warned of the danger of the proto-gunpowder mixture. The *Chu Chia Shen Phin Tan Fa*[1] (Methods of the Various Schools for Magical Elixir Preparations) is a collection made at some time during the Sung (i.e. after +960), but it assembled many recipes of much earlier dates. No name of principal author or compiler is given. Here in one place we find a 'Process in the Elixir Manuals for the subduing of Sulphur' (*Tan ching nei fu liu huang fa*[2]). It reads as follows:[b]

Take of sulphur and saltpetre (*hsiao shih*[3]) 2 oz. each and grind them together, then put them in a silver-melting crucible or a refractory pot (*sha kuan*[4]). Dig a pit in the ground and put the vessel inside it so that its top is level with the ground, and pack it all round with earth. Take three perfect pods of the soap-bean tree (*tsao chio*[5]),[c] uneaten by insects, and char them so that they keep their shape, then put them into the pot with the sulphur and the saltpetre. After the flames have subsided close the mouth and place 3 lb. of glowing charcoal on the lid; when this has been about one-third consumed remove all of it. The substance need not be cool before it is taken out—it has been 'subdued by fire' (*fu huo*[6]).

What this meant was that chemical changes had taken place giving a new and more stable product. As we remarked when we first discussed the passage, this operation seems to have been designed to produce potassium sulphate, and was therefore not in itself very exciting, but on the way Sun Ssu-Mo[7] (or whoever it was) stumbled upon a truly deflagrative mixture, later to lead to veritable explosions.[d]

The attribution is in fact a little difficult.[e] The process as it stands is anonymous, but the one before it carries the name of Huang San Kuan-jen[8] (His Excellency Huang Tertius), about whom nothing is known, and the one before that is given the name of the great Sui alchemist and physician Sun Ssu-Mo (+581 to +682).[f] If Sun was responsible for the soap-bean pods the date of the recipe would be about +650. The belief that he was is strengthened by another which

[a] Which sublimes as a black, grey or yellow substance (Durrant (1), p. 479; Multhauf (5), p. 108). The whole sequence could be demonstrated experimentally by Wang's collaborators Kuo Thung & Yuan Shu-Yü. What exactly would happen if all the constituents were heated together under various conditions remains a matter for speculation.

[b] *TT* 911, ch. 5, p. 11*a*, tr. auct. Fêng Chia-Shêng was the first to find this text, (1), p. 41, (4), p. 35, (6), p. 10. It is also quoted in Yen Tun-Chieh (20), p. 19.

[c] *Gleditsia sinensis* (CC 587).

[d] Vol. 5, pt. 3, pp. 137–8.   [e] Cf. Sivin (1), pp. 76–7.

[f] Neither of these concerned gunpowder or proto-gunpowder.

[1] 諸家神品丹法   [2] 丹經內伏硫黃法   [3] 硝石
[4] 沙罐   [5] 皂角   [6] 伏火   [7] 孫思邈   [8] 黃三官人

appears on the previous page,[a] attributed to the Holy Immortal Sun,[b] almost certainly Sun Ssu-Mo. This mixes an ounce each of saltpetre and sulphur with half an ounce of borax (tincal, sodium borate, *phêng sha*[3]), a combination which would surely be inflammable though hardly deflagrative; here again simple laboratory tests would settle its properties. Even if Sun himself was not responsible for these sulphur–saltpetre contiguities,[c] there were still two centuries to run before +850, and their rather archaic nature must place them then in the +8th or at latest early +9th century.

From this latter time there comes another instance of our theme, found in the *Chhien Hung Chia Kêng Chih Pao Chi Chhêng*[4] (Complete Compendium on the Perfect Treasure of Lead, Mercury, Wood and Metal)[d], compiled by Chao Nai-An[5] about +808.[e] In this work there is a 'Method of subduing Alum (or Vitriol) by Fire' (*Fu huo fan fa*[6]);[f] it involved mixing together 2 oz. each of saltpetre and sulphur with 0.35 oz. of dried aristolochia (*ma tou ling*[7]).[g] With the carbon present in such a form the preparation would have ignited suddenly, bursting into flames but not actually exploding—as indeed Fêng Chia-Shêng & Li Chhiao-Phing were able to show by actual experiment.[h] Thus one may say that the whole succession hangs together, from Ko Hung's use of saltpetre and sulphur mixtures (and probably some of the late Han alchemists had done it before him) through Sun Ssu-Mo and Chao Nai-An to the warnings of the mid +9th century, and then the application of the 'fire-drug' in war at the very beginning of the +10th, finally the printed compositions in the *Wu Ching Tsung Yao* dating from the early +11th.

Before examining these, however, we should take another look at the rather curious appellation 'fire-drug' (*huo yao*[10]), for it has a background the significance of which has already been hinted at (p. 6). The word *yao* originally meant a drug-plant or plant drug, but medicines of mineral and animal origin were always included in the Chinese pharmaceutical natural histories, from the *Shen Nung Pên Tshao Ching* onwards. Hence it came to mean for the alchemists any chemical substance, and the phrase could therefore be translated 'fire-medicine' or 'fire-chemical' equally well. It indicates clearly that those who first occupied themselves with it, indeed those who discovered it, were Taoist alchemists and

---

[a] P. 10*b*.

[b] From a *Sun chen-jen Tan Ching*[1], otherwise unknown, and not the same as Sun's *Tan Ching Yao Chüeh*[2] (Essentials of the Elixir Manuals, for Oral Transmission).

[c] Both Kuo Chêng-I (*1, 2*) and Wang Khuei-Kho & Chu Shêng (*1*) doubt the connection.

[d] *TT* 912. Chao's philosophical names were Chih I Tzu[8] and Chhing Hsü Tzu[9].

[e] A discrepancy in the stated datings has led some to prefer a time in Wu Tai or even Sung, but that would be too late to correspond with the military evidence.

[f] This was first noted by Fêng Chia-Shêng.

[g] *Aristolochia*, prob. *debilis* (R 585; CC 1559).

[h] Fêng Chia-Shêng (*1*), pp. 41–2, (*6*), p. 10.

[1] 孫眞人丹經	[2] 丹經要訣	[3] 硼砂	[4] 鉛汞甲庚至寶集成
[5] 趙耐菴	[6] 伏火礬法	[7] 馬兜鈴	[8] 知一子 [9] 清虛子
[10] 火藥			

physicians rather than military men.[a] Thus its appearance as a medicine in the *Pên Tshao Kang Mu* of +1596 was quite in character. Li Shih-Chen wrote:[b]

Gunpowder has a bitter-sour sapidity, and is slightly toxic. It can be used to treat sores and ringworm, it kills worms and insects, and it dispels damp *chhi* and hot epidemic fevers. It is composed of saltpetre, sulphur and pine charcoal, and it is used for (making) various (incendiary and explosive) preparations (*yao*[1]) for beacon-fires, guns and cannon (*fêng sui chhung (fo-lang) chi*[2]).[c]

There was thus no sharp line of distinction between medicines, drugs and chemicals;[d] and the technical term betrays the centuries of medico-alchemical work which preceded the adoption of the mixture in war—in China, that is to say, since in the Western world gunpowder appeared full-fledged. Indeed, when physiological alchemy (*nei tan*[3]) tended in China to replace chemical laboratory alchemy (*wai tan*[4]), the phrase *huo yao* was taken over in connection with the formation of 'inner elixirs' or enchymomas.[e] Lastly, the alchemists' secrecy injunctions of those centuries have often been remarked on, and this must have operated with particular force when a dangerous substance had been discovered.[f] And the relations of the Taoists with the military were quite close, as is shown, for example, by the authorship of the *Thai Pai Yin Ching*,[g] by the titles of many military manuals in the bibliographies of the dynastic histories, and by the names of military formations which often conformed to Taoist cosmological lore.

(ii) *The Sung formulae*

It would have been around the year +1040, during the life of William the Conqueror, that Tsêng Kung-Liang and his assistants were writing down the first gunpowder formulae to be printed and published in any civilisation, though evidence already given (pp. 80, 111) shows that the essentials of the mixture must have been known and used for at least a century previously. In the *Wu Ching Tsung Yao* there are three of these formulae, first for a quasi-explosive bomb to be shot off from a trebuchet (*huo (phao) yao*[5]), secondly a similar bomb with hooks attached so that it would fasten itself to any wooden structures and set them on fire (*chi li huo chhiu*[6]), and thirdly a poison-smoke ball which would attack the enemy chemically (*tu yao yen chhiu*[7]). To these we can add a formula

---

[a] The point was strongly made by Fêng Chia-Shêng (*1*), p. 31.
[b] Ch. 11, p. 60*a* (p. 78), tr. auct.
[c] On the last item see p. 369 below.
[d] All, so often, in the form of powders.
[e] This is fully explained in Vol. 5, pt. 5. It should be emphasised that this is the only context where the 'pyrial drug' has any meaning other than gunpowder. Cf. p. 7 above.
[f] Cf. Vol. 5, pt. 3, pp. 38, 74, 104, pt. 4, pp. 70, 259. An outstanding instance, of an oath sealed with blood, occurs in *PPT/NP*, ch. 4, p. 5*b*, tr. Ware (*5*), p. 75.
[g] These points were well put by Fêng Chia-Shêng (*1*), p. 45. Cf. p. 19 above.

[1] 藥　　　[2] 烽燧銃(佛狼)機　　　[3] 內丹　　　[4] 外丹
[5] 火砲藥　　　[6] 蒺藜火毬　　　[7] 毒藥烟毬

for another poisonous mixture probably intended to accompany a gunpowder-containing missile, which would provide the flame necessary to dissipate it as a smoke. It has been suggested that the relatively high content of sulphur betrays the origin of these projectiles from simple incendiaries,[a] but in fact the sulphur is not particularly prominent in them; carbonaceous materials other than charcoal form the main bulk, though charcoal itself is generally specified—what betrays the incendiary origin is surely rather the pitch and the various oils. Although we cannot deny the name of gunpowder to these compositions, it is still at a very low-nitrate level, deflagrative and incendiary rather than explosive. Later on we shall trace the rise in saltpetre content as the years and battles went by.

The first specification (Fig. 9) is called simply 'Method for making the fire-chemical' (*huo yao fa*[1])[b], and it lists the following ingredients:[c]

	oz.
Chin-chou[2] sulphur[d]	14
wo huang[3] (perhaps nodular sulphur)[e]	7
saltpetre (*yen hsiao*[4])	40
hemp roots (*ma ju*[5])	1
dried lacquer	1
arsenic (*phi huang*[6])[f]	1
white lead (*ting fên*[7])[g]	1
bamboo roots (*chu ju*[8])	1
minium (*huang tan*[9])[h]	1
yellow wax	0·5
clear oil	0·1
tung oil[i]	0·5
pine resin	14
thick oil	0·1
	82·2[j]

[a] Chao Thieh-Han (*1*), p. 10.
[b] *WCTY/CC*, ch. 12, p. 57*b*, 58*a*, *b*, tr. auct.
[c] The original figures are given in *chhêng*[10] (15 lb.), *chin*[11] (lb.), *liang*[12] (oz.) and *fên*[13] (tenths of an oz.), but we reduce them all to ounces for the sake of uniformity and simplicity.
[d] In Sung times this was the name of a city in Shansi, where there are indeed good sulphur deposits.
[e] This item is difficult, for the phrase 'nest yellow' cannot be found in any dictionary or reference work. It seems that 'chicken's nest yellow' (*chi kho huang*[14]) was a synonym of *hsiung huang*[15], realgar (arsenic disulphide), see *PTKMSI*, ch. 2 (p. 64); but we suspect that some kind of sulphur was meant here.
[f] Probably one of the sulphides rather than one of the oxides.
[g] Lead carbonate.
[h] Lead tetroxide, red lead.
[i] From *Aleurites fordii*.
[j] It will be noticed that the weight generally totals about 80 oz. (5 lb.).

[1] 火藥法	[2] 晉州	[3] 窩黃	[4] 焰硝	[5] 麻茹
[6] 砒黃	[7] 定粉	[8] 竹茹	[9] 黃丹	[10] 秤
[11] 斤	[12] 兩	[13] 分	[14] 雉窠黃	[15] 雄黃

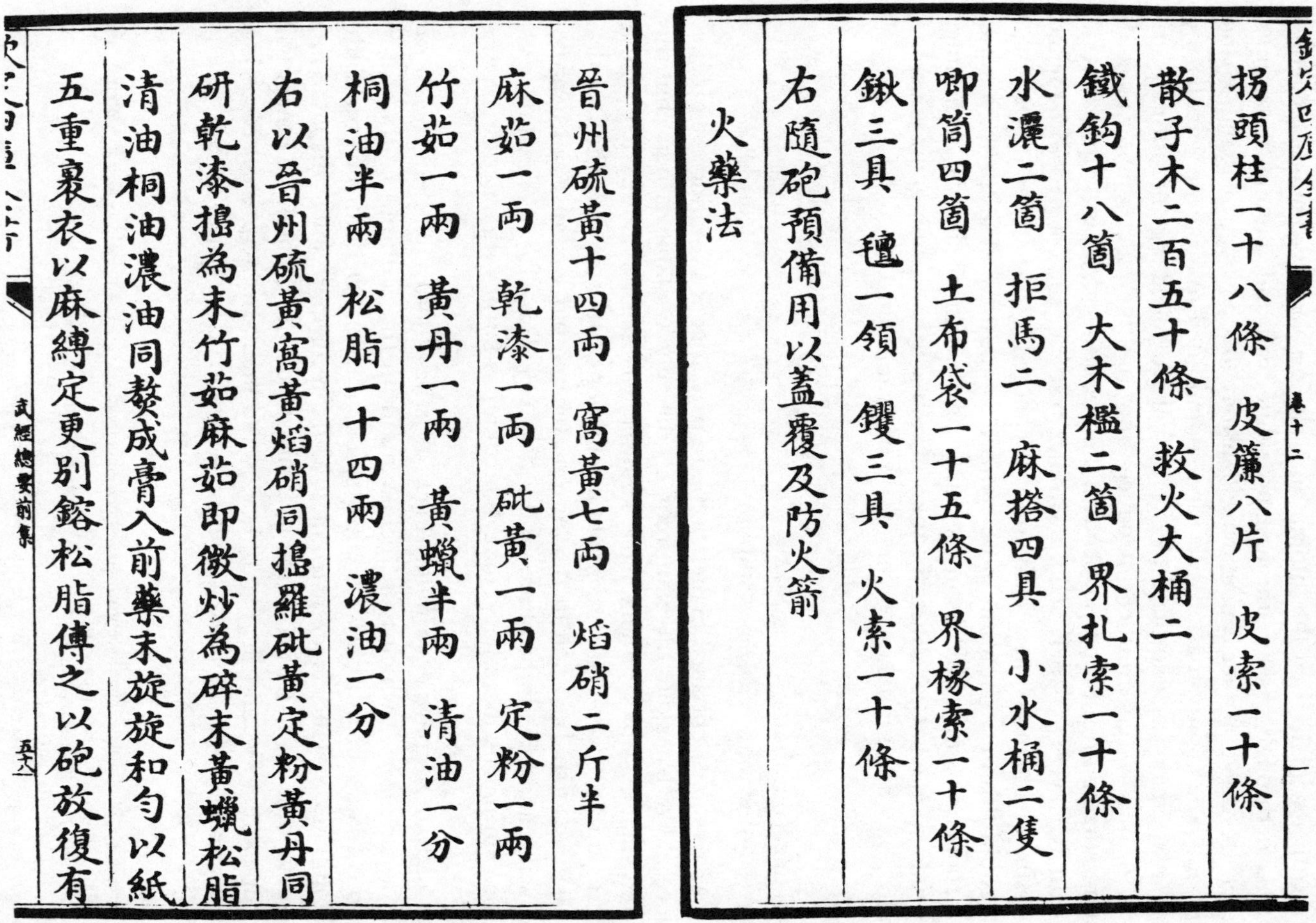

皮竇一十條　皮簾八片　拋頭柱一十八條

界扎索一十條　救火大桶二　散子木二百五十條

小水桶二隻　大木檻二箇　鐵鉤十八箇

界梯索二十條　拒馬二　水濺二箇

火索二十條　土布袋一十五條　即筒四箇

　耙二具　穗一領　鍬三具

右隨砲預備用以蓋覆及防火箭

火藥法

晉州硫黃十四兩　窩黃七兩　焰硝二斤半

麻茹一兩　乾漆一兩　砒黃一兩　定粉一兩

竹茹一兩　黃丹一兩　黃蠟半兩　清油一分

桐油半兩　松脂一十四兩　濃油一分

右以晉州硫黃窩黃焰硝同搗羅砒黃定粉黃丹同

研乾漆搗為末竹茹麻茹即微炒為碎末黃蠟松脂

清油桐油濃油同熬成膏入前藥末旋旋和勻以紙

五重裹衣以麻縛定更別鎔松脂傅之以砲放復有

Fig. 9. One of the earliest specifications of the gunpowder formula, pages from *WCTY*, ch. 12, p. 58*a, b* (+1044). The entry is headed 'method for making the fire-chemical' (*huo yao fa*)

The text then explains that the sulphur and saltpetre are to be pounded together and passed through a sieve, the arsenic, white lead and minium are ground together, the dried lacquer is pounded separately to a powder, the bamboo and hemp roots are slightly roasted and then comminuted to a powder, finally the yellow wax, pine resin, and the three sorts of oil are boiled together into a pasty mass. All the powders are then introduced into the thick soupy material, with constant stirring until evenly suspended. The resulting mass is wrapped into a ball with five layers of paper, tied up with hemp string, and covered with melted pine resin. The bomb is then ready to be discharged by a trebuchet (*phao*[1]).[a]

What did this mean in terms of percentage composition of saltpetre, sulphur and carbonaceous material? As in the other cases which we shall be examining, it depends upon what constituents (apart of course from the inorganic ones) are included. So we can tabulate the first formula as follows:

	%		
	N	S	C
all carbonaceous matter being taken	55·4	19·4	25·2
ditto, assuming that *wo huang* is sulphur	50·5	26·5	23·0

Some writers have obtained higher saltpetre percentages by noting the absence of charcoal,[b] or by taking the plant roots as the only available carbon,[c] but this is surely inadmissible. Others have followed the second alternative,[d] while others again have avoided percentage calculations.[e] In any case, the fact that there is no charcoal as such in the formula means that we must call it proto-gunpowder; and like all the others in the *Wu Ching Tsung Yao*, its function must have been primarily incendiary, though no doubt it burnt with a fierce deflagration.

The second gunpowder formula (Fig. 10) is for 'thorny fire-balls' (*chi li huo chhiu*[2] or *huo chi li*[3]),[f] which took their name from the calthrop[g] or the water-calthrop,[h] plants the fruits of which have spines or horns. Derivatively, the calthrop in military parlance is an instrument with four or more spikes disposed in a triangular form so that when three of them are on the ground the others will point upwards to wound the feet of horses and men.[i] In the present case (Fig. 17) the spikes were shaped like hooks or arrow-heads, designed to catch on to

---

[a] Nothing is said of a fuse, but there must have been one, for the artillerist to light before despatching the missile. The text says 'for attacking gates'.
[b] E.g. Chao Thieh-Han (*1*), p. 10.
[c] Ho Ping-Yü (priv. comm.).
[d] Arima (*1*), p. 43; Partington (5), p. 273; and ourselves formerly.
[e] Wang Ling (1); Davis & Ware (1), p. 524.
[f] *WCTY/CC*, ch. 12, p. 65*a*, *b*, tr. auct.
[g] *Tribulus terrestris*, R 364, Stuart (1), p. 441.
[h] *Trapa natans*, R 243; the *ling*[4] or *chi shih*[5].
[i] Cf. Wang Chung-Shu (1), p. 123 and fig. 156.

[1] 砲　　　[2] 蒺藜火毬　　　[3] 火蒺藜　　　[4] 菱　　　[5] 芰實

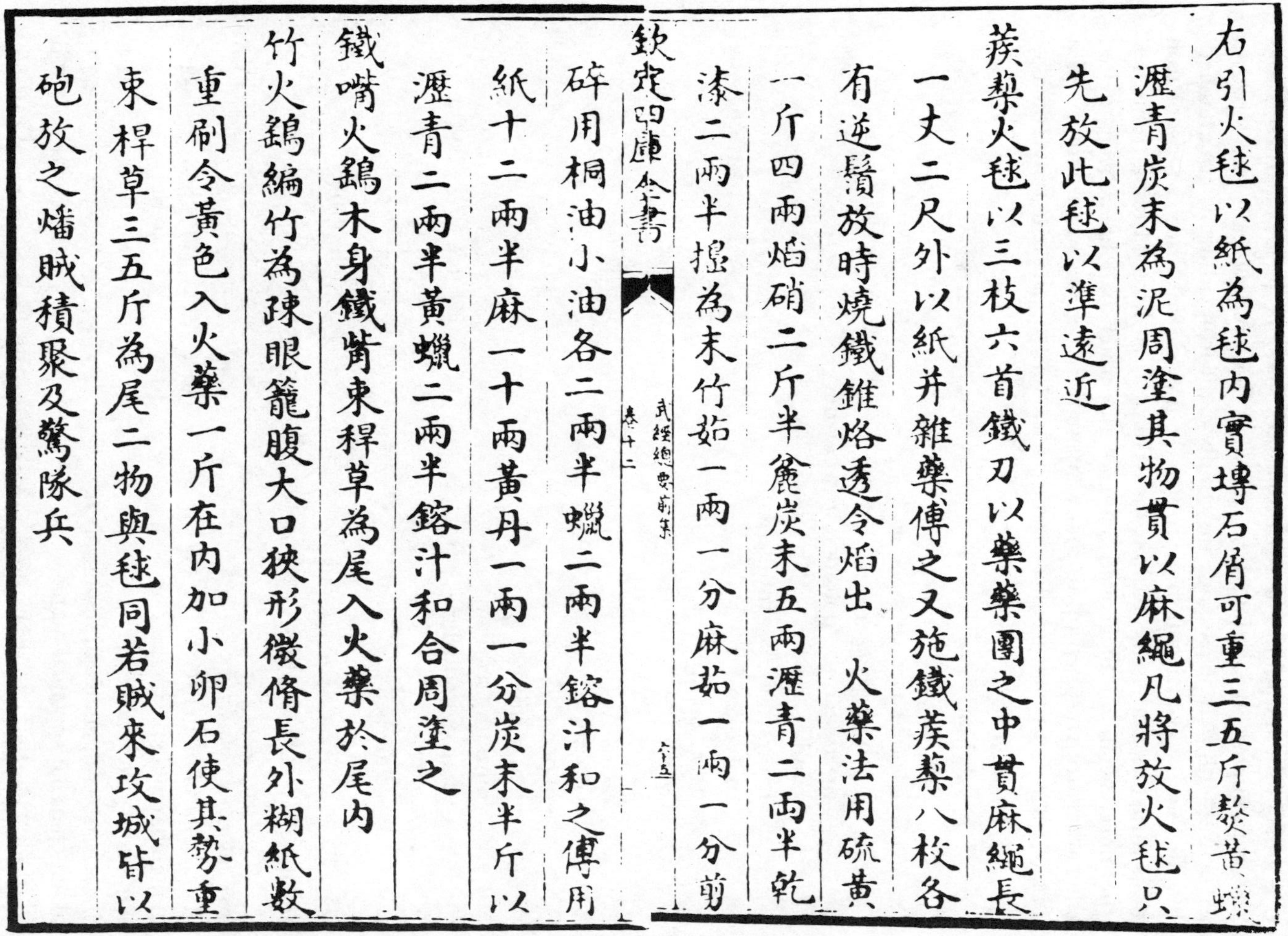

右引火毬以紙為毬內實塼石屑可重三五斤熬黃蠟
瀝青炭末為泥周塗其物貫以麻繩凡將放火毬只
先放此毬以準遠近
蒺藜火毬以三枝六首鐵刃以藥藥團之中貫麻繩長
一丈二尺外以紙并雜藥傅之又施鐵蒺藜八枚各
有逆鬚放時燒鐵錐烙透令焰出　火藥法用硫黃
一斤四兩焰硝二斤半麄炭末五兩瀝青二兩半乾
漆二兩半擣為末竹茹一兩一分麻茹一兩一分剪

欽定四庫全書　　武經總要前集　卷十二　　六十五

碎用桐油小油各二兩半蠟二兩半鎔汁和之傅用
紙十二兩半麻一十兩黃丹一兩一分炭末半斤以
瀝青二兩半黃蠟二兩半鎔汁和合周塗之
鐵嘴火鷂木身鐵嘴束稈草為尾入火藥於尾內
竹火鷂編竹為疎眼籠腹大口狹形微脩長外糊紙數
重刷令黃色入火藥一斤在內加小卵石使其勢重
束稈草三五斤為尾二物與毬同若賊來攻城皆以
砲放之燔賊積聚及驚隊兵

Fig. 10. Two further pages of the *Wu Ching Tsung Yao* of +1044 showing examples of the gunpowder formula, the earliest descriptions in any civilisation. *WCTY*, ch. 12, p. 65*a*, *b*. The text begins in the sixth column from the right and ends in the eleventh column.

objects and set them alight. The preamble says that the device has three hooks and six iron knives, inside the frame of which the gunpowder is enveloped in paper bound with hempen string, and finally eight iron calthrops are attached, each with fine backward-pointing prongs. At the time of firing a red-hot iron brand is thrust in, and as the ball begins to blaze it is shot off from a trebuchet. Furthermore, when all the constituents of the inner ball have been melted together, they are wrapped up in many layers of paper tied with hemp, and a coating of the second mixture plastered over it. Here are the listed ingredients:

Inner ball	oz.	
sulphur	20	
saltpetre	40	
coarse charcoal powder	5	
pitch (*li chhing*[1])	2·5	
dried lacquer, pounded to a powder	2·5	
bamboo roots	1·1	
hemp roots, cut into shreds	1·1	
tung oil	2·5	
lesser oil (*hsiao yu*[2])[a]	2·5	
wax	2·5	
	79·7	
Outer coating		
paper	12·5	
hemp (fibre)	10	
minium (red lead)	1·1	
charcoal powder	8	
pitch	2·5	
yellow wax	2·5	
	36·6	116·3

Calculating the percentages we find the following figures:

		%	
If inner ball alone considered	N	S	C
all carbonaceous matter being taken	50·2	25·1	24·7
taking charcoal specified alone	61·5	30·8	7·7
If outer covering included			
all carbonaceous material being taken	34·7	17·4	47·9
taking charcoal specified alone	54·8	27·4	17·8

Here Chao Thieh-Han and Ho Ping-Yü, omitting all the non-charcoal carbon, obtained approximately the result on the second line, while Partington preferred

---

[a] This is again a mysterious item, but probably edible oil from small beans of some kind as opposed to soya-beans.

[1] 瀝青   [2] 小油

the third, and Arima adopted the first, as we ourselves originally did. Since the function of the device was essentially incendiary, it may be that the third line is the best conclusion. Although the nitrate-content is in this case so low, the name of true gunpowder cannot be withheld because a considerable quantity of charcoal was present; and it would be reasonable to compare the figures on the first line with those established just above for the incendiary bomb of proto-gunpowder.

With the third formula we enter the field of chemical, or at least pharmaco-logical, warfare, in its medieval manifestation. It is called *tu yao yen chhiu*[1] (poisonous smoke bomb), and appears in the previous chapter,[a] which has a section entitled *huo kung*[2] (attack by fire).[b] According to the preamble about ordinary non-toxic smoke bombs, the inner core weighing 3 lb. is to be of a gunpowder composition thickly plastered over with about 1 lb. of yellow *hao* tinder (*huang hao i huai*[3]) as a wrapping.[c] But the poison-smoke bomb core, after being thoroughly mixed and made into a ball tied up with 12·5 ft. of hemp string, is to have a different coating, as specified in the table, almost identical with that of the preceding formula. At the time of firing, the ball is ignited by means of a red-hot iron brand pushed in, and then quickly shot off. The ingredients are as follows:

Inner ball	oz.
sulphur	15
blaze-solve (saltpetre, *yen hsiao*[4])	30
aconite (*tshao wu thou*[5])[d]	5
croton oil (*pa tou*[6])[e]	5
wolfsbane (*lang tu*[7])[f]	5
tung oil	2·5
lesser oil (*hsiao yu*[8])	2·5
charcoal powder	5
pitch (*li chhing*[9])	2·5
arsenic (*phi shuang*[10])[g]	2
yellow wax	1
bamboo roots	1·1
hemp roots	1·1
	77·7

[a] *WCTY/CC*, ch. 11, pp. 27*b*, 28*a*, tr. auct.
[b] Pp. 19*b* ff.
[c] This is from *Gnaphalium multiceps* (R 35; Stuart (1), p. 197), a composite allied to *Artemisia argyi*, the source of the celebrated moxa tinder (cf. Burkill (1), vol. 1, pp. 243 ff., 247; Lu Gwei-Djen & Needham (5), pp. 1, 170 ff.). Both these plants have hirsute leaves the hairs of which contribute to the tinder texture when dried, pounded, compacted, and again dried to evaporate the oil. This was an ingenious device, for the glowing tinder would have acted as a fuse, igniting the gunpowder only when it was nearing its destination.
[d] *Aconitum fischeri*, R 523.   [e] *Croton tiglium*, R 322.
[f] *Aconitum ferox* or *lycoctonum*, R 526.   [g] Arsenious oxide.

[1] 毒藥煙毬   [2] 火攻   [3] 黃蒿一裹   [4] 焰硝   [5] 草烏頭
[6] 芭豆   [7] 狼毒   [8] 小油   [9] 瀝清   [10] 砒霜

The recognition of gunpowder as a strategic war material already in the +11th century is reflected in the prohibitions of the export of the explosive and its raw materials to the Chhitan Tartars during the Northern Sung period. In modern terms one might say that the Sung government was apprehensive of the proliferation of gunpowder weapons.[a] For example, the *Sung Shih* tells us[b] that in +1067 the people of the Ho-tung[1] and Ho-pei[2] prefectures[c] were forbidden by edict to sell to foreigners any sulphur, saltpetre or *lu kan shih*[3].[d] Similarly, in +1076 an order was issued banning all private transactions in sulphur and salt-petre, for fear of their being smuggled out across the border.[e] This suggests the existence of a fairly large production by private enterprises, against which the government would have wanted to retain a monopoly.[f] Curiously enough, some six hundred years later the attempt to deny sulphur and saltpetre to 'for-eigners' was made again; in a time of troubles with the Miao people in the South-west, it was memorialised that the constituents of gunpowder should be with-held from them. But the enlightened Khang-Hsi emperor would not agree, for he knew that they depended on gunpowder for their livelihood by hunting game, and such an embargo would only make matters worse.[g]

Finally, Chang Yün-Ming (*1*) has brought forward evidence to show that the sulphur used in Chinese gunpowder in and after the +11th century was mostly not native, but rather that produced by the roasting of iron pyrites (*tzu-jan thung*[7] or 'fool's gold')[h], converting the sulphide to the oxide. There is a well-known illustration of the process in *Thien Kung Khai Wu* (+1637)[i] showing how the ore was piled up with coal briquettes in an earthen furnace with a kind of still-head to send over the sulphur as vapour, after which it solidified and crystallised.

---

[a] History repeated itself when in the +16th and +17th centuries the Papacy made great efforts to prohibit the export of gunpowder and metals to the Turks from Christendom. This was blithely ignored by the English and the Dutch, as indeed also by Venice and Genoa, more interested in trade than theology. Cf. Parry (1), pp. 225–6; Petrovič (1), p. 176.

[b] Ch. 186, p. 24*a*. Cf. Fêng Chia-Shêng (*1*), p. 53.

[c] Mod. Shansi and Hopei.

[d] Smithsonite (zinc carbonate), for making brass; cf. Vol. 5, pt. 2, pp. 195 ff.

[e] *Hsü Tzu Chih Thung Chien Chhang Phien*[4], Hsi-Ning reign-period, 8th year.

[f] Tshên Chia-Wu (*1*) has given us an interesting study of the trade between the Sung and the Liao (Chhi-tan Tartars). In +1084 the export of sulphur and saltpetre to them was so prosperous that it had to be heavily taxed. In +986 they had bought and translated many medical books. The sale of horses was prohibited in +1042 because they were reaching the Hsi-Hsia people through the Liao, and in +1056 all export of copper and iron to the Uighurs was stopped. But Taoists came and went freely, propagating a knowledge of wrestling—among other things.

[g] See *Thing Hsün Ko Yen*[5], p. 39*b*; *Ta Chhing Shêng Tsu Jen Huang Ti Shih Lu*[6], ch. 14, p. 5, ch. 106, p. 18. Tr. in Spence (1), p. 35.

[h] Cf. Vol. 5, pt. 2, p. 172; pt. 4, pp. 177, 198–9, 200.

[i] Ch. 11, pp. 5*a*, *b*, 6*a*, Ming ed. ch. 2, pp. 60*a*, *b*, 61*a*, *b*; Sun & Sun tr., pp. 208–10; Li Chhiao-Phing (2), pp. 296–7, 307. Judging from the description, a Hellenistic type of still-head was being used at this time, but that had not necessarily always been so. On the odysseys of still-heads, see Vol. 5, pt. 4, pp. 103 ff.

[1] 河東   [2] 河北   [3] 爐甘石   [4] 續資治通鑑長編
[5] 庭訓格言   [6] 大清聖祖仁皇帝實錄   [7] 自然銅

(8) FIRE-CRACKERS AND FIREWORKS

In July +1719, John Bell of Antermony, a Scottish gentleman and a medical graduate of Glasgow[a], set out for China from Moscow as physician to the embassy from His Most Czarish Majesty to the Court of Peking which His Excellency Leoff Vassilovitch Ismailov was undertaking. The account which Bell afterwards wrote of his travels is one of the classics of the kind. It is relevant here because of certain conversations which took place during the residence of the embassy in Peking. Let us quote:[b]

1st January 1721, the [Khang-Hsi] Emperor's general of the artillery, together with Father Fridelly,[c] and a gentleman, called Stadlin,[d] an old German, and a watchmaker, dined at the ambassador's. He [the general] was, by birth, a Tartar; and, by his conversation, it appeared he was by no means ignorant in his profession, particularly with respect to the various compositions of gun-powder used in artificial fire-works. I asked him, how long the Chinese had known the use of gun-powder? He replied, above two thousand years, in fire-works, according to their records; but that its application to the purposes of war, was only a late introduction. As the veracity and candour of this gentleman were well known, there was no room to question the truth of what he advanced on this subject.

The conversation then turned on printing. He said, he could not then ascertain, precisely, the antiquity of this invention; but, was absolutely certain, it was much ancienter than that of gun-powder. It is to be observed, that the Chinese print with stamps, in the manner that cards are made in Europe. Indeed, the connection, between stamping and printing, is so close and obvious, that it is surprising that the ingenious Greeks and Romans, so famous for their medals, never discovered the art of printing....

Next day, the ambassador and his train had a private audience of the Emperor, which lasted more than two hours, Khang-Hsi talking very affably and familiarly on many subjects, especially history, chronology, and inventions.[e]

The Emperor also confirmed most of the particulars, mentioned above, concerning printing and gun-powder. It is from the holy scriptures, most part of which have been translated by the missionaries, that the learned men, in China, have acquired any knowledge of the Western ancient history. But their own records, they say, contain accounts of transactions of much greater antiquity.[f]

Later, on the last day of January and the Chinese New Year on 1st February, the members of the embassy witnessed magnificent imperial firework displays,

---

[a] +1691 to +1780. Antermony is near Campsie in Stirlingshire.

[b] From vol. 2, p. 43, mod. ed., pp. 153 ff.

[c] This was one of the Jesuits, Xavier Ehrenbert Friede (Fei Yin[1]), an Austrian geographer and cartographer (+1673–+1743).

[d] A Jesuit lay brother from Switzerland, Francois Louis Stadlin (+1658 to +1740), maker and repairer of all kinds of mechanical instruments especially clocks and orreries. His Chinese name was Lin Chi-Ko[2].

[e] We quoted him on the 'load-stone' and the south-pointing carriage in Vol. 4, pt. 2, p. 288.

[f] This was an echo of the +18th-century chronological controversies to which we referred in Vol. 3, p. 173.

[1] 費隱   [2] 林濟各

The *Yuan Huang Ching*[1a] calls these spirits *shan sao kuei*[2]. It is quite all right for the common people to make bamboos burst in the fire (*pao chu*[3]) in the courtyards of their homes as a beacon-signal (*liao*[4]) for their families; but it is superfluous for rulers and officials (to carry out such) activities.

There speaks the sceptical Confucian scholar-bureaucrat, but the practice is fully established by his words. Indeed, it became universal in China at different times of the year, but especially at its beginning.[b] For example, Li Thien[5] (d. +1006)[c] tells us in his *Kai Wên Lu*[6] that each family would explode more than ten stems of bamboo on New Year's Eve.

Moreover, the custom of exploding bamboos continued long after the introduction of gunpowder fire-crackers. At the end of the +16th century, Fêng Ying-Ching[7], in his *Yüeh Ling Kuang I*[8] (Amplifications of the 'Monthly Ordinances')[d] remarked that

on the last night of the year bamboo is exploded in fires throughout the night until morning, in order to shake and arouse the Yang of the spring, and to remove and dissipate all evil influences (*hsieh li*[9]). Men in our time make a sport of it, and waste their money in attempts to outvie each other therein, so that the fundamental meaning of the matter is nearly forgotten.[e]

And in Thang and later poetry there are many references.[f]

As we often note elsewhere (pp. 11, 22, 40), one great difficulty in pursuing the history of science and technology in China is that certain things changed fundamentally while the terminology for them remained the same. An obvious example is the use of the phrase *huo chien*[11] (fire-arrow) first for incendiary arrows, and then for rockets (p. 147). But in other cases there does arise a change of appellation, and it does seem to mark a definite break in the continuity, in fact the appearance of something novel. We have already had the example of *shih yu*[12] (rock oil, natural petroleum or naphtha) giving place to *mêng huo yu*[13] (fierce fire oil), and it makes sense to assume that this new term meant only distilled petroleum or petrol, i.e. Greek Fire (p. 76). So with crackers; the term *pao chu*[14] (exploding bamboo) or sometimes *pao kan*[15] (exploding stem), is largely replaced after a certain time by *pao chang*[16] (exploding cracker)[g], and this coin-

---

[a] We have been unable so far to identify this book, the 'Original Yellow Canon'.
[b] Cf. de Groot (2), vol. 6, pp. 941 ff.
[c] According to Werner (4), p. 128, Li Thien was deified in later times as the god of fire-crackers.
[d] The *Yüeh Ling* itself is a text of perhaps the −5th century; cf. Vol. 3, p. 195.
[e] Ch. 20, p. 19a, tr. de Groot (2), vol. 6, p. 943, mod. auct.
[f] E.g. *Chhüan Thang Shih*, pt. 10, tsê 2, *Lai ku shih, tsao chhun*.[10] Many further quotations are given in Wang Ling (1), p. 163.
[g] The general meaning of *chang* is weapons of any kind, but in this binome it came to mean specifically gunpowder fire-crackers. Still, the choice of the word may have considerable significance in relation to the general history of gunpowder.

[1] 元黃經	[2] 山獟鬼	[3] 爆竹	[4] 燎	[5] 李畋
[6] 該聞錄	[7] 馮應京	[8] 月令廣義	[9] 邪厲	[10] 來鵠詩，早春
[11] 火箭	[12] 石油	[13] 猛火油	[14] 爆竹	[15] 爆竿
[16] 爆仗				

cides exactly with what we know of the origin and diffusion of gunpowder in China, i.e. it appears in the literature of the +12th century, and probably came into use in the +11th. Nevertheless, here again there was a tendency to use the old name loosely long after the new one had come in, so that when we meet with *pao chu* in Sung, Yuan, Ming or Chhing, it is likely, though not certain (since bamboos continued to explode), that gunpowder fire-crackers are meant. Of course, if the context is specified, it is easy to be sure.[a] Again, as we shall see (p. 133) *pao chang* and fireworks go together. The ancient name for these was *yen huo*[1] (smoke-fires), and this appellation assuredly goes back much earlier than gunpowder; yet after the application of the explosive mixture for 'feux d'ar-tifices' or true fireworks, the old name of 'smoke fires' still continued, indeed down to the end of the Chhing.[b] Coloured smoke and flame undoubtedly long preceded gunpowder fireworks, and so did gunpowder fire-crackers by some time also.

Positively the first appearance of the term *pao chang* seems to occur in that book which Mêng Yuan-Lao[5] wrote about +1148 describing life in Khaifêng during the first two decades of the century, before the fall of the Northern Sung capital to the Chin Tartars; he called it *Tung Ching Mêng Hua Lu*[6] (Dreams of the Glories of the Eastern Capital)[c]. Some time hard to fix between +1103 and +1120 the emperor visited the Pao-chin Lou[7] pavilion to see a great entertain-ment given by the army, including fireworks of sorts. 'Suddenly a noise like thunder (*phi li*[8]) was heard, the setting off of *pao chang*[9] (fire-crackers), and then the fireworks (*yen huo*[10]) began'.[d] These seem to have consisted almost wholly of dancers dressed in strange costumes moving about through clouds of coloured smokes, each act being separated by the resounding noise of fire-crackers. Thus the first act was called 'carrying the gong' (*pao lo*[11]), the second 'obstinate de-vils' (*ying kuei*[12]), the third 'the dancing judge of the ghosts' (*wu phan*[13]), and so on.[e] Gunpowder was certainly here for the fire-crackers, though not necessarily for the coloured smoke and flame.[f]

Another +12th-century reference comes in the *Hsi Hu Chih Yü*[14] of Thien I-Hêng.[15] Although this book was written about +1570 it was based on local

---

[a] Other names for gunpowder fire-crackers were *pien pao*[2] (whip bangs), *chhuan pao*[3] (linked bangs) and *hua pao*[4] (spark bangs).

[b] Then *hua phao*[16] (spark cannons) came also to be used; and especially *huo hsi*.[17]

[c] Cf. Balazs & Hervouet (1), pp. 150–2.

[d] Ch. 7 (p. 43), tr. auct.

[e] We need not enumerate all eight, but in the first there was fire-vomiting by the actors, while the fourth had 'dumb pantomime' and the eighth sword-juggling.

[f] Fêng Chia-Shêng (6), p. 74, fixes the period +1163 to +1189 as the first appearance of *yen huo*[18], when during imperial inspections of the Sung navy by Hsiao Tsung, smoke balls of five colours (*wu sê yen phao*[19]) were shot off from trebuchets. This could have been but a further stage in their development.

---

[1] 煙火　　　[2] 鞭爆　　　[3] 串爆　　　[4] 花爆　　　[5] 孟元老
[6] 東京夢華錄　[7] 寶津樓　　[8] 霹靂　　　[9] 爆仗　　　[10] 煙火
[11] 抱鑼　　　[12] 硬鬼　　　[13] 舞判　　　[14] 西湖志餘　[15] 田藝蘅
[16] 花炮　　　[17] 火戲　　　[18] 烟火　　　[19] 五色烟炮

records of the West Lake at Hangchow, and may therefore be worthy of credence here. It says:

In the 12th year of the Shun-Hsi reign-period (+1185), on New Year's Eve, there were
abundance of lanterns set out within the palace; and at the second watch of the night the
emperor rode in a small carriage to Kuan-ao Shan[1] hill outside the Hsüan-tê Mên[2] gate.
As the night wore on, they let off more than a hundred fireworks (*yen huo*[3]) attached to
frameworks, after which the emperor returned to his apartments.[a]

Still another mention arises out of a famous court case when in +1183 the great
philosopher Chu Hsi[4] impeached a provincial governor Thang Chung-Yu[5] for
counterfeiting *hui tzu*[6] paper money, and other alleged misdemeanours.[b] Among
the accusations was that since there was a man of Wuchow named Chou Ssu,[7]
who had a great reputation for making and managing fireworks, Thang Chung-
Yu asked him to come to the city, and spent several thousand ounces of silver
out of the public funds on the public performance of his displays.[c]

In the +13th century the references to gunpowder fire-crackers and fireworks
become more numerous. In +1275 Wu Tzu-Mu[9] wrote his often-quoted book
Mêng Liang Lu[10] (Dreaming of the Capital while the Rice is Cooking), a description of Hangchow towards the end of the Southern Sung, from about +1240
onwards.[d]

On New Year's Eve [he says] people all bought *tshang shu*[11e] and small dates, and there
were stalls selling fire-crackers (*pao chang*[12])[f] and fireworks (*yen huo*[3]) on frameworks, and
that sort of thing ...[g]

Inside the palace the fire-crackers (*pao chu*[13]) made a glorious noise, which could be
heard in the streets outside.... All the boats (on the lake) were letting off fireworks and
fire-crackers, the rumbling and banging of which was really like thunder. Ashore people
sat around braziers drinking wine and singing songs and beating drums. It was called
'Guarding the Year' (Shou Sui[14]).[h]

Interestingly, we read elsewhere that at the naval exercises following the ceremony of 'watching the bore' on the Chhien-thang River,[i] they practised firing off
smoke balls (cf. p. 123) from trebuchets, and shooting hundreds of *huo chien*[15],
now almost certainly rockets, and setting targets on fire (*shao hui*[16]), probably
with petrol flame-throwers.[j]

[a] Quoted in Morohashi's dictionary (Chinese ed., vol. 5, p. 1805), tr. auct.
[b] *Hui-An hsien shêng Chu Wên Kung Wên Chi*[8], ch. 18, pp. 17a ff., ch. 19, pp. 1a ff.
[c] This was first noted in Wang Ling (1), p. 165.
[d] Cf. Balazs & Hervouet (1), pp. 154-5.
[e] Roots of the composite *Atractylis ovata* (R 14), good for longevity; cf. Vol. 5, pt. 3, p. 11.
[f] Probably a misprint for the usual *chang*.
[g] Ch. 6, p. 6b (p. 181).   [h] Ch. 6, p. 7b (p. 182).   [i] Cf. Vol. 3, p. 483.
[j] Ch. 4, p. 7b (p. 163). Parallel accounts of these naval displays, which included coloured signal smokes and
smoke-screens, are in *Wu Lin Chiu Shih*, ch. 3, p. 11a (pp. 371-2) and ch. 7, p. 15a (p. 475).

[1] 觀鰲山	[2] 宣德門	[3] 煙火	[4] 朱熹	[5] 唐仲友
[6] 會子	[7] 周四	[8] 晦庵先生朱文公文集	[9] 吳自牧	
[10] 夢粱錄	[11] 蒼朮	[12] 爆杖	[13] 爆竹	[14] 守歲
[15] 火箭	[16] 燒燬			

The tale is continued in the *Wu Lin Chiu Shih*[1] and the *Tu Chhêng Chi Shêng*[2]. The former book (Customs and Institutions of the Old Capital) refers to events from about +1165 onwards, and was written by the eminent scholar Chou Mi[3] a century after that date.[a] He said:

At the festival of the Year Remnant (*Sui Chhu*[4]) from the 24th of the 12th month (*Hsiao chieh yeh*[5]) to the 30th (*Ta chieh yeh*[6]) ... there were many fire-crackers (*pao chang*[7]), some made in the shape of fruits or men or other things ... and between them there were fuses (*yao hsien*[8])[b] so arranged that when you lit one it set off hundreds of others connected with it.... Pipes and drums were played too, to welcome the spring....[c]

Bamboo crackers (*pao chu*[9]) were also let off.[d]

At the New Year on the West Lake many people went back and forth on boats with flags and picnics and singing ... and fireworks (*yen huo*[10]). Some of these were like wheels and revolving things, others like comets (*liu hsing*[11]), and others again shooting along the surface of the water (*shui pao*[12]), or flying like kites—too many to mention....[e] Young people competed in kite-flying...

Others let off fire-crackers on circular frames connected with long fuses, as an amusement....[f]

Chou Mi also mentions,[g] as a precious fragment from those times, the names of two citizens of Hangchow who were renowned for making and displaying fireworks (*yen huo*[10])—these were Chhen Thai-Pao[13] and Hsia Tao-Tzu.[14] The second book (The Wonder of the Capital) was written rather earlier, in +1235, by a Mr Chao[15], and it lists the same entertainments among skills (*chhiao*[16]) such as puppet-theatres and ball-play. There was the burning of fireworks (*shao yen huo*[17]), and letting off fire-crackers (*fang pao chang*[18]), and performing fire-plays (*huo hsi erh*[20]), whatever these were at that time.[h]

To these references we may add one or two from books very little known. Wang Chih[21], probably in the Sung, wrote a *Têng Wu Shih Phien*[22] (Records of a Journey up to the Cities of Wu, i.e. Chiangsu), and he recorded that at Hu-chhiu Shan[23], a hill resort near Suchow, there were at festival time masses of fire-crackers (*pao chang*[24]) that took four men to carry.[i] Again, in the *Nung Chi*[25], a

---

[a] Cf. Balazs & Hervouet (1), p. 155.
[b] The commoner phrase is *yin hsien*[19].
[c] Ch. 3, pp. 13*b*, 14*a* (p. 383), tr. auct.
[d] Ch. 3, p. 15*a* (p. 384).
[e] It is rather difficult to disentangle here the names of games from those of the various kinds of fireworks. Ch. 3, p. 1*b* (p. 375).
[f] Ch. 3, p. 3*b* (p. 376).
[g] Ch. 6, p. 30*b* (p. 465). There were shops where one could buy fireworks (ch. 6, p. 17*a*, p. 453), and even one which specialised in gunpowder fuses alone (ch. 6, p. 15*b*, p. 452). For a special display of fire-crackers in +1180 see ch. 7 p. 11*b* (p. 473).
[h] *Tu Chhêng Chi Shêng*, §7 (p. 97).
[i] In *KCCY*, ch. 50, p. 30*a*. The book is not in any of the dynastic bibliographies.

[1] 武林舊事	[2] 都城紀勝	[3] 周密	[4] 歲除	[5] 小節夜
[6] 大節夜	[7] 爆仗	[8] 藥線	[9] 爆竹	[10] 煙火
[11] 流星	[12] 水爆	[13] 陳太保	[14] 夏島子	[15] 趙氏
[16] 巧	[17] 燒煙火	[18] 放爆仗	[19] 引線	[20] 火戲兒
[21] 王穉	[22] 登吳社編	[23] 虎邱山	[24] 爆仗	[25] 農紀

book of agricultural prognostications, it is said that one can foretell the future by the sounds, pleasant or unpleasant, made by the fire-crackers.[a]

As for the continuing use of the term *pao chu*[1] long after gunpowder had become general, it was probably a matter of literary elegance as opposed to common speech, since the ancient term had been consecrated by centuries of scholarly writing. Some made a clear distinction, for example, Shih Hsiu[2], in his book on the neighbourhood of Shao-hsing in Chekiang, *Chia-Thai Kuei-Chi Chih*[3], dating from soon after +1205,[b] who says that on the last day of the year the sound of fire-crackers (*pao chu*[1]) is everywhere heard, but that there are people who mix sulphur with other chemicals to cause even more violent explosions, and these are *pao chang*. Yet when in +1380 Chhü Yu[4] wrote a poem on a picture of the immortal demon-quelling scholar Chung Khuei[5], it included the lines:

> At the sound of a burst of fire-crackers (*pao chu*[1])
> People ran away in every direction...[c]

which rather suggests the explosions of gunpowder. And many other examples of the elegant euphemism could be given.[d]

By the +14th and +15th centuries, fireworks were in full swing, gunpowder being now generally available, but there are very few detailed descriptions of them. The best, perhaps, were written in the close neighbourhood of +1593, one in Shen Pang's[9] *Yuan Shu Tsa Chi*[10] (Records of the Seat of Government at Yuan(-phing), i.e. Peking), and the other in Fêng Ying-Ching's[11] *Yüeh Ling Kuang I*[12] (Amplifications of the 'Monthly Ordinances').[e] In the first of these we read:[f]

Fireworks (*yen huo*[13]) are made in many sorts.

Those which give a loud noise are called 'resounding bombs' (*hsiang phao*[14]). Those which go up very high are called 'ascending fires' (*chhi huo*[15]). Those which give several explosions in mid-air when let off are called 'three breaking waves' (*san chi lang*[16]). Those which don't make much noise nor go up high, but rush round and round (*hsüan jao*[17]) twisting about on the ground, are called 'earth rats' (*ti lao shu*[18]). Those fireworks which

---

[a]　*KCCY, loc. cit.* The book is not in the bibliography of Wang Yü-Hu (*1*).
[b]　Also not in any of the dynastic bibliographies.
[c]　*Kuei Thien Shih Hua*[6], ch. 3, p. 7*a*.
[d]　So Fan Chhêng-Ta[7] (+1126 to +1193 in *Shih Hu Shih Chi*[8], ch. 30, p. 3*a, b*, ch. 23, pp. 2*a*, 9*a*. The poems are collected in Fêng Chia-Shêng (*1*), pp. 72–3.
[e]　These were the two books which appeared, disguised as Chinese authors, under the names of Wan-Shu and Tüch-Ling respectively, in Brock (*1*), p. 23. What they said must have been well known because they are the only two sources mentioned by Yang Fu-Chi[19] in his postface of +1753 to the *Huo Hsi Lüeh*, on which see p. 139 below.
[f]　Cit. *KCCY*, ch. 50, p. 30*a, b*, tr. auct. Mayers (6), p. 82, was aware of the passage, and gave an excerpt from it, but thinking it was of Thang date, put it seven centuries or so too early.

[1] 爆竹	[2] 施宿	[3] 嘉泰會稽志	[4] 瞿佑	[5] 鍾旭
[6] 歸田詩話	[7] 范成大	[8] 石湖詩集	[9] 沈榜	[10] 宛署雜記
[11] 馮應京	[12] 月令廣義	[13] 烟火	[14] 響炮	[15] 起火
[16] 三汲浪	[17] 旋遶	[18] 地老鼠	[19] 楊復吉	

are packed loose or tight are of two kinds (releasing) many or few sparks (flowers), plants, and shapes like men, are known as 'flower children' (*hua erh*[1]). Those which are enclosed in clay are called 'sand stone-rollers' (*sha thuo erh*[2]). Those which are enclosed in (layers of) paper are called 'spark (or flower) tubes' (*hua thung*[3]). Those which are enclosed in baskets are called 'spark (or flower) bowls' (*hua phên*[4]).

All these are varieties of fireworks (*yen huo*[5]), and a hundred of them or more may be skilfully collected together on to one single framework.

Here the first sort might well be what we should call maroons, and the second clearly rockets, but the fourth is the most interesting, as we shall see (p. 473). It appears again in the remarks of Fêng Ying-Ching, who says:

In Fukien there are fireworks called 'Chhin (Shih) Huang (Ti)'s hair-braid' (*Chhin Huang pien*[6]). From a single bomb (*huo phao*[7]) there burst forth all sorts of sparks and flowers, ground rats (*ti lao shu*), 'water rats' (*shui shu*[8]), etc. Hundreds of them are strung together (*chhuan*[9]) inside one tube, and come out at the same time. One man holds (the fuse) and sets them all off, which is a very wonderful technique.[a]

Still another type of firework is mentioned in a poem by Yang Hsün-Chi[10] (*fl.* +1465 to +1487), the *li thung*[11] or 'pear tube'; a kind of Roman candle (cf. p. 143) which, when lit, shot out sparks of different colours forming a radiating shape like the flowers of pear-trees.

The reason why the 'earth-rat' or 'ground-rat' has such importance is that it may well have been at the origin of rocket propulsion. We assume that it was a tube of bamboo filled with gunpowder, probably occluded by a hole in the node at one end, which was allowed to rush violently about on the floor or the ground. We have a certain record of it from +1264 on account of an incident which frightened an empress at an indoor firework display; and this may be relevant, as we shall see (p. 477 below), to the beginnings of rocketry both in peace and war. The incident is reported in the *Chhi Tung Yeh Yu*[12].[b]

When Mu-Ling[13] (another name for Emperor Li Tsung[14]) retired,[c] he prepared a feast in the Chhing-yen Tien[15] Palace Hall on the 15th day of the first month of the year in honour of (his mother), the Empress-Mother Kung Shêng[16]. A display of fireworks (*yen huo*[17]) was given in the courtyard. One of these, of the 'ground-rat' (*ti lao shu*[18]) type, went straight to the steps of the throne of the Empress-Mother, and gave her quite a fright. She stood up in anger, gathered her skirts around her, and stopped the feast. Mu-Ling, being very worried, arrested the officials who had been responsible for making the arrangements for the occasion, and awaited orders from the Empress-Mother. At dawn next day he went to apologise to her, saying that the responsible officials had been

---

[a] Cit. *KCCY*, ch. 50, p. 30*b*, tr. auct. These two sources were the main reliance of Martin (2), pp. 24 ff., but he also knew the events of +994, +1131 and +1232 (cf. pp. 148, 155, 171).
[b] Ch. 11, p. 18*a*, tr. auct.   [c] R. +1225 to +1264.

[1] 花兒 [2] 沙碢兒 [3] 花筒 [4] 花盆 [5] 煙火
[6] 秦皇辮 [7] 火炮 [8] 水鼠 [9] 串 [10] 楊循吉
[11] 梨筒 [12] 齊東野語 [13] 穆陵 [14] 理宗 [15] 清燕殿
[16] 恭聖 [17] 煙火 [18] 地老鼠

careless, and took the blame upon himself. But the Empress-Mother laughed and said, 'That thing seemed to come specially to frighten me, but probably it was an unintentional mistake, and it can be forgiven.' So mother and son were reconciled and just as affectionate as before.

Having now come this far, we can turn our attention to the debatable period of Liu Chhao, Sui and Thang, which preceded the development of gunpowder in the +10th and +11th centuries. Ever since the *Wu Yuan*[1] (Origins of Things), written by Lo Chhi[2] in the +15th century, it has been customary to follow his asseveration that Sui Yang Ti (r. +605–+616) invented and used 'miscellaneous gunpowder fireworks' (*huo yao tsa hsi*[3]).[a] But he had got his history wrong, for there is no contemporary evidence to substantiate this. Poems which have been called upon do not provide it, though they indicate other things, quite interesting in themselves. For example, Sui Yang Ti himself wrote:[b]

> Wheel of the Law turns, up in the sky
> Indic sounds ascend to the heavens,
> Lamp-trees shine with a thousand lights
> Flower flames open on the seven branches;
> Moon image freezes in flowing water
> Spring wind holds the night-time plums,
> Banderoles move on yellow-gold ground
> Bells come out from beryl estrade.

Monks chanting *sūtras*, and girls dancing, appear at the beginning and end of the stanza, but in between is a clear reference to the 'lamp-trees' (*huo shu*[4]) which were a custom of the age.[c] Trees and their branches, whether real or of brass or bronze, were made to support thousands of 'fairy lights' (as we might say) on ceremonial occasions. Ennin found them in +839 when he visited Yangchow at festival time.[d] But this has nothing to do with fireworks. Nor has the reply poem written by one of Sui Yang Ti's ministers, Chuko Ying[5]:[e]

> Light flashes as the lamp rotates,
> Peach blossoms drop from falling branches,
> Wreathing smoke moves round the buildings
> And the lake of the immortals reflects the floating lights.

Here the smoke may well have some relevance, but his first line probably referred to the 'pacing-horse lamp' (*tsou ma têng*[6]), i.e. the zoetrope.[f] And the last

---

[a] Ch. 14 (p. 30).
[b] *CHSK*, tsê 20, ch. 1, p. 5*b*, tr. Schafer (13), p. 259.
[c] Fang Hao (3).
[d] Reischauer (2), p. 71, (3), p. 128. We ourselves have experience of lamp-trees, since the custom continues in India and Sri Lanka till now. When I gave the first Wickramasinghe Lecture in Colombo in 1978, the opening ceremony conducted by the Mahanayake Thero in the chair, was the lighting of the brass lamp-tree.
[e] *CHSK*, etc., *loc. cit.* tsê 20, ch. 3.
[f] Cf. Vol. 4, pt. 1, pp. 123 ff., and Bodde (12), pp. 80–1. Cf. also J. F. Davis (1), vol. 2, p. 3.

[1] 物原   [2] 羅頎   [3] 火藥雜戲   [4] 火樹   [5] 諸葛穎
[6] 走馬燈

line recalls the Chung-Yuan[1] festival, when little paper boats with lights in them used to be liberated on the Chinese All Souls' Night in myriads.[a] In any case, fireworks were not in evidence at the court of Sui Yang Ti.

Lamp-trees, of course, continued on into the Sung and later. They were probably referred to in a poem by Chang Tzu-Yeh[2] of the mid +11th century, when, speaking of the Têng-Chieh[3] festival in the first month, he says:[b] 'Above and below the towers and terraces the fiery lights are like stars. As one leans on the parapet, the sparks seem to be flying high up in the heavens near the constellations of Tou and Niu.' But other mentions of sparks do suggest that the practice of using fine iron filings in combustibles to give silver spangles in the smokes may go back to the Thang at least. There is for instance a famous poem by Su Wei-Tao[4] (+648 to +705) which talks of *yin hua*[5], silver sparks, among the illuminations and the smokes.[c] A thousand years later, Fang I-Chih[6], in his *Wu Li Hsiao Shih*[7] encyclopaedia,[d] noticed this, and concluded, in his slightly muddle-headed way, that the trees of fire, the silver sparkles and the *pao* crackers all indicated gunpowder in Sui and Thang. But they did not.

What Sui Yang Ti undoubtedly had at his celebrations were huge bonfires (*huo shan*[8]) on which inordinate quantities of incense were burnt. We get a glimpse of this from a conversation between Thang Thai Tsung and his consorts about +630, reported in the *Thai-Phing Kuang Chi* under the heading of 'extravagant displays' (*shih chhih*[9]).[e] Apparently the emperor enquired about the illuminations of the halls and courtyards at the court of Sui Yang Ti, with the implication that he could probably do better. The empress then described the bonfires which he had had all over the palace, burning hundreds of cartloads of garroo or aloes-wood[f] and 200 *tan* of onycha perfume.[g] Night was turned into day, and the aroma (*yen chhi*[10]) could be smelt for several dozen *li*. Such extravagance helped to lose the empire for him. So Thang Thai Tsung desisted. But again this had nothing to do with fireworks properly so called.

Entering now into their pre-history, we can easily find that the ploy of smokes, together with the actual expression *yen huo*[11], goes much further back. There is a passage in the *Ching Chhu Sui Shih Chi*, already quoted in connection with fire-crackers, which shows this well. It says:[h]

---

[a] The fifteenth of the seventh month. Cf. Bodde (12), p. 62; Eberhard (31), p. 107; Bredon & Mitrophanov (1), p. 385 and Weig (1), p. 18.
[b] *Chang Tzu-Yeh Tzhu Pu I*, ch. 1, p. 15*a*, tr. auct.
[c] *Chhüan Thang Shih*, pt. 2, tsê 2. Wang Ling (1), p. 164 drew attention to this.
[d] Ch. 8, p. 26*a, b*.   [e] Ch. 236, pp. 8*b*, 9*a*.
[f] See Vol. 5, pt. 2, p. 141.
[g] Of molluscan origin, Vol. 5, pt. 2, pp. 138–9.
[h] P. 1*b*, tr. auct.

[1] 中元   [2] 張子野   [3] 燈節   [4] 蘇味道   [5] 銀花
[6] 方以智   [7] 物理小識   [8] 火山   [9] 奢侈   [10] 烟氣
[11] 烟火

According to Tung Hsün[1] of the Wei (in the Three Kingdoms period)[a] people nowadays make smoke-fires (*yen huo*) on Chêng-La[2] morning (the 8th day of the 12th month) and set up figurines of peach-wood (*thao shen*[3])[b] in front of the doors of their houses. They also plait rush garlands to hang between the pines and cypresses, and sacrifice a chicken and hang it on the door—all a ceremony to exorcise and get rid of the demons of epidemic disease.

Here then the smoke-fires have become fumigatory and demonifuge. Tung Hsün was a folklorist of the +3rd century, while the text itself, as we saw before, is of *c.* +550.

In each of the preceding centuries something relevant and interesting occurs. For instance, according to the dynastic histories,[c] in the year +493, a popular song circulated in Northern Wei to the effect that 'Red fire is spreading south, destroying the southern States'; and sure enough, in the same year there arrived a monk (*sha-mên*[5]) at Nanking who dispensed this fire. Its colour was redder than ordinary fire and also more ethereal (*sê chhih yü chhang huo erh wei*[6]). The monk performed cures of diseased persons with this 'holy fire', and the emperor of Southern Chhi tried to forbid it, but eventually the monk went away and the cult died out. Could he, one wonders, have come upon some natural ore of strontium and used it for his exhibitions of coloured flame and smoke?

Again, in the +4th century, there is a story[d] about an upright man of Kuei-chi named Hsia Thung[7]. His stepfather, Hsia Ching-Ning[8], engaged two sorceresses (*nü wu*[9]), Chang Tan[10] and Chhen Chu[11], to sacrifice to the ancestors; they chanted and danced, performing all kinds of tricks such as juggling and sword-swallowing, finally 'they spat fire, and were hidden from view by a great cloud, whence streams of light flashed like lightning'. Hsia Thung had strong Confucian objections to all this, and managed to stop the performance. But evidently it included smoke and flame, so it was yet another step on the road to fireworks. Beyond this, perhaps, we need not pursue the matter, since we would enter the whole field of *hsün*[12], hygienic fumigations, liturgical incense-burning, and war-smokes containing poisons, which went back far into the −1st millennium.[e] It is time to return to the end of the +2nd, with a few words on fireworks in their full development.

So let us take a brief look at fireworks in the eighteenth and nineteenth centuries; perhaps it will help us to interpret the long line of intermediate stages which led to them. And first, with respect to gunpowder fire-crackers, a preparation in

---

[a] The remaining fragments of his *Wên Li Su*[4], preserved in *TPYL*, ch. 29, are collected in *YHSF*, ch. 28, pp. 72 a ff.

[b] Cf. Bodde (25), pp. 127 ff.

[c] *Nan Shih*, ch. 4, p. 26 a, b, tr. de Groot (2), vol. 6, p. 951; also *Nan Chhi Shu*, ch. 19, p. 15 b.

[d] *Chin Shu*, ch. 94, p. 3 a, tr. de Groot (2), vol. 6, p. 1212.

[e] See Vol. 5, pt. 2, pp. 128 ff. esp. pp. 148 ff.; Needham & Lu Gwei-Djen (1), pp. 436–7. Cf. pp. 1 ff. above.

[1] 董勛   [2] 正臘   [3] 桃神   [4] 問禮俗   [5] 沙門
[6] 色赤於常火而微   [7] 夏統   [8] 夏敬寧   [9] 女巫
[10] 章丹   [11] 陳珠   [12] 燻

which the Chinese have remained pre-eminent to this day, there are vivid accounts of the process by Barbotin (1), Dyer Ball,[a] Weingart[b] and Tenney Davis.[c] The powder used is rather low in nitrate, with a percentage composition something like: N 66·6, S 16·6, C 16·8.[d] It is filled in to little tubes of stout paper tied together into hexagonal figurate bundles, with thin paper tubes containing gunpowder as the fuse for each one successively. It is interesting to learn from Huang Shang (1) that old book paper was considered the best for making fire-crackers, even during the present century, a circumstance which helps to account for the vast losses of old books sold away not for pulping but for the firework manufacturers.

There are many descriptions of fireworks in China (Fig. 11), from Louis Lecomte in +1696 to Dyer Ball in 1892, through John Bell (+1720), Pierre d'Incarville (+1763), John Barrow (+1794), A. Caillot (1818), J. F. Davis (1836) and the Abbé Huc (1855).[e] They all descant on the flares, gerbs, lances, drums, rockets, Roman candles, saxons, mines, maroons, Chinese fire and Chinese flyers, with many other works too tedious (as John Bell would have said) to rehearse.[f] Down to the middle of the nineteenth century, Chinese fire-works were generally considered much superior to those of Europe,[g] but one can see C. F. Ruggieri changing his mind on this between his two editions of 1801 and 1821.[h] If such a turning-point ever arrived at all, it would have been rather towards the beginning of the present century. But it is more interesting to turn to the consideration of the book which seems to be the only monument of civilian pyrotechny in all Chinese literature.

The *Huo Hsi Lüeh*[1] (Treatise on Fireworks) was written in his youth by Chao Hsüeh-Min[2], that scholar of scientific bent who was later to produce the *Pên Tshao Kang Mu Shih I*[3] (Supplementary Amplifications for the 'Pandects of Natural History' by Li Shih-Chen). This was begun in +1760, first prefaced in +1765, the prolegomena added in +1780, but not printed till 1871; the last date in the text being 1803. By contrast, Yang Fu-Chi[4] wrote the preface for the *Huo Hsi Lüeh* as early as +1753, so Chao must have been studying fireworks some time before. The work was not printed till 1833.[i] It shows that the state of the

---

[a] (1), pp. 239 ff.   [b] (1), pp. 166 ff.   [c] (17), pp. 111 ff.

[d] In recent times the chlorates of aluminium and potassium have been used, but this was a dangerous procedure. Chlorates were discovered by Berthollet in +1786, and much used for fireworks as by Cutbush in the eighteen twenties, but have since been replaced by perchlorates, which are much safer; cf. Davis (17), p. 59 (18).

[e] I cannot refrain from mentioning the splendour of the displays which I myself have witnessed in Peking from Thien-an Mên on several Mays and Octobers since the Second World War.

[f] The descriptions of Tissandier (7) are particularly interesting as he had had a scientific training.

[g] In +1751 one of the displays in London was a 'Collection of Fire Works in the Chinese manner', sub-scribed for by 'some Gentlemen curious of seeing the New Fire Works' (Brock (1), p. 57).

There seems to be no really good book on the history of fireworks, but Lotz (1) may be mentioned as well as Brock (1, 2).

[h] Cf. Brock (1), pp. 22–3.

[i] There is an excellent analysis of it by Tenney Davis & Chao Yün-Tshung (9).

[1] 火戲略   [2] 趙學敏   [3] 本草綱目拾遺   [4] 楊復吉

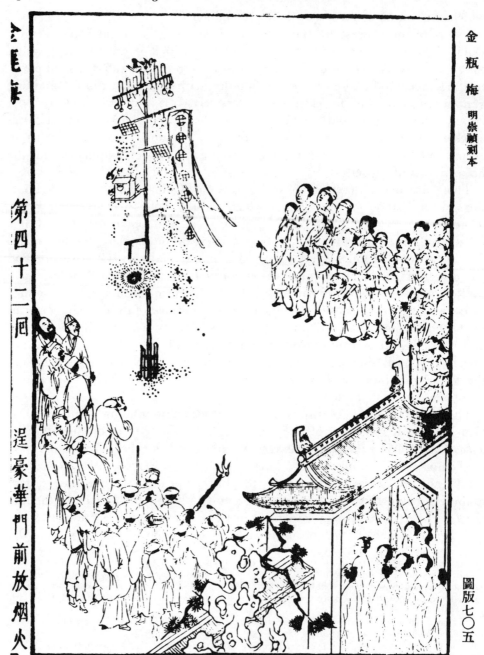

Fig. 11. Scene of a fireworks display, from ch. 42 of the Ming novel *Chin Phing Mei* (edition of +1628–43).

art was really more advanced than that of Europe at the time, and refers to a number of things not then known in the West, notably 'Chinese Fire' (to which we shall return in a moment), piped match fuses, conical containers for fountains forming chokes, coated grains of composition, etc. Chao gives a description of purifying saltpetre, which includes clearing by glue and turnip slices (cf. p. 105 above), and the making and drying of charcoal, which he calls the Yin soul (*pho*[1]) of fire.[a] He recalls the medieval alchemical experiments by his remark that 'by being mixed with the fat of soap-bean pods the gunpowder acquires slowness' (cf. p. 115 above).[b] 'Ground-rats' are mentioned,[c] and 'water-rats' too.[d]

The point about Chinese Fire is that it was nothing but filings of cast iron or steel reduced to a powder more or less fine, and this, when mixed with low-nitrate gunpowder and other combustibles, would upon firing yield flame and smoke containing an infinite number of silver sparkles. A favourite Chinese name was *thieh ê*[2] (iron moths), another *thieh sha*[3] (iron sand), a third *thieh hsieh*[4] (iron granules). Wrought iron will not do, so some carbon must be present.[e] 'Cast-iron', said Cutbush,[f] 'reduced to a powder more or less fine, is called iron-sand, because it answers to the name given to it by the Chinese. They use old iron pots, which they pulverise till the grains are no larger than a radish seed, and these they separate into sizes or numbers for particular purposes.' We have already seen that this procedure may go back to the Thang (p. 137). The simple secret was not revealed to Europe until d'Incarville (1) wrote his paper about it in +1763,[g] but though he mentioned rusting he did not clearly say that the iron-sand grains must be coated with tung oil or glue to prevent it.[h] The so-called 'brilliant fire' depended on powdered steel, but after 1860 the introduction of magnesium and aluminium led to enormously increased brilliancy.[i]

So much for Chinese Fire, but what about 'Chinese Flyers'? As Robert Jones well knew in +1765, the saxon or tourbillon depended on jet propulsion; it was a single tube pivoted half-way along its length, and made to revolve in the plane of

[a] Cf. Vol. 5, pt. 2, pp. 85 ff.   [b] Davis & Chao (9), p. 101.

[c] See pp. 135 ff. above and pp. 473 ff. below. Davis & Chao (9), p. 103.

[d] Davis & Chao (9), pp. 103–4. J. F. Davis (1), vol. 2, pp. 4–5, spoke of the little boats or water-rats which the Chinese made to skim on the surface of water by rocket jet propulsion. We shall return more compendiously to the important subject of rockets below (pp. 477 ff.), but here it may be mentioned that just in Chao Hsüeh-Min's time Chinese pyrotechnics were attracting some attention in Russia. In +1756 Larion Rossochin wrote a paper on Chinese fireworks, mostly rockets, which has been found and reproduced recently by Starikov (1).

[e] Cutbush (2).

[f] (1), p. 202. He also wrote a special paper on the technique (2).

[g] True, John Bate (1) in +1635 had used 'yron scales', no doubt in an attempt to reproduce the sparks which flew from blacksmiths' anvils, but this could have had no great success. Anon. (159) is an abstract of d'Incarville in English.

[h] See Brock (1), pp. 23, 152, 154, 189, 231, (2), p. 98; Tenney Davis (17), p. 57. D'Incarville explained that different grain-sizes would give the different effects of flowers desired. For the finest, a 'gunpowder' of 86·6 % nitrate was used, for the coarsest only 60·6 %. Cf. Fig. 12.

[i] Brock (1), pp. 154, 163.

[1] 魄   [2] 鐵蛾   [3] 鐵沙   [4] 鐵屑

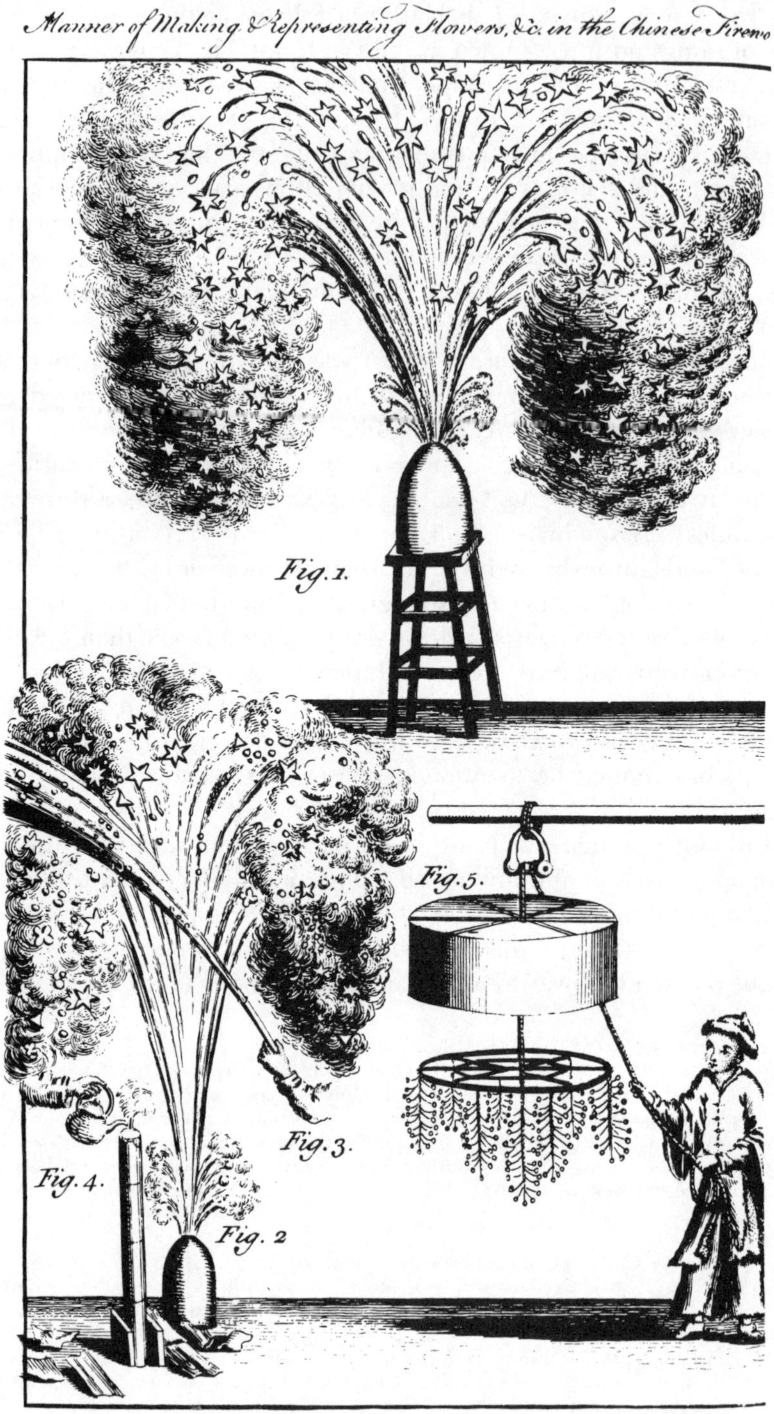

Manner of Making & Representing Flowers, &c. in the Chinese Fireworks

Fig. 1.

Fig. 5.

Fig. 3.

Fig. 4.

Fig. 2

Fig. 12. A representation of Chinese fireworks from an account by the Jesuit d'Incarville (1) in +1763. The plate is from the English abstract of the following year.

its axis by jets of fire projected through holes pointing in opposite directions at right angles to the axis. Furthermore, when revolving fast enough it could be projected into the air by two additional holes bored in the under surface of the tube.[a] That this should have been developed in China is quite interesting, for it was a principle parallel to the helicopter top which was so prominent there;[b] the direct ancestor of the helicopter rotor, and the godfather of the aeroplane propeller. The energy of this was derived from rotation given by a cord previously wound round the stem, or from the pull of a bowdrill spring which travelled with it; in the Chinese flyer the energy was also self-contained, chemically provided by the gunpowder filling, but it lasted a little longer. Moreover, this was no more than a rotary application of those 'water-rats' of which J. F. Davis wrote in 1836 that 'they also make paper figures of boats to float and move upon the water by means of a stream of fire issuing from the stern'.[c] And these in turn were simply derivatives of the rocket principle, which in its origins and warlike uses we shall study in detail presently (pp. 477ff.). In another way, of course, the Chinese flyer was a development of the steam aeolipile of Heron of Alexandria (though people in China could not have known about that),[d] a development most appropriate to the land of gunpowder; but besides it was in a sense the ancestor of the vertical take-off aircraft of the present day.

It is instructive to make some comparisons between the devices used in civilian pyrotechnics and those used in war. In the early stages of gunpowder weapons, as we shall see (p. 163) quantities of the mixture were probably enclosed in carton containers for use as bombs, and this persists down to the present day in maroons—cubical pasteboard boxes filled with gunpowder and exploded like extremely large crackers.[e] The fire-lance has left many descendants. 'Fire-clubbs', cylinders of low-nitrate composition shooting forth flame, even though the range was very short, were prominent in the European seventeenth century, and known to John Bate (+1635).[f] At the same date, John Babington spoke of 'a trunck of fire which shall cast forth divers fire-balls', so co-viative projectiles were known then too (cf. p. 42).[g] Even the word lance passed into civilian pyrotechny, though only for tubes of very small size filled with ammonium picrate or coloured fire compositions.[h] The larger cylinders came to be called 'Roman Candles', familiar to all, formerly 'star pumps' or 'pumps with starrs'; these add dextrin to the mixture, which is very low in nitrate, having a percentage composition about N 53·9, S 11·2, C 34·9.[i] The candles may be up to

[a] Cf. Brock (1), p. 187, (2), p. 116.
[b] See Vol. 4, pt. 2, pp. 580 ff.
[c] These had also been mentioned by d'Incarville (1).
[d] Cf. Vol. 4, pt. 2, pp. 226, 407, 576. For the background see Sarton (1), vol. 1, p. 208; Woodcroft (1), p. 72; Usher (1), 2nd ed., p. 392; and Drachmann (2).
[e] Tenney Davis (17), p. 104.
[f] Davis (17), p. 54; Brock (1), opp. p. 17, and pp. 32, 247, (2), pp. 111 ff.
[g] Brock (2), p. 112. This goes back to Biringuccio in +1540, if no further.
[h] Davis (17), p. 69; Brock (1), pp. 196, 226.
[i] Davis (17), p. 79; Brock (1), pp. 191 ff.

6 in. in diameter, and since they throw out individual pellets of combustible material, the co-viative principle is present, while as for size they may be compared with the eruptors (p. 263) rather than with hand-held fire-lances. Brock makes a weak attempt to derive the name, not found in England before +1769, from the Mardi Gras carnival of that time in Rome,[a] but we suspect it is much more likely that the place was really al-Rūm, i.e. Byzantium, and goes back almost to the time of Marcus Graecus and Ḥasan al-Rammaḥ, Mines or carton mortars[b] also clearly derive from the eruptor conception. As for rockets, the identity is complete, save that they are not armed, producing coloured stars instead and aimed at the zenith.[c]

When we review the whole history and prehistory of fireworks in China, it seems clear that the imparting of colours to smokes and flames is the backbone of the question. 'The diversity of colours indeed', wrote Barrow about +1797, 'with which the Chinese have the secret of cloathing fire seems one of the chief merits of their pyrotechny.'[d] Everyone noticed the same thing, Caillot, for example, in 1818: 'It is certain that the variety of colours which the Chinese have the secret of giving to flame is the greatest mystery of their fireworks.'[e] And so also Cutbush: 'The Chinese have long been in possession of a method of rendering fire brilliant, and variegated in its colours.'[f] But then comes the important point, not generally realised, that the gunpowder formula is by no means necessary for coloured smoke and fire. It was not incumbent upon the Chinese, therefore, to wait for the +10th century before producing some of these remarkable effects.

In pursuing these we can go back to the +14th century because the *Huo Lung Ching* contains recipes for military signal smokes of five colours,[g] and these are repeated word for word in the *Wu Pei Chih*.[h] Four of them include low-nitrate gunpowder,[i] but one does not. We can tabulate them as follows:

Blue-green	indigo (*chhing tai*[1])[j] + gunpowder.
White	white lead (carbonate)[k] + gunpowder.
Red	red lead (tetroxide) + saltpetre, pitch and resin.
Purple	cinnabar (*tzu fên*[2])[l] + gunpowder and hemp oil.
Black	lignite and soap-beans + gunpowder.

---

[a] (1), p. 193.   [b] Davis (17), p. 97.

[c] Davis (17), pp. 73 ff.; Brock (1), pp. 181 ff. Another connection with China is that the stationary jet or fountain of fire, the gerbe or wheatsheaf, was originally often called a 'Chinese Tree', especially when incorporating steel filings (*loc. cit.* p. 189). Cf. Brock (2), pp. 97–8.

[d] (1), p. 206.   [e] (1), p. 100.   [f] (1), p. 371.

[g] Pt. 1, ch. 1, p. 14*a*, *b*.

[h] Ch. 120, pp. 5*b*, 6*a*, tr. Davis & Ware (1), p. 527.

[i] The proportions of saltpetre vary from 47.6% to 71.4%, with an average of 66%.

[j] Also mentioned by d'Incarville (1), but somewhat doubtfully.

[k] Mentioned by d'Incarville as one of the chemicals used with the iron-sand for making silver stars or sparkles (flowers), but by his time camphor was also added.

[l] Again mentioned by d'Incarville as used with the iron-sand.

[1] 青黛   [2] 紫粉

Most of these, which would have depended on the formation of some kind of aerosol of the suspended particles, would have coloured the smoke but not the flame. This kind of thing could be done with any combustible; for example the *Wu Pei Huo Lung Ching* has two formulae for firework-like signals, the *san chang chü*[1] (thirty-foot chrysanthemum) and the *pai chang lien*[2] (hundred-foot lotus) which combine sulphur, charcoal powder and iron filings, with no saltpetre at all.[a] This would produce the silver sparkles in the smoke.[b]

When we come to the time of the *Huo Hsi Lüeh* (+1753) we find a variety of colorations by chemical substances, and now they tint the flame as well as the smoke.[c] For example, in all cases using low-nitrate gunpowder, the flame itself would be coloured:

Yellow	arsenical sulphides.
Violet	cotton fibres.
Green	verdigris (copper acetate),[d] and indigo.
Lilac-white	lead carbonate.
White	calomel (mercurous chloride).
Black smoke	pine soot and pitch.

The earliest use of a salt like copper acetate in Europe seems not to antedate Ruggieri in 1801,[e] but in later times powerful effects such as the red of strontium and the green of barium were introduced.[f] Chao Hsüeh-Min, however, also mentions compositions not involving gunpowder at all. Sulphur alone gives a blue light, and with copper sulphate it is a more intense blue,[g] saltpetre alone or with miscellaneous combustibles yields a violet light,[h] and sodium nitrate a yellow one. It is noteworthy that the intense blue-white light of the 'Bengal Fire', so well known in England a century ago, was produced only by saltpetre and sulphur with antimony sulphide.[i] Antimony, it seems, was first used in European pyrotechnics by Jean Appier in +1630,[j] but since China has the largest antimony deposits in the world, it would be strange if no alchemist in those parts used one of its ores in fireworks at some time or other.[k]

From what has now been said one can see how closely related were recreation-

---

[a] Ch. 2, pp. 30*b*, 31*a*.

[b] We speak here only of signal-smokes, but it is evident from the *Wu Ching Tsung Yao* that the Chinese were very familiar with the principle of smoke-screens in the +11th century. Yet apparently they were considered a military novelty in England in +1760 (Brock (1), p. 240).

[c] Davis & Chao Yün-Tshung (9), pp. 101–2, 106–7.

[d] Copper filings can also be used (Cutbush, 2).

[e] Brock (1), p. 23.

[f] Brock (1), pp. 198–9; Davis (17), pp. 64, 67; Cutbush (2).

[g] So do zinc filings, according to Cutbush (2). Since the Chinese were in possession of isolated metallic zinc from about +900 onwards, it would be strange if this was not tried too. See Vol. 5, pt. 2, p. 214.

[h] No doubt the potassium flame that Thao Hung-Ching saw so long ago.

[i] Davis (17), p. 64; Brock (1), p. 196; Cutbush (2).

[j] Davis (17), p. 55.

[k] Such as stibnite (antimony sulphide). See the discussion in Vol. 5, pt. 2, pp. 189 ff. on what metallic and other elements were available to the Chinese alchemists of the Middle Ages.

[1] 三丈菊   [2] 百丈蓮

al pyrotechnic coloured smokes and military signal smokes. To say much here about these latter would be to encroach too much on another sub-section in this volume, but it may just be worth while mentioning a fascinating passage on army beacon towers in the *Thung Tien*[1] (Comprehensive Institutes) of +812.[a] The groups of five beacon-towers all along the Han *limes* in the North-west are very familiar to those who have travelled along the Old Silk Road (cf. pt. 8*n*).[b] Tu Yu[2], however, recommends groups of three.[c] Each beacon-tower (*fêng thai*[3]) is provided with three raised fire-baskets (*chhai lung*[4]), each of which can be lit from below the battlements by a kind of incendiary fuse (*liu huo shêng*[5])[d] running up a tube (*huo thung*[6])[e]. If all is clear, one smoke-fire is lit, if danger seems nigh, then two, and if enemy troops are in sight, all three are to give their signal. The tower is provided with a flag and a drum, it has a fire-drill and moxa tinder, it is defended by a guard of six men with arrows and fire-arrows, crossbows and trebuchets, and they have adequate stores of food. Here was yet another demonstration of the characteristically Chinese skill in smoke-making.

Looking back over the whole subject, our conclusions must be that civilian pyrotechnics in the modern sense arose along with gunpowder and its warlike uses between about +850 and +1050.[f] But the colouring of smokes and flames by combining various chemicals with combustible substances, including sulphur and saltpetre, must have started a good while earlier, possibly in the Han or soon afterwards; and it would have derived from very ancient customs and processes of fumigation as such. The transition in Wu Tai and Sung times would thus have paralleled that which took place with regard to explosion itself, as the ancient decrepitating bamboos gave place to fire-crackers containing gunpowder. We may end, as we began, with the doings of the Scotsman in Peking. When John Bell in +1720 went with other gentlemen of the Russia-embassy to dine at the palace of the Khang-Hsi emperor's ninth son, they were magnificently entertained with stage-plays 'accompanied with musick, dancing, and a kind of comedy, which lasted most part of the day', though they could not understand any of the dialogue. Towards the end, a fight between heroes was interrupted by a spirit, who 'descended from the clouds, in a flash of lightning, with a monstrous sword in his hand, and soon parted the combatants, by driving them all off the stage; which done, he ascended in the same manner as he came down, in a cloud of fire and smoke.'[g] Nothing could have been more in the Chinese tradition.

---

[a] The passage is almost verbally identical with one in *TPYC*, ch. 5 (ch. 46), p. 2*a*, *b*, which would be some fifty years earlier. There are other examples of the same dependence.
[b] Cf. Vol. 1, Fig. 15 and Vol. 4, pt. 3, pp. 35, 37.
[c] Ch. 152 (pp. 801·2, 801·3).
[d] Nothing to do with gunpowder at this time of course.
[e] This term is interesting, because later on the expression applied solely to metal-barrel guns and hand-guns (cf. pp. 304, 306).
[f] It is interesting that Brock (1), p. 230, the technician, joined with Fêng Chia-Shêng (6), p. 12, the historian, in the conviction that no fireworks, properly so called, existed in the Sui and Thang.
[g] Bell (1), p. 143.

[1] 通典   [2] 杜佑   [3] 烽臺   [4] 柴籠   [5] 流火繩
[6] 火筒

(9) GUNPOWDER AS INCENDIARY

To suppose that gunpowder produced brisant explosions as soon as it was known would be a rather simple-minded mistake. In the days when low-nitrate compositions predominated it found employment mainly as an incendiary, setting on fire the enemy's wooden buildings, tents and other equipment. Perhaps therefore we should keep in mind the following five stages of gunpowder weapons:

(1) gunpowder as incendiary (projectiles launched from bows, crossbows, trebuchets and *arcuballistae*);

(2) gunpowder as flame-thrower (fire-lances and their variants);

(3) gunpowder as explosive (maroons or bombs launched from trebuchets or *arcuballistae*);

(4) gunpowder as retroactive propellant (rockets);

(5) gunpowder as forward propellant (barrel guns, cannon).

Here then we shall concentrate for a while on incendiary projectiles.

Earlier on (p. 130) we drew attention to the ambiguity of the expression 'fire-arrow' (*huo chien*[1]). The rough and relatively unlettered soldiers of Chinese antiquity and the Middle Ages made no terminological distinctions between the different types, and the scholars, lacking familiarity with the things themselves, made little distinction either. But we can see at least the following three types: (*a*) the early incendiary arrows using oil and sulphur and miscellaneous combustibles (*huo chien*[1]); (*b*) incendiary arrows using gunpowder (sometimes called *huo yao chien*[2], but often this was not specified); (*c*) rockets, generally with arrow warheads (again *huo chien*[1]). Now as we follow through the sequence of events we find that towards the end of the +10th century a change occurred in these techniques, and we suggest that this was the time when the first stage gave place to the second.

Something has already been said about fire-arrows in connection with incendiary warfare in general (p. 124). One of the earliest references must be that in the *Wei Lüeh*[3] by Yü Huan[4], which describes the attack on Chhen-tshang by the troops of Shu under Chuko Liang.[5] They used battering-rams and storming ladders, but the defenders, led by Hao Chao[6], shot *huo chien*[7] and burnt them all together with the soldiers on them.' This took place early in +229, a date at which there could have been no question of gunpowder. Then there is an account of a naval battle *c.* +425 between Tu Hui-Tu[8] and Lu Hsün[9], in which the latter's ships all went up in flames because of the *huo chien*[7] launched by the sailors of the Liu Sung admiral.[a] Another instance, some hundred years later, in +535, recounts how the soldiers of Wang Ssu-Chêng[10] shot *huo chien*[7] and burnt all the siege engines of the attacking army.[b] The *Thai Pai Yin Ching* of +759 has

---

[a] *Sung Shu*, ch. 92, p. 5b.   [b] *Pei Shih*, ch. 62, p. 7a.

[1] 火箭   [2] 火藥箭   [3] 魏略   [4] 魚豢   [5] 諸葛亮
[6] 郝昭   [7] 火箭   [8] 杜慧度   [9] 盧循   [10] 王思政

many references to *huo chien* from which we learn that oil was enclosed in a gourd (*yu phiao*[1]) and sent over attached to an arrow, presumably with some kind of fuse; this was useful for shooting upwards to attack watch-towers, or downward to burn siege equipment.[a] Similar projectiles (*huo shih*[2]) could be shot from crossbows (*nu*[3]) with a range of 300 paces.[b] And very similar arrangements are discussed in the *Wei Kung Ping Fa*[4] (Military Treatise of Li Wei-Kung), a +7th-century work by Li Ching[5], the fragments of which were recovered and published by Wang Tsung-I[6] a couple of centuries ago.[c]

On an earlier page (p. 85) we noted what Hsü Tung[7] had to say about 'flying fire' (*fei huo*[8]). Writing just about +1000, he remarked that it was of the nature of trebuchet 'bombs', probably incendiary, and incendiary arrows (*huo chien*)[d]. And it is exactly in his time that we enter a new phase of incendiary projectiles, marked by a wave of new inventions demonstrated to the emperor and his commanding generals. These, we suggest, involved the use of gunpowder as incendiary.

Almost as soon as the Sung dynasty had begun, in +969, Yo I-Fang[10] presented a new type of fire-arrow to the emperor, and was rewarded by a gift of silk.[e] In +970 one of the generals, Fêng Chi-Shêng[11], together with some other officers, presented another new model for fire-arrows; the emperor ordered it to be tested, and as it proved successful, gowns and silk were bestowed upon the inventors.[f] In +976 the King of Wu-Yüeh State sent as a present to the Sung emperor a band of soldiers especially skilled in the shooting of incendiary arrows.[g] Before the century was out there arose several opportunities of using the new devices in combat; for example, in +975 Thai Tsu employed fire-arrows and also incendiary bombs hurled from trebuchets against the last defenders of the Nan Thang State.[h] Then in +994 a force of 100,000 Liao troops besieged the city of Tzu-thung, and the population was greatly alarmed, but the officer in command, Chang Yung[12], ordered the trebuchets to play upon the enemy with stones while the new fire-arrows were shot off, whereupon the investing force retreated.[i]

---

    [a] *TPYC*, ch. 4 (ch. 35), p. 2*b*.   [b] *TPYC*, ch. 4 (ch. 38), p. 8*b*.

    [c] This must be a different work from the *Li Wei Kung Wên Tui*[9], which has been translated by Boodberg (5). It became one of the seven Sung military classics, though probably dating from the beginning of that dynasty, and we have not found anything in it about incendiary arrows.

    [d] *Hu Chhien Ching*, ch. 6 (ch. 53), p. 4*b*. Cf. Fêng Chia-Shêng (*1*), p. 46, (6), p. 73.

    [e] *Wu Li Hsiao Shih*, ch. 8, p. 26*a*; *Thung Ya*, ch. 35, p. 4*b*. Fang I-Chih's source is not clear; possibly Yo was one of the associates of Gen. Fêng.

    [f] *Sung Shih*, ch. 197, p. 1*b*.

    [g] *Ibid.* ch. 3, p. 11*b*.

    [h] See Fêng Chia-Shêng (*1*), p. 47, (6), p. 16. The evidence is later, from the *Chhao Yeh Chhien Yen*[13] (Narratives of Court and Country), as quoted in *San Chhao Pei Mêng Hui Pien*;[14] but we have mentioned already (p. 89) the probable presence of gunpowder weapons at the battles which extinguished Nan Thang. Hsü Mêng-Hsin's[15] quotation is in ch. 97, p. 5*b*.

    [i] *Sung Shih*, ch. 307, p. 3*b*.

[1] 油瓢	[2] 火矢	[3] 弩	[4] 衛公兵法	[5] 李靖
[6] 汪宗沂	[7] 許洞	[8] 飛火	[9] 李衛公問對	[10] 岳義方
[11] 馮繼昇	[12] 張雍	[13] 朝野僉言	[14] 三朝北盟會編	[15] 徐夢莘

At the beginning of the next century the inventors were again busy. In the 3rd year of the Hsien-Phing reign-period (+1000), a naval captain, Thang Fu[1], presented models for an incendiary arrow (*huo chien*[2]), a fire-ball (*huo chhiu*[3]) and a barbed fire-ball (*huo chi li*[4]), while at the same time a naval architect, Hsiang Wan[5], presented designs for warships.[a] They were rewarded with numerous strings of cash. Then in +1002 a military officer, Shih Phu[6], reported that he knew how to make better fire-balls (*huo chhiu*[7]) and fire-arrows (*huo chien*[8]). Accordingly his products were tried out, by imperial order, in tests watched by ministers of State and their assistants.[b] Thus in the decades preceding the first appearance of gunpowder formulae in the *Wu Ching Tsung Yao* there were many developments of something essentially new. Otherwise why all the fuss about tests and rewards? Surely these inventions were in fact connected with the use of gunpowder low in nitrate as an incendiary more controllable and more effective than any fire-producing mixtures previously available. It would have been much less haphazard than the old fire-arrows with oil and other combustibles, and the length of fuse could have been carefully adjusted to the estimated time of travel. The passages given in the previous paragraphs have sometimes been interpreted as signifying the first appearance of the rocket, though there is really no evidence for this, no indication that the projectiles flew off of themselves. But the use of 'rocket-composition' as an incendiary would be exactly what one would expect for this particular time; the gunpowder mixture had not been known for very long, and the properties of high-nitrate powder were still remaining for the future to discover.

Perhaps the most enigmatic text of about this time is that concerning the 'whip arrow, or javelin' and the 'gunpowder whip arrow, or javelin'(Fig. 13), which occurs in the *Wu Ching Tsung Yao* of +1044. The passage concerning it is very difficult to interpret, and we shall have to adopt the unusual course of giving two alternative translations.[c] The first is as follows:

The whip arrow (or javelin) called *pien chien*.[9]

Take a length of newly (-cut) green bamboo 10 ft. long, with a diameter of 1·5 in. (as the pole, *kan*[10]). The lower end is shod with iron (and fixed to the ground). A silk cord 6 ft. long is attached to the top end of it. Take also another piece of strong bamboo 6 ft. long to make the *pien chien*[9] itself, and give it a pointed head (*tsu*[11]). Check the junction of the two poles, and fix there a bamboo guide-hook (*chu nieh*[12]).

---

[a] *Ibid.* ch. 197, p. 2*a*; *Sung Hui Yao Kao*, ch. 185, p. 37*b*. Also *HTCTC/CP*, ch. 47, p. 15*b*.

[b] *HTCTC/CP*, ch. 52, p. 20*a*; cf. Fêng Chia-Shêng (*1*), p. 48, (*6*), p. 17. Shih Phu has a biography in *Sung Shih*, ch. 324, pp. 1*a* ff., but it does not tell us much about his technical interests. There are many references to him also in the former work, e.g. ch. 46, p. 6*a*, ch. 47, p. 15*a*, ch. 15, p. 4*a*, ch. 55, p. 9*b*, etc.

[c] *WCTY/CC*, ch. 12, pp. 60*b*, 61*a, b* (Ming ed. ch. 12, pp. 52*b*, 53*a*), tr. auct.

[1] 唐福	[2] 火箭	[3] 火毬	[4] 火蒺藜	[5] 項縮
[6] 石普	[7] 火毬	[8] 火箭	[9] 鞭箭	[10] 竿
[11] 鏃	[12] 竹槀			

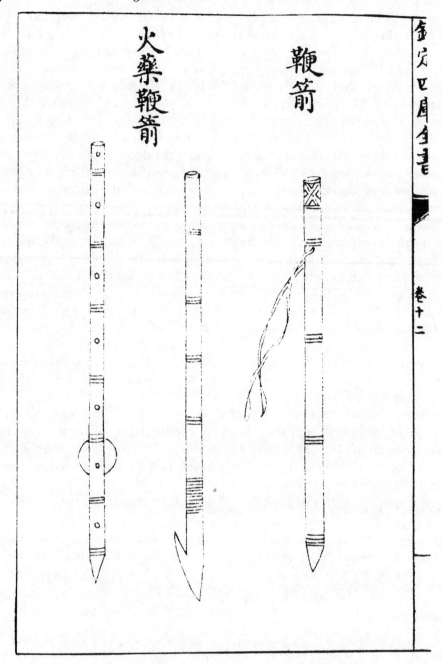

Fig. 13. The 'gunpowder whip-arrow' (*huo yao pien chien*), from *WCTY*, ch. 12, p. 60*b*.

[*Comm.* Some people call it a *pien tzu*[1].]

When the moment for shooting (*fang*[2]) comes, connect the javelin to the pole (through a loop of) the silk cord; then while one man shakes the pole and pulls it back, the other man holds the end of the javelin (aiming it), so that the pole hits against it (*chi*[3]) and propels it forth (*erh fa chih*[4]).

The advantage of the whip arrow (or javelin) is that it can shoot (far) upwards to hit the enemy above.

Then come the mysterious words about gunpowder.

But if there are low objects or (enemy) troops, then let off (*fang*[2]) the gunpowder whip arrow or javelin (*huo yao pien chien*[5]).[a] Make a container of the bast of birch bark, and put into it 5 oz. of gunpowder behind the javelin head. Light it and shoot it off (*fan erh fa chih*[6]).

Thus on this interpretation the main propulsive force was provided by the elasticity of the bamboo pole bent backwards by one of the soldiers, while the other one did the aiming. Nothing is said in detail of the fuse, but the gunpowder, presumably low in nitrate, was clearly acting as an incendiary. A drawing based on this view is given in Fig. 14 (*a*).

But there is another possibility, according to which the second pole acted more like an *atlatl* or throwing-stick. Let us look then at a second translation.

For whip arrows (or javelins, *pien chien*[7]) one must use a new and green bamboo 10 ft. long and of diameter 1.5 in., forming a long staff. At the back end an iron chain is attached, and to the other end a silk cord 6 ft. long is fastened. Another piece of strong bamboo is sharpened to form a whipjavelin, it is also 6 ft. long, and at a specific point in the middle a hook or projection (*nieh*[8]) is fitted.

[*Comm.* This is also called a *pien tzu*[9].]

At the time of shooting, the cord is hooked round the projection, attaching the javelin to the pole. One man wields the pole to give a force, while the other holds (and aims) the rear end of the javelin so that it can receive the impetus and fly forth. The benefit of this is the way it shoots high upwards, hitting the enemy with the accuracy of close-quarter weapons.

And the words about gunpowder follow on.

For letting off gunpowder arrows (or javelins, *huo yao chien*[10]) the bast of birch bark (*hua phi yu*[11]) is wrapped round forming a ball with 5 oz. of gunpowder placed inside it, behind the point (the shaft of the arrow passing through the middle). Upon igniting, the arrow is shot forth.

On this version, then, one soldier gave a strong turning movement to the bamboo pole, which pulled the silk cord with it, then as the second soldier aimed

---

[a] The text says only *huo yao chien*; the full phrase is taken from the caption in the illustration.

[1] 鞭子	[2] 放	[3] 激	[4] 而發之	[5] 火藥鞭箭
[6] 燔而發之	[7] 鞭箭	[8] 臬	[9] 鞭子	[10] 火藥箭
[11] 樺皮羽				

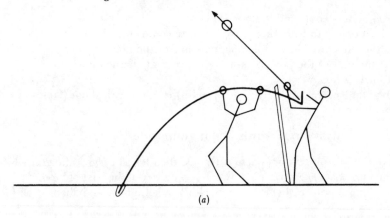

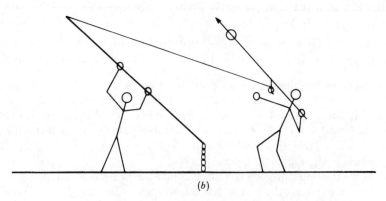

Fig. 14. Diagrams illustrating alternative reconstructions of the 'whip-javelin' and 'gunpowder whip-javelin'
of the *Wu Ching Tsung Yao*.

the javelin, the cord, doubtless with a ring at the end, slipped off the hook and
the projectile went on its way. The sketch in Fig. 14(b) attempts to explain the
mechanism. It would have been a somewhat sophisticated application of the
principle of the *atlatl, propulseur*, or throwing-stick, used by almost all peoples
from prehistoric times onwards to increase the range of their javelins.[a] We do not
feel able to decide as between the bent pole and the throwing-stick.[b]

[a] See for instance Singer *et al.* (1), vol. 1, p. 57; Kroeber (1), p. 643; Montandon (1), pp. 398 ff.; Heymann
(2); Pitt-Rivers (4), p. 132 and pl. 16; Underwood (1).
[b] Among its many obscurities the text gives no idea of the length of the iron chain, if there really was one. It
seems to say that the silk cord was attached to the chain, but *hsiao* must surely indicate the top of the pole.
Then the text states that the silk cord is attached to the 10 ft. pole, but the illustration (Fig. 13) shows it
attached near the rear end of what seems to be the 6 ft. javelin-arrow. And on the first interpretation the
purpose of the silk cord is unclear; perhaps it steadied the javelin, or held the pole back when sufficiently bent.

[1] 梢

But of course what really matters is the function of the gunpowder. An incendiary purpose for the *huo yao pien chien*[1] is clearly stated a few pages later on,[a] and birch-bark containers are mentioned again,[b] in the course of a long section entitled 'Methods for the Defence of Cities' (*Shou Chhêng chih Fa*[2]). But the later use of the words for designating rockets has impelled many to see in this description the earliest account of rocket propulsion. Wang Ling[c] was inclined to this, but Fêng Chia-Shêng[d] decided definitely against it. Unfortunately there is a certain ambiguity in the concluding words,[e] and this has impelled Li Ti (*1*) to defend the idea that the whip-javelin was a gunpowder rocket. He uses several philological arguments, first, that the verb *fang*[3] is used for the shooting, not *shê*[4]—but the former was already said of fire-arrows in the *Sung Shu*. What he says about *fan*[5] is more weighty; it means to boil or roast, or to cook meat offered to gods and spirits, i.e. a process rather than an action; so he would like to translate 'as it burns, it will fly forth'.[f] He wishes to differentiate the whip-javelin from the arrow carrying a gunpowder incendiary packet shot from a bow, the device we shall study next, since he doubts that the methods of propulsion stated would have carried the javelin any significant distance if it was not a rocket. Here opinions may differ. Our view is that the javelin-arrow was a javelin-arrow, with or without its payload of incendiary gunpowder, and therefore not a rocket. If it was really self-propelled, why did it need all that auxiliary equipment worked by two men?

One hardly ever comes across a reference to the *pien chien* in any of the kinds of literature of the following centuries. But it does appear in a poem written by Chang Hsien[8] in the middle of the +14th century. It is called Pei Fêng Hsing[9] (Affairs of the North Wind),[g] and in it a young man driving a cart near Chü-yung Kuan north of Peking meets a strange and fearful horseman, who carries with him a *pien chien* in a holder. Nothing more is said of it, but this gives us at least one literary reference which suggests that the technique still remained in existence.

We do think we have a clear and concise description of exactly the kind of thing that Thang Fu and his colleagues were introducing at the turn of the millennium, but naturally it comes from some three hundred and fifty years later.[h] In the *Huo Lung Ching* there is an illustration of the 'Fiery Pomegranate

---

[a] *WCTY/CC*, ch. 12, p. 73*a* (Ming ed. ch. 12, p. 64*b*).
[b] *WCTY/CC*, ch. 12, p. 74*b* (Ming ed. ch. 12, p. 65*b*).
[c] (*1*), p. 165.
[d] (*4*), p. 41, (*6*), pp. 23–4.
[e] They could mean either 'ignite it to set it off', or 'ignite it and set it off'. We prefer the latter.
[f] He feels that if just the single action of lighting had been meant, some expression like *cho huo*[6] or *jan*[7] would have been used. But we have not found these very commonly in texts of that time.
[g] *Yü Ssu Chi*[10], ch. 3, p. 9*a*.
[h] The passage is not in the *Wu Ching Tsung Yao* now, but there is a similar one in its *Wu Ching Yao Lan* version.

[1] 火藥鞭箭   [2] 守城之法   [3] 放   [4] 射   [5] 燔
[6] 着火   [7] 然   [8] 張憲   [9] 北風行   [10] 玉笥集

arrow shot from a bow' (*kung shê huo shih-liu chien*[1]), given here in Fig. 15, and a textual explanation.[a] It says:

Behind the arrow-head wrap up some gunpowder with two or three layers of soft paper, and bind it to the arrow shaft in a lump shaped like a pomegranate. Cover it with a piece of hemp cloth tightly tied, and sealed fast with molten pine resin. Light the fuse and then shoot it off from a bow.

The last sentence is amplified later on in the *Wu Pei Chih* as follows:[b]

You can use paper pasted and oiled to make the fuse (*yao hsien*[2]), which should lead into the front of the gunpowder ball.[c] The iron arrow-head must be sharp, with backward-pointing prongs. Light the fuse to start the fire, then release the arrow from the bow, and send it off. When it reaches the target, the fire caused in the protective matting or sails cannot be extinguished with water; so the device is of great advantage.

This last remark reminds us that the difficulty of putting out a blaze caused by a mixture with its own built-in oxygen supply was another signal advantage of gunpowder as an incendiary; a similarity in a way to the old Greek Fire, though that had depended on its physical property of liquidity. Primitive incendiary combustibles would not have been so hard to put out.

What was true of bows was also true of crossbows, as is shown by another passage in the *Wu Ching Tsung Yao*.[d] After explaining in detail the construction and operation of the three-spring or triple-bow *arcuballista* (cf. pt. 6 (*f*) above), it goes on to say that in the use of the San Kung Chhuang Tzu Nu[3] 'to all these bolts one can add gunpowder, but the amount, whether heavy or light, much or little, will depend upon the strength of the catapult'.[e] This will have been another version of the incendiary projectile.

When the Sung capital, Khaifêng, fell into the hands of the Chin Tartars in +1126 a great deal of war material was captured by them. Hsia Shao-Tsêng[4] afterwards wrote:

The palace eunuch Liang Phing-Wang[5] showed (the invaders) round the imperial palace, and told them of the toys and precious things contained therein, while Têng Shu[6] presented a complete list of queens, consorts, young princes, concubines and the like. A certain Li[7] handed over 20,000 fire-arrows (*huo chien*[8]), a (standard) model of the trebuchet for hurling projectiles filled with molten metal, and four-bow *arcuballistae*—all of which had been prepared by (Sung) Thai Tsung for the conquest of (Nan) Thang. In peacetime these officials had lived off the fat of the land, yet now what heartlessness to the country did they show![f]

---

[a] Pt. 1, ch. 2, p. 24*a*, *b*; Hsiang-yang-fu ed. p. 21*a*, tr. auct.
[b] Ch. 126, pp. 10*b*, 11*a*, tr. auct.
[c] This probably explains the antenna-like objects seen in the illustration; there is only one in *WPC*, drawn more clearly.
[d] Ch. 13, p. 7*a* in both editions, tr. auct.
[e] This passage was first noted by Wang Ling (1), p. 166, but he thought that some kind of rocket was meant.
[f] *Chhao Yeh Chhien Yen*, quoted by Hsü Mêng-Hsin in his *San Chhao Pei Mêng Hui Pien*, ch. 97, p. 5*b*, tr. auct.

[1] 弓射火柘榴箭   [2] 藥線   [3] 三弓牀子弩   [4] 夏少曾
[5] 梁平王   [6] 鄧述   [7] 李   [8] 火箭

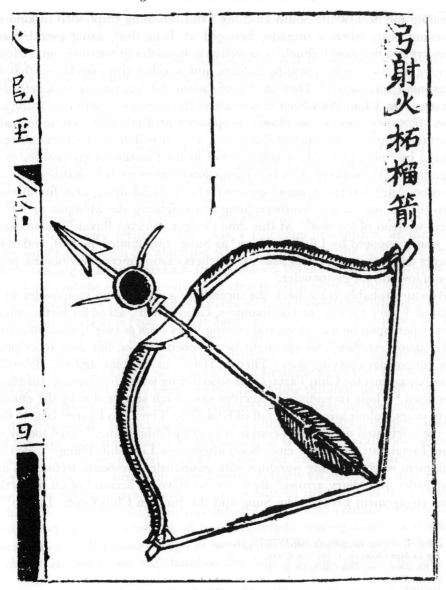

弓射火柘榴箭

Fig. 15. The true fire-arrow, the 'fiery pomegranate shot from a bow', from *HLC*, pt. 1, ch. 2, p. 24*a*.

Thus here again we have the incendiary arrows, at this time almost certainly containing gunpowder, or with arrangements to do so.

During the whole of the remainder of the Sung dynasty these weapons found much employment. In +1130, four years after the fall of the capital at Khaifêng, the Chin Tartars used them with much effect against Han Shih-Chung[1], com-

---

[1] 韓世忠

the illustration given in Fig. 16 says:

The 'iron-beaked fire kite' has a wooden body, an iron beak, and a bundle of straw as a tail. Gunpowder is enclosed in (front of) the tail.[a]

The 'bamboo fire kite' is made of a coarse bamboo basket framework, large in the belly and narrow at the mouth, with a rather elongated shape. Several layers of paper are pasted over the framework, and brushed (with oil) until the cover becomes yellow. 1 lb. of gunpowder is put inside, and some round stones are added to increase the weight. Then a bundle of straw weighing 3 to 5 lb. is tied on to form the tail.

These two things are (used) in the same way as the 'barbed fire-ball'. When the enemy comes to attack one's city wall, they are both launched from trebuchets (*phao*[1]). They will set fire to the equipment collected by the enemy, and strike terror into his troop formations.

In spite of the word 'kite', which could equally well be translated 'kestrel' or 'sparrow-hawk', these two projectiles were evidently meant to be thrown over like bombs, and the paper kite (which was certainly a Chinese invention) is not involved.[b]

This brings us to the subject of fire-balls, bombs and grenades in general, our discourse next in order. The distinction between the incendiary and the explosive is difficult to draw, since we need to know what we are never told, namely the proportion of saltpetre in the mixture. But as we shall see, certain items of terminology may inform us of the moment when the borderline was crossed.

Before going further, however, let us dispose of a projectile which was certainly not explosive, even though containing gunpowder. 'Fire-balls' (*huo chhiu*[2]) were made to be hurled from trebuchets towards the enemy. We have already described (p. 73) the 'igniter' or 'range-finding' fire-ball (*yin huo chhiu*[3]) used for ascertaining distances. But there was also a 'barbed' or hooked fire-ball (*chi li huo chhiu*[4]), intended for attaching itself to objects or structures (Fig. 17). The *Wu Ching Tsung Yao* says:[c]

The barbed, or calthrop, fire-ball has three sharp-edged six-pointed iron spikes, and is rolled up with gunpowder inside it. It has a hempen rope 2 ft. long (with a ring on the

[a] The point of the iron nozzle was presumably to direct the flames in a particular direction.
[b] This was discussed in Vol. 4, pt. 2, pp. 576 ff., whence we may recall the leaflet raid of +1232 (pp. 577–8) and the man-lifting kites of the +13th century (p. 589) which could have been used for spotting. It is true that a bomb of some kind is shown suspended over a city from a sort of flying windsock in Walter de Milamete's famous MS (cf. p. 287); James (2), pp. xxxiv, 154–5, fol. 77b, 78a. This representation would be evidence for +1327 in Europe, but the design is so fanciful and impractical that it must have been imaginary. There is a good deal more to be said about windsock aerostats or hot-air balloon dragon-standards than what we were able to put in Vol. 4 pt. 2, pp. 597–8, but it is not really relevant here.
[c] *WCTY/CC*, ch. 12, pp. 64a, 65a tr. auct. Cf. Arima (1), pp. 31–2. The *SKCS* edition writes *yao yao*[5] instead of *huo yao*[6], but this is an obvious misprint because all other texts say the latter, e.g. *Wu Ching Yao Lan*, ch. 12, pp. 60, 64; *Huo Lung Ching*, pt. 1, ch. 3, pp. 5a, b, 6a, b; Hsiang-yang-fu ed. pp. 27b, 28a; *Wu Pei Chih*, ch. 130, pp. 4a, b, 5a, b.

[1] 砲   [2] 火毬   [3] 引火毬   [4] 蒺藜火毬   [5] 藥藥
[6] 火藥

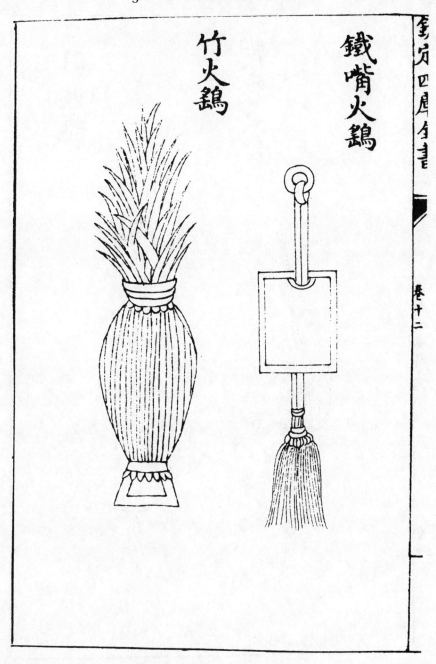

Fig. 16. The 'bamboo fire kite' (bird) and the 'iron-beaked fire kite' (bird); incendiary projectiles from *WCTY*, ch. 12, p. 64*b*.

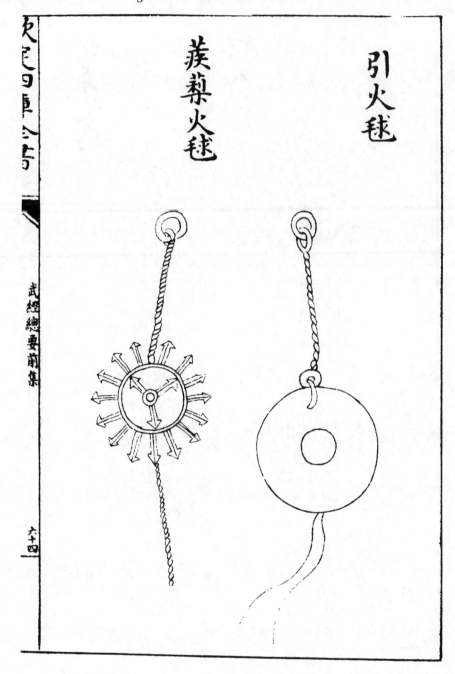

Fig. 17. The igniter or range-finding fire-ball and the barbed fire-ball, from *WCTY*, ch. 21, p. 64*a*.

end) threaded through it.[a] On the outside it is enveloped in paper, to which is applied various chemical substances.[b] It also has eight iron calthrops, each of which is provided with hooks (*ni hsü*[1]). When you want to let it go you set light to it by piercing it with a (red-hot) iron poker, so that smoke begins to come out.

The text continues by giving the gunpowder formula for use with this weapon; it is the second of those which we examined above (p. 122), and its nitrate-content did not exceed 50%.

A fire-ball with barbs, spikes or hooks (cf. p. 120) makes one naturally think of those clusters of radiating spikes, known as calthrops, which were scattered on a road or any piece of ground to deter the onset of cavalry. This principle was of course a very ancient one, going back to the −4th century with the *Mo Tzu* book.[c] But that was not quite what was at issue here—the barbed or hooked fire-ball was intended to attach itself to wooden buildings or to the sails of ships, and so set them on fire. It is interesting that exactly the same device was used later on in Europe, whether derivatively or independently, incendiary shot with hooks designed to catch on to rigging and sails, with destructive consequences.[d]

It was natural that incendiary bombs and grenades containing low-nitrate gunpowder should persist into the +17th century and later. Thus the *Ping Lu* of +1606 mentions some of these. For example, the 'flying fire-pestle' (*fei huo chhui*[2]) was simply a bottle-shaped wooden grenade, eight inches long and with sharp spikes or hooks protruding from its surface (Fig. 18). When thrown on to an enemy ship it would attach itself by these to sails, rigging or woodwork, and then the flames of the explosion or deflagration, even though strong enough only to break the casing, would set the craft on fire.[e] A rather simpler version was the 'flying swallow' (*fei yen*[3]), nothing but a tube of bamboo or carton containing gunpowder of 68·5% saltpetre,[f] yet also provided with hooks for attachment to the sails and structures of the enemy.[g] Such devices were the direct ancestors of the incendiary bombs of the present day.

(10) BOMBS AND GRENADES

We are now at the frontier between incendiary gunpowder and explosive gunpowder. The probability is that the *huo chhiu*[4] and *huo phao*[5] of the +10th and +11th centuries involved only low-nitrate mixtures, nevertheless very effective in

---

[a] Perhaps this acted in the same way as the sling in which the projectile rested at the end of the long arm of the trebuchet.
[b] This may well mean more gunpowder.
[c] Ch. 54, p. 15*b*. Then all the ancient military works, such as the *Liu Thao*, speak of it. Cf. Chang Hung-Chao (*1*), p. 426.
[d] Blackmore (*2*), p. 193.
[e] *PL*, ch. 12, pp. 59*b*, 60*a*.
[f] The composition is in this case actually given: in percentages, N 68·5; S 12·3; C 19·2.
[g] *FL*, ch. 12, p. 61*a*.

[1] 逆鬚　　　[2] 飛火槌　　　[3] 飛燕　　　[4] 火毬　　　[5] 火砲

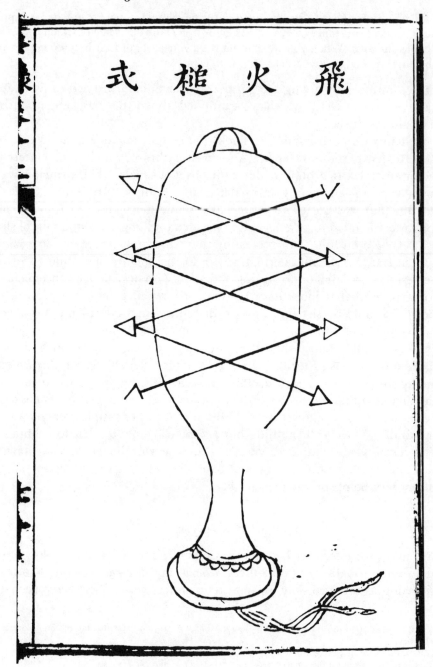

Fig. 18. The 'flying fire-pestle' (*fei huo chhui*) from *Ping Lu*, p. 76*a*.

setting fire to siege machines and towers on land or 'wooden walls' at sea. But now a new term, *phi li phao*[1], seems to mark the appearance of a new thing, the 'thunderclap bomb', for the first time truly explosive. It would have been something like a maroon, consisting of higher-nitrate gunpowder, enclosed in a weak case of bamboo, carton and the like; with the property of giving a loud bang when exploded, and therefore more suitable (unless combined with other things) for causing fright rather than serious injury to the enemy's horses and men. As we shall see, this weapon was characteristic of the conflicts of the +12th century. Following upon this, there was a further step to the *chen thien lei*[2] or 'heaven-shaking thunder-crash bomb', also identifiable as the *thieh huo phao*[3] or 'iron bomb', and also projected from trebuchets. Here for the first time brisant high-nitrate gunpowder was used, enclosed in a strong casing of metal, and thus calculated to cause serious injury to the enemy's troops upon detonation, a word we can now at last make use of. Broadly speaking, this development was characteristic of the +13th century. Its development had taken some two and a half centuries, since the first use of the term *huo phao* seems to have occurred in +1004, when Hsü Tung mentioned it in one of his discussions of attack by fire in the *Hu Chhien Ching*.[a]

There was one great advantage about the use of explosive projectiles, whether thin-walled or stout-walled, but so simple that it has not often been mentioned. When both sides were equipped with trebuchets, the stones hurled by the enemy could with relative ease be collected and used as ammunition to hurl back against them. But as Li Shao-I (*1*) has pointed out, maroons and bombs disintegrated, doing as much damage as possible in the process, and the fragments were not available for re-use in the opposite direction.

It is a matter of great interest that the 'thunderclap fire-ball, or bomb' already appears in the *Wu Ching Tsung Yao*; a fact which must surely mean that some of the Sung artisans of the first half of the +11th century already knew what would happen if one increased the percentage of saltpetre in the gunpowder mixture. The point was vital, since now for the first time a true explosion could be brought about. Here is the description (cf. Fig. 19):[b]

The thunderclap bomb (*phi li huo chhiu*[4]) contains a length of two or three internodes of dry bamboo with a diameter of 1.5 in. There must be no cracks, and the septa are to be retained to avoid any leakage. Thirty pieces of thin broken porcelain the size of iron coins are mixed with 3 or 4 lb. of gunpowder, and packed around the bamboo tube. The tube is wrapped within the ball, but with about an inch or so protruding at each end. A (gun)powder mixture is then applied all over the outer surface of the ball.

---

[a] Ch. 6, p. 4*b*, though even then in the author's commentary only.
[b] *WCTY/CC*, ch. 12, pp. 67*b*, 68*a*, 69*b*, tr. auct. There are parallel descriptions in *HLC*, pt. 1, ch. 3, p. 7*a, b*, and *WPC*, ch. 130, p. 6*a, b*, both abridged, otherwise essentially the same. *HLC* says 30 lb. of gunpowder, which would have been a much bigger bomb, but perhaps it was a misprint for 3 or 4. Cf. Okada Noboru (*3*).

[1] 霹靂砲   [2] 震天雷   [3] 鐵火砲   [4] 霹靂火毬

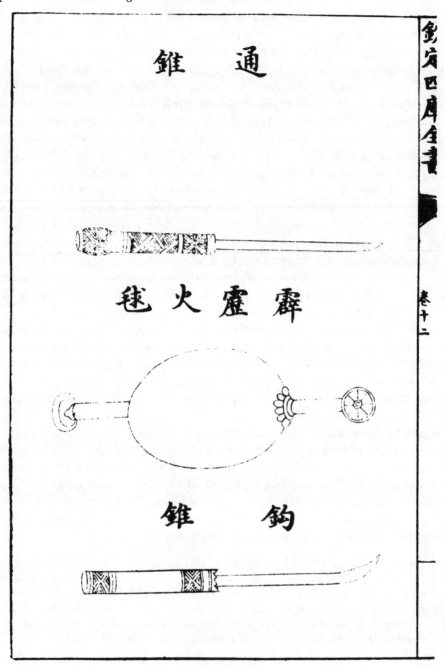

Fig. 19. The 'thunderclap bomb' (*phi li phao*, or *phi li huo chhiu*), type of the bomb with weak casing. From *WCTY*, ch. 12, pp. 67*b* ff. The other two objects illustrated are the red-hot iron brands used for igniting the projectile before it was hurled from the trebuchet.

[*Comm.* The gunpowder mixture for application around the outside is given under the fire-ball section.][a]

If the enemy digs a tunnel to attack the city, then a sap must be excavated so as to connect with it. A (long) red-hot iron brand is used to set off the thunderclap bomb, which produces a noise indeed like thunder. Bamboo fans are used to drive the smoke and flame down the tunnel, so as to stifle and burn the enemy's sappers.[b]

[*Comm.* The soldier setting off the bomb should suck some liquorice (*kan tshao*[1]) as a protection.][c]

Here several interesting points arise. Presumably the point of the unbroken bamboo was to act like a fire-cracker and add to the fearsomeness of the explosion.[d] Secondly, the nature of the covering is not stated, but as other descriptions will show, it was of carton or thick layers like a paper parcel. Thirdly, the sort of gunpowder applied round the outside is explicitly stated to have been of the fire-ball incendiary type, therefore low in nitrate, and it must have been mixed with some kind of gum to hold it in place. It would be extremely interesting to reconstruct and test the whole device.[e]

So far we have not found the thunderclap bomb referred to in battle descriptions before the end of the +11th century, but after that time they come thick and fast—perhaps a shortage of saltpetre delayed the general use of the weapon. One of the earliest concerns the valiant but unsuccessful defence of Khaifêng (Piencheng), the Sung capital, against the hosts of the Chin Tartars. One of the Sung commanders. Li Kang[3], left us an eye-witness account of the use of the thunderclap bomb. He wrote:[f]

First Tshai Mou[4] gave orders to all the officers and soldiers that (even) when the Chin troops came near the city, the trebuchets and *arcuballistae* were not to be used, and anyone who did so would be beaten; whereupon our men were very angry. I myself then took over the command, and ordered them to shoot off any such artillery as they should see fit, and those who attained their targets best were well rewarded. At night the thunderclap bombs were used, hitting the lines of the enemy well, and throwing them into great confusion. Many fled, howling with fright.

The thunderclap bomb was sometimes combined with the blinding lachrymatory smoke caused by finely powdered lime. Here the classical instance is the Battle of Tshai-shih[5], where in +1161 the Sung admiral Yü Yün-Wên[6] won a

---

[a] P. 65*a, b*; cf. p. 122 above.

[b] This is old stuff, going back to the −4th-century Mo Tzu book; cf. Vol. 4, pt. 2, pp. 137 ff.

[c] Cf. p. 125 above.

[d] Davis & Ware (1), p. 524, overlooked the specified integrity of the bamboo, and thought that as a hollow pipe it would make a roaring noise as the bomb flew through the air.

[e] Together with other forms of *huo chhiu*, it is referred to incidentally in the description of the trebuchet (*huo phao*[2]) on pp. 56*b*, 57*a*.

[f] *Ching-Khang Chhuan Hsin Lu*, ch. 2, p. 13*a, b*, tr. auct. See also *Sung Thung Chien Chhang Phien Chi Shih Pên Mo*, ch. 147, p. 10*a*. Knowledge of this incident in the West goes back to Mayers (6), pp. 89–90, but he could not make out what sort of a projectile it was.

[1] 甘草   [2] 火砲   [3] 李綱   [4] 蔡楙   [5] 采石

[6] 虞允文

great victory over the Jurchen Chin forces which were trying to cross the Yang-tze and invade the south. In his *Hai Chhiu Fu*[1] (Rhapsodic Ode on the Sea-eel Paddle-wheel Warships),[a] Yang Wan-Li[2] wrote as follows:[b]

In the *hsin-ssu* year of the Shao-Hsing reign-period, the rebels of (Wanyen) Liang[3c] came to the north (bank) of the River in force, intending to capture the people's boats, and hoisted flags indicating that they wished to cross over. But our fleet was hidden behind Chhi-pao Shan (island), with orders to come out when a flag signal was given. So a horseman was sent up to the top of the mountain with a hidden flag, and then when the enemy were in mid-stream suddenly the flag appeared; whereupon our ships rushed forth from behind (the island) on both sides. The men inside them paddled fast on the treadmills, and the ships glided forwards as though they were flying, yet no one was visible on board. The enemy thought that they were made of paper. Then all of a sudden a thunderclap bomb was let off. It was made with paper (carton) and filled with lime and sulphur. (Launched from trebuchets) these thunderclap bombs came dropping down from the air, and upon meeting the water exploded with a noise like thunder, the sulphur bursting into flames.[d] The carton case rebounded and broke, scattering the lime to form a smoky fog which blinded the eyes of men and horses so that they could see nothing. Our ships then went forward to attack theirs, and their men and horses were all drowned, so that they were utterly defeated.[e]

It would be interesting to know what the arrangements were which ensured that the lime would form an irritant fog without being wetted and slaked by the water.[f] The presence of quicklime in these thunderclap bombs has caused several writers to puzzle over analogies with the 'automatic fire' of Western antiquity (cf. p. 67 above), in which incendiary substances were supposedly ignited by the heat of slaking quicklime,[g] but in fact this was needless, since we know now that the thunderclap bomb contained explosive gunpowder. Probably the noise

---

[a] On these see Vol. 4, pt. 3, p. 416.

[b] *Chhêng Chai Chi*, ch. 44, pp. 8b, 9a, tr. auct. Cf. *Pi Chou Kao Lüeh*, ch. 1, p. 6a; *Chin Shih*, ch. 65, p. 16b; *Sung Shih*, ch. 368, p. 15a. We gave an account of this battle already in Vol. 4, pt. 2, p. 421, but reserved the text for the present place. For the historical background see Cordier (1).

[c] Fourth emperor of the Jurchen Chin, assassinated by his own generals after the defeat.

[d] Yang Wan-Li himself evidently did not know exactly how the thing worked. His actual words here were: 'I think most likely it was made of paper and contained lime and sulphur; when it hit the water, falling from the air, the sulphur began to burn, then it jumped up again with a noise like thunder, and the paper broke, liberating the lime and scattering it as a smoky fog.' Presumably the rebounding was the effect of the low-nitrate gunpowder on the outside, the fuse of which had been set just right to bring about the ignition when the projectile reached the water surface.

[e] Abbreviated, and not very good, versions of this text appeared later on in encyclopaedias such as *Wu Li Hsiao Shih*, ch. 8, p. 26a and *Ko Chih Ching Yuan*, ch. 42, p. 27b. Earlier Western writers, e.g. Romocki (1), vol. 1, pp. 43–4, knew only these, so it is not surprising that they ignored the gunpowder and evoked Greek Fire and the speculations about it. Lu Mou-Tê (1), p. 29, knew the original text, but thought that the thunder-clap lime bombs were shells fired from cannon.

[f] Artificial fogs in warfare are also mentioned in the +10th century. Under the Later Liang, when Chhien Liu[4] (cf. Vol. 4, pt. 3, pp. 320–1) was fighting a war in +918, his son Chhien Yuan-Kuan[5] made fireships which covered the enemy fleet in thick fog, and so turned the scales. See Chavannes (2), p. 202, translating *Chiu Wu Tai Shih*, ch. 133, pp. 4b to 8b.

[g] Notably Wang Ling (1), p. 169, who had the merit, however, of giving the translation of another passage from the *Chhêng Chai Chi*.

Chinese historians around 1800 were accustomed to say that this battle of +1161 was the first in which gunpowder was used—but now we know that it went back at least two centuries earlier. See, for example, Chao I's[6] *Kai Yü Tshung Khao*[7], ch. 30, and Liang Chang-Chü's[8] *Lang Chi Tshung Than*[9], ch. 5.

[1] 海鰍賦   [2] 楊萬里   [3] 完顏亮   [4] 錢鏐   [5] 錢元瓘
[6] 趙翼   [7] 陔餘叢考   [8] 梁章鉅   [9] 浪跡叢談

was as important here as the toxic smoke, and that required this higher-nitrate mixture.

Actually, the blinding effect of clouds of finely powdered lime was an old military technique in China; we have two accounts of it already in the same century. For example, in +1134, when a Sung garrison was shut up in Haochow by the Chin Tartars:[a]

Orders were given to the townspeople to transport (to the walls) jars of lime (*hui phing*[1])…. As before, the Chin soldiers erected (wooden) towers at the river mouth in order to attack the city; but from its ramparts projectiles of molten iron were sent over, together with the jars of lime, and stones (all from trebuchets) as well as arrows (from crossbows and *arcuballistae*).

Thus battered and confused, the enemy raised the siege. Then, in the following year, when the Sung general Yo Fei[2] was campaigning against the bandit chief Yang Yao[3],[b] we hear that[c]

the army also made 'lime-bombs' (*hui phao*[4]). Very thin and brittle earthenware containers were filled with poisonous chemicals, (powdered) lime, and iron calthrops.[d] In combat they were used to assail the enemy's ships. The lime formed clouds of fog in the air, so that the rebel soldiers could not open their eyes. They wished to make the same kinds of things themselves, but their potters were not able to produce them, so they suffered great defeats.

So here we get a little light on the nature of the vessels used to contain the lime when it was discharged in projectile form; an echo thus of the fragile bottles used in Arabic warfare for hurling over naphtha or distilled petroleum (p. 44 above). We can even trace the poison-gas effect a thousand years earlier, when in the Han period, about +178, the governor of Ling-ling, Yang Hsüan[5], was fighting a peasant revolt near Kueiyang. The *Hou Han Shu* says:[e]

The bandits were numerous, and Yang's forces very weak, so his men were filled with alarm and despondency. But he organised several dozen horse-drawn vehicles carrying bellows (*phai nang*[6]) to blow powdered lime (*shih hui*[7]) strongly forth, he caused incendiary rags to be tied to the tails of a number of horses, and he prepared other vehicles full of bowmen and crossbowmen. The lime chariots went forward first, and as the bellows were plied the smoke was blown forwards according to the wind (*shun fêng ku hui*[8]), then the rags were kindled and the frightened horses rushed forwards throwing the enemy lines into confusion, after which the bowmen and crossbowmen opened fire, the drums and gongs were sounded, and the terrified enemy was utterly destroyed and dispersed. Many were killed and wounded, and their commander beheaded.

---

[a] *San Chhao Pei Mêng Hui Pien*, ch. 165, p. 2b, tr. auct.
[b] We met him before, in Vol. 4, pt. 2, pp. 419 ff. in connection with his remarkable 22-wheeler paddle-boat warships.
[c] *Lao Hsüeh An Pi Chi*, ch. 1, p. 2a, tr. auct. The text dates from about +1190.
[d] Cf. p. 125 above.
[e] Ch. 68, p. 12a, tr. auct. Cf. Yang Khuan (1), p. 73.

[1] 灰瓶   [2] 岳飛   [3] 楊么   [4] 灰礮   [5] 楊璇
[6] 排囊   [7] 石灰   [8] 順風鼓灰

From about +1187 there comes a curious story, recorded by a scholar of the Jurchen Chin dynasty, Yuan Hao-Wên[1], some fifty years later. What he said was this:[a]

Towards the end of the Ta-Ting reign-period there lived north of Thaiyuan a certain hunter named Thieh Li[2]. One evening he found a great number of foxes in a certain place. So knowing the path that they followed, he set a trap, and at the second watch of the night he climbed up into a tree carrying at his waist a vessel of gunpowder (*huo yao kuan tzu*[3]). The coven of foxes duly came under the tree, whereupon he lit the fuse and threw the vessel down; it burst with a great report, and scared all the foxes. They were so confused that with one accord they rushed into the net which he had prepared for them. Then he climbed down the tree and killed them all (for their fur).

Here the bomb was in all probability a narrow-mouthed pottery vase or amphora; in any case it takes its place in the array of weak-walled containers.[b] It is interesting that such bombs could be used for hunting as well as for warfare.

But we must return to the thunderclap bombs. When Chao Shun[4] was conducting his successful defence of Hsiang-yang against the Chin Tartars in +1207[c] he found them a very useful weapon. Afterwards Chao Wan-Nien[5] wrote:[d]

In the evening he (Chao Shun) sent out a commando party of more than a thousand brave soldiers, and at midnight they went forward from Ho-thou to attack the enemy.... The artillerists held up their torches and shouted, while the soldiers on the city walls also shouted and beat drums while the thunderclap bombs were shot off. The (Chin) wretches were terrified and quite lost their senses, men and horses running away as fast as they could....

On the 5th day at 10 o'clock in the morning, the enemy collected themselves together, and again attacked the city.... Thereupon the (Sung) commander gave orders that the soldiers ... on the city walls should beat drums and raise their shouts, while at the same time more thunderclap bombs were hurled forth. The enemy cavalry were again frightened, and retreated....

On the evening of the 25th day, taking advantage of the rain and overcast sky, the commander urgently sent the officers Chang Fu[6] and Hao Yen[7] to prepare boats large and small, more than thirty in number, enough to carry 1000 crossbowmen, 500 trident spearmen, and 100 drummers, together with thunderclap bombs (*phi li phao*[8]) and gunpowder arrows (*huo yao chien*[9]).[e] They took cover by the river bank below the enemy's encampment.... Then at the stroke of a drum the crossbowmen let fly a volley, and immediately following this all the drums sounded and all the crossbows were fired. Simultaneously the thunderclap bombs and the fire-arrows were sent into the enemy's

---

[a] *Hsü I Chien Chih*, ch. 2, p. 1*b*, tr. auct.
[b] Cf. Fêng Chia-Shêng (*1*), pp. 41–2, 77–8, (*6*), p. 27.
[c] Cf. Franke (25).
[d] *Hsiang-Yang Shou Chhêng Lu*, pp. 13*b*–23*a*, tr. auct.
[e] By this date these could well have been rockets, but we must postpone till pp. 472 ff. our examination of the time of their first appearance.

[1] 元好問   [2] 鐵李   [3] 火藥罐子   [4] 趙淳   [5] 趙萬年
[6] 張福   [7] 郜彥   [8] 霹靂礮   [9] 火藥箭

camp. How many were killed and wounded in this attack could not be known, but men and horses were thrown into confusion and trampled upon each other. By the fifth night watch they were flying away in all directions. The (Sung) commander then ordered his men to retire, not even one being wounded....

On the 26th day, one of them, by name Fan Chhi[1], who had been captured, walked back and regained the lines, saying that the whole Chin force had been asleep when the attack took place, so that they had no time to mount their horses or to collect their baggage. Such was the confusion that the barbarian army lost two or three thousand dead or wounded, and more than eight hundred horses.

This vivid account suggests that the explosive character of the thunderclap bombs took its toll, even though their casings were quite weak, just as the cross-bow bolts, fire-arrows and close-quarter weapons certainly did. With this, then, we may proceed to gunpowder bombs with stronger casings.

On the way we may pause to notice that in the +13th century there are several references to 'signal bombs' (hsin phao[2]).[a] For example, in +1276, when A-Chu[3] was attacking Yangchow, these were fired as messages to troop detachments;[b] and there are other instances, including one of +1293, when the order was given to collect all those still in the stores in Chekiang.[c] Although they are called 'heaven-shaking' (chen thien[4]), they never seem to cause the slightest damage, so they were most probably carton bombs or maroons timed to explode in mid-air, and therefore belonging more to the phi li phao than the chen thien lei category.

From here onwards we have to adopt a method rather different from that used for the explosive projectiles with weak casings; for in the Wu Ching Tsung Yao of +1044 there is no mention of bombs or grenades with strong ones. We must therefore take a look at the battle accounts and other descriptions which deal with these cast-iron missiles. Then having considered these, we may say something of the literature from +1350 onwards, which has a good many specifications for explosive projectiles of both kinds, with casings strong as well as weak. By that time, of course, we are well beyond the time of arrival of gunpowder weapons in Europe, so we shall return to the much earlier development of the fire-lance, the rocket, and the metal-barrel cannon in China.

The story begins with the successful siege of Chhi-chou[5] by the Jurchen Chin forces in +1221. That dynasty was by now almost at the end of its tether, the rising Mongolian power in the north having taken their capital of Peking in +1215, since when they had set up at Khaifêng. Though menaced in their rear they continued to struggle with the Sung. On this occasion, as we can see from

[a] See Fêng Chia-Shêng (1), pp. 62–3.
[b] Chhien-Thang I Shih, ch. 9, pp. 4a, 5b; cf. Sung Shih, ch. 451, p. 4b.
[c] Kuo Chhao Wên Lei, ch. 41, p. 61b.

[1] 樊起   [2] 信砲   [3] 阿尤   [4] 震天   [5] 蘄州

the account of the siege, *Hsin-ssu Chhi Chhi Lu*[1], written by Chao Yü-Jung[2], who had himself been an eye-witness and participant,[a] the Sung division holding the fortified city seems to have had nearly everything—7000 incendiary gunpowder arrows for use with crossbows (*nu huo yao chien*[3]), and 10,000 to be shot from bows (*kung huo yao chien*[4]), 3000 barbed fire-balls (*chi li huo phao*[5]), and 20,000 large leather projectiles (*phi ta phao*[6]), presumably low-nitrate gunpowder in bags.[b] To these the somewhat later book *Hsing Chün Hsü Chih*[7] adds,[c] besides unspecified incendiary bombs (*huo phao*[8]) which could pass right over high obstructions, as also grenades (*shou phao*[9]), now—for the first time—true metal-barrel guns, or proto-guns (*huo thung*[10], literally 'fire-tubes'), a point to which we shall of course return.[d] The Sung soldiers had matting and wet clay as protection against the petrol flame-throwers and incendiary bombs and arrows of the Chin Tartars, who also used expendable birds (*huo chhin*[11]) to set the roofs of the houses within the city on fire.[e] Although the Sung artillerists would use more than 3000 incendiary bombs in a single day, there is no mention of explosive thunderclap bombs.[f] But now, again for the first time, comes something else new; the Chin army was provided with explosive bombs of cast iron (*thieh huo phao*[12]), and these they used to attack the defenders, which must mean that a detonating (*phao cha*[13]) high-nitrate gunpowder mixture had been reached at last, since nothing less would have burst the iron casing. 'Their shape was like that of a bottle-gourd' (*phao*[14]), says Chao Yü-Jung,[g] 'with a small opening,[h] and they were made from cast iron about 2 in. thick. Fêng Chia-Shêng suspected that the Sung troops were also equipped with these, but we do not know exactly what they were like.[i] In any case, it seems sure that we have to do here with an early appearance of the thunder-crash bomb or grenade (*chen thien lei*[15]), surpassing the thunderclap bomb (*phi li phao*[16]) because of the much greater strength of its casing, and the much greater damage that it would do when it burst.[j] And indeed Chao Yü-Jung does say that the sound was like thunder (*shêng ta ju phi li*[17]), and the effectiveness very great, shaking the walls of houses, and killing and wounding many people.[k]

---

[a] According to Fêng Chia-Shêng (*1*), p. 79, his biography is in *Sung Shih*, ch. 449, p. 24*b*. He had been a judge in Chhichow.

[b] *Hsin-Ssu Chhi Chhi Lu*, p. 3.   [c] Ch. 2, pp. 16*b*, 17*a*.   [d] Pp. 304 ff. below.

[e] *Hsing Chün Hsü Chih*, loc. cit.   [f] *Hsin-Ssu Chhi Chhi Lu*, p. 2.   [g] *Hsin-Ssu Chhi Chhi Lu*, p. 23.

[h] This was no doubt for filling, perhaps also to admit the fuse, but one wonders how the mouth was closed so that the casing would break rather than blow it open.

[i] (*1*), pp. 61, 67, 78.

[j] One officer, Chia Yung[18], was blinded in an explosion which also wounded half a dozen other men.

[k] *Hsin-Ssu Chhi Chhih Lu*, pp. 20–5. Goodrich & Fêng (*1*), p. 117 signalled this development, remarking very justly on the alacrity with which the Chin Tartar military took up new war devices and inventions, not least in connection with gunpowder.

¹ 辛巳泣蘄錄   ² 趙與袞   ³ 弩火藥箭   ⁴ 弓火藥箭   ⁵ 蒺藜火礮
⁶ 皮大礮   ⁷ 行軍須知   ⁸ 火砲   ⁹ 手砲   ¹⁰ 火筒
¹¹ 火禽   ¹² 鐵火砲   ¹³ 爆炸   ¹⁴ 匏   ¹⁵ 震天雷
¹⁶ 霹靂砲   ¹⁷ 聲大如霹靂   ¹⁸ 賈用

The first appearance of *chen thien lei* (thunder-crash bomb) as a technical term seems to occur just ten years later, in +1231, when the Chin Tartars were themselves in turn besieged in a city in Shansi by Mongol forces. A Chin general, Wanyen Ê-Kho[1], was in command at Ho-chung[2],[a] when his defences were overrun by the Mongolian army. So

he escaped in ships with three thousand of his men (down the Yellow River). The Mongols pursued them along the northern bank with clamour and uproar of drums, while arrows and stones fell like rain. Now several *li* away a Mongolian fleet came out and intercepted them, so that they could not get through. But the Chin ships had on board a supply of those fire-bombs called 'thunder-crash' missiles, and they hurled these at the enemy. The flashes and flames could distinctly be seen. The Northerners had not many troops on their barges, so eventually the Chin fleet broke through, and safely reached Tung-kuan.[b]

Thus the cast-iron explosive bombs were here used in a naval battle between the Chin and the Yuan.

The following year, +1232, saw the siege and capture of the Jurchen Chin capital, Khaifêng, by the Mongols. The Chin Tartars had held it only a little over a century since they had taken it in their turn from the Chinese dynasty of the Northern Sung. Now the investing troops of Ögötäi were commanded by the ferocious general Subotai[3], while the defence was organised by the more technically-mined Chhihchan Ho-Hsi[4]. In the *Chin Shih* we read:[c]

Among the weapons of the defenders there was the heaven-shaking thunder-crash bomb (*chen thien lei*). It consisted of gunpowder put into an iron container (*thieh kuan*[5]); then when the fuse was lit (and the projectile shot off) there was a great explosion the noise whereof was like thunder, audible for more than a hundred *li*, and the vegetation was scorched and blasted by the heat over an area of more than half a *mou*. When hit, even iron armour was quite pierced through. Therefore the Mongol soldiers made cowhide sheets to cover their approach trenches (*niu phi tung*[6]) and men beneath the walls, and dug as it were niches (*khan*[7]) each large enough to contain a man, hoping that in this way the (Chin) troops above would not be able to do anything about it. But someone (up there) suggested the technique of lowering the thunder-crash bombs on iron chains. When these reached the trenches where the Mongols were making their dug-outs, the bombs were set off, with the result that the cowhide and the attacking soldiers were all blown to bits, not even a trace being left behind.

Moreover, the defenders had at their disposal flying-fire spears (*fei huo chhiang*[8]). These were filled with gunpowder, and when ignited, the flames shot forwards for a distance of more than ten paces, so that no one durst come near.

These thunder-crash bombs and flying-fire spears were the only two weapons that the Mongol soldiers were really afraid of.

[a] Mod. Yung-chi.
[b] *Chin Shih*, ch. 111, p. 8*a*, *b*, tr. auct., adjuv. Wang Ling (1), p. 170. Cf. *Chin Shih*, ch. 7, p. 10*b*, *Yuan Shih*, ch. 115, p. 1*b*. Lu Mou-Tê (1), p. 32, knew of this battle, but interpreted the thunder-crash bombs as cannon.
[c] Ch. 113, p. 19*a*, tr. auct.

[1] 完顏訛可　　[2] 河中　　[3] 速不台　　[4] 赤盞合喜　　[5] 鐵礶
[6] 牛皮洞　　[7] 龕　　[8] 飛火槍

Thus in this graphic passage we can see the Chin Tartars using both explosive cast-iron bombs and incendiary gunpowder flame-throwers or fire-lances, this last an important development to which we shall shortly return (p. 220). It did not save them from the fall of the city and the virtual collapse of their dynasty. Destruction was due anyway for many other reasons, but we need not visualise the bombs as being so effective and reliable as modern weapons of the same kind; probably they often failed to go off, or even exploded prematurely. All the same, it was a famous defence, and worthy of note in any world military history.[a]

This passage has been the property of Western historians for nearly two and a half centuries. It would hardly be expected that eighteenth-century writers would have been very clear about the nature of the weapons used,[b] but in 1849 St Julien (8) gave a full translation of the passage.[c] He appreciated in principle the explosive character of the bombs,[d] but supposed the fire-lances to have been rockets.[e] Commenting on this, Reinaud & Favé (in whose long paper it was first published),[f] concluded that the bombs were essentially incendiary, though they did not rule out altogether a true explosive petard, the iron casing of which would shatter. As for the fire-lance, they accepted St Julien's interpretation of it as a rocket. Later on, Schlegel (12) understood the passage up to a point, but for a reason which will appear in a moment (p. 179), thought wrongly that the weapons were cannons, a mistake in which he was joined afterwards by Lu Mou-Tê,[g] though fully rectified by Pelliot (49, 59). Although Reinaud & Favé had been quite right in denying that the Khaifêng weapons were cannons,[h] their assertion that the propellant property of gunpowder was then quite unknown, is today more dubious, for we have already found *huo thung*[1] (metal-barrel guns or proto-guns) among the stores of Chhichow in +1221.[i] Perhaps the first person to state almost correctly the nature of both the Khaifêng weapons was Mayers in 1870, who gave quite a good translation.[j] Such are the vicissitudes of the history of technology.

Another eye-witness account comes from the pen of Liu Chhi[2], a scholar of the

[a] Background and general description in Cordier (1), vol. 2, pp. 231 ff., 236 ff.
[b] E.g. Gaubil (12), pp. 68 ff. in 1739; de Mailla (1), vol. 9, pp. 160 ff. in 1777.
[c] Using the closely similar version in the *Thung Chien Kang Mu*, pt. 3 (*Hsü*), ch. 19, p. 50b.
[d] But he thought that they raised themselves automatically into the air.
[e] Later on Romocki (1), vol. 1, pp. 47–9, erred in a similar way, accepting the rockets though regarding the bombs, more firmly than Reinaud & Favé, as truly explosive missiles.
[f] (2), pp. 288 ff.
[g] (1), p. 32.
[h] (2), p. 292.
[i] We shall explain more clearly the development of guns and proto-guns from fire-lances and co-viative projectiles on pp. 248 ff. below.
[j] It made more of fire than explosion, but his commentary shows that he understood well what was happening, (6), p. 91. Goodrich & Fêng (1), p. 117, were also fairly sound on the matter, but provided no details.

[1] 火筒   [2] 劉祁

Jurchen Chin realm. In his book of reminiscences, the *Kuei Chhien Chih*[1], he afterwards wrote:[a]

The army of the Northerners (the Mongols) then attacked the city (of Khaifêng) with their trebuchets.... The assault became more and more fierce, so that the trebuchet stones flew through the air like rain. People said that they were like half-millstones or half-sledgehammers. The Chin defenders could not face them. But in the city there were the kind of fire-missiles called 'heaven-shaking thunder-crash bombs', and these were at last used in reply, so that the Northern troops suffered many casualties, and when not wounded by the explosions were burnt to death by the fires that they caused....

All the people in the city were conscripted into a Home Guard called the Fang Chhêng Ting Chuang[2]. An order was issued to the effect that any man who remained at home would be summarily executed. Even the scholars and students in the Academy were drafted as soldiers. The students petitioned to have a University Guard formed, to be called Thai Hsüeh Ting Chuang[3]. But a discussion at court decided that the bookish gentlemen were too weak for the hard work involved in being bomb-throwing artillerists (*phao fu*[4]). So they appealed to the emperor himself, but his decision was that they should all be given desk jobs in the Ministry of the Interior (Hu Pu[5]), and thus in the end they were spared the painful labour of the artillerists....

This suggests that the Chin Tartar State could command a certain patriotism before it was overwhelmed by the Mongolian power.

A few years later, when the Chin State was at its last gasp a commander named Kuo Pin[6] found himself in +1236 defending a city called Hui-chou[7]. He commandeered all the metals that could be found, including gold and silver, copper and bronze, as well as iron, for making the explosive bomb-shells;[b] but it was all to no avail, and eventually the last pockets of resistance surrendered to the all-conquering Mongols.

Next in line, of course was the Southern Sung, and next we have to look at the warfare between them and the Mongols. In +1257, before the campaigns began, a meritorious official, Li Tsêng-Po[8], was gravely disturbed at the lack of preparedness in the arsenals of Ching[9] and Huai[10], near the border with the Mongols in the north. In his *Kho Chai Tsa Kao, Hsü Kao Hou*[11] he recorded his complaints,[c] and they concern us because of the fire-weapons he enumerated. He began by saying that the armour was rusty and the munitions decayed, and that repeated requests to the court brought no results. 'For every ten items we ask for, the Arsenals Administration[d] sends only one or two.'

---

[a] Ch. 11, p. 3a, b, tr. auct.
[b] *Chin Shih*, ch. 124, p. 15b, *Yuan Shih*, ch. 121, p. 10b, discussed in Fêng Chia-Shêng (1), p. 81.
[c] Ch. 5, p. 52a, tr. auct. The full text is given by Fêng Chia-Shêng (1), p. 66; cf. (6), pp. 21-2.
[d] The National Arsenals Administration had been set up in +1073 as the result of an important memorial by Wang Fang[12] two years earlier. He was the son of the famous politician Wang An-Shih[13], and himself a meritorious scholar and thinker. See Williamson (1), vol. 1, pp. 258-9, 276, vol. 2, pp. 251 ff.

[1] 歸潛志 [2] 防城丁壯 [3] 太學丁壯 [4] 砲夫 [5] 戶部
[6] 郭斌 [7] 會州 [8] 李曾伯 [9] 荆 [10] 淮
[11] 可齋雜槀,續稿後 [12] 王雱 [13] 王安石

As for the weapons for attack by fire [he went on], there are (or should be) several hundred thousand iron bomb-shells available. When I was in Chingchow they were making one or two thousand of them each month, and they used to despatch to Hsiang (-yang) and Ying(-chou)[a] ten or twenty thousand a time. Yet now at Chingchiang we have no more than 85 iron bomb-shells, large and small, 95 fire-arrows,[b] and 105 fire-lances (*huo chhiang*[1]). This is not sufficient for a mere hundred men, let alone a thousand, to use against an attack by the (Mongol) barbarians. The government supposedly wants to make preparations for the defence of its fortified cities, and to furnish them with military supplies against the enemy (yet this is all they give us). What chilling indifference!

This was not a very inspiring overture to a war against foreign enemies, but for us it does give an insight into the industry making explosive bombs.[c]

All the same, when it came to the crunch, the Sung armies seem to have been not too badly provided. Between +1267 and +1273 came the epic siege of Hsiang-yang on the Han River by the Mongols, about which we have already had something to say in connection with paddle-boat warships.[d] For it was by the use of a hundred of these that two gallant Sung officers, Chang Shun[2] and Chang Kuei[3], organised a relief convoy which successfully re-provisioned the city, though both were themselves killed, one on the way in and the other on the way out. We saved the gunpowder material until now, however, and here is some of it. The first passage refers to Liu Hsien-Ying[4], who was fighting on the side of the Mongols, attacking Fan-chhêng, the twin city across the river.

Outside Fan-chhêng [the stele inscription says][e] there were defences called Tung-thu-chhêng[5] (eastern earthen walls), and the commanding general ordered that these should be stormed. Having raised a scaling ladder, and being the first to climb on it, Mr Liu received a serious wound from a bomb-shell in the left thigh, but in spite of this he went on fighting fiercely, and the defences were taken....

After Hsiang-yang had been under siege for some time, there was a shortage of food inside. So 'Shorty' Chang, the Chief of Staff (Ai Chang Tu-Thung[6]), secretly organised a relief convoy of ships to bring in provisions.... But Mr Liu caught a spy and obtained intelligence as to the day when (the convoy) would come, so an attack was planned for the moment when the ships would be just half-way. Then bomb-shells were thrown with great noise and loud reports, and our (Mongol) army attacked (the ships) fiercely for a space of more then 30 *li*. On the ships they were up to the ankles in blood, and 'Shorty' Chang, the Chief of Staff, was captured alive.

---

[a] Mod. Chung-hsiang in Hupei.
[b] Possibly rockets by this date, cf. pp. 472 ff.
[c] Goodrich & Fêng (1), p. 118, were so impressed with this passage, which they gave in partial translation, that they said it was 'surprising, indeed almost incredible, information'.
[d] Vol. 4, pt. 2, pp. 423–4.
[e] From Liu Hsien-Ying pei[7], recorded in *Shan Tso Chin Shih Chih*[8], ch. 21, p. 29*b*, tr. auct. Mr Liu's biography is in *Yuan Shih*, ch. 162, p. 18*b*.

[1] 火槍　　　[2] 張順　　　[3] 張貴　　　[4] 劉先瑩　　　[5] 東土城
[6] 矮張都統　　[7] 碑　　　　[8] 山左金石志

This was certainly Chang Kuei, who commanded the vanguard of ships. But what is important for the historian is that both sides were now using iron bomb-shells. Mr Liu might have been burnt by a fireball, but would hardly have been seriously wounded, as by an iron bomb-fragment; while incendiary gunpowder might have set the Sung ships on fire, but would not have done so much damage to their men. Let us continue in the words of the Sung dynastic history. Early in the siege there were swimmers who left the cities to get salt and firewood, but many of these were captured, after which the siege was tightened, and a price put on the heads of dead Sung soldiers—then came the relief convoy of the two Changs in +1272. The *Sung Shih* says:[a]

Since the Han River was the only way of deliverance (for the garrison), one hundred (paddle-boat) ships were assembled at a point below Thuan-shan, and after a couple of days they entered Kao-thou-kang harbour. Then (after loading) they took up a rec-tangular formation, every ship being equipped with fire-lances (*huo chhiang*[1]), trebuchets and bombs (*huo phao*[2]), burning charcoal (*chhih than*[3]), large axes and heavy crossbows. When the night had worn on three quarter-hours by the water-clock, the fleet hoisted anchor and sailed out into the river using red lamps as signals. (Chang) Kuei led the van, and (Chang) Shun commanded the rearguard; so with a following wind they breasted the waves, making straight for the enemy ahead. When they got above Mo-hung-than, there were the ships of the northerners (the Mongols) stationed right across the river, with no gap where they could get through. So taking advantage of their arma-ment, they cut right through the iron cables (*thieh hsüan*[4]) and tore out several hundred stakes (*tsuan i*[5]), and sailing on they fought an energetic rearguard action for 120 *li* until dawn, when they reached the waters beside Hsiang-yang.

Here the historian will be interested in the bombs and the fire-lances, but one can see how Chang Kuei and his ship came to be captured.[b] This was, of course, the siege in which the Mongols employed the counterweighted, or 'Muslim' trebuchets for the first time (cf. pt. 6 (*f*) 5 above).

At the risk of a surfeit of battle accounts, just one more must be given before we return to technological description. In +1277 the Mongols mounted a great campaign against the remaining Sung resistance in Kuangsi; their army was led by a Uighur Muslim artillery general in the Mongol service named A-Li-Hai-Ya[6], while Ma Chi[7], the Sung general, attempted to oppose him. But he was outflanked and had to fall back upon the provincial capital Kweilin. After a siege of more than three months, the main Sung garrison gave in, but Ma and a

---

[a] Ch. 450, p. 3a, tr. auct. Other sources include *Chhi Tung Yeh Yü*, ch. 18, p. 12a, b; *Sung Chi San Chhao Chêng Yao*, ch. 4, pp. 5b, 6a; *Chao Chung Lu*, pp. 12a–13b; *Kuei Hsin Tsa Chih, Pieh Chi*, ch. 2, p. 39a, b.

[b] Lu Mou-Tê (1), p. 31 and Wang Ling (1), p. 170, both knew of this famous exploit, but both supposed that the weapons on the Sung ships were guns. Goodrich & Fêng (1), p. 118, got this right, but at first took the counterweighted trebuchets to be some kind of cannon. The texts are in Fêng Chia-Shêng (1), pp. 67, 70; cf. (6), p. 23.

[1] 火槍   [2] 火砲   [3] 熾炭   [4] 鐵絙   [5] 攢杙

[6] 阿里海牙   [7] 馬𡎴

event occurred so short a time before, in +1274, the drawing has a considerable authenticity.[a]

Neither caption nor narrative has anything to say about the *tetsuhō*[1],[b] but there are texts which do, notably the *Hachiman Gudōkun*[2] (Tales of the God of War told to the Simple), an anonymous work of the +14th century, which nevertheless agrees closely with the account of the battle in the scroll. Here we can read the following account of the incident:[c]

The commanding general kept his position on high ground, and directed the various detachments as need be with signals from hand-drums. But whenever the (Mongol) soldiers took to flight, they sent iron bomb-shells (*tetsuhō*) flying against us, which made our side dizzy and confused. Our soldiers were frightened out of their wits by the thundering explosions; their eyes were blinded, their ears deafened, so that they could hardly distinguish east from west. According to our manner of fighting, we must first call out by name someone from the enemy ranks, and then attack in single combat. But they (the Mongols) took no notice at all of such conventions; they rushed forward all together in a mass, grappling with any individuals they could catch and killing them.[d]

This was the expedition led by the Mongol general Hu-Tun[3] which landed at Hakata in Kyushu.[e] There is independent evidence from the Chinese side that iron bomb-shells were used in these engagements.[f] The Mongols did so again in the second invasion, in +1281, at Sekiura, under the command of the Chinese admiral Fan Wên-Hu[4], who had apparently asked Khubilai Khan for the services of Uighur or Muslim counterweighted trebuchet engineers, and been refused, the emperor seeing no use for them in naval warfare.[g] We need do no more than refer to the often-quoted parallel between these Mongol expeditions and the Spanish armada three hundred years later, both broken up by storm and gale not without energetic resistance by the island nations in question. At all events, the picture in Takezaki's scroll is a precious heritage for historians of technology.

Lastly, there are the words of a scholar who actually saw iron bomb-shells dating a couple of centuries back, dumped on the walls of a great city. This was

[a] Attention was drawn to it long ago by Arisaka Shozo (*1*), and it has since been discussed at length by Arima Seiho (*1*), pp. 86 ff; Nambo (*1*), p. 410. The picture was first reproduced with Western-language commentary by Goodrich & Fêng (*1*), opp. p. 118 and on p. 120 (but they surmised that the projectiles were solid iron cannon-balls) and by Wang Ling (*1*), p. 175·(but he took them to be explosive shells fired from cannon). There is a paper on the whole *makimono* by Fischer-Wierzuszowski (*1*).
[b] There is a lot of orthographic variation in the different Japanese texts and editions, but this is the most interesting form. In Chinese, pronounced *pao*, it meant a plane or a curry-comb, if *phao*, a brush.
[c] Gunsho Ruiji collection, ch. 13, pp. 328 ff. (ch. 1, p. 467). We offer grateful thanks to Dr Nakaoka Tetsuro for his interest and help with the translation.
[d] Cf. parallel accounts in *Taiheki*[5], ch. 39 (p. 881), a monastic chronicle of about +1370.
[e] For a general account of the two expeditions see the book of Yamada Nakaba (*1*). Khubilai thought of mounting a third one, in +1283, but it never came off.
[f] Documents assembled by Kho Shao-Min (*1*) in his *Hsin Yuan Shih*, ch. 250, pp. 6*a*, 8*b*, 9*a*, 11*b*. Fan's biography is in ch. 177, p. 18*b*.
[g] We have already seen examples of their use on shipboard, pp. 81, 89 above.

¹ 鐵鉋   ² 八幡愚童訓   ³ 忽敦   ⁴ 范文虎   ⁵ 太平記

in +1522. Ho Mêng-Chhun[1] afterwards wrote as follows:[a]

In the spring I was sent to Shensi, and there at Sian[b] on the city wall I saw some old cast-iron bomb-shells, of the kind that were known in former times as 'heaven-shaking thunder-crash' bombs. In shape they were like two bowls that could be joined together (*ho wan*[2]) to make a ball, and at the top there was a small hole the size of a finger. These things are not used by the army now, but I am sure that it was one of the weapons used by the Jurchen Chin people when defending Khaifêng (against the Mongols).

This was a satisfying observation, though one wishes that a few had been saved from the scrap-iron merchants for the benefit of military museums today. Curiously, it was this passage which gave rise to a classic misunderstanding. Ho Mêng-Chhun's words were supposedly quoted in Thang Shun-Chih's[3] *Pai Pien*[4] encyclopaedia of +1581, whence they found their way into the *Ko Chih Ching Yuan* of +1735,[c] and so to Schlegel (12) in 1902. Somewhere along the line the double bowls were corrupted to 'double rollers' (*ho tho*[5]) and this led to Schlegel's 'closed rollers', which dominated the literature for some time, since he was determined to prove that the heaven-shaking thunder-crash weapons were in fact cannon. It took Pelliot (49, 59), as so often, to put the matter right. The deceptiveness of the situation lay in the fact that just during the heyday of cast-iron bomb-shells, metal-barrel guns and cannon were in fact arising, as we see elsewhere (pp. 23, 170 above, pp. 304 below). But it takes many years to unravel these tangled skeins of history which the centuries have confused.

As for the inevitable comparison with Europe, we have now seen that it is possible to trace back the use of cast-iron bomb-shells and grenades in China to +1221, and they must have been coming into regular employment about the beginning of that century.[d] But the first date for hollow iron bomb-shells in Europe appears to be +1467, when they were used by the Burgundians in their wars.[e] Thus like all the other gunpowder weapons there is a lag of a couple of centuries at least between their first appearance in China and the earliest dates for them in the West. Judging from Romocki's account of the *Bellifortis* of Konrad Kyeser, written about +1410, the Europeans repeated the Chinese experience in having casings of different strengths for their bombs.[f]

It now remains only to take a look at the descriptions of bombs in the +14th century and later, seeing how the weak-casing devices (*phi li phao*) and the strong-casing ones (*chen thien lei*) had developed by those times. The best source is the *Huo Lung Ching* (Fire-Drake Manual) which refers to the techniques in use

---

[a] *Yü Tung Hsü Lu Tsê Chhao Wai Phien*, ch. 5, p. 11 b, tr. auct.

[b] Chhang-an, of course, in J/Chin times, as in Han.

[c] Ch. 42, p. 27 a, b.

[d] P. 169 above.

[e] Johannsen (3), p. 1464, (4), p. 273. Partington (5), p. 127 gives an earlier Byzantine reference for +1439. His other references are to cast-iron shells, which is rather a different matter.

[f] (1), vol. 1, p. 169, figs. 25–31.

[1] 何孟春   [2] 合椀(碗)   [3] 唐順之   [4] 稗編   [5] 合砣

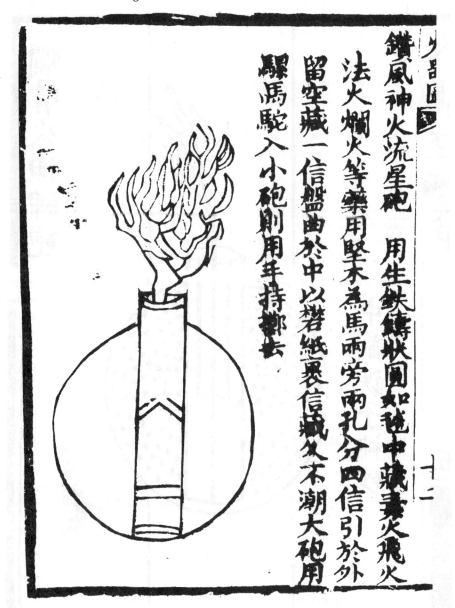

鑽風神火流星砲　用生鐵鑄狀圓如毬中藏毒火飛火
法火爤火等藥用堅末為馬兩旁兩孔分四信引於外
留空藏一信盤曲於中以捲紙裹信藏又不潮大砲用
騾馬駝入小砲則用羊持鄰去

Fig. 22. The 'magic-fire meteoric bomb that goes against the wind' (*tsuan fêng shen-huo liu-hsing phao*), from *HLC*, pt. 1, ch. 2, p. 7*a*, *b* and *HCT*, p. 12*b*.

Here the purpose of the core is not obvious, unless it was to act as a spout for the flames of the low-nitrate gunpowder pending the explosion of the high-nitrate mixture.[a] But the same principle is involved, that of mixing poisonous materials together with the saltpetre, sulphur and charcoal. Lastly, we may mention the 'dropping-from-heaven bomb' (*thien chui phao*[1]),[b] seen in Fig. 23. It is described as about the size of a bushel measure, and intended to be hurled up very high into the air, presumably by a trebuchet or an *arcuballista*, whence it should land on the enemy camp, preferably during a dark night. The enemy soldiers then fall to killing each other in their alarm. The sound of the explosion is like thunder, so a metal casing is to be supposed, and the bomb contains dozens of incendiary packets (*huo khuai*[2]) which are scattered in all directions.

Turning now to the weak-casing bombs deriving from the *phi li phao*, there is first the 'bee-swarm bomb' (*chhün fêng phao*[3]).[c] The description (see Fig. 24) says:

Bamboo strips are woven into the shape of a ball and pasted round with forty or fifty layers of thick paper, then dried in the sun. Afterwards it is wrapped up further in fifteen layers of oiled paper. Make an opening in it and fill it with 2 lb. of gunpowder, and half a pound of iron calthrops, putting in also several dozen flying-swallow poison-fire gunpowder (*fei yen tu huo*[4]) fire-crackers made of paper.[d] This bomb has a very strong power, for not only can it hit the enemy personnel (with the objects), but also when the flying-swallow fire comes forth it can stick to their persons and still burn. It can also set fire to the sails of enemy ships and burn fiercely; but it can be extinguished with water.[e]

This then was a projectile of fairly primitive type, primarily incendiary rather than bursting a metal case into fragments.[f] Our second example comes from a different source, the *Thien Kung Khai Wu* of +1637 (Fig. 25). It is the well-known

---

[a] The quantities are not always given in the composition formulae, only the constituents.

[b] *HLC*, pt. 1, ch. 2, p. 12*a*, *b*, Hsiangyang ed., p. 15*a*.

[c] *HLC*, pt. 1, ch. 2, p. 9*a*, *b*, Hsiangyang ed., p. 13*b*, tr. auct.

[d] The composition of this does not appear to be given in *HLC*, but it was probably similar to the others.

[e] A very similar type of bomb, the 'great bee-hive' (*ta fêng wo*[5]) is described in Chou Chhing-Yuan's[6] *Hsi Hu Erh Chi*[7] (Second Collection of Documents about West Lake at Hangchow and its Neighbourhood), ch. 17 (p. 335). This occurs in a piece of much interest, one of the few which describe gunpowder weapons (with illustrations) apart from the military compendia. It is entitled *Liu Po-Wên Chien Hsien Phing Chê Chung*[8], 'On the Pacification of Central Chekiang Province by the Able Officers recommended by Liu Po-Wên', and it refers to that remarkable military commander and scholar of scientific and technological interests, Liu Chi[9], whom we have already several times encountered (cf. pp. 25). He was of great assistance to Chu Yuan-Chang in conquering the empire for the Ming. Liu's campaigns in Chekiang were conducted between +1340 and +1350 against both the inland rebels under Hsü Shou-Hui[10] and the coastal pirates under Fang Kuo-Chen[11], who continued to rule the province, siding sometimes with the Yuan and sometimes with the Ming, until +1367. Liu was therefore acting at the time as a Yuan official. But the designs for bombs, fire-lances and rockets must certainly go back to +1340, so we should be grateful to Chou for preserving the document about +1620. For our knowledge of this interesting survival we are indebted to Dr Lin Yü-Thang in 1954.

[f] The projectile was evidently meant to be hurled like a grenade by means of a loop of rope used as a handle.

[1] 天墜砲	[2] 火塊	[3] 羣蜂砲	[4] 飛燕毒火	[5] 大蜂窠
[6] 周清源	[7] 西湖二集	[8] 劉伯溫薦賢平浙中		[9] 劉基
[10] 徐壽輝	[11] 方國珍			

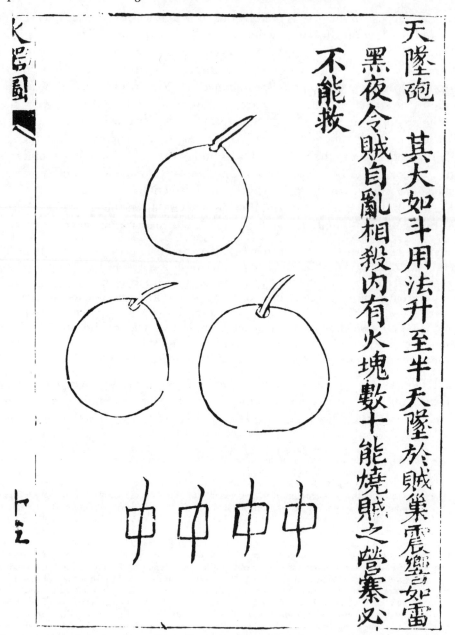

天隆砲 其大如斗用法升至半天隆於賊巢震響如雷

黑夜令賊自亂相殺內有火塊數十能燒賊之營寨必

不能救

Fig. 23. The 'dropping-from-heaven bomb' (*thien chui phao*), from *HLC*, pt. 1, ch. 2, p. 12*a* and *HCT*, p. 15*a*.

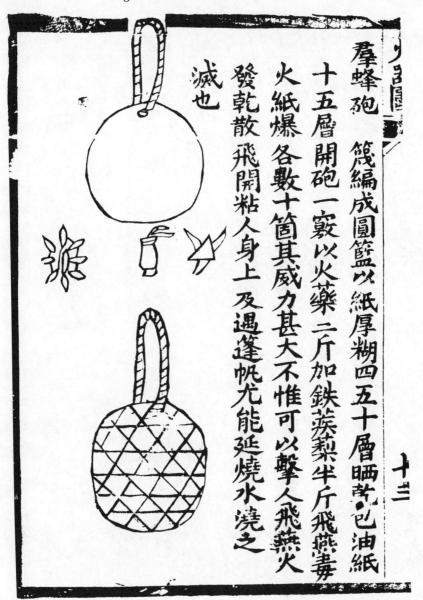

群蜂砲

篾編成圓籃以紙厚糊四五十層曬乾已油紙十五層開砲一窠以火藥二斤加鉄蒺藜半斤飛燕毒火紙爆各數十箇其威力甚大不惟可以擊人飛燕火發乾散飛開粘人身上及遇篷帆尤能延燒水澆之滅也

十三

Fig. 24. The 'bee-swarm bomb' (*chhün fêng phao*), from *HLC*, pt. 1, ch. 2, p. 9*a*, *b* and *HCT*, p. 13*b*. One of the class of weak-casing bombs.

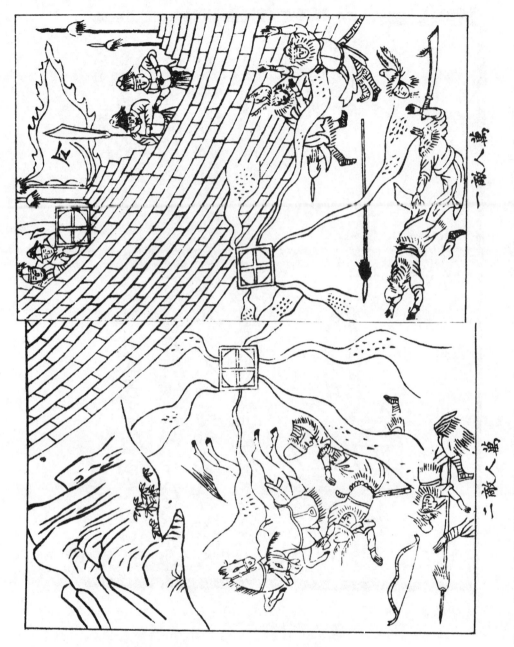

一敵人萬

二敵人萬

Fig. 25. Another similar bomb, the 'match for ten thousand enemies' (*wan jen ti*), described and illustrated by Sung Ying-Hsing in *Thien Kung Khai Wu* (+1637), ch. 15 (Excellent Weapons (Chia Ping) sect.) in ch. 3, pp. 36*b*, 37*a*. The picture is from another Ming edition used in the Kuo Hsüeh Chi Pên Tshung Shu series, pp. 265–6. Cf. Sun & Sun tr., p. 276.

passage on the bomb called 'match for ten thousand enemies' (*wan jen ti*[1]).[a] Sung Ying-Hsing says:

(When attacks are made upon) the walls of small cities in remote prefectures; if the available guns (*phao*[2])[b] are too weak to repulse the enemy, then bombs (*huo phao*[3]) should be suspended (i.e. dropped) from the battlements; if the situation continues to worsen, then the 'match for ten thousand' bomb should be employed. This recently developed weapon can be used according to the circumstances, and unlike the previous one, it can be thrown in any direction. The saltpetre and sulphur in the bomb, on being ignited (explode), and blow many men and horses to pieces in an instant.

The method is to use a dried empty clay ball with a small hole for filling, and in it are put the gunpowder, including sulphur and saltpetre, together with 'poison gunpowder' (*tu huo*[4]) and 'magic gunpowder' (*shen huo*[5]).[c] The relative proportion of the three gunpowders can be varied at will. After the fuse (*yin hsin*[6]) has been fitted, the bomb is enclosed in a wooden frame. Alternatively a wooden tub, coated on the inside with the sort of clay used for image-making, can be used. It is absolutely necessary to use the wooden framework or the tub in order to prevent any premature breakage as the missile falls (until the gunpowder explodes). When a city is under attack by an enemy the defenders on the walls light the fuse and throw the bomb down. The force of the explosion spins the bomb round in all directions, but the city walls protect one's own men from its effects on that side, while the enemy's men and horses are not so fortunate.[d] This is the best of weapons for the defence of cities. It is important that those who have charge of such affairs should realise that the understanding of gunpowder and the knowledge of the construction of fire-weapons come from human ingenuity, so that those concerned may have to take as much as ten years to master it all.

Sung Ying-Hsing was no army man, one feels, otherwise he would hardly have described such an archaic weapon with so much enthusiasm, but in the back-blocks it may well have been used still at the end of the Ming. A rather similar device described in the *Huo Lung Ching* two or three centuries earlier was the 'flying-sand magic bomb releasing ten thousand fires' (*wan huo fei sha shen phao*[7])[e] seen in Fig. 26. Here a tube of gunpowder was put into an earthenware pot containing quicklime, resin and alcoholic extracts of poisonous plants, all to be released by the explosion; this was thrown down from city-walls, recalling the lime bombs of Yü Yün-Wên's naval victory (p. 165 above).[f] Another of the

---

[a] Ch. 15, pp. 8*b*, 9*a*, 12*b*, 13*a*, Ming ed. ch. 15, pp. 34*b*, 35*a*, 36*b*, 37*a*, tr. auct. Cf. Sun & Sun (1), pp. 276–7; Li Chhiao-Phing (2), pp. 395–6.

[b] By this time the ancient word undoubtedly meant metal-barrel cannon, as is evident from the adjacent illustrations.

[c] The former composition is the same as that already given; the latter is in *HLC*, pt. 1, ch. 1, p. 6*a*, and included arsenicals, moxa, resin, croton and gingko leaf powder.

[d] There may have been some arrangement here similar to the saxons, tourbillons or Chinese flyers of the pyrotechnic world (cf. p. 141 above), but it is probably more likely that Sung Ying-Hsing had never seen one of these bombs go off, and was drawing on his imagination. Cf. Sun Fang-To (1).

[e] *HLC*, pt. 1, ch. 2, p. 6*a*, *b*; *PL*, ch. 12, p. 14*a*, *b*.

[f] Quicklime was still part of the standard equipment for fortress defence in 1840, when Jocelyn observed it in the course of his inspection of the Tinghai redoubts taken by the British on Choushan island; (1), p. 59.

[1] 萬人敵　　　[2] 砲　　　　　[3] 火砲　　　　[4] 毒火　　　　　[5] 神火
[6] 引信　　　　[7] 萬火飛砂神砲

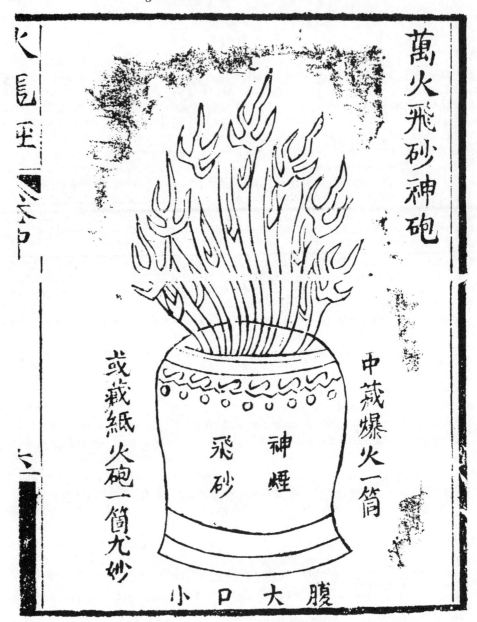

萬火飛砂神砲

火昆亞一汉口

或藏紙火砲一箇九妙

中藏爆火一箇

神煙飛砂

小口大腹

Fig. 26. The 'flying-sand magic bomb releasing ten thousand fires' (*wan huo fei sha shen phao*), from *HLC*, pt. 1, ch. 2, p. 6a. A weak-casing device reminiscent of Yü Yün-Wên's +12th-century lime bombs used in naval combat.

same kind was the 'wind-and-dust-bomb' (*fêng chhen phao*[1])[a] of Fig. 27. Many more could be discussed, but this should be sufficient to show that the parallel traditions of weak-casing *phi li phao* and strong-casing *chen thien lei* continued down to the latter part of the +17th century and the beginning of the Chhing.[b]

One gets an interesting sidelight on this from the book of Juan Mendoza, written in +1585 and translated into English by Robert Parke three years later. Speaking of the Chinese soldiers, he says:[c]

These footmen be marveillous full of policie, and ingenious in warlike or martiall affaires: and although they have some valor for to assalt and abide the enemie, yet doo they profite themselves of policies, devises and instruments of fire, and of fire workes. Thus do they use as wel by land in their wars as by sea, many bomes of fire, full of old iron, and arrowes made with powder and fire worke, with the which they doo much harm and destroy their enimies....

Here no doubt are references to fire-lances with co-viative projectiles, possibly also to rockets, as well as to the 'bombs' with casings weak or strong. In due course we shall discuss them all (pp. 220 and 472 below).

In fact their longevity was even greater than this. During the war of 1856–8, when the 'red-haired barbarians' (i.e. the British Navy) were attacking the Bogue Forts and the city of Canton,[d] Admiral Sir William Kennedy was a midshipman, and his subsequent account of the proceedings, written fifty years later, can be read in his rather jingoistic autobiography, full of period flavour. What he says is of considerable technological interest.[e]

The Chinese were fully prepared for us; the junks lay broadside on, with their guns run out on one side, springs on their cables to keep their broadside bearing, and 'stink-pots' at the mast-heads. These offensive weapons are worthy of description. The stink-pot is an earthenware vessel filled with (gun)powder, sulphur, etc. Each junk had cages at the mast-head, which in action were occupied by one or more men, whose duty it was to throw down these stink-pots on to the decks of the enemy, or into boats attempting to

---

[a] *HLC*, pt. 1, ch. 2, p. 11 *a, b*, Hsiangyang ed. p. 14 *b*. The vase shape of the container is worth noticing here, for it may have been this which gave to the earliest cannon-founders the idea of making guns in this form, thickening the walls round the explosion-chamber.

[b] There is an interesting illustration of bombs in the *Thai Tsu Shih Lu*[2], a work which we shall study more carefully later on (p. 398 below) because of the rich information it gives concerning the Chinese field artillery about +1620. The book recounts the exploits of Nurhachi (d. +1626), the Manchu prince afterwards regarded as the principal ancestor of the Chhing dynastic house. In Fig. 28 we see the siege of Ningyuan[3] in +1626, with Ming bombs bursting on the roofs of each of the Manchu assault ladders. Although they do not look as if they were doing much damage, this siege was in fact one of Nurhachi's few failures, and the city was held for the Ming by its gallant commanding general Yuan Chhung-Huan[4]. Note that by now the Manchus themselves are firing muskets from behind mobile ramparts.

[c] (1), ch. 6, p. 65 (+1588 ed.), p. 88 (Hakluyt Soc. ed.).

[d] This was the 'Arrow' war, so named from the pirate lorcha which precipitated it; the Anglo-French incursion which led to the treaty system. On the general background see Fairbank (4), pp. 243 ff.; Wakeman (2).

[e] Kennedy (1), p. 43. Later in his book he gives a translation of a proclamation by the Governor of Liang-Kuang which includes detailed instructions about the management and use of the stink-pots (pp. 65 ff., 67).

[1] 風塵砲   [2] 太祖實錄   [3] 寧遠   [4] 袁崇煥

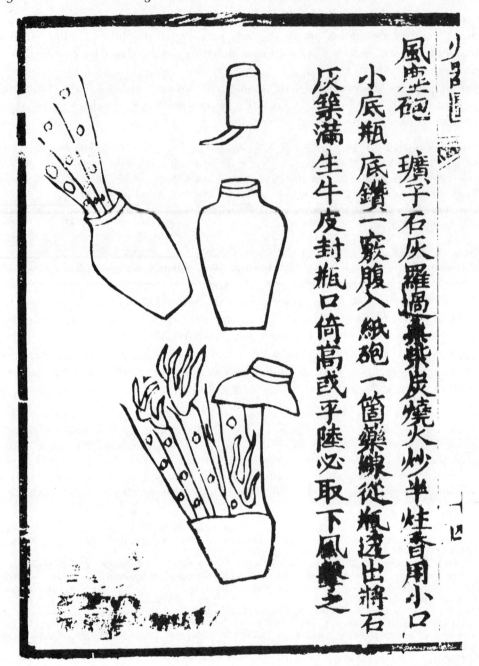

風塵砲　璣子石灰羅過共排炭燒火炒半炷香用小口

小底瓶底鑽一竅腹入紙砲一箇藥線從瓶透出將石

灰築滿生牛皮封瓶口倚高或平陸必取下風擊之

Fig. 27. The 'wind-and-dust bomb' (*fêng chhen phao*), from *HLC*, pt. 1, ch. 2, p. 11*a, b* and *HCT*, p. 14*b*.

Fig. 28. Bombs at the siege of Ningyuan by the Manchu prince Nurhachi in +1626. Although they do not seem to be very effective against the roofs of the Manchu assault ladders, they must have been by this time 'thunder-crash bombs' (*chen thien lei*), i.e. bombs with strong iron casings. The picture is taken from the *Thai Tsu Shih Lu*, a work which would not be expected to do much justice to Ming military technology, but in fact the siege had to be raised after a spirited defence by Yuan Chhung-Huan.

board; and woe betide any unlucky boat that received one of these missiles, for the crew would certainly have to jump overboard or be stifled.

From this description it is clear that the military technicians of the +11th century, or Yü Yün-Wen's men in the +12th, would have been quite at home with the poison-smoke weak-casing projectiles still used by Chinese forces in the middle of the nineteenth century. And from what we have seen on the composition of such bombs (pp. 123, 144, 167 above) we can fairly well imagine it too. Was not all this a chapter in the pre-history of tear-gas grenades? But we have no records that the medieval Chinese used it on civilian populations.

(11) LAND AND SEA MINES

We have now followed the fortunes of gunpowder with its continually rising nitrate content from its first uses as slow match and incendiary through the explosive weak-casing maroon to the strong-casing cast-iron bombs and grenades which gave true detonations. By +1277 the case of the 'enormous bomb' of Lou Chhien-Hsia (p. 176 above) makes it clear that something more of the nature of a land-mine was then available, and it is to these greater masses of explosive that we must now briefly turn our attention. It also seems probable that the thunder-crash bombs which were let down on chains to the Mongol sap trenches in the siege of +1232 were also larger than the customary iron bomb-shells lobbed over from trebuchets, in which case it would be more reasonable to speak of mines in that affray also. In any case, it looks as if the size of the infernal machines was growing steadily all through the +13th century. By the time that we get to the middle of the +14th we can find specific descriptions of mines in the *Huo Lung Ching*.

In early times the terms *phao* and *huo phao* evidently covered mines, though the name later adopted was 'ground thunder' (*ti lei*[1]). Several types of mine are described in the various versions of the 'Fire-Drake Manual' as well as in the *Wu Pei Chih*. For example, the 'invincible ground-thunder mine', *wu ti ti lei phao*[2] is clearly intended to be buried in places where the enemy is likely to pass. The *Huo Lung Ching* says:[a]

The mine, made of cast iron, is perfectly spherical in shape. It holds one peck or five pints of (black) powder, depending on its size. The 'magic gunpowder' (*shen huo*[3]), 'poison gunpowder' (*tu huo*[4]) and 'blinding and burning gunpowder' (*fa huo*[5]) compositions are all suitable for use (in this device).[b] Hard wood is used for making the wad (*fa ma*[6]), which carries three different fuses in case of defective connection, and they join at the 'touch hole' (*huo chhiao*[7]). The mines are buried in places where the enemy is

---

[a] *HLC*, pt. 1, ch. 3, p. 29a, b; tr. auct. Also in *Huo Kung Pei Yao*, ch. 3, p. 29a, b; *Huo Chhi Thu* (Hsiangyang-fu edition), p. 39b, and *Wu Pei Chih*, ch. 234, pp. 9b, 10a.

[b] Cf. p. 180 above.

[1] 地雷   [2] 無敵地雷砲   [3] 神火   [4] 毒火   [5] 法火
[6] 法馬   [7] 火竅

expected to come. When the enemy is induced to enter (the minefield) the mines are exploded at a given signal, emitting flames (and fragments) and a tremendous noise.

What exactly the triggering mechanism was we are not told, so one has to suppose that a long fuse was ignited by hand from an ambush or some sort of concealment just at the right time. The speed of transmission along the fuse would have had to be nicely calculated.

Another form of land-mine, but one using a firing device touched off by the enemy, is represented by the 'ground-thunder explosive camp' (*ti lei cha ying*[1]), one of the 'self-trespassing' (*tzu fan*[2]) type. It presumably derived its name from the fact that it was laid in the ground in large numbers in strategic positions, like the tents of an army encampment. The *Huo Lung Ching* says:[a]

These mines are mostly installed at frontier gates and passes. Pieces of bamboo are sawn into sections nine feet in length, all septa in the bamboo being removed, save only the last; and it is then bandaged round with fresh cow-hide tape. Boiling oil is next poured into (the tube) and left there for some time before being removed. The fuse starts from the bottom (of the tube), and (black powder) is compressed into it to form an explosive mine (*cha phao*[3]). The gunpowder fills up eight-tenths of the tube, while lead or iron pellets take up the rest of the space; then the open end is sealed with wax. A trench five feet in depth is dug (for the mines to be concealed). The fuse is connected to a firing device which ignites them when disturbed.

Although the text does not say so, the eight bamboo 'guns' are held together by two discoidal boards pierced by holes of just the right size, as can clearly be seen in the illustration (Fig. 29). From the specification they must have been buried at a slanting angle, probably pointing up the path, and as the picture says, the whole contraption is to be concealed by earth and grass.[b] As for the boiling oil, its purpose was presumably to harden the interior of the bamboo for its once-only function.[c]

The 'self-tripped trespass mine' (*tzu fan phao*[4]) operated in the same way (Fig. 30). Again the *Huo Lung Ching* says:[d]

It is made of iron or rock, or even porcelain or earthenware, with a cavity inside, very like the explosive mine mentioned above. Outside, the fuse runs through a series of

---

[a] *HLC*, pt. 1, ch. 3, p. 25*a*, *b*, tr. auct. Also in the Hsiangyang-fu edition, *Huo Chhi Thu*, p. 37*b*; *Huo Kung Pei Yao* ch. 3, p. 25*a*, *b*. See also *Wu Pei Chih* ch. 234, pp. 4*b*, 5*a*.

[b] This is reminiscent of the automatic crossbowmen in the tomb of Chhin Shih Huang Ti (pt. 6, (*e*) 2 above).

[c] Dr Clayton Bredt suggests, from recent experiments, that its purpose was rather to waterproof the bamboo and to kill boring insects. Freshly cut bamboo is immensely strong but very susceptible to insect attack, and the labyrinth of holes under 1 mm. in diameter is soon broken up by moisture, moulds and bacteria, especially if the tube is exposed to soil. Under such conditions it would soon be useless for firing anything. The bamboo borers are beetle larvae such as *Lyctus brunneus* and *Cyrtotrachelus longimanus* (R 55; Tu Ya-Chhüan *et al.* (*1*), p. 412·1).

[d] *HLC*, pt. 1, ch. 3, p. 26*a*; tr. auct. Also in the Hsiangyang-fu edition, *Huo Chhi Thu*, p. 38*a*, and *Huo Kung Pei Yao*, ch 3, p. 26*a*, *b*. See also *Wu Pei Chih* ch. 234, pp. 5*b*, 6*a*.

[1] 地雷炸營   [2] 自犯   [3] 炸砲   [4] 自犯砲

Fig. 29. The 'ground-thunder explosive camp land-mine' (*ti lei cha ying*). The eight bamboo 'guns' are held together by two discoidal boards pierced by holes of suitable size. From *HLC* (*HKPY*), ch. 3, p. 25*a*.

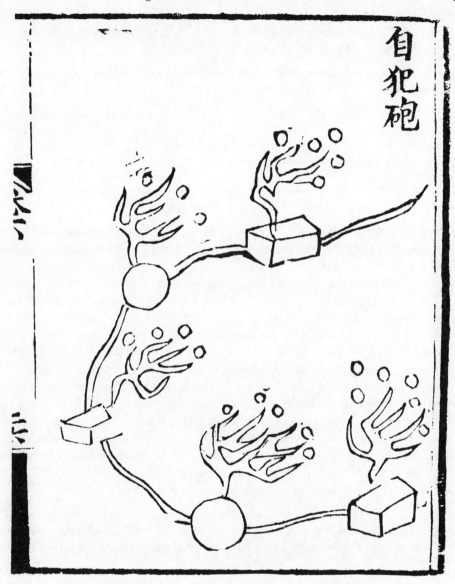

Fig. 30. The 'self-tripped trespass land-mine' (*tzu fan phao*). From *HLC* (*HKPY*), ch. 3, p. 26*a*.

'fire-ducts' (*huo tshao*[1]), which connect together several of these devices installed at strategic points. When the enemy ventures on to ground containing one of these mines, all the others are set to explode (quickly) one after the other.

Another rock-cut infernal machine was the 'stone-cut explosive land-mine' (*shih cha phao*[2]). Again the *Huo Lung Ching* says:[a]

This is a piece of rock carved into a spherical shape, and it can be of various sizes. Inside it is hollow, and contains explosive (black) gunpowder, which is packed in tight with a pestle to fill up nine-tenths of the space. A small section of bamboo is inserted for the fuse. The gunpowder is covered over with a piece of paper, above which is placed some dried earth, and a pound of clay above that in which the fuse is coiled round. For the defence of cities the land-mine is buried and hidden underground (at appropriate places), and this is what can be used for ground-thunder.

It says much for the labour-force available to the old Chinese military engineers, who were able to keep an army of stone-masons chipping away at such land-mines. But even where suitable lumps of rock were available, they would not be easily replaced once one had shot one's bolt, as it were.

Yet another apparatus, the 'Supreme Pole combination mine' (*Thai Chi tsung phao*[4]),[b] mounted a battery of little guns pointing in eight directions (*pa kua chhung*[5]),[c] which were set off by an automatic trigger mechanism. The case for them could be of wrought iron or hard wood, with ports for the muzzles, and it could be installed in unguarded camps or mounted passes, where a returning or advancing enemy would be likely to trip it (Fig. 31). This idea is old, probably of the early +14th century, because it occurs in the first stratum of the *Huo Lung Ching*,[d] but its specification persists in many later books.[e]

Nothing is given in the text to elucidate the firing device used to set off these mines.[f] But for the 'explosive mine' (*cha phao*[6]), the text of the Fire-Drake Manual mentions at least the type of ignition arrangement, though not describing it fully. The *Huo Lung Ching* explains the matter thus:[g]

The explosive mine is made of cast iron about the size of a rice-bowl, hollow inside with (black) powder rammed into it. A small bamboo tube is inserted and through this passes the fuse, while outside (the mine) a long fuse is led through fire-ducts. Pick a place where

---

[a]　*HLC*, pt. 1, ch. 3, p. 28*a*, *b*; tr. auct. See also the Hsiangyang-fu edition, *Huo Chhi Thu*, p. 39*a* and *Huo Kung Pei Yao*, ch. 3, p. 28*a*, *b*. Also in *Wu Pei Chih* ch. 234, pp. 7*b*, 8*a*. The illustration and description in ch. 122, p. 27*a* (*wei yuan shih phao*[3]) are also striking. A photograph of actual specimens is in Lo Chê-Wên (1).

[b]　For the significance of this name see Vol. 2, pp. 460 ff.

[c]　Again see Vol. 2, pp. 305, 312–13.

[d]　*HLC*, pt. 1, ch. 3, p. 30*a*, *b*; *HKPY*, *ibid.*; Hsiangyang ed. *HCT*, p. 40*a*.

[e]　*WPC*, ch. 134, pp. 22*b*, 23*a*; *PL*, ch. 12, p. 66*a*, *b*.

[f]　Liu Hsien-Chou (12) has devoted a special paper to the exploration of the firing and timing devices, which include, as we shall see, the joss-stick, the long-glowing composition, and the suddenly released flint-and-steel mechanism.

[g]　*HLC*, pt. 1, ch. 3, p. 27*a*, *b*; tr. auct. See also the Hsiangyang-fu edition, *Huo Chhi Thu*, p. 38*b* and *Huo Kung Pei Yao* ed. ch. 3, p. 27*a*, *b*. Also in *Wu Pei Chih*, ch. 234, pp. 6*b*, 7*a*.

¹ 火槽　　　² 石炸砲　　　³ 威遠石砲　　　⁴ 太極總砲　　　⁵ 八卦銃
⁶ 炸砲

Fig. 31. The 'Supreme Pole combination land-mine' (*Thai Chi tsung phao*), which had eight little guns pointing in all directions, set off by an automatic trip mechanism. From *PL*, ch. 12, p. 66*b*.

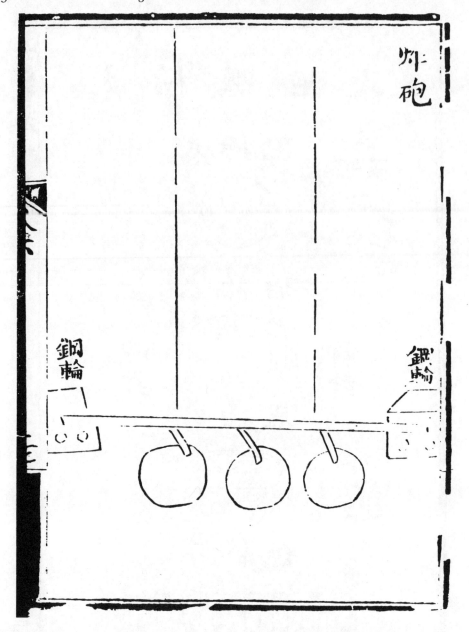

Fig. 32. An explosive land-mine (*cha phao*) set off by automatically operated steel wheels (*kang lun*) igniting tinder and thereby lighting the fuses by means of a spark from flint. *HLC* (*HKPY*), ch. 3, p. 27*a*.

the enemy will have to pass through, dig pits and bury several dozen such mines in the ground. All the mines are connected by fuses through the gunpowder fire-ducts, and all originate from a steel wheel (*kang lun*[1]). This must be well concealed from the enemy. On triggering the firing device the mines will explode, sending pieces of iron flying in all directions and shooting up flames towards the sky.

Thus from this it is clear that there was some arrangement of flint and steel, set in motion by the injudicious enemy, which directed sparks on to tinder and set light to the train of fuses. From the illustration (Fig. 32) it is clear that there could be at least two steel wheel systems either of which would activate the whole mechanism.

How exactly the arrangement worked was not revealed in a printed book until early in the +17th century. It consisted of a couple of the steel wheels, presumably serrated, fixed on a single axle and so placed as to rest on flints. A cord wound round a drum on the axle was attached to a weight at one end, and the mechanism kept in position by a pin. When the pin was removed by an unwary enemy stepping on a piece of board or plank attached to it, the pin released the weight, with the result that the wheels produced sparks by rubbing against the flints, thus lighting the fuses and setting off the mines. The earliest description of this steel-wheel firing device is in the *Ping Lu*,[a] and the same account is repeated in the *Wu Pei Chih*.[b] The illustration in the former (Fig. 33) shows the assembly rather diagrammatically, including the two wheels with their flints and weight-drive, the 'doorstep boards' (*huan pan*[2]) and the retaining pins (*chi chen*[3]) released by them. The picture in the latter is considerably more informative, and shows all the components both separately and assembled (Fig. 34).[c]

There is more than meets the eye in this set-up, especially when we remember that it goes back to the middle of the +14th century, certainly not later than +1360. Its two essential components, the flint-and-steel igniter and the weight-drive, both invite some thought, since they call to mind parallel devices in Europe, either of later date or not likely to have been known in China at the time. First, sparks struck off by steel on flint were a very ancient item in all civilisations, going back almost to the beginning of the iron age,[d] but their use in connection with gunpowder came in Europe much later than +1360.[e] The wheel-lock musket, which ignited its priming powder by a spark struck from a piece of iron pyrites and a steel wheel, does not go back further than the sketch by Leonardo da Vinci about +1500, and the first firm date for the actual thing is +1526. The flintlock musket, fired by the descent of a piece of flint and its impact on the steel pan-cover of the priming, was first mentioned only in

---

[a] *PL*, ch. 12, p. 61*b*–62*b*.
[b] *WPC*, ch. 134, p. 14*a*–15*b*.
[c] Several alternative mechanisms for releasing the wheels to act on the flints have been reconstructed by Liu Hsien-Chou (*12*). On Sea-mines and their ignition mechanisms, see Li Chhung-Chou (*3*).
[d] Cf. Vol. 4, pt. 1, p. 70.
[e] See Blackmore (*1*), pp. 19, 28; Reid (*1*), pp. 90, 96, 116–17; Partington (*5*), pp. 168 ff.

[1] 鋼輪　　　　[2] 桓板　　　　[3] 機針

+1547.[a] It has not so far been suggested that these were inspired by a previous Chinese practice, and perhaps the idea behind them was obvious enough, but it does remain true that one of those 'transmission clusters' in which techniques passed from China to Europe, did occupy the second half of the +14th century.[b] Secondly, the weight-drive is a rather curious device to find in China in the middle of the +14th century, for although it had successfully powered the first mechanical clocks of medieval Europe soon after +1300, it could hardly have passed East so quickly; and the indigenous hydro-mechanical linkwork-escapement clocks of China worked on a different principle, that of an inhibited vertically mounted driving-wheel using water or mercury from a constant-level tank.[c] On the other hand, there is evidence that the Hellenistic anaphoric clock was known and used by the medieval Chinese as well as the Arabs, so the weight-drive could have originated fairly easily from the anaphoric float, and it might have done so in both West and East independently.[d] One generally thinks of the weight-drive having come into China only with the mechanical clocks introduced by the Jesuits in the +17th century, but the present evidence suggests that it had been there for a long time already. Perhaps the secret, or 're-stricted', nature of its use helped to keep it dark. But the paradox remains that a flint-and-steel device was used in China for gunpowder ignition a century and a half or so before it found the same use in Europe; while on the other hand the weight-drive appeared in China for gunpowder ignition rather secretively half a century after it had begun to power European mechanical clocks, yet two and a half centuries before these came into China as the gifts of the Jesuit missionaries.[e]

As for explosive mines in Europe, there is not much evidence of them before the middle of the +15th century.[f] The first clearly recorded case of a plan for such an infernal machine seems to occur in +1403 in a war between Pisa and Florence, but whether it was actually practised is not quite clear.[g] The first design for anything similar to the firing device of the Fire-Drake Manual did not come, apparently, until +1573, when Samuel Zimmermann of Augsburg invented a contrivance for igniting fireworks or a mine at a distance by the use of

[a] Before this period, the priming powder or the charge itself was always ignited by a piece of slow-match, just as in the flame-thrower of +919 (cf. p. 81 above).

[b] Needham (64), pp. 61–2, 201. Among the great inventions coming at this time were the blast furnace for cast iron, block-printing, and segmental arch bridges.

[c] Vol. 4, pt. 2, pp. 446 ff., 469 ff.; cf. Needham, Wang & Price (1).

[d] Vol. 4, pt. 2, pp. 223, 466 ff., 532, 541. Liu Hsien-Chou (12) suggests that the well-windlass (cf. Vol. 4, pt. 2, p. 335) was the prototype. It must have been a very old observation that an empty bucket, let alone a full one, was liable to run away down the well if not checked at the well-head.

[e] All through the +16th century in China improvements were being made in land and sea mines, as for instance by Tsêng Hsien[1] about +1530. He was the military official who urged the recovery of the Ordos region from the Mongols, but was executed at the instance of opponents of his policy. See Wan Pai-Wu (1), p. 63; Liu Hsien-Chou (12).

[f] Romocki (1), vol. 1, p. 243.

[g] Partington (5), p. 172.

[1] 曾銑

flint-and-steel, springs and string.[a] To this clockwork was in time added, bringing about the time-bomb, once again by the use of the weight-drive.[b]

The weight-drive flint-and-steel mechanism was not the only one used in China for setting off infernal machines. Recipes existed for mixtures which would glow for long periods given an adequate supply of air, ready to ignite a fuse when brought mechanically into contact with it. For example, the *Wu Pei Chih* describes a device of this kind for use with a booby-trap (Fig. 35). In the 'underground sky-soaring thunder' (*fu ti chhung thien lei*[1]) the mines are placed three feet underground with the fuses leading to a point below a bowl containing a slow-burning incandescent material (*huo chung*[2]).[c] Lances or pikes with long handles are set up vertically above the bowl; then when the delighted enemy comes to appropriate the weapons the bowl is upset and the mine fuses ignited. The *Wu Pei Huo Lung Ching* contains a formula for making the slow-burning incandescent material, which it claims will burn continuously from 20 days to a month without going out. It is made of 1 lb. of white sandalwood powder (*hui mu*[3]), 3 oz. of iron rust (*huo lung i*[4] or *thieh i*[5], ferric oxide), 5 oz. of 'white' charcoal powder (*pai than mo*[6]), 2 oz. of willow charcoal powder, 6 oz. of the dried powdered flesh of 'red' dates (*hung tsao*[7]) and 3 oz. of bran. 'White' charcoal was simply charcoal whitened with quicklime. In a similar recipe given in the earlier *Ping Fa Pai Chan Ching* (c. +1590) it is simply given as 'charcoal powder'.[d] In principle the stuff was not unlike the glowing incense-powders used in temples, but without most of the fragrant constituents,[e] and the lime was doubtless added to keep the mixture dry.

Still other ignition and timing methods were used in sea-mines for naval warfare—one was the burning down of a joss-stick.[f] The *Huo Lung Ching* has an interesting specification for such a sea-mine.[g] It reads as follows:

The sea-mine called the 'submarine dragon-king' (*shui ti lung wang phao*[8]) is made of wrought iron, and carried on a (submerged) wooden board (*mu phai*[9]), [appropriately weighted with stones; see Fig. 37]. The (mine) is enclosed in an ox-bladder (*niu phao*[10]). Its subtlety lies in the fact that a thin incense(-stick) is arranged (to float) above the mine in a container. The (burning) of this joss-stick determines the time at which the fuse is ignited, but without air its glowing would of course go out, so the container is

[a] Zimmermann (1).

[b] Partington (5), p. 169.

[c] *WPC*, ch. 134, pp. 11*b*, 12*a*.

[d] P. 17*a*.

[e] Cf. Vol. 3, pp. 329 ff., Fig. 145, Vol. 5, pt. 2, pp. 134 ff.

[f] It was very natural that the Chinese sailors should have thought of this, for there can be no doubt that incense-sticks were used for the time-keeping of watches at sea from the early Middle Ages onwards at least. Cf. Vol. 4, pt. 3, p. 570.

[g] *HLC*, pt. 1, ch. 3, p. 24*a*, *b*; also in the Hsiang-yang edition, *Huo Chhi Thu*, p. 37*a*, and in the *Huo Kung Pei Yao* edition, ch. 3, p. 24*a*, *b*, tr. auct. The passage is somewhat corrupt, so all three versions of the text have to be used.

[1] 伏地衝天雷	[2] 火種	[3] 灰木	[4] 火龍衣	[5] 鐵衣
[6] 白炭木	[7] 紅棗	[8] 水底龍王砲	[9] 木牌	[10] 牛脬

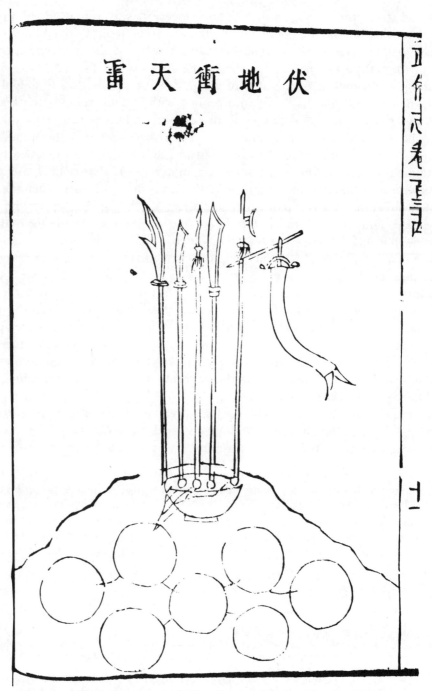

Fig. 35. A booby-trap called the 'underground sky-soaring thunder' (*fu ti chhung thien lei*). A stand of arms is set up with a land-mine underneath, and this is set off by the enemy whose step upsets a bowl of slow-burning incandescent material when he comes to take possession of the halberds, pikes and lances. From *WPC*, ch. 134, p. 11*b*.

connected with the mine by a (long) piece of goat's intestine (through which passes the fuse).

  [Comm. The saltpetre-saturated (fuse) can also come from a roughly made iron fish (as the floating container).][a]

 At the upper end the (joss-stick in the container) is kept floating by (an arrangement of) goose and wild-duck feathers, so that it moves up and down with the ripples of the water. On a dark (night) the mine is sent downstream (towards the enemy's ships), and when the joss-stick has burnt down to the fuse, there is a great explosion.

The illustration for this in all the editions (Fig. 36) is diagrammatic in the extreme, with no indicative lettering,[b] so it is necessary to look also at the picture in the *Thien Kung Khai Wu* of +1637, three centuries later, which does give a few identifications though still very badly (Fig. 37).[c] The only difference is that a lacquered leather bag replaces the ox-bladder,[d] while a cord pulled from the shore releases a flint-and-steel firing mechanism. The Chhing edition adds only artistic detail to the Ming picture, though it provides a graphic drawing of an underwater explosion. Also the name of the sea-mine has now changed, to the 'chaos-producing river-dragon' (*hun chiang lung*[1]).

 So far we have been elucidating Chinese practice of the +14th century. Apparently Europeans had not advanced so far at that time (if it is an advance to be able to blow up ships), for the first plan for sea-mines was presented to Queen Elizabeth by Ralph Rabbards in +1574.[e] When it came to the nineteenth century, the Chinese naturally improved their sea-mines by borrowing from Western practice.[f] In 1842, at the time of the Opium Wars, Phan Shih-Chhêng[2], the wealthy merchant-shipbuilder and technologist at Canton,[g] engaged an American naval officer J. D. Reynolds (Jen Lei-Ssu[3]) to conduct experiments with sea-mines as part of the modernisation of China's coastal defences, and Phan himself participated in the trials. Later he contributed a piece on them for the 1852 edition of the *Hai Kuo Thu Chih*[4] (Illustrated Memoir on the Occidental Maritime Nations) of Wei Yuan[5] & Lin Tsê-Hsü[6]; and this was the

---

 [a] Here again there is an echo of another Chinese medieval technique, namely the floating magnetic compass, where a shallow hollow iron fish took the place of the needle. Cf. Vol. 4, pt. 1, pp. 252–3. This one would have to have been rather deeper so as to take the joss-stick. On combustion clocks in general in Chinese culture see Vol. 3, pp. 329 ff.

 [b] The same applies to the description in *WPC*, ch. 133, p. 4b.

 [c] Ming ed. ch. 15, pp. 34a, 38a; Chhing ed. ch. 15, pp. 8a, 14b, 15a. The translations of Sun & Sun (1), p. 272 and Li Chhiao-Phing (2), p. 394, are not to be recommended.

 [d] In the text but not in the illustration; which shows a derivation from *HLC*.

 [e] Partington (5), p. 166.

 [f] The background to this will be found in the paper of Bauermeister (1), though it deals with designs only from +1787 onwards, and mentions no Chinese antecedents.

 [g] Phan was known to Beal as Poon Sse-Sing, and to Rondot as Pwann Sse-Ching. His common name among foreigners was Tinqua (derived from the office he held), and he was a descendant of the founder of the famous merchant Co-Hong in the city.

[1] 混江龍 [2] 潘仕成 [3] 任雷斯 [4] 海國圖志 [5] 魏源
[6] 林則徐

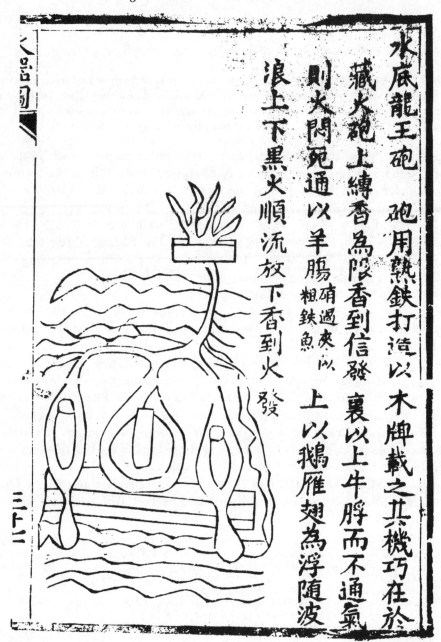

水底龍王砲

砲用熟鐵打造以木牌載之其機巧在於

藏火砲上縛香為限香到信發裏以上牛脬而不通氣

則火悶宛通以羊腸硝過爽以粗鐵魚上以鵝雁翅為浮隨波

浪上下黑火順流放下香到火發

Fig. 36. A sea-mine, the 'submarine dragon-king' (*shui ti lung wang phao*), from the mid +14th century, in *HLC* (*HCT*), p. 37*a*. For explanation, see text. The firing mechanism consists of a floated incense-stick which lights the fuse when it burns down, this last being contained in a length of goat's intestine, and connecting with the explosive charge which is floated at a certain depth submerged below.

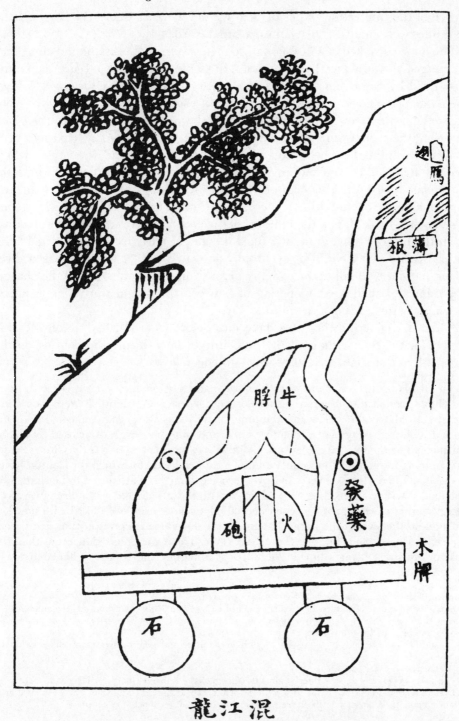

龍江混

Fig. 37. The more explicit diagram of the same device in *TKKW*, ch. 3 (Excellent Weapons sect.), p. 269. In the current facsimile Ming edition it is ch. 3 (ch. 15), p. 38*a*. The derivation from the *HLC* drawing is very evident, but the text says that the ox-bladder containing the explosive is replaced by a lacquered leather bag, and that the joss-stick gives way to a cord pulled from the shore to release a flint-and-steel firing mechanism.

source of the interesting paper of Beal (4) seven years later.[a] Twenty of these sea-mines were made at the Canton arsenal in 1843.[b]

The mine consisted of a hexagonal wooden waterproof brass-bound chest (tu^1) submerged in the water by means of adjustable iron weights (*thieh chui*[2]) and suspended by two chains or ropes (*hsüan shêng*[3]) from a buoy (*fou chhiu*[4]). Two openings in the cover enabled the charge of some 160 lb. of gunpowder to be inserted (*ju yao khung*[5]) and a third allowed water to enter a tube (*shui kuan khung*[6]) at the proper time. All three openings were sealed, two by 'charge covers' (*yao kai*[7]) and the third by a protective cover (*hu kai*[9]) for the water-tube, which incorporated a filter to avoid any danger of blockage. When the time came the mine was towed on the end of a rope (*yin shêng*[10]) from a boat silently approaching the enemy ship, or even by a swimmer or a diver, and then fastened to its anchor-cable. Upon the removal of the protective cover, the water, passing down the narrow tube, gradually filled a cylinder with accordion-pleated sides and raised its upper end; this eventually activated a lever which released three spring-hammers to fall upon as many percussion-caps and so set off the charge. The mine thus had a latent period of between thirty and thirty-five minutes, allowing for the escape of the mining party.[c]

Admiral Kennedy, whom we have met before (p. 189), had a considerable respect for the sea-mines used by the Chinese Navy during the fighting on the Canton River in 1856. He afterwards wrote as follows:[d]

To guard against fire-rafts and torpedoes[e] we made a boom across the River with spars and chains, connecting it with the shore on both sides. Some old junks were moored in mid-stream above and below the shipping; these junks were also connected with the shore, leaving a passage for a friendly vessel, and this space was also closed by chains which could be removed at pleasure.... All this was most necessary, as the Chinese were very cunning in the use of torpedoes and infernal machines, for which the Canton River was well adapted. Almost every night we received some kind attention in the shape of a junk loaded with combustibles, floated down with the stream and set on fire when close to us. Another clever apparatus consisted of one or more iron tanks filled with powder, and sunk to the level of the water, having on the outside wire springs connected with a trigger, so as to explode on touching a ship's side. These were more dangerous than the junks or fire-ships, being so low in the water as to require the utmost vigilance to detect

---

[a] A copy of the book had been found on one of the principal junks at Fatshan in June 1857. Chs. 92, 93 reprinted Phan's *Shui Lei Thu Shuo*.[8] Cf. Kennedy (1), pp. 77 ff.

[b] An interesting echo of all these proceedings reached Liang Chang-Chü, who wrote an account of them in his *Lang Chi Tshung Than* of 1845 (Impressions Collected during my Official Travels), ch. 5. Forty years later this was translated by Imbault-Huart (4).

[c] As the illustrations (pp. 3 a, 5 a) are very Western in character, and none too informative either, we do not reproduce them.

[d] Kennedy (1), p. 41.

[e] This was an old term for sea-mines; cf. the title of the paper of Imbault-Huart (4). The use of 'torpedo' in the modern sense did not come in till later, with the Whitehead self-propelled torpedo. The word must have been taken from the name of a genus of electric fishes which administer shocks to their prey.

[1] 櫝	[2] 鐵墜	[3] 懸繩	[4] 浮毬	[5] 入藥孔
[6] 水管孔	[7] 藥蓋	[8] 水雷圖說	[9] 護蓋	[10] 引繩

them. Our business was to sink or explode them before they got near enough to do us any harm, but this was not always possible; at times we managed to destroy some, and others drifted wide of the mark, but on one occasion they very nearly succeeded....

Chiao Yü and his +14th-century military and naval technologists would probably have been quite pleased if they had known that no less than five hundred years later their sea-mines, suitably improved, would have given so much bother to the heirs of modern science and the industrial revolution.

Now that we have become accustomed to the detonation of large masses of high-nitrate gunpowder, the moment has perhaps come to look again at accounts of arsenal explosions, of which several have come down to us from late Sung and Yuan times. For example, in Chou Mi's[1] *Kuei Hsin Tsa Chih*[2] we find note of the following incident.[a]

Chao Nan-Chung[3], when prime minister (*chhêng hsiang*[4]) ... reared four tigers at his private house in Li-yang, and kept them within a palisade near the gunpowder arsenal. On a certain day, while the gunpowder was being dried, a fire broke out and a terrible explosion followed. The noise was like the crash of thunder, the ground trembled, and many houses collapsed. The four tigers were killed instantly. This news spread from mouth to mouth among the people, and the incident was considered a frightening marvel.

Since Chao Khuei[5] (Chao Nan-Chung) died in +1266, the calamity would have occurred about +1260, just the time when the Southern Sung were defending their dynasty against the Mongols, a few years after Li Tsêng-Po had been complaining about the arsenal administration (p. 173 above) and a few years before the siege of Hsiang-yang (p. 168 above).

The *Kuei Hsin Tsa Chih*, written in +1295, goes on to report an even more alarming occurrence,[b] the destruction of an arsenal in +1280, just after the Mongols had taken over.

The disaster of the trebuchet bomb arsenal at Wei-yang[6] was still more terrible. Formerly the artisan positions were all held by southerners (i.e. the Chinese). But they engaged in peculation, so they had to be dismissed, and all their jobs were given to northerners (probably Mongols, or Chinese who had served them).[c] Unfortunately, these men understood nothing of the handling of chemical substances. Suddenly, one day, while sulphur was being ground fine, it burst into flame, then the (stored) fire-lances caught fire, and flashed hither and thither like frightened snakes. (At first) the workers thought it was funny, laughing and joking, but after a short time the fire got into the bomb store, and then there was a noise like a volcanic eruption and the howling of a storm at sea.

---

[a] Chhien Chi, p. 13b, 14a, tr. auct.
[b] *Ibid.* p. 14a, b, tr. auct.
[c] Under the Yuan dynasty there were three classes of citizens. First came the Mongols (Mêng-ku[7] or *pei jen*[8]), secondly the Arabs, Persians or Europeans hired to serve the Great Khan, people with coloured pupils in their eyes (*sê mu jen*[9]), and finally in the third class the Chinese (southerners, *nan jen*[10]).

[1] 周密　　　[2] 癸辛雜識　　[3] 趙南仲　　[4] 丞相　　　[5] 趙葵
[6] 維揚　　　[7] 蒙古　　　　[8] 北人　　　[9] 色目人　　[10] 南人

The whole city was terrified, thinking that an army was approaching, and panic soon spread among the people, who could not tell whether it was near or far away. Even at a distance of a hundred *li* tiles shook and houses trembled. People gave alarms of fire, but the troops were held strictly to discipline. The disturbance lasted a whole day and night. After order had been restored an inspection was made, and it was found that a hundred men of the guards had been blown to bits, beams and pillars had been cleft asunder or carried away by the force of the explosion to a distance of over ten *li*. The smooth ground was scooped into craters and trenches more than ten feet deep. Above two hundred families living in the neighbourhood were victims of this unexpected disaster. This was indeed an unusual occurrence.

This graphic description bears witness to what the principle of the 'sorcerer's apprentice' could do where explosives were concerned. It is interesting to reflect that it could not have happened in the Europe of that time, for Roger Bacon's first notice of gunpowder was then only a dozen years old, while nearly half a century had yet to pass before the first practical use of gunpowder in Western warfare.[a]

Similar stories continue down in Chinese literature until modern times. For example, it is recorded[b] that in +1363 a certain augur named Chang Chung[1] predicted a great disaster, but assured Chu Yuan-Chang, the future Ming emperor, that no harm would come to him personally. Sure enough, a month later the Chung-Chhin Pavilion caught fire, and a thunderous explosion followed when the bombs and powder stored there were touched off.

(12) BIZARRE DELIVERY SYSTEMS

We now approach the tubular fire-lance, ancestor of all metal-barrel guns and cannon, but before we can discuss it we must pause to take note of various peculiar methods for the delivery of incendiaries and explosives devised by the Chinese in medieval times. Although at first sight a modern reader may be inclined to dismiss them as fanciful, they are in fact of quite considerable interest because one can trace in several of the devices a clear progress in sophistication as time went by. One can also find the whole succession from incendiary to high-nitrate gunpowder plainly exposed in them.[c] Most began with the use of expendable animals, a method which must be almost as old as warfare itself. We

---

[a] Cf. p. 179 above, and p. 287 below.

[b] By Sung Lien, in his *Sung Hsüeh Shih Chhüan Chi* ch. 10 (p. 356); cf. Goodrich & Fêng Chia-Shêng (1), p. 121. Since the account was written only a decade or so later, there is every reason to accept the reality of the arsenal explosion.

[c] Even the most archaic of them survived in military books into the +17th century and later, but to what extent they were any use by then or whether anyone attempted to use them, is not so sure. Such was Chinese conservatism, however, that their description and illustration certainly persisted long after the introduction of relatively modern types of guns and cannon. But Chingiz Khan is said to have used the fire-bird technique when besieging J/Chin cities; Schmidt (1), Eng. tr. p. 50; Franke (24) pp. 199, 354.

[1] 張中

have already noted an example from ancient Hebrew history, Samson tying firebrands to the tails of foxes and driving them towards the enemy's cornfields.[a]

Now in what follows we shall trace the development of such systems from the beginning of the +8th century onwards. First there were the 'fire-birds' (*huo chhin*[1]), partridge-like creatures sent off to sit upon the enemy's thatched roofs and set them alight (Fig. 38);[b] they carried a walnut receptacle filled with burning moxa tinder, pierced with two holes, and tied round their necks. The illustration comes from *Wu Ching Tsung Yao* (+1044), but the text about them is verbally identical with a predecessor in the *Hu Chhien Ching*,[c] and that takes us back as far as +962. The picture and description is then repeated in practically all the subsequent military compendia.[d]

Another sort of fire-bird was represented by the 'apricot(-stone fire-)sparrows' (*chhiao hsing*[2]). They were smaller birds, and the burning moxa tinder was enclosed in a split apricot-stone attached to their legs; several hundred were to be let loose at one time to fly to the enemy's camps and granaries and set them on fire.[e] The *Huo Lung Ching* also has this,[f] but here comes an interesting surprise, for on the very next page this work of about +1350 has something very different, no less than a winged rocket, i.e. an artificial bird propelled by four rocket-tubes attached to feathered rods.[g] This is the 'magic-fire flying crow' (*shen huo fei ya*[3]), with its accompanying text and illustration repeated in later works. Here we must do no more than mention it, reserving its description for the sub-section on rockets, but it does show the continuity between the ancient techniques and the far more ingenious ones of later times.

The expendable animals continue with the 'fire-beasts' (*huo shou*[4]), deer, boars or other wild animals sent towards the enemy carrying on their heads burning moxa tinder in a gourd (*phiao*[5]) with four holes.[h] One of the earliest appearances of this text must be that in the *Hu Chhien Ching* (began in +962 and finished by +1004),[i] and there is a good illustration in the *Wu Ching Tsung Yao*.[j] This particular ploy seems not to have led to anything further, but it may be added that the birds were induced to leave by pricking their tails, while the animals were despatched towards the enemy by tying oil-soaked reeds to their tails and setting them alight.

[a] Judges 15, 4–5, cf. p. 66 above.
[b] *WCTY/CC*, ch. 11, p. 21a, b.
[c] *HCC*, ch. 6 (ch. 54), p. 5a. There is mention also in *TPYC* (+759), ch. 4, p. 8b.
[d] *HLC*, pt. 1, ch. 3, p. 16a, b; *HKPY*, ch. 3, p. 16a, b; *HCT*, p. 33a; *WPC*, ch. 131, pp. 10b, 11a.
[e] *WCTY/CC*, ch. 11, p. 22a, b.
[f] *HLC*, pt. 1, ch. 3, p. 17a, b. Also in *HKPY*, ch. 3, p. 17a, b; *HCT*, p. 33b; *WPC*, ch. 131, pp. 11b, 12a.
[g] *HLC*, pt. 1, ch. 3, p. 18a, b. Also in *HKPY*, ch. 3, p. 18a, b; *HCT*, p. 34a; *WPC*, ch. 131, pp. 12b, 13a.
[h] The object of these in all the cases must have been to supply air for the glowing tinder, as well as to allow some of it to come out on the thatch.
[i] *HCC*, ch. 6 (ch. 54), p. 5a.
[j] *WCTY/CC*, ch. 11, p. 24a, b. It appears again in *HLC*, pt. 1, ch. 3, p. 19a, b; *HKPY, ibid.*; *HCT*, p. 34b; and *WPC*, ch. 131, pp. 13b, 14a.

[1] 火禽　　　[2] 雀杏　　　[3] 神火飛鴉　　　[4] 火獸　　　[5] 瓢

Fig. 38. Expendable bird carrying an incendiary receptacle round its neck (*huo chhin*), a technique in use at least as far back as the +10th century. From *WCTY*, ch. 11, p. 21*a*.

Next came the 'fire-ox' (*huo niu*[1]), which moved, as we shall see, with the times.[a] He is not in the *Hu Chhien Ching*, but in +1040 the *Wu Ching Tsung Yao* shows him (Fig. 39) pounding away towards the enemy lines at a very un-cattle-like pace, but that is because a large tub of incendiary material is attached to his rump and burning.[b] Although the animal is provided with a couple of spears tied on, it cannot have caused much damage if the fire was put out, and the generals must have felt it undesirable to give the enemy such free supplies of beef. Very different was the situation after it became possible to attach a large bomb of high-nitrate gunpowder to the back of the animal, as we see (Fig. 40) in the later works.[c] It now had some real point, though it was still an expendable animal; this was the situation by the end of the +13th century, though our illustration comes from the *Wu Pei Huo Lung Ching* of about +1400. And it had changed its name, for it was now called the 'rolling thunder-bomb fire-ox' (*huo niu hung lei phao*[2]).

Another striking change which came over an ancient plan or tactic as time went by appears in the matter of the 'fire-soldier' (*huo ping*[3]). In the *Hu Chhien Ching* and the *Wu Ching Tsung Yao* he is a real person, a rider on a gagged horse who gallops round the enemy camp lighting and throwing in combustible materials (Fig. 41), then, if confusion results, an attack is made.[d] But by the +14th century he is replaced by a wooden human figure, also mounted on a horse, but stuffed behind his paper face and bamboo accoutrements with incendiary substances and carrying a large bomb timed to explode when it reaches the enemy lines (Fig. 42).[e] Here of course the animal had to be despatched by the age-old method of tying firebrands to its tail. The arrangement is now called the 'heaven(-shaking) thunder-bomb carried by the wooden man on the fire-horse' (*mu jen huo ma thien lei phao*[5]), or in the more succinct phrase of the *Wu Pei Chih* the 'wooden man on the live horse' (*mu jen huo ma*[6]).

A variant of this, rather less convincing in character, was the 'flying-carriage fire-dragon pushed along the ground' (*huo lung chüan ti fei chhê*[7]).[f] Here a bomb or land-mine was concealed within the body of a wooden figure of a winged dragon or similar animal mounted on a two-wheeled cart and pushed forward by two

---

[a] His traditional originator was Thien Tan[4], a general of Chhi, in the campaigns of −279 against Yen State. But there was no incendiary element about Thien's stampeding cattle, with sharp blades attached to their horns, except the bundles of oiled rushes tied to their tails, by which, when lit, they were set in motion.

[b] *WCTY/CC*, ch. 11, p. 25*a*, *b*. Also in *HLC*, pt. 1, ch. 3, p. 20*a*, *b*; *HKPY, ibid.*; *HCT*, p. 35*a*; *WPC*, ch. 131, pp. 17*b*, 18*b*.

[c] *WPHLC*, ch. 2, pp. 2*b*–3*b*; *HLC*, pt. 2, ch. 3, p. 17*a*, *b*; *HKPY*, ch. 3, p. 21*a*, *b*; *HCT*, p. 35*b*; *WPC*, ch. 131, pp. 18*b*, 19*b*, *b*, 20*a*. One of the two pictures in this last work shows a camouflage cover over the animal and its burden.

[d] *HCC*, ch. 6 (ch. 54), p. 5*a*; *WCTY/CC*, ch. 11, p. 23*a*, *b*. Also in *TPYC*, ch. 4, p. 8*b*.

[e] We illustrate from *WPHLC*, ch. 2, pp. 1*a*–2*a*. But there is a particularly long specification in *WPC*, ch. 131, pp. 15*b*–17*a*. Attack is to be made as soon as the explosion occurs.

[f] *HLC*, pt. 2, ch. 3, pp. 18*b*, 19*b*; *WPHLC*, ch. 2, pp. 4*a*–5*a*; *WPC*, ch. 132, pp. 2*b*, 3*a*.

[1] 火牛   [2] 火牛轟雷砲   [3] 火兵   [4] 田單   [5] 木人火馬天雷砲
[6] 木人活馬   [7] 火龍捲地飛車

Fig. 39. The 'fire-ox', another expendable animal, from *WCTY*, ch. 11, p. 25a.

火牛轟雷砲式

用守城

尾縛蘆葦灌
油然尾牛怒
而奔銳不可當

火砲安安蓋以
羽翎一以被面
一以便觀火一
搖動卽發此安
平火牛發法也

少膏絵

Fig. 40. The bomb-carrying fire-ox, from *WPHLC*, ch. 2, p. 2*b*.

Fig. 41. A 'fire-soldier' rider (*huo ping*), from *WCTY*, ch. 11 p. 23a. He gallops round the enemy camp, throwing in combustible incendiary materials.

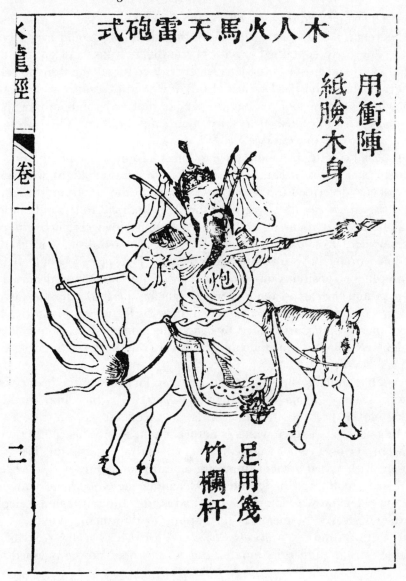

Fig. 42. Bomb-containing robot rider, intended to explode when carried into the enemy camp;
WPHLC, ch. 2, p. 2*a*.

soldiers. The wings had eye-holes through which they could look, and acted as a shield for them, but it is hard to believe that the device could ever have been effective. The body contained several of the different kinds of gunpowder (cf. p. 117 above), and in some versions spears projecting at the front were tipped with tiger-poison. The *Wu Pei Chih* also figures and describes wooden animals with a wheel at each foot, containing smoke materials or bombs with metal fragments, each to be pushed forward by a single soldier.[a] These were called 'wooden fire-beasts' (*mu huo shou*[1]).

Very special conditions would have to have pertained before anything of this kind could have been useful, and the same conclusion might well apply to another device described in the old books, namely the 'wind-and-thunder fire-rollers' (*fêng lei huo kun*[2]).[b] These were simply cylindrical rollers of bamboo and paper about a foot in diameter and three feet long, which were filled with poison-fire gunpowder,[c] iron fragments, and five cast-iron bombshells in each, then rolled down from above into the enemy camp (Fig. 43). No doubt defenders occupying higher positions on slopes had from time immemorial hurled down loose rocks and logs upon their enemies, but for the fire-rollers to have been effective it would have been desirable to induce the foe to encamp at the head of a valley, for example, surrounded by grassy declivities on nearly all sides, and that might well have been difficult.[d] However, this peculiar firearm is worth recording.

Expendable animals carrying incendiaries go far back in the history of most civilisations. They are mentioned in the *Arthaśastra*,[e] which implies the early centuries of the present era; and birds in particular appear in Ḥasan al-Rammāḥ's books, *c*. +1280, which is natural enough since his connections with China were so close.[f] As for frightening figures of dragons and the like, we have to contend with a vast ancient literature on automata.[g] Firdawsī, for example, about +1020, took from the legends of Alexander the Great a story about iron horses and riders on wheels that he had had made, and sent against elephants, which they destroyed by means of the naphtha within them.[h] As late as +1463 Roberto Valturio figured a *machina arabica*, a great dragon figure, which shot forth arrows from guns in its mouth.[i] But all this need not be pursued further here.

---

[a] *WPC*, ch. 131, pp. 14*b*, 15*a*.
[b] *Huo Lung Ching*, pt. 1, ch. 3, p. 10*a*, *b*; *HCT*, p. 30*a*; also in *HKPY*, ch. 3, p. 10*a*, *b*; copied in *WPC*, ch. 130, p. 13*a*, *b*.
[c] Cf. p. 123 above.
[d] The texts do not always say that the cylinders have to be rolled down from a higher position, but it is fairly self-evident.
[e] Shamasastry (1), p. 434 (§405); Partington (5), p. 210.
[f] Partington (5), pp. 200–1. Cf. p. 41 above.
[g] Adumbrated briefly in Vol. 4, pt. 2, pp. 156ff. and Vol. 5, pt. 4, p. 488.
[h] See Elliot (1), vol. 6, pp. 475–80.
[i] Partington (5), p. 164.

[1] 木火獸   [2] 風雷火滾

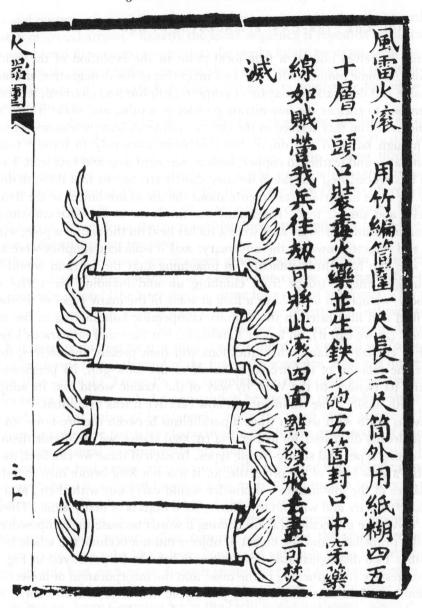

風雷火滾　用竹編筒圍一尺長三尺筒外用紙糊四五

十層一頭口裝毒火藥並生鉄少砲五簡封口中穿藥

線如賊當我兵往如可將此滾四面黙發戒毒盡可焚

滅

火器圖八

三一

Fig. 43. 'Wind-and-thunder fire rollers' (*fēng léi huo kun*), from *HLC* (*HCT*), p. 30 *a*.

(petrol for flame-throwers).[a] What exactly it referred to at that time remains uncertain, perhaps a fire-lance with co-viative projectiles, perhaps a true barrel-gun; after all, these last are now attested by archaeological evidence for *c.* +1290 (cf. p. 293 below), so the fully developed form was already 'in the air'. The general upshot is that the passage from the fire-lance to the hand-gun and the bombard was one of slow stages with many shades of meaning and distinction, no sharp break occurring at any time.

Let us now review the passages in Chinese literature which mention the use of fire-lances in warfare. For many years past the *locus classicus* has been the *Tê-An Shou Chhêng Lu*[1], the account of the famous siege of Tê-an by the Chin Tartars in +1132, when Chhen Kuei[2] successfully held it for the Sung. The text says:[b]

We also used bomb gunpowder (lit. fire bomb powder, *huo phao yao*[3]) and long poles of bamboo to make more than twenty fire-lances (*huo chhiang*[4]). Also striking lances (*chuang chhiang*[5]) and swords with hooks at the ends (*kou lien*[6]), many of each. It took two men to handle each one. These things were got ready to use from the ramparts whenever the assault towers with their flying bridges (*thien chhiao*[7]) approached the city.

And a further reference occurs in Chhen Kuei's biography in the dynastic history, which tells how[c]

(Chhen) Kuei, with sixty men, carrying fire-lances (*huo chhiang*[8])[d], made a sally from the West Gate, and using a fire-ox (*huo niu*[9]) to assist them,[e] burnt the flying bridges, so that in a short time all were completely destroyed. So (Li) Hêng[10] pulled up his stockades and went away.[f]

Here the nuance seems to be that the fire-lances were used not so much to oppose invading soldiers who got on to the city wall as to set light to the wood-work of the enemy's siege equipment. Still, they could well have been used in close-quarter combat too.

But did these events, occurring in the early years of the +12th century, really constitute the first appearance of the fire-lance upon the stage of warfare? It has been customary to think so. But in 1978 an unusually important discovery was made by Clayton Bredt in the Musée Guimet in Paris.[g] There he found a painted silk banner from Tunhuang, doubtless one of Paul Pelliot's acquisitions, which

---

[a] Cf. pp. 170 ff. above.
[b] *Shou Chhêng Lu*, ch. 4, p. 6a, tr. auct. Cf. p. 8a.
[c] *Sung Shih*, ch. 377, p. 6a, tr. auct. Cf. Fêng Chia-Shêng (*1*), pp. 68–9, (*6*), pp. 22–3.
[d] The wood and metal radicals are interchangeable. It should be possible to find out in what texts at what date the metal radical first began to replace the wood one, but how much light that would throw on the history of technology we would hesitate to say. And the same applies to the stone and fire radicals in *phao*[11,12], about which there has been some debate in the past.
[e] Here we see the use of an expendable animal, as described on p. 211 above.
[f] Li Hêng was one of the chief generals on the Jurchen Chin side.
[g] Dr Bredt kindly reported it to one of us (W. L.) in a letter of 27 January that year.

[1] 德安守城錄   [2] 陳規   [3] 火砲藥   [4] 火鎗   [5] 撞鎗
[6] 鈎鎌   [7] 天橋   [8] 火槍   [9] 火牛   [10] 李橫
[11] 砲   [12] 炮

dates from the middle of the +10th century,[a] and which may well contain, as he himself said,[b] 'the earliest representation of a Chinese pyrotechnic weapon'. The painting[c] portrays the assault of the demon Māra[1] and his cohorts on the meditating Buddha, seeking to distract him from his attainment of understanding of the nature and mechanism of the universe, and to prevent his enlightenment (Fig. 44). Although the figures in the painting are all supernatural beings a number of them are dressed in military armour and bear weapons—reflex bows, and straight double-edged swords, in particular—which are reasonably accurate depictions of late Thang arms. The weapon which is important for us here, however, is a fire-lance, namely a long pole with a cylinder at its end from which issue flames that shoot forward. They do not go upward like a torch, as if there were no pressure behind them, but rather blaze forwards as if from a flame-thrower containing rocket composition—which was exactly what the fire-lance was. The figure holding this is a devil with a head-dress of three serpents (Fig. 45), and he is pointing it to the left about the level of the upper part of the Buddha's halo. Just below him to the left there is another devil with a serpent entangled in his eyes and mouth, who seems to be about to throw a small bomb or grenade from which flames are already coming out. Many other features of great interest occur in the painting, but we cannot discuss them here.[d]

If the dating is right, and there is no reason for doubting it, the implication can only be that the fire-lance originated about +950 in the Wu Tai period not long before the Sung, i.e. some thirty years after the time when we concluded (p. 85 above) that the gunpowder mixture was first used in war, namely to make a kind of slow-match for a petrol flame-thrower (+919). The transition to a low-nitrate gunpowder flame-thrower would thus have been extremely natural, but it remains remarkable that (so far as we can see) there is neither mention nor illustration of a fire-lance in +11th-century books such as the *Hu Chhien Ching* or the *Wu Ching Tsung Yao*. For the next picture of the device we have to await +1350 or so, the time of the *Huo Lung Ching*, though then and thereafter there are dozens of varieties, which lasted well into the musket era. Perhaps the fire-lance was kept extremely secret throughout the rest of the +10th century and all through the +11th;[e] perhaps gunpowder could not then be produced in quantity, and other uses therefore had the preference. But the iconographic evidence of the Buddhist banner seems incontrovertible.

[a] According to the archaeological expertise of Dr Robert Jéra-Bézard and his colleagues of the Museum staff.
[b] In what follows we have kept as far as possible to the wording of Dr Bredt's own letter.
[c] MG 17.655, no. 6 in the Vandier–Nicolas Catalogue. It was no. 315 in the 1976 Exhibition on the Old Silk Road.
[d] Dr Bredt makes the point that the Chinese Buddhist frescoes and paintings have hitherto been quite insufficiently combed for technological details, which is indeed undeniable.
[e] Examples of the secrecy there was in those times have already been given on p. 93 above.

[1] 魔羅

In the following year fire-lances figured largely again in an engagement on the canals surrounding the city of Kuei-Tê[1] in Honan, where the Chin troops defeated a large force of Mongols. In +1233 a detachment of the Chung Hsiao Chün[2] (Loyal and Filial Army) prepared to evacuate Kuei-Tê in the knowledge that the Mongols were coming, and intended to retreat to Tshai. But the commander, Phuchha Kuan-Nu[3], drew up a plan of attack on the positions which he guessed that the Mongols would occupy. The text continues:[a]

On the fifth day of the fifth month they sacrificed to Heaven, secretly prepared fire-lances, and embarked 450 (Chin) soldiers outside the south gate, whence they sailed first east and then north. During the night they killed the (enemy) guards outside the dykes, and reached the Wang family temple. Then they got to the north gate and waited, but the Mongols, fearing a defeat, retired in part to Haüchow. However, at the fourth night watch there came an attack, in which at first the Chin troops gave way; but (Phuchha) Kuan-Nu divided his small craft into squadrons of five, seven and ten boats, which came out from behind the defences and caught (the Mongols) both from front and rear, using the fire-spouting lances (*huo chhiang thu ju*[10]). The northern army could not stand up to this and fled, losing more than 3500 men drowned. Finally their stockades were burnt, and our force returned.

Here the five-minute flame-throwers seem clearly to have been used as close-combat weapons, as well perhaps as incendiaries for wooden defence works. The same account gives details of how they were made in Phuchha Kuan-Nu's time, but we postpone it for a moment in order to dwell first on other battle relations.

By +1257 we have the complaints of Li Tsêng-Po about the inefficiency of the Sung arsenals administration (already translated p. 174 above), in which he mentions fire-lances. Then two years later there comes an important statement about inventions made in the arsenal at Shou-chhun-fu[11], which suggests that the 'boys in the back room' were more active than the clerks in the issuing office.[b]

---

researches of the latter raised an interesting point, however; it may just be that the flying-fire lances of +1232 shot out arrows as well as flames. Sun noticed that the texts of two modern authors who tried to improve the dynastic history of the Mongols used a slightly different phrase—*phên huo chien thung*[4], i.e. tubes which spurted arrows as well as fire. This would fix an early date for the appearance of arrows as co-viative projectiles (cf. p. 230 below). The two books concerned were the *Mêng Wu Erh Shih Chi*[5] (History of the Mongols) by Thu Chi[6] (1912), ch. 29, p. 12*a*, and the *Hsin Yuan Shih*[7] (New History of the Yuan Dynasty) by Kho Shao-Min[8] (1922), ch. 122, pp. 3*b*, 4*a*. But neither of these works has authority as high as the *Chin Shih* and the *Yuan Shih* themselves, so unless their authors had access to old records not generally available, their military nomenclature is not very sure. One point that makes us suspicious is that we have not found the word *phên*[9] in any of the contemporary texts, though it does frequently occur in the later military compendia from the *Huo Lung Ching* onwards. So the matter must remain in doubt. But whatever the weapons were, they were not rockets.

The American National Aeronautics and Space Administration (NASA) Museum and Exhibition at Clear Lake City, Texas, in a very laudable effort to do justice to Chinese priority, displays a painting showing the 'rockets' of +1232 being let off by soldiers from basket launchers (cf. p. 488 and Fig. 198). So once again the mistake is perpetuated.

[a] *Chin Shih*, ch. 116, p. 12*a*, *b*, tr. auct.   [b] *Sung Shih*, ch. 197, p. 15*b*, tr. auct.

[1] 歸德   [2] 忠孝軍   [3] 蒲察官奴   [4] 噴火箭筒   [5] 蒙兀兒史記
[6] 屠寄   [7] 新元史   [8] 柯紹忞   [9] 噴   [10] 火槍突入
[11] 壽春府

One of these was the 'box-and-tube crossbow' (*kan thung mu nu*[1], i.e. the magazine crossbow) which was very convenient and steady (for loading and firing), and could also be used at night (because the projectiles fell into place automatically).[a] The other was the 'flame-spurting lance' (*thu huo chhiang*[2]). A bamboo tube of large diameter was used as the barrel (*thung*[3]), and inside it they put a bundle of projectiles (lit. a nest of eggs, *tzu kho*[4]). After ignition, and when the blazing stream of flame was ending, the bundle was shot forth as if it was a trebuchet projectile (*phao*[5]), with a noise that could be heard more than 150 paces away.[b]

Presumably the bundle disintegrated as it flew, sending the objects, whether fragments of metal or pottery, pellets or bullets, in all directions. In any case this is one of the earliest references we have to co-viative projectiles,[c] but we doubt whether the invention was really new in +1259.[d]

The epic story of the relief of Hsiangyang by Chang Shun and Chang Kuei has already been given (p. 174–5 above), and it will be remembered that all the ships of their fleet were provided with fire-lances to repel boarders. That was in +1272. Four years later we have an account of a battle of the Sung against the Mongols which emphasises the lance aspect of the fire-lance. Bayan (Po Yen[6]) when invading the Sung territory, ordered one of his officers, Shih Pi[7], to attack Yangchow; the garrison commander, Chiang Tshai[8], led a sortie to surround the Yang-tzu bridge, but suffered a great defeat. Then

Chiang Tshai returned with his men by night, but thrice Shih Pi was victorious. At daybreak Chiang Tshai, seeing that Shih Pi's troops were few, pressed an attack, but Shih Pi resisted furiously. Two (Sung) cavalrymen rushed at him to pierce him with fire-lances, but he so defended himself with his sabre that to left and right every man fell; and he himself personally killed more than a hundred....[e]

This shows that the lance or pike to which the barrel of the weapon was attached was a very real arm in itself, and could be used when the flames and projectiles of the latter were all spent. It may be of significance that the word *chhiang*[9] here is once again (cf. p. 222) written with the metal rather than the wood

---

[a] This has been discussed already in Vol. 5, pt. 6, Sect. 30 *e* (2), iv above.

[b] I.e., some 250 yards.

[c] Brock (1), p. 232, describes a Roman candle mortar throwing out discrete balls of half-combusted rocket composition, and elsewhere (pp. 206 ff.) discusses 'shells' containing a 'filling charge' and shot off (or up) by a 'lifting charge'. But we agree with Partington (5), pp. 246–7, that the Chinese projectiles were solid objects, as in the many later cases where the expression *tzu kho* occurs.

[d] The fire-lance was so unlike anything used in modern times that Lu Mou-Tê (1), p. 30, may well be excused for supposing that the Shou-chhun arsenal was making guns with truly propellant gunpowder. Von Lippmann (22), in (3), vol. 1, p. 133, understood it better, but believed that the projectiles were only 'Brand-satzklümpfchen' i.e. balls of combustible and incendiary material; we doubt whether the text will bear this interpretation. Where he went far wrong, however, was in his view that Chinese firearms never got independently beyond this stage.

[e] *Yuan Shih*, ch. 162, p. 11 *b*, tr. auct.

---

[1] 匣筒木弩 [2] 突火槍 [3] 筒 [4] 子窠 [5] 砲

[6] 伯顏 [7] 史弼 [8] 姜才 [9] 鎗

radical, but it would be quite dangerous to deduce from this that there were no metal tubes in use before +1276.

Lastly we return to the precious description of the making of fire-lances about +1230, written a century or so later, at a time when they were certainly still in use. The *Chin Shih* says:[a]

The method of making (fire-)lances was to take (thick) 'imperial yellow' paper and to make it into a tube (with walls composed of) sixteen layers, about two feet long. It was then filled (with a mixture of) willow charcoal,[b] iron in the form of powder, porcelain fragments, sulphur, arsenious oxide (*phi shuang*[1]), and other things.[c] It was then bound with cords to the end of the lance. Each soldier carried with him, hanging down (from his belt) a small iron fire box (of glowing tinder).[d] At the appropriate time during combat he lit (the fuse), and the flames shot forth from the lance head more than a dozen feet. After the composition had burnt out the tube was not damaged. When Pien-ching (i.e. Khaifêng) was being besieged (in +1126) these (fire-lances) were used a great deal,[e] and they still are today.

Here the omission of the essential constituent, saltpetre, may or may not have been deliberate, but we can be quite sure that it was present. The enemy was probably blinded and confused by the sparkling of the 'Chinese iron' (cf. p. 141 above). The tubes seem rather short, but that was perhaps the best length that paper would stand; and the use of paper is doubtless surprising in itself for those who do not know that in suitable conditions a paper like carton can be made, strong and hard enough even to be used for protective combat armour.[f] Perhaps in the northern part of the country there was not enough bamboo available for tubes of the necessary calibre. At all events, this passage forms a good transition to the types of fire-lances which we are now in a position to examine. And as we proceed to do this, it may be well to recall that the fire-lance had become the possession of the Arabs by about +1280 (as the book of Ḥasan al-Rammaḥ, p. 42 above, shows). With three centuries of Chinese experience behind it, this can hardly be conceived as anything but derivative.

It is possible also to find references to the fire-lance in poetry. For example, Chang Hsien[3] (*fl.* +1341) wrote a number of poems on military subjects, and one of them is entitled Fu Yang Hsing[4] (On Soldierly Proceeding[5] at Fu-yang).[g] It goes as follows:

---

　[a]　Ch. 116, p. 12*b*, tr. auct.
　[b]　Lit. 'ash', as often in these texts.
　[c]　A noteworthy expression.
　[d]　*Chün shih ko hsüan thieh kuan tshang huo*[2].
　[e]　Of course the writers may have been thinking more of +1232.
　[f]　See Vol. 5, pt. 1, pp. 114 ff. and pt. 8, Sect. 30 *l* (2), iii.
　[g]　In his *Yü Ssu Chi*[5] (Jade Box Collection), ch. 3, p. 27*a* (p. 765·1). Fu-yang was in Shantung province. The poem was first noted by Wang Ling (1), p. 172, but he interpreted it as referring to bombards, which the context shows can hardly be right.

　[1]　砒霜　　　　　　[2]　軍士各懸小鐵鑵藏火　　　　　[3]　張憲　　　　[4]　富陽行
　[5]　玉笥集

The general mounted his horse, and flourished his gold-painted whip,
On the mountain-top the white flags were like birds flying.
Westwards came cavalry in thousands, crowding like bees,
And the sound of the drums echoed from all four quarters.
The Tartar town was defended around by a palisade
And the Chin people lurked behind great wooden stakes
Boasting that five hundred fierce commanders guarded it withal.
When the iron gate (of the entrance) did not open
Fire-barrels (*huo thung*[1]) were used to attack and burn it.
Soon the strength of the Tartars (*hua yao*[2]) ebbed, and their soldiers fled,
From north to south of the town there were puddles of blood,
And the clouds were red for the space of ten *li* afar
As the flames set by the flying fire-crows (*fei huo ya*[3]) enveloped the town.[a]
Our brave general drinks a cup, and forbears from chasing the enemy,
Letting his men just pillage the homes and household goods
Of the three hundred contumelious barbarian families.

Here the focal point is evidently flame-throwers rather than metal-barrel guns. But if the reference is indeed to the wars between the Chin Tartars and the Mongols, as it seems to be, the poem must refer to some episode before +1234, the date of extinction of the Jurchen Chin dynasty. Chang Hsien need not have been giving an eye-witness account, but rather depicting an assault of the previous century which was part of the heritage of Yuan tradition.

The prototypic fire-lance is the weapon named in the books *li hua chhiang*[4] (pear-flower spear),[b] and consisted simply of the tube of low-nitrate gunpowder attached to the business end of a lance or pike. According to the descriptions:[c]

A pear-flower tube[d] is bound tightly to the end of a long spear, and ignited when face to face with the enemy. As it burns (the flames) shoot forth as far as several dozen feet.[e] Anyone that gets in the way [of its chemic force] is inevitably burnt to death; and after the fire has ended, you can still use the spear-point to pierce the enemy through. It is the best of all fire-weapons. [In the Sung, Li Chhüan[5] always used it in his heroic exploits in Shantung, where there was a saying that with twenty (loyal) pear-flower spearmen no enemy would be left in the world.][f] [This technique was for some time lost, but Hsü

---

[a] We take this as a reference to incendiary birds (cf. p. 211 above). But it could be purely and simply poetic, or alternatively it could just be a reference to incendiary arrows (p. 154 above) or even rockets (p. 502 below).

[b] The term 'pear-flower' was probably a reference to the flame thrown out, expanding like a flower.

[c] *HLC*, pt. 2, ch. 2, p. 24a; *Wu Pei Huo Lung Ching*, ch. 2, p. 30b; *WPC*, ch. 128, pp. 3b, 4a; *CHTP*, ch. 13, p. 63a, b, quoted verbatim in *STTH*. The portions of text in square brackets are from *WPC*, and *CHTP* adds the portion in italic square brackets. Cf. Davis & Ware (1), p. 524.

[d] This could have been a kind of firework like a Roman Candle; evidently the material of which it was made needed no further explanation.

[e] This may have been a slight exaggeration.

[f] This comes from *Sung Shih*, ch. 477, p. 18b (copied in *Pa Pien Lei Tsuan* and other places); but what Li Chhüan's wife Yang Miao-Chen[6] (herself a military commander at times) said to one of his officers Chêng Yen-Tê[7] was slightly different: 'After using the pear-flower fire-lance for twenty years past, there is no enemy left in the land.' But now times had changed, and new alliances had to be sought, etc.

[1] 火筒   [2] 花猺   [3] 飛火鴉   [4] 梨花鎗   [5] 李全
[6] 楊妙眞   [7] 鄭衍德

Kuo[1], a Repayment Clerk in the Administrative Commission office, re-discovered it and successfully tested it, so that it was again brought into standard use.][a]

None of the sources says what the material of the tube was, but bamboo is the most likely, rather than paper or leather (Fig. 46).

Li Chhüan (b. *c.* +1180) was an intriguing and enigmatic figure, a military adventurer who rose to the command of very substantial forces, part brigands, part rebel guerrillas, and for some twenty years played off the Sung, the Chin Tartars and the Mongols against one another. His army was known as the Red Jackets (Hung Ao[3]). In +1213 he joined forces with a similar character, Yang An-Kuo[4], and became his brother-in-law. Shantung province was at this time debatable land, and first Li Chhüan won and held it as a kind of Lord of the Marches for the Sung, then later went over to the Mongol side and died besieging Yangchow on their behalf in +1231. He was closely associated with the use of the fire-lance in combat.

The 'pear-flower spear' appears again in the section on spear play in Chhi Chi-Kuang's[5] *Chi Hsiao Hsin Shu*[6] (New Treatise on Military and Naval Efficiency) of +1560.[b] He says that the best school of practice with the long spear comes from the Yang tradition (presumably referring to Yang An-Kuo[4], d. +1215); it surpasses the Sha[7] school of short-spearmanship and the Ma[8] school of pikemanship. Chhi does not positively say that all pear-flower spears have flamethrower tubes attached to them, nor do his illustrations show this, but he does refer to earthquake and thunder in connection with them, so he may have been wrapping something up. On the other hand it may be more likely that the later simple spear acquired that name from the fact that it had once been a fire-lance, before these were abandoned in favour of hand-guns and muskets. Otherwise one could hardly explain the nomenclature.

Parallel to the fire-lance was that other Sung device called the 'fire-tube' (*huo thung*[9]). One account of it, in the *Hsing Chün Hsü Chih*[10] of about +1230, describes it as a short section of large-diameter bamboo, so it was most probably also essentially a flame-thrower, differing from the fire-lance mainly in that it was held directly in the hand, often being provided with a wooden handle or 'tiller' (cf. Figs. 48, 50, 61) not attached to the head of a spear. Co-viative projectiles were already in the fire-lance by that same date, the time of Phuchha Kuan-Nu, as we have just seen (pp. 221, 226 above), and the 'flame-spurting lances' (*thu huo*

---

[a] This seems somewhat garbled. Hsü Kuo was in fact a scholar-official sent north by the Sung government in +1223 as Commissioner (in succession to Chia Shê[2]) to hold together the quasi-bandit irregulars under Li Chhüan in Shantung, and keep them loyal to the Sung against the Chin Tartars. After his assassination in +1225 it was not long before Li had to give in to the Mongols and enter their service with his troops. What the story probably means, therefore, is that Hsü Kuo transmitted various forms of know-how regarding fire-lances from Li's military engineers to those of the Sung in the south.

[b] Ch. 10, p. 1*b*.

[1] 許國   [2] 賈涉   [3] 紅襖   [4] 楊安國   [5] 戚繼光
[6] 紀效新書   [7] 沙   [8] 馬   [9] 火筒   [10] 行軍須知

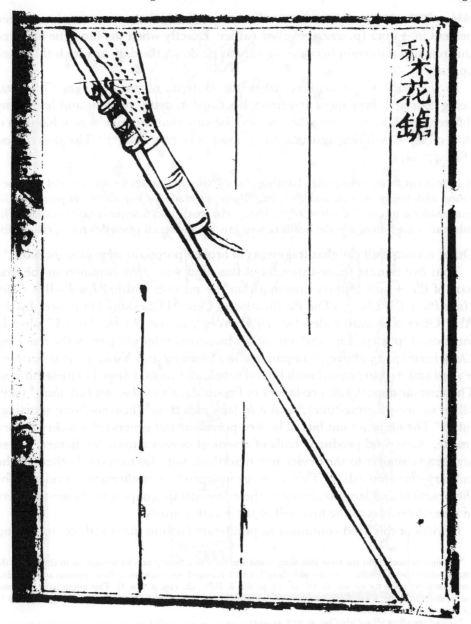

Fig. 46. The prototypic fire-lance called the 'pear-flower spear' (*li hua chhiang*). A tube of low-nitrate gunpowder attached to the front end of a lance or pike, it would act as a three-minute flame-thrower. From *HLC*, pt. 2, ch. 2, p. 24*a*.

magically efficient fire-spurting tube' (*tu lung phên huo shen thung*[1]).[a] It consists of a bamboo tube emitting poisonous flames, and was recommended, as usual, for defending city-walls. Then come the co-viative projectiles again, in their most highly divided small-particle form. The 'empyrean-soaring sand-tube' (*fei khùng sha thung*[2]) sends out flame and sand from a bamboo tube, with the intention of causing blindness when it gets into the eyes of the enemy.[b] Another similar weapon described also in the Fire-Drake Manual is the 'orifices-penetrating flying-sand magic-mist tube' (*tsuan hsüeh fei sha shen wu thung*[3]),[c] which spurts out flame, sand, poisons, sal ammoniac and many other chemicals (Fig. 48).[d] With a favourable wind, it is said, the mixture will exert its effects several *li* away, and if soporific drugs are added the enemy will not awaken so that they can easily be attacked. However that may be, the use of fire-lances primarily for generating smoke-screens or poison-smokes must go back rather a long way, for the *Wu Lin Chiu Shih* (written in +1270 concerning events around +1170), speaks of imperial army drill demonstrations in which smoke lances (*yen chhiang*[4]) were used.[e]

It will have been noticed from the preceding paragraphs that arsenical compounds were often ejected with the flames of fire-lance flame-throwers, and that was only one aspect of a tendency to mix poisonous substances with gunpowder which runs through all the formulae in Chinese texts from the +11th century onwards (cf. pp. 118, 123, 125). An almost contemporary experience of this can be found in the book of Mildred Cable and Francesca French,[f] two British missionaries who lived and worked in Sinkiang during the days of the warlords Ma Pu-Fang[6] and Ma Chung-Ying. In 1930 they were called upon to treat some soldiers whose wounds had been caused by 'fire-arrows' (more probably fire-lances) discovered in the old disused arsenal at Hami. 'These wounds were septic, and the flesh was charred as though burned by some chemical.' The suggestion of Cable and French that it was due to phosphorus is highly improbable, but mercury, often as the sulphide or one of the chlorides, was commonly used in such compositions in addition to arsenic,[g] and that could well have led to the effects described.[h]

And now we come to another great turning-point, the first appearance of the metal barrel. It is of cardinal importance that this occurred in connection with a flame-thrower and co-viative projectiles, not with high-nitrate gunpowder and

---

[a] *WPC*, ch. 129, p. 5*a, b*.
[b] *WPC*, ch. 129, pp. 7*a* to 8*a*.
[c] *Huo Lung Ching* pt. 1, ch. 2, p. 34*a, b*, *Huo Chhi Thu*, p. 25*a* and *WPC*, ch. 129, pp. 8*b*, 9*a, b*.
[d] The word *hsüeh*[5] used here is the same as that for the acupuncture-points, but the meaning was probably just the eyes, nose and ears. See Lu Gwei-Djen & Needham (5), pp. 13, 52.
[e] Ch. 2, p. 2*a* (p. 358). Cf. also what has been said above concerning coloured smokes in the sub-section on fireworks (pp. 144ff. above).
[f] (1), p. 241.
[g] Cf. *HLC*, pt. 1, ch. 1, p. 13*b*; *WPC*, ch. 119, p. 20*a, b*, ch. 128, p. 18*a*.
[h] As Davis & Ware (1), p. 525 suggested.

[1] 毒龍噴火神筒   [2] 飛空砂筒   [3] 鑽穴飛砂神霧筒
[4] 煙槍   [5] 穴   [6] 馬步芳

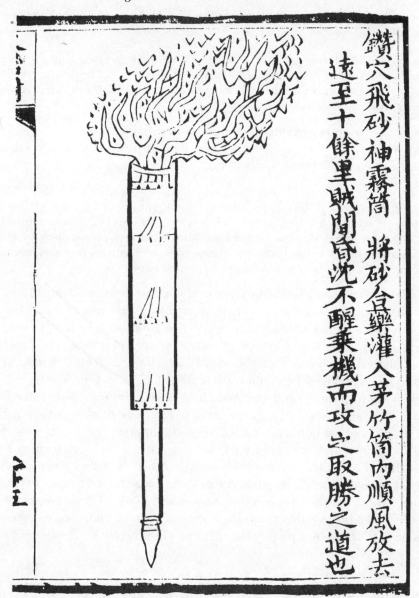

鑽穴飛砂神霧筒　將砂合藥灌入茅竹筒內順風放去遠至十餘里賊聞骨沈不醒乘機而攻必取勝之道也

Fig. 48. A fire-lance designed to send out sand or blinding dust together with sal ammoniac and poisonous chemicals along with the flames (cf. Figs. 26, 27). It was named the 'orifices-penetrating flying-sand magic-mist tube' (*tsuan hsüeh fei sha shen wu thung*); *HLC*, pt. 1, ch. 2, p. 34*a*, *b* and *HCT*, p. 25*a*.

an occlusive shot. The date is very difficult to fix, but the fact that the following description occurs in the first part of the *Huo Lung Ching* puts it in the earliest stratum of the weaponry there assembled, so that it must be before about +1350, and was probably as old as +1200, if we take into account all the other historical data we have. The name of the device was the 'bandit-striking penetrating gun' (*chi tsei pien chhung*[1]),[a] and the Fire-Dragon Manual describes it as follows (Fig. 49):[b]

The barrel is made of iron, 3 ft long, with a stock or handle 2 ft long, and the weapon is used by foot-soldiers. It has a range of 300 paces.[c] The enemy can be shot with pellets (at a distance) or struck with the gun itself (at close quarters), and the device is very useful because of this dual function.

We shall soon see other examples of metal-barrel flame-throwers.

Next we may consider a bottle-shaped flame-thrower called the 'phalanx-charging fire-gourd' (*chhung chen huo hu-lu*[2]). The *Huo Lung Ching* (Fig. 50) says:[d]

Its shape is like a bottle-gourd, and the interior forms the (ignition) chamber of the gun (*chhung*[3]) holding lead pellets and 1 *shêng* of 'poison gunpowder' (*tu huo*[4]).[e] The stock, made of hard wood, is 6 ft long. In action it is wielded by one brave soldier, in between men holding 'fire shields',[f] right in the front line. When enemy positions are charged with a detachment of such weapons, they cause panic among both men and horses. It is an efficient weapon for cavalrymen as well as foot-soldiers.

None of the sources tells us what the gourd itself was made of, but rather than being carved out of wood it is perhaps more likely that it was fashioned from metal. In this case we should very much like to know whether its bore was uniform inside or whether it followed the outer shape. This raises the question of the vase-shaped or 'ampulliform' bombards, which we shall have to discuss later on (p. 289); a significant common trait between China and the West.

Besides flame, posion, sand, porcelain fragments and metal pellets, the flame-throwers were also used to discharge arrows. This was another step on the way to the true gun or cannon, and it has the special interest that the earliest depiction of bombards in Europe (+1327, see Figs. 82, 83) shows them firing off arrows. But the Chinese illustrations usually draw the tubes with parallel or divergent straight sides rather than the bulbous vase-shaped forms of Walter de Milamete, though these were certainly known and used in China too, as we shall duly see (p. 329 below). Since nothing is ever said (in so many words) about the occlusion of the bore by a plug at the rear end of the arrow,[g] one supposes that it

---

[a] Two things are noteworthy about this name. First, *pien* was the ancient term for stone acupuncture needles, so the idea of the penetration of the projectile here had a very long literary background (cf. Lu Gwei-Djen & Needham (5), pp. 1, 70–1). Secondly the term *chhung*, afterwards universally used for the true gun and cannon, is seen now coming in.

[b] *HLC*, pt. 1, ch. 2, p. 18*a*, *b*. *Huo Chhi Thu* ed. p. 18*a*; tr. auct.

[c] This would mean 1500 feet or 300 yards, so it looks like a considerable exaggeration. The flames reached only 20 or 30 feet at most (p. 228 above). The pellets would hardly have had high velocity either.

[d] *HLC*, pt. 1, ch. 3, p. 14*a*, *b*; *Huo Chhi Thu* ed. p. 32*a*; *WPC*, ch. 130, p. 25*a*, *b*; tr. auct.

[e] See p. 180.

[f] We shall return to this subject shortly (p. 414 below).

[g] But some versions of the pictures seem to make an attempt to show what may have been a plug or wad, e.g. the illustration of the 'single-flight magic-fire arrow' in *HLC*, pt. 1, ch. 2, p. 25*a* (Fig. 51). And in other cases some kind of wad may be implied.

[1] 擊賊砭銃　　　[2] 衝陣火葫蘆　　　[3] 銃　　　[4] 毒火

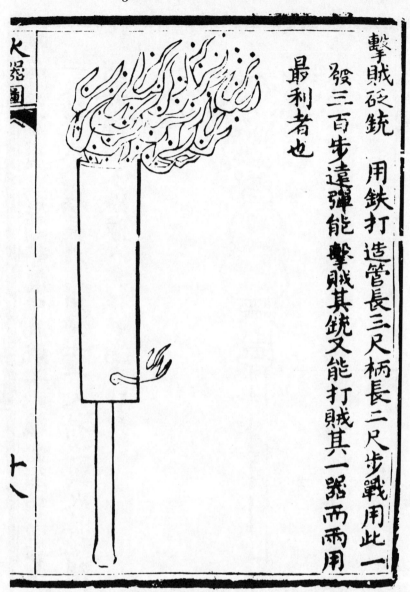

Fig. 49. The first of all metal barrels, not for high-nitrate gunpowder and a bore-filling projectile, but for a low-nitrate flame-throwing fire-lance and small co-viative missiles. The 'bandit-striking penetrating gun' (*chi tsei pien chhung*), in *HLC* (*HCT*), p. 18a.

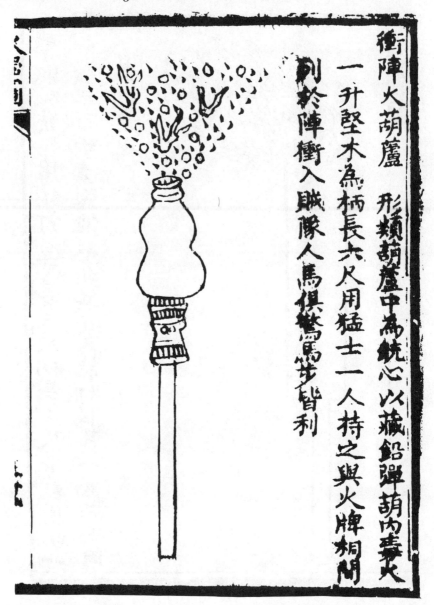

衝陣火葫蘆

形類葫蘆中為銃心以藏鉛彈葫內春火

一升堅木為柄長六尺用猛士一人持之與火牌相間

列於陣衝入賊隊人馬俱驚寫等皆利

Fig. 50. An ampulliform fire-lance with lead pellets as co-viative projectiles, the 'phalanx-charging fire-gourd'
(*chhung chen huo hu-lu*), in *HLC* (*HCT*), p. 32*a*.

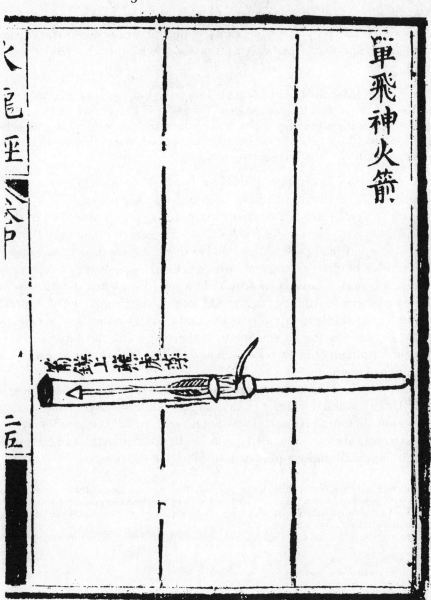

Fig. 51. Arrows as co-viative projectiles; the 'single-flight magic-fire arrow' (*tan fei shen huo chien*), from *HLC*, pt. 1, ch. 2, p. 25*a*. One can see here what may have been an attempt to depict a plug or wad filling the bore at the base of the arrow; if so the device may have approximated to a true gun (a 'proto-gun', cf. p. 251), but there is no evidence that the gunpowder was filled in only behind it.

was just shot out by the force and rush of the rocket-composition burning in the enclosed space (cf. p. 480).[a] One can see this situation, for instance, in the 'single-flight magic-fire arrow' (*tan fei shen huo chien*[1]). The *Huo Lung Ching* says (Figs. 51, 52):[b]

Use a barrel 3 ft long cast from high-grade bronze and designed to take only a single arrow. Put 0.3 oz. of 'blinding gunpowder' (*fa yao*[2])[c] as charge into the barrel before firing, whereupon the arrow is sent flying like a fiery serpent, with a range of between 200 and 300 paces.[d] It can pierce the heart or the belly when it strikes a man or a horse, and can even transfix several persons at once.

In the *Huo Kung Pei Yao* edition of the Fire-Drake Manual the caption indicates that the arrow-head should be tipped with bear or tiger poison.[e]

Then the 'magical (fire-)lance arrow' (*shen chhiang chien*[3]) described in the *Huo Lung Ching*, the *Ping Lu* and the *Wu Pei Chih*,[f] discharged amidst the flames not only an arrow but lead pellets as well. These were contained in some kind of wooden tube or holder (*mu sung tzu*[4]) which conceivably acted as a plug or wad. The weapon is said to have been acquired or developed when the Ming sent an expeditionary force to Annam;[g] this would refer to the campaigns of +1406 and +1410 under the able generals Chang Fu[5] and Chhen Chhia[6]. The special point about the device was that it was made of the very hard and heavy iron-wood (*thieh-li-mu*[7]), probably the barrel as well as the tiller.[h] Again, the arrow is said to have a range of about 300 paces.

The 'awe-inspiring fierce-fire *yaksha* gun' (*shen wei lieh huo yeh-chha chhung*[8])[i] of the Fire-Drake Manual (Fig. 53) discharged multiple arrows together with strong poisoned flames.[j] Its barrel was bound with rough cloth and many turns of iron wire, and its arrows came from a cradle ejected at the same time. This once again may well imply a plug or wad filling up the bore.[k]

---

[a] On the other hand the ranges given, up to 300 paces (500 yards), suggest (if they are not optimistic exaggerations) a primitive gun with fully occluded bore rather than co-viative projectiles sent out with the flames of a fire-lance. With that one would expect at most only a tenth of such a distance. Evidently we are here trembling on the verge of the true gun.

[b] *HLC*, pt. 1, ch. 2, p. 25 *b*; *Huo Chhi Thu* ed. p. 21 *b*; tr. auct. Cf. *WPC*, ch. 126, pp. 15 *b*, 16 *a*. Davis & Ware (1), p. 533.

[c] Cf. p. 180 above.

[d] Again this may be an exaggeration.

[e] *Huo Kung Pei Yao*, ch. 2, p. 25 *a*, *b*.

[f] *HLC*, pt. 1, ch. 2, p. 23 *a*, *b*; *Huo Chhi Thu* ed. p. 20 *b*; *Huo Kung Pei Yao* ch. 2, p. 23 *a*, *b* ch. 11, pp. 37 *a*–38 *b*; and *WPC*, ch. 126, pp. 9 *b*, 10 *a*.

[g] Cf. p. 311 below.

[h] We have come across this before, in Vol. 4, pt. 3, p. 416, in connection with shipbuilding. It may have come from palm-trees of Kuangtung and Annam such as *Sagus rumphii* or *Arenga engleri*; or it may have been the *Mesua ferrea* of Kuangsi, or the hemlock-spruce *Tsuga sinensis*.

[i] This name was derived from the ogres of Buddhism (and India) called *yakṣa*, supposed to devour human beings at night.

[j] *HLC*, pt. 1, ch. 2, p. 19 *a*, *b*; *Huo Chhi Thu* ed. p. 18 *b*.

[k] The text says: *yung chien mu chhê wei fa-ma*[9]; use a block of hard wood to make a common cradle.

[1] 單飛神火箭	[2] 法藥	[3] 神槍箭	[4] 木送子	[5] 張輔
[6] 陳洽	[7] 鐵力木	[8] 神威烈火夜叉銃		[9] 用堅木車爲法馬

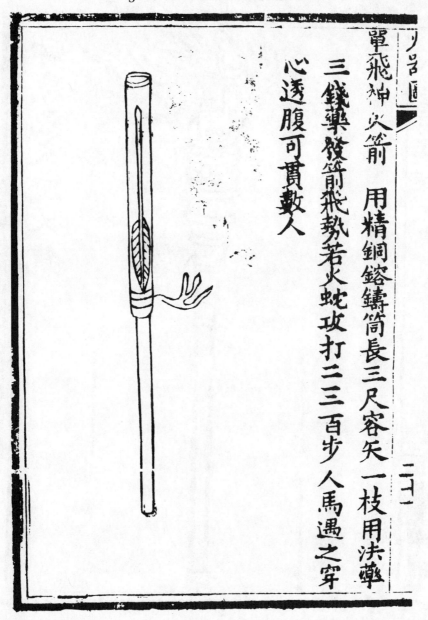

單飛神火箭　用精銅鎔鑄筒長三尺容矢一枝用決藥

三錢藥發筒前飛勢若火蛇攻打二三百步人馬遇之穿

心透腹可貫數人

Fig. 52. The picture of the same device (Fig. 51) from *HCT*, p. 21*b*; there is no sign of any wad. But the description of the penetrating power of the arrow suggests propellance rather than co-viative discharge. On the other hand, some versions indicate that the arrow-head was tipped with tiger-poison, so that its speed of flight need not have been great for it to do much damage.

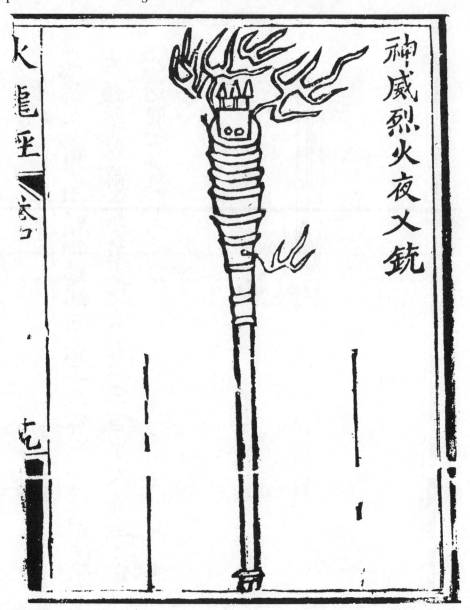

Fig. 53. Another fire-lance discharging arrows, the 'awe-inspiring fierce-fire *yaksha* gun' (*shen wei lieh huo yeh-chha chhung*), from *HLC*, pt. 1, ch. 2, p. 19*a*. The 'cradle' which held the arrows can be seen in the illustration, but it is not clear whether it occluded the bore, or whether, if it did, the gunpowder was filled in only behind it. It may have been a proto-gun, but its name is not very significant in this regard.

The 'lotus-bunch (*i pa lien*[1]) emitted flame and many small arrows, but its chief interest lies in the fact that it was the only other one of the whole series of arrow-firing flame-throwers which had a bamboo barrel (Fig. 54).[a] This was 2 ft 5 in. long, with all the septa removed except the end one (which was protected by clay), and a metal ring at the mouth. Outside it had to be bound with many layers of hempen cord, and cloth soaked in a saturated solution of alum to protect against the fire. Here there is no sign of a plug or wad.

The fire-tube or spurting-tube naturally took some considerable time to be reloaded after firing (when it was feasible to do this at all), so for dealing with enemy soldiers at close quarters the fire-lance remained the weapon of preference—just as in the West bayonets were fixed to muskets four hundred years later, from about +1650 onwards.[b] The prototypic 'pear-flower lance' (*li hua chhiang*[2]) described above (pp. 229 ff.) was soon elaborated in various ways. For instance, the number of barrels could be increased, and indeed the fire-lance (*huo chhiang*[3]) as such, of the military compendia, had twin flame-thrower tubes (Fig. 55). When one of these was burnt out, a fuse automatically ignited the second, thus prolonging the flames, and after that the halberd-like blade, knives and hooks of the lance-head came into play.[c]

We can also illustrate these fire-lances from a little-known source, the *Chhê Chhung Thu*[4] (Illustrated Account of Muskets, Field Artillery and Mobile Shields), written by Chao Shih-Chên[5] as an appendix to other military tractates in +1585. As Fig. 56 shows, he depicts[d] 'ten types of weapons for use by soldiers accompanying field-guns'. Two of these fire-lances have three barrels each, and are called 'the three-eyed lance of the beginning of the dynasty' (*kuo chhu san yen chhiang*[6]), and 'the miraculous triple resister' (*san shen tang*[7]) respectively. The double-tube one is also said to derive from the same time. That would take the prototype back to +1368, just the period of the *Huo Lung Ching*. The inscription says that in the *kêng-tzu* year (+1360) a Taoist of the Kung-tê Ssu[8] temple, more than a hundred years old at the time, first made the designs for these weapons, and transmitted them to posterity—a statement which links up in an interesting way with the Taoist connections described in Chiao Yü's preface (p. 28 above).

Of course various types of fire-lances also discharged co-viative projectiles. Even with bamboo tubes this was possible, as in the case of the 'winged-tiger gun' (*i hu chhung*[9]), a fire-lance sending forth lead pellets (*chhien tan*[10]) as well as

---

[a] *HLC*, pt. 2, ch. 2, pp. 26*a*, *b*, 27*a*; *Wu Pei Chih*, ch. 129, p. 6*a*, *b*. Cf. Davis & Ware (1), p. 529.
[b] Cf. Reid (1), pp. 144, 147, 172.
[c] *HLC*, pt. 2, ch. 2, p. 23*a*, *b*; *WPC*, ch. 128, pp. 2*b*, 3*a*. Cf. Davis & Ware (1), p. 524. The idea here was the same as that which generated the 'axe-pistols' and 'mace-pistols' in the European +16th and +17th centuries; cf. Blackmore (4), pp. 36–7, 39.
[d] Pp. 4*b*, 5*a*.

[1] 一把蓮　　[2] 梨花槍　　[3] 火槍　　[4] 車銃圖　　[5] 趙士楨
[6] 國初三眼鎗　[7] 三神攛　　[8] 功德寺　　[9] 翼虎銃　　[10] 鉛彈

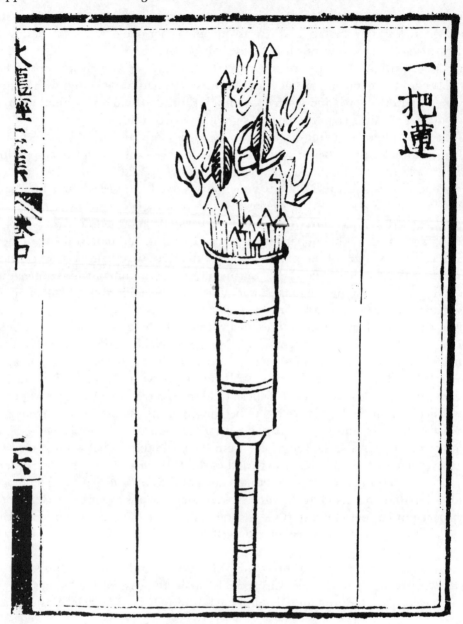

Fig. 54. A bamboo-tube fire-lance emitting many arrows along with the flames; the 'lotus-bunch' (*i pa lien*) from *HLC*, pt. 2, ch. 2, p. 26*a*. No sign of any wad or cradle.

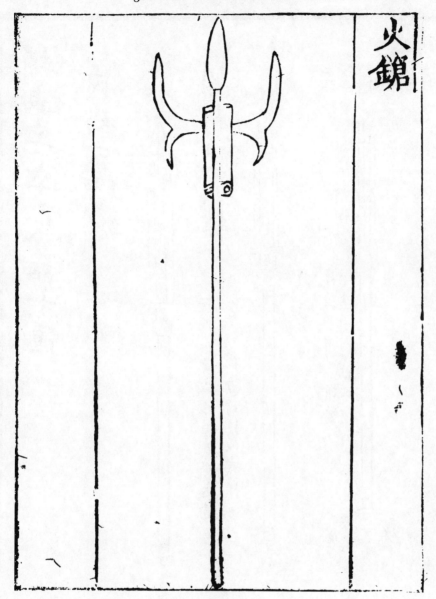

Fig. 55. Fire-lance with two tubes, the second of which ignited automatically when the first one had almost burnt out. *HLC*, pt. 2, ch. 2, p. 23*a*.

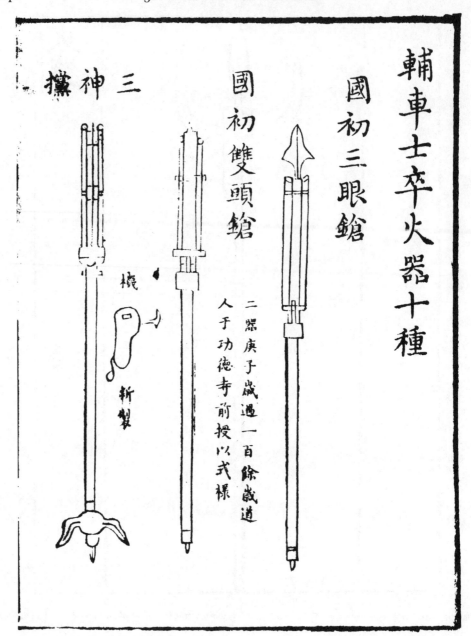

輔車士卒火器十種

國初三眼銃

國初雙頭銃

三神攛

二器庚子歲過一百餘歲道
人于功德寺前授以式樣

機

斬製

Fig. 56. Fire-lances from the *Chhê Chhung Thu* by Chao Shih-Chên (+1585). Double and triple tubes went back to the beginning of the dynasty, just the time of the first version of the *Huo Lung Ching*.

flames from three large barrels just behind the spear-point.[a] The 'wasps-nest of lead pellets' (*chhien tan i wo fêng*[1]) however, a fire-tube rather than a fire-lance, was carried on a bandolier, and shot out several hundred lead pellets at one firing from a metal tube.[b] This would have been a rather later stage of development.

Sometimes fire-lances approximated very closely to guns, with parallel-sided tubular barrels. For example, the 'horse-felling fire-serpent magically efficient cudgel' (*tao ma huo shê shen kun*[2])[c] in the Fire-Drake Manual and the *Wu Pei Chih*, is described as follows:[d]

It is made of wrought iron in the form of a hollow tube, which holds lead pellets and magic-fire gunpowder [mixed with poison-gunpowder]. It is 3 ft long and is fixed to a wooden stock 4 ft long. In practice it is held by a soldier to bring down horses in the front line of an attack. [Another way is to have two parallel tubes of iron, one like a musket for the lead balls, the other for the fire-lance flames; this is very useful in combat.]

The word *kun* (cudgel) may remind us of the names 'fire-stick' and 'thunder-stick' used outside China and Europe to denote light Western fire-arms.[e] Curiously, there was also a weapon looking very like a sword, and called the 'thunder-fire whip' (*lei huo pien*[4]).[f] This name must have been derived from a sword-like object described and illustrated in the *Wu Ching Tsung Yao*,[g] the 'iron whip' (*thieh pien*[5]), seemingly articulated, and presumably used for bashing the enemy about.[h] But the 'thunder-fire whip' was not like this, it was essentially a rigid fire-tube in the shape of a sword, made of bronze or iron and tapering to a small muzzle. It was 3 ft 2 in. long, with a gunpowder charge of 5 in., a small hole through the barrel-'blade' for the fuse, and a wooden hilt 4 in. long. Only a particularly strong man could wield it, and it discharged three lead balls as big as coins in diameter. This is a striking instance of a 'skeuomorph'.[i] In the history of technology there has always been a tendency for the shapes of objects in a new material to imitate the shapes which they had in older ones; so here, where the artisans were dealing with a proto-gun, they felt impelled to make it take the shape of that familiar weapon, the sword.

Somewhat analogous to this was the 'vast-as-heaven enemy-exterminating

---

[a] *HLC*, pt. 1, ch. 2, p. 17*a*, *b*; *Huo Chhi Thu* ed. p. 17*b*.

[b] *PL*, ch. 12, p. 34*a*, *b*; *WPC*, ch. 123, p. 21*a*, *b*. Cf. Davis & Ware (1), p. 534.

[c] Known as the 'horse-felling fire-cannon magic cudgel' (*tao ma huo phao shen kun*[3]) in the *Wu Pei Huo Lung Ching*.

[d] *HLC*, pt. 1, ch. 2, p. 29*a*, *b*; *Huo Chhi Thu* ed. p. 23*b*; *Wu Pei Huo Lung Ching*, ch. 2, p. 14*a*, *b*; *WPC*, ch. 128, pp. 10*b*, 11*a*; tr. auct. Words in square brackets are in the *WPC* text only. Cf. Davis & Ware (1), p. 528.

[e] But when the word 'stick' occurs in early European literature, it usually means the handle or tiller on which the hand-gun was mounted, as in *Stangenbüchse*; cf. Partington (5), p. 147.

[f] *HLC*, pt. 1, ch. 2, p. 31*a*, *b*; *Huo Kung Pei Yao*, ch. 2, p. 31*a*, *b*; *WPC*, ch. 128, pp. 14*b*, 15*a*.

[g] *WCTY/CC*, ch. 13, p. 14*a*, *b*.

[h] Articulation is not distinctly stated, but the drawing stands next to that of a war-flail. Cf. Vol. 4, pt. 2, pp. 70, 461 and Fig. 374.

[i] Cf. Vol. 2, p. 468.

[1] 鉛彈一窩蜂　　　　[2] 倒馬火蛇神棍　　　　[3] 倒馬火砲神棍
[4] 雷火鞭　　　　　　[5] 鐵鞭

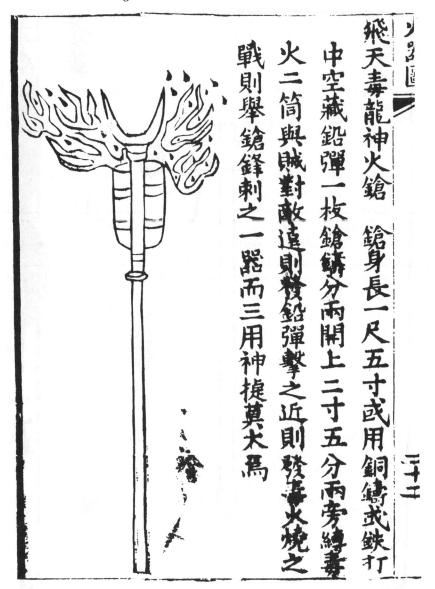

飛天毒龍神火鎗　鎗身長一尺五寸或用銅鑄武鐵打
中空藏鉛彈一枚鎗鏽分兩開上二寸五分兩旁綁毒
火二筒與賊對敵遠則噴鉛彈擊之近則發毒火燒之
戰則舉鎗鋒刺之一器而三用神搋莫犬焉

Fig. 58. Fire-lance and true gun combined in one weapon, the 'sky-soaring poison-dragon magically efficient fire-lance' (*fei thien tu lung shen huo chhiang*), from *HKPY*, ch. 2, p. 27 *a*, *b*, *HCT*, p. 22 *b*. The central barrel fired only one bore-occluding lead shot, after which poisoned flames issued from the two fire-lance tubes, and eventually the bifid lance could be used.

Fig. 59. A bamboo gun or cannon wrapped with raw-hide and rattan (preserved from China in British Museum (Museum of Mankind) no. 9572). It is said to show signs of considerable use, but is still in good condition (photo Clayton Bredt, 1).

seen in the 'magic-mechanism ever-conquering fire-dragon halberd' (*shen chi wan shêng huo lung tao*[1]), in which the chief difference is that the blades curve outwards like antennae rather than forwards like horns—but still only one lead ball is fired from the central barrel.[a] The fact that both these combinatory devices belong to the oldest stratum of the *Huo Lung Ching* must place them in the early part of the +14th century, if not some distance back in the +13th, remembering always that (as we shall see) the oldest small cannon of standard type belongs to about +1290.

The metal-barrel gun did not fully replace the bamboo-barrel proto-gun immediately because of the portability and wide availability of the latter. In fact the bamboo barrel was not eliminated until long after the introduction of the musket soon after the early part of the +16th century (cf. Fig. 59). Here is a description of a bamboo-barrel gun by Mao Yuan-I:[b]

Bamboo fire-lance (*chu huo chhiang*[4]).

Take a piece of cat-bamboo (*mao chu*[5])[c] 3 ft long, and drill through the septa (to get a barrel) like that of the iron 'bird-beak' musket (*niao chhiang*[6]). At the breech (the solid end is protected by) a layer of firmly compressed clay one or two inches thick; and just above it a touch-hole (*huo mên*[7]) is drilled for firing by a fuse or slow-match (*yao hsien*[8]). On the outside the tube is tightly bound with iron wire or hempen string, and sealed with earthenware sealing compound, ashes and lacquer. The inside of the barrel should be cleaned with a cleaning compound (*thang yao*[9]), and then loaded with 0·16 oz. of propellant (*chih hsing*[10]) gunpowder followed by a single lead bullet. One aims and shoots (with this weapon). It has the double advantage of being light and portable, and able to kill (*pi*[11]) a man instantaneously.

---

[a] *HLC*, pt. 1, ch. 2, p. 28a, b; *Huo Chhi Thu* ed. p. 23a; *WPC*, ch. 128, pp. 9b, 10a. Other examples include the *chia pa chhung*[2] (stock-clasping gun) of *HLC*, pt. 1, ch. 2, p. 21a, b; *Huo Chhi Thu* ed. p. 19b; *Huo Kung Pei Yao*, ch. 2, p. 21a, b; which has two barrels on each side of a trident staff, but we should not be able to interpret it if we did not know from *Ping Lu*, ch. 12, p. 34a, b that each barrel fires but one single lead shot. Here it is called *chia pa chhung*[3] (rake-handle-clasping gun).

[b] *WPC*, ch. 128, pp. 5b, 6a; tr. auct.

[c] A strong bamboo stem of large diameter frequently used in shipbuilding (cf. Morohashi dict., vol. 10, p. 690).

[1] 神機萬勝火龍刀　　　[2] 夾欄銃　　　[3] 夾耙銃　　　[4] 竹火鎗
[5] 猫竹　　　[6] 鳥鎗　　　[7] 火門　　　[8] 藥線　　　[9] 盪藥
[10] 直性　　　[11] 斃

the battlements providing fresh fire-lances for the defenders. When one remembers that each of these would probably burn out in five minutes or so, it would obviously have been very desirable to organise constant supplies, and mobile racks like these would have been very useful.

The use of the fire-lance continued to be recommended throughout the +16th century; as an example one could take the *Shen Chhi Phu Huo Wên*[1] (Miscellaneous Questions and Answers arising out of the Treatise on Guns) written by Chao Shih-Chên[2] in +1599.[a] Once again the *li hua chhiang*[3] makes its appearance, but now alongside all kinds of more modern things, such as mobile armoured shields for field-guns, bullet-moulds and muskets, and even a kind of primitive machine-gun.[b] The fire-lance was not yet quite dead. Indeed, the forms of it which projected arrows had been quite prominent in the successful operations of Chhi Chi-Kuang's 'new model army' against the Sino-Japanese pirates on the south-eastern coasts in the fifties and sixties of the century.[c]

There was still a place for the spurting-tube (*phên thung*[9]) as late as +1643, in the *Huo Kung Chhieh Yao*[10] of Chiao Hsü[11] and Adam Schall von Bell. The illustration[d] shows it as a fire-barrel, with a handle 4 ft long; it was certainly of bamboo bound with wire and string, and it emitted an arrow or lead shot as well as flames.[e] From the gunpowder formula given for it, which was rather high in nitrate,[f] one may guess that the projectile was no longer co-viative, but the weapon may have burst quite often and could hardly have been used more than a few times. By a striking coincidence, this date is the same as that for the last recorded use of fire-lances in the West, at the siege of Bristol, in the English Civil War.[g] Another bamboo-barrel proto-gun was the 'invincible bamboo general' (*wu ti chu chiang-chün*[12]) described by Ho Ju-Pin in the *Ping Lu* of +1606.[h] The barrel was fortified by a winding of iron wire, and the weapon fired a single stone ball; from the illustration (Fig. 61) it is hard to tell whether this completely filled

---

[a] Fig. 56 is taken from his *Chhê Chhung Thu*, some dozen years earlier in date.

[b] The same is true of Wang Ming-Hao's[4] *Huo Kung Wên Ta*[5], written a year or two earlier (c. +1598) and preserved in the *Huang Ming Ching Shih Shih Yung Pien*[6] of Fêng Ying-Ching[7] (+1603), ch. 16, p. 51a (pp. 1287–1318). Wang includes much on fire-lances, along with accounts of bombs, mines and sea-mines, breech-loading culverins and cannons small and large, muskets, rockets and rocket-launchers. Elvin (2), pp. 94 ff. was the first to draw attention to this piece, but we cannot associate ourselves with his estimate of Ming gunpowder technology. Wang Ming-Hao was also the author of important military books such as the *Têng Than Pi Chiu*[8].

[c] See Huang Jen-Yü (5), pp. 168, 179, 180, with references.

[d] Ch. 1 (Thu), p. 19b (p. 32).

[e] Ch. 1, p. 26a, b.

[f] Ch. 2, p. 10a, b. The percentage composition was: N, S, C; 74 : 4 : 22.

[g] Partington (5), p. 5.

[h] *PL*, ch. 12, p. 33a, b; *Phing Phi Pai Chin Fang*, ch. 4, p. 22a, b, description, pp. 20aff. Cf. *WPC*, ch. 123, pp. 9a–11b; Davis & Ware (1), p. 533.

¹ 神器譜或問   ² 趙士禎   ³ 藝花鎗   ⁴ 王鳴鶴   ⁵ 火攻問答
⁶ 皇明經世實用編   ⁷ 馮應京   ⁸ 登壇必究   ⁹ 噴筒
¹⁰ 火攻挈要   ¹¹ 焦勗   ¹² 無敵竹將軍

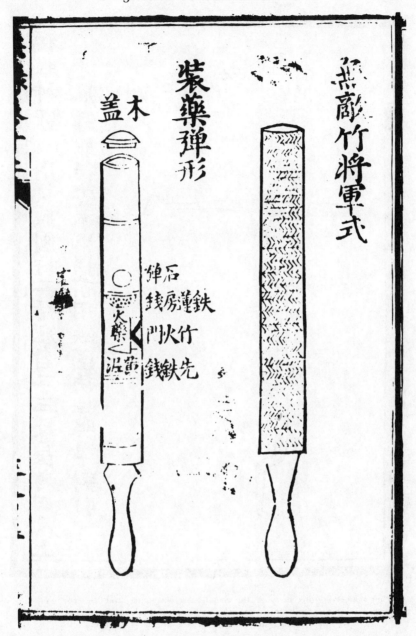

Fig. 61. A hand-held bamboo-barrel proto-gun, the 'invincible bamboo general' (*wu ti chu chiang-chün*), from *PL*, ch. 12, p. 33*a*. It fired a single stone ball. The wooden barrel-cap prevented the powder from getting wet.

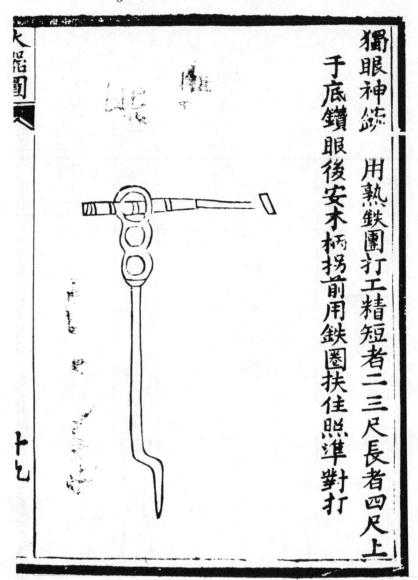

Fig. 62. Another connecting-link between the metal-barrel fire-lance and the true gun, the 'one-eyed magically efficient gun' (*tu yen shen chhung*), from *HCT*, p. 19*a*. It was a kind of gingall or swivel-gun, provided with a wooden tiller and fired from a pole with several rings as rests.

the bore, but it probably did.[a] The shape of the whole is sophisticated enough, but one would rather have been somewhere else when it was being let off.[b]

However, it remains true that the majority of fire-lances and fire-tubes that discharged projectiles, whether co-viative or not, described in the *Huo Lung Ching* and later books such as *Ping Lu* and *Wu Pei Chih*, had metal barrels. How closely they could approach to the true gun, arquebus or musket, can be seen from our final example, the 'one-eyed magically efficient gun' (*tu yen shen chhung*[1]), a kind of gingall,[c] fired with the help of a rest or support. The *Huo Lung Ching* says:[d]

This is made of wrought iron by a skilled smith. It can be as short as two or three feet, or else four feet or more. A hole is drilled underneath (i.e. at the back of) the gun, so that a wooden tiller can be attached to it. In front the gun is supported by an iron ring, which also serves the purpose of taking a better aim at the target.

The illustration (Fig. 62) shows the support, very reminiscent of the forked rests which were standard in Europe later on for matchlock muskets.[e]

Believe it or not, the fire-lance lasted down to our own times on the rivers and round the coasts of the South China Sea.[f] Cardwell, who got to know well the passenger-carrying and cargo junks of that region in the thirties, as also the pirate ships which preyed upon the traffic, has a remarkable picture (Fig. 63) of the fire-lances used for the defence of the junks.[g] They were, he said, a kind of Roman candle composed of a mixture of tow, wax, gunpowder and other ingredients, pressed in alternate layers into a length of hollow bamboo bound with rattan. Upon ignition at the muzzle, the tube was aimed at the attacking craft with the object of setting it on fire, or driving the helmsman from his post, by means of the cataract of sputtering fire and burning wads of tow, which could also do great damage to the pirate's sails. Many junks carried a good supply of these incendiary tubes. Another picture (Fig. 64), from a Japanese source, shows a passenger junk from Wuchow or Shao-chow, with fire-lances protruding from the bulwarks outboard ready to repel bandits whether in boats or on the river-bank.[h]

[a] The illustration specifies two iron coins, one below the gunpowder charge and one on top of it. The latter could have acted as a wad for a ball of less diameter than the bore.

[b] Elvin (2), p. 95, knew about this, and its seven advantages, from *Huo Kung Wên Ta* (p. 1302), but mistook it for some kind of mortar.

[c] This is a word not to be found in most military histories, but Hobson-Jobson knew it as a term for swivel- or wall-pieces (of ordnance) though unable to trace its origin. The editor of the second edition, however, felt able to derive it from Ar. al-Jazā'il, a 'heavy Afghan rifle fired from a fixed rest'. Ball (1), p. 44, considered it a musket from 6–14 ft in length, resting on a stand or tripod like a telescope.

[d] *HLC*, pt. 1, ch. 2, p. 20*a*, *b*; *Huo Chhi Thu* ed. p. 19*a*; *Huo Kung Pei Yao*, ch. 2, p. 20*a*, *b*; tr. auct.

[e] Cf. Reid (1), p. 61, and Fig. 180 below.

[f] Narratives of the Opium Wars in the eighteen-forties sometimes describe weapons that may have been fire-lances. Thus Ouchterlony (1), p. 262 speaks of long brass tubes, wound round with silk and catgut, found in a captured Chinese fort.

[g] (1), pp. 788, 794.

[h] Thanks are due to Mr Rewi Alley for this document.

[1] 獨眼神銃

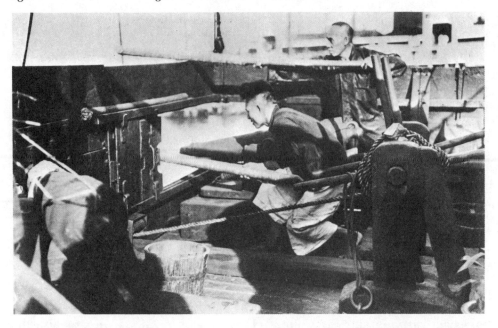

Fig. 63. Two long fire-lances, still used in the thirties in the South China seas (photo. Cardwell).

Fig. 64. Muzzles of two fire-lances projecting from the side of a Cantonese river-going passenger junk (1929).

Fig. 65. Drawing from the Arabic Rzevuski MS. of about +1320, showing on the left a soldier with a fire-tube held in the hand, and on the right another soldier with a naphtha flask or incendiary bomb in his right hand and a proto-gun or fire-tube in his left. After Partington (5), p. 207.

Was there now, one may well ask, anything similar to the fire-lance in Europe? There was indeed, and we can learn a good deal by following its fortunes. From the +14th to the +17th century we can recognise it under a variety of names deriving from Latin *tromba*, a trumpet.[a] The trombe we have already met with;[b] it was a metallurgical blower and mine ventilator, with a cascade of water descending into a closed space, through the outlet of which the air carried down blew forth in a continuous stream.[c] This was as old as the +8th century, and supplied the Catalan bloomery furnaces.[d] In the +14th 'trumba' was the name used for a bombard, particularly its fore-part, corresponding to the later usage of muzzle and chase.[e] But what we are looking for comes under the name 'trump', or *trompe à feu*,[f] and it was just as fearsome as the earlier weapons of the same kind in China.[g]

There are fire-lances in the book of Hasan al-Rammāḥ, *c.* +1280, just as one would expect if such Arabic circles were the means of transmission of Chinese fire-weapons westwards, and some of them may have had co-viative projectiles, for there is mention of 'Roman Candles' throwing out 'chick-peas' and incendiary balls of burning materials.[h] The fire-lances appear again in the Arabic Rzewuski MS of about +1320 (cf. p. 43 above), and in the drawings (Fig. 65) as

---

[a] It is curious that there was no Chinese parallel to this in some term derived from *la-pa*[1] (nothing to do with Lat. *labarum*, the imperial standard), or *hao chio*[2]. The Western name no doubt arose because of the snoring noise made by the tubes when giving out their flames.

[b] Vol. 4, pt. 2, pp. 149, 379.

[c] The principle was just the opposite of the familiar filter-pump. One can feel it in shower-baths today.

[d] Cf. Needham (32), p. 11.

[e] Partington (5), pp. 117–19; e.g. +1340 and +1376, +1379. Cf. Blackmore (2), p. 216.

[f] Hence *tromba di fuoco* (It.), *turonba* (Tk.) and *troumpa* (Byz., Gk). Cf. Kahane & Tietze (1), p. 449.

[g] A related group of words came from Lat. *troncus* or truncus, a tree-trunk or headless body. A trunk was a wooden support for a cannon, sometimes on wheels, cf. Partington (5), p. 182; Tout (1), p. 685. A truncke was a land-mine (p. 199 above); cf. Partington, *op. cit.* p. 166; Romocki (1), vol. 1, p. 275, fig. 65. The word 'trunnion' has the same origin—two cylindrical metal projections cast on a cannon to give an axis for elevation.

[h] See Partington (5), pp. 200 ff.

[1] 喇叭   [2] 號角

Fig. 66. An early appearance of the fire-lance in Europe, from a Latin MS. studied by Reinaud & Favé (1), fol. 199. The date would be about +1396.

well as the text.[a] Their first appearance in Western Europe occurs in the Latin MS. studied by Reinaud and Favé,[b] datable at about +1396; here we have drawings of a fire-lance used by a horseman (Fig. 66), another borne at the end of a chariot-pole, and another held by a dismounted knight.[c] The weapon is described again in the *De Re Militari* of Roberto Valturio, about +1460;[d] but for the most detailed account we have to wait for the *Pirotechnia* of Vanoccio Biringuccio of +1540.[e]

Biringuccio gives detailed specifications for fire-lances, 'tongues of fire', he says, 'to be tied on the ends of lances, like squibs'.[f] They are to be made of carton-paper 'in the form of rockets', and contain, just as in so many of the Chinese formulae, gunpowder plus x, y and z, for example pitch, sulphur, salt, iron filings, crushed glass, arsenic and other poisons. When lighted, they send

[a] Partington, *ibid.* p. 207.
[b] (1), pp. 213 ff., 217–18, 279–80. Lat. MS. Bib. Roy. 7239, done between +1384 and +1444, most probably +1395/6; Italian.
[c] Pls. 8, 10, 11, in their book.
[d] Reinaud & Favé (1), pp. 224, 226; Partington (5), pp. 146, 164.
[e] Tr. Smith & Gnudi (1). Cf. Brock (1), p. 30; Partington (5), p. 61; Reinaud & Favé (1), pp. 170, 229, pl. 14 fig. 1.
[f] Bk. 10, ch. 7, Smith & Gnudi tr., p. 433.

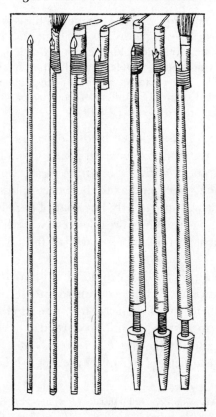

Fig. 67. Illustrations of fire-lances from the *Pirotechnia* of Biringuccio (+1540). Bk. 10, ch. 7, p. 433.

out 'a very hot tongue of flame more than two *braccia* long,[a] full of explosions and horror', and they are as useful at sea as on land (Fig. 67). Parallel with this, in his chapter on fireworks, Biringuccio describes trunks or *trombe di fuocho*, cylinders like Roman candles for the projection of fire-balls.[b] It was the custom, too, in the +16th and +17th centuries, for state processions to be headed by men like Jack-in-the-green holding 'clubbs' which spouted forth fire in a continuous stream; this happened on the occasion of Anne Boleyn's coronation in +1533, and is illustrated on the title-page of John Bate's book of +1635.[c]

But the *pièce de resistance* of fire-lances in late medieval Europe was the defence of Malta by the Knights of St John against the Turks in +1565.

The trumps [wrote Bradford][d] were hollowed-out tubes of wood or metal secured to long poles. Like the pots of wildfire[e] they were filled with an inflammable mixture,

---

[a] I.e. two or three yards.
[b] Bk. 10, ch. 10, Smith & Gnudi tr. pp. 441–2; cf. Brock (1), p. 30.
[c] Brock (1), p. 32, and opp. p. 17.
[d] (1), pp. 97–8, cf. pp. 105, 120; (2), p. 241.
[e] I.e. incendiary grenades containing saltpetre, sulphur and various carbonaceous combustibles.

except that it was made more liquid by the addition of linseed oil or turpentine. 'When you light the trump', wrote one authority,[a] 'it continues for a long time snorting and belching vivid furious flames, and large, and several yards long.' The trump derived its name from the harsh snoring sound it made when alight. A smaller version was attached to the head of a pike. This often had an ingenious mechanism whereby, when it was almost burnt out, it fired two small cylinders of iron or brass which were loaded with (ordinary) gunpowder, and discharged lead balls.[b]

Such were the *trombe de fuego* mentioned by di Correggio, writing only a couple of years later.[c] The doctored gunpowder used in the grenades and fire-lances was enhanced by more saltpetre, with the addition of sal ammoniac, sulphur, varnish, camphor and pitch,[d] very similar to the earlier Chinese compositions, and its anti-personnel effect was apparently like that of napalm. The opinion of the Victorian military engineer, Whitworth Porter, was that these trumps must have constituted 'a most formidable obstacle to the advance of any storming party'.

After the siege of Malta, anything would be an anticlimax, so it may suffice to say that the fire-lance in Europe continued in use down to the middle of the +17th century, when it was replaced by more modern guns and artillery. Diego Ufano (1) described it in his military treatise of +1613, and so did the pyrotechnists Appier and Thybourel a few years later.[e] Its final appearance seems to have been at the siege of Bristol in the English Civil War in +1643.[f]

---

[a] Bosio (1), vol. 3, pp. 561–2, a word-for-word translation from the old Italian.

[b] 'As if they were wheel-lock muskets', said Bosio, so they seem not to have been co-viative, but it is hard to be sure. This weapon, combining as it did the fire-lance and the gun, is extraordinarily reminiscent of the Chinese triple-function devices described on pp. 248, 251 above, and it is hard to believe that there could have been no connection between them.

[c] Eng. tr. by Balbi, p. 79.

[d] Porter (1), vol. 2, p. 97.

[e] Appier & Thybourel (1), p. 58 in +1620; Appier (1), p. 164 in +1630. Cf. Partington (5), pp. 176–7.

[f] Partington, *op. cit.* p. 5. Perhaps the existence of the fire-lance till this time could illuminate certain literary allusions otherwise hard to explain. For example, in the version of 'Tom o'Bedlam's Song' written by Giles Earle in +1615 the madman says:

> With an hoste of furious fancies
> 　　Whereof I am commander
> With a burning speare, and a horse of aire
> 　　To the wilderness I wander;
> By a knight of ghostes and shadowes
> 　　I summoned am to tourney
> Ten leagues beyond the wide world's end—
> 　　Methinks it is no journey...

There are several other versions of this, as in Percy's *Reliques* (+1765), vol. 2, p. 370. Tom was one of the 'Bedlam Beggars', so named after the Bethlehem Mental Hospital in London, founded in +1547 after the suppression of the abbeys (complete by +1540) which had previously harboured the psychologically deranged.

Similarly there was the *Knight of the Burning Pestle*, one of the comedies by Francis Beaumont & John Fletcher, printed in +1613; it was (like *Don Quixote*) a burlesque on knight errantry (Bowers, 1). Here the Grocer Errant had a burning pestle on his shield, reminiscent of the 'clubbs' mentioned on p. 261 above. And in *Amadis de Gaul*, a prose romance printed early in the +16th century, there had been a 'knight of the burning sword' (Hattaway, 1). Without overlooking other aspects of this symbolism, one cannot but draw attention, hitherto rather neglected, to the presence of fire-lances among European weaponry in the +16th and early +17th centuries.

When we survey the origins and development of the fire-lance in the Western world, we are at once impressed by the fact that it seems to have started there with no antecedents. The bombard was in Europe by +1327, and the fire-lance very probably accompanied it since there are several illustrations before the end of the same century. In Europe one cannot trace any long prior development similar to that which takes the fire-lance back in China to the middle of the +10th century.[a] This is surely circumstantial evidence that both weapons came to the West already fully fledged as it were, after which the cannon had a long development yet to undergo, while the fire-lance was probably very similar in the mid +17th century to what it had been like in the mid +14th. And it is interesting that in Europe, just as in China, it was still found useful down to that date, only then succumbing to the new and more efficient firearms of the time.

Another point well worth emphasising here is that the metal barrel did not have to await the coming of the true gun and cannon in China; on the contrary it was specified for many types of fire-lance, where the design was that of a close-quarters incendiary flame-thrower, even when combined with co-viative projectiles. We shall find that the same is true for those large-scale flame-throwers mounted on carriages or trunks, and sending out co-viative objects, even including proto-shells. It is to a brief examination of these that we must now turn.

(14) THE ERUPTOR, ANCESTOR OF ALL CANNON

So far all the weapons of fire-lance type which we have been considering were wielded by a single combatant, or else stacked in a mobile trolley which could be manoeuvred by several men. But when we come to 'fire-lances' with large-diameter tubes mounted on frames, like the *arcuballistae* of old (cf. pt. 6, (*f*), 3 above), we have to turn over a new page. Several of these are described and illustrated in the military compendia from +1350 onwards, but their character is so archaic that they must surely belong, at any rate in their earliest forms, to the previous century. Let us look at a few examples.

To begin with, there was the 'multiple bullets magazine eruptor' (*pai tzu lien chu phao*[1]). As we know, the term *phao* originally meant the trebuchet, and the stone projectile, or later the bomb, which was hurled from it, while later still it came to mean in common parlance any kind of cannon; but there was an intermediate phase when the gunpowder was low in nitrate, and the projectile did not fit the bore. It was for this gargantuan fire-lance that we felt the need to coin

---

[a] Not much is heard of Greek Fire petrol flame-throwers after the +12th century in the Byzantine region, and whether they were made use of in the later Crusades is uncertain; so there is no reason for the belief that the European fire-lances were derivative from them. The new factor was essentially gunpowder, and the existence of that in Europe before +1300 is hard to substantiate. Cf. p. 272 below.

[1] 百子連珠砲

the word eruptor, and we use it here. Of this magazine eruptor the *Huo Lung Ching* says:[a]

It is made of cast bronze, and measures 4 ft 5 in. in length. It contains 1·5 *shêng*[b] of 'blinding fire' gunpowder[c] which sends forth (flames) from the muzzle. At the side of the barrel a beak(-shaped tube) is cast on, rather more than a foot in length, and it is filled with a hundred or so lead balls. A frame of hard wood is made for the carriage, and on it the eruptor can be rotated in all directions. First the magazine is held horizontal, but when it is turned vertically the lead bullets all fall down into the firing chamber, and are spewed forth at the enemy soldiers one after the other, hitting them and preventing them from assaulting one's camp. One such eruptor can resist as many as fifty determined soldiers of the opposite side.

From the illustration (Fig. 68) one can see that the bronze tube was provided with a tiller (*yen wei*[3]), and the axis on which it was turned to aim is visible underneath the barrel. From the text we visualise that the magazine was filled while the eruptor was on its side, then immediately after ignition the barrel was turned so that the magazine pointed upwards, allowing the projectiles to slip down and be shot forth with the flames. It would seem quite certain here that the diameter of the balls must have been much smaller than that of the barrel; assuredly they were co-viative.

Perhaps the greatest surprise of this genre is that the eruptors could toss over shells. They must have popped out like the 'stars' from Roman candles or 'pumps', each one lighting the 'blowing charge' of the next one beneath it before leaving the tube,[d] but clearly they were capable of landing on the top of city walls in sieges. Moreover, in some cases they carried 'bursting charges' as well as 'lifting charges', for they would explode when they got to their destination.[e] For example, there was the 'flying-cloud thunderclap eruptor' (*fei yün phi-li phao*[4]). The text reads:[f]

The shells (*phao*[5]) are made of cast iron, as large as a bowl and shaped like a ball.[g] Inside they contain half a pound of 'magic' gunpowder (*shen huo*[6]).[h] They are sent flying towards the enemy camp from an eruptor (*mu phao*[7]); and when they get there a sound like a thunder-clap is heard, and flashes of light appear. If ten of these shells are fired successively into the enemy camp, the whole place will be set ablaze and his men will be thrown into confusion. [You can use any of the kinds of gunpowder in the shells—

---

[a] *HLC*, pt. 2, ch. 2, p. 6a, b; *WPC*, ch. 122, p. 13a, b; tr. auct. Cf. Davis & Ware (1), p. 528.
[b] The *shêng*[1] was a liquid and cereal measure often translated as pint, though perhaps better as gill; here it might be equivalent to lb., or rather less.
[c] On this translation of *fa yao*[2] cf. p. 180 above, and *HLC*, pt. 1, ch. 1, pp. 7a, 8a.
[d] See Brock (1), pp. 192–3.
[e] *Ibid.* p. 211.
[f] *HLC*, pt. 2, ch. 2, p. 8a, b; *WPC*, ch. 122, pp. 18a, 19a; tr. auct. Cf. Davis & Ware (1), p. 530.
[g] Cf. what has been said on pp. 163, 176 above regarding the *chen thien lei*[8] or thunder-crash cast-iron bombs used early in the +13th century.
[h] Formula in *HLC*, pt. 1, ch. 1, p. 6a.

---

¹ 升   ² 法藥   ³ 燕尾   ⁴ 飛雲霹靂砲   ⁵ 砲
⁶ 神火   ⁷ 母砲   ⁸ 震天雷

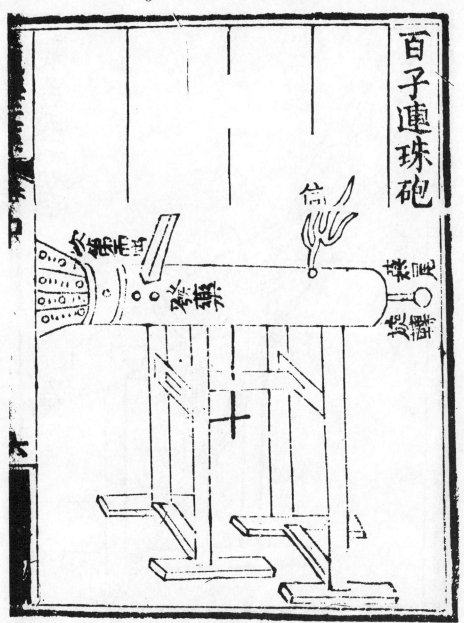

Fig. 68. An eruptor, i.e. a large fire-lance on a frame. This one, the 'multiple bullets magazine eruptor' (*pai tzu lien chu phao*) is taken from *HLC*, pt. 2, ch. 2, p. 6a. The magazine is filled with lead shot when it is on its side, then when the tiller is turned round on its axis they are fed into the barrel and issue forth along with the flames.

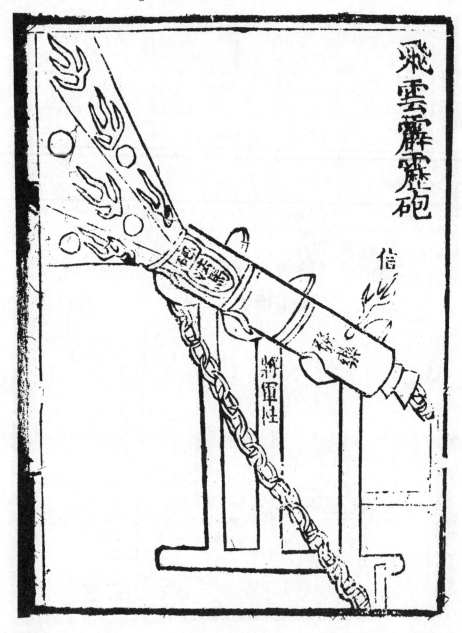

Fig. 69. An eruptor which fired proto-shells, i.e. gunpowder-containing cast-iron bombs; the 'flying-cloud thunderclap eruptor' (*fei yün phi-li phao*), from *HLC*, pt. 2, ch. 2, p. 8*a*. The proto-shells evidently did not fit the bore, but some kind of wad or cradle for them is shown (the *fa-ma*). The belching forth of the low-nitrate gunpowder must have been sufficient to send them on their way.

blinding powder (*fa yao*[1]), flying powder (*fei huo*[2]), violent powder (*lieh huo*[3]), poison powder (*tu huo*[4]), bruising and burning powder (*lan huo*[5]), and smoke-screen powder (*shen yen*[6]), according to the circumstances.][a]

These proto-shells can be seen in the illustration (Fig. 69), which shows well enough that they did not fill the bore. Underneath, the rotating axis which allowed of aiming the eruptor in different directions is called the 'general's column' (*chiang-chün chu*[7]).

Here it would be natural to ask when the shell, i.e. the cannon-ball which itself carries a charge of gunpowder exploding on impact, and is therefore essentially a propelled bomb, arose in the history of European warfare. The answer points to the early decades of the +15th century, because while the 'dracon' of Konrad Kyeser in his *Bellifortis* of *c.* +1405 is only a bomb, the shell is clearly present and described in the anonymous *Feuerwerkbuch* of about +1437. After Valturio's *De Re Militari* of +1460 shells become commonplace, but a good deal of time must have passed before they became reliable and effective.[b] Valturio's shells were clearly intended to burst, wrote Partington, but it needed probably a century more before the difficulties about the fuse were fully overcome.[c] From the passage just given, the shells from the eruptors of the *Huo Lung Ching* also burst on reaching their target. If the second part of this work is dated in the +16th century, developments in China and Europe were going on simultaneously, but we have already mentioned our conviction that the fire-lances and eruptors were archaic devices, to be placed before +1350 and indeed before +1290, so that the proto-shells here described may really have been among the first of their kind.

Other eruptors used shells designed to spread poison-smokes among the defenders of a city wall. The 'poison-fog magic-smoke eruptor' (*tu wu shen yen phao*[8]) is thus described in the *Huo Lung Ching* (Fig. 70);[d]

If blinding gunpowder (*fa huo*[9]), flying gunpowder (*fei huo*[10]), poison gunpowder (*tu huo*[11]) and spurting gunpowder (*phên huo*[12]) are filled into a shell (*phao*[13]) and fired at the top of a city wall, fire will break out and smoke will spread in all directions as the shell explodes. Enemy soldiers will get their faces and eyes burnt, and the smoke will attack their noses, mouths and eyes. If the right moment is chosen, no defenders can withstand such an attack.

The description of the 'heaven-rumbling thunderclap fierce fire eruptor' (*hung thien phi-li mêng huo phao*[14]) is more explicit about the poisons used in the smoke-shell.[e] These include wolf dung,[f] sal ammoniac, arsenical salts, soap-bean

---

[a] The formulae for the first, second, fourth and fifth of these are all in *HLC*, pt. 1, *loc. cit.* ff. The passage in square brackets is in *WPC* only.

[b] Cf. Partington (5), pp. 149, 157, 164–5.   [c] Hime (1), pp. 192 ff., 195, 202.

[d] *HLC*, pt. 2, ch. 2, p. 9a, b, tr. auct. Cf. *WPC*, ch. 122, pp. 23b, 24a. Davis & Ware (1), p. 529.

[e] *PL*, ch. 12, p. 15b; *WPC*, ch. 122, pp. 21b, 22a.

[f] This produced a particularly heavy smoke, and was therefore used in the signals system of the Ming forts along the northern border, but in the end it became very hard to get, especially in the south; Serruys (2), p. 19.

[1] 法藥   [2] 飛火   [3] 烈火   [4] 毒火   [5] 爛火
[6] 神煙   [7] 將軍柱   [8] 毒霧神烟砲   [9] 法火   [10] 飛火
[11] 毒火   [12] 噴火   [13] 砲   [14] 轟天霹靂猛火砲

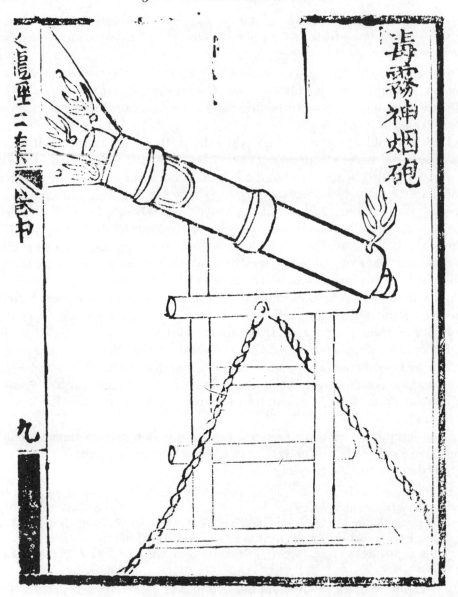

Fig. 70. Other eruptors used proto-shells to give forth clouds of poisonous smoke when fired so as to reach the enemy's city walls. This one is the 'poison-fog magic-smoke eruptor' (*tu wu shen yen phao*), depicted in *HLC*, pt. 2, ch. 2, p. 9a.

powder, pepper and croton oil, among other things, and from the name one would expect that some petrol came in somewhere. The illustration (Fig. 71) shows no discrete bombs or shells, but the text is clear that they were present and contained the poison-smoke ingredients.

After what we have seen for fire-lances, it would be only natural to find eruptors designed to shoot forth arrows as well as flames. Such a missile projector

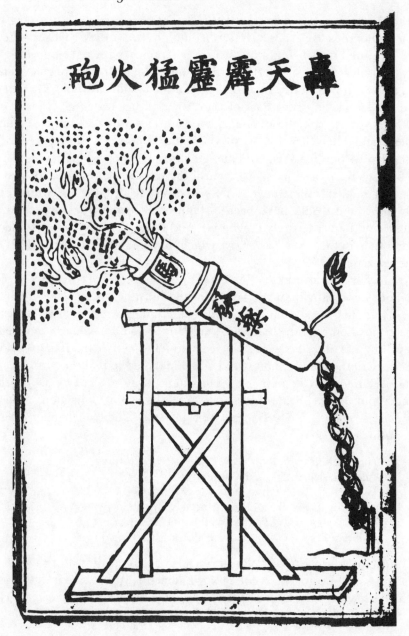

砲火猛靂霹天轟

Fig. 71. Another smoke-shell eruptor, the 'heaven-rumbling thunderclap fierce fire eruptor' (*hung thien phi-li mêng huo phao*), from *PL*, ch. 12, p. 15*b*. Arsenic, pepper and croton oil were constituents of the smoke, and no proto-shells are shown, but the wad or cradle for them is there.

was the *chiu shih tsuan hsin shen tu huo lei phao*[1] (nine-arrow heart-piercing magic-poison thunderous fire eruptor) described in the *Wu Pei Chih*.[a] This was designed to fire off nine arrows simultaneously, each tipped with tiger-hunting poison, from a cast bronze barrel 3 ft 8 in. long, mounted on a framework with arrangements for varying altitude and direction of aim. The illustration (Fig. 72) shows that the tiller was of iron. Part of the text is rather obscurely worded, but it seems to say that: 'sometimes one uses a cloth bag (or bags) full of "flying gunpowder", and when they (the arrows) are loaded like this, it has the advantage that the arrows don't shake about and get into confusion.' This can hardly mean that the bags were used like shells, but if the bags were attached to each arrow like sausages, they might have done something to occlude the whole bore, in which case there would have been an approach to the true cannon, with the projectiles no longer entirely co-viative. And indeed the projector is referred to now and then in the text as a *chhung*[2], which may be significant in understanding how it worked.

It seems fairly clear that in all these strange weapons the co-viative projectiles were more important than the flames of the burning gunpowder, for it would have been difficult to station enough of them in the protective lines of a camp or defensive position, and the hand-held fire-lances would have been more effective for repelling assaults. So we really seem to have here a final stage before the appearance of the true cannon with its ball matched to its bore.

There seem to be references to eruptors in poetry too. Chang Hsien[3], who was writing about +1341, has in his *Yü Ssu Chi*[4] (Jade Box Collection) a poem entitled *Thieh Phao Hsing*[5], which might be translated 'The Iron Cannon Affair'.[b] It starts in this way:

> The black dragon[c] lobbed over[d] an egg-shaped thing
> Fully the size of a peck measure it was,
> And it burst, and a dragon flew out with peals of thunder rolling.
> In the air it was like a blazing and flashing fire.
> The first bang was like the dividing of chaos in two,
> As if mountains and rivers were all turned upside down....

This must surely refer to a shell sent forth from an eruptor, but the rest of the poem shows that people were not very frightened of it, because it did little harm and 'its bark was worse than its bite'. But it would seem that in certain circumstances eruptors could have been more fearsome weapons.

---

[a] Ch. 127, pp. 8*b*, 9*a*, tr. auct.
[b] Ch. 3, p. 27*b* (p. 765·1). The word *phao* in the title is only a variant of the more usual *phao*[6].
[c] Black probably because of the black smoke emitted with the flames. Cf. Wang Ling (1), p. 172.
[d] We translate thus because the verb used is *to*[7], to fall or to let drop, suggesting a mortar-like trajectory.

[1] 九矢鑽心神毒火雷砲   [2] 銃   [3] 張憲   [4] 玉笥集
[5] 鐵礮行   [6] 礮   [7] 墮

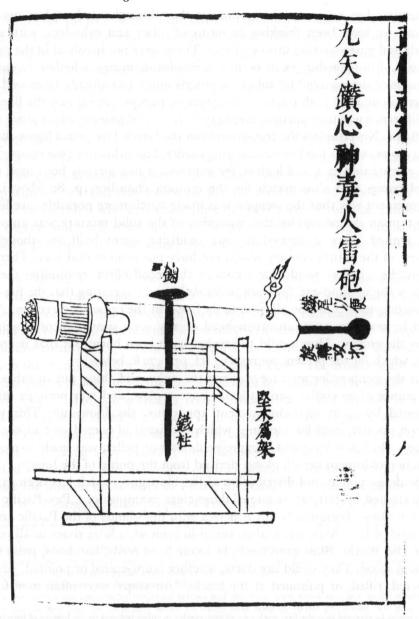

Fig. 72. An eruptor with a barrel of cast bronze and a tiller of iron, designed to shoot forth nine arrows simultaneously along with the flame and smoke. From *WPC*, ch. 127, p. 8*b*. The arrows were contained in cloth bags.

and history of the blow-gun. The reason for the connection was that although today only the Saisiat and Tsou folk retain the crossbow, it has a tube of bamboo at the head to guide the arrow-dart as it is sent forth;[a] and Thang was therefore led to suggest that the crossbow itself in Asia was the product of a marriage between the simple bow and the blow-gun.[b] However this may be, there can be no doubt that references to the blow-gun occur in ancient Chinese literature. For example, Tso Ssu[1] in his *Wu Tu Fu*[2] (Ode on the Capital of the Wu Kingdom) spoke in *c.* +270 of the 'cinnamon-tree arrows shot from tubes' (*kuei chien shê thung*[3]);[c] and in the *Chu Phu*[4] (Treatise on Bamboos) about +460 Tai Khai-Chih[5] referred to the *yün tang* bamboos as useful for shooting-tubes (*yün tang shê thung*[6]).[d] Again, Fan Chho[7], in the *Man Shu*[8] of *c.* +862 in the Thang (Monograph on the Southern Barbarians, i.e. Minority Peoples) mentioned the *pai chi chu*[9] bamboos which were useful for making blow-guns (*chhui thung*[10]). As would be expected, the mentions become rarer in more recent literature. But enough is there to show that the blow-gun was quite widespread among the people of South China and Thaiwan in ancient times,[e] and therefore that the inclusion of projectiles in tubes, when the gunpowder mixture at last became known, was something which had already had a very long history behind it.

There is yet another matter on which something must be said before we can leave the realm of fire-lances, eruptors and co-viative projectiles. More recent centuries have also known volleys of complex and discrete objects—what was the difference then between co-viative projectiles and chain-shot? The answer is that after the +17th century the fragments were always put together in some sort of casing which fitted the bore of the cannon or gun; leaving them free amidst the erupting gunpowder was a much earlier stage of evolution.[f]

'Chain-shot' itself, for instance, consisted of two cannon-balls joined together by a chain or iron bar, which, when fired from a gun, rotated at great speed through the air, smashing the spars and rigging of an enemy ship and clearing her upper deck of men.[g] Since the balls issued from the muzzle in succession there was generally no need for a casing. But 'case-shot' always had this.[h] In +1644 Mainwaring described it as 'made of any kind of old iron, stones, musket-bullets or the like, which we put into cases to shoot out of our great ordnance'. These cases were made preferably of wood, fitting the bore, or simply canvas

---

[a] This is still often poison-tipped, linking up with a wide area of practice both in the Old World and the New. Probably it was the very weakness of the propulsive force which led to the intensification of the effect caused by a hit.

[b] Cf. pt. 6, (*e*) above on the slur-bow.   [c] *Wên Hsüan*, ch. 5, p. 6*a*; tr. von Zach (6), vol. 1, p. 60.

[d] P. 4*a*; tr. Hagerty (2), p. 395. Both these scholars took the words *shê-thung* to be the name of a species of bamboo. This may well be, but from the commentators it can be seen that the argument is not affected.

[e] It has persisted among the tribal minority peoples of the South-west till contemporary times, and was observed among the Semang people of the Leichow peninsula in Kuangtung early in the present century by Imbert (1).

[f] We are grateful to Prof. Robert Maddin of the University of Pennsylvania for raising this point.

[g] Kemp (1), p. 150.   [h] *Ibid.* p. 143.

[1] 左思   [2] 吳都賦   [3] 桂箭射筒   [4] 竹譜   [5] 戴凱之

[6] 篔簹射筒   [7] 樊綽   [8] 蠻書   [9] 白箕竹   [10] 吹筒

Fig. 73. 'Langrage', or fragments of old metal enclosed in canvas bags, fired from a small cannon on the deck of a junk in the South China seas in the thirties (photo, Caldwell, 1).

bags which would do so. 'Canister-shot' was usually the same thing, put in cylindrical tin boxes, while 'grape-shot' was a number of iron balls bound together in a receptacle with canvas sides and circular cast-iron plates at top and bottom. Finally 'langrel' or 'langrage' consisted of iron bolts, nails, jagged fragments and any old metal pieces, enclosed in a thin cloth bag to fit the bore of the gun; it was a favourite weapon of privateers attacking merchant-ships.[a] In fact, in Chinese waters as recently as the thirties of the present century, merchant junks responded against pirates with just the same coin, as is seen in one of the photographs of Caldwell (Fig. 73). So, to sum it up, all the varieties of case-shot belonged to the era of true guns and cannon when the projectile always fitted the bore and had high-nitrate gunpowder behind it, while the co-viative projectiles were simply mixed with the low-nitrate gunpowder of the fire-lance, spurting-tube or eruptor, and issued together with the flames, obviously with much less force behind them, and consequently a much less range. In fact it was an earlier chapter in the story.

It is noteworthy that the sharp distinction which we draw between the co-viative projectiles of fire-lances and eruptors (even when their barrels were made of metal), and the full application of the propellant force of gunpowder upon projectiles that fitted the bore or calibre of the barrel, would have been fully appreciated in the +14th century by Chiao Yü himself. For in the earliest stra-

---

[a] *Ibid.* p. 465. For an eye-witness account of case-shot used in the English Civil War (+1648) See Temple (1).

tum of the *Huo Lung Ching*[a] there is a brief discussion of the composition of shells filled with combustible material calculated to set the enemy's works on fire (*huo tan yao*[1]). Here we find the remark that 'the size of the (incendiary or poisonous) shell must be just right to fit the bore of the iron tube; i.e. the gun or cannon (*nai yao yü thieh thung ho thang khou*[2])'. Chiao Yü would certainly have been quite clear about the great divide in this story.

(15) GUNPOWDER AS PROPELLANT (I): THE FIRST METAL-BARREL BOMBARDS AND HAND-GUNS[b]

In modern times the cannon has been commonly known in Chinese as *phao*[3,4] or *huo phao*[5]. But as we have noted earlier (pp. 11, 22) these two terms originally referred to the trebuchet[c] which, from antiquity onwards, hurled large pieces of stone,[d] and then later on incendiary bombs, and finally explosive bombs, into the cities or camps of the enemy.[e] The very word *phao* for trebuchet was actually a homophone of the verb *phao*[7], meaning 'to throw'. The phrase *huo phao*[5,8] seems to have appeared first in connection with the conquest of the kingdom of Southern Thang by the Sung army in +975.[f] From then on it recurs constantly in accounts of medieval battles, at first for gunpowder bombs with weak casings, then later on for bombs with strong casings, e.g. cast iron.[g] The fact is that when it was first introduced, probably in the late +10th century, it was essentially a new technical term, and as such it appears in the *Wu Ching Tsung Yao* towards the middle of the following century.[h] In just the same way we can trace other technical terms back to their starting-points—for example, *pao chang*[9] for gunpowder fire-crackers (as opposed to bamboo ones) to +1148 (cf. p. 131 above), and *chhung*[10] for hand-guns to a date which we shall shortly see (p. 294) somewhere in the +13th century.

---

[a] Pt. 1, ch. 1, p. 11*a, b*.

[b] A difference between British and American usage needs signalising here. While in American English the term hand-gun is still applied to all pistols and revolvers (even of the most modern types), in British English it designates only those earliest bombards which were small enough to be wielded by a single man holding the wooden handle or tiller which projected from their rear end.

[c] An exactly similar step in the evolution of technical terminology occurred in Europe, for Burtt (1) tells us that the word 'gun' was formed unquestionably from *mangona*, i.e. the mangonel, or trebuchet as we usually call it. Mangonels are even called guns in some +14th-century poems. Similarly, 'cannon' came from *canna*, a reed or tube, again closely paralleling the word *thung*, to give *huo thung*[6].

[d] And, as we have often seen (e.g. p. 163), the projectile itself was also called *phao*—causing no small difficulty sometimes.

[e] From Vol. 4, pt. 1, pp. 319, 323, it will be remembered that Chinese chess (*hsiang chhi*[11,12]) has a piece called *phao*[4] equivalent to the knight in European chess. This is generally thought of in artillery terms, but since 'combat' chess (as opposed to the earlier divinatory star-chess) became widely popular already in the Thang period, it must originally have meant the trebuchet, and only afterwards the cannon.

[f] P. 89 above; cf. Fêng Chia-Shêng (6), p. 16.

[g] See pp. 192 ff. above.   [h] *WCTY/CC*, ch. 12, pp. 56*b* ff.

[1] 火彈藥	[2] 乃要與鐵筒合堂口	[3] 炮	[4] 砲
[5] 火炮	[6] 火筒	[7] 拋	[8] 火砲	[9] 爆仗
[10] 銃	[11] 相棊	[12] 象棊	

Now by chance it happened that this last period was also the heyday (if a comparatively short one) of the most highly developed form of pre-gunpowder artillery, the counterweighted trebuchet ('the Muslim *phao*', *hui-hui phao*[1]). Earlier on (pt. 6, (*f*) 5) we had a good deal to say about the confusions which this caused for later writers, confusions only resolved in our own time. The siege of Hsiang-yang[2] and Fan-chhêng[3] by the Mongols between +1269 and +1273 provided the chief occasion of stumbling; and even today unwary historians[a] are liable to maintain that the *hui-hui phao* was a metal-barrel cannon. Paradoxically, this thing may quite possibly have come into existence by that time, but the *hui-hui phao* or counterweighted trebuchet was definitely something else. The loud crashing noise made by the projectiles as they demolished houses and made fortifications crumble, burying themselves deep in the ground, accounts easily for the idea that gunpowder was involved, yet neither fire nor explosions are ever mentioned in the descriptions.

Another confusing feature was that particular designations for projectile-propelling machines got carried over from the trebuchet era to the cannon era. This was the case, for example, with the 'crouching-tiger *phao*' (*hu tun phao*[4]). We see it in the Ming edition of the *Wu Ching Tsung Yao* as a trebuchet with a triangular frame (Fig. 74)[b] so that was what it looked like in +1044;[c] but by the time we get to +1350 (or +1412), we find the name applied in the *Huo Lung Ching*[d] to a small metal-barrel cannon weighing 36 lb. and provided with spikes for sticking into the ground to attenuate the recoil effect (Fig. 75).

Similarly with the two bombards substituted by the editors of the Chhing edition of the *Wu Ching Tsung Yao* (without any explanation) for two of the trebuchets formerly illustrated. These are both called 'mobile trebuchet carriages' (*hsing phao chhê*[5]), but in none of the available editions is there any text concerning them. The preceding pages describe and illustrate a curious shielded counter-weighted trebuchet (*thou chhê*[6]), designed for stationing at the head of a sap in siege warfare, and give no help, nor does the following one, which deals with a mobile bridge (*hao chhiao*[7]) for crossing moats or other water obstacles. However, the two trebuchets are quite usual projectile-throwers, only mounted on wheels, the first with four, the second with two. But then, instead of the first trebuchet illustration (Fig. 76)[e] the editors give a picture of a long-barrelled bombard jacked up to a high elevation so that the barrel superficially resembled the trebuchet arm (Fig. 77).[f]

---

[a] Such as Chang Chou-Hsün (*1*).
[b] Ch. 12, p. 45*a*; cf. the Chhing ed. ch. 12, p. 52*a*.
[c] Assuming that the illustrations in the +1510 edition were fairly accurate reproductions of the oldest drawings.
[d] *HLC*, pt. 1, ch. 2, p. 3*a*, *b*; *HCT* ed. p. 10*b*, as figured here. *Huo Kung Pei Yao* ed. ch. 2, p. 3*a*, *b*.
[e] *WCTY/CC* (Ming ed.), ch. 10, p. 14*a*.
[f] *WCTY/CC* (Chhing ed.), ch. 10, p. 13*a*.

[1] 回回砲   [2] 襄陽   [3] 樊城   [4] 虎蹲砲   [5] 行砲車
[6] 頭車   [7] 壕橋

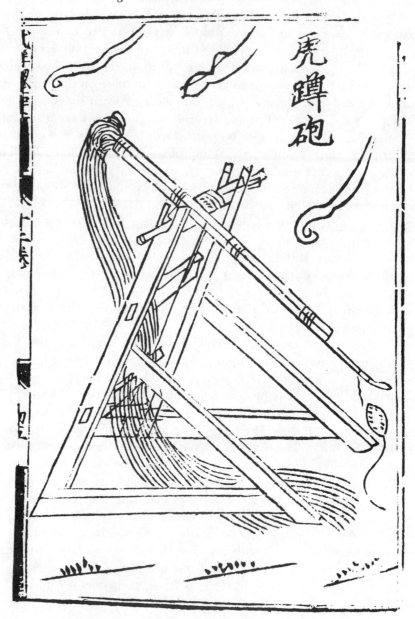

Fig. 74. The *hu tun phao* ('crouching-tiger' trebuchet) as it was in +1044, from the *WCTY* (Ming ed.), ch. 12, p. 45*a*. A detail of men pulled down suddenly on the ropes to the left, thereby sending the projectile, whether stone or bomb, into its trajectory from the pocket of the sling on the right.

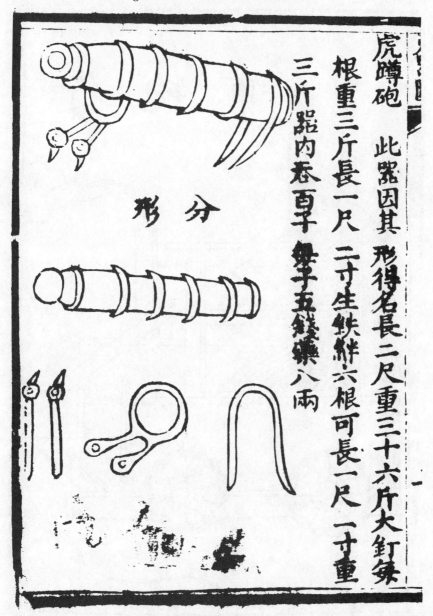

虎蹲砲　此器因其形得名長二尺重三十六斤大釘四根重三斤長一尺二寸生鐵絆六根可長一尺一寸重三斤器內吞百子銃子五錢藥八兩

形分

Fig. 75. The same name (*hu tun phao*) applied to a small metal-barrel cannon, 36 lb. in weight, from the *HLC*, pt. 1, ch. 2, p. 3*a* (*HCT* ed., p. 10*b*), therefore about +1350. Note the four anti-recoil pins to be stuck in the ground, showing that the muzzle, contrary to appearances, must be pointing to the right. Note also the bands encircling the barrel, on which cf. p. 331 below.

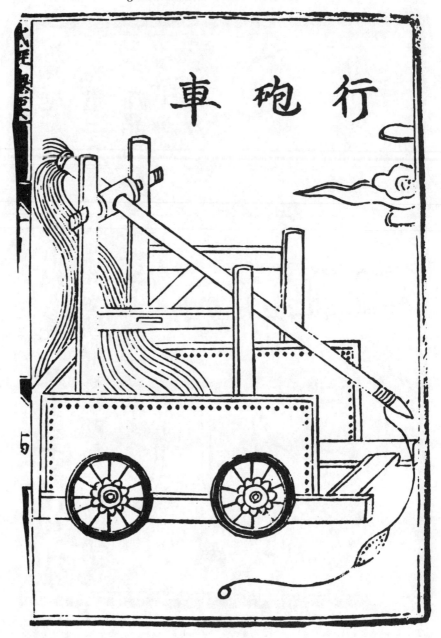

車砲行

Fig. 76. The *hsing phao chhê* (mobile trebuchet carriage) as it looked in +1044. The picture is from the *WCTY* (Ming ed.), ch. 10, p. 14*a* using the original copy in the library of Dr Hsü Ti-Shan at Canberra.

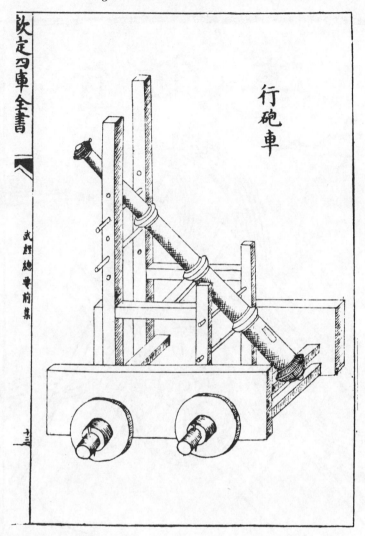

Fig. 77. In the Chhing edition (ch. 10, p. 13*a*), the arm of the trebuchet is replaced by a long-barrelled bombard jacked up to a high elevation, but the weapon still has four wheels, and bears the same name.

Adjacent to this a parallel substitution took place. Where before there was a trebuchet on a two-wheeled barrow-like carriage (Fig. 78)[a], also called *hsing phao chhê*, we now see another bombard with a long thin barrel carried on a two-wheeled barrow (Fig. 79)[b]. But it has had a slight change of name, becoming 'bombard on a high-fronted carriage' (*hsien chhê phao*[1]), and in its elevation slanting like the kind of perspective drawing of the mobile bridge on the opposite page.

[a] *WCTY/CC* (Ming ed.), ch. 10, p. 14*b*.　　[b] *WCTY/CC* (Chhing ed.), ch. 10, p. 13*b*.

[1] 軒車砲

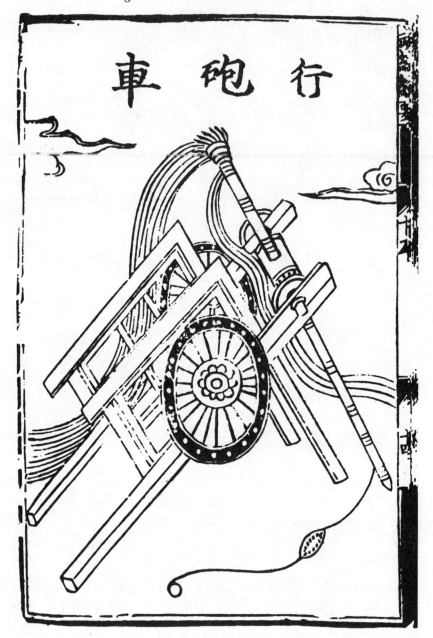

車砲行

Fig. 78. The name was also applied to a trebuchet borne on two-wheeled barrow-like carriage, as we see from
the Ming edition, ch. 10, p. 14b (again from the Hsü Ti-Shan library original).

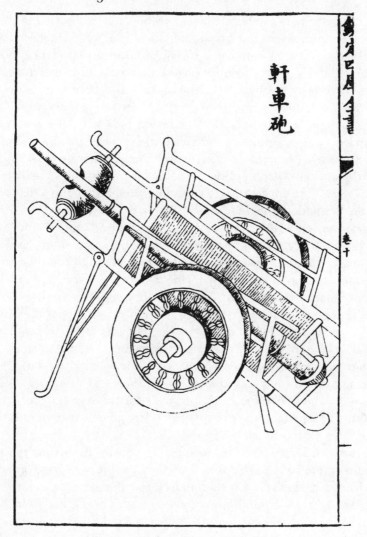

Fig. 79. The Chhing edition of *WCTY*, which might represent the situation about +1650, shows again a long-barrelled bombard on a two-wheeled carriage, but the weapon has undergone a slight change of name, becoming now *hsien chhê phao* (bombard on a high-fronted conveyance). Again ch. 10, p. 13*b*. But there is no relevant text for either of the bombard illustrations, so that they must be considered interpolations inserted at some time between +1350 and +1650.

After all, it was natural enough that the term *huo phao*[1] should have continued in common parlance for the metal-barrel bombard and hand-gun after the term *chhung*[2] had been appropriated for application to them (p. 248 above). In fact the more firearms developed the more natural it was, for the longer the barrel became, as in slings, culverins and muskets, the more reminiscent it was of the arm of the trebuchet; and the more often shells were fired from cannons the more

[1] 火砲   [2] 銃

reminiscent they were of the bombs which trebuchets had hurled in the old days of the +12th century. Here too a characteristic of the traditional scholars is very relevant, their predilection for using the most antique expressions possible because of the greater literary elegance one obtained thereby; we saw good examples of this already in connection with crackers and fireworks (p. 131 above). At the same time there was the tendency (often remarked upon in earlier volumes) to use professional wood-block artists (*hua kung*[1]) for making illustrations, men who knew nothing about what they were drawing, and probably rather despised it as banausic.[a] These two features can be seen quite well in ch. 101 of the military section of the *Thu Shu Chi Chhêng* encyclopaedia (+1726) entitled *chhê chan*[2] (chariot-fighting), a heading itself archaic to a degree, but one which could be made to include any military device on wheels.[b] Most of the chapter is concerned with references in the ancient *Shu Ching*[3] (Book of Documents) and *Shih Ching*[4] (Book of Odes) of the -1st millennium, and commentaries on them, but the illustrations at the end include a mobile windlass, a battering-ram and a mobile tank-like shield. Finally, a quite reasonable bombard on four wheels is given (Fig. 80), the 'subduing and burying cannon' (*mai fu chhung*[5]), though it would have been more appropriate in +1326, four hundred years before. But in the last illustration (Fig. 81) a climax of bewilderment is reached, for although the artist seems to have been trying to draw a mobile counterweighted trebuchet, the caption says 'the wonder-working long-range awe-inspiring cannon' (*wei yuan shen chhung*[6]).[c] Such was the conservatism of the scholars, and the indifference of the artists—fortunately not mirrored in the military compendia, which were clearly intended (like the pharmaceutical natural histories)[d] for practical use.

Nevertheless, this present sub-section differs from almost all the preceding ones in that concrete archaeological evidence is available in support of the texts. To put the matter in a nutshell, several hundred specimens of metal-barrel cannon, large and small, as also hand-guns, have survived in China from the +14th century (even indeed the +13th) and are preserved mostly in Chinese museums. In considering this we have always to bear in mind that the earliest date for bombards in Europe is +1327, the year of the two illustrations in the Oxford MS. of Walter de Milamete's book *De Nobilitatibus, Sapientis et Prudentiis Regum* (On the Majesty, Wisdom and Prudence of Kings),[e] both showing vase-shaped

---

[a] Cf. Vol. 4, pt. 2, pp. 48–9, 373, Vol. 5, pt. 4, pp. 70–1, etc.

[b] The encyclopaedia editors were in fact mixing up archaeology, ancient history and popular technological explanation.

[c] Unlike some of the other pictures, this is one of those which have no accompanying text. *TSCC, Jung chêng tien*, ch. 101 (*chhê chan pu*), *hui khao*, p. 14b.

[d] See Sect. 38 in Vol. 6, pt. 1.

[e] Ed. James (2).

[1] 畫工　　　　　[2] 車戰　　　　　[3] 書經　　　　　[4] 詩經　　　　　[5] 埋伏銃

[6] 威遠神銃

埋伏銃圖

Fig. 80. Another bombard on a four-wheeled carriage, the *mai fu chhung* ('subduing and burying cannon'),
from *TSCC, Chhê chen pu* in *Jung chêng tien*, ch. 101, p. 14*a*. From all else that we know, this would have
been more appropriate for +1326 or +1426 rather than +1726.

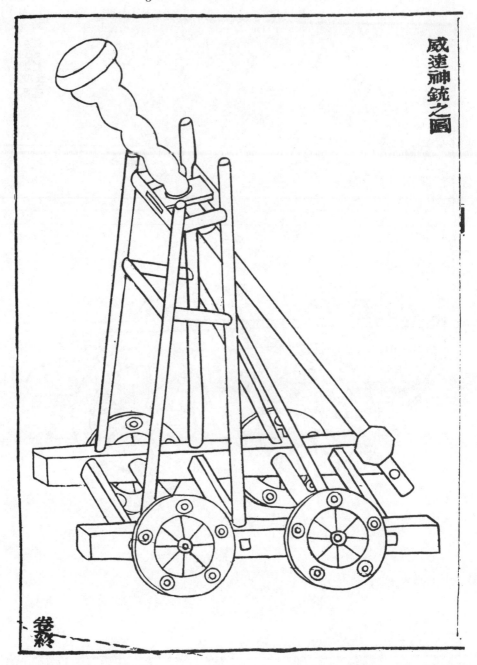

Fig. 81. A further illustration from *TSCC, ibid.* p. 14*b*. One cannot tell whether it was intended to be a counterweighted trebuchet (*hui-hui phao*) or a high-elevation bombard; at any rate, the name given is *wei yuan shen chhung* ('wonder-working long-range awe-inspiring cannon'). In works of this general kind, the scholars were very conservative, and the artists indifferent to what they were drawing, but such a situation is far different from what pertained in the professional military compendia.

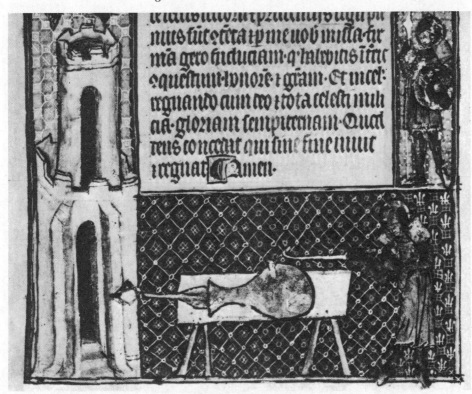

Fig. 82. The oldest illustration of a bombard in Europe, a page from the Bodleian MS., datable at +1327, of Walter de Milamete's *De Nobilitatibus, Sapientis et Prudentiis Regum* (On the Majesty, Wisdom and Prudence of Kings). A figure in armour on the right is gingerly applying a red-hot rod to the touch-hole of a vase-shaped cannon, out of the muzzle of which appears an arrow. Everything goes to show that the bore of the cannon was uniform, but it was thought wise to strengthen it by thickening the walls over the explosion chamber. The 'carpenter's bench' support for the bombard is worth noting, in view of what we see in Figs. 88, 106, 155 below.

bombards both of which are firing arrows (Fig. 82, 83).[a] Some specimens of European cannon or hand-guns rather later than this in the +14th century are also preserved in Western museums; but the difference is that many of the Chinese ones are self-dated by inscriptions, either cast or incised. Let it not be thought, as some amateurs of Chinese art objects might be tempted to suppose, that these dated inscriptions could be forgeries;[b] on the contrary, the low estimate in which technology was held by the traditional scholar-officials meant that no possible kudos could be gained by anyone in dating a bombard earlier than it really was.[c] We made the point at the beginning of our work, when

[a] This is the iconographic evidence, but in order to get into a picture the thing itself must have been known in Europe at any rate a few years earlier. Yet Partington (5), p. 101, could not adduce any textual evidence older than +1326, the date of a Florentine decree.

[b] The question of modern copies is of course another matter. Chinese museums habitually make them, for simultaneous display in several locations, but expert examination easily distinguishes them from the original. Cf. on this Arima (1), p. 134.

[c] Besides, no one in China before very modern times had the slightest idea of the comparative history of gunpowder and firearms. For a striking example of the disdainful, almost contemptuous, attitude of the Confucian scholars towards inventors and technologists, see Vol. 4, pt. 2, pp. 39 ff.

Fig. 83. Another, similar, page from Walter de Milamete's MS. One of four knights on the left is again applying a heated rod to the rear end of a bombard lying on a sort of table with an arrow-head again visible at the muzzle.

speaking of scientific texts we said: 'One may feel confident that these have never been intentionally interfered with, partly because the Confucian scholars considered them too unimportant, and partly because until modern times it would never have occurred to any Chinese scholar that the slightest interest attached to placing of scientific knowledge or a technical process earlier than its proper date.'[a]

From here onwards, the first thing to do will be to present a list of the earliest Chinese bombards and hand-guns now known, adding some commentary on the most interesting and important pieces; after which we may take a glance at some of the textual evidence for their use during the +13th and +14th centuries. Lastly we can have recourse to the descriptions and illustrations of bombards and hand-guns in the military compendia, which, sometimes irrespective of their date, belong clearly to the archaic period of artillery.

In 1962 Arima Seihō was able to list twenty-eight early Chinese examples of

[a] Vol. 1, p. 77. On the other hand, there have been clear forgeries in Europe, for example a mortar dated +1322 but self-dismissing on account of its completely incongruous decoration; cf. Arima (1), pp. 345–6. Another, purporting to be of +1303, is equally unacceptable; cf. Partington (5), p. 98.

these weapons, seven of which were of the +14th century.[a] His oldest specimen dated from +1372, with four others from +1377 and two from the years immediately following. But in 1957 Chou Wei had listed six others,[b] some decidedly earlier in date, for the oldest was of +1332, and two others of +1356 and +1357. It was in the former year, according to Arima, that the Koreans obtained their first bronze cannon from China, and it may be that the transmitter was a Chinese merchant named Li Khang[1].[c] Again, in 1957, an artillery exhibition which was mounted in Peking displayed several early bronze cannons, and three of these were afterwards described in detail by Wang Jung (1).[d] But the climax of the series so far was the bronze gun of about +1288 reported and described by Wei Kuo-Chung (1) seven years ago; we shall return to it when discussing Table 1, which lists most of those known that date from before the end of the Yung-Lo reign-period (+1424).

From the illustrations (Fig. 85, 88, 92, 93) it is already possible to sketch one or two characteristics of the successive periods. The early metal-barrel handguns or cannons tend to have a muzzle of blunderbuss type, the later ones are plain or with a single fillet beading;[e] but in nearly all cases the wall is made bulbous at the base (or closed breech end), i.e. intentionally thickened, with the bore remaining the same, at the part where the propellant explosion was to take place, and this was in fact called *yao shih*[5] (the gunpowder chamber).[f] Behind this they all have a hollow projection into which a wooden tiller or handle could be fitted. Towards the end of the +14th century the vent or touch-hole was elaborated to include a priming-pan, the hinged lid of which has in a few cases been preserved.[g] The bulbous strengthening (or reinforce) of the barrel (or chase) at the breech end (cf. Fig. 84, 90a, b, 91, 93)[h] brings up the question of the vase-shaped or bottle-shaped character of many of the early bombards both in East and West, indeed a significant common trait, and we shall return to it presently (p. 329). Later on, in the +15th century and the +16th, the chase or barrel of cannons was strengthened by very rugged rings or bands included as part of the casting, as we shall see (p. 331). This form continued into the era of breech-loading cannon with removable powder-chambers held in place with wooden wedges (cf. p. 365 below).

[a] (1), pp. 137–9.
[b] (1), p. 270 and pl. 83.
[c] Boots (1), p. 20, took this from the dynastic history *Koryŏ-sa*[2], providing no reference and giving the impression that the date was +1392; but that cannot be right, as on the following page he says that an arsenal for guns and bombards was established in Koryŏ in +1377. A later writer, Yu Sŏngnyong[3], in his *Su-a Manjip*[4] (Essays from the Western Cliff), of *c.* +1605, puts the date at +1372, which is also quite possible; cf. Cipolla (1), p. 105.
[d] One of +1332, the +1351 example, and one of those of +1372.
[e] For the explanation of the technical terms used in this paragraph and later see the diagram in Blackmore (2), p. 216 and accompanying text.
[f] Arima (1), p. 112.
[g] Arima, *op. cit.* p. 129.
[h] Arima, *op. cit.* p. 112. Naturally all these archaic fire-arms were muzzle-loaded.

[1] 李旽　　　[2] 高麗史　　　[3] 柳成龍　　　[4] 西崖文集　　　[5] 藥室

Table 1. *Early Chinese hand-guns and cannon* (to approximately the end of the Yung-Lo reign-period, +1424)

Year	Provenance and where preserved	Length overall cm.	Dimensions muzzle bore diameter cm.	Weight[a] kg.	Metal	Inscription[b]	References
c. +1288	Pan-la-chhêng-tzu in A-chhêng hsien, Heilungchiang. Provincial Museum	34	2·6	3·55	bronze		Wei Kuo-Chung (1); Fig. 84
+1332	National Histor. Mus. Peking	35·3	10·5	6·94	bronze	I	Wang Jung (1); Goodrich (25); Needham (82); Arima (1), p. 134; Figs. 85, 86, 87, 88
+1332	Thaiyuan Provincial Museum	26·5	2·3		bronze		Chou Wei (1), p. 270, pl. 83
c. +1334	Sian, Shensi	47·5	10·5	1·78	cast iron		Chao Hua-Shan (1)[c]
c. +1338	Rotunda Museum, Woolwich	32	2·2		bronze		H. Blackmore (p.c.); Fig. 89
c. +1340	Ta-ming (Yuan capital)	21·5	2·6		bronze		Arima (1), pp. 153 ff.
	Ta-ming (Yuan capital)	31·5	2·6		bronze		Arima (1), pp. 153 ff.
	Ta-ming (Yuan capital) Arima Collection						Arima (1), pp. 153 ff.
+1351	Shantung, Nat. Milit. Museum, Peking	43·5	3	4·75	bronze	I	Wang Jung (1); Goodrich (25); Needham (82); Figs. 90a, b, 91
+1356 +1357	Thaiyuan Provincial Museum						Chou Wei (1), p. 270, pl. 83
+1356 +1357	Nanthung Museum, Chiangsu	several hundred bombards made for Chang Shih-Chhêng's 'Chou' dynasty		302·7[d] 211·8	cast iron	{ I I }	Wang Jung (priv. comm.); Goodrich (24); Han Kuo-Chün (1)
+1372	Nat. Milit. Museum, Peking	36·5	11	15·75	bronze	I	Wang Jung (1); Goodrich (25); Figs. 92, 93
+1372	Harvard-Yenching Inst. Mus. Peking	45·7	2·54	–	cast iron		Goodrich (25)
+1372	Huhehot Museum, Inner Mongolia						Anon. (211)
+1372	Thaiyuan Provincial Museum	43	2	–	bronze		Chou Wei (1), p. 270, pl. 83
+1372	Arima Collection of Kuroda Genji					I	Arima (1), pp. 110–1, 137
+1372	Thaiyuan Provincial Museum	44·6	3·9	2·04	bronze	I	Chou Wei (1), p. 270, pl. 83
+1372	Provincial Museum, Nanking					I	Goodrich (15, who saw another in the grounds of Academia Sinica; Needham (82)).

Date	Location				Material	No.	References
+1377	Tho-kho-tho, Inner Mongolia	42	2·2	—	bronze		Li I-Yu (1)[e]
+1377	Tho-kho-tho, Inner Mongolia	44·3	1·9	2·1	bronze	I	Li I-Yu (1)
+1377	Tho-kho-tho, Inner Mongolia	44	2·1	2·14	bronze	I	Li I-Yu (1)
+1377	Tho-kho-tho, Inner Mongolia	42	2·1	1·95	bronze	I	Li I-Yu (1)
+1377	Tho-kho-tho, Inner Mongolia	36	1·9	—	bronze		Li I-Yu (1)
+1377	Tho-kho-tho, Inner Mongolia	27	2·3	—	bronze		Li I-Yu (1)
+1377	Tho-kho-tho, Inner Mongolia	38·5	1·9	—	bronze[f]		Li I-Yu (1)
+1377	Arima Collection	32·2	2·1	2·2	bronze	I	Arima & Kuroda (1); Arima (1), pp. 112, 137, 141
+1377	Arima Collection	43	2	2	bronze	I	Arima & Kuroda (1); Arima (1), pp. 112–13, 137, 141
+1377	Military Weapons Museum, Berlin	44	2	—	bronze	I	Arima & Kuroda (1); Arima (1), pp. 113–14, 137
+1377	Thaiyuan Provincial Library	44	2	—	bronze	I	Arima (1), pp. 114–15, 137
+1377	Thaiyuan Provincial Museum	101·6	21·6	>150	cast iron[g]	I	Sarton (14); Bishop (14); Goodrich (24); Read (4); Needham (80); Chou Wei (1); p. 270, pl. 83; (Figs. 94a, b)
+1377	Huhehot Museum, Inner Mongolia	44	2	2·1	bronze	I	Tshui Hsüan (1)
+1377	Huhehot Museum, Inner Mongolia	43·5	2	2·1	bronze	I	Tshui Hsüan (1)
+1378	Collection of Lo Chen-Yü	barrel broken off		3·5	bronze	I	Arima & Kuroda (1); Arima (1), pp. 115–16, 137
+1378 (approx.)	Kuangtung Provincial Mus. (from Kao-yao Hsien)	36	2·3	1·1	bronze	I	Ku Yün-Chhüan (1)
		30	2	1	bronze	I	
+1379	National Histor. Mus. Peking	26·5	2	1	cast iron	I	Arima (1), pp. 116–17, 137; Goodrich (15)
+1379		25·4	—	—	bronze	I	Tshui Hsüan (1)
+1379	Huhehot Museum, Inner Mongolia	44·5	2	1·9	bronze	I	
+1379	Tho-Kho-Tho, Inner Mongolia	44·2	2·1	2·1	bronze	I	Li I-Yu (1)
+1379	Sui-yuan Museum	43·2	2·1	2·1	bronze	I	Goodrich, in Goodrich & Fêng Chia-Shêng (1), p. 122[h]
+1409	Collection of Prince Chichibu	35	1·5	2·27	bronze	I	Kuroda (1); Arima (1), pp. 118–19, 137
+1409	Collection of Fujiwara Teiki	35·5	1·5	2·5	bronze	I	Kuroda (1); Arima (1), pp. 119–20, 137
+1409	Rotunda Museum, Woolwich	61	—	—	bronze	I	H. Blackmore (p.c.); Figs. 95, 96 Okada Noboru (p.c.), but date is hard to be sure of
+1412	Rotunda Museum, Woolwich	—	—	—	brass or bronze		
+1414	Kuroda Collection	36	1·4	2·2	bronze	I	Kuroda (1); Arima (1), pp. 120–1, 137

Fig. 86. Another view of the dated gun or bombard of +1332 (orig. photo.). The inscription can be seen in both these pictures.

All these labellings may be compared with that on the crossbow or *arcuballista* trigger mechanism of bronze depicted and described above (pt. 6, 30 (*e*) 2, (*f*) 3). For reasons already given, there can be no doubt of their authenticity.

The finds reported by Li I-Yu (*1*) from inside the east gate of the old city of Tokoto in Inner Mongolia, at the junction of the Black River with the Yellow R., are interesting for a number of reasons, such as the large pile of bronze-ball ammunition that was with them. But several of the Ming examples of hand-bombards (+1377, +1379) have inscriptions showing that they were intended for gunnery practice. Thus one says:

Hand-gun made on a fortunate day in the 10th year of the Hung-Wu reign-period, for Training Officer (Chiao Shih[1]) Shen Ming-Erh[2] and Instructor (Hsi Hsüeh Chün Jen[3]) A Tê[4], at the Left Naval Station, for teaching the troops, wt. 3 catties, 8 *liang*.[a]

Another mentions by name Training Officer Chu I[5] and Instructor Shang Shih-San[6] belonging to an Assault Guard Unit (*hu pên tso wei*[7]). A third gives the names of the makers, Artisan Hsü Chhêng[8] and Apprentice Military Artisan

---

[a] Tr. auct.

[1] 教師   [2] 沈名二   [3] 習學軍人   [4] 阿德   [5] 祝一

[6] 尙十三   [7] 虎賁左衛   [8] 民匠徐成

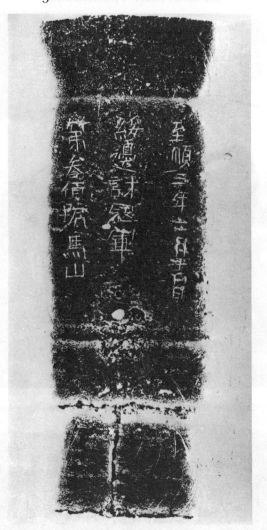

Fig. 87. Rubbing of the inscription on this gun, specifying the third year of the Chih-Shun reign-period. Translation on p. 296.

Wang[1] at the Yuanchow[2] Arsenal, working in this case for Local Commander and Battalion Judge Ho Hsiang[3]. What a strange and unexpected form of immortality it was to have one's name and title inscribed on a bronze gun which archaeologists six centuries later would uncover.[a]

It is not really possible as yet to pinpoint the origin of the true metal-barrel

---

[a] Names of gunners came down, of course, in song and story too. For example, the famous novel *Shui Hu Chuan*[4] (Stories of the River-Banks) tells how Sung Chiang[5] managed to lure and capture Ling Chen[6], the greatest artillerist of his age. The work was first collected from older plays and tales just about this time. The incident is related in Hui 54.

[1] 習學軍匠王   [2] 袁州   [3] 何祥   [4] 水滸傳   [5] 宋江
[6] 凌振

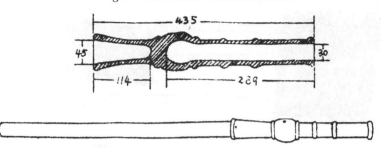

Fig. 91. Cross-section of the gun of +1351, with a reconstruction showing how the wooden tiller was fitted into the recess at the end opposite the muzzle. This recess will have been evident in all the early guns so far depicted.

Fig. 92. A gun dated +1372 with a blunderbuss muzzle (*ta wan khou thung*) after Wang Jung (*1*).

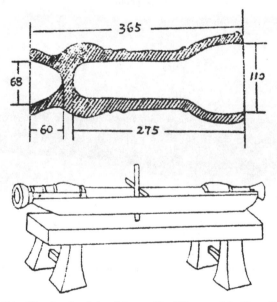

Fig. 93. Cross-section of the Ming bombard dated +1372 (the fifth year of the Hung-Wu reign-period), with a suggestion from the *Ping Lu* as to how it was mounted (cf. Fig. 106 below). Such a 'Mr Facing-both-Ways' device could certainly have doubled the rate of fire, but might not have been very comfortable for gunners standing behind it. From Wang Jung (*1*).

Fig. 94a. A cast-iron mortar or bombard dated +1377 with two pairs of trunnions, in the Provincial Historical Museum at Thaiyuan in Shansi (orig. photo.).

Fig. 94b. Another view of the same mortar (orig. photo.). The touch-hole is very clear.

Fig. 95. Chinese bronze gun conserved in the Rotunda Museum at Woolwich (Class II, 261). The inscription dates it as made in the 7th year of the Yung-Lo reign-period (+1409). Length *c.* 2 ft. Photo. Blackmore.

gun. But we have seen that metal barrels were first introduced for fire-lances and eruptors, quite a long time probably before single bore-fitting projectiles made full use of gunpowder's propellant force. We also noted that the term *huo thung*[1] (fire-tube) goes back at least as far as Thang times (p. 221) when it meant only a fuse in a tube for lighting signal-fires on the roofs of outpost-towers. Then, from the beginning of the +13th century we shall remember the *Hsing Chün Hsü Chih* and its references to *huo thung*[1] about +1230; here the difficulty is to be sure whether 'tube' or 'barrel' meant real barrel-guns or simply the tubes of fire-lances and eruptors (proto-cannon).[a] When we get to +1288 we do really meet with the metal-barrel gun, and under its subsequent name of *chhung*[2], in the affairs of Li Thing in the far north, making surprise attacks on the camps of Nayan (p. 294). Accordingly we can only suppose at present that about the middle of the +13th century would be the time of origin of the weapon, and it may be hoped that further study of the literature, together with fortunate archaeological finds, may help in due course to make the conclusion more precise.

The next item on the agenda would be the battle accounts which describe the use of the earliest fire-arms, but since by the +14th century these were becoming so widespread all over the Old World, we may be content with but a few examples. By +1353 the Yuan Mongol forces were using 'fire-tubes' (*huo thung*[3]) against the armies of Chang Shih-Chhêng,[b] and they were firing *huo tsu*[4], lit. 'fire-barbs' or 'javelin-heads'; but by now these can hardly have been either the fire-arrows (*huo chien*[5]) of antiquity, nor yet the rockets so prominent later— rather they were arrows shot from guns, exactly as we see in the famous pictures

---

[a] P. 221 above.
[b] Cf. Goodrich & Fang Chao-Ying (1), vol. 1, pp. 99 ff.

[1] 火筒   [2] 銃   [3] 火筩   [4] 火鏃   [5] 火箭

Fig. 96. The inscription on this gun, occupying the rear section of the barrel.

of Walter de Milamete (Figs. 82, 83).[a] In +1358 and the following year Lü Chen[1], one of Chang's generals, successfully defended the city of Shao-hsing against a siege train commanded for Chu Yuan-Chang by Hsü Ta[2][b] and Hu Ta-Hai[3]. From the *Pao Yüeh Lu*[4] written soon afterwards by Hsü Mien-Chih[5],[c] we know that cannon and hand-guns (*huo thung*[6]) were liberally used by both sides,[d] firing not only stone balls (*shih chhiu*[7]) but iron ones too (*thieh tan wan*[8]).[e] Actual cannon-balls of those times have been recovered by the Japanese archaeologists excavating the Mongol summer capital of Shang-tu[9] at Dolon Nor.[f] Both were of stone, 3 and 4 in. in diameter,[g] and since the palace was destroyed by fire in +1358 they are not likely to be much later. Then the term *huo chhung*[10] comes in again during the internecine strife among the Mongol generals at the close of the dynasty, as when Ta-Chha-Ma-Shih-Li[11] was fighting (and defeating) Polo Timur (Po-Lo Thieh-Mu-Erh[12]) near Peking.[h] The latter was loyal to the last Yuan emperor, Shun Ti (Toghan Timur), but to no avail; in the fifties and sixties both Mongols and Chinese fought among themselves in kaleidoscopic alliances until finally Chu Yuan-Chang won everything. By the end of that time quite large cannon were coming into use, like the 'bronze general' (*thung chiang-chün*[13]) used by Hsü Ta in +1366 when attacking Chang Shih-Chhêng's capital at Suchow.[i] Beyond this point we need hardly pursue the story.

Yet there are a few matters of interest in the late +14th century which refuse to be passed over in silence.[j] For example, we hear, as we rarely do, of a gunner in person; his name was Yang (Yang Phao-Shou[14]) and he deserted from the Mongol side to that of Chu Yuan-Chang in +1356. He was then put in charge of a detachment of soldiers armed with hand-guns (*chhung shou*[15]) in the engagements leading to the defeat of Chhen Yu-Liang[16], one of the provincial rulers who resisted the rise of the Ming—this was in +1363.[k] The campaign depended much on firearms. Têng Yü[17] defended Nanchhang successfully with hand-guns

---

[a] Fêng Chia-Shêng (6), p. 75.

[b] Cf. his biography in *Ming Shih*, ch. 125, p. 3b.

[c] Goodrich & Fang Chao-Ying (1), vol. 2, p. 1396.

[d] Franke (23) tr., pp. 9, 23, 33, 35 for the Chou side, pp. 18, 43 for the Ming side.

[e] *Ibid.* p. 43. Cf. Fêng Chia-Shêng (6), p. 75.

[f] Harada Yoshito & Komai Kazuchika (2), pp. 24, 67, fig. 21.

[g] Weighing 21·2 and 55·6 oz. respectively. Associated objects were, a Sung coin, a stele of +1322, and various tiles, some of cobalt blue, others ornamented with leaves and stems. Cf. p. 292 e.

[h] Fêng Chia-Shêng (6), p. 75; on Polo Timur see Cordier (1), vol. 2, p. 362. His biography is in *Yuan Shih*, ch. 95, p. 9b and ch. 207, p. 2b.

[i] Fêng, *loc. cit.* Cf. also *Phing Wu Lu*, p. 40a, and Wang Jen-Chün (1), *Ko Chih Ku Wei*, ch. 5, p. 11b.

[j] The best account of the period from the present point of view is that of Goodrich & Fêng Chia-Shêng (1), pp. 120–1.

[k] *Sung Hsüeh Shih Chhüan Chi Pu I*, ch. 3 (p. 1347). Cf. pp. 30–1 above.

[1] 呂珍	[2] 徐達	[3] 胡大海	[4] 保越錄	[5] 徐勉之
[6] 火筒	[7] 石球	[8] 鐵彈丸	[9] 上都	[10] 火銃
[11] 達札麻識理	[12] 李羅帖木耳	[13] 銅將軍	[14] 楊砲手	[15] 銃手
[16] 陳友諒	[17] 鄧俞			

(*huo chhung*[1]) in +1362,[a] and Yü Thung-Hai[2] used them to great effect in the famous naval Battle of the Po-yang Lake in the following year.[b] So did the Ming forces in the struggle of +1371 against yet another provincial potentate, Ming Shêng[3], the ruler in Szechuan, whose father had been self-appointed like the others. Chu Yuan-Chang's fleet advanced up the Yangtze defying the enemy's *thieh chhung*[4] and firing its *huo phao*[5] and *huo thung*[6], all now certainly guns, and in a final battle around Chhêngtu routed by the same means the elephants which Shu had imported.[c] Then in +1387 came hostilities with Ava-Burma, in preparation for which Mu Ying[7], the general on the frontier, was ordered to make ready no less than a couple of thousand *huo chhung*[8] hand-guns, and have his arsenals working night and day manufacturing gunpowder for them.[d] In the following year the Shan Burmese attacked, with a large elephant corps, but once again the Chinese artillery was more than a match for them, and their prince Ssu-Lun-Fa[9] was heavily defeated and fled.[e] So much for the last years of the +14th century.

Let us not forget, however, the cultures neighbouring to China. Korea, for example, got bombards as well as Europe during the +14th century. From +1356 onwards that country was much harassed by Japanese *wo khou*[10] pirates, and the Koryŏ king, Kongmin Wang[11], sent a special envoy to the Ming court appealing for a supply of firearms. Strictly speaking, of course, the Ming had not begun, but Chu Yuan-Chang[12] seems to have treated the request kindly and responded in some measure.[f] The *Koryŏ-sa* mentions a certain type of bombard (*ch'ong t'ong*[13]) which could send arrows from the Nam-kang[14] hill to the south of the Sun-ch'on Sa[15] temple with such force and velocity that they would penetrate completely into the ground together with their feathering (Figs. 97, 98).[g] In +1372 one Li Khang[16] (or Li Yuan[17]) came from South China to Korea, a saltpetre expert (*yen hsiao chiang*[18]), perhaps a merchant, and he was befriended by the courtier Choi Muson[19]. He asked him confidentially about the secrets of his mystery, and sent several of his retainers to learn his arts from him. Choi became the first Korean to manufacture gunpowder and gun barrels, all depending on Li Khang's transmission.[h] We also hear of a royal inspection of a new fleet in +1373,

---

[a] *Ming Shih Lu*, Thai Tsu sect. 12.
[b] *Ming Shih*, ch. 133, p. 4a; Dreyer (2).
[c] *Ming Shih*, ch. 129, pp. 12a ff.; *Phing Hsia Lu*, p. 19a; *Ming Shih Lu*, Thai Tsu sect. 67.
[d] *Yünnan Chi Wu Chhao Huang*, pp. 35 ff.; *Ming Shih Lu*, Thai Tsu sect. 182.
[e] *Ming Shih*, ch. 126, pp. 19a, b; *Ming Shih Lu*, Thai Tsu sect. 189.
[f] *Koryŏ-sa*, ch. 44, cf. Arima (1), pp. 227.
[g] *Ibid.* ch. 81, Arima, *loc. cit.* This recalls vividly the bombards in Walter de Milamete's picture (Figs. 82, 83).
[h] See Arima (1), p. 231. Cf. also p. 289 above.

[1] 火銃	[2] 俞通海	[3] 明昇	[4] 鐵銃	[5] 火砲
[6] 火筒	[7] 沐英	[8] 火銃	[9] 思倫發	[10] 倭寇
[11] 恭愍王	[12] 朱元璋	[13] 銃筒	[14] 南岡	[15] 順天寺
[16] 李亢	[17] 李元	[18] 焰硝匠	[19] 崔茂宣	

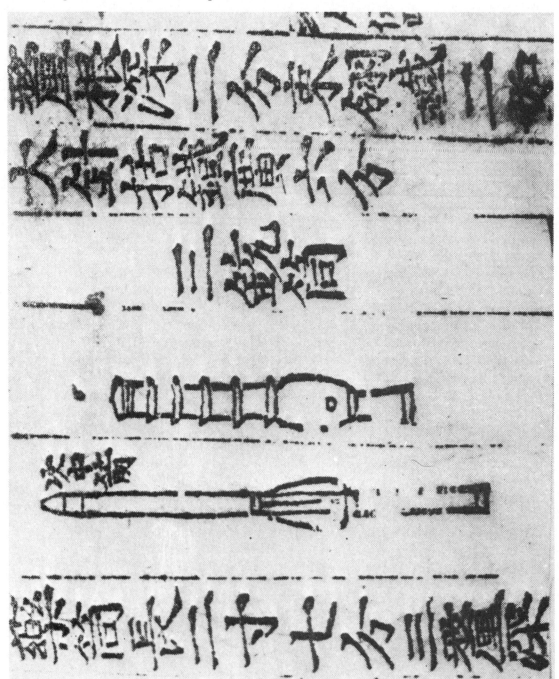

Fig. 97. A page of the Korean *Kukcho Orye-ŭi* (+1474) showing an early type of hand-gun (*chhung thung*) and the bolt-like arrow with metal fins ... From *Boots (1), pl. 91 b*.

Fig. 98. Arrow of the same kind but larger, over 9 ft long, with metal head and fins, shot from a similar type of gun. Seoul Museum. From Boots (1), pl. 21 a.

including tests of guns with larger barrels for shooting incendiary arrows against the pirate ships.[a]

Then in +1373 a new mission, led by Sang Sa-on[1] was sent to the Chinese capital asking for urgent supplies of gunpowder.[b] The Koreans had built special ships for repelling the Japanese pirates, and these needed gunpowder for their cannon. In the following year another request was made to the Ming emperor after the military camps at Happo[2] were set ablaze by Japanese pirates, with over five thousand casualties. At first Thai Tsu was reluctant to supply powder and arms to the Koreans, but in the middle of +1374 he changed his mind, and besides supplying what was sought, he sent military officers to inspect the anti-pirate ships built by the Koreans. The Koryŏ-sa records the first systematic manufacture of hand-guns and bombards in Korea in +1377, saying that the arsenal was directed by a 'Fire-Barrel Superintendent' (Huo Thung Tu Chien[3]),

[a] Koryŏ-sa, ch. 44; cf. Arima (1), p. 228.
[b] Hui-ch'an Ryŏsa, ch. 6; cf. Arima (1), p. 229. One has to remember also that this was just before the disintegration period of the Koryŏ kingdom, and the establishment of the Chosŏn kingdom under a new dynasty in +1392.

[1] 張子溫   [2] 合浦   [3] 火桶都監

whose new post was established at the suggestion of Choi Muson.[a] After this the Korean artillery never looked back, and played a quite important part in the campaigns of the +16th century.[b]

During the following couple of centuries the knowledge of firearms was a 'restricted' item in the Ming dynasty. Hence scholars were not sufficiently acquainted with guns and cannon to deal with them adequately in their writings. Such a lack of knowledge on military affairs and weaponry under the Ming is clearly demonstrated by the compilers of the official history in the +17th century. The sub-section on military writings in the bibliographical chapters of the *Ming Shih*, for example, contains fifty-eight titles but misses out more than half the list of Ming military works mentioned by Chiao Hsü[1] in his preface to the *Huo Kung Chhieh Yao*[2] in +1643.[c]

Chang Thing-Yü[3] in the *Ming Shih* spoke thus about firearms:[d]

What the ancient people called *phao* was a trebuchet for hurling stone projectiles. At the beginning of the Yuan (dynasty) a *phao*[4] (introduced) from Western parts (*hsi yü*[5]) was used to attack Tshai-chou[6], a city held by the Jurchen Chin Tartars. This was the first use (of the *phao* as a form) of fire(-arm), but the technique was not handed down and the weapon was little used. Afterwards, the technique of the *shen chi chhiang phao*[7] was acquired (*tê*[8]) during the conquest of Annam in Chhêng Tsu's time, and he set up a special establishment (*shen chi ying*[9]) so that the army could learn how to make them.[e] They used bronze, brass, and copper, in layers, as also iron[f]....[g] They were made in several different sizes, the largest being placed on carriages, the smaller ones on frameworks, and the smallest on posts or other supports. Their chief value was for defence, but they could be used also in the field.

This was all a muddle. The first sentence was right enough, but the reference to the counter-weighted or 'Muslim' trebuchet (*hui hui phao*[11]), which the writer evidently thought was a cannon, was misguided in another way as well, for the Jurchen Chin State was extinguished in +1234 and the Yuan dynasty did not begin till +1280.[h] Understandably perhaps in view of the fluctuating terminol-

---

[a] Ch. 18; cf. Arima (1), p. 230.

[b] Cf. p. 289. Many museum specimens of old Korean hand-guns and bombards are illustrated by Boots (1).

[c] *Ming Shih*, ch. 99, pp. 8b to 10a.

[d] Ch. 92, p. 10a; tr. auct. This dynastic history was commissioned in +1646 but not finished until ninety years later. We say Chang Thing-Yü because he was chief editor, but the monograph on military affairs was assuredly not written by him. Perhaps Chiang Chhen-Ying[10] was responsible, but we do not really know the name of the writer.

[e] Repeated in *Ming Hui Yao*, ch. 61 (p. 1188), which says +1410. Chhêng Tsu was the posthumous temple title of the Yung-Lo emperor, r. +1403 to +1424. The invasions of Annam took place between +1405 and +1410 (cf. Cordier (1), vol. 3, pp. 33 ff.).

[f] *Shêng shu chhih thung hsiang chien*[12], difficult to make sense of metallurgically. Cf. Vol. 5, pt. 2, p. 208.

[g] Here the text must be faulty for a few words.

[h] There can be no doubt that the former misunderstanding planted a seed of error which when grown up entangled many later historians (cf. p. 277 above). The true date for the appearance of counterweighted trebuchets in China was more like +1270 (cf. pt. 6 (f), 5 above).

[1] 焦勗	[2] 火功挈要	[3] 張廷玉	[4] 礮	[5] 西域
[6] 蔡州	[7] 神機槍礮	[8] 得	[9] 神機營	[10] 姜宸英
[11] 回回砲	[12] 生熟赤銅相間			

ogy already described (p. 248), the confusion then becomes worse confounded, for the *shen chi*[1] (magical trigger machine) was the term used for a musket, while a *chhiang*[2,3] was a fire-lance,[a] with or without co-viative projectiles, and a *phao*[4] could be a cannon but not usually a hand-gun. The writer was also weak in metallurgy, for otherwise he could hardly have written as he did about the materials used. After this, the concluding sentences clearly refer to bombards and early cannon (known since +1320 or so), hand-guns (from +1290 at least), and even muskets with their forked supporting sticks. In sum, the writer had no clear idea of what he was talking about, and we must try to unravel what we can from his words.

The fourth sentence in the passage was liable to convey the false impression that something called a *shen chi chhiang* was the first barrel-gun. Chao I, for example, interpreted it in this way in +1790,[b] and he was followed by many Westerners such as Mayers[c] and H. A. Giles,[d] who all supposed that metal-barrel guns had first entered China as a Vietnamese invention. Arima tried to save the situation by rendering the same sentence: 'When it came to the time of the Ming emperor Chhêng Tsu, (new) techniques of (using) the *shen chi chhiang* were developed during the conquest of Annam.'[e] And indeed it is true that the *Huo Lung Ching* describes the *shen chhiang chien*[5] 'magical (fire-)lance arrow' (Fig. 99) as *tzhu chi phing An-Nan chih chhi yeh*[6], i.e. 'this is the very weapon (used in) the subduing of Annam'.[f] It was in fact a fire-lance made of ironwood, which sent out an arrow and a number of lead bullets as co-viative projectiles. However, one cannot exclude the meaning that 'this is the very weapon (acquired during) the conquest of Annam'. Moreover, one can find a statement in the *Phing Phi Pai Chin Fang* (+1626) that during the Yung-Lo reign-period (+1403 to +1424) when Annam was conquered, the Annamese were found to be skilful in making this type of fire-lance, whereupon the Ming emperor ordered it to be copied and manufactured.[g] Thus what came up from Vietnam was only one of the many fire-lances described in Chinese sources.[h]

Without wishing to diminish in any way the ingenuity of the Annamese mili-

---

[a] Originally of course just a spear.

[b] *Kai Yü Tshung Khao*, ch. 30, p. 16a, b.

[c] (6), p. 94.   [d] (1), p. 21.   [e] (1), pp. 168–9.

[f] *Huo Chhi Thu*, p. 20b; other references are given on p. 251 above. But it is noteworthy that the illustration and caption appears in the earliest stratum, i.e. +1412 (*HLC*, pt. 1, ch. 2, p. 23a, b).

[g] Ch. 4, p. 32b.

[h] Cf. pp. 240 ff. above. Nevertheless Chinese historians continued to talk about the origin of the arquebus or musket from Annam at this time, e.g. Ling Yang-Tsao[7] in his *Li Shao Phien*[8] of +1799, ch. 40 (p. 649).

This Annamese connection may perhaps have been the origin of the statement in Tavernier (1), first Engl. tr. bk. 111, p. 187 that the people of Asem (Assam) were those who had 'formerly invented Guns and Powder, which spread itself from Asem to Pegu [in Burma], and from Pegu to China, from whence the invention has been attributed to the Chineses'. This passage was reproduced by Gait (1), p. 92, in the context of the expedition of Mirgimola (Mir Jumlah) to Assam in +1663. For these references we are indebted to Dr Anthony Butler of St. Andrews.

[1] 神機   [2] 槍   [3] 鎗   [4] 礮   [5] 神槍箭
[6] 此即平安南之器也   [7] 淩揚藻   [8] 蠡勺編

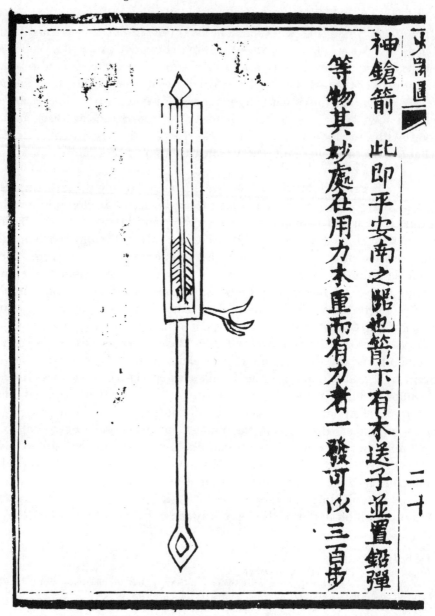

Fig. 99. The 'magical (fire-)lance arrow' (*shen chhiang chien*) in *HCT*, p. 20*b*, the description of which has misled some writers into supposing that guns originated in Annam. It was in fact a fire-lance made of iron-wood, and sent out an arrow along with some lead bullets as co-viative projectiles. But it might be described as a proto-gun because it is said to have shot the projectiles 300 paces, and to have had a wooden wad behind the arrow.

tary technicians, it is a fact that the Southern Sung people had been moving further and further south under pressure from Mongolian troops, from the middle of the +13th century onwards.[a] Remnants of the Sung soldiery could have crossed over into Annam to escape the onslaught of the Mongols, and some of them could have brought along their fire-lance designs.[b] The Annamese could well have developed their *shen chhiang chien* from these, using the very hard local wood. The Chinese of the Ming did not regard scientific and technical inventions and discoveries as a matter of national prestige; no controversies arose paralleling that concerning the invention of the calculus, or the discovery of Neptune, in Europe. On the contrary, the Chinese were always ready to acknowledge the foreign origin of any product or device, often using words like *hu*[1] and *yang*[2] before the names of objects. So the Annamese origin of the *shen chhiang chien* would have caused them no difficulty.

Chhi Chi-Kuang[3] mentions in his *Lien Ping Shih Chi*[4] (Treatise on Military Training) of +1568 that the 'crouching-tiger cannon' (*hu tun phao*[5]) was already put into service at various points along the Chinese border at the very beginning of the Ming dynasty (+1368).[c] This is only what one would expect, and throws useful light on the general development. The *Lien Ping Shih Chi* itself is listed in the bibliographical chapters of the *Ming Shih*, so it appears that the editors did not have time to read all the books they listed. Surely it was the lack of knowledge of firearms on the part of the compilers of the military monograph of the *Ming Shih* that made the passage so muddled. To understand more clearly about the earliest bombards and cannon we have to turn once again to the military compendia.

But first let us follow further the *Ming Shih* text as it deals with developments in the first half of the +15th century. The writer, whose terminology now becomes more consistent and comprehensible, tells us that in +1412, just after the campaigns in Annam, and the same year that saw the first printing of the *Huo Lung Ching*, an imperial edict ordered the stationing of batteries of five cannon (*phao*[6]) at each of the frontier passes as a kind of garrison artillery. In +1422 Chang Fu[7], one of the generals who had been victorious in Vietnam, petitioned that the system be extended to the northern frontiers, e.g. in Shansi, and this was done, though great secrecy was enjoined. In +1430 another officer, Than Kuang[8], suggested that hand-guns (*shen chhung*[9]) should be supplied to all frontier guard towers and fortified villages.[d] So also in +1441 two border comman-

---

[a] It will be remembered from p. 209 above that the Chinese, especially southerners, were regarded in Yuan times as third-class citizens.
[b] In more recent times we have seen remnants of the Chinese Kuomintang Army establishing themselves outside the Chinese borders in Indo-China, Burma and Thailand.
[c] *Lien Ping Shih Chi* (Tsa Chi), ch. 5, p. 20a (p. 235).
[d] The reference for all these three events is *Ming Shih*, ch. 92, p. 10b.

[1] 胡   [2] 洋   [3] 戚繼光   [4] 練兵實紀   [5] 虎蹲砲
[6] 礮   [7] 張輔   [8] 譚廣   [9] 神銃

ders, Huang Chen[1] and Yang Hung[2], began to establish arsenals for these weapons (*shen chhung chü*[3]) near the frontiers, but now the emperor Ying Tsung stepped in and forbade the decentralisation of gun foundries on security grounds. Next, in +1448, a Warden of the Marches, Yang Shan[4], petitioned for leave to make more double-ended bronze bombards (*liang thou thung chhung*[5]), which must have been the same as the rotating double bowl-muzzle bombards (*wan khou chhung*[6]) which we shall encounter presently (Fig. 106)[a]—and presumably he got it.[b] Then comes another interesting sidelight on southerners, for in +1450 an official named Chiang Chhao[7] recommended the manufacture of triple-barrel iron fire-lances[c] (*san thung thieh huo-yao chhiang*[9]) such as were used with shields (*huo san*[10])[d] by his military colleague Phing An[11], and made especially well by Shih Ao[12] and the people of Ying-chou[13] in Kweichow province. Tested and approved.[e] Arrow-firing guns were still in use, for in +1464 Fang Nêng[14] reported gratifying results in border combats with 'nine-dragon guns' (*chiu lung thung*[15]) which shot nine arrows at a time when ignited only once.[f] The nervousness about frontier arsenals continued all through the century, for in +1496 it was again insisted that the centralised Ministry of Works (Kung Pu[16]) alone should manufacture guns and cannon, and send them out to the various border units.[g] Finally, with relation to +1529, the text begins to tell about the Frankish cannon (*fo-lang-chi*[17]) or Portuguese breech-loading slings, and that is another story which for the moment we must postpone.[h]

For a clear picture of the various types of early bombards and cannon we have to fall back on the *Huo Lung Ching*. The Fire-Drake Manual describes them as follows. The 'crouching-tiger cannon' (*hu tun phao*[18], Fig. 75)[i] we have already had occasion to mention several times (pp. 21, 277). The text says:

---

[a] These were found also in Europe. The Munich Latin MS. CLM 197, of about +1442, has an illustration of two guns pointing opposite ways on a carriage. It was the earliest solution, perhaps, of the problem of achieving repeating fire.

[b] *Ming Shih*, ch. 92, p. 11a.

[c] Or guns, for the word *chhung*[8] is also used in the same breath, and they shot something like 300 paces. They may have been some kind of ribaudequin, but there are plenty of examples of fire-lances with several barrels in earlier centuries (p. 243 above).

[d] On shields used with firearms see p. 414 below.

[e] Again *Ming Shih*, ch. 92, p. 11a.

[f] This was the case in Europe too; cf. Partington (5), pp. 101, 144, 154. The *Livre de Cannonerie* of about +1430 describes guns and bombards that shoot arrows (like Walter de Milamete's), and so does the Latin MS. BN 7239 of c. +1450.

[g] *Ming Hui Yao*, ch. 61 (p. 1189).

[h] Cf. pp. 369 ff. below.

[i] *HLC*, pt. 1, ch. 2, p. 3a, b; *Huo Chhi Thu*, p. 10b; tr. auct. A more detailed description of the *hu tun phao* cannon is given in the *Ping Lu*, ch. 12, pp. 8b, 9a, b. Also in *WPC*, ch. 122, pp. 14a–16a, tr. Davis & Ware (1), p. 534. The text there describes the lead shot as still co-viative, except for the large lead ball that fills the muzzle.

---

[1] 黃眞	[2] 楊洪	[3] 神銃局	[4] 楊善	[5] 兩頭銅銃
[6] 碗口銃	[7] 江潮	[8] 銃	[9] 三筩鐵火藥鎗	
[10] 火傘	[11] 平安	[12] 師翺	[13] 應州	[14] 房能
[15] 九龍筒	[16] 工部	[17] 佛狼機	[18] 虎蹲砲	

This is so called because of its shape. It measures 2 feet in length and weighs 36 catties.[a] Each of the (iron) staples (used to pin down the cannon in position) weighs 3 catties and measures 1 ft 2 in. in length. The six cast-iron bands (for strengthening the barrel) each measures 1 ft 1 in. and weighs 3 catties. The barrel holds 100 bullets, each weighing 0·5 oz. (5 chhien[1]) and 8 oz. of (gun-)powder.

Contrary to natural expectation from the figure, this small bombard must have fired to the right, as otherwise the staples would not have acted to deaden the recoil effect. The balls may have been placed in a bag, like langrage, otherwise the gun would have been but an eruptor. This belongs to the earliest stratum of the *Huo Lung Ching*, i.e. by +1350; and the bands round the barrel were the forerunners of the much more prominent strengthenings of later date.[b]

Adjacent to this is a weapon called the 'long-range awe-inspiring cannon' (*wei yuan phao*[2], Fig. 100).[c] The text says:

Each weighs 120 catties[d] and measures 2 ft. 8 in. long. The touch-hole is 5 in. from the base and 3·2 in. from where the belly begins. The diameter of the bore at the muzzle is more than 2·2 in. Above the touch-hole there is a movable lid to protect (the priming powder) from rain. This cannon does not give a great bang nor much recoil. With 8 oz. of gunpowder use one large lead ball weighing 2 catties, or 100 small lead bullets (in a bag), each weighing 0·6 oz. (6 chhien). Firing is done very conveniently by hand.

Here we see the very model of the bombards already illustrated which had walls thickened round the explosion chamber, but we can be sure that the bore was uniform in diameter all through. Here again is the type of early vase-shaped or bottle-shape bombard often mentioned (p. 236 above, p. 330 below). In the +14th-century illustrations, the weapon is not provided with sights, but by +1600 they are there, cast on.[e]

[a] I.e. 21·6 kg. The Ming catty (lb.) equalled just under 0·6 kg.
[b] This was actually a very important weapon, and an example of how Chinese artisans could get things right at an extremely early stage. The crouching-tiger guns were still being used in the armament of the very sophisticated Sino-Korean fleet which was the decisive factor in the repulsion of the Japanese invasion attempts of the +1590s. These light-weight cannon seem to have been the earliest successful attempt to produce a built-up cast-iron gun, anticipating the methods of Armstrong and Whitworth by about five centuries. The bands or rings were probably shrunk on while red-hot, like the tyres of wheelwrights (and they may have been of wrought iron with its greater tensile strength), while the prongs again were cast separately. One would very much like to know whether the method of making malleable cast iron, so long supposed to have been a +17th- or +18th-century European invention but now recognised as having started in Warring States and Han times (cf. Sect. 36), was used for these cannon. In any case the Chinese had an age-long experience of cooling iron castings (cf. Needham, 32), and in the present case consummate control of size for accuracy of fit was essential. The result indicates an amazing mastery of the process by the foundry-men under mass-production conditions, where the problems of quality control are notoriously difficult. They certainly did not have the method introduced by Dahlgren about 1858 for cooling cannon castings by having water-tubes in the core so as to ensure that all parts cooled at a uniform rate. It seems they did not need it. For much of this commentary we are indebted to Dr Clayton Bredt.
[c] *HLC*, pt. 1, ch. 2, p. 2a, b; *Huo Chhi Thu*, p. 10a; tr. auct; cf. *WPC*, ch. 122, pp. 11b, 12a, b, as also *Ping Lu*, ch. 12, p. 5b, 6a, b. This must not be confused with the 'long-range awe-inspiring general' (*wei yuan chiang-chün*[3]), a title given by the Chhing emperor to a cannon made by Tai Tzu[4] in +1676. See p. 409 below.
[d] I.e. 72 kg.
[e] The breech one is labelled *chao mên*[5], and that at the muzzle is marked *chao hsing*[6].

[1] 錢　　　[2] 威遠砲　　　[3] 威遠將軍　　　[4] 戴梓　　　[5] 照門
[6] 照星

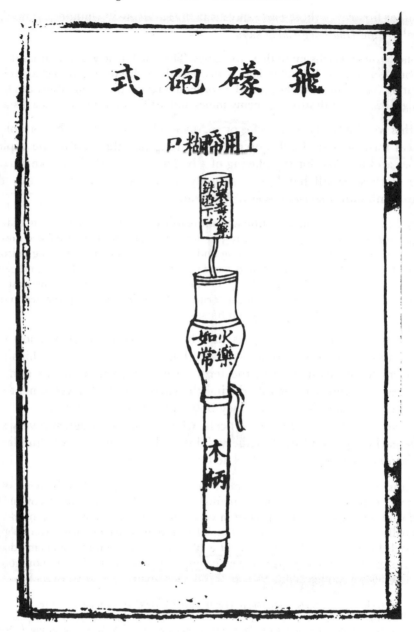

Fig. 101. A small hand-gun that fired a shell, the 'flying hidden-bomb gun' (*fei mêng phao*), from *HLC*, pt. 1, ch. 2, p. 10*a*, and *PL*, ch. 12, p. 19*b*, whence the picture. The short barrel was of iron, the tiller of wood, and the strong-casing canister, which contained poisonous substances as well as gunpowder, was fired lit by the propellant explosion.

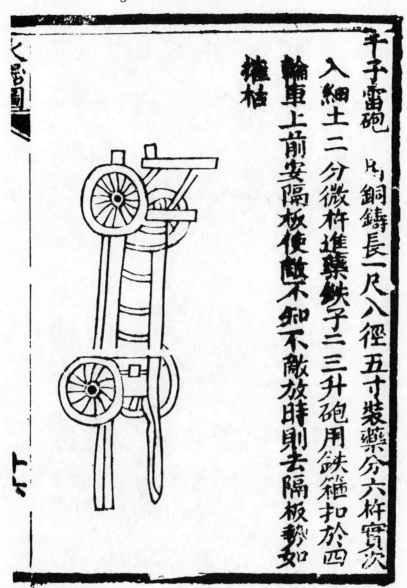

千子雷砲　用銅鑄長一尺八徑五寸裝藥分六杵實次

入細土二分微杵進藥鐵子二三升砲用鐵箍扣於四

輪車上前安隔板使敵不知不敵放時則去隔板勢如

摧枯

Fig. 102. Perhaps the earliest piece of field artillery, the 'thousand-ball thunder cannon' (*chhien tzu lei phao*), from *HCT*, p. 16*a*, which makes it date from between +1300 and 1350. It is noteworthy that the cannon is not vase-shaped, showing that better metallurgy had rendered the thickening of the wall over the explosion chamber unnecessary.

Fig. 103. A European field piece of about a century later, from a German Firework Book of $c.$ +1450, for comparison of the four-wheel carriage (photo. Tower of London Armouries). Note also the incendiary bombs being fired from crossbows by the soldiers on the right of the picture.

This was now something like grape-shot or langrage (cf. p. 275 above). Many other forms of firearm were also mounted on two-wheel barrows, for instance (Fig. 104) a seven-barrel ribaudequin called the 'seven-stars gun' (*chhi hsing chhung*[1]).[a] Similar cannon, perhaps rather more developed, occur in the later strata of the *Huo Lung Ching*, such as the 'barbarian-attacking cannon' (*kung jung phao*[2]; cf. Fig. 105); simply a mobile artillery piece carried on a two-wheel carriage.[b] But its markedly vase-shaped form is clearly seen in the illustration. Other gun- or cannon-bearing vehicles occur in the literature, notably the 'great effective mobile gun-carriage' (*ta shen chhung kun chhê*[3]) described by Lü Khun in +1607,[c] and the 'gun-carriage' (*chhung chhê*[4]) mentioned by Chiao Hsü in +1643.[d] But by that time we are almost in the modern period.

The *Ming Shih* mentions among many other types a 'wine-cup muzzle cannon' (*chan khou phao*[5]) and a 'bowl-size muzzle cannon' (*wan khou phao*[6]).[e] We have already drawn attention (p. 297 above) to the convenience of the bowl-shaped mouth for holding a projectile that could be a little larger in diameter than the bore itself. One of the most curious of these types was a 'Mr Facing-Both-Ways' or Janus-like weapon which consisted of two guns pointing in opposite directions and mounted on a pivoted support (Fig. 106). According to Ho Ju-Pin's description:[f]

The 'double bowl-mouthed gun' (*wan khou chhung*[9]) consists of (two) guns set on a movable support pivoted (so that it can rotate horizontally) on a (wooden) bench. Thus there are two heads (muzzles) pointing away from one another. Immediately after firing the first gun the second is (rotated into position and) fired, each one being muzzle-loaded with a large stone projectile. If the gun is aimed at the hull of an enemy ship below the water-line, the cannon-balls shoot along the surface and smash its side into splinters (so that it sinks). It is a very handy weapon.

This was evidently one of the earliest solutions of the problem of accomplishing repeating fire,[g] but one would rather not have been one of the gunners standing in the background while the sergeant was firing off the front-pointing component. Loading and re-loading would also have been rather slow, unless the barrels were replaceable, and several kept in reserve.[h]

---

[a] *HLC*, pt. 1, ch. 2, p. 15a, b; as in the *Huo Kung Pei Yao* ed.; *Huo Chhi Thu*, p. 16b. Cf. *WPC*, ch. 124, pp. 16b, 17a.

[b] *HLC*, pt. 2, ch. 2, pp. 10a, b, 11b; *WPC*, ch. 123, pp. 24b, 25a. The grapnel anchors are for fixing the bombard to minimise recoil.

[c] *Shou Chhêng Chiu Ming Shu*, pp. 14b, 15a.   [d] *Huo Kung Chhieh Yao*; Thu sect. p. 20a, b (p. 33).

[e] Ch. 92, p. 12a. Both of these appear elsewhere with the words *chhung*[7] or *chiang-chün*[8] (general) substituted for *phao*. The exact term chosen probably depended on the size of the piece.

In the late Ming it was recommended that one of the bowl-mouthed cannon should be mounted on each observation-tower of the northern frontier defences; *WPC*, ch. 97, pp. 14b, 15a; cf. Serruys (2), p. 19.

[f] *Ping Lu*, ch. 12, p. 10b, tr. auct.

[g] A further step was the 'cartwheel guns' (*chhê lun phao*[10]), which had 36 of them radiating from a centre like the spokes of a wheel; *WPC*, ch. 123, pp. 23b, 24a, tr. Davis & Ware (1), p. 535. A single mule could carry one of these on each side—but still some of the barrels pointed at the gunner.

[h] A closely similar device appeared in Europe too, c. +1386, with two *capita* or *testes* (Tout (1), p. 635).

[1] 七星銃   [2] 攻戎砲   [3] 大神銃滾車   [4] 銃車   [5] 盞口砲
[6] 碗口砲   [7] 銃   [8] 將軍   [9] 碗口銃   [10] 車輪砲

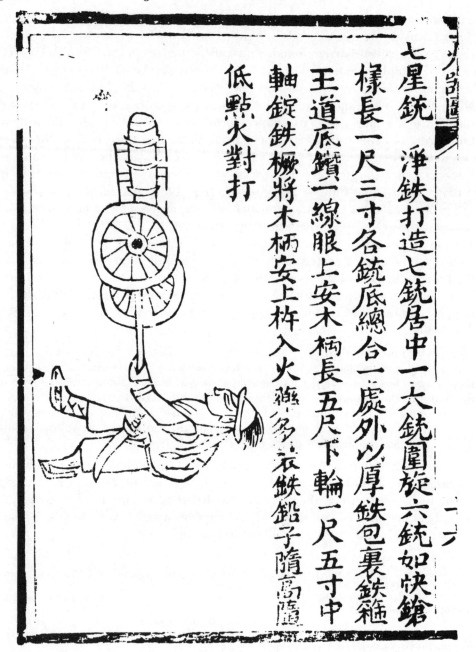

七星銃　淨鉄打造七銃居中一六銃圍旋六銃如快鎗
樣長一尺三寸各銃底總合一處外以厚鉄包裹鉄箍
王道底鑽一線眼上安木柄長五尺下輪一尺五寸中
軸錠鉄橛將木柄安上杵入火藥多衣鉄鉛子隨高隨
低點火對打

Fig. 104. A seven-barrelled ribaudequin carried on two wheels, the *chhi hsing chhung* from *HCT*, p. 16*b*.
Although only two auxiliary barrels are shown, it looks as if six smaller ones surrounded the central large one.

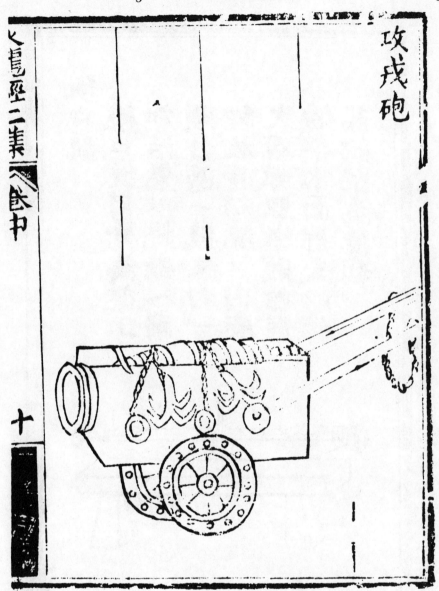

Fig. 105. The 'barbarian-attacking cannon' (*kung jung phao*) depicted in *HLC*, pt. 2, ch. 2, p. 10*a*, One can see that a certain degree of vase shape is still present, and that the recoil is modified by grapnel anchors. A two-wheeled barrow carries the artillery piece.

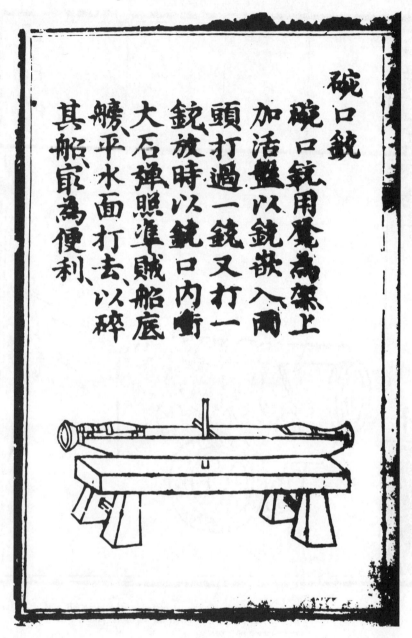

碗口銃

碗口銃用蘷為架上
加活盤以銃嵌入兩
頭打過一銃又打一
銃放時以銃口內衝
大石彈照準賊船底
艛平水面打去以碎
其船冣為便利

Fig. 106. The double-ended blunderbuss-mouthed gun supported on a 'carpenter's bench' (*PL*, ch. 12, p. 10*b*). This *wan khou chhung* was intended for naval warfare, and rotating the two barrels about a central point doubled the rate of fire.

In the illustration the bulge over the explosion-chamber can be well seen, and this raises the question of the vase- or bottle-shaped character of the earliest bombards both in East and West, a question we can no longer postpone. The form undoubtedly arose because the first gunners felt that they ought to strengthen the metal wall of the tube at the point where the propellant explosion was going to take place, even though, as we have seen (p. 315) the bore was uniform throughout.[a] Some have contrasted the vase-forms with the straight-sided cylindrical uniform-diameter tubes developing from the bamboo stem into all the later muskets and field pieces, but when one realises that whatever the shape externally, the bore always remained uniform, any seeming contradiction disappears. Antique forms exactly like vases occur in the military compendia, for example the 'eight directions over-aweing wind-fire cannon' (*pa mien shen wei fêng huo phao*[1])[b] shown in Fig. 107. We are told that it is 3 ft long, can be pointed in any direction, takes two men to work it (one to aim and one to fire), and sends forth a lead ball which can pass right through several men or shatter the sides of enemy ships to sink them, with a range of more than 200 paces.

Another vase-shaped bombard is the 'boring-through-mountains and smashing-up-places thunder-fire cannon' (*chhuan shan pho ti huo lei phao*[2]).[c] It is described as made of bronze, 4 ft long, and firing a packet of 3 pint measures of lead balls or a single iron cannon-ball as large as a couple of rice-bowls put together.[d] A similar but more stumpy bottle-shaped gun, rather like a mortar, is the 'flying, smashing and bursting bomb-cannon' (*fei tshui cha phao*[3]),[e] which fires cast-iron bomb-shells containing calthrops as well as gunpowder (Fig. 108); fuses running to the touch-holes through bamboo tubes enable the artilleryman to make a quick getaway.[f]

But it is when we come to the stack of bottle-guns called the 'nine ox-jar battery' (*chiu niu wêng*[7]) that we see the shape in its most characteristic form (Fig. 109).[g] Each one is 5 ft long and 1 ft in diameter,[h] nine being fastened to a frame

---

[a] Longitudinal sections have been given by many writers, e.g. Kuroda (*1*); Arima & Kuroda (*1*); Arima (*1*) and Wang Jung (*1*). Cf. Figs. 88, 93 above.

[b] *HLC*, pt. 2, ch. 2, pp. 23*b*, 24*a*, *b*; *Wu Pei Huo Lung Ching*, ch. 2, p. 18*b*; *WPC*, ch. 133, pp. 5*b*, 6*a*.

[c] *HLC*, pt. 2, ch. 3, pp. 26*b*, 27*a*; *Wu Pei Huo Lung Ching*, ch. 2, p. 29*a*.

[d] We say a packet, but it must be remembered that no sharp line of distinction existed between the co-viative projectiles of the fire-lances and eruptors on the one hand, and the truly propelled cannon-balls of the later guns and cannon on the other. The co-viative principle persisted sometimes into the stage of metal-barrel weapons, with strong walls.

[e] *WPC*, ch. 122, p. 26*a*, *b*. Cf. Davis & Ware (1), p. 530.

[f] Perhaps these forms were inspired by the vase-shaped pottery vessels with narrow base and mouth (*hsiao khou hsiao ti*[4]) containing lime and other offensive substances (cf. pp. 123 ff above) which had been used for the projectiles called 'wind-and-dust bombs' (*fêng chhen phao*[5]). These are described and illustrated in *HLC*, pt. 1, ch. 2, p. 11*a*, *b*; *Huo Chhi Thu*, p. 14*b*. Cf. Fig. 27 above.

It is curious to see that Fêng Chia-Shêng (6), pp. 74–5 says that the Mongols used 'iron fire-vases' (*thieh huo phing*[6]) at the siege of Baghdad in +1258, while the English also had these in +1327, i.e. the date of the Milamete MS. Yet the former must have been iron bombs while the latter were undoubtedly bombards—a world of difference.

[g] *HLC*, pt. 2, ch. 3, pp. 14*b*, 15*a*, *b*; *WPC*, ch. 131, p. 6*a*, *b*.

[h] The text does not say at which point along the length.

[1] 八面神威風火砲　　　[2] 穿山破地火雷砲　　　[3] 飛摧炸砲
[4] 小口小底　　　[5] 風塵砲　　　[6] 鐵火瓶　　　[7] 九牛甕

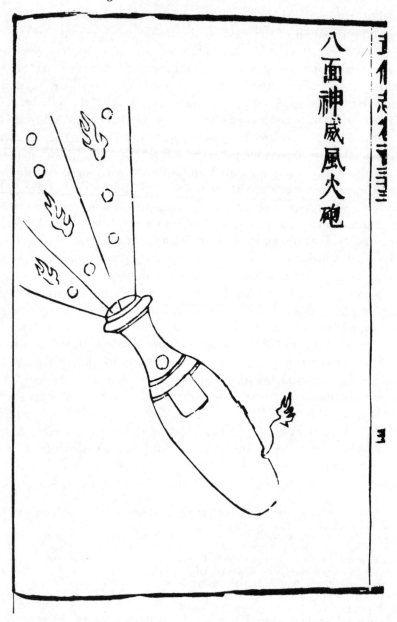

Fig. 107. The vase shape once again, seen in the 'eight directions over-aweing wind-fire cannon' (*pa mien shen wei fêng huo phao*), from *WPC*, ch. 133, p. 5*b*. The bore has to be visualised as uniform throughout. No information about the mounting is given, save that the weapon could be aimed in any one of the eight directions. This must have meant the four cardinal points and the four intermediate angles, as in the Eight Trigrams of the *I Ching* (cf. Mayers (1), p. 357, no. 247). But in the absence of universal joints, this may have been a slight exaggeration.

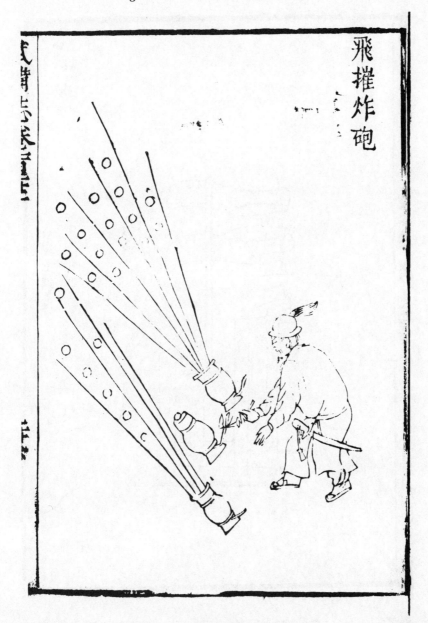

Fig. 108. Vase-shaped mortars firing cast-iron bomb-shells containing calthrops as well as gunpowder, the *fei tshui cha phao* from *WPC*, ch. 122, p. 26*a*.

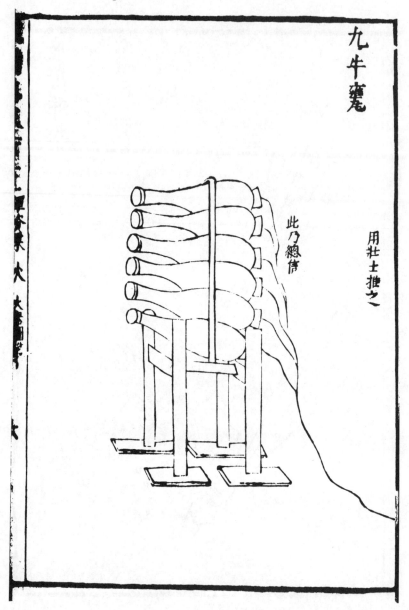

Fig. 109. Stack of bottle- or vase-shaped guns very reminiscent of those in Walter de Milamete's pictures (Figs. 82, 83). *WPC*, ch. 131, p. 6*a* actually calls it the 'nine ox-jar' battery (*chiu niu wêng*), alluding to the shape (though only six are shown). Each gun propelled a stone cannon-ball, and all of them were fired by a single fuse.

with bands or hoops. The stone cannon-balls each have the sizeable weight of 20 catties;[a] they are sent forth in a volley with a thunderous noise and, it is said, a range of more than 10 *li*. The captions of the illustration add that a single fuse is distributed to all the bombards, and that it takes very strong soldiers to move them about, which may well have been true as no wheels are shown.

Here the striking thing is that we have a shape almost exactly like that of the bombards of Walter de Milamete (Figs. 82, 83)[b] which doubtless also had a bore of uniform diameter; and the similarity is so remarkable that one feels oneself in the presence of a palpable transmission from China. Of course it is true that the pictures and the text belong to the late +16th or early +17th century, but the thing itself is so archaic, and the conservatism of Chinese writers and artists so extreme, that we are greatly tempted to date these bottle-guns, or smaller versions of them, in principle, to the neighbourhood of +1300.[c] It is at least very curious that both in East Asia and Western Europe the first bombards should have developed exactly the same pear-shaped form.[d] And the long prior evolution of firearms in the former part of the world, unparalleled in the latter, suggests strongly that this form was first arrived at there.

In this connection we may recall from p. 170 the iron firearms (*thieh huo phao*[1]) shaped like bottle-gourds (*phao*[2]) which Chao Yü-Jung used in the defence of Chhichow in +1221. It might be attractive to see in these the first metal-barrel bombards or hand-guns, but perhaps we should retain our previous interpretation of them as simply cast-iron gunpowder bombs. Chinese bottle-gourds, so often seen as recipients for medicines, or for life-elixirs when carried by the God of Longevity, always have a constricted waist, and this does recall the forepart of the earliest bombards, though the bulge ahead of it, or above it, is not present, and would have had no obvious purpose, in a cannon.[e] Possibly its function in the bombs may have been to make their fracture easier, as in the corrugations of a modern grenade. Still, the true nature of the bottle-gourd weapons of the early +13th century may be left an open question for the present.

There are, it is true, two gourd-shaped weapons in the *Huo Lung Ching*, and

[a] *C.* 12 kg.

[b] Cf. pp. 284, 289 above. Burtt (1) says that the oldest reference of any kind to gunpowder in Europe after +1327 is that of the Exchequer Pipe Rolls regarding the Battle of Creçy, in +1346. It is interesting that in +1353 he found references to 'gunnis cum sagittis et pellotis' (arrows and plugs), as also 'gunnis cum telar' (i.e. tillers for aiming).

[c] As we said at the outset, the numerous fire-weapons have to be judged on the basis of developmental logic as well as literary sequence. Later authors, always seeking for completeness, tended to describe and illustrate devices by their time long obsolete.

[d] Unfortunately not a single example of these has been preserved, either in East or West. The nearest approach is the Loshult gun (Fig. 110), named from the place in Skåne, Sweden, where it was found. It is a hand-gun made of bronze, *c.* 30 cm. long and with a bore of 3·6 cm. But it is trumpet- rather than pear-shaped. Blackmore (1), p. 5; cf. Reid (1), p. 54.

[e] Here it is worth recalling that the bottle-gourd, *Lagenaria siceraria* (= *vulgaris*, R 62; CC 178–9; Anon (*109*), vol. 4, pp. 364–5); was absolutely uncharacteristic of Europe but quite common in China. This might be relevant to any Chinese antecedents of the Milamete bombards.

[1] 鐵火炮   [2] 匏

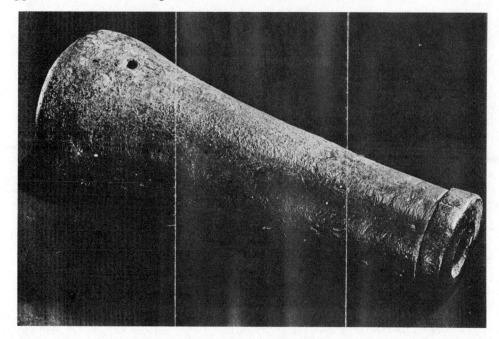

Fig. 110. The vase-shape in Europe again, a hand-gun of bronze found at Loshult in Skåne, Sweden, and considered to be of the +14th century. Now in the National Museum, Stockholm (no. 2891); photo. Blackmore. This early gun is trumpet- rather than pear-shaped, however.

one we have come across already. It is the 'phalanx-charging fire-gourd' (*chhung chen huo hu-lu*[1]), and we translated the passage concerning it on p. 236 (Fig. 50) above; it is unquestionably a form of fire-lance emitting lead bullets as co-viative projectiles along with the flames of poison-gunpowder.[a] The other one we have not so far mentioned; it was called the 'cavalry-opposing enemy-burning fire-gourd' (*tui ma shao jen huo hu-lu*[2]).[b] This was, it seems, a real gourd, strengthened with chemicalised and lacquered cloth, and filled with saltpetre, sulphur and carbonaceous materials. It was used as a kind of flame-thrower with a range of 30–40 ft. The formula seems decidedly archaic so in spite of its relatively late appearance the device may be very old. But neither of these items throw any light on the present question.

What, one may ask, led to the disappearance of the bombards with vase and bottle shapes, pear-like or gourd-like? It must have been metallurgical development leading to better cast iron and steel, less liable to crack and split, better able to withstand the strains of the boring process when this was done.[c] But the anxiety about wall strength long persisted, and we meet for a century or so more

[a] *HLC*, pt. 1, ch. 3, p. 14*a*, *b*; *Huo Chhi Thu*, p. 32*a*; *Huo Kung Pei Yao*, ch. 3, p. 14*a*, *b*; *WPC*, ch. 130, p. 25*a*, *b*.
[b] *HLC*, pt. 2, ch. 3, pp. 11*b*, 12*a*, *b*; *WPC*, ch. 130, p. 20*a*, *b*.
[c] A machine for boring cannon is illustrated and described in *WPC*, ch. 131, pp. 7*a*, *b*, 8*a*.

[1] 衝陣火葫蘆   [2] 對馬燒人火葫蘆

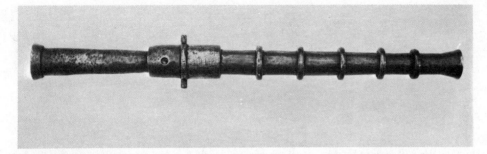

Fig. 111. The next phase of gun and cannon construction, hooped, ridged or ringed barrels. Small Chinese pivot-mounted iron cannon in the Tower Armouries collection, London, probably of the +16th or +17th century (Class XIX, no. 114). See the catalogue of Blackmore (2), no. 204 B (2), p. 154.

with barrels fortified by hoops or rings often cast on (Fig. 111, 112a, b, 113, 117, 118, 119).[a] It is to these that we must now turn.

For example, there is in the Tower of London Armouries a small Chinese pivot-mounted iron cannon cast in one piece, the chase having five rings projecting at intervals and intrinsic to the casting.[b] There is an inscription, unfortunately too worn to read. The wooden butt or tiller is preserved,[c] slender and slightly curving, with a whorl finial. Somehow or other it found its way to Benin in Africa, where it was captured by a British expeditionary force in 1897. This piece (Fig. 111) is dated by the curators as of the +18th or +19th century, and for this there may well be evidence, but judging by the form alone the +16th and +17th would be a better guess.[d] Similarly hooped, ribbed, ridged or crocketed, is a small three-barrelled signal-gun, only 18 cm. long (Fig. 112a, b),[e] the date and provenance of which are unknown. But it is very like two signal-guns of about +1600 or earlier from Korea described and illustrated by Boots,[f] though rather better made. He also gives a picture[g] taken from the *Wu I Thu Phu Thung Chih*[1] (Illustrated Military Encyclopaedia) of +1791, which shows a rider standing in the saddle and holding up just such a signal-gun. For these firearms the expression *hsin phao*[2], which will be remembered from p. 169 above, was retained. Every watch-tower in the northern defence system of the Ming had to be provided with at least one of these.[h]

[a] The *hu tun phao* (Fig. 75, pp. 279, 314 above) was an early example of this kind.
[b] Catalogue of Blackmore (2), no. 204, p. 154. Length 3 ft 3 in., bore 1·1 in. (2·8 cm.). As the entry says, the rings in such pieces as these may be vestigial in function, but we can make a guess about their origin.
[c] 1 ft 5 in. long.
[d] Many Korean examples have been illustrated, as by Boots (1), pl. 20, 22 A, B; Pak Hae-ill (1), pl. 1 B, C, pl. 5 B. Two of these last are self-dated by inscriptions as of +1555 and +1557. This was in the heyday of the Yi dynasty, which had started in +1392, and before the Japanese invasion of Hideyoshi (cf. p. 469). See Figs. 113, 114, 115, and, for Chinese parallels, Figs. 116, 117, 118, 119.
[e] In the collection of Mr Howard Blackmore, to whom we offer thanks for permission to reproduce the photographs.
[f] (1), pls. 24B, 26.   [g] *Ibid.* pl. 24A, from ch. 4.
[h] See Serruys (2), pp. 31, 68, citing a document of +1436.

[1] 武藝圖譜通志   [2] 信砲

Fig. 112*a*. Hand-held three-barrelled signal-gun with two bands or ribs on the barrel, probably of the +17th century. Length overall 7·125 in. (18 cm.). Photo. Blackmore, in whose collection the object is. Probably Korean in origin.

Fig. 112*b*. Another view of the three-barrelled signal-gun to show the muzzles. Cf. Boots (1), pl. 24*a* for a rider using one of these, and pls. 24*b*, 26 for Korean illustrations of similar signal-guns, but not so nicely made.

There is a suggestion in the literature, not at all to be despised, that the strengthening hoops or rings on the barrels of these guns and cannon from the late +14th century onwards were originally derived from the model of the nodes of the bamboo tubes which formed the earliest fire-lance barrels. This attractive idea is found in Horváth (2) and several other writers.[a] It is noteworthy that the earliest cast bronze hand-guns and bombards of Table 1 often had these rings, though there was no great necessity for them. Then, when the practice came in

---

[a] One of us (W. L.) has always maintained this.

Fig. 113. Korean +16th-century bombards in the Seoul Museum, from Boots (1), pl. 20. Some of the ridges or bands of these cast-iron weapons were prolonged so as to take rings for lifting into position.

Fig. 114. A page from the *Kukcho Orye-ŭi* depicting a mortar shooting a stone ball from a blunderbuss-like muzzle, as is implied by the name *chhung thung wan khou*. From Boots (1), pl. 22a.

Fig. 115. A similar mortar, with two ridges or hoops, in the Seoul Museum. From Boots (1), pl. 22*b*.

of building up barrels from forged longitudinal bars of low-carbon iron, the hoops would have acquired a very important function.[a] In later times the rings became rather exaggerated, but they seem to have had a dual origin, first skeuomorphic,[b] later technological. At all events, it was acute of Bernal to

[a] Banks (1), a naval surgeon, writing in 1861, reported finding, in the Taku Forts after capture, a number of large Chinese cannon of dual character. The inner part with the bore was made of longitudinal bars welded together and bound with hoops, also welded, but outside this was a layer of cast iron, 2·75 in. thick at the muzzle. These weapons were all more than 9 ft long, and in two cases the cast-iron layer had broken away at the muzzle, revealing the inner structure. Banks was all the more surprised because this double structure was one of the techniques used at the time for making Armstrong-Whitworth guns, but it could hardly have been derivative.

[b] Cf. Vol. 2, p. 468.

Fig. 116. A bronze cannon of about +1530, showing how the rings have flattened out so as to occupy the greater part of the barrel. Photo, Nat. Historical Museum, Peking.

remark that 'the very name of the "barrel" of a cannon indicates its primitive construction from iron staves hooped together'.[a]

Chhi Chi-Kuang says that in former times a large cannon weighing about 1050 catties[b] and known as the 'great invincible general' (*wu ti ta chiang-chün*[1]) was used, and later the weapon was modified but the name retained (Fig. 119).[c] Ho Ju-Pin refers to it as the 'great general cannon' (*ta chiang-chün chhung*[2]), and this is what he wrote about it:[d]

<hr>

[a] (1), p. 237. This holds good for English, but not perhaps for other European languages. Yet another possible origin might be sought in the fact that the arms of trebuchets, especially when composite, were strengthened with bands or hoops of iron wire (Vol. 5, pt. 6, (*f*), 5). We have already noted certain iconographic parallels between cannon-barrels and trebuchet-arms (pp. 280ff.).

[b] I.e. 630 kg., more than twice as heavy as any of those listed in Table 1.

[c] *Lien Ping Shih Chi* (Tsa chi sect.) ch. 5, pp. 13*b*–15*b* (pp. 226–9). He gives an illustration of it on a roofed two-wheeled gun-carriage, but by his time it had become a breech-loader with removable blocks. Cf. Huang Jen-Yü (5), pp. 179, 180.

[d] *Ping Lu*, ch. 12, pp. 7*a*, *b*, 8*a*; tr. auct. According to Thang Shun-Chih the *ta chiang-chün* was also called the 'thousand bullets cannon' (*chhien tzu chhung*[3]) which would imply some kind of grape-shot instead of a single cannon-ball; *Wu Pien*, ch. 5, p. 10*a*.

[1] 無敵大將軍   [2] 大將軍銃   [3] 千子銃

Fig. 117. Chinese bronze cannon of the Chhing period, dating towards the end of the +17th century, outside one of the gates of the Imperial Palace. Photo. Nat. Historical Museum, Peking.

Among the large firearms there is none that is greater than the 'great general gun'. Its barrel (used to) weigh 150 catties, and was attached to a stand made of bronze weighing 1000 catties. It looked rather like the *fo-lang-chi*[1] cannon. Yeh Mêng-Hsiung[2][a] changed the weight of the gun to 250 catties and doubled its length to 6 feet, but eliminated the stand, and now it is placed on a carriage with wheels. When fired it has a range of 800 paces. A large lead shell weighing 7 catties is called a 'grandfather shell' (*kung*[3]) and the next shell of medium size weighing 3 catties is a 'son shell' (*tzu*[4]), while a smaller shell weighing 1 catty is a 'grandson shell' (*sun*[5]). There are also 200 small bullets each weighing 0·3 to 0·2 oz. (contained in the same shell) and called 'grandchildren bullets' (*chhün sun*[6]), while the saying is that the 'grandfather' leads the way and the 'grandchildren' follow (*kung ling sun shang*[7]). They are supplemented with iron and porcelain fragments previously boiled in cantharides beetle (*pan mao*[8]) poison. The total weight of the projectile is some 20 catties. A single shot has the power of a thunderbolt, causing several hundred casualties among men and horses. If thousands, or tens of thousands, of (this

[a] Flourished in the second half of the +16th century.

[1] 佛狼機   [2] 葉夢熊   [3] 公   [4] 子   [5] 孫
[6] 羣孫   [7] 公領孫尙   [8] 班毛

Fig. 118. Chinese ribbed +17th-century cannon in the Rotunda Museum at Woolwich Arsenal. Photo. Okada Noboru.

weapon) were placed in position along the frontiers, and every one of them manned by soldiers well trained to use them, then (we should be) invincible. This weapon is indeed the ultimate among all firearms. At first its heavy weight caused some doubt as to whether or not it was too cumbersome; but if it is transported on its carriage then it is suitable, irrespective of height, distance or difficulty of terrain. During the 6th year of the Thien-Sun reign-period (+1462) 1200 gun-carriages were made. They included carriages for the 'large bronze cannon' (*ta thung chhung*[1]). During the 1st year of the Chhêng-Hua reign-period (+1465) 300 different 'great general (guns)' were manufactured and 500 carriages for cannon were made. This was an excellent strategy in using Chinese expertise to keep the barbarians under control.

Cannon of essentially this type went on being produced in China until the early 19th century.[a]

With this then we may leave the first phase of gunnery in China, postponing the 'Frankish culverin' (*fo-lang-chi*[2]) and later developments for a short while. We need only add that for those interested in the logistic organisation of Chinese artillery in the +15th century there are plenty of passages which detail the number and kind of guns in each unit, the amounts of powder and shot with which

[a] In the Tower of London Armouries there is an example, ribbed in this way, 10 ft 6 in. long and with a calibre of 7.5 in. (19 cm.). The Chinese inscription, recording its casting in 1841, during the Opium Wars, is translated in the catalogue of Blackmore (2), no. 207, p. 155.

[1] 大銅銃   [2] 佛狼機

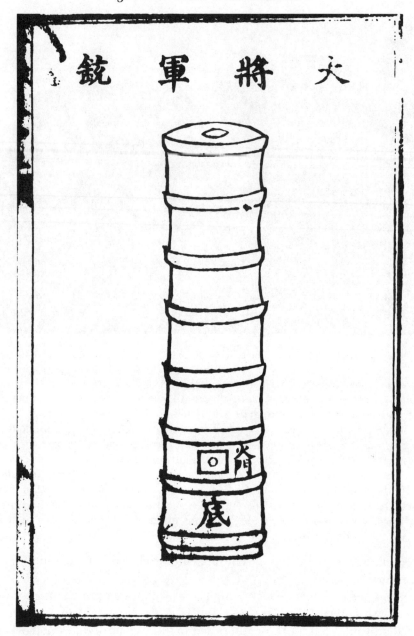

Fig. 119. The 'great general cannon' (*ta chiang-chün chhung*) as shown in *PL*, ch. 12, p. 8*a*. With its marked ribs, hoops or rings, it was carried on a four-wheeled vehicle; and many examples of this weapon were made around +1465.

they were supplied, and the total weight involved. For example, we read that one battalion (*ying*[1]), consisting of 40 batteries or units (*tui*[2]), was equipped with 3600 'thunderbolt shells' (*phi-li phao*[3]), 160 'wine-cup muzzle general cannon' (*chan khou chiang-chün phao*[4]), 200 large and 328 small 'continuous bullet cannon' (*lien chu phao*[5]) presumably firing grape-shot,[a] 624 hand-guns (*shou pa chhung*[6]), 300 small grenades (*hsiao fei phao*[7]), some 6·97 tons of gunpowder, and no less than 1,051,600 bullets of about 0·8 oz. weight each. This was quite some fire-power, and the total weight of the weaponry was reckoned to be 29·4 tons.[b]

Clayton Bredt (1) made a good point when he remarked that China had a very considerable priority over Europe in the making of cannon from cast iron. It was not until the second half of the +16th century that they could be used with safety there. But as we see from Table 1, cast iron had been used for cannon on a grand scale already in +1356 and the following year by Chang Shih-Chhêng's ill-fated 'Chou' dynasty;[c] and then further examples are extant from +1372 and +1377, not to speak of +1426 and later. This was only natural since, as is well known,[d] the art of iron-casting had been mastered in China from the −4th century onwards, though it did not reach Europe until towards the end of the +14th century. All this does not mean that iron was not used for ordnance in the West till then; but it was wrought iron, forged and welded. European iron cannon of late +14th-century date were built up of hammer-welded low-carbon iron bars and billets, bound together tightly with wrought-iron hoops as if for strengthened wooden barrels.[e] Earlier European castings were always of bronze, just as the earliest Chinese hand-guns and cannon also were. Indeed, even after iron-casting was thoroughly understood, guns of really high quality continued to be made of cast bronze, in Europe just as much as in China, down to the early nineteenth century and the advent of steel handling on an adequate scale.[f]

---

[a] Unless these were still eruptors, like that described on p. 263 above; but that would seem perhaps unlikely.

[b] *Wu Pien* (Chhien Pien), ch. 3, p. 95*a*, *b*. Note how in a passage like this, the significance of the word *phao* still has to be delicately adjusted to the context.

[c] Cf. pp. 295–6 above.

[d] See especially Needham (31, 32, 60, 72).

[e] A good example of this built-up construction is to be seen in the hand-guns excavated at Castle Rising in Norfolk, now in the Tower of London Armouries. They are not easily datable, but probably belong to the first half of the +15th century. Cf. Blackmore (2), no. 17, p. 55, the big gun from the *Mary Rose*, +16th century, and no. 196, p. 151. See Figs. 120, 121.

[f] Brass was also used in China as well as in the West. In 1959 two of us (J. N. and W. L.) had a long discussion with the late Prof. J. R. Partington on the meaning of certain statements in Sung Ying-Hsing's *Thien Kung Khai Wu* (+1637). Sung's knowledge of artillery was distinctly limited, as one can see from what he says about the heavy cannon of that time (ch. 15, p. 7*b*, tr. Sun & Sun (1), p. 271 and Li Chhiao-Phing (2), p. 393, both metallurgically misleading). But elsewhere he says that for casting Frankish culverins one should use brass (*shu thung*[8]), for hand-guns and signal guns one uses bronze, or some such alloy (*shêng shu thung*[9]), possibly gun-metal, and for large cannon such as the 'cup-mouthed great general' (*chan khou ta chiang-chün*[10]) one must use iron (ch. 8, p. 4*a*, tr. Sun & Sun (1), p. 165 and Li Chhiao-Phing (2), p. 230, again both metallurgically misleading). One reason why we then knew that *shu thung*[8] must mean brass was because in the previous

[1] 營   [2] 隊   [3] 霹靂砲   [4] 盞口將軍砲   [5] 連珠砲
[6] 手把銃   [7] 小飛砲   [8] 熟銅   [9] 生熟銅   [10] 盞口大將軍

Fig. 120. The Boxted Hall cannon, c +1450. Here the hoops have not been cast in, but were fixed on afterwards. Photo. Blackmore.

Fig. 121. Similar banding on Sultan Mehmet's cannon, now on the European bank of the Bosphorus at Istanbul. Photo. Bredt.

Detailed descriptions of gun-founding metallurgy are not lacking for the +15th and +16th centuries in China.[a] Iron from Fukien or Shansi was considered the best. If not poured directly into the mould from a kind of cupola furnace, a variety of the co-fusion method[b] was used to get a more steely iron, 5–7 parts of cast being combined with 1 of wrought. For some purposes[c] the billets of this were forged into long bars and four of them welded together to make a barrel, then held tightly by hoops of iron carefully forged on. Two or more of these barrels could be combined by these straps into a ribaudequin. Such methods paralleled the early use of iron for guns in Europe, but already by the mid +14th century good examples of cast-iron cannons were being made in China.

We might know more about gun-founding in the +15th and +16th centuries if it had not been regarded by the bureaucracy as so 'restricted' or 'top secret'. In the dynastic history itself we have the following passage:

'The casting of guns and cannon is done in the Nei Fu[11] palace compound, and it is forbidden to disclose any of the secrets of the techniques and designs'.[d]

The close association of gun-founding with the royal prerogative is just what we find in the early centuries of the art in Europe too. The compound in question was part of the Palace Treasury, superintended by eunuchs, and one would like to know more about its exact relations with the Arsenals Bureau (Chün Chhi Chü[12]) and the Ministry of Works (Kung Pu[13]).[e] But we must not pursue this subject further here.

chapter (ch. 14, p. 7b) Sung Ying-Hsing said that for 'four times melted (i.e. refined) shu thung' one takes 7 parts of thung[1] and 3 parts of chhien[2]. Since a copper–lead mixture would be quite useless for guns, Sung must have meant zinc here (wo chhien[3] or pai chhien[4]); cf. Vol. 5 pt. 2, p. 184. Li Chhiao-Phing (2), p. 349 muffed this, saying tin, but Sun & Sun (1), p. 247 got it right. Cf. also Vol. 4, pt. 2, p. 145 and Vol. 5, pt. 2, p. 208.

Sung Ying-Hsing also illustrated four small cannon (ch. 15, pp. 15b, 16a, b). The first two were vase- or bottle-shaped, with a tiller, and therefore quite archaic in his time. One was called the 'eight directions (i.e. pivot-mounted) hundred-bullets gun' (pa-mien chuan pai-tzu lien-chu phao[5]), probably for firing some kind of grape-shot; while the other was a 'magically effective smoke gun' (shen yen phao[6]) looking most suspiciously like an eruptor. Yet it was also labelled 'general gun' (chiang-chün phao[7]). In the Ming edition a certain iconographic similarity with the Huo Lung Ching two centuries earlier can be detected. The third was called a 'large magically-effective over-aweing cannon' (shen wei ta phao[8]), and the fourth a 'nine-arrow heart-piercing cannon' (chiu shih tsuan-hsin phao[9]). All these were almost three hundred years out of date, and no more need be said of them here.

[a] For example in Ping Lu, ch. 12, p. 1a, b. But they need a good deal more investigation than they have yet received.
[b] See Needham (32), pp. 26 ff. (72).
[c] As for the making of the hu tun phao[10], p. 315 above.
[d] Ming Shih, ch. 72, p. 30a.
[e] See Hucker (6, 7).

[1] 銅	[2] 鉛	[3] 倭鉛	[4] 白鉛
[5] 八面轉百子連珠砲	[6] 神烟砲	[7] 將軍砲	[8] 神威大砲
[9] 九矢鑽心砲	[10] 虎蹲砲	[11] 內府	[12] 軍器局
[13] 工部			

(16) FROM DEFLAGRATION TO DETONATION[a]

(i) *The rise in nitrate content*

The moment has now arrived when we must pause and look back over the way we have come.[b] We have been able to show that there was no great break in continuity between the oldest slow-burning incendiary weapons, through the quick-burning ones such as the petrol flame-throwers, to the deflagration of the first gunpowder mixtures. Sulphur and charcoal were both of very ancient use; it was the addition of saltpetre which gave a new turn to the story. But for a long time the compositions seemed to be no more than better incendiaries; then it was found that they would explode when placed in weak-walled containers, still later that the explosions could be strong enough to break cast-iron bombs and grenades into fragments.[c] In due course, explosive mixtures were produced of sufficient power, when used in land-mines and the like, to give brisant explosions capable of destroying fortifications and breaking down city gates. Eventually the true propellant property of optimally fast-burning[d] within the bore of the metal-barrel hand-gun or bombard was attained, capable of sending on its way a projectile, whether simple or composite, which perfectly occluded the charge. We saw, too, that the barrel, first of bamboo and only later of metal, long preceded the true gun, for advantage was taken of the incendiary property of gunpowder compositions to make those five-minute flame-throwers the fire-lances and eruptors, ancestors respectively of the gun and the cannon.

All this implies, upon analysis, one single thing more outstanding than any other, a slow and persistent rise in the nitrate content of the mixtures. It is only reasonable to ask how far the available figures bear out this interpretation. The first thing to do is to tabulate the data we have for the periods preceding any possible Western influence (Table 2). Here the important sources are the *Wu Ching Tsung Yao*[e] of +1044, and the *Huo Lung Ching*[f], first printed in +1412 but containing material which represents the situation about +1350. To have an

---

[a] As noted on p. 110 above, contemporary explosives chemists do not use the term 'detonation' in connection with gunpowder, reserving it for substances the rate of burning of which reaches supersonic speeds. Some of these have oxygen built into the molecule itself, like trinitro-toluene, but others do not, like lead azide (PbN_6), or silver fulminate, mentioned elsewhere in this Section. A mixture of acetylene and oxygen can also detonate in the strict sense. Therefore the title of this sub-section might preferably be 'From Burning, through Deflagration, to Explosion'.

[b] Cf. particularly pp. 117, 248 above.

[c] It is possible that a gunpowder charge inside a brittle cast-iron container might be not much less effective in certain circumstances than a detonating charge (in the modern sense), which would produce more air shock but much smaller fragments less effective against hard structures.

[d] Cf. p. 484 below.

[e] See p. 20 above, where the exact references are given.

[f] See p. 25 above. The section on gunpowder compositions is common to all the versions. In the *Huo Lung Ching Chhüan Chi* (the Nanyang edition) it is in pt. 1, ch. 1, pp. 6a–15a, and this places it firmly in the oldest stratum of the book. In the *Huo Chhi Thu* or Hsiangyang edition it is in ch. 1, pp. 3b–9a; and in the *Huo Kung Pei Yao* version it is in ch. 1, pp. 6a–15a. The *Wu Pei Huo Lung Ching* gives half-a-dozen composition figures not in the other versions, which state only the constituents, but the proportions are so archaic that it must have been copying earlier sources.

Table 2. *Early Chinese gunpowder compositions*

composition	name	nature	N	percentage S	C	other constituents
Wu Ching Tsung Yao[a]						
gunpowder	*huo yao fa*[1]	weak explosive	50·5	26·5	23·0	arsenic,[b] lead salts, dried plant materials, oils, resin
thorny fire-ball	*chi li huo chhiu*[2]	hooked incendiary projectile inner ball only	50·2	25·1	24·7	pitch, dried plant materials, oils
		outer covering incl.	34·7	17·4	47·9	arsenic,[b] plant poisons, wax, oils, dried plant materials
poison smoke-ball	*tu yao yen chhiu*[3]	incendiary with toxic smoke inner ball only	39·6	19·8	40·5	
		outer covering incl.	27·0	13·5	59·5	
Huo Lung Ching						
magic gunpowder	*shen huo yao*[4]	incendiary with toxic smoke	(28·6)	(21·4)	(50·0)[c]	arsenic sulphides,[b] plant poisons, faeces
poison gunpowder	*tu huo yao*[5]	strong explosive, with toxic smoke	77·5	9·3	13·2	arsenic, plant poisons
violent gunpowder	*lieh huo yao*[6]	probably incendiary, with toxic smoke	proportions not specified			arsenic, plant and insect poisons
flying gunpowder	*fei huo yao*[7]	mild incendiary, with toxic smoke	(12·3)	(57·5)	(30·2)	arsenic, plant and insect poisons
blinding gunpowder	*fa huo yao*[8]	strong explosive, with toxic smoke	(74·7)	(17·3)	(8·0)	lime,[d] arsenic sulphides, and dried plant materials
smoke gunpowder (or, bruising and burning gunpowder)	*yen huo yao*[9] *lan huo yao*[10]	tetoxic smoke composition, with co-viative projectiles	proportions not specified but nitrate probably about 60%			sal ammoniac,[e] bits of broken porcelain, iron filings, plant and insect poisons, urine and tung oil[f]
against-the-wind gunpowder	*ni fêng huo yao*[11]	probably incendiary, with toxic smoke	proportions not specified			insect poisons, wolf faeces, dolphin oil and bone
flying-in-air gunpowder	*fei khung huo yao*[12]	use uncertain	proportions specified only for the additives			camphor,[e] resin, realgar
rising-by-day gunpowder	*jih chhi huo yao*[13]	rocket propellant	50·1	(5·0)[g]	44·9	none
rising-by-night gun-powder	*yeh chhi huo yao*[14]	rocket propellant	76·9	3·9	19·2	none
			(45·5)	(9·0)	(45·5)[h]	(none)
spattering gunpowder	*phên huo yao*[15]	fire-lance composition	57·1	5·7	37·2[h]	fine sand (cf. p. 234 above).
explosive gunpowder	*pao huo yao*[16]	strong-casing bomb filling	91·3	6·9	1·8[h]	none
projectile gunpowder	*phao huo yao*[17]	weak-casing bomb filling	50·0	30·0	20·0	arsenic sulphides

Table 2 (*Cont.*)

composition	name	nature	percentage N	percentage S	percentage C	other constituents
fizzing-in-the-water gunpowder	*shui huo yao*[18]	use uncertain	proportions not specified			quicklime (cf. p. 166 above), sal ammoniac[e] and dried plant materials
fire-projectile gunpowder	*huo tan yao*[19]	incendiary ball for projection from fire-lance or eruptor, just fitting its iron barrel[i]	—	39·0	61·0	camphor,[e] resin
five-league fog	*wu li wu*[20]	incendiary, with lachrymatory smoke	27·8	27·8	44·3	arsenic, sawdust, resin, human hair, chicken, wolf and human excreta
soul-hunting fog	*chui pho wu*[21]	strong explosive, with toxic smoke	83·3	8·3	8·4	arsenic sulphides, animal poisons
poison-smoke ball	*yen chhiu tu yao*[22]	weak explosive, with toxic smoke	36·4	36·3	27·3	arsenic, plant poisons, wax, tung oil
magic fire	*shen huo*[23]	incendiary composition	26·6	66·7	8·6	sal ammoniac and iron filings
magic smoke	*shen yen*[24]	explosive, or rocket propellant, with toxic smoke	69·6	17·4	13·0	arsenic, camphor, calomel, and calcium-magnesium silicates

N.B. Besides these there are half-a-dozen more compositions, mostly for coloured signal-smokes, blue-green, red, purple, white and black; these have already been considered on p. 144 above. The average nitrate content is 66%.

[a] As noted above (p. 120), Arima (*1*), p. 43, worked out figures rather higher, but it depends on what assumptions are made. We think ours are better, but the general argument would not be greatly affected.
[b] Arsenic (and mercury too) were often constituents of early European gunpowders, e.g. in the *Bellifortis* of +1405; cf. Partington (5), p. 149.
[c] Figures in brackets are those derivable only from the *Wu Pei Huo Lung Ching* version.
[d] Cf. pp. 167 ff. above.
[e] Sal ammoniac (ammonium chloride) and camphor were often used also in early European gunpowder compositions; cf. Partington (5), pp. 144, 160.
[f] The presence of solid discrete constituents here shows that the mixture was intended for a fire-lance.
[g] Estimated, as no figure is given.
[h] Some editions give unreasonably small quantities of sulphur for these compositions.
[i] Cf. p. 221 above. This was definitely not gunpowder as it contained no saltpetre, only combustibles.

[1] 火藥法 [2] 蒺藜火毬 [3] 毒藥烟毬 [4] 神火藥 [5] 毒火藥 [6] 烈火藥 [7] 飛火藥 [8] 法火藥
[9] 烟火藥 [10] 爛火藥 [11] 逆風火藥 [12] 飛空火藥 [13] 日起火藥 [14] 夜起火藥 [15] 噴火藥 [16] 爆火藥
[17] 砲火藥 [18] 水火藥 [19] 火彈藥 [20] 五里霧 [21] 追魄霧 [22] 烟毬毒藥 [23] 神火 [24] 神烟

adequate perspective in mind we must recall the dates of $+1290$ and $+1327$ for the first bombards in China and in Europe respectively; then (as we shall see) the Portuguese or 'Frankish' breech-block cannon were in China by about $+1511$, and the Jesuit period, which so greatly intensified East–West relationships, was well under way by $+1600$.

This list of about two dozen different gunpowder compositions in the Hsiang-yang version of the *Huo Lung Ching* (the *Huo Chhi Thu*) is identical with that in the first part of the large Nanyang edition (*Huo Lung Ching Chhüan Chi*), and that of the *Huo Kung Pei Yao* version also. Indeed the last two are exactly identical down to the number of words on the page, and are more carefully printed than the Hsiang-yang version, though it includes a number of corrections. The list, without the smoke formulae, is repeated in the *Ping Lu* ($+1606$), but new inclusions increase the number of gunpowder compositions to about three dozen. The *Wu Pei Chih* ($+1628$) incorporates all the gunpowder compositions in the *Ping Lu* and the smoke formulae omitted by the latter, giving a total of more than forty compositions. The only new gunpowder mixture it gives is a 'lead (bullet) gunpowder formula' (*chhien chhung huo yao*[1]) comprising 40 oz. of saltpetre, 6 oz. of sulphur and 6·8 oz. of charcoal.[a] The explosive is used here as a charge of black powder. The *Wu Pei Huo Lung Ching*, containing more than twice the number of specifications of gunpowder constituents than that in the other versions, is much the most comprehensive, but the date of compilation of this text is even later than the *Wu Pei Chih*. Yet it looks as if the list of some two dozen gunpowder compositions in the *Huo Lung Ching Chhüan Chi*, the *Huo Chhi Thu* and the *Huo Kung Pei Yao* versions does indeed represent the knowledge of the mid $+14$th century incorporated into later military compendia.

In order to get the most out of these figures, it is desirable to plot them in a graph. This we did first using the three-vertex method of Fisher (1),[b] but later found it more convenient to use triangular graph-paper as in Figs. 122 and 123.[c] Here the two important reference-points are that for equal proportions of the three constituents, represented by a dot, and that for the approximately theoretical composition ($75:13:12$), represented by a triangle.[d] Indications are also given for a typical rocket composition, and for the lowest limit of blasting powder, i.e. the 'slowest' of explosives.[e] Now at once we see that the points for this period of early Chinese experimentation are scattered all over the map, the *Wu Ching Tsung Yao* figures ranging from 27 to 50 % nitrate, but the *Huo Lung Ching* ones covering a range from 12 to 91 %. Half-a-dozen of them attain the region of

[a] I.e. $N:S:C:75\cdot7:11\cdot4:12-9$, almost the theoretical nitrate level. This was not remarked upon by Davis & Ware (1), p. 526, who gave no percentages. *WPC*, ch. 119, p. 21 b.

[b] Cf. Needham (12), p. 71.

[c] We are indebted to Dr Peter Gray, at that time (1953) also a Fellow of Caius, for suggesting the use of this, and for advising us in general upon our entry into the subject.

[d] Cf. Mellor (1), p. 707.

[e] Berthelot (13), vol. 2, p. 311; Marshall (1), vol. 1, p. 74; Partington (5), p. 327.

[1] 鉛銃火藥

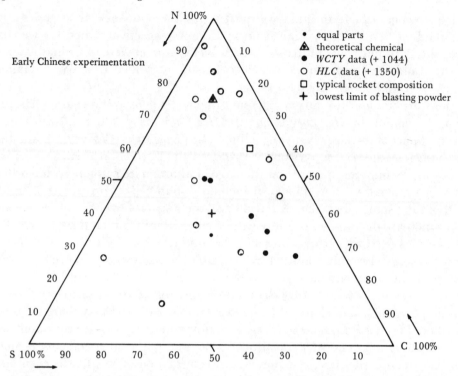

Fig. 122. The composition of gunpowder portrayed on triangular graph-paper; nitrate at the upper vertex, sulphur at the lower one on the left, and carbon at the lower one on the right. The points are all drawn from early Chinese experimentation, and their wide scatter is evident. Time +1000 to +1350.

the theoretical composition for maximal explosive force. What this must imply is many decades, even centuries, of trial and error, starting no doubt from the simplest scheme of equal proportions[a] and slowly finding its way up towards the most effective nitrate admixture. The low-nitrate mixtures would have been difficult, though not impossible, to get to explode, the high-nitrate ones would have been difficult, but not impossible, to make to burn in fire-lances or rockets. Many must have been the disappointments, and many also the dangerous, even fatal, accidents encountered on the way, though on the whole history is silent about them.[b] Of course we are well aware that the percentage proportions are only a part of the picture; much depends on the conditions of firing, and much on the physical character of the mixture. For example, the pressure situation, and the degree of confinement. All forms of gunpowder can be induced to burn quietly on an open surface,[c] but when enclosed in containers, even of paper or carton, will explode with a loud report.[d] This must have been an early discovery in China, and from the evidence already given we can place it pretty

[a] Brock always believed this, cf. (1), p. 17.
[b] But cf. pp. 112, 209 above.   [c] Even cordite will do this.
[d] For this effect the nitrate content probably has to reach 50% or more. Cf. Foley & Perry (1).

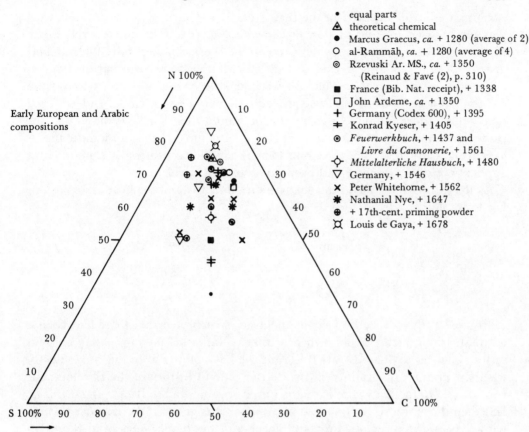

Early European and Arabic compositions

- • equal parts
- ⚠ theoretical chemical
- ● Marcus Graecus, *ca.* + 1280 (average of 2)
- ○ al-Rammāḥ, *ca.* + 1280 (average of 4)
- ◉ Rzevuski Ar. MS., *ca.* + 1350
  (Reinaud & Favé (2), p. 310)
- ■ France (Bib. Nat. receipt), + 1338
- □ John Arderne, *ca.* + 1350
- + Germany (Codex 600), + 1395
- ╪ Konrad Kyeser, + 1405
- ◎ *Feuerwerkbuch*, + 1437 and
  Livre du Cannonerie, + 1561
- ◇ *Mittelalterliche Hausbuch*, + 1480
- ▽ Germany, + 1546
- ✕ Peter Whitehorne, + 1562
- ✳ Nathanial Nye, + 1647
- ⊕ + 17th-cent. priming powder
- ⏣ Louis de Gaya, + 1678

Fig. 123. A similar triangular graph for the early European and Arabic compositions. It can be seen that they are all clustering round the value of about 70% nitrate, i.e. quite close to the theoretical value of some 75%. This suggests that by the time that gunpowder became known to the Arabs and Europeans, the optimum mixture was already known. Time from +1280 to +1700.

safely in the middle or second half of the +10th century. Equal proportions would have come in the first half, and nitrate contents high enough to burst cast-iron or other metal containers would have been arrived at towards the end of the +12th. Then in the following century came the application of the full propellant force of high-nitrate gunpowder in the first metal-barrel hand-guns and bombards.

Now when we plot the earliest figures from Arabic and European sources[a] in just the same way we come upon a remarkable difference.[b] Almost without

[a] Among primary sources one may mention Anon. (157, 158); Whitehorne (1); Nye (1); Sprat (1); Anon. (160); Turner (1); Robins (1); Muller (1); Watson (1).

[b] Data in secondary sources can be found in Partington (5), pp. 42 ff., 102, 144, 148–9, 154, 157, 204, 253, 316, 323, 324–7, 338; Sarton (1), vol. 3, p. 1700; Hime (1), pp. 149 ff., 168–9, 218; Reinaud & Favé (1), p. 166 (*Livre du Cannonerie*), 180 (Amiot); (2), pp. 310–11; Marshall (1), vol. 1, pp. 26–7, 74; Spak (1), pp. 62, 66, 157. Comparable late figures for China are in Davis & Ware (1). pp. 526–7; Rondot (2); as well as Reinaud & Favé above.

exception[a] all are clustered in the region of high efficiency, between the lowest blasting-powder line and the low 80 % level (Fig. 123). This must surely mean that the constitution of gunpowder came to the West as it were fully-fledged; and just as we find no long period of experimentation there with carton bombs, fire-lances, eruptors and the like, so we find little or no uncertainty about the most suitable composition. It was already known before it came. Indeed if we reflect upon the common Western name of 'gunpowder' itself, we may well conclude that it arose there solely in connection with guns. Is this not a mute philological indicator that the preceding four and a half centuries of experimental applications of the mixture had been done somewhere else?

As the nitrate rises, so do the gas pressure maxima and the heat of explosion. For example:[b]

% KNO_3	max. gas pressure (bar.)	heat of explosion [k]/g
80	98	3·05
75	92	2·87
70	84	2·71
68	78	–

In the +17th century Europeans themselves suspected that there had been a gradual rise of nitrate content in past times. Nathaniel Nye in +1647 chose a series of figures to demonstrate this,[c] and we have plotted it in Fig. 124.[d] Lastly we show another diagram to illustrate the general history of the development (Fig. 125). On the left we indicate the range of values in the *Wu Ching Tsung Yao*, and then the extraordinary spread of compositions given in the *Huo Lung Ching*, with the lower figures of Marcus Graecus and the higher ones of al-Rammāḥ sandwiched between them. There follows, on the right-hand side of the plot, a depiction of the way in which the diverse uses crystallised out—some 40 to 65 % for blasting powder, some 55 to 70 % for rocket compositions, fire-lances and Roman candles, with 65 to 85 % for propellant and other explosions or detonations.

All this, as we are well assured, is rather schematic.

The inflammability of gunpowder [wrote Partington,[e]] is not greatly affected by the mixture ratio. The propulsive force depends mainly on the burning rate and the volume of

---

[a] It is true that the equal parts formula is also given in the *Feuerwerkbuch* of +1437 and its French translation the *Livre du Cannonerie* of +1561, contemporary with Whitehorne (1). It comes down from the *Liber Ignium* probably. But there never seem to be any instructions what to do with it.

Then there is a composition supposedly used at Amiens in +1417 containing the ratio 27·2 : 26·1 : 46·7, but the descriptions do not well agree (cf. Partington (5), pp. 148, 324). Further research would be necessary here.

Again, Reinaud & Favé (1), p. 166, give the proportions 20 : 40 : 40 for one of the preparations listed in the *Livre du Cannonerie*, but its use is not described; presumably, like the former, it was an incendiary.

[b] Figures from a recent paper by Hahn, Hintze & Treumann (1).

[c] See Ffoulkes' edition of de Gaya (1), as well as Nye (1) itself.

[d] Ayalon (1), pp. 25–6, 42, also perceived the general rise, and cited Hime (1), pp. 168–9, whose figures, though rather widely scattered, do certainly show it.

[e] (5), p. 328. Inflammability is very hard to assess or quantify.

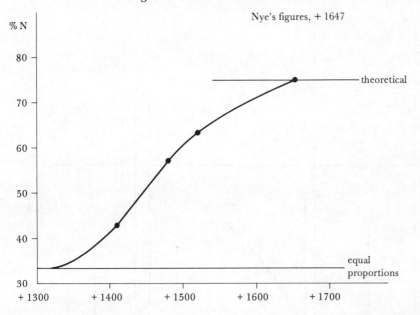

Fig. 124. The rise in nitrate content over the centuries; figures given in the book of +1647 by Nathaniel Nye.

gas, both of which do depend on the mixture ratio. The right mixture for military gunpowder was found only after many trials over a considerable period of time; and even today more stress is laid on the method of manufacture than on the mixture ratio.

For example, there was the matter of 'corning'; a form of granulation first attained by sieving away the impalpable powder, so that the oxygen of the air could gain better access to the particles, reinforcing the built-in oxidising capacity arising from the nature of the mixture itself.[a] This seems to have been first done in the West at Nürnberg about +1450.[b] A seventeenth-century writer summed up everything well when he wrote:

The whole Secret of the Art [of making gunpowder] consists in the proportion of the Materials, and the exact mixture of them, so that in every the least part of *Powder* may be found all the Materials in their just proportion; then the Corning or making of it into Grains; and lastly the Drying and Dusting of it.

Then after mentioning the various proportions recommended by such authors as John Baptist da Porta, Bonfadini and Jerome Cardan, he continued:

Indeed there is so great a Latitude, that Provided the Materials be perfectly mixt, you make good *Powder* with any of the proportions above mention'd; but the more Peter you allow it, it will still be the better, till you come to observe eight Parts.[c]

---

[a] Partington (5), pp. 154, 328.
[b] Räthgen (1), pp. 77, 109 ff.
[c] Anon. (160), in Sprat (1). He would have meant between 72 and 78 %.

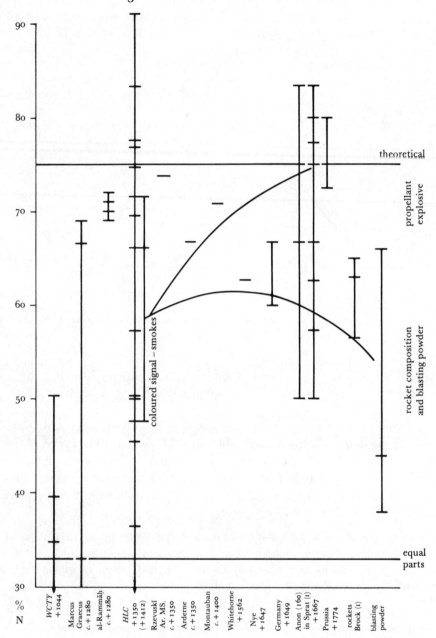

Fig. 125. Nitrate percentages derived from mixture specifications given from +1044 to the present day. One can see how the propellant–explosive usage at some 72% gradually differentiates itself from the rocket-composition blasting-powder usage at some 55%.

Some idea of the effect of grain size can be seen from the following figures.[a]

compressed grains average size (mm.)	whirling height in *éprouvette* when exploded[b]	
	70% KNO$_3$	75% KNO$_3$
0·75	51·5	65·6
1·75	31·3	38·2
2·60	16·0	30·0
3·10	14·0	19·4
3·75	10·1	16·6

Thus we have a family of descending curves, tending towards linearity for the larger grain sizes.

Remembering now the introduction of the Portuguese culverin (if we may so call it)[c] about +1511, it remains to look at the figures given in the military compendia after that date, for it is obvious that European experience of black-powder compositions would have come with it. Something of the sorting-out process in China can be seen in Fig. 126, where rocket compositions cluster in the neighbourhood of 60 % nitrate while explosive ones surround the theoretical value of 75 %. Eighteen proportions of this kind are given[d] in Thang Shun-Chih's *Wu Pien*[1], compiled about +1550, and by this time powder suitable for the 'bird-beak' muskets[e] as well as bombards and cannon was also specified. This was certainly the earliest Chinese book to give particulars about the gun-powder used for the arquebus, which had been introduced by way of Japan in +1548.[f] Only one formula appears in the *Chi Hsiao Hsin Shu*[2] of Chhi Chi-Kuang, written some ten years later, but at 75·7:10·6:13·7 it was very close to the theoretical mixture established by chemists. After the beginnings of the Jesuit mission there came out, in +1598, the *Shen Chhi Phu*[3], devoted primarily to mus-kets, but Chao Shih-Chên's two formulae were rather higher in nitrate.[g] As already noted, the *Ping Lu*[4] of Ho Ju-Pin (+1606) reproduced, apart from its twenty or so new mixtures, all the figures of the *Huo Lung Ching*[5], with the excep-tion of the coloured signal-smokes, which in Mao Yuan-I's *Wu Pei Chih*[6] of +1628 were re-placed and again recorded. How far some of the archaic formulae were actually still used at this time remains rather uncertain. The theoretical percentage of nitrate appeared again in Hui Lu's contemporary *Phing Phi Pai Chin Fang*[7], together with three others none of which were new. Finally the *Huo Kung Chhieh Yao*[8], which Chiao Hsü wrote in +1643 in collaboration with the Jesuit John Adam Schall von Bell (Thang Jo-Wang), carried fourteen composi-

---

[a] From Hahn, Hintze & Treumann (1).
[b] See p. 552 below.
[d] Ch. 5, pp. 63 *b*–78 *a*.
[f] But see p. 440 below.

[c] See pp. 367ff. below.
[e] See pp. 432 ff. below.
[g] 80·7 to 83·3 %.

[1] 武編   [2] 紀效新書   [3] 神器譜   [4] 兵錄   [5] 火龍經
[6] 武備志   [7] 洴澼百金方   [8] 火攻挈要

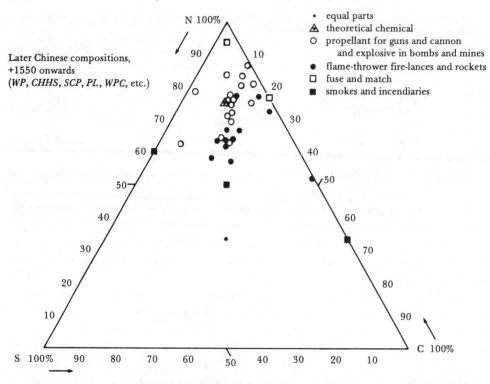

Fig. 126. A third triangular graph for the later Chinese compositions, from about +1550 onwards. Here too they cluster in the neighbourhood of 75% nitrate, showing that the optimum mixture was well known there also.

tions covering the whole range of possibilities from 33·3 to 86·4% nitrate, but mostly in the propellant region of 70 to 80%.[a]

Among all these late mixture formulae, two things are noteworthy. First there is the fact that the old Chinese predilection for high nitrate contents, in the eighties and even nineties, persisted alongside the proportions characteristic of European practice, which doubtless came in after +1511, together with the Portuguese breech-loader and the bird-beak muskets. Such high figures can be found even in the book where the Jesuit was joint author. But often the proportions approached the theoretical value closely, for example the *Ping Lu* (+1606) gives two gunpowder compositions for musket (*ta chhung*[3]) and pistol (*hsiao chhung*[4]) at 75·1% and 71·4% respectively.[b] Partington hit the nail on the head when he

---

[a] Besides all the books mentioned in this paragraph, a wealth of gunpowder compositions is also to be found in the *Chin Thang Chieh Chu Shih-erh Chhou*[1] (Twelve Suggestions for Impregnable Defence), written by Li Phan about +1630. Some of these were noted by Arima (*1*), p. 221. Other formulae, very near the theoretical, were also recorded in the *Ping Chhien*[2] (Wai Shu section), composed by Lü Phan & Liu Chhêng-Ên in +1675. This 'Key to Military Art' belongs to the next dynasty, of course, after the end of the Ming.

[b] Ch. 11, pp. 6*b*, 7*a*; 13, pp. 24*b*, 25*a*.

[1] 金湯借箸十二籌   [2] 兵鈐   [3] 大銃   [4] 小銃

wrote that 'the development from the compositions given in the *Wu Ching Tsung Yao* (+1044) to the modern gunpowder of the *Wu Pei Chih* (+1628) could have resulted from Chinese experiments rather than from the import of European information'.[a] From the data given in Fig. 122 we can now be sure that this was in fact precisely what happened. Secondly, in Fig. 126 it can be seen that Chinese experimentation continued, involving curious mixtures sometimes without carbon, sometimes without sulphur; these probably had no great future before them.

The nomenclature of the weapons and purposes mentioned in these late formula lists does not call for much remark, apart from the arquebuses, muskets and breech-loading cannon which we shall be considering in the following subsections. Many of the names we have encountered already, like 'swarm of bees' for fire-lances with co-viative projectiles (*hsiao i wo fêng*[1]), or 'river-dragon' (*ching chiang lung*[2]) for a sea-mine, which are still in the lists. But there is also in *Wu Pien* an incendiary bomb called by the colourful name of the fruit, *li-chih phao*[3]; while the *Huo Kung Chhieh Yao* has a land-mine with the explanatory appellation 'foot-tripped, buried and lying-in-wait powder' (*mai fu tsou hsien (huo) yao*[4]).[b]

It is interesting that throughout this series of books the old belief in the value of mixing poisonous or opprobrious substances with the gunpowder persisted, not excluding the last one where a Jesuit was co-author. These included arsenic, mercury, lead and copper in various forms, sal ammoniac, camphor,[c] borax, quicklime, plant and animal poisons,[d] and the excreta of man and beast. No surprise need be occasioned by this, for Leonardo da Vinci himself had been interested in attacking the enemy by sulphurous smokes,[e] fumes of burnt feathers, sulphur and arsenic,[f] and even toad and tarantula venoms mixed with rabid saliva and conveyed by bombs.[g] This would have been about +1500. Faith in arsenicals continued at least till +1580, with von Senfftenberg (1),[h] and mercury figured still in the smoke-balls of Appier and Thybourel in +1620 and +1630.[i] Such was the pre-history, probably mercifully inefficient, of chemical warfare.

Before leaving the subject of percentage proportions in China, it may be noted that information of value can sometimes be gained from records of bulk purchases by the Arsenals Administration for the preparation of the gunpowder needed. We have come across this kind of thing before, in relation to the require-

---

[a] (5), p. 274.   [b] Ch. 2, p. 11*a, b*.

[c] In the European Middle Ages a mixture of saltpetre, sal ammoniac and camphor was named 'sal practica' (with many spelling variations), and commonly added to gunpowder, because of a vague idea that it gave more 'volatility' to the mixture; cf. Marshall (1), vol. 1, p. 25; Partington (5), p. 155, etc.

[d] On Chinese arrow poisons see the studies of Bisset (1, 2).

[e] Partington (5), p. 175; McCurdy (1), vol. 2, p. 198.

[f] McCurdy (1), vol. 2, pp. 201, 210.

[g] *Ibid.* vol. 2, pp. 217–19.   [h] Partington (5), pp. 170, 183.

[i] On Jean Appier and François Thybourel see Partington (5), pp. 176–7. The plant and other poisons continued down into the report of Amiot (2), Suppl., as late as +1782; cf. Partington (5), p. 253.

[1] 小一窩蜂   [2] 淨江龍   [3] 荔枝砲   [4] 埋伏走線藥

ments of the Chinese Mints for metals and alloys with which to issue currency at different times.[a] Thus some details regarding the gunpowder manufactured in the State workshops in the early +17th century can be found in Ho Shih-Chin's[1] *Kung Pu Chhang Khu Hsü Chih*[2] (What Officials ought to know about the Factories and Storehouses of the Ministry of Works, +1615).[b] It says that 300,000 *chin* of gunpowder (about 150 tons) were made annually for fire-lances and cannon. The making of the fire-lance gunpowder required 100,312 lb. 8 oz. of saltpetre, 19,687 lb. 8oz. of sulphur and 30,000 lb. of willow charcoal; while the making of gunpowder for cannon took 106,875 lb. of saltpetre, 20,625 lb. of sulphur and 22,500 lb. of willow charcoal. This implies the following N : S : C proportions: for fire-lance gunpowder 66·9 : 13·1 : 20·0, and for cannon gunpowder 71·2 : 13·7 : 15·0.[c] The cost for each item is given.[d] The text also states the cost of gunpowder manufacture for the 'bird-beak gun', i.e. the arquebus, but unfortunately without giving the proportions in the specification. It is interesting too that 200,000 lead balls are to be made not only for some kind of chain-shot cannon (*lien chu phao*[3]) but also as co-viative projectiles for the fire-lances (*pa chhiang*[4]).[e]

Pai wên pu ju i chien[5] says the Chinese proverb, which could be Englished in the words 'a thousand explanations are not as good as once seeing for oneself'. Accordingly we resolved to view the ignition of a number of powders made up with varying proportions of nitrate, thus elucidating what seems to be a historical sequence by actual experiment. Here we were very fortunate in gaining the co-operation of the staff of the Royal Armament Research and Development Establishment at Fort Halstead in Kent,[f] who prepared and let off for us more than a dozen mixtures. The results are shown in the accompanying Tables and photographs.

Table 3 gives the compositions examined,[g] and Table 4 the times of burning, with the phenomena observed.[h] The same volume of powder was taken for each

---

[a] See Vol. 5, pt. 2, p. 216 for a Thang example.
[b] In the *Hsüan Lan Thang Tshung Shu Hsü Chi* collection.
[c] These figures are included in Fig. 126.
[d] Ch. 8, pp. 4*a*–6*a*.
[e] *Ibid.* pp. 1*b*–2*b*. Each year 5000 of these were made. The chapter gives a wealth of information on the natures and quantities of many kinds of fire-weapons.
[f] Our warmest thanks are due to Mr Cliff Woodman, Dr Nigel Davies, Dr John Robertson and Mr Philip Seth. For our introduction we are most grateful to Mr Howard Blackmore, then Deputy Master of the Armouries at H.M. Tower of London. The experiments were done on 20 Feb. 1981.
[g] The potassium nitrate used was of pyrotechnic grade, dried at 70°C. for 24 hr., the sulphur was of laboratory reagent grade, and the charcoal was prepared from the wood of the alder buckthorn (*Frangula alnus*). All the reagents were sieved through a B.S. No. 120 sieve to remove any lumps and ensure that the powders were free-flowing. The sieved reagents were weighed into a dust-tight container which was then tumbled rapidly in a Turbula mixer for 20 minutes, then the mixtures were sealed into anti-static plastic bags. The commercial preparation (no. 1) was probably more perfectly mixed because it was ground in an edge-runner mill.
[h] A colour film was taken at a speed of 500 frames per sec., and later studied at a speed twenty times slower.

[1] 何士晉   [2] 工部廠庫須知   [3] 連珠砲   [4] 靶鎗
[5] 百文不如一見

Table 3. *Compositions studied at Fort Halstead (1981) R.A.R.D.E.*

Experiment no.	Percentage composition			Notes
	KNO$_3$	sulphur	charcoal	
1	75	10	15	commercial, corned (ICI, Ardeer)
2	75	10	15	lab. preparation (fine powder) electric fuse
3	90	–	10	electric fuse
4	70	10	20	electric fuse
5	63	27	10	electric fuse
6	42	42	16	electric fuse
7	42	16	42	no ignition
8	42	16	42	slow-match cord
9	33	33	33	slow-match cord
10	50	50	–	electric fuse
11	50	–	50	electric fuse
12	54	23	23	electric fuse
13	81	9	10	electric fuse
14	81	9	10	hand-pressed candle electric fuse

test[a] and ignited as a free unconfined heap. Under these conditions it was at once clear that all the compositions blazed; but some more quickly and fiercely than others.[b]

An initial burst of flame (Fig. 127) was seen only when the nitrate exceeded 60 %, and the maximum speed of this occurred when the composition approached most closely to the theoretical. At lower proportions there was simply a column of flame, sometimes continuing for a good while, and illustrating well the incendiary effects for which gunpowder found its first uses (Fig. 128). The more the nitrate was reduced the more difficult ignition became, and slow-match sometimes had to be substituted for an electric spark. No compositions under 33 % saltpetre were tried, but as low as 12 % it would be extremely hard to ignite, and the sulphur would just burn to sulphur dioxide (SO_2). Explosive effects appeared only when the powder was confined, as in a carton tube, and if this was open at one end the fire-lance effect was clearly seen. For that a slow burning-rate was quite all right, but the great problem of the first cannoneers was how to avoid on the one hand burning too slow and lacking adequate propulsion, or on the other burning too fast and bursting the gun. Thus the rate of burning, and hence of gas production, and hence of pressure rise within the confined space, and hence of imparting motion to the projectile, had to be just right.

Broadly speaking, then, the experiments bore out the deduction from the

[a] A loose pile ranging between 100 and 200 gm.
[b] The termination of burning was hard to assess, and the filming in each case was not continued longer than about 4 to 8 sec.; except in the case of no. 14, when it was run at 100 frames per sec. and continued for nearly 20 sec.

Table 4. *Observations on the compositions of Table 3*

% KNO$_3$	Exp. no.	Time of burning in secs.		Notes
		initial burst of flame	length of blaze	
81	13	0·32	3·04	red flash, strong column of flame with blue-black smoke
75 (comm.)	1	0·16	1·12	big puff, dull thud, white smoke, fast-burning flame, many incandescent particles (prob carbon) thrown out
75 (lab.)	2	0·16	2·4	big puff, white smoke, flame of longer duration
70	4	0·48	>2·48	sideways gush, strong column of flame, with white smoke
63	5	0·56	2·8	strong flame column, with sparks, blue smoke towards the end
54	12	—	3·76	slow to ignite, no burst, strong flame column, with reddish-brown streaks (prob. nitric oxide) in the white smoke
42 (high S)	6	—	>2·88	no burst, slow start, weak fire, little bluish smoke, flame column persisting
42 (high C)	8	—	>2·56	hard to ignite, no burst, strong initial flame then weakening to thin column, with sparks, little bluish smoke, burning for a long time
33	9	—	>3·76	hard to ignite, no burst, strong flame column with reddish-brown streaks
90 (no S)	3	—	7·52	no burst, steady gush of flame, blue smoke (K colour)
50 (no C)	10	—	>4·32	no burst, weak yellow flame with many detached flames, hardly any smoke
90 (no S)	11	—	4·24	no burst, spasmodic flame building up slowly, blue smoke
Confined in 0·5 in. diameter carton candle				
81	14	—	16·0	no burst, steady flame like flame-thrower, little smoke, no marked explosion
75	—	—	—	loud explosion, burning continuing after bursting of tape seal
66	—	—	—	definite explosion, then flame-thrower effect after bursting of tape seal

Fig. 127. Fort Halstead experiments. No. 5 at 63% nitrate gives an initial burst of flame for 0·56 sec., then a strong column of flame and smoke with sparks for 2·8 sec.

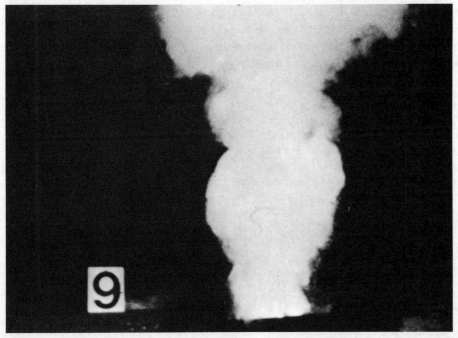

Fig. 128. Fort Halstead experiments. No. 9 at 33% nitrate was hard to ignite and gave no burst, then burnt with a strong flame column and reddish-brown streaks for as long as 4 sec.

history of gunpowder in China, between the mid +9th century and the mid
+14th, that gradually more and more nitrate was used in the mixtures.

This is indicated also by the literature, which records how others beside
ourselves have at sundry times and places been moved to try experiments. Thus
Lassen (1) found that a powder of N : S : C : 35 : 35 : 30 composition would throw
a ball from an iron tube like a +14th-century hand-gun only about 40 ft, even
if the weight of the charge equalled that of the projectile; but it was very hard
to ignite, and burnt but slowly. This was no more than a fire-lance effect. At
66·5 : 11 : 22·5 Williams (1) had many mis-fires, the gases issuing through the
touch-hole, and sometimes the ball just dropped from the muzzle; even when it
did fly, it failed to penetrate an iron sheet 18 ft away. Foley & Perry (1), using
powders ranging from 66·5 to 69·2 % nitrate, could get fire-cracker explosions
like those that Roger Bacon knew, and rocket effects also, but at 41 : 29·4 : 29·4
there was nothing but smoky combustion. Tried in simulated hand-gun conditions, the former mixtures only displaced or ejected a paper wad in the barrel,
and the 41 % composition refused to burn at all. These results, together with
many indications from European +15th- and +16th-century writings, support
our conviction that the nitrate-content gradually increased as time went on,
between the +9th and the +13th centuries.

(ii) *Powder manufacture and powder theory*

Descriptions of powder-making in China can be found in several texts, such as
the *Chi Hsiao Hsin Shu* of +1584, which gives a composition of 75·7 % saltpetre,
10·6 % sulphur and 13·7 % carbon for gunpowder used in the 'bird-beak gun',
the arquebus.[a] The account of the making of the gunpowder reads:[b]

Making of gunpowder: (Each round requires) 1 oz. of saltpetre, 0·14 oz. of sulphur and
0·18 oz. of charcoal made from willow wood.

Take altogether 40 oz. of saltpetre, 5·6 oz. of sulphur, 7·2 oz. of willow charcoal[c] and
three cupfuls of water, and grind (the ingredients) until they become extremely fine—
the finer the better. The best method is to pound and grind the saltpetre, the sulphur
and the charcoal separately into powder. Then put them together according to the right
proportion in a wooden mortar containing two bowlfuls of water. The ingredients are
pounded with a wooden pestle, and a stone pestle is never used for fear of (a spark
causing) fire. They should be pounded thousands of times; if they become dry during the
process a bowlful of water should be added, and pounding continued until they come

---

[a] This seems to be the earliest Chinese record extant on a gunpowder formula for the arquebus.

[b] Ch. 15, p. 9a, b, tr. auct. A passage almost identical is contained in *WPC*, ch. 124, pp. 8b, 9a. Words
enclosed in square brackets are in the latter version only.

[c] Great attention has always been paid to the kind of wood from which the charcoal was prepared. Baddeley (1) in 1857 mentioned willow (*Salix* spp.), alder, and 'black dogwood', as did Marshall (1), vol. 1,
pp. 67 ff. Gray (1) identified dogwood and black dogwood as alder buckthorn (*Frangula alnus*), but true alder
(*Alnus glutinosa*) and beech (*Fagus sylvatica*) have also been used. The first of these (which has the lowest spontaneous ignition temperature, and the highest and most even porosity) is employed today for evenly burning
fuses, the second for most commercial powders, and the third for those where precise burning is not needed.

(very) fine. (The gunpowder) is removed (from the mortar) when it is half dry, and then dried under the sun. Eventually it is broken up into pieces each as big as a small pea.

This powder is wonderfully good because it has been ground and pounded so fine for so long. [If pure water is used and changed it will take away any alkali from the saltpetre.] The pounding process is the same sort of thing as making the best kind of ink.

If you have added water more than a dozen times you may test (the powder) by setting light to a pinch of it on a piece of paper; should it burn off without damaging the paper you must not dare to put it into a gun. Or you may burn a few tenths of an ounce in the palm of the hand, then if the hand is not warmed it can be used in guns. But if it leaves behind black or white spots and the hand feels a sensation of heat, it is not of good quality. Water should again be added, and the pounding and grinding continued until the tests succeed.

There is much more than meets the eye in this rather deadpan passage. In the first place it conjures up a scene of workers all grinding away manually with pestles—but the Chinese were much more sophisticated than that. From an earlier disquisition it will be remembered that the pedal-operated trip-hammer (*tui*[1]) for cereal pounding can be traced back well into the Chou period,[a] and that water-powered trip-hammers (*shui tui*[2]) worked by lugs on the horizontal shaft of a water-wheel appear as early as the Han.[b] By this time too circular-trough edge-runner mills (*nien*[3]) were also known and used,[c] and roller-mills (*kun nien*[4]) developed soon afterwards,[d] doubtless derived from the simple stone hand-roller (*shih kun*[5]).[e] Machinery for pounding and grinding therefore went back a very long way in China, and though at present there is no means of knowing when it was first used for the mixing of gunpowder, we should be likely to find this point before the time of the *Wu Ching Tsung Yao*, i.e. by +1000 at least. That roller-mills (*nien*[3]) were used in Chinese gunpowder manufacture is in any case certain from another passage in the *Wu Pei Chih*, which mentions them by name.[f] It also alludes to the use of strong distilled alcohol (*shao chiu*[8]) for purifying and drying the powder.[g] All this got into the account of Amiot in +1782, who spoke of rollers grinding the wetted paste on marble slabs, then the drying, then the corning.[h]

In Europe, from the early +14th century onwards, horse- or water-powered vertical stamp-mills with hardwood (*lignum vitae*) pestles,[i] substituted for the

---

[a] See Vol. 4, pt. 2, pp. 51, 183–4, Figs. 358, 359.

[b] *Ibid.* pp. 390 ff., Fig. 617.

[c] *Ibid.* p. 199, Figs. 453, 454.

[d] Vol. 4, pt. 2, p. 178, Fig. 456. To be carefully distinguished from rolling-mills, where the substance is made to pass between two adjacent rollers; cf. *ibid.* pp. 122, 204.

[e] Water-power was applied to these also, at least from the beginning of the Sung, where we meet with them under the names of *shui nien*[6] and *shui wei*[7]; cf. Vol. 4, pt. 2, p. 403.

[f] Ch. 119, p. 10a.

[g] One pound for each three pounds of gunpowder.

[h] (2) Suppl.; cf. Partington (5), pp. 253–4.

[i] Cf. Forbes (8), p. 69; Gille (14), fig. 581 from the *Mittelalterliche Hausbuch* of *c.* +1480.

[1] 碓　　　　[2] 水碓　　　　[3] 碾　　　　[4] 輥碾　　　　[5] 石㲲

[6] 水碾　　　　[7] 水礧　　　　[8] 燒酒

water-powered horizontal trip-hammers, but in general the principle was the same.[a] The first powder-mill in Europe is attested for +1431, earlier references being doubtful.[b] Double roller-mills work to this day upon the moistened powder.[c]

The passage concludes with an interesting account of rough tests for the goodness of the gunpowder made.[d] The use of the palm of the hand must obviously be related to the testing of saltpetre itself by the same means.[e] At first sight the tests seem rather self-contradictory, but this is not really so. If the powder is weak because of poor admixture or other fault, it will burn away with a slow flame and no explosion, therefore the paper is perhaps browned but not damaged; if it is good it will go off in a flash and the explosion will blow a hole in the paper. The palm of the hand is much more solid, but a good puff and pop will be so rapid that little or no heat will be felt;[f] if on the other hand there is a slow burning there will be a sensation of heat, and a residue will be left,[g] which again will show that the powder is not good.[h] Amiot mentions this text.[i]

Next it is interesting to observe that the classical Chinese theory of medical prescriptions and elixir formulae was applied by the military theorists to different gunpowder compositions. Just as in medicine and pharmacy, the various components of a formula were looked upon as 'princely' (chün[1]), 'ministerial' (chhen[2]), and tso shih[3], which has the sense of auxiliary efficacious official envoy, hence 'adjutant'.[j] In the oldest of the pharmaceutical natural histories, the Shen Nung Pên Tshao Ching[4] (−1st century) the drugs in the first category are those with the largest minimal lethal dose, while those in the third are the extremely powerful and toxic ones. The Wu Pien (c. +1550) says:[k]

Saltpetre is the prince and sulphur the minister; their mutual dependence (hsiang hsü[5]) is what gives rise to their usefulness. The nature of saltpetre is to go forwards and that of sulphur to spread out sideways, so that is why the two can act together without contradicting each other. 'Ash' (charcoal) is their adjutant, being able to follow (substances of) the same category (thung lei[6]).[l]

    [a] A contemporary stamp-mill for powder is shown in Davis (17), p. 44, fig. 19. Cf. Marshall (1), vol. 1, pp. 23–4.
    [b] Köhler (2), vol. 1, p. 37; Partington (5), p. 328.
    [c] A contemporary example with ten-ton wheels is shown in Davis (17), p. 46, fig. 20. Today black powder is used almost wholly in fireworks.
    [d] Apparatus for more exact quantitative measurement of its properties arose later, and we shall briefly consider some types of it below (p. 548) in connection with the idea of gunpowder engines.
    [e] Cf. pp. 105–6 above.
    [f] Cf. Davis (17), p. 47.
    [g] This trial was widely used in the West too; cf. Marshall (1), vol. 1, p. 27.
    [h] Perhaps the oldest European test is that in the Feuerwerkbuch of c. +1437, on which see Hassenstein (1), p. 64 and Partington (5), p. 155.
    [i] (2) Suppl.; cf. Partington (5), p. 272, following WPC, ch. 119, p. 10a.
    [j] See Vol. 6, pt. 1, p. 243.
    [k] Ch. 5, p. 61b; tr. auct.
    [l] On this concept, see Vol. 5, pt. 4, pp. 305ff., 316ff., and Needham (83, 84); Ho Ping-Yü & Needham (2).

    [1] 君   [2] 臣   [3] 佐使   [4] 神農本草經   [5] 相須
    [6] 同類

Again, the *Ping Lu* (+1606) gives a theory of the substances that went to compose gunpowder:[a]

The nature of the chemicals (*yao*[1]) used in attack by fire is as follows. Among the principal substances saltpetre and sulphur are the princely ones, charcoal is the ministerial one, the various poisons are adjutants (*tso*[2]), and those constituents that produce *chhi*[3] are the envoys (*Shih*[4]).[b] One must know the suitability of the ingredients before one can master the wonderful (effects) of attacks with incendiaries and explosives. Now the nature of saltpetre is to be linear (*chih*[5]);[c] the nature of sulphur is to radiate (*hêng*[6]);[d] and the nature of charcoal is to take fire (*jan*[7]).[e] That which is straight by nature governs impact at a great distance, so for propulsion we take nine parts of saltpetre to one part of sulphur. That which goes sideways by nature governs explosion, so for detonation we take seven parts of saltpetre to three parts of sulphur.[f] Charcoal from green willow is most sharp in nature, charcoal from dried fir is slow, while that from the leaves of the white mountain bamboo (*jo yeh*[13]) is particularly fiery.

The *chhi* of realgar is high, causing the flame to rise[g]...; the *chhi* of arsenic (*phi huang*[14]) is placid, but its fire is toxic.[h] If iron pellets and sharp porcelain fragments previously prepared by roasting with urine (*chin chih*[15]) and its sediment (*yin hsiu*[16]) and sal ammoniac, hit one of the enemy, they will cause his flesh to rot until the bone shows.[i] Wild aconite (*tshao wu*[17]), croton oil (*pa tou*[18]), and parts of the thunder-god vine (*lei thêng*[19]),[j] roasted together with a small quantity of (dried) sea-horse (*shui ma*[23]),[k] can be used on dragon-lances as a poison, which kills if it draws blood.[l] Dolphin (*chiang tzu*[24]) (oil),

[a] Ch. 11, pp. 3*b*–4*b*. Also given in *Wu Pei Huo Lung Ching* ch. 1, pp. 4*b*–5*b*; tr. auct.
[b] Cf. what was said about 'sal practica' on p. 353 above.
[c] *Comm.*: 'for forward motion, saltpetre plays the major role'.
[d] *Comm.*: 'for explosion sideways, sulphur plays the major role'.
[e] *Comm.*: 'the fire is different depending on the charcoal, and of this there are many different kinds, such as that from the white mountain bamboo (*jo*[8]), that from the willow (*liu*[9]), that from the fir (*shan*[10]), that from the catalpa tree (*tzu*[11]), and that from the chive (*hu*[12])'. These plants are, in the order named: (*a*) *Sasa albomarginata*, R757, CC2084, (*b*) *Salix babylonica*, R624, CC1697, (*c*) *Cryptomeria japonica*, R786a, CC2137, (*d*) *Catalpa ovata*, R98, CC1260, (*e*) *Allium scorodoprasum*, R672, CC1824.
A recent account of the woods used for gunpowder charcoal in Europe is that of Gray, Marsh & McLaren (1). Weeping willow was prominent, but the preference was for alder buckthorn, *Frangula alnus* (wrongly called black dogwood), the alder itself (*Alnus glutinosa*), and the beech (*Fagus sylvatica*).
[f] Counting one part of charcoal in both cases, this would mean 81·8% nitrate in the first case and 63·6% in the second.
[g] *Comm.*: 'magical fire (*shen huo*[20]) uses realgar as the princely constituent'.
[h] *Comm.*: 'in poison fire (*tu huo*[21]) arsenic plays the princely role'.
[i] *Yin hsiu*[16] means literally silver rust, but a commentary in the *Wu Pei Huo Lung Ching* says that it refers here to the sediment of urine, and that *chin chih*[15] refers to urine itself. The commentary of our text says that these are used in bruising and burning gunpowder (*lan huo yao*[22]).
[j] These plants are, in the order named (*a*) *Aconitum uncinatum*, R527, (*b*) *Croton tiglium*, R822, CC857, (*c*) *Tripterygium wilfordii*, CC826.
[k] *Hippocampus*, spp., usually known as *hai ma*[25] (R190). The reason for its presence here is not obvious, as *PTKM* records only its use in difficult parturition and as an aphrodisiac.
[l] *Comm.*: 'the poison can be used with rockets and fire-lances, and will kill an enemy instantly if hit or wounded by the point'.

[1] 藥	[2] 佐	[3] 氣	[4] 使	[5] 直
[6] 橫	[7] 然	[8] 箬	[9] 柳	[10] 杉
[11] 梓	[12] 蒿	[13] 箬葉	[14] 砒黃	[15] 金汁
[16] 銀銹	[17] 草烏	[18] 芭豆	[19] 雷籐	[20] 神火
[21] 毒火	[22] 爛火藥	[23] 水馬	[24] 江子	[25] 海馬

Szechuan varnish (*chhou chhang shan*[1]), and the arum (*pan hsia*[2]),[a] are used for anointing the spearheads of fire-lances, and anyone wounded by these will be struck dumb.[b]

As for tung oil, soap-bean powder, and resin, they are used for burning (enemy) food stores and attacking (enemy) camps. Human hair, molten iron and croton oil are used to smash leather-covered siege-engines and hide screens.[c] Smoke from (burning) wolf faeces, which looks red both in the daytime and on dark nights, can be used for sending warning signals. The ashes of dolphin meat (when added to gunpowder) intensify the flame in a head-wind, this being their unusual property. There are also other substances like 'petrol' (*mêng huo yu*[5]), the flame of which intensifies in the presence of water and will burn wet objects; and oil from the 'nine-tailed fish' (*chiu wei yü*[6]) which causes flames to go against the wind so that none can escape from them.[d] Such substances are difficult to obtain, but the commanding officer of an army should (at least) be aware of their existence.

This passage as a whole is indeed a precious résumé of traditional theorising at the end of the medieval centuries, contemporary with the upsurge of modern scientific method applied to gunnery at that other end of the Old World where the Scientific Revolution was now taking place. Some of the reasoning is quite Aristotelian in character, just as it was in Europe down to the time of Robert Boyle; and the pharmaceutical classification came straight from the *Pên Ching* of the −1st or −2nd century. The remarks on the varying properties of different kinds of charcoal were quite acute, and must have resulted, like so much else, from practical experiment. Then comes the traditional enthralment with poisons, reasonable enough when they were applied to arrow-heads and spear-points, much less so when incorporated in gunpowder compositions.[e] The text ends with miscellaneous customary lore on siege warfare, smoke-signals and flame-throwers, much of it by this time surely obsolete.

It is not perhaps surprising that a very similar passage occurs in the *Thien Kung Khai Wu* of +1637, where Sung Ying-Hsing introduces a further idea, namely that saltpetre is very Yin and sulphur very Yang.[f] One can easily see

---

[a] *Chiang tzu* was a synonym for the dolphin (*chiang thun*[3], R 176). The plant named in the second place is *Orixa japonica*, a poisonous member of the Rutaceae, related to the oranges, R 353, CC 915; and the third is *Pinellia tuberifera*, very poisonous, R 711, CC 1929, Steward (2), p. 500.

[b] *Comm.*: 'this is also used in spattering gunpowder (*phên huo yao*[4])'.

[c] I.e. shields covered with animal hides as protection.

[d] *Comm.*: 'this fierce fire oil is produced in Champa (mod. South Vietnam), and the nine-tailed fish is found in Thailand'. The *Wu Pei Huo Lung Ching* version makes no mention of the latter, but says only 'fish-oil produced in Borneo'. These were probably all remote echoes of the trade in Greek Fire, on which see pp. 86 ff. above. On Champa cf. Vol. 4, pt. 3, p. 487, and on Thailand, Vol. 5, pt. 4, p. 136.

[e] All organic substances would be largely decomposed by the burning or explosion, but a smoke heavily charged with lead or arsenic would have been quite toxic, though the effects would not necessarily have been rapidly apparent. On Chinese arrow-poisons see again Bisset (1, 2).

[f] Ch. 15, pp. 5*b*, 6*a* (pp. 258–9), tr. Sun & Sun (1), p. 268; Li Chhiao-Phing (2), pp. 389 ff. Sung was rather sceptical about how far the statements of the military technologists were based on experiment. Another passage (ch. 11, p. 6*a*) reads as follows: 'Of the components of gunpowder, sulphur is pure Yang, and saltpetre is pure Yin; when these two essences come together the result is noise and change. This is a mystery wrought by Chhien and Khun (the two *kua* of the *I Ching* corresponding to Yang and Yin). This is a marvel of Nature.' Tr. auct., adjuv. Sun & Sun (1), p. 210; Li Chhiao-Phing (2), p. 297.

[1] 臭常山   [2] 半夏   [3] 江豚   [4] 噴火藥   [5] 猛火油
[6] 九尾魚

that this was derived from the classical theory of the nature of thunder, which went back far into antiquity, at least as early as the beginning of the Han.[a] Here our Chinese authors came close to a conception much agitated in the Europe of the +16th and +17th centuries, that of the 'aerial nitre'.[b] Giving rise to a large literature, it had more to do with the explanation of thunderstorms, and ultimately with the discovery of oxygen, than with the chemistry of saltpetre, but it played a considerable part in the thought of the time.

Actually, the men of the Scientific Revolution had been anticipated by Chu Hsi[1], writing in the latter half of the +12th century. As Huang Jen-Yü noticed,[c] he regarded thunder as due to the sudden expansion of intolerably compressed *chhi*, and analogised it explicitly with gunpowder explosions. His words were:[d] 'Thunder is just like our present day fire-crackers (*pao chang*[2]); most probably (the *chhi*) is densely compressed, and when this attains its climax, then it bursts forth and dissipates, scattering in all directions.' As we have seen so often,[e] the word 'thunder' was applied to so many gunpowder weapons from the end of the +10th century onwards, that it is not surprising that the 'aerial nitre' should appear in Neo-Confucian dress.

What is rather extraordinary is that a text of closely similar wording to that in the *Ping Lu* can be found in the *Huo Kung Chhieh Yao* of +1643, written by Chiao Hsü and Adam Schall von Bell. The section bears the title 'Huo-Kung Chu Yao Hsing Chhing Li-Yung Hsü-Chih'[3] (What one ought to know about the Profitable Use of the Natures and Relationships of the various Chemicals used in Attacks by Fire).[f] The exposition follows the *Ping Lu* passage closely, with much of the commentary incorporated into the text, all the ideas being essentially the same. We hear of realgar, croton oil, soap-bean powder, the wolf dung, the dolphin and the sea-horse, the petrol and even the nine-tailed fish. These are among the many adjutant (*tso*[4]) ingredients. This is the book, we remember, in which Thang Jo-Wang the Jesuit is described as the transmitter or instructor (*shou*[5]), Chiao Hsü as the compiler (*tsuan*[6]), and Chao Chung[7] as the editor (*ting*[8]). From this it has been concluded by some that the Jesuit was the responsible writer, with the others just taking down what he said. But if this were so one would hardly expect to find passages of such highly traditional ideology. Presumably Schall von Bell found nothing to object to in them. The bringing of modern science into China was necessarily a slow business, and the Jesuits only partly effected it. We prefer our usual course of regarding von Bell's title as partly an honorific one; Chiao Hsü must have been a true collaborator, and not someone writing to dictation. Hence the medieval account of the nature of gunpowder's constituents.

---

[a] Cf. Vol. 3, pp. 480 ff.   [b] See Debus (9, 10) and Multhauf (5), p. 332.
[c] (5), p. 202.
[d] *Chu Tzu Chhüan Shu*, ch. 50, p. 47*b*, tr. auct.
[e] Pp. 192, 203, 213 above.   [f] Ch. 2, p. 8*a, b* (pp. 28–9).

[1] 朱熹   [2] 爆仗   [3] 火攻諸藥性情利用須知   [4] 佐
[5] 授   [6] 纂   [7] 趙仲   [8] 訂

At this point mention may be made of an interesting piece by Mao Yuan-I[1], the famous author of the *Wu Pei Chih* which we so often quote. It is entitled *Huo Yao Fu*[2] (Poetical Dissertation on Gunpowder) and would be well worth a translation in full, epitomising as it does the traditional thinking about the mechanism of the explosive mixture. The nature of saltpetre is to expand vertically (*shu*[3]) while sulphur expands horizontally; saltpetre is the prince, with sulphur and charcoal as the ministers, and even poisonous substances are brought in as adjutants. It could show very clearly how Chinese technologists thought of explosive phenomena in the early years of the +17th century.[a]

It remains only to say a few words about the time of the Opium Wars, when the Chinese were busy catching up with the gunnery developments, modern for that day, which had been made by the European nations. Thus in 1843 Chhen Chieh-Phing[4], Admiral of Fukien, memorialised that the remaining gunpowder-mills (*nien*[5]) worked by man-power should be done away with, and animal-power or water-power, seven times more effective, universally substituted.[b] He also had something to say on the preparation and purification of saltpetre (cf. p. 94 above), recommending oxhide glue for the clearing. Rondot (2) knew this text when he visited some Chinese arsenals in 1849; there he found that the nitrate percentage of the powder made was equivalent to that of the best French product (75.5 %). Ting Kung-Chhen[6], who was one of the leading gunnery and powder experts of the time, observed this too.[c] Rondot found, rather to his surprise, a large Chinese chemical laboratory and works organised and equipped by Phan Shih-Chhêng[7,d] where saltpetre was prepared and recrystallised in bulk, and alcohol and nitric acid distilled. Some of this was used for making silver fulminate detonator caps, which had been produced in China since 1842.[e]

---

[a] Perhaps the nearest Western parallel to Mao Yuan-I's essay would be the pages which Sir Thomas Browne consecrated to the nature of gunpowder in his *Pseudodoxia Epidemica* (commonly called 'Vulgar Errors') of +1646. They occur in bk. 2, ch. 5, para. 5 (Sayle ed., vol. 1, pp. 271 ff.). 'Now all these (constituents)', says Browne, 'although they bear a share in the discharge, yet they have distinct intentions, and different offices in the composition. From Brimstone proceedeth the piercing and powerful firing.... From Small-coal ensueth the black colour and quick accension.... From Salt-petre proceedeth the force and the report; for Sulphur and Small-coal mixed will not take fire with noise, or exilition; and Powder which is made of impure and greasie Petre hath but a weak emission, and giveth a faint report. And therefore in the three sorts of Powder the strongest containeth most Salt-petre....'

[b] The memorial he submitted is to be found in *Hai Kuo Thu Chih*, ch. 91, pp. 8b–11b. He appears in European accounts as Ching Ki-Pimm. He also recommended the use of vine charcoal instead of that made from pine or fir.

[c] *HKTC*, ch. 91, pp. 11b–15a. On him, see Chhen Chhi-Thien (1); Huang Thien-chu, Tshai Chhang-Chhi & Liao Yuan-Chhüan (1).

[d] Pwann Sse-Ching (or Tinqua) to Europeans; cf. Chhen Chhi-Thien (1), pp. 36 ff., pp. 40 ff., 56 ff., (2), pp. 8–9, and p. 205 above, where we discussed the attention he gave to sea-mines, and his employment of an American expert to assist in their construction.

[e] See Davis (17), pp. 400 ff., 405, 412. Silver fulminate had first been prepared by Berthollet in +1788, but on account of its excessive sensitivity it was soon replaced for military purposes by mercuric fulminate. One of us (J. N.) always remembers the nervosity which accompanied a visit he made to a Chinese fulminate factory during the second world war in his capacity as Adviser to the Arsenals Administration. Cf. p. 56 above.

[1] 茅元儀   [2] 火藥賦   [3] 豎   [4] 陳階平   [5] 碾
[6] 丁拱辰   [7] 潘仕成

The memorials of Chhen and Ting both urged that lessons should be learnt from European methods of powder manufacture. Then after the foundation of the famous Kiangnan Arsenal (Chiang Nan Chi-Chhi Chih-Tsao Chü[1]) near Shanghai in 1865, and the establishment of a Translation Bureau (Fan I Kuan[2]) within it two years later, an American book by Watt on powder-making was translated into Chinese by John Fryer (Fu Lan-Ya[3]) with the title *Chih Huo Yao Fa*[4] (Procedures in Gunpowder Manufacture).[a] But it is probable that neither Chhen nor Ting, nor Fryer and his associates, had any idea of how old gunpowder really was in Chinese history, nor that China had been the land of its birth. And now we must retrace our steps to the last years of the +15th century in order to follow the later development of artillery and musketry.

(17) THE LATER DEVELOPMENT OF ARTILLERY

From this point onwards we find ourselves in the presence of a great wave of influence back from Europe upon China. If Chinese culture had been left entirely to itself it is possible that the same developments would have occurred, according to that slow and steady progress which the whole of its history had manifested.[b] But now the new economic system of capitalism was arising in Europe in strength, and innovation as well as invention[c] was getting full rein;[d] thus it came about that superior forms of light cannon originating in the West spread rapidly everywhere over the Old World.[e] We deal with them now (and their heavier congeners too) because improved hand-guns such as the arquebus and the musket reached China only some forty or fifty years later.

Here the key invention was that of breech-loading. Rather than waste a lot of time ramming the charge and the projectile down the muzzle, and probably a wad as well, it was much more convenient to have a separate container for all these, shaped rather like a beer-mug with an appropriate handle, and placed in position in a cavity arranged to receive it at the breech of the cannon, then wedged into place with a transverse wooden billet.[f] This replaceable cylinder was known as the chamber or culasse. A drawing of the whole system is given in Fig. 129.[g]

---

[a] Cf. Bennett (1), p. 118.　　[b] Cf. Needham (59), (64), p. 414.

[c] See Schumpeter (1, 2). It was not only a matter of the new, but of the adoption and mass application of the new.

[d] This was what vitiated the otherwise meritorious book of Cipolla (1). To show that the full-rigged ship (+1500 onwards, cf. Vol. 4, pt. 3, pp. 512, 594–5, 606, 611, 697–7), with its broadside of up-to-date guns, soon outclassed the ocean-going junk was one thing. To ignore completely that the former was based upon capitalist applications of invention and scientific knowledge, while the latter still had only traditional bureaucratic feudalism as its background, was quite another.

[e] The spread of European artillery pieces among all the States of Eastern, South-eastern and South Asia has been well described by Boxer (11), Gibson-Hill (1) and Crucq (1). They were greatly sought after. So also were the sulphur and saltpetre from China to be used in them; see Tomé Pires' 'Suma Oriental' (+1515), tr. Cortesão (2), vol. 1, pp. 115, 125.

[f] This invention was particularly important at sea, since it avoided running the guns back and forth for loading.　　[g] Reid (1), p. 113.

[1] 江南機器製造局　　　[2] 繙譯舘　　　[3] 傅蘭雅　　　[4] 製火藥法

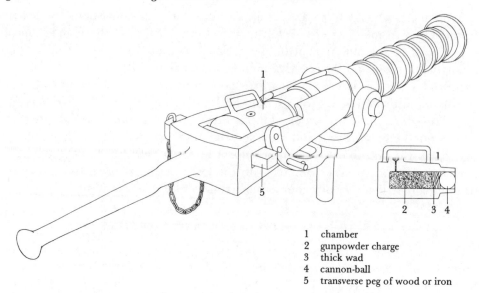

1   chamber
2   gunpowder charge
3   thick wad
4   cannon-ball
5   transverse peg of wood or iron

Fig. 129. The key invention of breech-loading; a separate container (the chamber or culasse) shaped rather like a beer-mug, with an appropriate handle, contained the propellant charge, the wad and the projectile. It was wedged into place by a transverse wooden billet. As many of these as was convenient were prepared beforehand, then quickly fitted into place, thus increasing the rate of fire. Drawing from Reid (1).

There has been much disagreement about the date when breech-loading artillery appeared in Europe. Reid may not be far off the mark when he concludes[a] that the evidence points to some time not long before 1372. Räthgen[b] said +1380 or +1398, and Köhler[c] chose +1397, but these were all German datings, and the Burgundians seem to have had the device as early as +1364.[d] England comes in with a picture of +1485 referring to 'port-pieces' of +1417, and Portugal with 'versos' (*berços*) in +1410.[e] The design lasted for several centuries, but it could never be made satisfactorily airtight, and the serious loss of gas resulting naturally decreased the propellant force.[f] Only in 1809 was the problem

---

[a] (1) p. 59.   [b] (1), pp. 58 ff., 181.   [c] (1), Vol. 3, pt. 1, p. 282.
[d] Bonaparte & Favé (1), vol. 3, pp. 130–2. They were called *veuglaires* from a maker named Vögler, hence *fuggeler bussen*.   [e] See Partington (5), pp. 110, 112, 115, 121, 224.
[f] There were alternatives, especially screw-in breech-blocks, such as Leonardo da Vinci sketched in his *Codex Atlanticus* about +1500 (cf. McCurdy (1), vol. 2, opp. p. 206); but they were not much taken up, probably because of their slowness and awkwardness. The great Turkish cannon of +1464 in the Tower of London has them for its 2 ft bore (Blackmore (2), p. 172, no. 242 and pl. 3), but they found little use till after +1770, and never got to China. Movable breech screw-plugs came in from +1593, and pivoted chambers from +1680, but again not in China. On the whole development see Blackmore (1), pp. 58–9, 62, 64; Ffoulkes (2), pp. 94, 98. Screw-in breech-block chambers generally had sockets for handspikes.
    We should like to know a great deal more than we do about the artillery which was used on both sides at the finally successful Turkish siege of Constantinople in +1453. Several of the frescoed churches of northern Moldavia in Rumania (Arbore, Humor, Moldoviţa, Suceviţa) have paintings depicting this, which range in date from +1503 to +1595. We reproduce what are perhaps the best pictures (Fig. 130) done in +1537 at Moldoviţa. On the right are seen the muzzles of three Turkish field-guns, but four other cannons can be made out on the city's ramparts, two confronting the land battery on the right, and two firing at an assault by naval vessels on the left, apparently with devastating results. All are clearly painted with dragon-scales, in accordance with the appellation so often given to artillery pieces. An enlargement (Fig. 131) shows the battery more clearly, but not clearly enough, unfortunately, to make out the breech arrangements.

Fig. 130. A fresco painting from the exterior wall of the church at Moldoviţa in Moldavia, Rumania. Though done in +1537, it depicts the siege of Byzantium eighty years earlier. Some artillery is visible on the ramparts, as well as crossbowmen in the tower on the right. Orig. photo.

solved when S. J. Pauly invented the cartridge, first of many varieties to come.[a]

When, early in the +16th century, the breech-loader entered China, it got the name of *fo-lang-chi*[1,2], the 'Frankish culverin'. But although we occasionally use this translation ourselves, culverin is not the right word,[b] yet unfortunately there is no satisfactory or well-recognised one. In +15th- and +16th-century English

According to Runciman (3), pp. 66–7, 108, 116–17, 119, 126, the Turks, on the whole, took artillery a good deal more seriously than did the Byzantines. In the city they had few cannon, and if fired from the walls the recoil shook and damaged them, moreover there was a saltpetre shortage (p. 94). The Turkish bombardment continued for six weeks, but in circumstances difficult for the gunners, since their cannon lacked proper mountings (pp. 97–8). However, Sultan Mehmet II was advised from +1451 onwards by a Jewish physician Jacopo of Gaeta, who knew something about guns; and in the following year they were joined by Urban, a cannon-founder from Hungary, who manufactured at least one monster some 27 ft long (pp. 77–8). It is particularly interesting, in view of their later failure to adopt modern science and technology, that at this time, just before the rise of the Scientific Revolution, the Turks should have been more open to advanced military technology than the Greeks. Cf. Fig. 121.

For those who would like to pursue the matter further, the best collection of texts, translated and annotated, is that of Pertusi (1).

[a] A Swiss artillerist, originally a wagon-maker, b. +1766. See Reid (1), p. 188; Blackmore (1), p. 66. Associated with this was the introduction of the fulminate percussion-cap, worked out by Alexander Forsyth (+1768 to 1843) in the Tower of London in +1799 (Blackmore (1), pp. 45 ff.).

[b] Its worst feature is that it was generally used to imply muzzle-loading cannon.

[1] 佛郎機   [2] 佛朗機

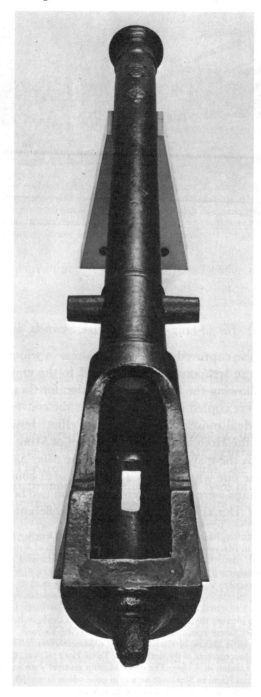

Fig. 133. Portuguese breech-loader of *c.* +1520 bearing the national arms. The chamber and the iron tiller are missing. Overall length 8′2″. Photo. Tower of London Armouries. Blackmore (2) catalogue, p. 139, no. 178.

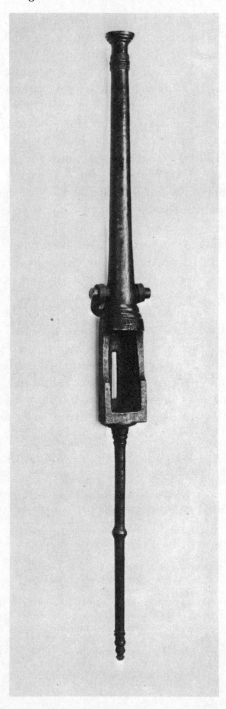

Fig. 134. An African copy of a Portuguese breech-loader, date uncertain. The chamber is missing but the iron tiller is intact. Photo. Tower of London Armouries. Blackmore (2) catalogue, p. 170, no. 239.

Tshung-Chien[1] (+1574),[a] and the *Huang Ming Shih Fa Lu*[2] (Ming Political Encyclopaedia) of Chhen Jen-Hsi[3] (+1630); but both of these say that it was a
lower War Ministry official, Ho Ju[4], who got hold of the guns in +1522, and that
later on copies were cast at the capital by two Westernised Chinese, Yang San[5]
(Pedro) and Tai Ming[6].[b] However, when in +1519 the famous philosopher
Wang Yang-Ming[7] (d. +1529), then Governor of Chiangsi, was putting down
the revolt of a prince named Chu Chhen-Hao[8], he used, or intended to use,
fo-lang-chi cannon.[c] In his collected works there is a piece[d] in which he says that
his friend Lin Chün[9], army commander against the prince, had his bronze-
founders cast *fo-lang-chi chhung*[10] at this time—consequently the weapon was
known in China, at least in Fukien and Chiangsi, before +1522. Moreover, there
had been another rebellion in the same province twelve years earlier, when
Huang Kuan[11] was prefect, and it had been put down largely by a volunteer
officer named Wei Shêng[12], who attacked the brigands with more than a hundred *fo-lang-chi*, and destroyed them.[e] Therefore the Frankish breech-loaders
were a fairly familiar weapon in the south as early as +1510.

If this is the case, it cannot have reached China directly from the Portuguese,
because Malacca did not fall until +1511.[f] Pelliot thought it most probable that
the guns came up from Malaya before Chinese people had ever met anyone from
Portugal,[g] in which case the word *chi*[15] really meant 'machine' from the first, i.e.
'the engine of the Farangi, or Franks', and then the syllable stayed on in the
transliteration of the name for the people.[h] As Pelliot put it: 'on avait connu les
canons *fo-lang-chi* avant les étrangers *Fo-lang-chi*'. At all events there was a perva-
sive association of *fo-lang-chi* breech-loaders with southern regions, as witness

[a] Ch. 9, p. 9b.
[b] Dates of +1530 (9th month) and +1533 (8th month) are both given for this in *Ming Shih Lu*.
[c] See Goodrich & Fang Chao-Ying (1), pp. 1412–13.
[d] *Wang Wên Chhêng Kung Chhüan Shu*, ch. 24, p. 12a.
[e] The evidence for this comes, it is true, from the *Fukien Thung Chih*[13] of Chhen Shou-Chhi[14] (+1771 to
1834), ch. 267, p. 10a, which was compiled long afterwards, but he used local manuscript records, and there is
no reason for doubting his account.
[f] The first Portuguese ship to touch at a Chinese haven was commanded by Jorge Alvares and the year was
+1514. The first Portuguese diplomatic contact was that of Tomé Pires and began in +1517. See Vol. 4, pt. 3,
pp. 507, 534.
[g] If so, things must have happened rather quickly, as the first Portuguese visit to Malacca was only in
+1509. One wonders whether other sources ought not to be looked for—Spanish, or even English? On the
+1514 contact see Chang Thien-Tsê (1), pp. 35 ff.
[h] This designation of Europeans was widespread all over Asia at the time, derived, no doubt, from Arabs
talking of Frankistan. For example, the *Yuan Shih* (ch. 40, p. 6a) already uses the phrase Fu-lang[16] for the
Marignolli embassy (cf. Vol. 1, p. 189); and this was easily assimilated to the old Thang term for Byzantium—
Rūm (New Rome) → Fröm → Fu-lin[17] (cf. Vol. 1, pp. 186, 205). The Farangs also generated the name for
cloisonné work (see Sect. 35), which was of Western origin, *fa-lan*[18] (later *fa-lang*[19]). A closer parallel to the
fo-lang-chi breech-loaders comes from the fact that Bābur, the first Mogul emperor (r. +1526 to +1530) used
the names *firingihā* or *farangī* for cannon of Frankish design, though made in India (Partington (5), pp. 219, 234,
279). Cf. Chang Wei-Hua (1).

[1] 嚴從簡	[2] 皇明世法錄	[3] 陳仁錫	[4] 何儒	[5] 楊三
[6] 戴明	[7] 王陽明	[8] 朱宸濠	[9] 林俊	[10] 佛狼機銃
[11] 黃琯	[12] 魏昇	[13] 福建通志	[14] 陳壽祺	[15] 機
[16] 柛郎	[17] 拂秣	[18] 發藍	[19] 琺琅	

the *Yüeh Shan Tshung Than*[1] (Collected Discourses of Mr Moon-Mountain), i.e. Li Wên-Fêng[2], who was writing about +1545. In the course of this book of memorabilia, he notes that the design came originally from abroad, and in his time only the Cantonese gun-founders could make them as well as the foreigners could.[a]

It is often said that the earliest Chinese description of the *fo-lang-chi* breechloader occurs in the *Chhou Hai Thu Pien* of +1562, and this may be true, but when one takes a closer look one finds that Chêng Jo-Tsêng was quoting a much earlier memorandum, written in fact by Ku Ying-Hsiang[3], the scholar we met with long ago as a distinguished mathematician.[b] When Ku was Acting Superintendent of Foreign Trade at Canton in +1517 he became an eye-witness of the arrival of a fleet under Fernão Peres de Andrade, which brought the first Portuguese ambassador to China, the ill-fated Tomé Pires.[c] What he said about the breech-loading cannon must therefore have been written long before, probably about +1525 or +1530.

The report, which Chêng Jo-Tsêng says did not get into the *Ming Hui Tien*, is given in his *Chhou Hai Thu Pien*;[d] it speaks of two Portuguese vessels carrying the *Capitão-mor* (*Chia-pi-tan-mo*[4]), i.e. the ambassador, Pires, surrounded by tall men with prominent noses and deepset eyes wearing white head-cloths like Muslims. The Viceroy of the two Kuang provinces, Chhen Hsi-Hsien[5], came to examine them, and the party was sent up to the capital, where it stayed in the Hostel for Foreign Tribute Missions (Hui Thung Kuan[6]) for a year, but the Chinese were upset because the Westerners did not know the proper customs of civilised intercourse, and the embassy ended in failure. Actually, what was much more significant were the depredations of other Portuguese captains, and the bitter complaints of the ousted Rajah of Malacca. Then follows the passage about the guns (Figs. 135, 136):[e]

This cannon (*chhung*[7]) is made of iron,[f] and measures five or six feet in length. It has a large belly and a long barrel. At the bulge there is a long cavity, into which five smaller chambers (*chhung*[7])[g] can be inserted in rotation, and these contain the gunpowder for firing.[h] The gun is wrapped on the outside with wooden staves and fastened with iron hoops to ensure that it does not split.[i] Four or five of these cannon are concealed behind a ship's bulwarks on each side, and if an opposing ship comes near, one single shot,

---

[a] This text was first noted by Parker (7).
[b] Vol. 3, pp. 51–2.
[c] Vol. 4, pt. 3, pp. 534–5. The full details are in Cortesão (2), Pelliot (53) and Chang Thien-Tsê (1), pp. 42 ff.
[d] Ch. 13, pp. 31 a, b, 32 a.   [e] Tr. auct.
[f] *Ming Shih* later on (ch. 92, p. 11 a) says bronze.
[g] Notable here is the failure to invent a new term for what was clearly a new thing. We have often come across this misfortune before, as in Vol. 4, pt. 2, p. 465. Cf. Needham (2), pp. 215–16 (27). It was characteristic of medieval and traditional science and technology. See pp. 11, 130 above.
[h] 'And the projectile too', he might have added.
[i] This we doubt; perhaps it was a mistake of Ku's. After all, he was not a gunner himself.

[1] 月山叢談   [2] 李文鳳   [3] 顧應祥   [4] 加必丹末   [5] 陳西軒
[6] 會同舘   [7] 銃

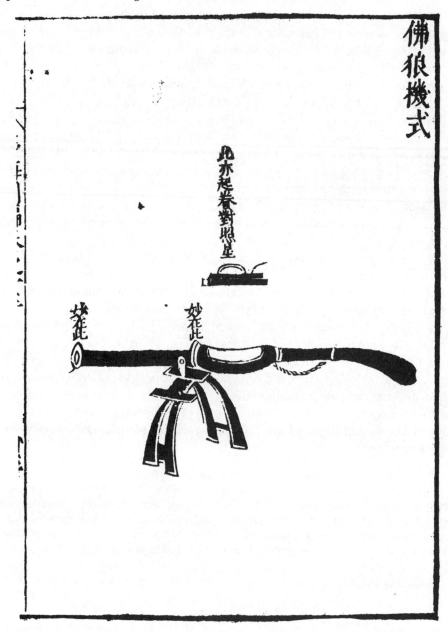

Fig. 135. The first Chinese illustration of a 'Frankish culverin' (*fo-lang-chi chhung*), from *CHTP*, ch. 13, p. 33*a*. One chamber or culasse is also shown. This book came out in +1562, but the relevant quotation came from a report of +1525 or so. The small cannon is mounted with its trunnions supported on a swivelling pivot.

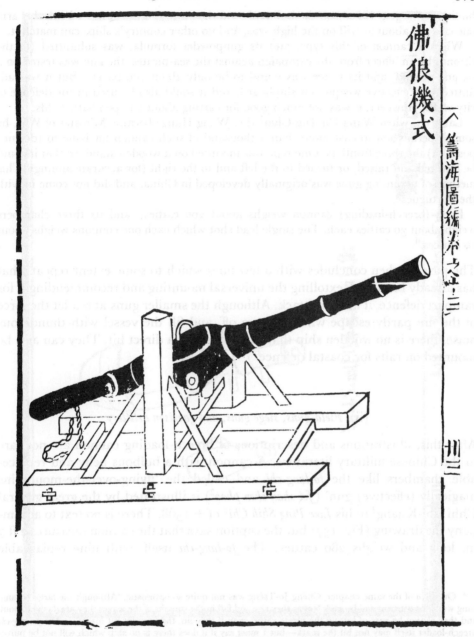

佛狼機式

〔籌海圖編卷之十三〕

三

Fig. 136. Another illustration from the same work (ch. 13, p. 33 b). The mounting is more elaborate but the same principle of swivelling trunnions pertains.

chambers, also appears in this work.[a] Another artillery piece rather smaller than the *fo-lang-chi* but faster to fire was called the 'cannon-rivalling gun' (*sai kung chhung*[1]) and is described in the *Ping Lu*[b] of +1606 with a diagrammatic illustration. Before long the breech-loading principle was extended to quite heavy guns, like the 'invincible general' (*wu ti ta chiang chün*[2]) illustrated and described by Chhi Chi-Kuang (Figs. 140, 141).[c] This weighed 1050 catties, and was carried into position on a kind of barrow. Here a good new term was at last found for the chambers, *tzu chhung*[3]. The range for grape-shot was over 200 ft.

Cannon of this name we have already come across (p. 338), but like all the largest ones they were muzzle-loading. Let us look at another one in the *Chhou Hai Thu Pien*, that called the 'bronze outburst cannon' (*thung fa kung*[4]),[d] Fig. 142. Chêng Jo-Tsêng says:[e]

Each of these weighs 500 catties or thereabouts, and fires 100 lead shot, each weighing about 4 catties. It is a powerful weapon for assaulting city-walls, as also for attacking the enemy when tens of thousands of them are gathered in massed formations. The stone cannon-balls are as large as a small peck measure, and any object struck by them must inevitably disintegrate. Walls will be penetrated, houses in their path will crumble, trees hit by them will fall, and from any men or animals that get in the way blood will flow in streams. If fired at a mountain-side, the balls will bury themselves several feet deep. Not only are the cannon-balls not to be withstood, but objects which are struck by them will ricochet and strike other objects—even parts of the human body like limbs and trunks thrown about in this way will also cause damage.

Not only are the cannon-balls so powerful and frightening, but after the priming-powder (*i*[7]) is ignited, the gas (*chhi*[8]) coming from (the explosion) is poisonous, the blast can blow people to death, and even the earthquake-like noise can kill. Hence before letting off a bronze outburst cannon it is necessary to dig a trench in which the gunner can take cover before lighting the fuse. Then, as the fire, the gas, and the roar all go upwards, he is protected from injury and death.[f]

Of course it is always necessary to guard the gun with a detachment of brave soldiers so as to prevent the misfortune of the enemy capturing it. But if you are not attacking strong defensive works, nor getting out of a dangerous situation, you do not need to use this (great siege cannon).

---

[a] *Tsa Chi*, ch. 5, pp. 16b, 17a, with two pages of description following. Fig. 138.

[b] Ch. 12, p. 28a, b.

[c] *Tsa Chi*, ch. 5, pp. 13b–16a.

[d] The caption of the illustration has *kung*[5] without the fire radical, but properly *kung*[6] meant any great piece of ordnance.

[e] Ch. 13, pp. 34b, 35a, tr. auct. The same picture appears in *HLC*, pt. 2, ch. 2, p. 2a, with text on pp. 2b, 3a, b, identical with that translated here. It is also in *WPC*, ch. 122, pp. 4b, 5a, b.

[f] One remembers having come across this curious procedure before, and in fact it is (derivatively) in the *Thien Kung Khai Wu* (+1637), ch. 15, p. 7b (Sun & Sun tr., p. 271, Li Chhiao-Phing tr., p. 393, both misunderstanding in different ways). One wonders whether it is not a relic of the ever-present danger of these early big guns bursting, and killing the gunners.

[1] 賽熕銃   [2] 無敵大將軍   [3] 子銃   [4] 銅發熕   [5] 貢
[6] 熕   [7] 藝   [8] 氣

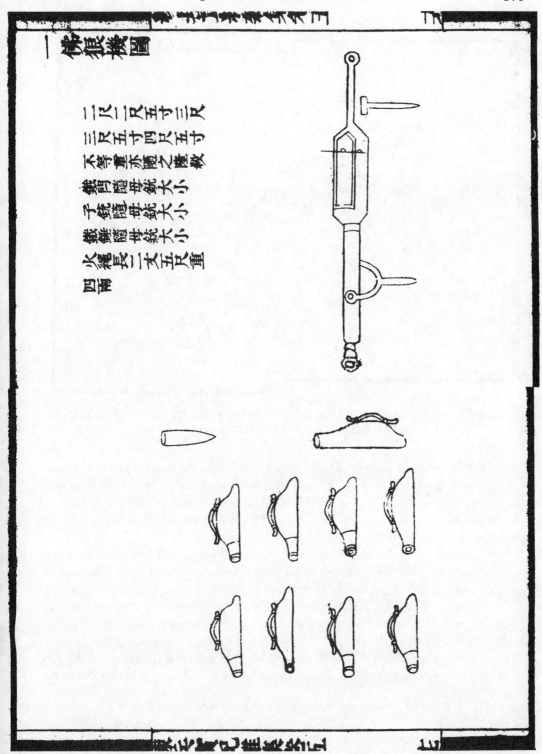

Fig. 138. A Frankish culverin shown in the same book (ch. 5, pp. 16b, 17a), together with nine culasses to fit into it.

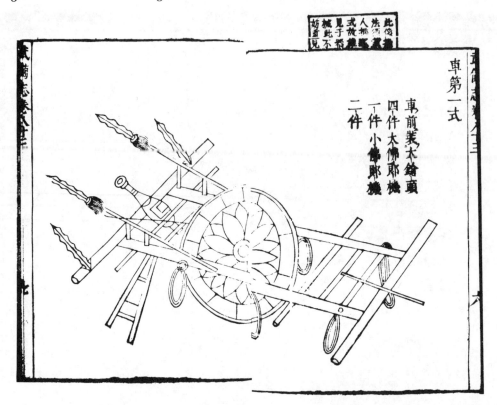

Fig. 139. One of Chhi Chi-Kuang's vase- or bottle-shaped breech-loaders (cf. Fig. 137) mounted at the front of an assault wheelbarrow (*WPC*, ch. 83, pp. 6*b*, 7*a*) accompanied by four spears. The text says that there were three such cannon, one large and two small, but only a single one is shown.

And the text goes on to say that this weapon could also be used on board ships at sea, if the vessels were large, and part of a fleet; it was also good for defending the gates of cities or encampments. The design was derived from the countries of the Western-ocean barbarians (Hsi-Yang Fan Kuo[1]) in the Chia-Ching reign-period (+1522 to +1566).

The passage further adds that just as the first bronze outburst cannons were developed from foreign examples, so Chinese ingenuity (*chhiao ssu*[2]) produced a smaller version of the *fo-lang-chi* breech-loaders, and called it the 'lead-and-tin gun' (*chhien hsi chhung*[3]), presumably because of the shot it fired. One of these is in the Tower of London (Figs. 143*a*, *b*);[a] it has a swivel mounting though hardly larger than a musket.

By +1605, when Ho Ju-Pin was writing his *Ping Lu*, even the terminology for cannon was reflecting Western usage, as we can see in Table 5, where 'serpen-

---

[a] Kindly provided by Mr Howard Blackmore. One of the photographs of Caldwell (1) shows such a weapon actually in use (Fig. 144). Cf. Fig. 73 on p. 275.

[1] 西洋番國   [2] 巧思   [3] 鉛錫銃

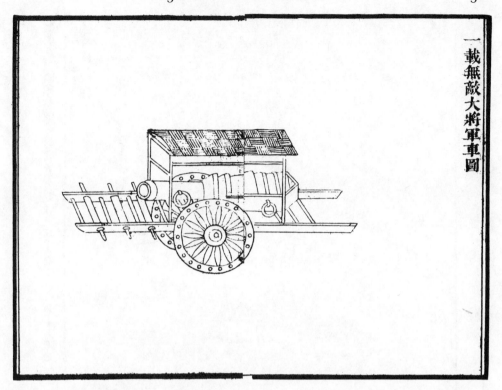

一載無敵大將軍車圖

Fig. 140. The culasse breech-loader applied to larger cannon; the 'invincible general' (*wu ti ta chiang-chün*) on its two-wheel carriage. *LPSC (TC)*, ch. 5, p. 14*a, b.*

tine', 'falconet' and 'saker' had their counterparts in Chinese. Illustrations too, now often show clear influence from the West, e.g. the field-gun with its trunnions (*chan chhung*[1]),[a] the heavy garrison piece (*shou chhung*[2]),[b] and the siege gun ornamented in very European style (*kung chhung*[3]), Fig. 145.[c] Variations in elevation are shown by the pictures in Figs. 146, 147, with the quadrant and plumb-bob at the cannon's mouth, set in the howitzer case at 60°, as the inscription says.[d] The carriage here resembles closely those of late +16th-century cannon in the West.[e] Finally, the 'tiger-cat mortar' (*fei piao chhung*[4]) is illustrated (Fig. 148) in the act of bombarding a city, which with its church towers and crenellated walls seems likely to have come out of some Western gunnery book.[f]

[a] *PL*, ch. 13, p. 6*a.* [b] *Ibid.* p. 22*a.* [c] *Ibid.* p. 13*b.* [d] *Ibid.* p. 2*a, b.*
[e] Blackmore (2), p. 12. When we come down as late as 1844, one can find in the *Hai Kuo Thu Chih*, ch. 87, p. 23*a*, a good drawing of a muzzle-loading ship's gun, like those which defended the 'wooden walls of Old England' in Nelson's time, complete with wheels, swabs, wedges and slow-match—but we do not reproduce it here.
[f] *PL*, ch. 13, p. 14*b*. But Yi Yangson's name has come down to us as the inventor of a bomb-throwing mortar which did good service against Hideyoshi's Japanese at the siege of Kyongju in Korea *c.* +1593; see Hulbert (2), vol. 1, pp. 407 ff.

[1] 戰銃　　　[2] 守銃　　　[3] 攻銃　　　[4] 飛彪銃

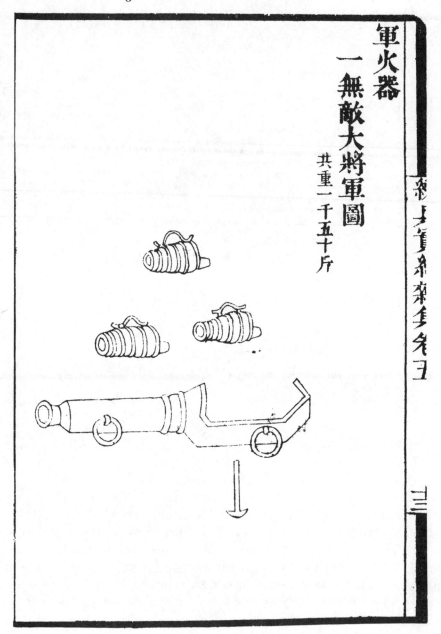

軍火器

一無敵大將軍圖

共重一千五十斤

Fig. 141. Three chambers or culasses for the same (*LPSC* (*TC*), ch. 5, p. 13*b*).

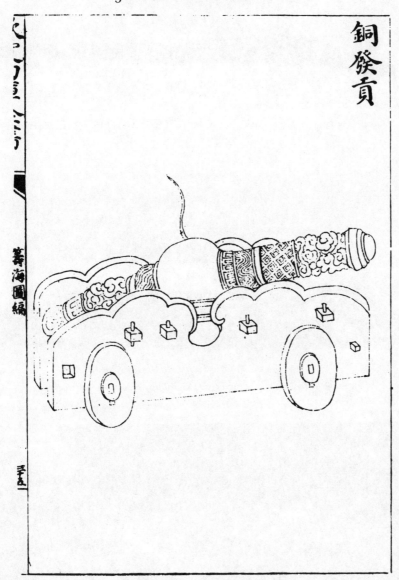

Fig. 142. A large artillery piece of the +16th century, the 'bronze outburst cannon' (*thung fa kung*), muzzle-loading. From *CHTP*, ch. 13, p. 35*a*. Cf. *WPC*, ch. 122, p. 4*b*.

Fig. 143a. Chinese musket-size breech-loading gun, in the Tower Armouries (photo. Blackmore). Overall length 8 ft.

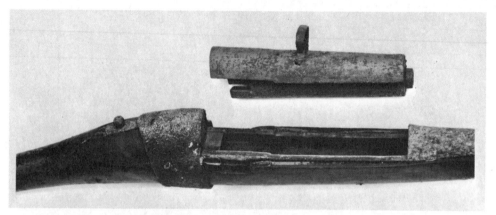

Fig. 143b. Close-up photograph of the same gun, with chamber removed.

Fig. 144. Breech-loading cannon (*fo-lang-chi chhung*) of musket size actually in use on board a junk in the South China seas in the thirties. Photo. Caldwell (1), p. 792.

Table 5. *Artillery pieces described in the* Ping Lu (*+1606*)

name		weight in catties		range in paces		ref.
		projectile	powder charge	horizontal	upwards (howitzer-style)	
Field-guns						
demi-serpentine[a]	*pan shê chhung*[1]	9–17	eq.	550–650	5500–6180	} 13/8b–9b
large serpentine	*ta shê chhung*[2]	18–25	eq.	700–900	6800–7270	
extra-large serpentine	*pei ta shê chhung*[3]	26–40	eq.	980	7190	13/10b, 11a
small Frankish sling	*hsiao fo-lang-chi*[4]	1		350	2900	
large Frankish sling	*ta fo-lang-chi*[5]			400	4000	
Siege-guns						
flying tiger-cat mortar	*fei piao chhung*[6]					13/14b and 20a–21a
falconet	*ying shun chhung*[7]	9–13	2/3 wt	500	3540	13/16b, 17a
pouncing-owl cannon[b]	*hsiao cho chhung*[8]	14–18	2/3 wt	600	4390	13/17a
demi-saker	*pan chen chhung*[9]	46		100	4620	} 13/17b–18b[c]
larger saker	*ta chen chhung*[10]	50		950	4730	
extra-large saker	*pei ta chen chhung*[11]	60		1000	4650	
roaring-tiger cannon	*hu hsiao chhung*[12]	60–100				13/18b–20a
Defence-guns[d]						
demi running-hog cannon	*pan thuan chhung*[13]	6–12				} 13/23b–24b
large running-hog cannon	*ta thuan chhung*[14]	12–18				
extra-large running-hog cannon	*pei ta thuan chhung*[15]	19–25				
leaping-tiger cannon	*hu chü chhung*[16]	26–50				

[a] This name for a small cannon should not be confused with the similar name for the lever in arquebuses that brought the slow-match to the touch-hole; cf. pp. 455 ff. below.
[b] The word 'saker' originally meant a kind of hawk; here *chen* is the serpent-eagle or poison-falcon, *Spilornis cheela* (R317; Chêng Tso-Hsin (2), vol. 2, p. 104), so the translation seems appropriate.
[c] Here there seem to be many printing errors.
[d] Said in the text to be of Western origin.

1 半蛇銃   2 大蛇銃   3 倍大蛇銃   4 小佛郎機   5 大佛郎機   6 飛彪銃   7 鷹隼銃   8 梟啄銃
9 半鵰銃   10 大鵰銃   11 倍大鵰銃   12 虎哮銃   13 半豘銃   14 大豘銃   15 倍大豘銃   16 虎距銃

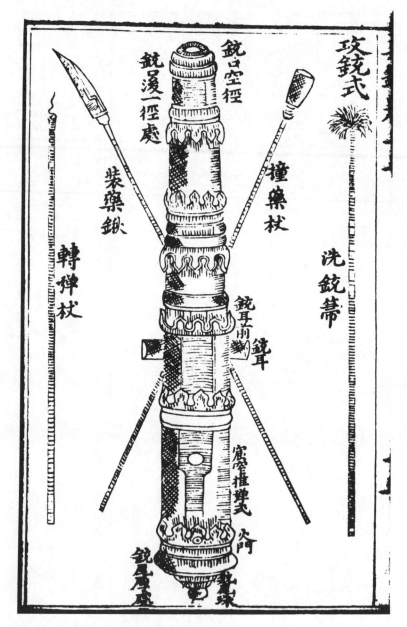

Fig. 145. Muzzle-loading siege gun ornamented in very European style, with swabs and other impedimenta,
PL, ch. 13, p. 13*b*.

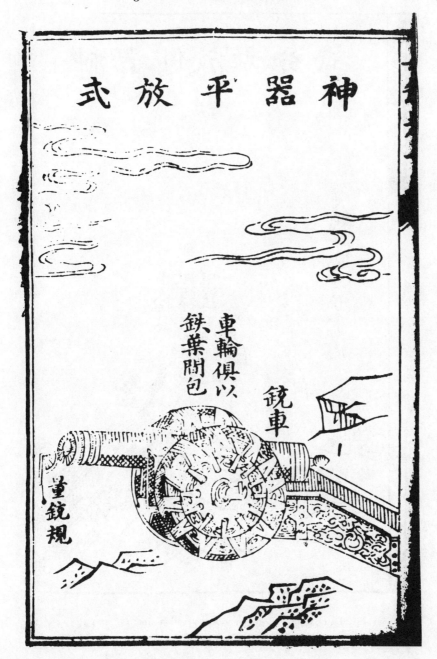

Fig. 146. European-type field-gun at low elevation, from *PL*, ch. 13, p. 2*b* (+1606).

Fig. 147. European-type field-gun at high elevation, from *PL*, ch. 13, p. 2*a*.

Fig. 148. The 'flying tiger-cat mortar' (*fei piao chhung*), no doubt copied from some Western gunnery book, since it is in the act of bombarding a town with church towers and crenellated walls. *PL*, ch. 13, p. 14*b*. And indeed the heading of the following page says 'Details of the Western Methods of casting large and small Cannon'. Note the quadrant, protractor and plumb-bob at the muzzles of the cannon in all these three pictures.

Nevertheless, Chinese artillery continued to impress Westerners quite favourably. In +1596 Jan van Linschoten wrote:[a]

> All the Townes in that Countrie are walled about with stone Walles, and have Ditches of water round about them for their Securitie; they use no Fortresse nor Castles, but onely upon every Gate of the Towne they have strong Towers, wherein they place their Ordnance for the defence of ye Towne. They use all kinde of armes, as Calivers, etc.

It was rather acute for an observer at this time to realise that there were no castles in China since for centuries there had been no military feudal aristocracy, but only centres of population and administration held for the king. After the passage from Juan de Mendoza (+1585) quoted on p. 54, he goes on to say that Friar Gerrardo saw some 'ill-wrought' pieces of artillery, but

> it was given them to understande that in other provinces of the kingdome there be that bee very curiously wrought and faire, which may bee of such that the Captain Artreda did see; who in a letter which he wrote unto King Phillip, giving him to understande of the secreets of this Countrie, amongst which he saide: 'the Chinos doe use all armour as wee doe, and the artillerie which they have is excellent good' ... I am of that opinion, for that I have seene vessels there of huge greatnesse, and better made than ours, and more stronger.[b]

Three hundred years later, after that long a period of capitalist enterprise and production, the disparity between Western and Chinese artillery became considerably greater.

Mention of the quadrant and plumb-bob at the mouth of the cannon in Fig. 146 leads us on to say something about the beginnings of external ballistics. As is generally known, the earliest Western speculations about the path of a projectile supposed it to move in a straight line for a while, before succumbing finally to the influence of gravity and then falling downwards in an equally straight line, not unlike the course of the mortar shell seen in Fig. 148. This was the conception in the days of Nicolo Tartaglia (+1537, +1546).[c] But Galileo (+1638) and Torricelli (+1644) proposed a parabolic trajectory,[d] and this eventually became more like a hyperbola when the resistance of the air was fully taken into account, as by Newton (+1674) and later mathematicians.[e]

In East Asia ballistics was pursued more in Japan than in China, but the connections were close. The famous Inatomi family of gunsmiths[f] left many MS books still extant on the theory and practice of gunnery, notably one of +1607 to +1610 in twenty-nine large volumes.[g] The most outstanding member of the family, Inatomi Naoie, recorded a tradition that Sasaki Shyō-huziro had first learnt the art in China, and then transmitted it to his grandfather Inatomi

---

[a] (1), vol. 1, pp. 130–1. The passage was largely a borrowing from Mendoza (1), p. 342 (Hakluyt Soc. ed., vol. 2, p. 288), whose translator said arcabuses or 'hargabushes'.
[b] (1), vol. 1, pp. 128 ff., ch. 15 of bk 3.
[c] See Hall (1), pp. 37 ff.   [d] Ibid. pp. 89 ff.
[e] Ibid. pp. 123 ff., 140 ff.   [f] Cf. p. 470 below.
[g] Itakura & Itakura (1), p. 83.

Sagami-no-kami Naotoki; this would take us back to +1500 or earlier, certainly before the arrival of either the Turkish or the Portuguese musket (cf. pp. 440 ff. below) in China and Japan, and would suggest that the Chinese hand-guns of the +15th century, probably with serpentines,[a] had begun the affair.[b] By +1618 trajectories were being studied by Shimizu Hidemasa, who visualised a slow rise followed by a slow fall.[c] Then from +1659 onwards[d] the parabolic trajectory was proposed, first in the *Kaisan-ki*[1] (Book of Improved Mathematics) by Yamada Shigemasa[2],[e] then in the remarkable work of Nozawa Sadanaga[3], the *San Kyūkai*[4] (Mathematics in Nine Chapters) of +1677.[f] This book, which accompanied the illustration of the curve with complicated quadratic equations, was the first Japanese treatise to explain physical phenomena using mathematical formulae. There may have been some Jesuit or other Western influence here,[g] but Nozawa's view of the world was at least as much Chinese, based on the Yin-Yang theory, decimal metrology,[h] and the standard pitch-pipe dimensions.[i] The *Suan Fa Thung Tsung*[5] (Systematic Treatise on Arithmetic) of Chhêng Ta-Wei[6] (+1592)[j] had been translated into Japanese only two years before Nozawa's own writing, and he was probably strongly influenced by it. Lastly there was the extension of Yamada's work by Mochinaga Toyotsugu[7] & Ōhashi Takusei[8], the *Kaisan-ki Kōmoku*[9], which continued to speak of gunshot parabolas.

After the *rangaku* (Dutch learning) period had opened, Shizuki Tadao[10] produced in +1793 the *Kaki Happō-den*[11] (On the Firing of Guns and Cannon) translated from the relevant parts of J. Keill's *Inleidinge tot de Waare Natuur- en Sterrekunde*.[k] This continued the parabolic interpretation. But gradually more advanced ideas became prevalent, as in the *Kikai Kanran*[12] (Survey of the Ocean of Pneuma) by Aoji Rinsō[13] (1825), the first work on modern science in Japanese, including besides physics much astronomy, meteorology and ballistics.[l] There was an extension of this in 1851 by Kawamoto Kōmin[14] entitled *Kikai Kanran Kōgi*[15], and the theory of projectile motion in this was studied by Mikami Yoshio (25).[m] But we need not pursue the story into the modern period further, and must return to the +17th century.

[a] Cf. p. 459 below.   [b] Itakura & Itakura (1), p. 89.   [c] Itakura & Itakura (1), p. 85.
[d] Note that this was after the closure of the country to foreign influences, and during the time when the production of fire-arms was steadily decreasing (cf. pp. 469 ff. below).
[e] See Mikami Yoshio (22). Having regard to possible contacts, this must almost certainly have been independent of Galileo.
[f] On this see Itakura (1).
[g] But it must have been very small and second-hand, since the Christian religion was outlawed in +1616. All the Latin Jesuits had been expelled, and the Dutch *rangaku* influence had hardly begun. Indeed, the year +1720 may be regarded as the beginning of it (cf. Fujikawa Yu (1), p. 56).
[h] Cf. Vol. 3, pp. 82 ff.   [i] Cf. Vol. 4, pt. 1, pp. 171 ff.   [j] Cf. Vol. 3, pp. 51-2.
[k] Leiden, 1741. On the translation and its ballistic content see Mikami Yoshio (23).
[l] Cf. Vol. 4, pt. 2, p. 531.
[m] He has also considered the contributions of Koide Shūki[16] (1847) and Ikebe Harutsune[17]. See Mikami (24, 26).

[1] 改算記   [2] 山田重正   [3] 野沢定長   [4] 算九回   [5] 算法統宗
[6] 程大位   [7] 持永豐次   [8] 大橋宅清   [9] 改算記綱目   [10] 志筑忠雄
[11] 火器發法傳   [12] 氣海觀瀾   [13] 青地林宗   [14] 川本幸民   [15] 氣海觀瀾廣義
[16] 小出修喜   [17] 池部春常

European cannon, in the era of nascent capitalism, were indeed now making all the running. In +1600 or soon after, late in the Wan-Li reign-period, Chinese artillerists obtained a cast-iron cannon larger than any hitherto known, from some European ship. The *Ming Shih* says:[a]

At this time, a ship arriving from the West (Ta Hsi-Yang[1]) brought an enormous cannon, which got the name of the 'red (-haired) barbarian gun' (*hung i phao*[2]). It measured over 20 ft long, and weighed as much as 3000 catties. It could demolish any stone city-walls, and its earthquake-like roar could be heard for several dozen *li* around.

During the Thien-Chhi reign-period (+1621 to 7) the (old) name of 'great general' (*ta chiang-chün*[3]) was given to it, and officials were sent to pay honour to it.[b]

During the Chhung-Chên reign-period (+1628 to 43) the grand secretary (*ta hsüeh shih*[4]) Hsü Kuang-Chhi[5] requested the emperor to issue an edict commissioning Westerners to fabricate weapons of this kind.

It will be remembered that Hsü Kuang-Chhi was a great friend of the Jesuits,[c] so this text immediately plunges us into the strange story of the apostles of Christianity engaging in gun-founding for the Chinese governments of the day.

It began in a relatively small way, with the Jesuits marginally involved;[d] because from +1620 onwards the danger of Manchu incursions and border fights[e] caused the Peking government, urged by Hsü Kuang-Chhi and other officials, to look with favour on the idea of inviting Portuguese artillery detachments north from Macao to oppose the Manchus.[f] The first group of these gunners set out, with some cannon, in +1621, but failed to get through; the second, consisting of gunnery instructors, arrived in Peking in the spring of the following year.[g] Urgent invitations, however, continued, and the colourful Jesuit João Rodrigues (Lu Jo-Han[6])[h] went with others to Kuangchow early in +1628 to arrange for a larger detachment, then accompanied it himself as interpreter. It was commanded by an artillery captain, Gonçalvo Teixeira-Correa (Kung-Sha Ti-Hsi-

---

[a] Ch. 92, p. 11*b*, tr. auct. Often afterwards quoted, as by Ling Yang-Tsao in his *Li Shao Phien* of +1799, ch. 40 (p. 650).

[b] In Taoist folk-religion any device or machine of almost miraculous potency was something which should receive veneration; analogous perhaps to Indian *puja* addressed to tools and instruments. This went against the grain of Confucian officialdom, but they generally played along with popular feeling.

[c] We have often discussed him and his work, cf. Vol. 1, p. 149, Vol. 3, pp. 52, 110, 447, and Vol. 4, pt. 2, *passim*. The Jesuits called him 'Doctor Paul'. According to Matteo Ricci's account and Hsü's biography in *Ming Shih*, ch. 251, p. 15*a*, they discussed together not only astronomy, mathematics and calendrical science, but also the modern firearms of the West.

[d] A good brief account is that of Cooper (1), pp. 334 ff.

[e] We shall see something of this at closer range from the gunnery point of view in a few moments (pp. 398 ff. and Figs. 152 to 155 below).

[f] Already in +1557 a force of Portuguese gunners and musketeers had helped the Governor of Kuangchow to suppress an uprising of pirates and dissident soldiers (Cooper (1), p. 335; Videira-Pires (1), pp. 698 ff.).

[g] Unfortunately, one of the artillerymen, João Correa, lost his life, together with two Chinese gunners, when one of the cannons blew up in +1624.

[h] He was always known as Rodrigues Tçuzzu, or Interpreter Rodrigues, partly because of his exceptional linguistic ability, and partly to distinguish him from other Jesuits of the same name. Tçuzzu comes from Jap. *tsuji* = *thung shih*[7]. He was really a long-standing member of the Japan Mission, but had been exiled to Macao.

[1] 大西洋   [2] 紅夷砲   [3] 大將軍   [4] 大學士   [5] 徐光啓
[6] 陸若漢   [7] 通事

Lao[1]), and took with it ten field-guns, but saw little fighting, as the Manchus thought discretion the better part of valour, and retreated. Substantial reinforcements under Pedro Cordeiro and Antonio Rodrigues del Campo arrived in +1630, but did not stay long.[a] Teixeira and his men, however, served under Sun Yuan-Hua[2], the Governor of Têngchow in Shantung, who had studied mathematics and gunnery with Hsü Kuang-Chhi, but both he and Rodrigues were caught there in a mutiny of troops in +1632, and the former was killed though the latter escaped. Afterwards Rodrigues wrote a eulogy of his friend the artillery captain, entitled *Kung-Sha Hsiao Chung Chi*[3] (Memoir of the Loyal and Gallant Gonçalvo).[b] At some point during their stay at Têngchow, a Korean embassy headed by Chŏng Tuwŏn[4] came through, and Rodrigues presented him with many scientific and technical books, including one *Explanation of Western Cannons*. We do not have the Chinese title, but quite probably it was the *Hsi-Yang Huo Kung Thu Shuo*[5] (Illustrated Treatise on Western Gunnery), which had been written by Chang Tao[6] and Sun Hsüeh-Shih[7] in +1625 or just before, in connection with the earlier expeditionary force of Portuguese artillerymen.[c] Rodrigues also gave to Chŏng a pair of quick-firing guns of some kind.[d] Finally he got back to Macao, and died there later in the same year.[e] All this goes to show two things, the intense interest which Chinese and Koreans both took in European gunnery developments at this time, and the natural, if regrettable, connection of the Jesuits with it.

What then happened followed inevitably from the new superiority of European armaments—the Jesuits were the most learned and scientific Westerners available, so they were 'drafted' into service.[f] In +1636, in the last decade of the Ming dynasty, Johann Adam Schall von Bell (Thang Jo-Wang[8])[g], the Director of the Astronomical Bureau, was called upon to advise about the fortifications of Peking, and had to do so again in +1643, though hardly any action was taken.[h] Then in +1642 he was visited by the Minister of War, Chhen Hsin-Chia[9], who invited him to set up a bronze cannon-foundry in the capital, and in spite of all

[a] This was partly because of growing bureaucratic nervousness at having so many armed Westerners around (in this connection cf. Vol. 4, pt. 3, p. 534), and partly because of the commercial interests of the Kuangchow merchants, who profited greatly by the Portuguese trade and wanted no weakening of the city of Macao, already subject to attacks by the Dutch. They actually paid the return travel expenses of the force.

[b] Pfister (1), p. 25* (add.).

[c] This work is now extremely rare, if extant at all; on it see Pelliot (55). It may be no coincidence that the two Chinese Christian officials who were sent down to Macao twice (in +1621 and 1622) to expedite the Portuguese artillery detachments were named Michael Chang and Paul Sun respectively. They may well have been the authors concerned. See Cooper (1), pp. 335–6 and Pfister (1), p. 12* (add.).

[d] *Kukcho Pogam*, ch. 3, pp. 65 a ff. For the probable nature of these weapons, cf. pp. 424, 461 below.

[e] On all these episodes see Boxer (12).

[f] We can be brief in this relation because we discussed the matter fairly fully in Vol. 5, pt. 3, pp. 240–1. The facts can be followed further in Pfister (1), p. 165; Bornet (1, 2, 3); Väth (1), pp. 111 ff., 370; and Duhr (1), pp. 60 ff. Occasional references occur, for example, in Rémusat (12), vol. 2, p. 220.

[g] Often discussed in Vol. 3, esp. pp. 447 ff.

[h] Schall von Bell (1), pp. 34, 90.

[1] 公沙的西勞　　[2] 孫元化　　　[3] 公沙効忠紀　　[4] 鄭斗源　　　[5] 西洋火攻圖說
[6] 張燾　　　　　[7] 孫學詩　　　[8] 湯若望　　　　[9] 陳新甲

expostulations this was what he had to do.[a] The arms desired were like sakers[b] and all he could do was to get their size reduced from 75-pounders to 40-pounders; of these twenty were cast that year, and 500 smaller ones in the year following. It was at this time that he collaborated with Chiao Hsü[1] in producing the book *Huo Kung Chhieh Yao*[2] (Essentials of Gunnery), an admirable work, which we quote from time to time. Schall von Bell survived both the end of the Ming, and a wave of severe persecution also, not dying till +1666, at which time he handed over his astronomical position to another Jesuit, the Belgian, Ferdinand Verbiest (Nan Huai-Jen[3]).

It must not be supposed that the Ming metal-workers were incapable of designing and casting good cannon themselves. In 1952 when in Shenyang, I visited the home of a former warlord, Thang Yü-Lin[4], and found outside two big guns, the larger one about 12 ft long and of 5 in. bore. It had on it the following inscription, which I copied:

Great General Pacifying Manchuria. Cast for the Regional Commander-in-Chief and High Commissioner for Military Affairs in Liaotung, Wu Chüan-Tzu[5]. Arsenal Superintendent and Regional Commander, Sun Ju-Chi[6]. Staff Officer in charge, Wang Pang-Wên.[7] Chief bronze-founder Shih Chün-Hsien.[8] Made on a fortunate day, in the 12th month of the 15th year of the Chhung-Chên reign-period.

That was +1642, and the day cannot have been so fortunate, for only two years later the Manchus captured Peking, and the cannon was probably used by them during the ensuing century.

What happened to Schall von Bell happened also to Verbiest—a decade later, the identical play was acted over again. Wu San-Kuei[9], the powerful general who had joined his army with the Manchu troops of Dorgon[10] in +1644 to capture the capital from the Ming,[c] and then served the Chhing dynasty loyally for nearly thirty years, especially by his successful campaigns against the remnants of the Southern Ming in Yunnan and Burma,[d] became in the end disaffected, and set up a standard of revolt in Kweichow and Hunan in +1673. He pro-

---

[a] Schall von Bell (1), pp. 63 ff., 80 ff.   [b] Cf. Table 5, p. 385 above.

[c] Actually the Ming had already fallen, and the last emperor had committed suicide, so the invaders were liquidating a great peasant uprising under Li Tzu-Chhêng[11], who had proclaimed a Ta Shun[12] dynasty. This has always been regarded as a classic case of class interest prevailing over national feeling.

[d] The Southern Ming were also capable of casting good cannon, and one of them, dredged up from Kaitak Bay in 1956, now stands beside the Central Government Offices in Hongkong (Fig. 149). The inscription gives the names of the three generals who ordered the casting, which was directed by a colonel, Hsiao Li-Jen,[14] and delivered to the commander of the ordnance depôt, Ho Hsing-Hsiang.[15] The date was the 6th month of the 4th year of the Yung-Li reign-period (positively the last that the Ming ever had), i.e. +1650. This was twelve years before the last extinction of the line, when the Ming Pretender, Chu Yu-Lang[13], was executed at Kunming. What service the gun saw before being sunk in the sea we do not know. From Goodrich (23) we learn that another Southern Ming cannon cast in the same year has been found near Hongkong, and the name of Hsiao Li-Jen is on it too, so he must have been some kind of Master-General of Ordnance for that remnant dynasty in its last days. A little-known study of these guns and their inscriptions is that of Lo Hsiang-Lin (6).

For the photograph of the Kaitak cannon we are much indebted to Mr J. Cranmer-Byng.

[1] 焦勗	[2] 火攻挈要	[3] 南懷仁	[4] 湯玉麟	[5] 吳捐資
[6] 孫如激	[7] 王邦文	[8] 石君顯	[9] 吳三桂	[10] 多爾袞
[11] 李自成	[12] 大順	[13] 朱由榔	[14] 蕭利仁	[15] 何興祥

Fig. 149. Southern Ming cannon cast in +1650, dredged up from Kaitak Bay in 1956, and now standing beside the Central Government Offices in Hongkong. See Lo Hsiang-Lin (6). Photo. John Cranmer-Byng.

claimed himself emperor of a new dynasty, the Chou[1], in +1678, but died of dysentery that same year. It was therefore perhaps not surprising that Verbiest, who had been re-equipping the Peking Observatory with splendid bronze instruments from +1669 to +1673,[a] should receive a summons in +1675 to set up another cannon-foundry, this time for the Manchus.

Let us listen to the elegant account of another Jesuit, Louis Lecomte (Li Ming[2]) written twenty years or so later.[b]

After the Emperor had tryed many feveral ways to no purpofe, he faw plainly that it was impoffible to force them [i.e. the troops of Wu San-Kuei] from the places where they had entrenched without ufing his great Artillery: but the Cannon which he had were Iron, and fo heavy that they dared not carry them over such fteep Rocks, as they muft do to come to him. He thought Father *Verbiest* might be affiftant to him in this matter; he commanded the Father therefore to give directions for cafting fome Cannon after the *European* manner. The Father prefently excufed himfelf, faying that he had lived his whole life far from the noife of War, that he was therefore little inftructed in thofe affairs. He added alfo that being a Religious, and wholly employed in the concerns of another World, he would pray for his Majefty's good fuccefs; but that he humbly begged that his Majesty would be pleafed to give him leave not to concern himfelf with the warfare of this World.

The Fathers Enemies (for a Missionary is never without fome) thought that now they had an opportunity to undermine him. They perfuaded the Emperor that what he commanded the Father to do, was no ways oppofite to the will or intention of the Gofpel: and

---

[a] See Vol. 3, pp. 451 ff., Figs. 189–92.
[b] (1), pp. 368–9.

[1] 周   [2] 李明

that it was no more inconvenient to him to caſt Cannon than to caſt Machines and Mathematical Inſtruments, eſpecially when the good and ſafety of the Empire were con-cerned: that therefore without doubt the reaſon of the Fathers refuſal was becauſe he kept Correſpondence with the Enemy, or at leaſt becauſe he had no reſpect for the Emperor. So that at laſt the Emperor gave the Father to underſtand, that he expected obedience to his laſt Order, not only upon pain of loſing his own Life, but alſo of having his Religion utterly rooted out.

This was to touch him in the most ſenſible part, and he was indeed too wiſe to ſtand out for a nicety or a ſcruple at the hazard of loſing all that was valuable. I have already aſſured your Majeſty [he said] that I have very little underſtanding in caſting Cannon; but ſince you command me I will endeavour to make your Workmen underſtand what our Books direct in this Affair. He took therefore upon himſelf the Care of this Work, and the Cannon was proved before the Emperor, and found to be extraordinary good. The Emperor was ſo well pleaſed with the Work, that he pulled off his Mantle, and in the preſence of the whole Court gave it to Father *Verbiest* for a token of his Affection.

All the Pieces of Cannon were made very light and ſmall, but ſtrengthned with a ſtock of Wood from the mouth to the breech, and girt with ſeveral bands or Iron; so that the Cannons were ſtrong enough to bear the Force of Powder, and light enough to be carried thro' any, even the worst, Roads. This new Artillery did every way anſwer what they propoſed from it. The Enemy were obliged to leave their Intrenchments in diſorder, and ſoon after to Capitulate; for they did not think it poſſible to hold out against thoſe any longer, who could deſtroy them without coming themſelves into reach.

It seems that the Manchu artillery had about 150 cannon, but (as Lecomte says) many were too heavy for a mountain campaign, so Verbiest was called upon to cast a lot of smaller ones. Having duly organised the foundry he cast twenty in the first month, then 320 during the rest of the year.[a] On a previous occasion we could not help commenting adversely on the Christian ceremonies that Schall von Bell carried out in his foundry,[b] but now Verbiest did not hesi-tate to bless the guns liturgically with asperges and incense, giving to each one the name of a saint, and inscribing it accordingly. He was awarded the title of Deputy Minister of Public Works (Kung Pu Shih Lang[1]) for his pains. By an extraordinary coincidence, two of his guns are still preserved in the Tower of London,[c] having been captured at the Taku Forts in 1860 (Fig. 150). One has a legible inscription, which runs as follows:

General of Holy Authority. Cast in the 28th year of the Khang-Hsi reign-period [+1689]. It takes 1 catty, 12 *liang*, of powder as charge, and fires an iron ball weighing 3 catties, 8 *liang*.[d] Height of the sight 6 *fên*, 3 *li*.[e] Official in charge, Nan Huai-Jen.

[a] See Bosmans (2, 4) and Pfister (1), pp. 347 ff. What is interesting here, as Dr Clayton Bredt points out, is that Verbiest seems simply to have made improvements on the long-established Chinese tradition of producing remarkably light-weight cast-iron ordnance, rather than introducing imported Western types. All the *Huo Lung Ching* and *Chhou Hai Thu Pien* weapons weighed very much less than Western guns of comparable calibre. These Chinese 'minions' continued in use right into the nineteenth century, as late as 1875 (cf. Bellew, 1). Verbiest's chief modification seems to have been the lengthening of the barrel.

[b] Vol. 5, pt. 3, p. 240.

[c] See Blackmore (2), pp. 153–4, no. 203, pl 42 a, b.

[d] About 1·66 lb. and 4·5 lb. respectively.　　[e] About 0·882 in.

[1] 工部侍郎

Fig. 150. One of Ferdinand Verbiest's field-guns, set on a mounting of about 1910 style, preserved at the Tower of London. Blackmore (2) catalogue, p. 153, no. 203 and pl. 42 *a*, *b*. Another of these cannon bears the date of +1689.

Officials supervising, Fo Pao and Shih Ssu-Thai. Artisan, Wang Chih-Chhen. Crafts-men, Li Wên-Tê and Yen Nai.

Since this was the year after Verbiest's death, his foundry must have gone on producing a whole series of cannon designed by him. Each has a solid trail fitted with a hinged traversing lever and elevating screw.[a] Here we cannot refrain from reproducing an imaginative drawing of Ferdinand Verbiest aiming and firing one of his guns (Fig. 151)[b] under the admiring gaze of assorted mandarins and artillerymen. Verbiest too seems to have written a treatise in Chinese on can-non and cannon-founding, but the title is not known and the text seems to have perished.[c]

Leaving now the exploits of the Jesuits as cannon-founders, we must retrace our steps a little to look at some quite remarkable drawings which have come down to us portraying the state of artillery in China in the second decade of the +17th century. They are battle pictures contained in the *Thai Tsu Shih Lu Thu*[2] (Veritable Records of the Great Ancestor (of the Chhing Dynasty) with Illustra-tions), first written in +1635.[d] This was Nurhachi[3], who fought the Ming from +1609 onwards, especially after +1616 when he proclaimed himself emperor of a Later Chin (Hou Chin[4]) dynasty, recalling that the Manchus traced their descent in part from the Jurchen Chin Tartars.[e] His first invasion of China was in +1618, and he continued at war until he died in +1626.

When one studies the pictures in the book it is clear at once that the Manchus are generally drawn as mounted archers wielding bow and sword, with the guns all on the side of the Ming; but towards the end the Manchus are using firearms too.[f] A characteristic study of the field-guns is that of Fig. 152, which shows Nurhachi's cavalry taking a Ming battery from the rear.[g] The eleven guns shown are mounted on two-wheeled barrows, the handles of which form the trails, and in front of each there is a shield, presumably of metal.[h] Three are

---

[a] These may have been later additions. Another of Verbiest's guns is preserved in the Hakozaki Shrine on Kyushu in Japan.

[b] It is the frontispiece of the second volume of the popular book of Caillot (1), published in 1818.

[c] Du Halde (1), vol. 2, p. 49; van Hée (17); Pelliot (55), p. 192; Pfister (1), p. 359. It was not known to Cordier (8), but Dr Hsi Tsê-Tsung tells us (priv. comm.) that its title was *Shen Wei Thu Shuo*[1] and its date +1681.

[d] The text is in Chinese, but the pictures have Manchu captions also. No writers are known by name, but they must have been official historians living very near the dates of the events described. The bibliography is complicated (see Hummel (1), pp. 598–9), and there are several versions of the text, while some sets of pictures were re-drawn by Mên Ying-Chao[5] in +1781. We use the MS. of +1740, reproduced in facsimile by the North-east University at Mukden in 1930. How exactly faithful the illustrations in this are to the MS. of +1635 one cannot know, but they have no flavour of the late +17th or +18th centuries.

[e] The name Chhing[6] was not adopted till +1636.

[f] Something should perhaps be allowed for Manchu self-congratulation in that they felt they had conquered troops better armed than themselves. But from Chhen Wên-Shih (1) we learn that the first 'modern' cannon was not cast by the Manchus until +1631, i.e. after the death of Nurhachi. Cf. *Li Shao Phien*, ch. 40 (p. 650).

[g] One of the Four Princes (Beile) is shown in the right-hand bottom corner with his staff. The caption says that the artillery belonged to the forces of Kung Nien-Sui[7], one of the Ming generals.

[h] We shall say something more of shields on pp. 414 ff. below. Here they are generally painted with lion-mouths, suns, etc.

[1] 神威圖說   [2] 太祖實錄圖   [3] 弩爾哈赤   [4] 後金   [5] 門應詔
[6] 清   [7] 龔念遂

Fig. 151. An imaginative reconstruction of Ferdinand Verbiest, in his Jesuit robes, aiming and firing one of the field-guns cast for the Chhing dynasty under his directions. Manchu officials look on. From Caillot (1), the frontispiece of vol. 2 (1818).

Fig. 152. A drawing from *TTSLT* (no. 2*a, b*) showing Nurhachi's cavalry taking a Ming battery from the rear. The eleven field-guns shown are mounted on two-wheeled barrows, the handles of which form the trails, and most of the gunners are either dead or fleeing. Normally a shield, presumably of metal, protects them. Note between each field-gun a double-barrelled bird-beak musket with prongs protruding beyond the muzzles as supports. The battery is described as one of those under the command of the Ming general Kung Nien-Sui.

partly or wholly overturned, and the gunners are dead or fleeing. Between each gun there are double-barrelled bird-beak muskets with prongs[a] at the front end as supports;[b] six of these can be seen, but none in use. One gets the general impression that the Chinese artillery was good when emplaced, but rather lacking in mobility.

Another picture (Fig. 153) shows a frontal attack on a Ming battery[c] by Manchu archers, both mounted and on foot, with Nurhachi himself commanding in the right-hand bottom corner. Again there are the field-guns and the shields to protect the gunners,[d] but besides these one can see five more guns simply resting on the parapet of the entrenchment, with a bombardier just about to fire one off at the bottom on the left.[e] The priming-pans of the cannon are carefully drawn in, and twelve of the bird-beak muskets may be noted, this time single-barrelled.[f] Double-barrelled muskets appear again, however, in Fig. 154, where the front line of Khang Ying-Chhien's[1] men is firing six of them, while he himself is indicated commanding behind.[g] The musketeers have quilted armour, but not the swordsmen with round shields.

The two-wheeled barrow-carriage was not the only way in which field-guns were mounted at this time, for Fig. 155 shows another frontal attack on a battery by the Manchu cavalry,[h] and here the guns are all attached to what we can only call 'carpenter's bench trolleys'. These trestles seem at first sight to have wheels at the end of each of their splayed legs, but a more careful look suggests that they were simply round flat feet, in which case the mobility was very poor.[i] Two of these trestles have overturned in the combat.[j] This curious type of carriage appears again in other illustrations, such as that depicting Nurhachi's siege of Liaoyang, which fell in +1621. Here they are all mounted on the flat ground between the city-wall and the moat, and in several cases the gunners can be seen applying their match (Fig. 157).[k] One could hardly get a better insight into

[a] Anyone wishing to see a photograph of such prongs in contemporary use may find it in Stone (1), p. 265, fig. 328, who calls them 'A-shaped rests'. The example comes from the Lamut, a Tungusic people in Siberia. And the Chinese army still had them on its muskets in 1860 (Fig. 156).

[b] On muskets see the following sub-section, pp. 429 ff.

[c] Under the general Phan Tsung-Yen[2], who can be seen in person in the left-hand top corner. Nurhachi's men are opposed by a few Ming archers, who do not seem to be doing anything however.

[d] One gun-carriage is already overturned.

[e] This lack of any form of carriage or mounting appears also in another drawing, which depicts the death of the Ming general Liu Thing[3] in +1619. On the whole campaign of this year see the paper of Huang Jen-Yü (6).

[f] Two of them can be seen firing, in the top right-hand corner of the picture.

[g] Besides these, there are thirteen bird-beak muskets to be seen.

[h] The Chinese were here commanded by a general named Ma Lin[4], and this may well be part of the battle of +1619 in which he was killed.

[i] Dr Clayton Bredt, however, is sure that they were wheels, and that most, if not all, of these field-guns were leather-wrapped.

[j] Besides the field-guns, nine bird-beak muskets, some double-barrelled, are to be seen.

[k] As in most of the other drawings, the field-guns are all breech-loaders, though no spare chambers ever appear.

[1] 康應乾   [2] 潘宗顏   [3] 劉綎   [4] 馬林

Fig. 153. Another drawing from the same work, *TTSLT* (no. 4*a*, *b*). Manchu archers, both mounted and on foot, are attacking frontally a Ming battery commanded by the general Phan Tsung-Yen, who is himself seen in the top left corner, while Nurhachi is depicted opposite at the bottom on the right. Besides the field-guns with their shields and the pronged muskets, several guns are simply resting on the parapet of the entrenchment, with an artillery-man about to fire one off with a brand in the left bottom corner.

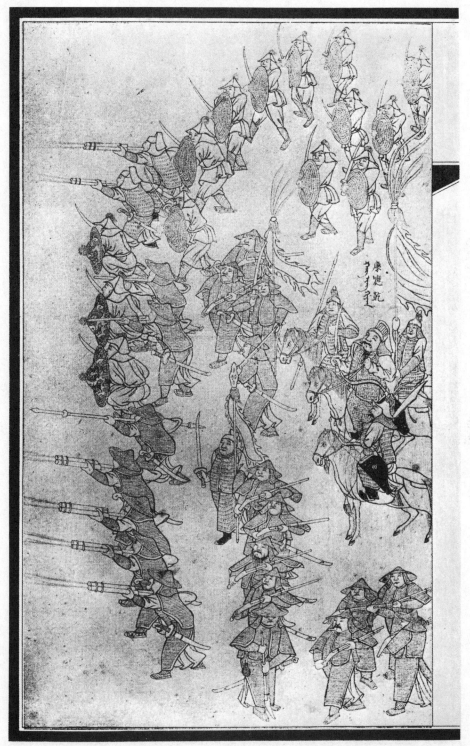

Fig. 154. A group of Ming musketeers firing off their guns, while their commander, Khang Ying-Chhien, is to be seen behind them on the right. At the top there are more lightly armoured swordsmen with round shields. From *TTSLT* (no. 6).

Fig. 155. Another mounting for field-guns in early +17th-century China; they are carried upon 'carpenter's bench' supports (cf. Figs. 82, 88, 106). These may have had wheels, but it is more likely that the legs simply ended in round flat feet, so that the mobility was very limited. In this picture (*TTSLT*, no. 3), the Manchu cavalry is overrunning a Ming battery commanded by Ma Lin, and it may be the very battle of +1619 in which he was killed. In this illustration the pronged muskets are again visible, and a couple of the trestle mountings have been overturned in the melée.

Fig. 156. Prongs still in use on muskets in 1860, a drawing from *Hutchings' California Magazine* for June of that year, p. 535. Ref. courtesy of Michael Rosen. The prongs are shown erroneously, however; they were evidently intended to help aiming when firing over a parapet or on the ground, and should therefore curve in the same direction as the butt (cf. the *TTSLT* illustrations). The present artist was not the only one who fell into this mistake, for it was also made in the illustration of Allom & Wright (1), vol. 1, opp. p. 87 in 1843, depicting a military guard-station at Thung-chang-fu on the Grand Canal. Cf. Vol. 4, pt. 3, Fig. 718 above.

Fig. 157. 'Carpenter's bench' trestle mountings seen again in a picture of Nurhachi's siege of Liaoyang in
+1621. The Ming artillery is deployed on the flat glacis outside the city wall; four of the seven guns are being
fired, and five musketeers are also to be seen. From *TTSLT* (no. 9).

the artillery of China in the early +17th century than from these drawings.[a] The world of learning has perhaps been unduly dazzled by the cannon of the Jesuits, so that the real achievements of the indigenous artillery have been somewhat overlooked.

All through the +16th and +17th centuries artillery was very prominent in the Chinese culture-area. One can see this from the many memoirs of adventures and narrow escapes in those troublous times, especially when the Manchus were fighting the remnants of the Ming, and both were in arms against the popular leader Li Tzu-Chhêng[1] and the tyrant of Szechuan Chang Hsien-Chung[2]. They constitute a whole genre of literature. For example, Shen Hsün-Wei[3] went with his father to Szechuan in +1642 at the age of five, then later his father was martyred by the tyrant, and he spent the rest of his youth escaping from manifold dangers, as he tells in his *Shu Nan Hsü Lüeh*[4] (Records of the Difficulties of Szechuan),[b] by which he meant something equivalent to our own +17th-century phrase: 'battle, murder, sudden death and other inconveniences'. In this book there are many references to gunpowder, gunfire and cannonades.[c] Another writer, Huang Hsiang-Chien[5], who described a decade of peregrinations escaping from combat zones (+1641 to 51), speaks in his *Huang Hsiao Tzu Wan Li Chi Chhêng*[6] of 'hearing the noise of cannon, and seeing the distant fire and smoke'.[d] In another place, he says that 'the sound of gunfire was like thunder, shaking the very mountains and valleys'.[e] Similar descriptions come in Pien Ta-Shou's[7] *Hu Khou Yü Shêng Chi*[8] (Life Regained out of the Tiger's Mouth)[f] of +1645, a book so called because after having devastated the tombs of Li Tzu-Chhêng's ancestors in order to stop his conquests, he actually fell into the hands of one of his commanders, but managed to escape therefrom.[g]

Nor was the age lacking for inventors, such as Ong Wân-Ta[9], who presented improved firearms in +1546,[h] while in the same year Chang To[10] offered prototypes of four-barrel and ten-barrel guns made of bronze, and capable of a range up to 700 paces.[i] In +1596 the judge Hua Kuang-Ta[11] presented further gunpowder-weapon inventions made by his father.[j] There were also great artillery generals such as Chhen Lin[12], who was prominent during the second invasion of Korea by the Japanese in +1597, and fought some decisive naval battles

---

[a] Further information on the early Manchu use of artillery can be obtained from Tanaka Katsumi (1). He fixes the first use in +1628, and says that it was very prominent in +1644/5, but much less so in the war against Koxinga (Chêng Chhêng-Kung[13]), who apparently made little use of field-guns. Tanaka noted that the artillery arsenals were always under the Eight Chinese Banners.

[b] Cf. Struve (1), pp. 346, 362.

[c] E.g. pp. 4a, b, 35a, b, 39b.

[d] P. 4b.   [e] Fu Chuan, p. 3b.   [f] P. 6b.

[g] Cf. Hummel (2), p. 741.

[h] *Ming Shih*, ch. 92, p. 11b.

[i] *Ibid.*   [j] *Ibid.*

¹ 李自成   ² 張獻忠   ³ 沈荀蔚   ⁴ 蜀難敍略   ⁵ 黃向堅
⁶ 黃孝子萬里紀程   ⁷ 邊大綬   ⁸ 虎口餘生記   ⁹ 翁萬達
¹⁰ 張鐸   ¹¹ 華光大   ¹² 陳璘   ¹³ 鄭成功

during their withdrawal.[a] A typical Fukien warship of this time (paralleling those that fought the Spanish Armada at the other end of the Old World) carried one heavy cannon (*ta kung*)[b], one mortar (*hu tun phao*)[c], six large culverins (*fo-lang-chi*)[d], three falconets (*wan khou chhung*)[e], and sixty fire-lances (*phên thung*)[f], doubtless to repel boarders or set fire to the enemy's sails and rigging, and finally a number of *shen chi chien*[1], probably arrows shot from guns.[g] Another Chinese gunner officer who distinguished himself in these campaigns was Lo Shih[2], who successfully defended Chiang-hua against the Japanese, and repulsed an attack by 500 of their ships upon the port-town of Phu-khou using shore-based artillery.[h]

The following century also produced some remarkable inventors. We may give the life-story of just one, Tai Tzu.[3] His biography runs as follows:[i]

Tai Tzu, whose other name was Tai Wên-Khai[4], was a Chhien-thang man from Chekiang.[j] His remarkable ingenuity appeared even while he was still young. He himself made a gunpowder weapon which could hit (a target) at more than a hundred paces away.

In the beginning of the Khang-Hsi reign-period (+1673) Kêng Ching-Chung[5] rebelled in Chekiang, and Prince Giyešu (Chieh-Shu[6])[k] led a government army south to overcome the uprising. Tai Tzu as a simple commoner or private scholar joined this army, and presented a design for a rapid-fire machine-gun (*lien chu huo chhung*[7]).[l] Its shape was like that of a balloon-guitar (*phi-pha*[8]). The gunpowder and lead balls (*huo yao, chhien wan*[9]) were all contained within the back of the gun, which was opened and closed by means of a wheel mechanism (*chi lun*[10]). There were also two parts fitting into each other like male and female. If one lever was pulled the gunpowder and lead bullets fell automatically into the barrel, whereupon the other mechanism followed suit and moved all together (*sui chih ping tung*[11]). The flint was struck, the spark came out, and the gun fired off accordingly. After twenty-eight rounds, the magazine had to be refilled with bullets. The design was in principle similar to that of the guns of the Westerners (*chi kuan chhiang*[12]). But the weapon was not at that time widely used, and the prototype was kept at Tai's home. This was still in existence during the Chhien-Lung reign-period (+1736 to 95).

When some Westerners presented 'coiled intestine (helical screw) bird guns' (*phan

---

[a] Lo Jung-Pang, in Goodrich & Fang Chao-Ying (1), vol. 1, pp. 167, 173.
[b] Cf. p. 378 above.   [c] Cf. p. 277 above.   [d] Cf. p. 367 above.
[e] Cf. p. 321 above.   [f] Cf. p. 232 above.

[g] It is not generally known that in +1588 Sir Francis Drake was still firing arrows from his muskets. Even as late as +1693 improvements to this system were still being canvassed. See Blackmore (1), p. 12.

[h] If Parker (9) was right, the Japanese were distinctly backward in fire-weapons at this time, still using carton 'thunderclap bombs' thrown from trebuchets.

[i] *Chhing Shih Kao*, ch. 505 (Lieh chüan, 292, I shu 4), pp. 5*b*, 6*a*; quoted by Chu Chhi-Chhien, Liang Chhi-Hsiung & Liu Ju-Lin (1), pp. 90–1 [*CCL* pt. 7], tr. auct.

[j] This is to say that he was born near Hangchow.

[k] Kêng was a Chinese, but Giyešu was a Manchu, with the princely title of Khang Chhin Wang.[13] The former died in +1682, the latter in +1697.

[l] Here follows a long paragraph about Tai's career, which we place at the conclusion of the passage.

[1] 神機箭   [2] 羅世   [3] 戴梓   [4] 戴文開   [5] 耿精忠
[6] 傑書   [7] 連珠火銃   [8] 琵琶   [9] 火藥鉛丸   [10] 機輪
[11] 隨之並動   [12] 機關槍   [13] 康親王

chhang niao chhiang[1]) Tai Tzu copied a number of these at the request of the emperor. Ten of his make were presented to the Western officials.[a]

Tai was also commissioned to design and make a 'mother-and-son' cannon (*tzu mu phao*[2]). It fired a projectile which burst and sent forth other projectiles that all fell down upon the enemy (*mu sung tzu chhu to erh sui lieh*[3]). It was rather like a Western mortar (*cha phao*[4]). The emperor, accompanied by all his ministers, watched a demonstration of it, and honoured the device with the name 'Awe-inspiring Far-reaching General' (*Wei Yuan Chiang-Chün*[5]).[b] The name and title of the inventor and maker was inscribed on the back of the cannon. When later the emperor personally commanded in the campaign against Galdan[6],[c] this weapon was among those used to defeat the enemy.

Because of (his part in the expeditionary force of Giyešu against Kêng), when national authority was restored over the territory, Tai Tzu was given the title of 'Acting Circuit Instructor' (*Tao Yuan Ta Fu Shih*[8]). Returning (to the capital) he had an interview with the Khang-Hsi emperor, who recognised his literary ability and examined him on the poem 'Dawn Audience in Springtime'. So he was given a post in the Han-Lin Academy as Expositor (*Shih Chiang*[9]), and (then), together with Kao Shih-Chhi[10],[d] was seconded to the Nan Shu Fang[11] (as one of the emperor's secretaries), and later to the Yang Hsin Tien[12].[e] Tai was expert in astronomy and mathematics, but when the *Lü Lü Chêng I*[13] (Collected Principles of Acoustics and Music)[f] was being edited, his views were not in agreement with those of Nan Huai-Jen[14] (Ferdinand Verbiest) and the other Westerners. So everybody envied him, and was at the same time jealous of him.

Unfortunately there was a person named Chhen Hung-Hsün[15], who had been a foster-son of Chang Hsien-Chung[16],[g] but switched his allegiance and became an official under the Chhing. This man accused (Tai) falsely, and it came to blows, so the matter was taken to court, giving Tai's enemies the opportunity of vilifying him; thus he lost his office, and was exiled to Kuan-tung. Later he was pardoned and went home, where he stayed at Thieh-ling for the rest of his life.

Thus was a remarkable talent wasted. How striking it was that when the Khang-Hsi emperor saw that he was literate, and called him into his direct service, all he could think of was to examine him in poetry. His scientific and technical ability was evidently considered quite secondary, and even so it got

[a] I hesitate to write 'ambassadors' though that is what *shih chhen*[7] should mean, because in those days there were no resident envoys at the Chinese court. But it could be a reference to the Russian embassy of +1693 headed by the Dutchman E. Ysbrandts Ides (cf. Vol. 4, pt. 3, p. 56); or some other mission of those times. If not, it must mean some of the Jesuits, who held many scientific offices under the crown. Cf. p. 366 (*f*).

[b] Not a very original name; cf. p. 315 above.

[c] This was the Bushktu Khan of the Sungars (part of the Eleuths), a tribal people like the Kalmuks or Western Mongols. He had conquered Sinkiang by +1679, and then fought against the Khang-Hsi emperor from +1689 till his death in +1697.

[d] Poet and calligrapher of note (+1645 to 1703), who spent many years as one of the Khang-Hsi emperor's private secretaries.

[e] The South Library and the Hall of Healing for the Soul were literary institutions at the imperial court.

[f] This was eventually issued as part of the *Lü Li Yuan Yuan*[17] (Ocean of Calendrical and Acoustic Calculations), and finished in +1713.

[g] The ferocious tyrant of Szechuan already mentioned (+1605 to 47).

[1] 蟠腸鳥槍	[2] 子母礮	[3] 母送子出墜而碎裂	[4] 炸礮
[5] 威遠將軍	[6] 噶爾丹	[7] 使臣	[8] 道員劄付師	[9] 侍講
[10] 高士奇	[11] 南書房	[12] 養心殿	[13] 律呂正義	[14] 南懷仁
[15] 陳宏勳	[16] 張獻忠	[17] 律曆淵源	

him into trouble in due course. One is very much reminded of the story of Ma Chün[1], the +3rd-century engineer and inventor, which we told earlier on.[a] Even in our own time and in the Western world, four centuries after the Scientific Revolution, the only avenue of promotion in technical services is all too often from 'blue-collar' practice to 'white-collar' paper-work.[b]

If we look at Tai's inventions in order, we see that the first must have been some kind of quick-firing machine-gun. It was a time when people everywhere were trying to make devices of this kind—for example, in Samuel Pepys' Diary for 3 July 1662 we read that the attention of the Royal Society was drawn to a 'rare mechanician' who claimed to be able 'to make a pistol shooting as fast as it could be presented, and yet to be stopped at pleasure, and wherein the motion of the fire and bullet within was made to charge the piece with powder and bullet, to prime it, and to bend the cock'.[c] But the problem was not practically resolved till +1718, when James Puckle developed his breech-loading gun with a revolving set of chambers which could fire sixty-three shots in seven minutes.[d] Thereafter the line led straight to the multi-barrel 'pepper-box' pistols and revolving 'coffee-mill' guns of Ethan Allen (1837) and others,[e] thence to the Gatling gun of the American Civil War (1862)[f] and the Maxim gun of 1883.[g] Chinese antecedents for Tai Tzu's efforts are easy to find, for we have already described (pt. 6 (e), 2, iv) the magazine crossbow, widespread in +16th-century Ming use, as also (pp. 263–4) the magazine eruptor, which may well have been common considerably earlier, indeed back to +1410 or even +1350. All the same, we should very much like to have further details about Tai Tzu's guitar-shaped machine-gun.[h]

The second of his exploits is more difficult to pin down, but it could have been some kind of screw-chamber breech-loader.[i] If it was a variety of musket, as one might at first sight suspect from the name 'bird-gun',[j] a screw of one sort or another was evidently involved.[k] Here rifling would not come altogether amiss.

[a] Vol. 4, pt. 2, pp. 39 ff.
[b] Tai Tzu's engineering skill became legendary. A century or more later, Ling Yang-Tsao averred that Verbiest had tried to cast cannon for a year without success, while Tai Tzu, when called upon by the emperor, succeeded in eight days; Li Shao Phien, ch. 40 (p. 650). They were certainly contemporaries and knew each other, so Tai probably knew Verbiest's cannon-foundry too. Earlier, Ling says that Tai made fo-lang-chi breech-loaders, which is not at all impossible, but he ends by garbling the third exploit of the shell-firing mortars.
[c] (1), Everyman ed., vol. 1, p. 271, noted by Hall (5), pp. 358–9. Cf. Birch (1), vol. 1, p. 396.
[d] Cf. Reid (1), pp. 161 ff.   [e] Ibid. pp. 205–6.   [f] Ibid. pp. 221 ff.
[g] Ibid. pp. 230–1, 245.
[h] Another forerunner was the magazine musket (lien tzu chhung[2]) described by Chhi Chi-Kuang in +1560 (see CHHS, ch. 15, p. 12a, b, and later PL, ch. 12, p. 30b). An iron side-tube was arranged to feed lead bullets into the barrel. But Chhi regarded the gun as complex and unreliable, so he only included it, he said, for the sake of completeness.
[i] Cf. p. 366 (f) above.   [j] Cf. p. 432 below.
[k] Cf. Vol. 4, pt. 2, p. 121 and Fig. 416, from STTH, Chhi yung sect., ch. 8, p. 6a (+1609). As Horwitz (6) noted, this was an early illustration of the screw-thread in China, for the screw was not indigenous there (Vol. 1, pp. 241, 243). See also Fig. 171 below.

[1] 馬鈞   [2] 連子銃

The rotational stabilisation of a projectile's flight by endowing it with a spin, due to spiral grooves contrived inside the barrel, may go back to Leonardo,[a] and in any case began to be fairly frequently used by gunsmiths from about +1500 onwards.[b] A number of examples have survived from the second half of the +16th, and from the following, century. These however were sporting guns, and general military use did not come in until the American War of Independence, from the late +18th century onwards.[c] Still, it is not at all impossible that rifling was what interested Tai Tzu at this point; the text says 'bird', not 'bird-beak', so it might well refer to the use, rather than to the shape of the cock or butt, hence perhaps the presentations to the ambassadors or officials, to please them in their fowling.[d]

The third and last of his designs was fairly clearly a shell-firing cannon, for the projectile burst and released other projectiles, falling down like the shower of sparks from a firework rocket. Shells had been known in Europe since the +15th-century *Feuerwerkbuch*, probably of +1437, and they are also described in Valturio's *De Re Militari* of +1460.[e] Moreover, we met with them already in China in connection with eruptors (p. 264, cf. p. 317), which would take them back to the +15th, if not the +14th century; and they were only a logical development from the 'thunder-crash' bombs with iron casings (p. 170 above), which were older still. It was only to be expected therefore that people in China should by this time (late +17th century) have been experimenting with shells. They finally came into their own in a memorandum addressed to the emperor by Lin Tsê-Hsü[1] in 1846, entitled *Cha Phao Fa*[2].[f] But it is interesting (and certainly not generally known) that shells or shrapnel of some kind were used by Khang-Hsi's artillery in the war against the Eleuths at the end of the seventeenth century.

Mention of the period of the Opium Wars[g] reminds us that an important gun-founding invention was made at this time by a pioneering Chinese engineer, Kung Chen-Lin[3], some thirty years before its adoption in the West.[h] This was

---

[a] Cf. Partington (5), p. 175. In the Codex Atlanticus.

[b] Reid (1), pp. 112–13, 143. Certainly by +1540 (Blackmore (1), pp. 14–15).

[c] Reid (1), pp. 167, 209.

[d] Ling Yang-Tsao, in his *Li Shao Phien*, ch. 40 (p. 650), dates the event at +1676, in which case the embassy would have been the Russian one headed by the Rumanian (Moldavian) scholar, Nikolaie Spătarul Milescu. On this see Cordier (1), vol. 3, p. 271, and Vol. 4, pt. 3 above, p. 149, with refs.

[e] Partington (5), pp. 149, 157, 164–5.

[f] *Hai Kuo Thu Chih*, ch. 87, p. 6*a*, *b*, abridged tr. Chhen Chhi-Thien (1), p. 17. This date was intermediate between Mercier's use of improved shells at the siege of Gibraltar in +1782 and Shrapnel's eponymous invention of 1852. Cf. Reid (1), p. 182.

[g] A number of Chinese cannon of this period (1841), cast under the superintendence of the Governors Yen Po-Thao[4] and Liu Hung-Ao[5], are preserved inside the western wall of the campus of Amoy University; they have been described by Chêng Tê-Khun (18, 19). It is not generally known that Wang Thao[6], the famous collaborator of James Legge, wrote a book rather later on, the *Tshao Shêng Yao Lan*[7] (Important Factors for Gaining Victories) on cannon, cannon-founding and boring, the manufacture and use of gunpowder, etc.

[h] Chhen Chhi-Thien (1), p. 43. A biography of Kung is given in Chu Chhi-Chhien, Liang Chhi-Hsiung & Liu Ju-Lin (1) [*CCL*, pt. 7], pp. 94 ff.

---

[1] 林則徐   [2] 炸礮法   [3] 龔振麟   [4] 顏伯燾   [5] 劉鴻翔
[6] 王韜   [7] 操勝要覽

her broadside bearing on the battery which annoyed her, shortly silenced its fire, and having fulfilled the object of examining the sea defences of the place, made sail for her point of rendezvous.

And again, with reference to the bombardment of the batteries of 'Ko-lang-soo':[a]

The engagement was a fine spectacle, but beyond the picturesqueness of the scene afforded no point worthy of comment, save that it furnished strong evidence of the excellence of the Chinese batteries, upon which the fire of the seventy-fours, though maintained for fully two hours, produced no effect whatever, not a gun being found disabled, and but few of the enemy killed in them when our troops entered. The principle of their construction was such as to render them almost impervious to the effects of horizontal fire, even from the 32-pounders of the seventy-fours, as, in addition to the solid mass of masonry, of which the parapets were formed, a bank of earth bound with sods had been constructed on the outer face, leaving to view only the narrow mark of the embrasure.

(iii) *Shields, 'battle-carts' and mobile crenellations*

In Figs. 152 and 153, taken from the *Thai Tsu Shih Lu Thu*, we have already seen pictures of the shields, presumably of iron, which protected the men who worked the field-guns, mostly on the Ming side, during the first quarter of the +17th century. But shields adapted to the uses of fire-weapons did not begin with guns and light cannon, they began with fire-lances (cf. pp. 236 ff. above).

This we know from an item called the 'mysteriously-moving phalanx-breaking fierce-flame sword-shield' (*shen hsing pho chen mêng huo tao phai*[1])—a rather enigmatic description the meaning of which will in a moment be clear. In the *Huo Lung Ching* we read:[b]

The apparently automotive fierce-flame-spouting[c] shield for use with cutlass-wielding soldiers to destroy enemy formations, is covered with fresh ox-hide. In it are concealed thirty-six (fire-lance) tubes, containing magical gunpowder, poisonous gunpowder,

[a] (1), pp. 174–5.
[b] Pt. 1, ch. 3, p. 2a, b. The same illustration and description occurs again in *WPC*, ch. 129, p. 11a, b. The only words added there appear in square brackets.
[c] A strict interpretation of these words would mean Greek Fire petrol (p. 86 above), but one might hesitate to insist on this. They could of course imply that shields of a similar kind had been used with petrol flame-throwers two or three centuries earlier. On the other hand, these devices may have been in use much later than we usually think; after all, we have seen a depiction (Fig. 8) of the 'fierce fire oil machine' in an encyclopaedia of +1609.
In general we can hardly talk of shields without recalling the armoured iron-clad roofed-over spike-studded combat 'turtle-ships' (*kuei chhuan*[2]) of Admiral Yi Sunsin[3], so successful against the Japanese invasion of Korea in +1592; cf. Vol. 4, pt. 3, pp. 683 ff. and Fig. 1050, as well as Pak Hae-ill (1). In (2), pp. 33–4 he figures (from a contemporary painting on a porcelain jar) the prominent animal heads at the bows of these ships, and suggests that 'sulphurous fumes' were poured forth from them. One wonders whether they were not rather relict Greek Fire petrol flame-throwers? If this was so, the 'siphon' of the Byzantine navy would have lived again. In any case, a contemporary Japanese source, the *Kōrai Sensenki*[4], records that incendiary rockets (*hiya*[5]) were fired from the turtle-ships, and did much damage to the Japanese war-vessels.

[1] 神行破陣猛火刀牌   [2] 龜船   [3] 李舜臣   [4] 高麗舩戰記
[5] 火矢

blinding gunpowder, and bruising and burning gunpowder[a] [six tubes of each]. Coiled slow-match is held by each man in the formation (to light the fire-lances as may seem best). When two opposing armies are confronting one another, at the sound of a maroon signal, the shields are rolled forward into action, and when they spout fire, the flames shoot 20 or 30 ft forward. One group of men in armour on the left hand work the shields, while another group on the right wield their cutlasses. They aim to decapitate the enemy soldiers, and to cut off the legs of their horses (during the confusion caused by the flame-throwers). One single one of these shields is in itself worth ten brave soldiers.

This may be considered fairly archaic, but the fact that it comes in the oldest stratum of the book means that it must belong at least to +1412, and most probably to +1350 or before. Fig. 158 shows the usual tiger face but no obvious means of movement, yet the verb *kun*[1], 'rolled', used in the text, indicates that the whole weapon was mounted on some kind of mobile stand, probably a two-wheeled barrow pushed by the fire-lance operators.

Actually the mobile shield had a long history going back before the time of guns and cannon if not of fire-lances. Wei Shêng[2], a Sung general already mentioned (p. 157) made in +1163 many hundreds of shielded vehicles, pushed or drawn by hand, which could be parked in defensive arrays to protect encampments and strategic positions. Some of these carried trebuchets hurling stones or weak-casing gunpowder bombs, while others bore multiple-bolt *arcuballistae* with two or three springs. Others again could transport personal armour and munitions.[b]

Indeed one can trace the laager tactic far back in Chinese military history. The 'deer-horn cart camp' (*lu chio chhê ying*[4]) may perhaps be as old as the Warring States time with Wu Chhi[5] (d. −381); but it was certainly used in the +3rd century under Ma Lung[6].[c] There may well have been also some relation with the 'mobile city-walls' (*hsing chhêng*[7]) mentioned in the Mohist military chapters, as also with the watch-towers mounted on carts which patrolled the frontiers of the Han.

By the +16th century there had been a great development of shields, though they were still made of wood. The celebrated general Chhi Chi-Kuang[8] (+1528 to +1588) based much of his tactics on what we may call 'battle-carts' which carried protective screens and could be formed into a laager, as seen in Fig. 159 taken from his *Lien Ping Shih Chi* (+1571).[d] These large two-wheeled carts were

---

[a] On these forms of gunpowder see p. 180 above.

[b] The passage (*Sung Shih*, ch. 368, pp. 15*b*, 16*a*) is given in translation by Průsek (4), pp. 255–6, who, as a Czech himself, was inspired by the resemblance to the *Wagenburg* of the Hussites (pp. 276–7); cf. p. 421 below. Unfortunately, he translated *phao*[3] as 'pièce', presumably 'd'artillerie', instead of trebuchet. However, the main thrust of his paper was against true guns or cannon at that time.

[c] *Chih Shu*, ch. 57, p. 3*a*; Yang Hung (*1*), pp. 92–3; and Boodberg (5), pp. 2, 6, translating from a source supposedly Thang. The name obviously derived from the fact that all sides bristled with defences when the vehicles formed a circle or a square.

[d] *Tsa chi* sect., ch. 6, pp. 8*b*, 9*a*.

[1] 滾   [2] 魏勝   [3] 砲   [4] 鹿角車營   [5] 吳起
[6] 馬隆   [7] 行城   [8] 戚繼光

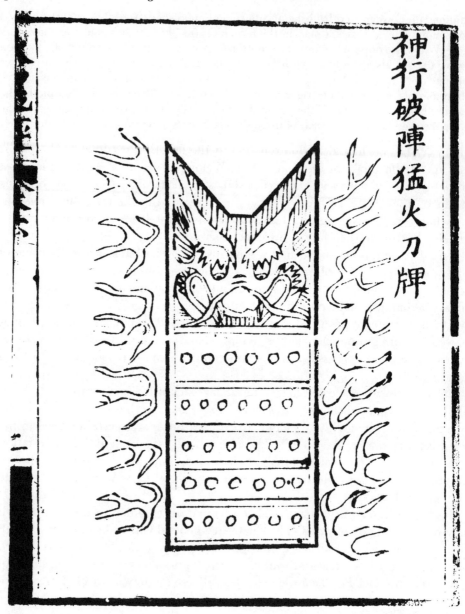

神行破陣猛火刀牌

Fig. 158. A mobile shield for fire-lances, from *HLC*, pt. 1, ch. 3, p. 2*a*, therefore at least as early as +1350 and probably used long before that. It is the 'mysteriously moving phalanx-breaking fierce-flame sword-shield' (*shen hsing pho chen mêng huo tao phai*). The fire-lances are stated to contain the usual rocket-composition (low-nitrate) gunpowder, but the title suggests a Greek Fire flame-thrower behind the shield. This last must have been mounted on wheels, and was certainly accompanied by swordsmen on either side.

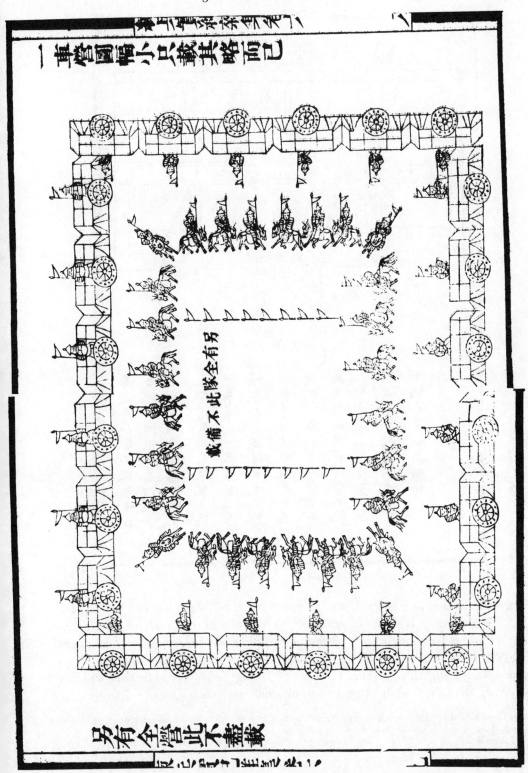

一車營圖幅小只載其略而已

載備不此隊全有另

另有全營此不盡載

Fig. 159. Chhi Chi-Kuang's +16th-century laager or *Wagenburg*, from *LPSC* (*TC*), ch. 6, pp. 8*b*, 9*a*. Each battle-cart was two-wheeled and had shield-screens which could be folded out to form a continuous defence, through which muskets or breech-loading cannon could be fired.

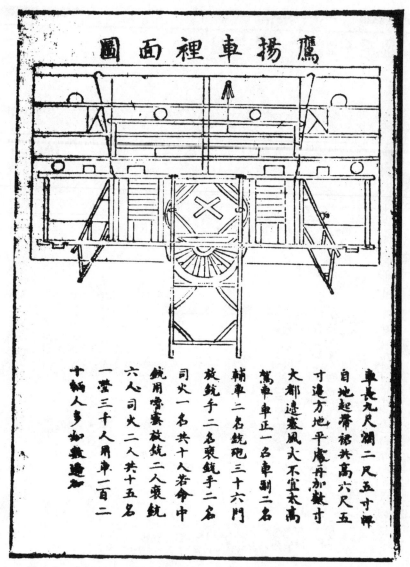

Fig. 160. A single two-wheeled battle-cart, from *Chhê Chhung Thu*, p. 6*b* (+1585).

drawn by mules, and carried screens which could be folded out to a length of 15 ft, thus forming a continuous battlement, only the hinged ends of each allowing for the ingress or egress of defending soldiers whether on foot or mounted. Twenty men were assigned to each battle-cart, ten of them manoeuvring it into place and firing the muskets or breech-loading *fo-lang-chi* culverins which it carried,[a] the other ten forming an assault team with close-combat weapons.[b]

[a] In many of the pictures which we reproduce it can be seen how the trail of a field-gun developed from the shafts of the two-wheel barrow on which it was originally borne.

[b] A good account of Chhi's methods is given by Huang Jen-Yü (5), pp. 179 ff.

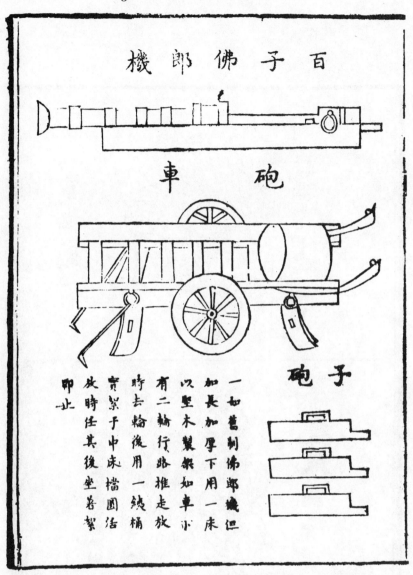

機郎佛子百

車砲

砲子

即時實時有以時去如
止放結去二堅緊長舊
放緊于輪木子加制
時于中行緊中厚佛
任中床路架床下郎
其床擋推如檔用機
後檔圓走車圓一但
坐圓活放小活床
著緊後推用
緊用一總
一總桶
總桶

Fig. 161. Breech-loading cannon with ridged barrel, the two-wheeled barrow on which it was mounted, and three of its culasses or chambers; from *CCT*, p. 3*b*. Intended to form part of the Wagenburg of battle-carts.

A closer view of one of these battle-carts is given (Fig. 160) in a slightly later book, the *Chhê Chhung Thu*[1] (Illustrated Account of Muskets, Field Artillery and Mobile Shields) dating from about +1585. In Fig. 161 we see a breech-loader, the barrow on which it was carried, and three of the chambers (here called *tzu phao*[2]) for the charges and projectiles. Finally, Fig. 162 shows two of the screens, with some men ready to shoot off their muskets, while others get ready to fight

---

[1] 車銃圖   [2] 子砲

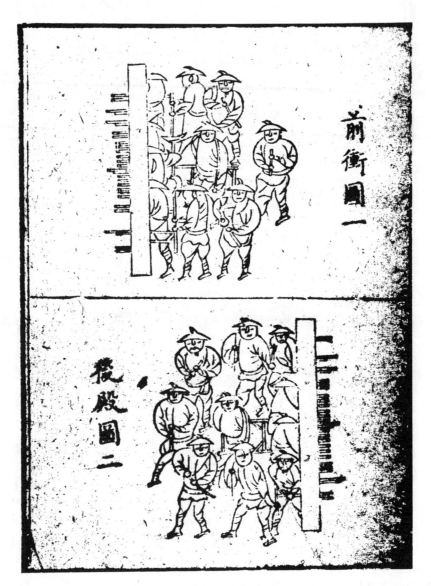

Fig. 162. Screens with soldiers behind them ready to fire off their muskets, or preparing to fight with swords; from *CCT*, p. 7*b*.

with swords.[a] We have already seen one or two illustrations from Chao Shih-Chên's[1] book, in Fig. 56 above, for fire-lances were still in use in his time, contemporaneously with the siege of Malta. The book by Chhen Phei[2] must belong to this same date, as one can see from its title *Huo Chhê Chen Thu Shuo*[3] (Illustrated Account of the Formations in which Mobile Shields can be used with Guns and Cannon).[b]

A rather different tank-prototype can be seen in some of the illustrations in the *Thai Tsu Shih Lu Thu* of +1635, namely mobile ramparts borne on two wheels and pushed by two men using four poles, with a platform on which a couple of musketeers could fire through crenellations (Fig. 163). This picture shows the defeat of the troops of Tung Chung-Kuei[4] by Nurhachi's men, but the former are not visible, and the platforms are in the act of being overtaken by the Manchu cavalry and mounted archers. By now the Manchus too are using muskets.[c]

Lastly a word must be said about the 'rapid thunder gun' (*hsün lei chhung*[6]) described in the *Shen Chhi Phu*[7] of +1598, also due to Chao Shih-Chên. Five barrels were fixed through a round shield, with a rotating stock in their midst which would bring the serpentine into five separate positions for touching off each barrel in turn (Fig. 164).[d] This ribaudequin itself is seen in an accompanying picture (Fig. 165). While firing, this multiple matchlock musket rested on the handle of an axe fixed in the ground, and the end of the stock took the form of a pointed spear, so that the soldier could defend himself with these weapons if the worst came to the worst. The barrels were only about 2 ft long, and each one was provided with its own fore-sight and back-sight. A model reconstruction is seen in Fig. 166.[e] This arrangement is very reminiscent of the discoidal pistol-shields used by Henry VIII's bodyguard and now in the Tower of London Armouries (Fig. 167), each shield having only one matchlock barrel at the centre.[f]

This discussion of shields has taken us into the territory of small arms such as muskets, as well as that of field-guns and artillery. We must now turn our attention briefly to the former subject in particular.

---

[a] The laager tactics of the +16th-century Chinese armies are irresistibly reminiscent of the methods of the great Hussite general Jan Žiska in Bohemia and Germany during the first half of the +15th. The 'battle-wagons' which carried cannon as well as folding ramparts of oak, and linked up to form a defensive square, circle or triangle upon the word of command, brought the Hussites and Taborites great success for many years; cf. Oman (1), vol. 2, pp. 361 ff.; Delbrück (1), vol. 3, pp. 497 ff; Denis (1). But the inspiration for the *Wagenburg* seems to have come from Russia, where the *goliaigorod* or movable city had been known and used long before. If this was so, could it not have been originally Mongolian? Perhaps further research will show that the ideas of Žiska and Chhi Chi-Kuang had both the same root.

Hall (2) says that the MS of the 'Anonymous Hussite' dates not from +1430 as usually thought, but consists of two parts, dating from +1470 and +1490 respectively, therefore after Taccola and Fontana. But this does not affect the present argument.

[b] In Wang Ming-Hao's *Huo Kung Wên Ta* of *c.* +1598 too, much is also said of gun-carriages and mobile shields forming laagers (pp. 1306 ff.).

[c] Another illustration in ch. 6 also shows these platforms, depicting the defeat of the Ming troops of Chhen Tshê[5].

[d] P. 22b.   [e] By Mr S. Videau of Brisbane, Australia.   [f] Reid (1), p. 107; Blackmore (4), p. 14.

[1] 趙士禎   [2] 陳棐   [3] 火車陣圖說   [4] 董仲貴   [5] 陳策
[6] 迅雷銃   [7] 神器譜

Fig. 163. Mobile crenellated rampart platforms used by the Manchu troops (musketeers now as well as archers) in their defeat of the Ming brigade under Tung Chung-Kuei, c. +1620. From *TTSLT*, (no. 8).

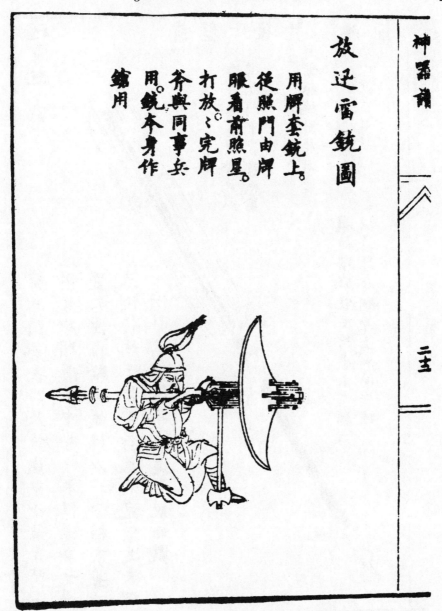

放迅雷銃圖

用腳套銃上。
從照門由砰
眼看前照星。
打放〈完砰
芥興同事兵
用銃本身作
鏡用

神器譜

Fig. 164. The 'rapid-fire thunder gun' (*hsün lei chhung*), a five-barrelled ribaudequin fired by a Ming gunner protected by a shield. From *Shen Chhi Phu*, p. 22*b* (+1598).

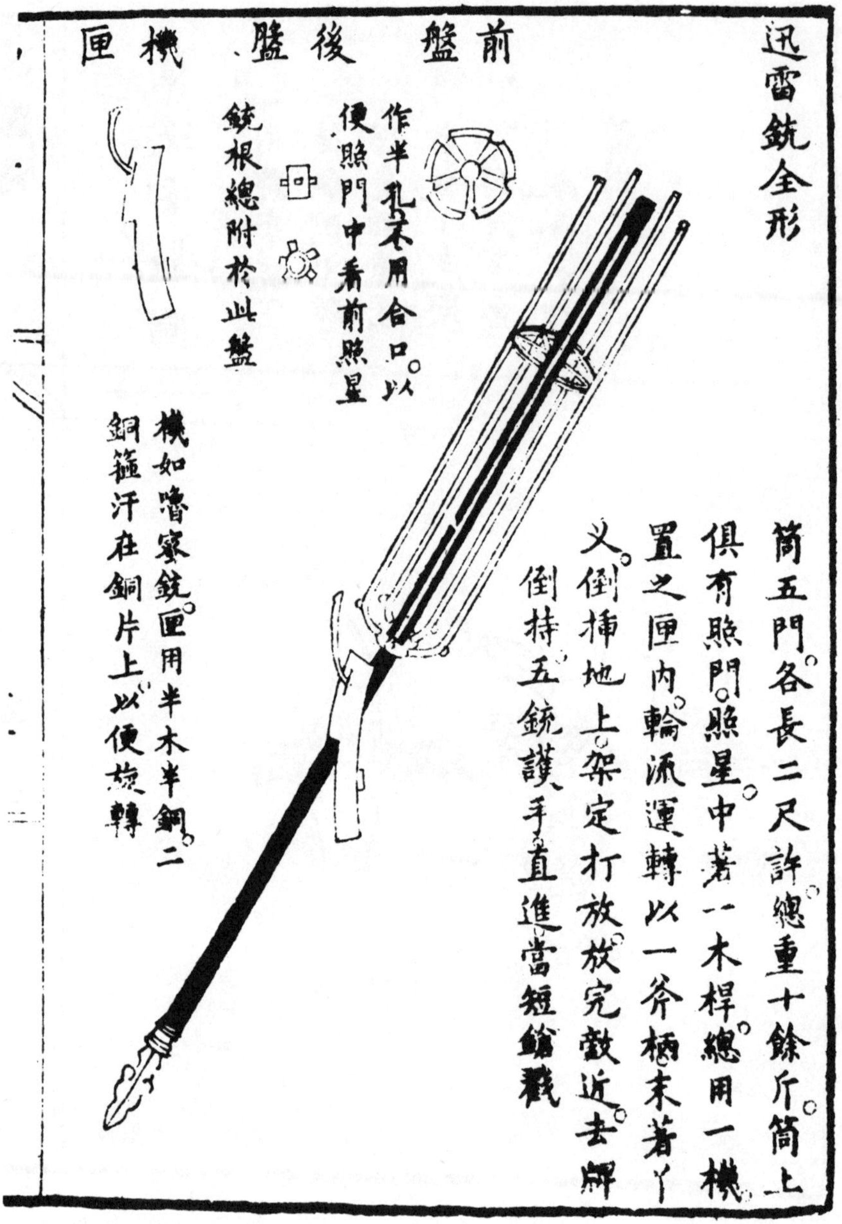
迅雷銃全形
前盤
後盤
機匣
作半孔不用合口以便照門中看前照星
銃根總附於此盤
筒五門各長二尺許總重十餘斤筒上俱有照門照星中著一木桿總用一機置之匣內輪流運轉以一斧柄末著丫叉倒插地上架定打放放完斂近去牌倒持五銃護手直進當短鎗戳
機如魯密銃匣用半木半銅二銅箍汗在銅片上以便旋轉

Fig. 165. Detailed view of the same weapon, also from *SCP*. A rotating stock in the midst of the barrels permitted the serpentine to touch off each barrel in turn.

Fig. 166. Model of this ribaudequin and shield made by S. Videau of Brisbane.

(18) LATER DEVELOPMENTS IN HAND-GUNS; THE ARQUEBUS AND THE MUSKET

(i) *Matchlocks, wheel-locks and flintlocks*

Before taking a brief view of the later history of portable fire-arms in China it is well to be clear about what happened in Europe, between the simple hand-guns of the +14th and +15th centuries, and the era of the cartridge and percussion-cap from the end of the +18th century onwards. This was the period of the arquebus and the musket. The simplest hand-guns, which go back in China, as we have seen (p. 294) into the last decades of the +13th century, consisted of nothing but the muzzle-loaded barrel, the touch-hole for the slow-match, and a socket with a wooden handle (the tiller) fixed into it. They must have been exceedingly difficult to hold, aim, and ignite effectively, all at the same time; so that from about +1400 a Z- or S-shaped lever (the serpentine)[a] was pivoted in the stock (which now began to be shaped to fit the shoulder) in such a way as to bring the burning slow-match (which it held in its jaws) near to the touch-hole (or priming-pan, filled with gunpowder, just above it).[b] The oldest depiction of this dates from +1411. All the technical terms are a little vague, partly because

---

[a] See Blackmore (1), p. 9; Reid (1), pp. 58–9. This was the origin of all triggers.
[b] Hence the idiomatic expression 'a flash in the pan'—just that and nothing more.

Fig. 167a. An example of parallel thinking in the +16th century; a shield of wood faced with steel and containing a matchlock pistol at the centre, said to have been carried by the bodyguard of Henry VIII, c. +1530 to 1540. Photo. Tower of London Armouries.

their contemporary use fluctuated like that of all pre-modern terms, and partly because of the variable uses of modern gun historians; but 'arquebus' belongs somewhere here, though in fact the name arose, with all its variants (such as hackbut, hargabush, etc.) from the German *Hakenbüchse*, i.e. a hand-gun with a hook-like projection or lug on its under surface, useful for steadying it against battlements or other objects when firing.[a]

---

[a] It thus had nothing to do with Latin *arcus* or *arcuballista*, as some +17th-century etymologists thought.

Fig. 167b. Rear view of the pistol shield (photo. Tower of London Armouries). Cf. Reid (1), p. 107; Blackmore (4), p. 14.

Although the match was thus held in a holder, and moved to ignite the charge at will, the true matchlock[a] had not yet been born. The ingenuity of the lock-smiths was still needed to fashion it,[b] and this they effected from about +1475 onwards. The smouldering match was now held by some kind of vice at the head of a curved lever called the cock[c] (which could face either forwards or backwards according to the design), and which was connected with the trigger by a series of

---

[a] See Blackmore (1), pp. 10 ff. (4), pp. 12 ff.; Reid (1), pp. 60–1, 122, 134–5; Pollard (1), pp. 6 ff.

[b] On the locksmith's art see Vol. 4, pt. 2, pp. 263 ff. Hall (5), p. 354 well emphasised the close relation of gun-locks to door- and coffer-locks, with their frame-plates, springs, levers and fixing-pins. One would like to know more about the social relations of musketeers and locksmiths at that time.

[c] Hence the idiomatic expression 'going off at half-cock', for forcible–feeble actions.

detents, working in many different ways, but generally including springs, sear levers,[a] tumblers, notches, lugs and the like. By about +1575 a trigger-guard was often added. The earlier form was known as a snap-matchlock because the cock was forced down on to the priming-powder by a spring; the latter, safer form, or sear-matchlock, had a spring which held the cock back until the trigger was pulled and a catch dislodged. In the +16th century matchlock muskets were heavy, weighing up to 20 lb., and had to be fired from a rest, i.e. a wooden pole with a Y-shaped piece of iron at its upper end.

The matchlock system lasted on a very long time, as we shall see, but it was deeply unsatisfactory if only because of the problem of keeping the slow-match (generally a hempen rope saturated with saltpetre) glowing in damp or wet weather. This was therefore thrown aside in favour of striking a spark each time from flint and steel. Two systems came into existence almost but not quite contemporaneously, first the wheel-lock from about +1530 onwards,[b] and then the flintlock from about +1550 onwards. In the wheel-lock[c] a V-shaped spring was connected by a chain to the spindle of a wheel, which, on being released, rotated very like that in a modern cigarette-lighter, and struck sparks from a piece of iron pyrites held in the jaws of the cock.[d] Since there was no small danger that one would lose the key for winding the mechanism up again, a rack-and-pinion device was presently developed to give a self-spanned wheel-lock.[e] But the wheel-lock system was always rather costly, sometimes including as many as fifty separate components.

As for the flint-lock musket,[f] it did away with the wheel, and arranged for the cock to hold a flint (in its screw-jaws) which came down upon a piece of steel, and ignited the powder in the priming-pan by the sparks so struck off.[g] We can now use the word 'musket' without hesitation[h] because it came into use first about +1550, and lasted on into the nineteenth century.[i] The piece of steel (also called hammer, battery or frizzen) could be in one unit with the priming-pan cover, and knocked aside by the cock as it descended; or it could be separate from it, and these latter types have today the name of snaphance muskets,[j] though in former times this term applied to both. These arrangements, like those

---

[a] Sear or sere implies locks and bolts, from OF *serre, serrer*.

[b] Allegedly +1517, but however this may be, Leonardo da Vinci was making drawings of flint-and-steel gun-locks as well as tinder lighters by about +1500. Of course the idea took time to spread. See Blair (1).

[c] See Blackmore (1), pp. 19 ff. (4), pp. 29 ff.; Reid (1), pp. 92 ff.; Pollard (1), pp. 19 ff.

[d] By a screw-tightened vice, interesting because it was a second application of the screw principle in fire-arms. The first had been the screw-in breech (p. 436). A third seems to have appeared in China (cf. p. 410 above).

[e] The rack-and-pinion appears again, also in a different context (p. 446 below).

[f] See Blackmore (1), pp. 28 ff., (4), pp. 61 ff.; Reid (1), pp. 116–17, 125 ff., 128, 141–2, 146, 148; Pollard (1), pp. 32 ff.

[g] As Blackmore (4), p. 61 has pointed out, the French word for any gun, *fusil*, comes from the Ital. *focile*, meaning a piece of steel for striking sparks from flint, cognate with Lat. *focus*.

[h] The name originated from Ital. *moschetto* (sparrow-hawk), a word allied to mosquito, from Lat. *musca*, a fly.

[i] They were still being made in Napoleonic times, e.g. 1810 (Reid (1), pp. 168–9), and for tripwire spring-guns even later (*ibid.* p. 185).

[j] From Dutch *snaphaan*, Ger. *Schnapphahn*, the action of a pecking hen, so similar to the motion of the cock as it descended.

of the wheel-lock, were a considerable help against rain or mist, and as the flintlock was a good infantry weapon, it completely displaced the matchlock by +1725 or so. In a similar way, the wheel-lock had been the limiting factor for cavalry pistols and carbines, as was seen in the English Civil War of the mid-seventeenth century.[a]

Of course all the weapons we have been discussing were muzzle-loaded, and the application of the breech-loading principle to portable firearms came about only slowly.[b] True, Leonardo had sketched an arquebus with a screw-on chamber in the Codex Atlanticus, but the idea was not very practical, and the first breech-loaders were wheel-lock guns of about +1650. In these the barrel unscrewed, hence the name 'turn-off' muskets. Only in China, so far as we know, were chambers like those of the *fo-lang-chi* culverins applied to portable guns (p. 380 and Figs. 143, 144). All this meant that the matchlock and even the two spark-producing types were very slow in firing, one round in 5 to 15 minutes depending on the skill or clumsiness of the musketeer.[c] Of course this could be compensated by having large numbers of them, like the 10,000 at the Battle of Nagashino[1] in Japan in +1575.[d]

(ii) *The musket in China and Japan*

What happened with artillery now happened with portable firearms too; the improved devices of the Western world were transmitted, and in more ways than one, to the Chinese culture-area. The usual view has been that matchlock muskets of Portuguese origin were acquired by the Chinese military from Japanese pirates (*wo khou*[2]) on their coasts about +1548. Indeed, Chhi Chi-Kuang says so clearly.[e] These raids were very severe throughout the forties and fifties of the century; in +1546 and +1552 much of Chekiang was devastated, in +1555 Suchow and Nanking were besieged, and in +1562 there was much fighting in Fukien.[f] Evidently we need to take a close look at how the Japanese themselves had acquired these muskets.

It seems historically established that in the year +1543 two Portuguese adventurers were shipwrecked on the island of Tanegashima[3], just south of the southernmost tip of the Japanese mainlands.[g] We can identify the name of one of

---

[a] Among the many later elaborations of the flintlock was the single-set or hair-trigger, set by pushing it forward, so that an extremely slight pull would set the levers and springs in motion to bring down the cock and fire. Pistols of this kind, used in the celebrated duel of 1804 between the two American politicians Aaron Burr and Alexander Hamilton, explain what then happened (Lindsay, 2).

[b] See Blackmore (1), pp. 58 ff. (4), pp. 147 ff.   [c] Nef (1), p. 32.

[d] Cf. Perrin (1), pp. 19, 98; Turnbull (1), pp. 156 ff.

[e] *Lien Ping Shih Chi (Tsa Chi)*, ch. 5, p. 22 b (p. 239). This was in +1568. Cf. Huang Jen-Yü (5), pp. 165, 250. Lang Ying, in *Chhi Hsiu Lei Kao*, ch. 45 (p. 662) makes the same statement, perhaps rather earlier, say +1558.

[f] Cordier (1), vol. 4, pp. 60–1.

[g] The best account is that of Arima (1), pp. 615 ff. Cf. Boxer (7), pp. 26–8; Perrin (1), p. x. The earliest Portuguese version, that of Fernão Mendes Pinto in +1614 (see Vol. 4, pt. 3, p. 535) is considered to be partly fictional.

[1] 長篠   [2] 倭寇   [3] 種子島

them, Kirishitadamōta[1], as Christopher da Mota;[a] the name of the other remains only in Japanese, Murashukusha[2].[b] The matchlocks which they carried, explained and demonstrated,[c] interested greatly the lord of the island, Tokitaka[3]; and largely through his efforts only a few years passed before the Japanese smiths were able to make such muskets themselves.[d] At first they were called by the name of the island, but soon the expression *teppo*[6] (*thieh phao*[6]) became universal. All this information comes to us partly from an almost contemporary source, the *Teppō-ki*[7] (Record of Iron Guns)[e] written by a monk, Nampo Bunshi[8], in +1606 though not printed till +1649.[f] Exactly a hundred years earlier than this last date the great unifier Oda Nobunaga[9] had ordered 500 matchlock muskets for his army,[g] and in +1560 at the siege of Marune[10] a Japanese general had been killed by a matchlock bullet.[h] The weapon was thus by this time well established in the country.[i]

The two Portuguese have earned a certain immortality, but one should not assume that their guns were the first which the Japanese had ever seen or known. There is some evidence that hand-guns in small numbers reached Japan from China during the +15th century;[j] and a well-authenticated case is the presentation of a *thieh phao*[6] to a feudal lord in +1510 by a Buddhist priest recently returned from China.[k] By this time a hand-gun would be far more likely than a cast-iron bomb. It is interesting that the lord in question was Hōjō Ujitsuna[11], a doughty warrior and general but also a great proponent and establisher of peace throughout his extensive domains.[l]

But how sure is it that the muskets of the Chinese were derived from Japan?

---

[a] Or perhaps 'the Christian', since his chief given name seems to have been Antonio.

[b] The Deshima Museum at Nagasaki identifies three names: Antonio da Mota, Francisco Zeimoto and Antonio Peixoto.

[c] By the aid of a Chinese interpreter, known to us only as Gohō.[4]

[d] The first of these was probably Yasaka Kinbei Kiyosado[5], an artisan metal-worker in the service of Tokitaka himself.

[e] Tr. Kikuoka Tadashi (1). Cf. the book of Hara Tomio (2).

[f] See Louis-Frédéric (1), pp. 172–3; Boxer (7), *loc. cit.*; Arima (1), p. 617. A full-page illustration of a musket from this book is reproduced by Arima, *op. cit.* fig. 273, p. 635.

[g] It will be remembered that after the seemingly endless wars between the feudal principalities during the Age of Strife (Senjoku Jidai[12], +1490 to +1600), the country was unified by Oda Nobunaga[9] (+1534 to 82) and Toyotomi Hideyoshi[13] (+1536 to 98), on whom see Dening (1). It was Tokugawa Ieyasu[14] (+1542 to +1616) who entered into this heritage with the founding of the Tokugawa Shogunate from +1603 onwards; on him see Sadler (1). Hara Tomio (1) has maintained, not implausibly, that the musket was a fundamental instrument of this unification.

[h] Sadler (1), p. 53.

[i] We shall have something more to say further on (p. 467) about the century of the musket in Japan. There is a brief account in Sugimoto & Swain (1), pp. 170ff. See also Okamura Shōji (1).

[j] See the studies of Nakamura Kenkai (1, 2, 3).

[k] The story comes in the *Hōjō Godai-ki*[15] (Chronicles of the Hōjō Family through Five Generations), in *Shiseki Shūran*[16], vol. 5, ch. 26, pp. 58–60. Cf. D. M. Brown (1), pp. 236–7.

[l] Papinot (1), p. 169; +1487 to +1541.

[1] 喜利志多佗孟大　　[2] 牟艮叔舍　　[3] 時堯　　[4] 五峯
[5] 八板金兵衛清定　　[6] 鐵砲　　[7] 鐵砲記　　[8] 南浦文之
[9] 織田信長　　[10] 丸根　　[11] 北條氏綱　　[12] 戰國時代　　[13] 豐臣秀吉
[14] 德川家康　　[15] 北條五代記　　[16] 史籍集覽

Quite a lot of information is available about the events of +1548, and Chêng Jo-Tsêng, who was in a position to know, wrote only a dozen years later that they came direct from the Western barbarians (Hsi Fan[1]) rather than the Japanese ones (Wo I[2]).[a] In the year in question the Governor of Chekiang and Fukien was the Censor Chu Wan[3] (+1494 to +1550), and under him he had an energetic young military commander, Lu Thang[4] (+1520 to c. +1570). At a certain point this brigadier attacked a pirate lair at Shuang-hsü-kang[5] near Tinghai, and reduced it, capturing many persons, including eleven Portuguese who had good muskets.[b] They were merchants, but not above a bit of smuggling when convenient. Chu thereupon ordered the volunteer officers Ma Hsien[6] to get the smiths to copy these guns, and Li Huai[7] to have the proper powder for them made, in which matter they succeeded well, so that the new weapons were even superior to the foreign ones. Yet it is also recorded that in the same year there came an official Japanese embassy or tribute mission led by the Buddhist cleric Sakugen Shuryō[8] (+1501 to +1579);[c] and Chu Wan gave them very hospitable treatment. So the question can hardly be settled, though we suspect that Chêng Jo-Tsêng knew what he was talking about.

There is another passage of Chêng Jo-Tsêng which may be worthy of more notice than it has yet received.[d] In his work of +1566, the *Chiang-Nan Ching Lüeh*[9] (Military Strategies South of the River) he wrote:[e]

Our first emperor Thai Tsu (Chu Yuan-Chang), because of his remarkable military accomplishments, gained control of the whole Middle Kingdom.[f] He possessed every sort of fire-weapon in existence from past to present, and kept them in his armouries. Every year when the Magically effective Weapons Brigade (Shen Chi Ying[10]) held its exercises, the names and appearances of most (of the weapons on display) were quite unfamiliar to the onlookers. And yet they were only several hundred types (out of what were stored in the armouries). People nowadays all say that the *fo-lang-chi* cannon[g] and the bird-beaked musket both came from foreign ships (i.e. from the Portuguese adventurers and the Japanese pirates respectively). But I once heard the Adjutant-Commander (Tshan Chiang[11]) Chhi Chi-Kuang[12] say that when formerly he held a garrison post in Shantung he excavated a pit and found a *fo-lang-chi* cannon. The date could be checked from the inscription (on the barrel) showing the year and month when it was cast and kept (in the armoury) of the Yung-Lo emperor Chhêng Tsu (r. +1403 to 24). Again, he also found (some) bird-beaked muskets in the garrison armoury. So these things must have been already possessed by the Middle Kingdom before the time of the Japanese pirates.

[a] *Chhou Hai Thu Pien*, ch. 13, p. 39a.
[b] The Dominican Gaspar da Cruz gave an extended account of the incident, translated in Boxer (1), pp. 194 ff., but says nothing about muskets. Law and administration seem to have interested him more than arms.
[c] The mission was quite sizeable, in four ships with more than 400 men.
[d] It was noted by Wang Ling (1), p. 175, but without reference.
[e] Ch. 8, p. 3a, tr. auct.   [f] Cf. p. 30 above.   [g] Cf. p. 369 above.

[1] 西番	[2] 倭夷	[3] 朱紈	[4] 盧鏜	[5] 雙嶼港
[6] 馬憲	[7] 李槐	[8] 策彥周良	[9] 江南經略	[10] 神機營
[11] 參將	[12] 戚繼光			

This is a rather precious relic of a conversation between the great geographer and the great general. There is no reason for doubting that Chhi Chi-Kuang found an old disused cannon in Shantung, and as we know from Table 1, there are plenty of inscribed artillery pieces from the Yung-Lo emperor's time still extant today; but Chêng Jo-Tsêng must have understood him wrongly in supposing it to have been a breech-loader. On the other hand, although matchlock muskets could hardly have existed in the Yung-Lo reign-period, there is a suggestion here that in fact they were known in China before the Japanese–Portuguese contact, and indeed this will shortly appear.

From the first coming of the musket into southern China, it was dubbed the 'bird-gun' (*niao chhung*[1]) or 'bird-beak gun' (*niao tsui chhung*[2]). Sometimes the word *chhung*[1] was reserved for guns with the shorter barrels, and those with longer ones were called *chhiang*[3].[a] The term bird-beak must have been derived, one would think, from the pecking action of the cock that held the match, paralleling the term 'snaphance' which developed in the West (cf. p. 428 above).[b] But there is some authority for the view that the reference to birds arose because of the use of muskets as fowling-pieces.[c] It has also been supposed that the stocks, which tended in China to be short like pistol-grips (cf. Fig. 168), might have been likened to bird-beaks—but the cock derivation is probably the right one.

The first illustrations and descriptions appeared in two books printed in the same year (+1562), Chêng Jo-Tsêng's *Chhou Hai Thu Pien*[d] and Chhi Chi-Kuang's *Chi Hsiao Hsin Shu*.[e] Fig. 168, taken from the former, shows the general view, and Fig. 169 the lock, of the bird-beak gun. The stock is called *mu chia*[4], the spring and trigger *kuei chhêng*[5],[f] while the forward-falling cock is a 'dragon-head' (*lung thou*[6])[g] with the spring acting on its other end (*lung wei*[7]) after release from a sear lever (*kou*[8]).[h] Figs. 170 and 171 give further details. The touch-hole is called *huo mên*[10] and the fence or pan-guard[i] *huo mên kai*[11], while the ramrod is the *chha*

---

[a]　As by Sung Ying-Hsing in *Thien Kung Khai Wu*, p. 438 below.

[b]　Davis & Ware (1), p. 536, got this right. Mayers (6), p. 98 had supposed that the name referred to the flared mouths of blunderbusses.

[c]　See Chhi Chi-Kuang in *Lien Ping Shih Chi (Tsa Chi)*, ch. 5, p. 22*b*, and Mao Yuan-I in *Wu Pei Chih*, ch. 124, p. 6*b*.

[d]　Ch. 13, pp. 36*a* ff. Similar illustrations with a different text are in *WPC*, ch. 124, pp. 2*a* ff. Chêng quotes Chhi at one place (p. 38*b*) because the latter had been writing a couple of years earlier.

[e]　Ch. 15, pp. 9*a* ff., with rather clear illustrations.

[f]　This usage arose from the old Neo-Confucian identification of *kuei*[5] (anciently 'devil') with all forms of compression and contraction (cf. Vol. 2, p. 490). *Shen*[9] (anciently 'spirit') comprised all forms of dispersion and relaxation.

[g]　This is the term later adopted in Chinese for a water-tap. One would have expected 'bird-beak' rather than 'dragon-head'.

[h]　From the diagrams in Figs. 169 and 170 it looks as though these locks were of the type known as snap matchlocks. Here the trigger is quite separate from the cock, and pulls back a horizontal sear which has previously retained its toe (or tail). The long-armed U-shaped spring which then drives the cock down can be well seen in both diagrams, though the artists did not draw the whole mechanism very clearly.

[i]　Cf. Blackmore (1), pp. 30, 32.

[1] 鳥銃　　　[2] 鳥嘴銃　　　[3] 鎗　　　[4] 木架　　　[5] 鬼撐
[6] 龍頭　　　[7] 龍尾　　　[8] 勾　　　[9] 神　　　[10] 火門
[11] 火門蓋

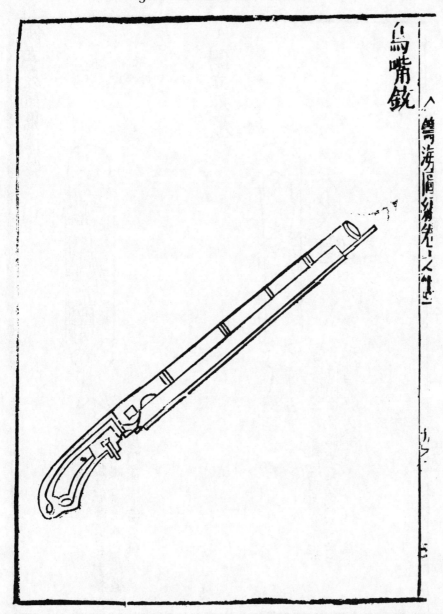

鳥嘴銃

Fig. 168. The 'bird-beak musket' (*niao tsui chhung*), an illustration of +1562, from *CHTP*, ch. 13, p. 36*b*.

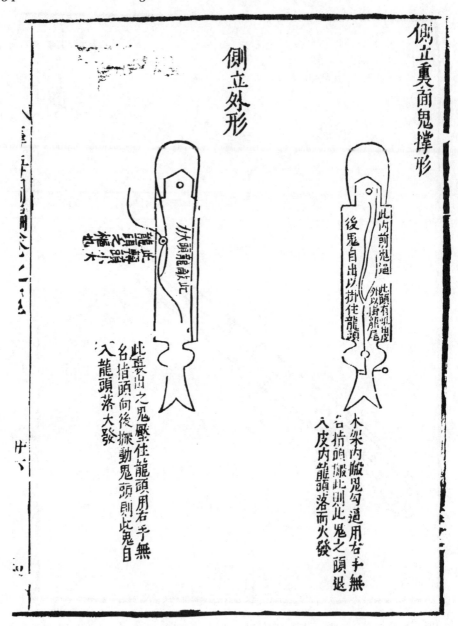

Fig. 169. The lock of the bird-beak matchlock musket, with its system of springs, from *CHTP*, ch. 13, p. 36*a*.

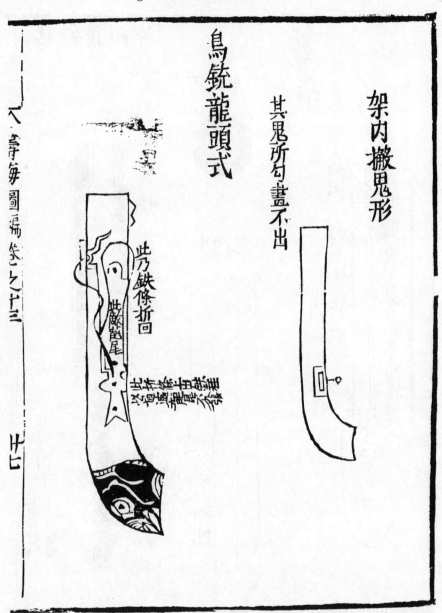

Fig. 170. Another drawing of the same (*CHTP*, ch. 13, p. 37*a*). For explanation see text.

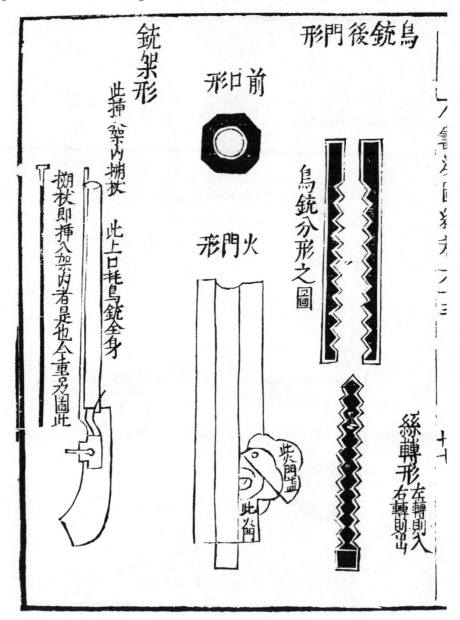

Fig. 171. Depiction of the breech-screw of the bird-beak matchlock musket, from *CHTP*, ch. 13, p. 37*b*. In Fig. 416 (p. 121) of Vol. 4, pt. 2, we gave the similar picture from the *San Tshai Thu Hui* encyclopaedia of +1609, calling it the first Chinese diagram of a continuous screw or worm with its male and female threads, but *CHTP* is just over forty years earlier. The matter has some importance because although tangent-plane helicoidal structures were known and used in ancient and medieval China, the screw as such was one of the rare mechanical devices not indigenous to that civilisation.

shuo chang[1].[a] The most curious feature is the screw, or 'turning coiled silk-thread' (*lo ssu chuan*[2]) used for plugging the breech (*hou mên*[3]).[b] The illustration is similar to that in the *San Tshai Thu Hui*[4] encyclopaedia (+1609) which we gave at an earlier stage,[c] but pre-dates it by almost half a century. It is important because the screw was one of the rare mechanical devices not indigenous to Chinese civilisation,[d] and its appearance here shows the careful way in which the musket was being copied. In flintlocks it found still another use,[e] but the question of flintlocks in East Asia must be postponed for a few pages.[f]

Old Chinese matchlock muskets are rare in Western collections, but we can show one (Fig. 172*a*, *b*) which is in the Maidstone Museum in England.

In +1637 Sung Ying-Hsing gave an intelligent layman's account of matchlock muskets and their making. He said:[g]

The bird-gun (*niao chhung*[7])[h] is about three feet long, an iron pipe containing gunpowder, and inset in a wooden holder so that it can be conveniently grasped by the hands. To make the bird-gun tube, an iron rod of the size of a chopstick is used as the cold core, and three strips of extremely red-hot iron are forged and welded together around it longitudinally.[i] Then the interior of the barrel wall is highly smoothed by drilling with a four-edged steel reamer of the diameter of a chopstick, so that frictional resistance to the discharge is minimised. The bore at the back end is larger than that at the muzzle,[j] so that it can hold the gunpowder.

Each bird-gun is loaded with about 0·12 oz. of gunpowder[k] and 0·2 oz. of lead or iron bullets. No fuse is used to ignite the gunpowder. [*Comm.* Except in South China sometimes.][l] but instead a touch-hole (*khung khou thung nei*[8]) is filled with gunpowder, and a slow-match made of ramie or hemp (*chhu ma*[9]) is employed for the ignition. The musketeer holds the bird-gun in his left hand, and points it at the enemy, then with his right

[a] The fore and back sights above the barrel were *chhien hsing*[5] and *hou hsing*[6].
[b] See *CHHS*, ch. 15, p. 11*b*, and even more explicitly *WPC*, ch. 124, p. 2*a*. Cf. Blackmore (1), p. 9. It is curious because the Japanese gunsmiths consistently avoided using screws (Blackmore, p. 17); and this might be an argument for direct acquisition from the Portuguese. It is also curious because the screw-thread shown is left-handed, and you turned it to the right, not the left (as our common usage is), to screw it out.
[c] Vol. 4, pt. 2, Fig. 416.
[d] Tangent-plane helicoid structures were, however, known and much used. Cf. Vol. 4, pt. 2, pp. 119 ff.
[e] Apart from the screw-plug of the breech, and the screws which held the tang or tangs of the barrel in place on the stock, the screw came in again as a vice to keep the flint tight in the jaws of the cock. Cf. Blackmore (1), pp. 17, 43 (4), p. 23.
[f] See p. 465 below.
[g] *TKKW*, ch. 15, p. 8*a*, *b*, tr. auct., adjuv. Sun & Sun (1), pp. 272, 276 and Li Chhiao-Phing (2), pp. 394–5.
[h] We cannot accept the term 'bird-pistol' used by Sun & Sun, nor the 'fowling-piece or sporting-gun' proposed by Li Chhiao-Phing. Enemy soldiers are distinctly mentioned; furthermore, the illustration (Fig. 173) shows them.
[i] Sun & Sun interpreted this as three sections of tube welded together end on, but this cannot be right.
[j] One wonders whether he meant to say that the wall was made thicker at that point, as it had been in the earliest vase-shaped cannon (cf. p. 287 above), thus strengthening the explosion chamber.
[k] The word actually used here is *hsiao*[10], nitre, saltpetre, but probably Sung was only using it in order to avoid saying gunpowder so often, according to the 'principle of elegant variation'.
[l] One wonders whether this means that a serpentine was employed in the south at times, or even a free match rather than the developed matchlock.

[1] 揷搠杖	[2] 螺絲轉	[3] 後門	[4] 三才圖會	[5] 前星
[6] 後星	[7] 鳥銃	[8] 孔口通內	[9] 苧麻	[10] 消

Fig. 172a. The barrel of a Chinese bird-beak matchlock gun now in the Maidstone Museum (photo. Tower of London Armouries). The inscription says: 'You press the trigger as if it was a crossbow, but the impact is better than that of any arrow.'

Fig. 172b. Rear part of the barrel, and trigger, of the Maidstone Museum gun (photo. Tower of London Armouries). The three medallions from above downwards indicate the three characteristic Chinese felicities, *fu*, blessedness or happiness, *lu*, rise to high official position, and *shou*, longevity. But here there are in the order *fu*, *shou*, *lu*, the first giving the character, the second symbolised by a crane-bird (*hao*) and the third symbolised by a deer (*lu*) because of the pun on its name.

hand he pulls the iron trigger (*fa thieh chi*[1]) (of the gun-lock), thus bringing the glowing hemp to the top of the (touch-hole filled with) gunpowder. The gun then fires.

Sparrows and other birds when struck by the bullets within a distance of 30 paces are shattered all to fragments; but those more than 50 paces away are killed without being destroyed. At a distance of some 100 paces[a] the force of the bird-gun is almost exhausted.

The long bird-gun (*niao chhiang*[2]) has a further range, about 200 paces. It is constructed in the same manner as the shorter bird-gun, but its barrel is of greater length, and it needs about double the amount of gunpowder.

More professional, one might say, were the comments of the general Chhi Chi-Kuang,[b] who remarked that the only sound way to make musket barrels was to

---

[a] I.e. 500 ft, rather under 200 yds.
[b] In *WPC*, ch. 124, p. 4*b* (+1621). Chhi's accounts of muskets can be found in *CHHS*, ch. 15, pp. 9*a*–11*b* (+1560) and *LPSC/TC*, ch. 5, pp. 21*b*–23*a* (+1568).

[1] 發鐵機　　　[2] 鳥鎗

鳥銃

Fig. 173. The bird-beak matchlock musket in use, a picture from *TKKW*, ch. 3 (ch. 15), p. 35 *b* (Ming ed.).
The fleeing enemy are probably Miao tribesmen.

Ssu-Ma also said: 'I have received great favour by being looked after all through three reigns[a] and I have often worried that I could see no way to express my gratitude. I should be delighted if I could get an opportunity to spread the design (of this weapon), so as to add to the military power of the Imperial Court.' He then explained to me the technique of making (the Turkish musket).

After this I disbursed funds and employed smiths to manufacture this weapon. I showed the product to (To) Ssu-Ma and it met with his approval. During my younger days I often observed musketeers in combat not being able to recharge their powder and shot in time, and as a result being taken advantage of by the enemy. I therefore deliberated over something between the Western musket and the *fo-lang-chi* cannon, and so made the 'gripped-lightning musket' (*chhê tien chhung*[1]).[b] Similarly, as an intermediate between the musket and the 'three-eye gun' (*san yen chhung*[2])[c] I made the 'fast thunder gun' (*hsün lei chhung*[3]).[d] I think that on the battlefield, besides the larger firearms like the 'third general' (*san chiang-chün*[4]), the *fo-lang-chi* (breech-loading) cannon, and the 'thousand-*li* thunder' (*chhien li lei*[5]), among the small firearms nothing has a greater range and does more destruction than the Rūm (Turkish) musket, and next to it is the Western (Portuguese) musket.

All this requires a little commentary. We have often come across Byzantium before, under its early medieval Chinese name of Fu-Lin[6], transliterating Frōm and Hrom, the names that Eastern Rome had acquired in passing further east across the Old World.[e] In the Thang it replaced the earlier name Ta-Chhin[7] which the Han people had used for Roman Syria.[f] Then after the Arab conquests got under way, and Asia Minor as well as the Great City (after +1453) had become Rūm, the transliteration changed again, becoming Lu-Mi[8], as in the passage just given. Many masters in Islam were from Rūm; one thinks at once of opposite extremes—Jalāl al-Dīn Rūmī (+1207–73), perhaps the greatest of mystical theologians and poet-philosophers,[g] and Muḥammad al-Rūmī, doubtless a Turk, who cast a howitzer of enormous size for the Mughal emperor Humāyūn in India in +1548 (again that fateful year).[h]

The surmise of Chao Shih-Chên's grandfather that the Turkish culture was the origin of all muskets, though at first sight bizarre, is not to be lightly dis-

---

[a] This would mean the Lung-Chhing and Chia-Ching reign-periods as well as Wan-Li, therefore taking us back as far as +1522.

[b] Clearly this was a breech-loader with a number of replaceable chambers, cf. pp. 380ff. Fig. 143.

[c] Often used as a signal-gun, cf. p. 331 above, and Fig. 112. It was a triple-barrel gun used on horseback in northern China.

[d] We have come across this before (p. 421, Fig. 165) in connection with shields for infantrymen and gunners.

[e] Vol. I, pp. 186, 205. Among the many Byzantine embassies there was one in +1371 led by Nicholas Comanos (Nieh-Ku-Lun[9]) who was accompanied back some part of the way by a Chinese ambassador named Phu La[10]. We have often wondered whether part of his mission did not concern the know-how of gunpowder weapons (by then real cannon) which might be acquired from the Chinese. See *Ming Shih*, ch. 326, p. 17b, and Vol. I, p. 206.

[f] Vol. I, p. 174.

[g] Born at Balkh, but his ancestors must have been from Rūm.

[h] Partington (5), p. 220.

---

[1] 掣電銃	[2] 三眼銃	[3] 迅雷銃	[4] 三將軍	[5] 千里雷
[6] 拂林	[7] 大秦	[8] 魯迷	[9] 捏古倫	[10] 普剌

missed. At the least we can say that the Turks had matchlocks about as early as the Europeans did. Remembering (from p. 425 above) that the true musket developed from the match-holding serpentine arquebus a little before +1475 in the West, we find philological arguments that this change had occurred among the Turks by +1465.[a] Other evidence puts its beginnings in the Egyptian Mamlūk kingdom under al-Ashraf Saif-al-Dīn Qāyt Bey (r. +1468 to 95), with dates such as +1489 and +1497.[b] Matchlocks were prominent in the battles of +1514 between Turks and Mamlūks. After all, if we are right (cf. pp. 573–6) in regarding the Islamic culture-area as the principal way-station between China and Europe in the transmission of all gunpowder weapons, an early Turkish expertise in portable firearms would be natural enough. So perhaps Chao Hsing-Lu was not so far wrong after all.

Those who have examined the background of the gunnery and musketry of the Ottoman Turks derive it from the energetic early development of firearms in the Balkan kingdoms, and city-States like Dubrovnik. Influence in this southeastward direction is considered more likely, and certainly better documented, than any north-westward influences from the Arab world.[c] It seems that the Turks had cannon, directed by an artillerist named Ḥaydār, at the battle of Kossovo, already in +1389.[d] The arquebus would have reached them perhaps as early as +1425,[e] and it can still remain an open question whether the Turkish locksmiths were not the first in the field as regards matchlocks. At any rate, the Ottomans soon became the most advanced Muslim nation in respect of firearms,[f] however much aided by Hungarian and other European gun-founders, and by the middle of the +15th century had the most powerful artillery in the Western world.[g] What is more, the men of al-Rūm then supplied muskets to many countries east and south—to Egypt, Ethiopia, Gujerat, Sumatra and, significantly for us, the Khanates of Central Asia and Turkestan.[h] This was where we came in.

The question now at issue is when exactly the Ottoman Turkish muskets reached China. Chao Shih-Chên's grandfather must have been born about +1500, and the Turfan–Hami campaign would have happened during his youth, so that the Lu-Mi muskets could have become known to the Chinese military before +1530. The Ottoman embassies of +1524 and +1526 could both have introduced them too,[i] even that of +1543–4, and all this would have been well

---

[a] Ayalon (1), p. 143 (appendix by P. Wittek). The reason is that the purely Turkish word *tüfek* or *tüfeng*, the later name for musket, can be traced back to that date. Originally *tüwek* meant a blow-gun. There is evidence for some kind of hand-gun at the siege of Byzantium by Mehmet II in +1453, but no clue as to what exactly it was. Probably it had a serpentine.

[b] Partington (5), pp. 208–9.

[c] This might support a view that Western, as well as Arabic, developments came direct from China in the +13th century, afterwards passing back to the Turks.

[d] Petrović (1), pp. 172, 175. The formerly accepted opinion that the Turks had no firearms before the time of Mehmet I (+1413–21) cannot be right.

[e] Petrović (1), pp. 186, 191.   [f] Inalcik (1), p. 210.

[g] Petrović (1), pp. 190–1, 194.   [h] Inalcik (1), pp. 202, 208–9.

[i] The embassy guards might well have had them.

before +1548. It was a remarkable chance that Chao should have known the Pa and To families of sinified Turks by the end of the century, and that he should have been able to learn from them at first hand the details of the Turkish musket. Perhaps the most likely conclusion is that there were two introductions of the matchlock to China, first from Turkey by way of the Muslims of Sinkiang, forming a tradition known only to restricted circles in the north and north-west;[a] and secondly, in the south and south-east, a little later, either from the Japanese pirates or directly from the Portuguese merchant-adventurers.

The *Shen Chhi Phu* also gives a picture of the Turkish match-lock and describes how it was operated by the musketeer. Of this Rūm musket (*Lu-mi chhung*[1]) it says:[b]

The musket weighs about 7 or 8 lb., or (sometimes) 6 lb., and is about 6 to 7 feet long. The holding mechanism of the cock (*lung thou*[2]) is situated inside the stock. On pressing (the trigger), the cock falls, and after ignition it rises again. A steel knife is attached to the end of the stock, so that (the musket) can be used as lance if the enemy should get too near.[c] Or it can be used for defence against cavalry. At the time of firing, one hand should hold the grip in front, and the end of the stock should rest against the arm-pit. When firing one should only squeeze (the trigger), not pull it, and the body and hands should be still. (In the Rūm musket) the touch-hole is slightly further away from the place where the eye takes aim (than in the case of the Japanese bird-beak gun), and hence the smoke and flame developed when the musket is fired affects the eye and startles the musketeer less. This is one way in which (the Rūm musket) is superior to the Japanese bird(-beak) gun. It uses 0·4 oz. of powder and a lead shot weighing 0·3 oz.

The illustration of the Turkish musket (Fig. 174) also shows the match, made up of four strands of cord; and two sorts of copper dispensers containing the gunpowder. The larger one (*yao kuan*[3]) carried the propellant charge, the smaller one (*fa yao kuan*[4]) provided the priming powder. Each bottle had an elongated neck, and that of the larger one was pierced by a sliding copper diaphragm, or 'cut-out', so that it would dispense just the right amount. The operation was first to remove the wooden stopper with the teeth, then to place a finger over the opening of the container and turn it upside down. When the powder filled the

---

[a] Such a transmission inevitably calls to mind another, which occurred in the opposite direction, and about the same time, namely the gift of the technique of inoculation against smallpox (variolation), from the Chinese to the Ottoman Turks through Central Asian intermediaries. This must have happened in the +16th or +17th century, because the technique was passed on further West by the celebrated Lady Mary Wortley Montagu early in the +18th. On this whole subject see Needham (85). It is rather a piquant fact that the West provided the Chinese with instruments of death in the shape of a more effective war weapon, while they for their part presented the West with a life-giving technique in the form of an extremely beneficial medical invention. And these transmissions must have been approximately contemporaneous.

[b] P. 11a, tr. auct. This translation is taken from the main right-hand caption of Fig. 174. A representation of the *Lu-mi chhung*, with a brief description, is also found in *HLC*, pt. 2, ch. 2, pp. 13b, 14a.

[c] This is the first mention so far of a bayonet. The name is said to derive from the town of Bayonne in south-western France, where they would have first been made, but it was not much before the last quarter of the +17th century. Thus this Turkish arrangement was rather advanced. See Reid (1), p. 124; Blackmore (1), p. 36, (4), p. 19.

[1] 嚕蜜銃   [2] 龍頭   [3] 藥罐   [4] 發藥罐

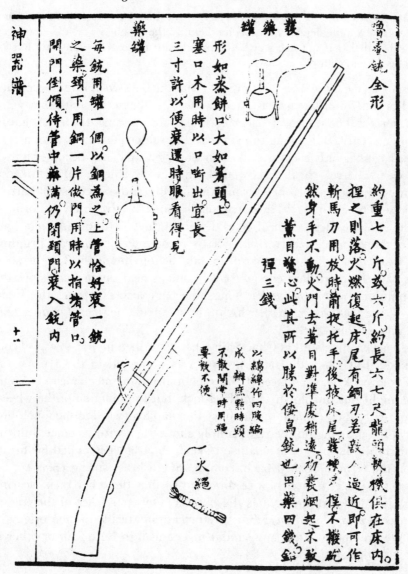

神器譜

魯密銃全形

藥罐

形如蒸餅口大如箸頭上
塞口木用時以口断出宜長
三寸許以便裏還時眼看得見

發藥罐

約重七八斤致六斤約長六七尺龍頭軟機俱在床內。
捏之則落火燃復起火若敲八逼近即可作
斬馬刀用放時前挺托手後披床尾費螺尸撲不撒矼
然身手不動火門去着目對準虐稍遠勿裝烟起亦不敢矼
薰目驚心此其所以滕於倭鳥銃也用藥四錢鉛
彈三錢

以綿線作四腰編
成一辨燕點時用繩
不敢閉當時用繩
要敢不便

每銃用罐一個以銅為之上管恰好裹一銃
之藥頸下用銅一片做門用時以指拽管口。
開門倒傾侍管中藥滿仍閉頸門裹入銃內

十二

火繩

Fig. 174. The 'musket of Rūm' (Byzantium, i.e. Turkish), from *SCP*, p. 11*a* (*Lu-mi chhung*). For translation see text. The picture also figures two powder-containers and a slow-match.

neck the diaphragm was set so as to prevent further flow, and then the charge was poured into the barrel down the muzzle. As for the priming, it was just shaken in until the pan was full.

Other diagrams in the *Shen Chhi Phu* explain the various component parts of the Turkish musket in detail. Of the sights the caption says:[a]

---

[a] P. 12*b*, tr. auct. The back-sight was a small plate with a hole, the fore-sight a pin.

The back-sight (*chao mên*[1]) and the fore-sight (*chao hsing*[2]) are essential for the musket, since accurate aiming depends entirely on them. In the Japanese gun a U-shaped sight is employed, but that is nothing worth compared with these. The sights are fixed on by horseshoe-shaped clamps.

Of the stock (*chhung chhuang*[3]) we are told that the best wood is mulberry, and the second-best is from the tamarisk tree (*ho liu*[4])[a], while in the south they generally use *chhou mu*[5].[b] The ramrod (*shuo chang*[6]), used for pressing down the charge and the bullet, is carried in a long tube in the stock under the barrel; after action it is wrapped round with a cloth soaked in boiling water and becomes a swab to clean out the barrel. It can be of wood for most of its length, but its forked head, just fitting the bore, must be of iron. Another illustration shows the barrel[c] and the screw-in breech-plug.[d] The bore must of course be absolutely uniform, otherwise the gun is useless. A third gives details of the touch-hole (*huo mên*[7]) and its pivoted copper fence (*huo yen kai*[8]).[e] It is recommended that the priming-pan (*shêng yao chhih*[9]) should be rather deep, but the touch-hole itself small, so as not to dissipate the propellant energy (*chhi*[10]), and as near the breech as possible, so as to minimise recoil (*hou tso*[11]). The same page also illustrates the V-shaped spring of the lock (*hsüan chi*[12]), which must be made neither of copper nor iron but of unquenched steel.[f]

The Turkish musket that Chao Shih-Chên described in +1598 was of course a matchlock. But when the same firearm was described in the *Wu Pei Chih* of +1628 the cock was replaced by a rack-and-pinion mechanism.[g] A cursory glance might put this down as a wheel-lock, but the function of the wheel here was not to produce a spark, but to move the match forward to the touch-hole. In Fig. 175 the drawing on the right simply shows this, with its fence, but that on the left shows the trigger mechanism. When the trigger is pulled for firing, the rack on the right is pushed back, compressing a brass spring (*ping tzu*[15]), and rotating the wheel in a clockwise direction so that the other rack, bearing the match, goes forward and ignites the powder. On the release of the trigger the two racks (*hsia kuei*[16] and *shang kuei*[17]) return automatically to their original positions. We are not aware of any similar mechanism in European or other Asian matchlock muskets.

---

  [a]　*Tamarix chinensis* (R 260).
  [b]　This wood is rather difficult to identify, but it may be the sweet oak, *Quercus glauca* (R 614; Chhen Jung (*1*), p. 203).
  [c]　P. 11*b*. Octagonal exteriorly.
  [d]　As in Fig. 171 above, for the Japanese bird-beak matchlock musket.
  [e]　P. 12*a*.
  [f]　One limb is called *kuei*[13] and the other one *fa kuei*[14].
  [g]　Ch. 124, p. 11*a*.

[1] 照門	[2] 照星	[3] 銃床	[4] 河柳	[5] 紬木
[6] 搠杖	[7] 火門	[8] 火眼盖	[9] 盛藥池	[10] 氣
[11] 後坐	[12] 旋機	[13] 軌	[14] 發軌	[15] 搠子
[16] 下軌	[17] 上軌			

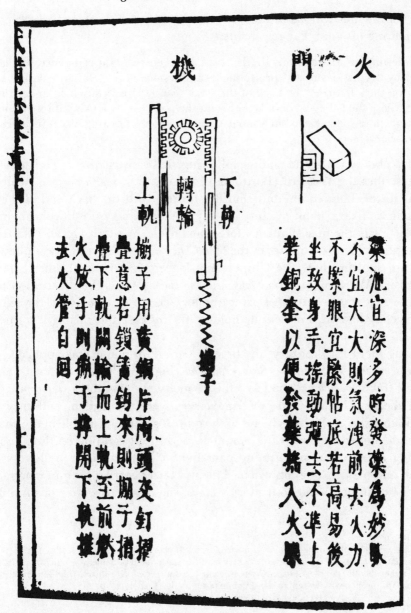

機　　　門火

上軌　　轉輪　　下軌

去火放火管自回　　　疊下軌闗輪而上軌至前籤　　搬子用黃銅片兩頭交釘擢　　疊意若鎖簧鈎來則掤于拥

藥池宜深多貯發藥為妙眼
不宜大大則氣洩前去火力
不緊眼宜際帖底若高易後
坐致身手搖動彈去不準上
著銅奎以便發藥擠入火眼

Fig. 175. The cock of the Turkish musket replaced by a rack-and-pinion mechanism in *WPC*, ch. 124, p. 11*a*. On the right the touch-hole with its fence, on the left the trigger mechanism embodying spring as well as rack-and-pinion.

One of the illustrations in the *Shen Chhi Phu* shows a Turkic musketeer with a turban firing his gun. The text reads:[a]

After inserting the match firmly (in the cock), kneel down on the right knee, and hold the musket by the peg (*tho shou*[1])[b] projecting from the stock in the left hand, with the elbow resting on the left knee. The back of the stock (*hou wei*[2]) is firmly held under the right armpit. Close the left eye, and take aim with the right eye by looking through the back-sight at the fore-sight. Keep one's mouth shut, hold one's breath, aim at the enemy and squeeze the trigger.

Fig. 176 shows a Muslim soldier following these instructions. The same book also describes the Western (Portuguese) musket (*Hsi-yang chhung*[3]),[c] which was similar to the Turkish one but slightly shorter, with the back end of the stock bent almost like a hook, as in Fig. 177.[d] A drawing of a European firing a musket, one of the miscellaneous foreigners of the Western Ocean countries, as the caption says, also appears in the *Shen Chhi Phu*; see Fig. 178.[e] It is followed by a picture showing how the Chinese musketeers improved the firing position when using Western guns (*kai fang hsi-yang chhung*[4]).[f] As the illustration in Fig. 179 shows, it was a matter of using the fork-clamp rest-peg with the left hand and kneeling like the Turk, while holding the musket and raising it to the right eye.[g]

During the +16th century, the breech-loading principle was applied to these muskets, as we see from the *Shen Chhi Phu* (Fig. 181).[h] The 'gripped-lightning musket' (*chhê tien chhung*[6]) had several six-inch-long chambers (*tzu chhung*[7]) into which the lead bullets and powder charges were pre-loaded, and they easily fitted in to a slot at the breech end of the gun. A small touch-hole in the chamber then came under the cock with its match, permitting ignition on the pressing of the trigger; there was no priming pan. Each chamber took a bullet 0·2 oz. in weight, and 0·25 oz. of gunpowder. The musketeer carried his gun ready loaded with one chamber, while four more were borne in a leather bag of suitable size

[a] P. 19*a*, tr. auct.

[b] This is said to be three inches long and made of wood. It clamped on to the stock (Fig. 177). Its presence suggests that the barrels were considerably lighter than European models. Something similar is used today in modern target shooting by the 'off-hand' system, using a palm-rest which transmits the weight of the barrel down through the elbow to the hip, in the standing position. See Trench (1), p. 292.

[c] P. 13*a*.   [d] P. 13*b* gives details of the barrel and the stock.

[e] P. 21*a*.   [f] P. 21*b*.

[g] We have not come across any Chinese illustrations of the staffs with Y-shaped heads that Western musketeers used for supporting and steadying their guns (p. ■■ above). But in *WPC* (ch. 123, pp. 2*b*–3*b*) there is a picture (Fig. 180) of a musketeer aiming his gun and firing through an iron ring attached to a slanting wooden staff with an iron ferrule held by a second soldier. This is called the 'successive rotation detachment musket system' (*tsao-hua hsün huan phao*[5]), and it is explained that while one man is finding the required elevation with the sights and firing accurately with this aid, four or five more are loading and awaiting their turn.

A version of the picture was given by Mayers (6), p. 97, but he misunderstood the purport of the entry's title.

[h] P. 14*a*.

[1] 托手	[2] 後尾	[3] 西洋銃	[4] 改放西洋銃	[5] 造化循環砲
[6] 掣電銃	[7] 子銃			

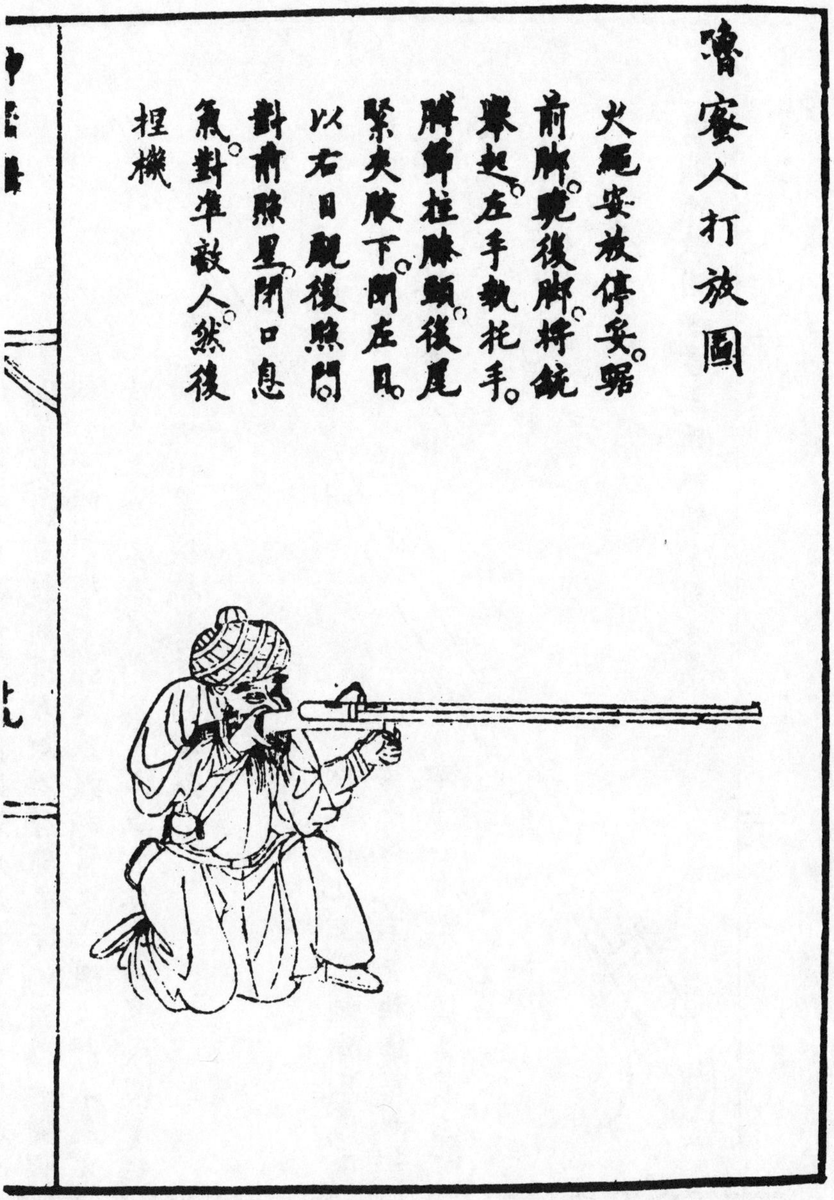
嚕密人打放圖
火繩安放停妥。踞
前脚。跪後脚。將銃
舉起。左手執托手。
脚節拄膝頭。後尾
緊夾腋下。開左眼。
以右目覻後照門。
對前照星。閉口息
氣。對準敵人。然後
捏機

Fig. 176. A turbaned Muslim soldier kneeling and firing his gun (*SCP*, p. 19*a*). His left hand holds the peg, a downward-pointing handle which was clipped on to the barrel and stock; this is shown separately in the next illustration.

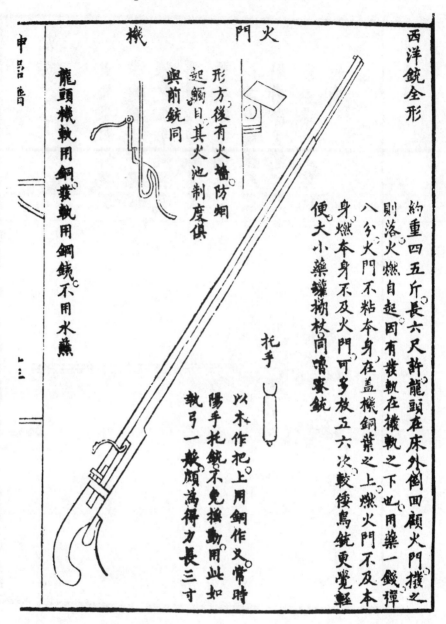

Fig. 177. The *Hsi-yang chhung* or European musket (*SCP*, p. 13a). At the left on the top the spring system of the trigger is seen, then to its right the touch-hole and fence, finally below the barrel to the right, the clamp-peg.

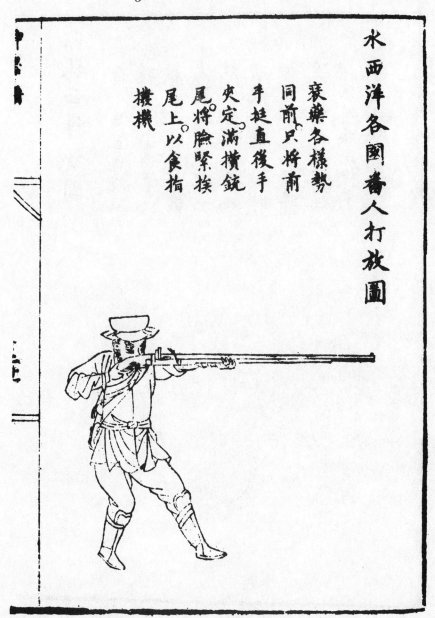

衰藥各樣勢
同前只將前
手挺直後手
夾定滿攢銃
尾將臉緊挨
尾上以食指
攪機

水西洋各國番人打放圖

Fig. 178. A European firing his musket from the standing position (*SCP*, p. 21 *a*). The caption reads: 'A foreigner from one of the miscellaneous Western-Ocean sea countries letting off his gun.'

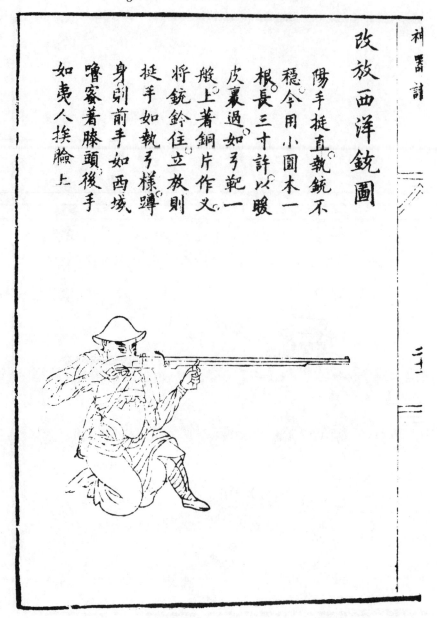

改放西洋銃圖

陽手挺直執銃不
穩令用小圓木一
根長三寸許以暖
皮裏過如弓靶一
般上著銅片作义
將銃鈴住立放則
挺手如執弓樣蹲
身別前手如西域
嚕蜜著膝頭後手
如夷人挨臉上

補器註

二十

Fig. 179. A Chinese musketeer combining the kneeling position of the Turk, with the fork-clamp rest-peg, while using a Western musket (*SCP*, p. 21*b*).

Fig. 180. The ring-rest for muskets (from *WPC*, ch. 123, p. 2*b*).

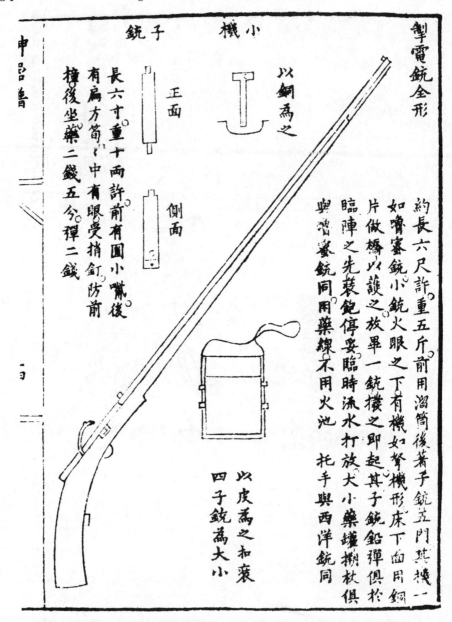

Fig. 181. A breech-loader musket depicted in *SCP*, p. 14*a*. Above on the left are two of the chambers and a sketch of the bronze or copper trigger-spring, while below on the right is a drawing of the leather bag in which four more of the chambers were carried. Cf. Figs. 143*a, b,* 144.

and shape slung round him. The lock, the two powder-dispensers and the ramrod were all the same as those of the Turkish musket, but the peg-rest for the left hand was as in the modified European musket. On the other hand, a bronze or copper trigger guard (*hu chhiao*[1]) was now provided.

Another breech-loading musket was the 'sons-and-mother gun' (*tzu mu chhung*[2]) described in the *Ping Lu* of +1606 (Fig. 182).[a] The barrel (the 'mother') was 4 ft 2 in. long, while the chamber (the 'son') was but 7 in. long. The bore of the two was carefully made identical, and then one 'son' after another could be inserted at the breech and fired off. The number of chambers carried by each musketeer was the same as in the previous case. Since, as we saw on p. 442, Chao Shih-Chên claimed to have invented breech-loading muskets himself, this one of Ho Ju-Pin's would seem to be derivative—but of course there may well have been many such inventors.[b]

So much for matchlocks. But what about the hither side, as it were, of the history of the matchlock musket? Nothing has yet been said about the simple serpentine in China, the pivoted S-shaped lever which brought the glowing match to the touch-hole (cf. p. 425 above). Now the late Ming military compendia preserve a family of archaic hand-guns, both single-barrelled and multi-barrelled, which in some cases show what appears to be a serpentine. We have something to say about these first, and then we must turn to what one might call the further side, and consider the flintlock musket in East Asia.

The *Wu Pei Chih* describes a long hand-gun with a barrel weighing 18 catties, and 4 ft 4 in. long, attached to a handle 1 ft 9 in. long and bent in scroll-shaped curves.[c] This is called the 'large blowing-away-the-enemy lance-gun' (*ta chui fêng chhiang*[4]).[d] It had sights and was operated by two soldiers using a tripod support; with a blunderbuss muzzle it would fire a lead ball weighing 0·65 oz. with a range of more than 200 paces. No serpentine is shown in the illustration (Fig. 184), but it probably had one because an alternative name is given: 'match-holding lance-gun' (*chih huo-shêng chhiang*[5]).

The rest of the series consisted in multiplying the number of barrels. Thus the 'triple-victory magically effective contraption' (*san chieh shen chi*[6]) had three barrels rotating on a central shaft so that they could be fired off in turn.[e] Each one had a fore-sight, but there was only a single back-sight, fitted on the handle itself, which ended in a curve as before. But now for the first time we see what

---

[a] Ch. 12, pp. 11*a*–12*b*. Note the plug-bayonet (*chhung tao*[3]) which could be fixed in the muzzle for close combat.

[b] Cf. Fig. 183*a*, *b*, *c*, an English example of +1537 in the Tower of London Armouries.

[c] *WPC*, ch. 125, pp. 9*b*, 10*a*. Unlike the tillers of old, which fitted as a male peg into a female socket, this handle was the socket and the barrel extended 5 in. into it.

[d] On the usage of retaining the term *chhiang* for a long gun, and calling a short one *chhung*, cf. p. 432 above.

[e] *WPC*, ch. 125, p. 15*a*. No explanation is given for the curious apparatus depicted on the left of the illustration. Perhaps it was some kind of holster for carrying the gun, though it looks like a bow-case.

[1] 護橋　　　　[2] 子母銃　　　　[3] 銃刀　　　　[4] 大追風鎗　　　[5] 執火繩鎗
[6] 三捷神機

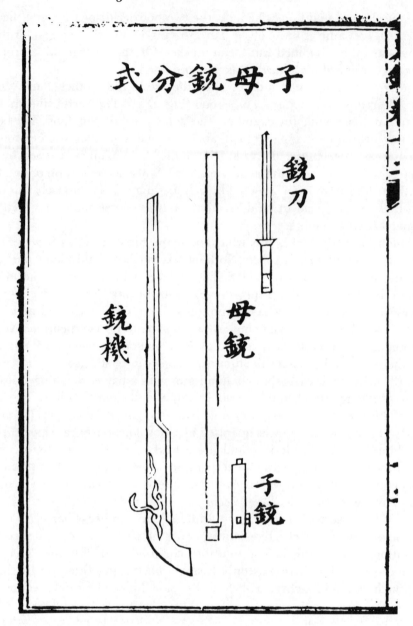

Fig. 182. Another breech-loading musket, from *PL*, ch. 12, p. 12*b*. One of the chambers is seen below on the right, and a bayonet for fixing in the muzzle is to the right above.

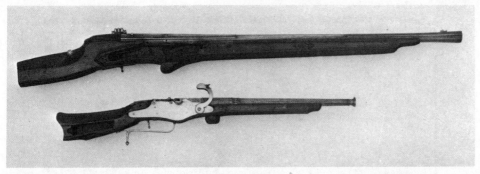

Fig. 183a. Above, a breech-loading arquebus of Henry VIII's time; below, a breech-loading carbine of about the same date. Photo. Tower of London Armouries. Note in both cases the peg or palm-rest for the left hand, analogous to the Chinese examples just seen (Figs. 176, 177, 179).

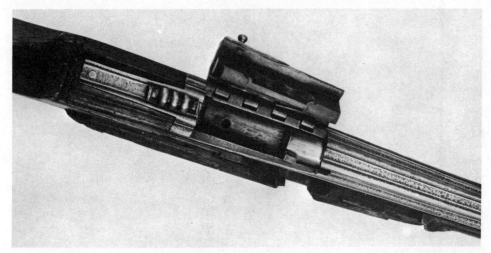

Fig. 183b. The breech-loading arquebus of the previous picture with the space for the culasse. Photo. Tower of London Armouries.

Fig. 183c. The same breech-loading arquebus, showing the culasse extracted. Photo. Tower of London Armouries. All these three illustrations by courtesy of Howard Blackmore.

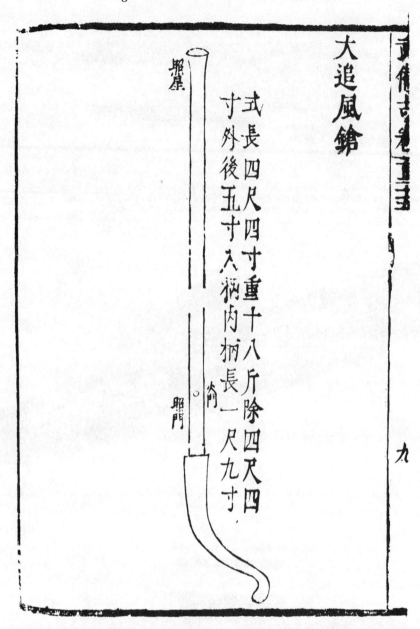

Fig. 184. The origin of the serpentine. An archaic weapon in *WPC*, ch. 125, p. 9*b*; the 'large blowing-away-the-enemy lance-gun' (*ta chui fêng chhiang*). No serpentine is shown, but it probably had one because of its alternative name, the 'match-holding lance-gun' (*chih huo-shêng chhiang*).

looks extremely like a serpentine, so arranged as to bring the match down to the touch-hole of each of the three barrels one after the other (Fig. 185).[a] The principle was then extended to five barrels, and these came in two forms, the rotary,[b] and a set arranged in a row.[c] The 'five thunder-claps magically effective contraption' (*wu lei shen chi*[1]) was quite similar to the three-barrel gun just described. The barrels (now called *chhung*[2]) were all 1 ft 5 in. long, and the whole apparatus weighed 5 catties. The sights were arranged in the same way, and it is explained that the gun should be held in the left hand for aiming, whereupon the forefinger of the right hand should bring down the serpentine upon the touch-hole of each barrel in turn.[d] One can see the slow-match (*huo shêng*[3]) held in a copper tube at the business end of the serpentine (Fig. 186), through which it must have been fed as it burned away, and a whole length of it is shown dangling.

Just how practical any of these devices were remains a little uncertain, but the principle was extended to seven and ten barrels. The 'seven stars gun' (*chhi hsing chhung*[5]) consisted of six barrels 1 ft 3 in. long turning around a longer and larger barrel of pure iron, and all attached to a wooden handle 5 ft long. The barrels were bound with iron straps, and the whole set-up mounted on two wheels 1 ft 5 in. in diameter, approximating it to a small field-gun (Fig. 104). No serpentine is shown, but perhaps we may deduce it from the other weapons in the series. What is important here, however, is that this design goes back not only to the *Wu Pei Chih*, but to the *Huo Lung Ching*, and to the earliest stratum of that too,[e] so that we are dealing with a firearm of the mid +14th century, certainly well before +1400. Finally, the 'sons-and-mothers hundred-bullets gun' (*tzu mu pai tan chhung*[7]) consisted of ten small wrought-iron barrels each 5 in. long, surrounding a larger barrel 1 ft 5 in. long, all attached to a wooden handle. Each barrel fired several dozen small lead bullets, and we are told that it takes a strong man to wield it (Fig. 188).[f] Again the serpentine is not shown, but there must have been some way of igniting the barrels one after the other.

Could the serpentine have been in fact a Chinese invention? As we know from p. 425, it would have to have reached Europe by +1410, and that probably means the last quarter of the previous century. When we look at the possible

---

[a] The same device appears in a weapon called the 'sword lance-gun' (*chien chhiang*[6]); *WPC*, ch. 138, p. 8a. This had but one barrel, but was provided with a close-combat blade at the stock end.

[b] *WPC*, ch. 125, p. 14a.

[c] This was the 'row-of-five lance-gun' (*wu phai chhiang*[4]) in *WPC*, ch. 125, pp. 15b, 16a (Fig. 187). Each barrel, said to be of pure iron, fired 4 or 5 lead bullets. No serpentine is shown. We encountered similar arrangements in the fire-lance period, cf. p. 243 above.

[d] The explanation is clear, but it doesn't look as if the artist quite understood how the serpentine worked. On the deficiencies of the old Chinese technical illustrators cf. Vol. 4, pt. 2, pp. 369 ff.

[e] *HLC*, pt. 1, ch. 2, p. 15a, b, also *Huo Chhi Thu*, p. 16b and *Huo Kung Pei Yao*, ch. 2, p. 15a, b. The *WPC* reference is ch. 124, pp. 16b, 17a.

[f] *HLC*, pt. 2, ch. 2, pp. 14b, 15a, repeated in *WPC*, ch. 125, pp. 13b, 14a. This is not the oldest stratum of the *Fire-Drake Manual*, but the weapon clearly belongs to the +15th century rather than the +16th.

[1] 五雷神機   [2] 銃   [3] 火繩   [4] 五排鎗   [5] 七星銃
[6] 劍鎗   [7] 子母百彈銃

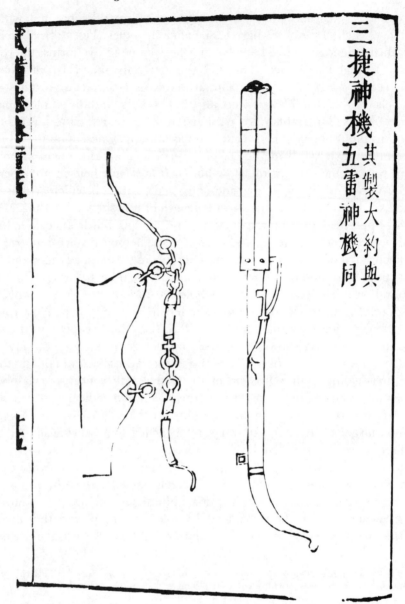

Fig. 185. Gun with three barrels rotating on a central shaft so that each one could be fired in turn by a slow-match held in a serpentine. The 'triple-victory magically effective contraption' (*san chieh shen chi*) from *WPC*, ch. 125, p. 15*a*. The unexplained object on the left must have been some kind of holster for carrying the weapon.

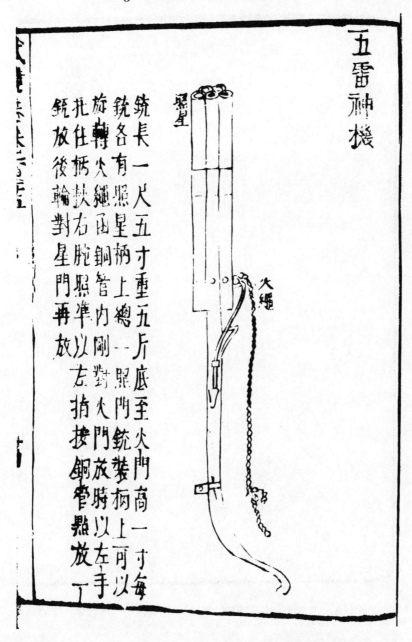

五雷神機

銃長一尺五寸重五斤底至火門高一寸每
銃各有照星柄上總一照門銃裝柄上可以
旋轉火繩函銅管內剛對火門放時以左手
扡仕柄扶右腕照準以左拾按銅管黗放一
銃放後輪對星門再放

火繩

照星

Fig. 186. The five-barrelled version of the same arm (*WPC*, ch. 125, p. 14*a*). Here the serpentine is quite clearly seen, with the slow-match (*huo shêng*) dangling from it.

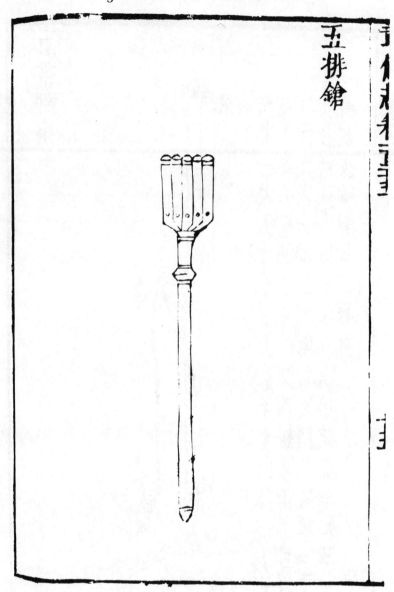

五排鎗

Fig. 187. Gun with the five barrels aligned in a row (*WPC*, ch. 125, p. 15*b*).

子母百彈銃

Fig. 188. A ten-barrelled ribaudequin, the 'sons-and-mothers hundred-bullets gun' (*tzu mu pai tan chhung*), from *HLC*, pt. 2, ch. 2, p. 14*b*, repeated in *WPC*, ch. 125, p. 13*b*.

means of transmission we find some that are too early to fit the case,[a] and others too late.[b] But the voyages of the fleets under the admiralty of Chêng Ho[1], which certainly brought a knowledge of the most up-to-date things the Chinese had to the attention of technicians and potentates in all the Indian and Arabic lands, would just qualify ($+1405-33$);[c] and in $+1403$ the Timurid court of Shāh Rūkh at Samarqand, which was in touch with China,[d] saw the arrival of the Spanish ambassador Ruy Gonzalez de Clavijo, who picked up a lot of information there.[e] Other Europeans, like the Bavarian Johann Schiltberger, were present in the Timurid service at the same time, and eventually got home safely too.[f] Moreover, the embassy of Nicholas Comanos from Byzantium to China in $+1371$ (cf. p. 442 (e) above) would not really have been too soon.[g] More important perhaps than any of these is the remarkable but little-known fact that between $+1330$ and $+1430$ there was a slave-trade from Mongolia to Italy which brought many hundreds, even thousands, of 'Tartar' servants from north-east Asia to Europe.[h] We speculated long ago on the technology which some of them must have brought with them, whether in textiles or in firearms.[i] We have also recognised 'clusters' of transmissions from China to Europe, in the $+12$th, $+13$th and $+15$th centuries, but especially the $+14$th century;[j] and it may be that the serpentine lever belongs with these lists. After all, it was the simplest possible improvement on the original hand-gun, with its slow-match being flourished about, and once the method of control (for that was essentially what it was) had passed to Europe, it would have been just like the locksmiths of the West[k] to take the further step of inserting springs, levers, detents and tumblers, between the trigger and the touch-hole. Early Renaissance sophistication would thus have added valuable safety devices to a system which had had its origin at the other end of the Old World.

One important consideration touching the first appearance of the serpentine, which further suggests a Chinese origin, is the fact that the trigger as such,

---

[a] For example, the travels of the Franciscan friars in Mongolia and Cathay centre on the period from $+1230$ to $+1300$, which is too soon (cf. Vol. 1, pp. 189, 202, 224).

[b] Among the travellers Nicolo Conti centres on $+1430$, Athanasius Nikitin $+1468-74$, and Hieronimo di Santo Stefano $+1496$ (Yule (2), vol. 1, pp. 124, 174 ff., 179; Cordier (1), vol. 3, p. 94). The Portuguese voyages ($+1415$ to 98) are of course much too late; cf. Vol. 4, pt. 3, pp. 503 ff.)

[c] See Vol. 4, pt. 3, pp. 489 ff.

[d] The exchange of embassies between the Timurid dynasty and the Chinese emperor in $+1414$ and $+1419$ is only just too late; cf. Dunlop (10), Maitra (1).

[e] Yule (2), vol. 1, pp. 173-4, 264 ff.

[f] Ibid. p. 174.

[g] Of course we should like to know something about the musket in Byzantium before $+1453$, but all is silence.

[h] See Olschki (6).

[i] Vol. 1, p. 189.

[j] One may mention, besides gunpowder itself, the mechanical clock, the blast-furnace for cast iron, block-printing, segmental arch bridges and summit-canal lock-gates. The three-component assembly for the inter-conversion of rotary and longitudinal motion came in the $+15$th century. See Vol. 4, pt. 2, p. 383, and Needham (64), pp. 61-2, 119 ff., 201.

[k] Or even of the Turks.

downward-hanging as we know it, was so age-old in that country.[a] As we saw (pt. 6 (e), 2, ii above), China of the Warring States period (−5th century) was almost certainly the home of the cross-bow, and by the time that this became the standard weapon of the Han armies (−1st to +2nd centuries) it had acquired a bronze trigger-mechanism of beautiful and intricate construction (see K.P. Mayer, 1). The crossbow was introduced to the Western world probably twice, before the +5th century and again during the +10th.[b] Of course there were trigger-mechanisms of various kinds in the Greek and Roman proto-artillery of catapults, onagers, *arcuballistae* and the like, but they seem almost always to have released the bowstrings from a holding-claw by a transverse lever working up or down, or around a pivot.[c] This is true even of the hand-held *gastraphetes*. In other words, the triggers did not operate from below, through a stock in which they were pivoted, as the Chinese triggers did. Hence their relevance to the serpentine.

We can now return to the later trans-matchlock territory, namely that of the flintlock musket. First coming into use about +1550, it displaced the matchlock slowly but steadily in the Western world, gaining complete dominance from about +1725 onwards; but it was destined to obsolescence itself a century later when mercuric fulminate was successfully confined in percussion-caps and used to detonate the cartridges with the charges that propelled the bullets.[d] This was the work primarily of Alexander Forsyth in the first decade of the nineteenth century, but the man with the best claim to be the inventor of the little top-hat-shaped copper cap was Joshua Shaw in +1822.[e]

By and large there seems to have been no flintlock period either in China or Japan, the former because of military conservatism, the latter because of the Sakoku[1] closure of the country to all outside influences between +1636 and 1853. As we have noted already (p. 37), very few military compendia were produced during the Chhing period, and it would therefore be difficult to say on what occasions flintlock weapons came to the attention of the Chinese; at any rate in 1841 Wei Yuan[2] described and eloquently recommended them,[f] telling how the flint (*huo shih*[3]) was held in the screw-vice or 'jaws' of the cock.[g] But

---

[a] This point was made by our friend Prof. Yoshida Mitsukuni during a symposium in Kyoto on 5 Oct. 1981. Cf. Allen (1), pp. 78 ff., 110.

[b] The oldest European illustrations of crossbow-triggers are in the *Book of Ezekiel* by Haimo of Auxerre (Bib. Nat. Lat. MS. 12302; Blackmore (5), p. 174, fig. 72a), a late +10th-century work; and in the Catalan version of *The Four Riders of the Apocalypse*, +1086 (Cathedral Library MS, Burgo de Osma; Blackmore (5), p. 176, fig. 73).

[c] Cf. Marsden (1), pp. 6, 11, 34–5; (2), pp. 47–8, 102, 179, 180–1, 219–20 and 261.

[d] Among many accounts that of Blackmore (4), pp. 124 ff. is one of the shortest and clearest. Flintlock mechanisms still lingered on for some decades after 1800. On the nature, history and use of the fulminates see Davis (17), pp. 400 ff.

[e] It was Forsyth who had the idea of using fulminate as priming, and Shaw who devised the mass-produced metal caps.

[f] *Hai Kuo Thu Chih*, ch. 91, pp. 1a ff.

[g] As Waley (26), p. 53 remarked, 'percussion-guns' were just coming in at this time, so it was rather late to recommend flintlocks, though the Westerners still used them to some extent.

[1] 鎖國   [2] 魏源   [3] 火石

still in 1860, at the time of the Anglo-French war against China, matchlocks were in regular use on the side of the defenders (Fig. 156).[a] As for Japan, there is evidence that the Dutch +1636 presented a dozen new flintlock pistols to the Shogun,[b] and that certain provincial samurai tried out flintlock guns with satisfaction on board the Dutch ship *Breskens* in +1643[c]—but the matter went no further, perhaps for a reason which we shall mention a few pages hence.

It is rather extraordinary that the flintlock musket did not catch on in China, for we have seen (pp. 198 ff.) that flint-and-steel had been used from the middle of the +14th century onward for the automatic igniting of land-mines and other infernal machines.[d] The striking of sparks from flint and steel must have been known in China much earlier, quite probably as far back as steel itself, and that would mean the −3rd century.[e] Opinions about the beginning of this method of fire-making in the West have differed a good deal; it was certainly known to Pliny[f] and the Romans of early Christian times, but some put it much earlier than that.[g] There was therefore no reason at all why the advantages of flintlock muskets should not have been appreciated in East Asia. They just were not.[h]

So one may say that the matchlock musket was superseded only by the percussion-cap[i] and cartridge rifle in the second half of the nineteenth century. The first significant modernisation of Chinese armed forces is generally said to have been due to Li Hung-Chang[1], who in 1864 equipped his army with 15,000 foreign-made rifles.[j] But the troops of the Taiping revolutionaries under Li Hsiu-Chhêng[2] had acquired several thousand similar small-arms already two years before.[k] At the same time Chhêng Hsüeh-Chhi[3] was organising 'foreign arms

---

[a] The prong supports at the muzzle end are seen again here, just as they were in the pictures of early +17th-century battles (Figs. 152 and 155). Chou Wei (*1*), p. 336, wrote in 1957 that the matchlock musket was still being used in modern times by the border tribesfolk in western and south-western China. And Tibetan matchlocks still used by soldiers of the Muli king are illustrated in Rock (2), pl. 12.

[b] Caron & Schouten (1), p. xxxiv.

[c] Montanus (1), pp. 352–3.

[d] The steel wheel (*kang lun*[4]) method, *HLC*, pt. 1, ch. 3, p. 27*a*, *b*; *PL*, ch. 12, pp. 61*b*–62*b*; , ch. 134, pp. 14*a*–15*b*, and 6*b*, 7*a*. The flint (*huo shih*[5]) is distinctly mentioned. Further analysis in Liu Hsien-Chou (*12*).

[e] Cf. Needham (*32*), pp. 24 ff.

[f] *Nat. Hist.* XXXVI, 138. See also Harrison (4), p. 219; Neuburger (1), p. 234; Feldhaus (1), col. 319.

[g] Forbes (7) put steel in the Mediterranean area as old as −700, which may well be right, and flint-and-steel by −300; see (15), p. 9. This makes rather absurd the suggestion of Pollard (1), p. 37, smiled on by Perrin (1), p. 70, that the flintlock principle was brought back from Japan by the Portuguese. Leonardo da Vinci had had the idea already (cf. p. 428).

[h] Unless we should see a reference to them in something called an 'automatically closing touch-hole' (*tzu pi huo men*[9]) which Wang Ming-Hao dismissed about +1598 as a 'strange novelty of the day' (*Huo Kung Wên Ta*, p. 1291).

[i] From Ting Shou-Tshun[6] in 1844 we learn that the technical term for the flintlock was *tzu lai huo chi*[7], while that for the percussion-cap system was *tzu lai huo yao*[8] (*Hai Kuo Thu Chih*, ch. 91).

[j] See Kuhn (1), pp. 308 ff. Li's memorial of 1863 on Western musketry and gunnery is translated in Têng Ssu-Yü & Fairbank (1), pp. 70 ff.

[k] The famous 'Taiping Rebellion' lasted from 1851 to 1864; cf. Curwen (1); Kuhn (1).

---

¹ 李鴻章   ² 李秀成   ³ 程學啓   ⁴ 鋼輪   ⁵ 火石
⁶ 丁守存   ⁷ 自來火機   ⁸ 自來火藥   ⁹ 自閉火門

companies' (*yang chhiang tui*[1]),[a] and in 1863 the cavalry of the Nien rebels (*Nien fei*[2]) was routed by machine-guns (*lien huan chhiang phao*[3]) of some kind or other.[b] This was the time when China began to set up her own arsenals, among which the An-chhing Ping Kung Chhang[4], founded by Tsêng Kuo-Fan[5] and directed by the engineers Hsü Shou[6] and Hua Hêng-Fang[7] (1862)[c] continued to make matchlock muskets, but also began the making of percussion-caps for rifles.[d] The famous Kiangnan Arsenal (Chiang-Nan Chi-Chhi Chih-Tsao Chü[8]) was founded in 1865, but it did not produce satisfactory modern rifles until 1871.[e] Ten years earlier one of the censors, Wei Mu-Thing[9], had memorialised his conviction that China ought to copy Western firearms without hesitation. He claimed that 'the vaunted European weapon technology was, after all, a legacy of China herself'. He asserted that 'it was the Mongols of the Yuan dynasty who had introduced gunpowder and firearms to Europe, though they had afterwards been greatly improved there by extraordinary skills multiplied in a hundred ways'.[f] Many other scholars said the same, for example Wang Jen-Chün[10] in his *Ko Chih Ku Wei*[11] (Scientific Traces in Olden Times).[g] How right they were; and even more right would they have been if they had ventured to claim the preceding dynasty, the late Sung, as the time of transmission.

Lastly we may take up a point touched on a couple of paragraphs ago, the failure of Japan to adopt flintlock muskets—it was because they almost abandoned muskets altogether. There was a period, the hundred years before the Sakoku closure of +1636, when fire-arms were very prominent in Japanese strategy and tactics, but after the turn of the century controls increasingly strict were brought in, and the activities of the gunsmiths diverted in other directions.[h] We mentioned already (p. 429) the ten thousand musketeers under Oda Nobunaga[12] at the Battle of Nagashino[13] in +1575 when he defeated Takeda Katsuyori[14],[i] only a few years before (+1567) another lord of the same clan,

[a] Liu Kuang-Ching (1), p. 426.
[b] *Ibid.* p. 471. See also Têng Ssu-Yü (3), pp. 170 ff.
[c] Cf. Vol. 4, pt. 2, p. 390.
[d] Kuo Ting-Yi & Liu Kuang-Ching (1), p. 519.
[e] *Ibid.* p. 521.
[f] *Chhing-Tai Chhou-Pan I-Wu Shih-Mo* (Thung-Chih r.p.), Anon. (*212*), ch. 2, pp. 35 a ff. On these collections see Hummel (2), p. 383.
[g] See ch. 2, pp. 25a, 27b, 28a.
[h] Earlier they had shown much skill and originality, producing a match-protector (Fig. 189) for matchlocks in rainy weather (cf. Perrin (1), p. 18), as also a helical mainspring (cf. the Turkish musket detail in Fig. 175), and an adjustable trigger-pull (Gluckman (1), p. 28). It is even said that old +17th-century matchlock barrels preserved in armouries were converted to percussion-cap rifles after Cdr Perry's time (1853), and then again to bolt-action rifles for the Russo-Japanese war of 1905 (Kimbrough (1), pp. 464–5; Perrin (1), p. 67).
[i] Turnbull (1), pp. 156 ff. The musketeers were commanded in part by a notable soldier, Honda Tadakutsu[15] (+1548 to +1610).

[1] 洋槍隊	[2] 捻匪	[3] 連環槍礮	[4] 安慶兵工廠	[5] 曾國藩
[6] 徐壽	[7] 華蘅芳	[8] 江南機器製造局		[9] 魏睦庭
[10] 王仁俊	[11] 格致古微	[12] 織田信長	[13] 長篠	[14] 武田勝賴
[15] 本多忠勝				

Fig. 189. Japanese match-lock match-protectors in use in rainy weather, a picture by Utagawa Kuniyoshi (1855) in the collection of S. Yoshioka, Kyoto. After Perrin (1), p. 18.

Takeda Harunobu[1], had recommended the musket as the most important weapon of the future.[a] Indeed it was much used in the Korean expeditions of Hideyoshi,[b] but especially after the Chinese armies flooded in to support the Koreans, frantic letters were sent home by Japanese commanders asking for urgent reinforcements of muskets and musketeers.[c] The last important engagement in which muskets were used was the siege of Hara[2] in the Shimabara[3] Rebellion of +1637, an uprising of Christian peasant-farmers and landless samurai.[d]

Then came the period of firearm control and almost abolition. The first step was taken by Hideyoshi himself in +1586, when he announced that he needed as much iron as possible to make a giant Buddha image, and all farmers, samurai and monks had to 'volunteer' to surrender their guns for this purpose.[e] Then, after Ieyasu[5] had won the Battle of Sekigahara[6] and established the Tokugawa shogunate in +1603, came the licensing of the two great gunsmith centres, Nagahama[7] on Lake Biwa and Sakai[8] near Osaka.[f] An office of a Commissioner of Guns (Teppō Bugyō[9]) was set up, and he cleared no orders except those from the central government, but the gunsmiths got an annual salary, whether they made any muskets or not. By +1625 the monopoly was complete, but the orders were reduced to a minimum.[g] Export of guns was also forbidden.[h] And so matters continued until the arrival of Commodore Perry in 1853 jolted the Japanese shogunate into the modern world, and set the stage for the Meiji[10] Restoration of 1867. The firearm suppression policy of Tokugawa times probably accounts for the fact that the *Honchō Gunkikō*[11] (Investigation of the Military Weapons of the Present Dynasty), written by Arai Hakuseki[12] (+1656 to +1725) and published in +1737, has only one brief chapter on guns and cannon.[i]

Five reasons have been given for this singular story, all convincing enough.[j]

[a] Brown (1), p. 239. The given name of Takeda Harunobu (+1521–1573) was by this time Takeda Shingen[4], for he had changed it in +1551.

[b] From +1592 to 1598. The translation by Pfizmaier (107) of the *Chōsen Monogatari*, an account of these campaigns, though now very old, remains of much interest.

[c] Examples in Brown (1), pp. 239, 241, 244; cf. Perrin (1), pp. 30–1.

[d] Murdoch (1), vol. 2, p. 658.   [e] Murdoch (1), vol. 2, p. 369.

[f] A graph showing the continuously decreasing production of firearms from Sakai during the +17th century will be found in Itakura (1), p. 143.

[g] The best account of all this is in Arima (1), pp. 657ff., 667ff., 670–1, 676–7.

[h] Though Richard Wickham of the Hon. East India Co. managed to get out a few for Siam in +1617 (Pratt (1), vol. 1, pp. 243–4, 265).

[i] Cf. Waterhouse (1), p. 95. But as time went on, there was considerable uneasiness about the policy. In 1808 the eminent *rangaku* scholar Satō Nobuhiro[13] (+1769 to 1850) published a book on the use of small-arms and even made some inventions himself; while in 1828 there were experiments with flintlock guns. Only the year before Perry a third scholar, Sakuma Shōzen[14] (1811 to 1864) deplored the parlous condition of Japanese shore batteries. See Tsunoda Ryosaku (2), pp. 568, 615. There was also Murakami Sadahe[15] (1808 to 1872) whose school of gunnery had, according to the study of Iwasaki Tetsushi (1), considerable influence. And at the term of the century Honda Toshiaki had advocated the making of gunpowder; see Keene (1), p. 162.

[j] See Perrin (1), pp. 24ff., 33ff.

[1] 武田晴信   [2] 原   [3] 島原   [4] 武田信玄   [5] 家康
[6] 關原   [7] 長濱   [8] 境   [9] 鐵砲奉行   [10] 明治
[11] 本朝軍器考   [12] 新井白石   [13] 佐藤信淵   [14] 佐久間象山   [15] 村上定平

First, muskets and gunnery interfered with the age-old feudal class-relationships of Japan. The lords (*daimyō*[1]), armed retainers (*samurai*[2,3]), professional soldiers (*bushi*[4]) and knights errant (*ronin*[5]) were accustomed to look down on the local worthies (*ji-samurai*[6]), yeomen (*goshi*[7]), peasant-farmers (*ashigaru*[8]) and the artisans and merchants (*heimin*[9]). Putting weapons in the hands of the common people which would enable them to kill at a distance the finest lord or knight in the country, was an affront to all right-thinking feudal values. As the Governor of Izu, Matsudaira[10], said at the time of the Shimabara Rebellion, 'firearms destroy the difference between soldiers and peasants'.[a] Musketry also interfered with feudal knightly customs such as the single combat of champions (cf. pt. 6, c (2) above), and it had the effect of transferring skill from the field commander to the gunsmith and the arsenal mechanic. No wonder the Japanese military aristocrats, so different from the non-hereditary bureaucratic élite of China, which for the most part of two thousand years could successfully keep down the military in a subordinate place, intensely disliked both musketry and gunnery.

Secondly, there was a great mystique of swords in Japan, as opposed to firearms. The 'privilege of name and sword' (*myōji taitō*[11]) was forbidden to peasant-farmers and merchants. Sword-play, involving as it did elegant body-movement, was esteemed as an aspect of aesthetics.[b] In contrast the musketeer's motions were uncouth or humdrum, and that remarkable MS. work, the *Inatomi-ryu Teppō Densho*[12],[c] depicts the figures illustrated clad only in the *fundoshi*[13] or characteristic Japanese apron-loincloth, as if to emphasise their plebeian or ugly, unadorned and unaccoutred, nature.[d] Third, the warrior class was much more numerous relatively in Japan than in the Western world (perhaps eight per cent of the population as against 0·6 per cent in England), so the prejudice against firearms was more able to find a voice in public policy. This may go some way to explain why gunpowder successfully played its part in overthrowing occidental feudalism while it could not easily do so in Japan—apart from all the other factors such as the city-State tradition of Europe, and the burghers and merchant-adventurers who had for centuries been waiting in the wings.[e]

---

[a] Murdoch (1), vol. 2, p. 658.

[b] Of course, such martial arts, with a strong aesthetic element, were practised in China too (cf. Lu Gwei-Djen & Needham (5), p. 303), often as part of Taoist self-cultivation, but somehow they never came into conflict with the serious public business of suppressing rebellions, establishing new dynasties, or repelling invasions.

[c] Cf. p. 390 above.

[d] This work, which has 32 illustrations, was produced in +1595 for one of the Inatomi family, famous for gun-making. A copy of +1607 is in the New York Public Library, Spencer Coll. MS. 53. Perrin (1) has reproduced several pages of it, pp. 43 ff.

[e] Seventeenth-century group portraits such as the 'Honorable Company of the Musketeers of Antwerp' sipping their wine from conical glasses, are deeply symbolic here.

[1] 大名　　　　[2] 士　　　　　[3] 侍　　　　　[4] 武士　　　　　[5] 浪人
[6] 地士　　　　[7] 郷士　　　　[8] 足輕　　　　[9] 平民　　　　　[10] 松平伊豆の守
[11] 名字帶刀　　[12] 稻富流鐵砲傳書　　[13] 褌

Fourthly, there was a great wave of xenophobia in Japan after its first contacts with Westerners. As we know, Christianity was illegal after +1616. The English gave up their factory in +1623, the Spaniards were expelled in +1624, the Portuguese in +1638, and the Dutch were confined to Deshima Island from +1641 onwards. Obviously firearms were (and always had been) something essentially foreign. And fifth, the Japanese could close their country completely because they were, as a single political entity, more isolated geographically than any country on the Eurasian continent could be. Historically they had always been isolated too, far more so than England from Europe or Ceylon from India.

In a brilliant and stimulating book, Noel Perrin (1)[a] has used this history to demonstrate that over a period of time a certain people did succeed in 'putting back the clock' of military technology, or at least in stopping its hands. He argues that this was a successful instance of the 'selective control of technology', and that it ought to inspire us with the conviction that the atomic arms race is not inevitable, nor the holocaust of nuclear warfare either, which no one can win. He claims that a 'no-growth' community is perfectly compatible with prosperous and civilised life,[b] and that human beings are less the passive victims of their own knowledge and skills than most people in the modern world suppose.[c] The history of Tokugawa Japan demonstrates, he feels, that men *can* give up a new and dreadful weapon. With much of this argument we deeply sympathise, but Japan was a very special case; the decision to abandon firearms was essentially a feudal and anti-democratic one, which could only work because it was possible to isolate the whole country indefinitely from the rest of the world.

Today no people is an 'Ilande unto itselfe' (in John Donne's famous words); orbiting satellites keep watch on everybody, the trade nexus links all communities, telecommunications connect all parts of the globe, and for good or ill the world is one. Rather do we feel that the mastery of nuclear energy as well as atomic weapons, of laser beams, space flight and computerisation, is something that is set before the human race to achieve. What we know, we cannot unknow. Nor can we refuse new knowledge. But we *can* decline to use. At an earlier stage[d] we quoted the words of the +8th-century *Kuan Yin Tzu*[1] book, where the writer[e] was talking about many wonders of Art and Nature—how to induce thunder in winter, how to restore the dead to life, how to make images speak, and how (strangely appropriate in the present context) to make exceedingly sharp swords. 'Only those who have the Tao', he said, 'can perform such actions—and better still, not perform them, though capable of performing

[a] To whom we are indebted for much of the material of the preceding paragraphs.
[b] Though this rather contradicts his enthusiasm for the technical progress of Tokugawa Japan, described in Tuge Hideomi (1).
[c] Cf. what Roszak (1, 2) has had to say on the 'technological imperative'.
[d] Vol. 2, p. 449.   [e] Probably Thien Thung-Hsiu[2].

[1] 關伊子   [2] 田同秀

them.'[a] Knowing but refraining, this is the lesson that humanity must at all costs learn, for the price is survival, continued existence, itself, no less.[b]

(19) GUNPOWDER AS PROPELLANT (II); THE DEVELOPMENT OF THE ROCKET

Now at long last we come to the problem of the rocket. It is a peculiarly difficult one for many reasons, not least because a device changed fundamentally while a name did not. 'Fire-arrow' (huo chien[2]), as we have seen (pp. 11 ff.), was a term applied in Thang times and much earlier to the incendiary arrow; but in the days of the Mongolian dynasty, the Yuan, it had come to mean the rocket. Nobody noticed the change, and no-one gave a thought to the difficulties which in the course of centuries it would cause for historians of technology. Thus rockets were certainly in use in warfare by about +1280, but that is just the time when Ḥasan al-Rammāḥ[c] was calling them 'arrows of China' (sahm al-Khiṭāi), which implies that they had already been known and used there for some time previously (p. 41 above). Their presence in Marcus Graecus, at a roughly similar date, is rather less certain; his 'ignis volantis in aere' may have been rockets, but were much more probably fire-lances.[d] At another point earlier on (pp. 153, 226) we were driven to the conclusion that the rocket is almost certainly not described in the Wu Ching Tsung Yao of +1044; the 'gunpowder whip-arrow' (huo yao pien chien[3]) was rather an incendiary javelin projected by two men. Yet the rockets are present in full force by +1340, so it is somewhere in those three centuries that we have to look to find their origin. We believe that it is to be sought essentially in the 'ground-rat' or 'earth-rat' (ti lao shu[4]), a firework first used for scaring troops and upsetting cavalry, then applied, with stick (the arrow shaft) and balance-weight, to long-distance trajectories.[e] But exactly when?

---

[a] Wei yu Tao chih shih nêng wei chih; i nêng nêng chih erh pu wei chih![1] (Kuan Yin Tzu, p. 20a; Wên Shih Chen Ching, ch. 7, p. 1b).

[b] To decide what to refrain from will of course necessitate great judgment. The Tokugawa Japanese knew, but refrained, in our view for the wrong reasons, and under conditions unrepeatable today. But Perrin's admiration for them was not wholly unjustified. Nor are we maintaining that pacifist reason and feeling have always been justified; later we may say something on war as an instrument of human progress. Meanwhile it may be noted that the history of pacifist philosophy in China has been told by Tomkinson (1).

Today, when many find it hard to distinguish 'terrorists' from 'freedom-fighters', we are witnessing an unprecedented 'democratisation', or better, universalisation, of sophisticated explosives and highly developed firearms. One can only hope that it is a phase which will give way to the just, equitable and healthy society of the future.

[c] Cf. Reinaud & Favé (2), pp. 314 ff.; Partington (5), p. 203.

[d] We shall remember the case of the fei huo chhiang[5] (p. 225 above) which must mean 'flying-fire spears' and not 'flying fire-spears'. It does not therefore attest the presence of rockets in +1232, as has so often been thought (e.g. von Romocki (1), vol. 1, pp. 46 ff.; Feldhaus (1), col. 853), though by that time they may well have existed (cf. p. 512 below). Davis (10) got it right.

[e] See pp. 477 ff. below. Of course, it may have originated as a recreational firework.

---

[1] 惟有道之士能爲之, 亦能能之而不爲之   [2] 火箭   [3] 火藥鞭箭
[4] 地老鼠   [5] 飛火鎗

Many things make this search difficult. For example, there was the secrecy generally surrounding arsenal and military supplies (cf. pp. 24, 93);[a] and there happens to be a dearth of battle accounts between +1100 and +1300 which mention rockets or anything similar. They do not seem to have been used in the wars between the Sung and the Chin Tartars which culminated in the fall of Khaifêng (+1126). Yet the fire-lance, as we have noted (p. 223) was already in use by +950, and over the centuries the strong backward pressure on the arms of the wielder, the recoil, must have become well known. Moreover, during fights a chance sword-cut which hacked off the haft of a fire-lance would have released its flame-throwing tube to fly swishing backwards, perhaps up into the air.[b] And there is another close connection here, in that fire-lances were occasionally rocket-propelled (cf. p. 484 below). We shall suggest that the rocket originated, as it were, from the tube of the fire-lance filled with gunpowder, but detached from its handle, and therefore free to travel in whatever direction chance dictated.

In these circumstances the best plan will be to describe first the several types of rocket weapon at the time when they first come fully into the limelight, and then to look again at their history with a view to sketching out as far as we can their probable origin and development. Here it will be desirable to follow the most logical order of arrangement possible, and this we try to do in the following sub-sections. Such order cannot be found in the military compendia of the Yuan and Ming themselves, where the weapons are all jumbled up with juxtapositions which are sometimes quite confusing; each text and each illustration have to be carefully studied before one can decide to what genus and species the weapon in question belongs.[c]

(i) *The* ti lao shu[4] *(ground-rat or earth-rat) in military use*

This contraption we met with at a much earlier point when speaking of civilian firework displays (p. 134), concluding that it was a tube of bamboo filled with low-nitrate gunpowder and having a hole in the septum at one end, then lit and allowed to rush violently about all over the floor or the ground, in a rudimentary form of rocket-propulsion. The thing could just as easily be made by floats to skate over the surface of water, when it was called *shui shu*[5]; and it took other

---

[a] To take a concrete case, gunpowder weapons are completely excluded from Wang Ying-Lin's *Yü Hai* encyclopaedia, though compiled as late as +1267.

[b] We owe this point to Dr Nigel Davies of RARDE.

[c] There is quite a literature on the origin and development of rockets in China, but most of it is misleading when not positively wrong, as for instance the paper of Strubell (1).

At an earlier point (p. 108) we drew attention to the possible significance of the fact that in the Germanic languages gunpowder is called *kraut*, normally a vegetable drug, like *yao*[1] in *huo yao*[2]. Now we find that the Dutch word for rocket is *vuurpijlen*, as if it was a direct translation from *huo chien*[3], i.e. fire-arrow. It was Winter (5), p. 10, who drew attention to this. Such strange similarities are at least worth meditating.

[1] 藥   [2] 火藥   [3] 火箭   [4] 地老鼠   [5] 水鼠

forms also, as we shall see (p. 514 below). Within the military realm we find it mostly enclosed in weak-casing bombs which released a dozen or more of these mini-rockets to annoy the enemy's horsemen—and foot soldiers too. This was perhaps the most primitive form and first appearance of jet-propulsion in warfare.

Perhaps the type-specimen is the 'water-melon bomb' (*hsi kua phao*[1]),[a] and significantly it appears in the oldest stratum of the *Huo Lung Ching* (Fig. 190), which would date it to the first half of the +14th century at least. Here we translate the relevant passage:[b]

The 'water-melon bomb' is the most efficacious weapon for defending city-walls, best used from a high position when (the enemy) is below. Inside the bomb there are one or two hundred small (iron) calthrops, and fifty to sixty 'fire-rats' (*huo lao shu*[2]). [On the surface of each fire-rat tube three little hooks are fastened, and each such tube has a fuse going to it. All these are put into the bomb first before it is filled with gunpowder, and this should be packed in it loosely, not pressed down. The bomb is now sealed, two layers of hempen cloth with twenty layers of strong paper being glued over it, after which it is dried in the sun. The circumference of the bomb is divided into three parts, and three small holes are bored to take in three fuses. Another hole is bored directly at the top, and a small two-inch long bamboo tube is put in. A fuse goes right into the bomb through this, to make the bomb explode evenly, and the four fuses are connected together (at the top).]

When the enemy appears below the city wall the main fuse is lit, then when the burning reaches the point of junction with the four subsidiary fuses, one throws it down into the midst of the enemy. The four fuses are necessary to prevent the flame going out as the bomb is dropped. [At the moment of explosion, even the coating can cause some damage, but in a trice the calthrops are scattered all over the ground, while the fire-rats rush about in confusion, burning the soldiers. Thus the attackers can only run away, and as they do so the calthrops hurt their feet and injure them when they fall over. They never dare to come back beneath those city walls again.][c]

Thus it would seem that each fire-rat had its own fuse and was not just ignited by the flames of the main explosion. The illustration is instructive, first because it shows inside each mini-rocket a rectangle which we think indicates the bored cavity that gives equal burning; and secondly because the three hooks on each fire-rat are clearly shown. These evidently were designed to attach themselves to the clothing and accoutrements of men and horses, causing lesions and other damage as they burnt themselves out.

Another projectile of similar type was the 'rumbling-thunder bomb' (*hung lei phao*[3]), also in the oldest stratum of the *Huo Lung Ching*.[d] It was more like a grenade in size,[e] and contained poisons as well as gunpowder, but it had its

[a] *HLC*, pt. 1, ch. 2, p. 8*a*, *b*; *HKPY*, *ibid.*; Hsiangyang ed., *HCT*, p. 13*a*.

[b] Passages in square brackets belong to the enlarged version in *WPC*, ch. 122, pp. 24*b*, 25*a*, *b*, and *PL*, ch. 12, pp. 16*a*, *b*, 17*a*.

[c] Tr. auct.

[d] *HLC*, pt. 1, ch. 2, p. 13*a*, *b*; *HKPY*, *ibid.*; Hsiangyang ed., *HCT*, p. 15*b*.

[e] Because it used sun-dried mule droppings as the spherical moulds or matrices round which to wrap the cloth and paper, after which they were broken up and taken out through the fuse hole.

¹ 西瓜砲   ² 火老鼠   ³ 轟雷砲

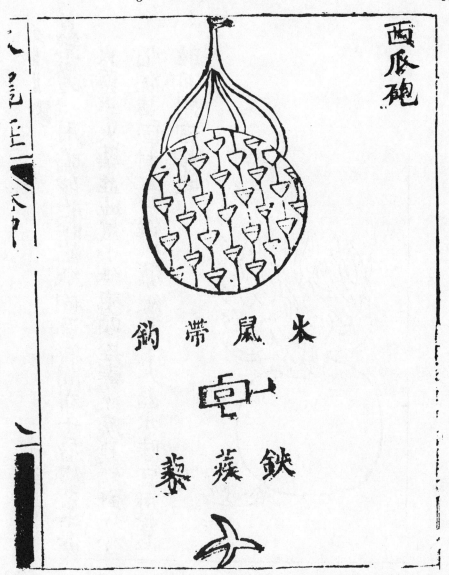

Fig. 190. The origin of rockets; the 'water-melon bomb' (*hsi kua phao*), containing a number of hooked 'earth-rats or ground-rats' (*ti lao shu*) or 'fire-rats' (*huo lao shu*), i.e. mini-rockets. From *HLC*, pt. 1, ch. 2, p. 8a.

ground-rats (*ti shu*[1]) made of carton tubes, and its iron calthrops, so it was quite similar in conception (Fig. 191). If carefully made, it 'caused the enemy to become dizzy and disheartened', and could be used either on land or afloat. In the beginning these weapons could have had an element of real surprise, since an enemy would expect attack from above or horizontally rather than at ground level from objects originally thought of as toys.

[1] 地鼠

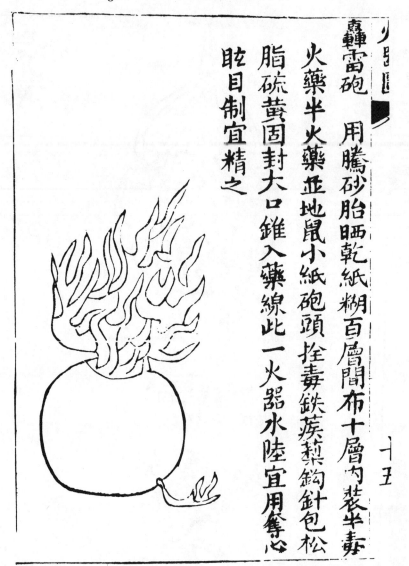

轟雷砲　用騰砂胎晒乾紙糊百層間布十層內裝半毒

火藥半火藥並地鼠小紙砲頭拴毒鐵蒺藜鈎針包松

脂硫黃固封大口錐入藥線此一火器水陸宜用籌心

眩目制宜精之

Fig. 191. Another bomb containing mini-rockets along with calthrops and poisons (*HLC*, pt. 1, ch. 2, p. 13*a*, here from *HCT*, p. 15*b*). It also belongs to the oldest stratum of the *Huo Lung Ching*.

Rather larger was the 'bandit-burning vision-confusing magic fireball' (*shao tsei mi mu shen huo chhiu*[1]), but again on the same principle, yet not described till the +17th century.[a] It has a clay matrix, over which were pasted many layers of paper with persimmon glue to make a casing, then filled with gunpowder, calthrops, ground-rats and fire-crackers, with the addition of 'flying sand' and 'magic smoke' composition. The *Huo Lung Ching* says:

[a] *HLC*, pt. 2, ch. 3, pp. 3*a*ff.; *WPC*, ch. 130, pp. 8*a*, *b*, 9*a*.
[1] 燒賊迷目神火毬

Troops carry these bombs in bags made of oiled string. In combat the soldiers light them, and throw them into the enemy's position or camp; as they explode, the (iron) calthrops are thrown about underfoot, causing wounds, while the ground-rats rush in all directions and get into the enemy's armour, hopping and bouncing up and down, so as to bring about alarm and confusion. Opportunity should be taken of this to press the attack by fire, using guns and bombards. In this way the troops of the enemy never fail to be defeated.[a]

Finally the ground-rats occur again, this time fitted with sharp little spikes, in a device called the 'fire-brick' (*huo chuan*[1]), though very different from our meaning of the term.[b] It was just a bomb (Fig. 192) made in elongated rectangular shape and filled with individually fused mini-rockets amidst the gunpowder. On ignition the brick was hurled into the enemy's camp to set it alight and sow confusion.[c]

(ii) *Rocket arrows*

The classical 'fire-arrow' (*huo chien*[2]) is shown in the *Wu Ching Tsung Yao* (+1044)[d] with the explanation that it is sent on its way from bow or crossbow, the amount of gunpowder attached to it depending on the strength of the bow.[e] Therefore it is clearly an incendiary arrow using a low-nitrate composition. But the *huo chien*[2] in the *Huo Lung Ching* is entirely different, for it is a perforating shock-weapon rocket-propelled. The name might remain the same, but the device was something entirely different.

Some time between +1150 and +1350 it occurred to someone who had seen ground-rats leave the ground and fly a short distance through the air, that if such a tube were attached to a feathered stick, i.e. the arrow-shaft, it would propel it with sufficient force to enable one to dispense with bow and crossbow altogether. This was a fundamental discovery. The oldest stratum of the *Huo Lung Ching* says:[f]

One uses a bamboo stick 4 ft 2 in. long, with an iron (or steel) arrow-head 4·5 in. long [smeared with poison; and some smear that on the rocket-tube too.] Behind the feathering there is an iron weight (*thieh chui*[3]) 0·4 in. long. At the front end there is a carton tube bound on to the stick, where the 'rising gunpowder' (*chhi huo*[4]) is lit [and it is oiled to prevent its getting wet.] When you want to fire it off, you use a frame shaped like a dragon, or else conveniently a tube of wood or bamboo to contain it [or launcher boxes of different kinds].

---

[a] Tr. auct.
[b] *HLC*, pt. 2, ch. 3, pp. 6b, 7a; *WPC*, ch. 130, pp. 18a, b, 19a, b.
[c] One of the two specifications stipulates also poisonous smoke-producing material. Two centuries later (c. +1565) the fire-brick is mentioned again by Chhi Chi-Kuang (*Chi Hsiao Hsin Shu*, ch. 18, p. 26a; *Lien Ping Shih Chi, Tsa Chi*, ch. 5, p. 29b), but now classed with obsolete weapons no longer made in the arsenals. Wang Ming-Hao still talks about it too at the end of the century (*Huo Kung Wên Ta*, p. 1296).
[d] Ch. 13, p. 3a, b.
[e] Even the nock on the end of the arrow is depicted.
[f] *HLC*, pt. 1, ch. 2, p. 22a, b; *HKPY, ibid.*; Hsiangyang ed., *HCT*, p. 20a. Sentences in square brackets come from *WPC*, ch. 126, pp. 4b, 5a. Tr. auct.

[1] 火磚   [2] 火箭   [3] 鐵墜   [4] 起火

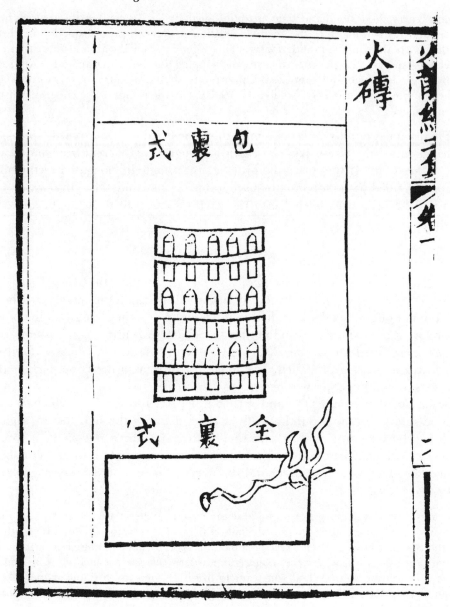

Fig. 192. The 'fire-brick' (*huo chuan*) bomb, filled with the mini-rockets bearing sharp little spikes. From *HLC*, pt. 2, ch. 3, p. 6*b*.

The illustration (Fig. 193) shows two launching cylinders, one with a dragon head. Very significant is the mention of the balance-weight at the tail; it must soon have been obviously necessary to make up for the weight of the rocket-tube, and as the gunpowder burnt away it would have added force to the rocket's velocity. This was a second aspect of the invention. A passage from the *Wu Pei*

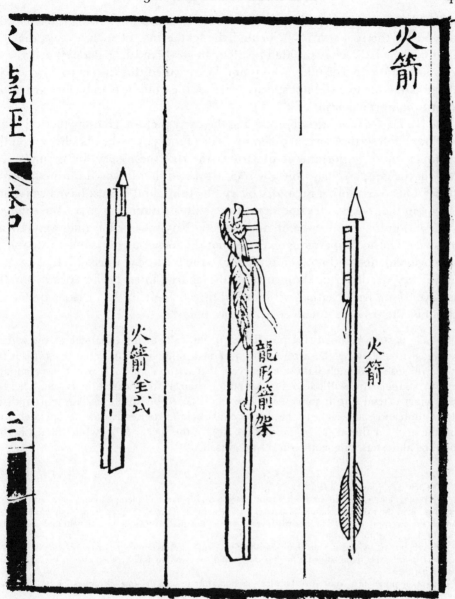

Fig. 193. The oldest illustration of rocket-arrows, from *HLC*, pt. 1, ch. 2, p. 22*a*. Although this must date from the neighbourhood of +1350, there is good reason to think that the rocket-arrow had been known and used at least a century and a half earlier. Here we see also two launching cylinders, the middle one with a carved dragon head.

Chih spells it out more clearly.[a] It says: 'An iron weight (*thieh chui*[1]) is fixed at the rear end (of the rocket-arrow), behind the feathering, of such a mass that the fulcrum of the balance is situated just four finger-breadths (*ssu chih*[2]) away from the mouth of the rocket-tube.' Davis and Ware called this the centre of gravity;[b] unfortunately the text did not specify whether the point was to be forward of the rocket-tube's orifice or aft of it.

The *Wu Pei Chih*, besides reproducing the early picture, gives further information. First, it describes several different kinds of rocket war-heads.[c] But secondly, and much more important, it illustrates the drill necessary for boring out a cavity in the gunpowder within the rocket-tube so that it would burn equally as it flew.[d] This was another great discovery, the third, and it must have been made early on in the rocket's development. In one illustration (Fig. 194)[e] the cavity is shown within the rocket-tube; in another (Fig. 195)[f] the drill is diagrammatically drawn.[g] The accompanying text says that the rocket-arrow is most valuable in land engagements, and not at all inferior to the bird-beak musket (cf. p. 432). It can be very deadly. But the centre of the charge must be bored out, for the 'fuse-eye' (*hsien yen*[3]),[h] either with an awl or a bow-drill; the artisans prefer the latter, but the result is not so good. It goes on:

If the hole is straight-sided (i.e. parallel with the walls of the tube) the arrow will fly straight; if it is slanting the arrow will go off at a tangent. If the hole is too deep the rocket will lose too much flame at the rear, if it is too shallow it won't have enough strength, so the arrow will fall to the ground too soon. If the rocket-tube is 5 in. long, the cavity must extend into it some 4 in. The shaft has to be absolutely straight, and the (rocket-tube and end-weight of the) arrow must balance perfectly when suspended 2 in. from the neck, or throat (*ching*[8] i.e. the nozzle), of the rocket-tube, while the feathering should be almost as long as the rocket-tube itself.[i]

[a] *WPC*, ch. 127, p. 12a, tr. auct.
[b] (1), p. 532. Chinese engineers from Thang times onwards had had plenty of experience with counter-weighting, as in the hydro-mechanical link-work escapement of clocks (Vol. 4, pt. 2, pp. 459–60; Needham, Wang & Price (1), pp. 50–1). The steelyard, or balance of unequal arms, was both ancient and prevalent in China (Vol. 4, pt. 1, pp. 24 ff.).
[c] Ch. 126, pp. 5b, 6a, b, 7a. Thus there was the 'flying knife rocket-arrow' (*fei tao chien*[4]), the 'flying spear' (*fei chhiang chien*[5]), the 'flying sword' (*fei chien chien*[6]) and the 'swallow-tail' (*yen wei chien*[7]). We refrain from reproducing them.
[d] This is the principle of 'concentric burning', used in order to keep the area of combustion surface as near as possible constant. Cf. Anon. (161), Vol. 1, pp. 580–1, Vol. 2, pp. 363–4.
[e] From *Ping Lu* (+1606), ch. 12, p. 44a, equivalent exactly to *WPC*, ch. 126, p. 2b.
[f] *WPC*, ch. 126, p. 3a; *PL*, ch. 12, p. 46b.
[g] This was the 'thorn' of early European rocket-makers. Kyeser was perhaps the first to mention it (+1405), and of course it is in Schmidlap (+1591) and many others. Cf. Ley (2), pp. 60 ff., 63.
[h] The technical term at that time for the cavity.
[i] *WPC*, ch. 126, p. 3b, 4a, tr. auct. adjuv. Davis & Ware (1), p. 532. The passage is a good deal older than might be supposed, for it is verbally identical with what the great general Chhi Chi-Kuang said in his *Chi Hsiao Hsin Shu* of +1560 (ch. 15, p. 14a, b).

[1] 鐵硾   [2] 四指   [3] 線眼   [4] 飛刀箭   [5] 飛槍箭
[6] 飛劍箭   [7] 燕尾箭   [8] 頸

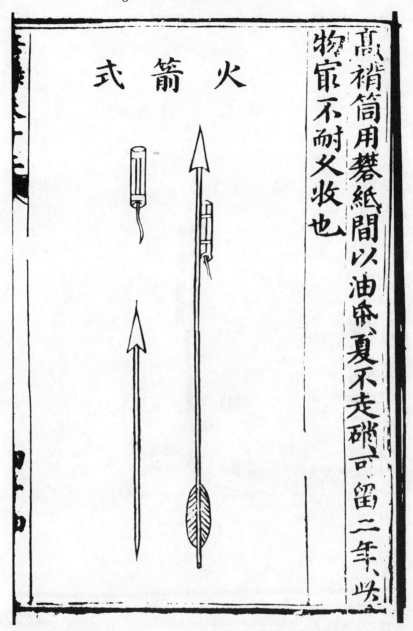

火箭式

高梢筒用碁紙間以油帚夏不走碩可留二年此
物宜不耐久收也

Fig. 194. A picture of the rocket-arrow from *PL*, ch. 12, p. 44*a*, important because it shows the cylindrical cavity within the rocket tube which was needed for even and equal combustion during flight.

Fig. 195. Diagrammatic drawing of the drill or 'thorn' for making the cavity in the rocket tube. From *WPC*, ch. 126, p. 3*a*.

Here comes the fourth part of the invention. By +1300 at least the rocket-makers must have known that it was desirable to constrict the orifice of the rocket-tube in such a way as to increase the flow-velocity of the issuing gases, and therefore the retro-active force of the combustion. This was the principle of the 'choke',[a] later called in Europe the Venturi[b] 'waist' or nozzle.[c] Finally, a description is given of large-diameter rocket-arrows (ta thung huo chien[1], weighing as much as two catties, with a range of some 300 paces (500 yards), and again a drilling apparatus is illustrated.[d]

In Chhi Chi-Kuang's[2] time (+1550 to 80) the rocket-arrow was much prized as a war weapon.[e] It would fly into the enemy's rear as well as his front line, to left or to right, keeping everyone in alarm, not knowing where it was going to strike—and the launching side would of course not know either, since accurate aiming was distinctly difficult if not impossible. Hence the tendency to release flocks of rocket-arrows at the same time, from the launching-frames which we shall described in due course (pp. 486 ff.); as also the practice of poisoning the arrow-tips to make a direct hit much worse. It was said to be as potent as the hand-gun, penetrating wooden planks an inch thick and piercing metal breast-plates. As for the drilling of the rocket cavity,[f] it was recommended that the boring tool be frequently wetted with water to reduce the friction which was capable of igniting the composition, and that a drill should be discarded for re-sharpening after half-a-dozen borings. Apparently rockets deflagrated or exploded quite often in the making, so directions were given for dispersing the work and the stores of powder among separate buildings. Great care was taken in making and rolling the strong carton case of the rocket-tube, but sometimes iron tubes were employed, especially for the constricted end (the choke) whence the gases escaped.

Exactly what kind of gunpowder was used for the rockets of the Yuan and Ming is not very clear, but the *Huo Lung Ching* lists several compositions the names of which would have been appropriate.[g] For example, there was 'flying gunpowder' (*fei huo yao*[3]), 'wind-opposing gunpowder' (*ni fêng huo yao*[4]), 'flying-in-the-air gunpowder' (*fei khung huo yao*[5]), 'day-rising gunpowder' (*jih chhi huo yao*[6]) and 'night-rising gunpowder' (*yeh chhi huo yao*[7]). But while the text gives

---

[a] Cf. Brock (1), p. 183.

[b] See Rouse & Ince (1), pp. 134 ff., 189; Biswas (3), pp. 272 ff., 305.

[c] Giovanni-Battista Venturi (+1746 to 1822) was a hydrodynamician very little noticed by historians of science, in spite of his important book (Venturi, 1) Cf. Anon. (161), vol. 1, pp. 206–7, 248–9. Hence the Venturi flow-meter, and a device embodied in most of our gas 'geyser' water-heaters.

[d] *WPC*, ch. 126, pp. 8b, 9a.

[e] Cf. *Chi Hsiao Hsin Shu*, ch.15, pp. 14a, b, 15a; ch. 18, p. 28a; *Lien Ping Shih Chi, Tsa Chi*, ch. 5, pp. 27b, 28a, b, 29a, b, 30a.

[f] Today a conical 'spindle' is used, on top of which the packing of the gunpowder is done; cf. Brock (1), p. 183.

[g] *HLC*, pt. 1, ch. 1, pp. 6a–11b; parallel texts in *HKPY* and *HCT*.

[1] 大筩火箭　　[2] 戚繼光　　[3] 飛火藥　　[4] 逆風火藥　　[5] 飛空火藥
[6] 日起火藥　　[7] 夜起火藥

many constituents of each of these, including saltpetre, actual quantities are listed only for two of them (the last-named)[a], and then the sulphur is so low as to cast doubt on the validity of the percentages. Perhaps the original quantities were all removed as a security measure before the book was printed. But we know (p. 351 above) that the nitrate must have been in the neighbourhood of sixty per cent to work a successful rocket.[b]

The technical affinities between the fire-lance and the rocket have already been pointed out (p. 472), and one might therefore well expect to find some attempt at combining the two. This indeed occurs, under the name of the 'tiger-catching-up-with-the-sheep rocket-arrow' (*i hu chui yang chien*[1]).[c] The explanation says that this is a five-foot-long shaft (Fig. 196)[d] with a trident at the business end and two rocket-tubes just behind it. At the rear end there are two more gunpowder tubes secured to the shaft, but these are fire-lances, not rockets, and are ignited automatically as the rocket is nearing the end of its course, said to attain 500 paces (830 or so yards).[e] It can set light to the enemy's wooden defences and ships; one man can use it yet a hundred men will be terrified of it, especially if poison is applied to the trident.[f] Verily, recondite is the craft of this weapon,[g] says the text—but on the principle of the survival of the fittest it can hardly have been all that effective. Still, a flock of them could have been rather a nuisance. Such was what could really be called the 'flying fire-lance'.[h]

Thus it would appear, looking back, that in spite of its seeming simplicity four distinct inventions had to be combined in the development of effective rocket flight. First, there was the basic idea of applying a ground-rat tube to a projectile, and secondly the balancing of the whole to give the arrow an adequate range. Thirdly there was the drilling of an internal cavity to promote equal areas of combustion surface, and fourthly the addition of a waist, throat or choke, in fact a Venturi constriction,[i] to accelerate the flow-velocity of the discharged gases, thus increasing the propulsive reaction.[j]

At some time during the +14th or +15th centuries it occurred to some ingenious Chinese artificer that if a rocket could be made to go, it could also be made

[a] *Chhi huo yao* often appears in the accounts of rocket-arrows in the military books.

[b] The standard rocket composition is 63·6 : 22·7 : 13·6 (Brock (1), p. 188).

[c] *HLC*, pt. 2, ch. 2, p. 22*a*, *b* (not the oldest stratum); *WPC*, ch. 127, pp. 3*b*, 4*a*. Nevertheless, on intrinsic grounds, this weapon could be considered rather old, quite probably developed soon after the rocket itself.

[d] From *WPC*; that in *HLC* is identical, save that the former has (more logically) 'two tigers' (*erh hu*[3]).

[e] Standard rocket ranges at the end of the +16th century are usually given as 600–700 paces, or about 1000 yards (*Huo Kung Wên Ta*, p. 1293).

[f] The second part of this sentence is only in the longer *WPC* version.

[g] *Ta yu hsüan miao*[2]. Could one not suspect a Taoist echo here? Cf. p. 117 above.

[h] Cf. pp. 171, 225 above.

[i] Among rocket engineers this is often called a Laval convergent-divergent nozzle, after the Swede Carl de Laval who introduced it for gas turbines in 1889. Cf. Baker (1), p. 18.

[j] The arrow-shaft itself can hardly be counted as an invention, but presumably the stick of later rockets must derive from it, and therefore indirectly from the shaft of the even more ancient fire-lance. Modern pyrotechnists say simply that the stick 'balances and directs the flight' (Brock (1), p. 183). Spinning, fins and wings, ultimately, it seems, took over this function.

[1] 一虎追羊箭   [2] 大有玄妙   [3] 二虎

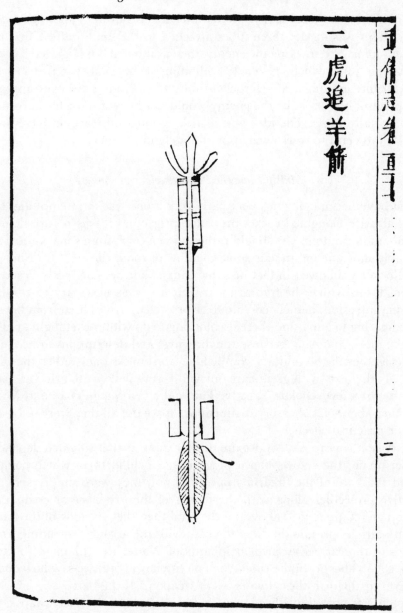

二虎追羊箭

Fig. 196. A manifestation of the technical affinity between the fire-lance and the rocket, the two combined in one device. The 'tigers-catching-up-with-the-sheep rocket-arrow' (*erh hu chui yang chien*), from *WPC*, ch. 127, p. 3*b*. Two rocket-tubes are placed behind the trident, but that is not the only warhead, for two fire-lances are carried just ahead of the feathering.

to come, at least theoretically. Hence the 'flying powder tube' (*fei khung sha thung*[1]).[a] This was in fact three tubes attached to the same staff. A first rocket-tube sent it forward towards the enemy, then as it burnt out it ignited a charge in the leading tube which expelled a blinding lachrymatory powder over the enemy, before igniting a return rocket-tube and so sending the contraption back to its point of origin. Thus the enemy would not know from what direction the attack actually came. The idea was a striking one, but it would have involved great skill to get it to work even approximately in practice.[b]

(iii) *Multiple launchers and wheelbarrow batteries*

It must have become obvious very early that if one was to attempt any kind of aim at all with elongated rocket-propelled projectiles it was no use flourishing them about at random, one should rather launch them from some kind of frame, preferably movable on an axis so as to allow of some choice of trajectory (Fig. 197). Rocketry followed in fact just this course, and we can easily describe the different forms which the frames took.[c] But first it is necessary to eliminate a confusing intrusion, namely co-viative or projected arrows fired from fire-lances approximating to guns, and therefore nothing to do with rocket flight at all. This is all the more confusing because the drawings and descriptions are completely mixed together in the military compendia, and unless one studies the pictures and reads the texts with great care one will certainly come to grief, as has happened to not a few scholars already. The soldiers of Sung, Yuan and Ming did not bother about classificatory distinctions, as we do; all they were interested in were the practical effects.

The reason why we say 'approximating to guns' is that so much depended on whether or not the arrows had a plug or wad behind them which completely blocked the bore of the firearm's barrel. If not, they were simply shot out as co-viative projectiles along with the flames of the fire-lance at comparatively short range (cf. pp. 236 ff. above); if they did then they partook of the nature of cannon-balls, as presumably was the case with the arrows protruding from the muzzles of the early European bombards of Walter de Milamete (+1327, cf. pp. 10, 287–8 above). In the sub-section on fire-lances (13) we saw how difficult it can be to distinguish these two types of weapon.[d] If the barrel was of wood or bamboo it was probably a fire-lance, if of bronze or iron it was perhaps a proto-

---

[a] *WPC*, ch. 129, pp. 7*b*, 8*a*. Hsü Hui-Lin (1) deserves credit for having taken notice of it. There was a model of it in the National Military Museum in Peking in 1964.

[b] A closely similar 'come-and-go' rocket occurs in the Sibiu MS. of Konrad Haas, dating from about +1560 (von Braun & Ordway (1), p. 11). The question of how derivative from the Chinese sources this could have been might admit of a wide solution.

[c] It is surely needless to emphasise the great role played at the present day by all forms of launchers, whether for military uses or for space-flight. Cf. Humphries (1), p. 140 and opp. p. 150.

[d] Davis & Ware (1), p. 533, called them all guns or bombards, but they were not very sensitive to the distinction we have to make here.

[1] 飛空砂筒

Fig. 197. Oblong-section rectangular rocket-launcher, with all the rockets ignited and sent on their way by one fuse, the *shen huo chien phai* (*HLC*, pt. 2, ch. 3, p. 2*a*).

gun. If there was no vase-shaped bulge, indicating a thicker wall for the explo-sion-chamber, then it was a fire-lance; if one is mentioned or illustrated, then it was probably a kind of early gun. If the range is said to be short, it was a fire-lance; if it was some 500 yards or more, as often stated, then it was more likely to be a gun. This is why long ranges are so confusing, because they do not neces-sarily imply rocket-propulsion, as some have thought.[a]

Among the fire-lances or proto-guns we have already described, the two sim-plest cases involved only one arrow,[b] but there was another which shot three at a time,[c] and yet another which discharged many.[d] To these we can now add sever-al more, the 'triple tiger-halberd' (*san chih hu yüeh*[5]) delivering three arrows,[e] the 'sevenfold tube arrow' (*chhi thung chien*[6]) sending out seven,[f] the 'nine-dragon arrows' (*chiu lung chien*[7]) shooting nine at a time,[g] and the 'hundred-aimed bow-like arrow-shooter' (*pai shih hu chien*[8]), letting off ninety-six from six tubes at one ignition.[h] All these are relevant to the present discussion only because they are scattered disorderly in the books among the true rocket-launchers, to which we must now turn. It is significant that none of the projectile arrows in these quasi-guns ever show rocket-tubes.

The most succinct means of surveying the launchers is tabulation, and this is done in Table 6, passing from the simplest to the most complicated. We must remember that all the data come from books written just before and after +1600, but it may be assumed that the simpler forms would go back one or two centu-ries before that time.[i] Broadly speaking, three materials were used for the laun-chers, basketry (cf. Fig. 198), bamboo tubing, and woodwork. All were provided with internal grids or frames to hold the individual rocket-arrows apart (Fig. 199),[j] and there was a marked tendency to make the launchers more or less

---

[a] Bows and crossbows are of course not at issue here at all.

[b] The 'single-flight magic-fire arrow' (*tan fei shen huo chien*[1]), and the 'magical (fire-)lance arrow' (*shen chhiang chien*[2]). See p. 240 above and Figs. 51, 52. The former was of cast bronze, and does have something that might have been a wad, while a long range and great impact force are noted. The latter was of ironwood, yet it also has something that could have been a wad, and again a long range is mentioned. This last was the weapon associated with the expeditions of +1406 and +1410 against Annam. See also p. 240 above. One can only call these weapons quasi-guns, leaving open the exact shade of difference between fire-lances and true guns. Much would depend on the tightness of the wad—and the long ranges may have been exaggerations.

[c] The 'awe-inspiring fierce-fire yaksha gun' (*shen wei lieh huo yeh-chha chhung*[3]); see p. 240 and Fig. 53 above.

[d] The 'lotus bunch' (*i pa lien*[4]); see p. 243 and Fig. 54 above.

[e] *HLC*, pt. 1, ch. 2, p. 26*a, b*; *WPC*, ch. 127, pp. 4*b*, 5*a*. Here a bulge over the explosion-chamber is mentioned, but not shown in either drawing. It is called an 'iron gun' (*thieh chhung*[9]) with three barrels, but there is no indication of any plugs or wads. Since this is the oldest part of the *Huo Lung Ching*, this quasi-gun may well go back to the beginning of the +14th century.

[f] *WPC*, ch. 127, p. 7*a, b*. The arrows were to be tipped with poison.

[g] *WPC*, ch. 127, p. 8*a*. This has no text.

[h] *WPC*, ch. 127, pp. 10*b*, 11*a*. The tubes were of carton, and it is significantly said that fire should be reserved until the enemy is quite near. We refrain from reproducing any of these.

[i] It will be seen that two items are in the *Huo Lung Ching*, but neither in the oldest stratum.

[j] From time to time there is mention of arrow-lengths (nos. 4, 8), poison applied to the tips (nos. 4, 12), and tail-end balance-weights (nos. 7, 9), etc. but we need not go into further detail. Also the usual romantic names are in the Table, so we omit them here.

[1] 單飛神火箭   [2] 神鎗箭   [3] 神威烈火夜叉銃   [4] 一把蓮
[5] 三隻虎鉞   [6] 七筒箭   [7] 九龍箭   [8] 百矢弧箭   [9] 鐵銃

Table 6. *Types of rocket-launchers*

	Nature	Name	Chinese name	HLC	WPC	PL
1	Basket-work rocket-launcher (conical)	Rocket-arrow firing basket	*huo lung chien*[1] (Fig. 198)		126/16b, 17a	
2	Basket-work rocket-launcher (cylindrical)	Mr Facing-both-ways rocket-arrow firing basket	*shuang fei huo lung chien*[2]	2/2/21a, b	127/2b, 3a	
3	Basket-work rocket-launcher (cylindrical)	Forty-nine simultaneously fired rocket-arrows	*ssu-shih-chiu shih fei lien chien*[3]		127/9b, 10a	
4	Portable bamboo rocket-arrow carrier or quiver with sling	Small bamboo rocket-arrow tube	*hsiao chu thung chien*[4] (Fig. 202)		126/14b, 15a	12/49b, 50a
5	Bamboo 3-arrow rocket-launcher	Magical mechanism rocket-arrows	*shen chi chien*[5]		126/7b, 8a	
6	Bamboo 5-arrow rocket-launcher	Five-tigers-springing-from-a-cave rocket-arrows	*wu hu chhu hsüeh chien*[6] (Fig. 199)		127/5b, 6b	
7	Smaller bamboo 5-arrow rocket-launcher	Lesser five-tigers, etc., rocket-arrows	*hsiao wu hu chien*[7]		127/6a, b	
8	Shield with racks for rocket-arrows	Tiger-head fire shield	*hu thou huo phai*[8]		129/12a, b, 14b, 15a, b	
9	Square-section rectangular rocket-launcher	Pack of 100 tigers running together	*pai hu chhi phen chien*[9] (Fig. 206)		127/11b, 12a	
10	Oblong-section rectangular rocket-launcher	Magical rocket-arrow block (or screen)	*shen huo chien phai (or phing)*[10] (Fig. 197)	2/3/3a, b	129/16a, b	
11	Elongated rectangular double-ended rocket-launcher (cf. 2)	Covey of hawks catching rabbits	*chhün ying cho thu chien*[11]		127/14b, 15a	
12	Elongated slightly flared rectangular rocket-launcher	Long-serpent enemy-destroying rocket-arrows	*chhang shê pho ti chien*[12] (Fig. 203)[a]		127/13b, 14a	
13	Flared octagonal rocket-launcher	Leopard pack unexpectedly scattering	*chhün pao hêng pên chien*[13] (Figs. 200, 201)		127/12b, 13a	
14	Flared octagonal rocket-launcher	Wasp's nest	*i wo fêng*[14][b]		127/15b, 16a	12/55b
15	Wheelbarrow rocket-launcher	Fire-frame combat-vehicle	*chia huo chan chhê*[15] (Figs. 204, 205)		132/9a, b	
16	Wheelbarrow launcher battery	Battery of fire-frame combat-vehicles	*lien lo chan chhê*[16] (Fig. 207)		132/10a, b	

[a] This, and the 'wasp's nest', are much spoken of in the *Huo Kung Wên Ta* towards the end of the +16th century (pp. 1294, 1303).

[b] This we shall come across again in our historical quest (p. 514 below).

To illustrate the danger of relying on terminology alone, a 'smaller wasp's nest' (*hsiao i wo fêng*) is illustrated and described in *WPC*, ch. 128, pp. 17b, 18a, b. It has obviously nothing to do with rocket-launching frames, but one might take it for a double-tube rocket-arrow itself if it were not too long (12 ft), and shaped like a spear or lance with a ferrule. Showers of sparks looking like wasps are said to come out of the 1 ft 3 in. wooden or bamboo canisters, causing blindness among the enemy, and the flames reach forward 30–40 ft (which is said to be the range). The weapon is thus a lachrymatory fire-lance, though several kinds of co-viative projectiles are described also. There follows a gunpowder composition corresponding to 67.8:19.1:13.1; but with additions of mercury and mercuric sulphide (cf. p. 344 above), and doubtless plant drugs also.

[1] 火籠箭 [2] 雙頭火籠箭 [3] 四十九矢飛簾箭 [4] 小竹筒箭 [5] 神機箭 [6] 五虎出穴箭 [7] 小五虎箭
[8] 虎頭火牌 [9] 百虎齊奔箭 [10] 神火箭牌(屏) [11] 羣鷹逐兔箭 [12] 長蛇破敵箭 [13] 羣豹橫奔箭 [14] 一窩蜂
[15] 架火戰車 [16] 聯絡戰車

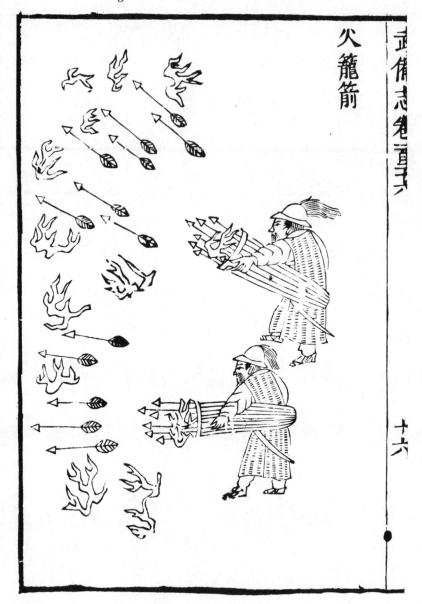

Fig. 198. Conical rocket-arrow launchers made of basketwork (*WPC*, ch. 126, p. 16*b*).

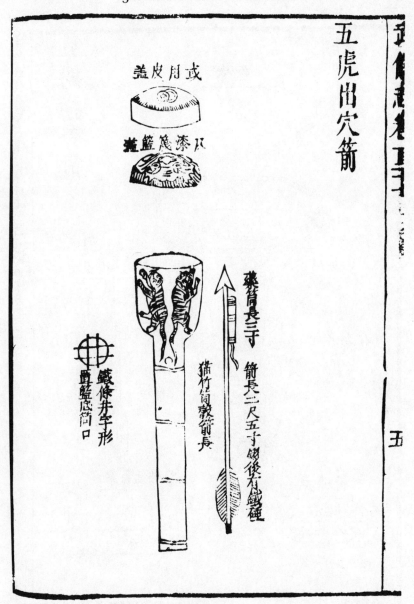

Fig. 199. Bamboo or wooden rocket-launcher with internal grid to keep the arrows apart
(*WPC*, ch. 127, p. 5*b*).

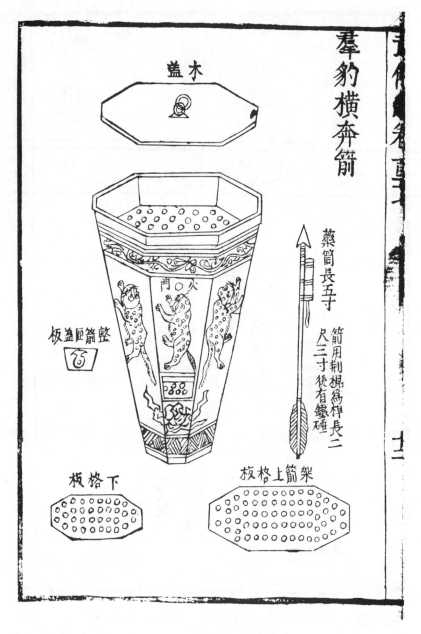

Fig. 200. Splayed conical rocket-launcher with internal diaphragm to keep the arrows apart
(*WPC*, ch. 127, p. 12*b*).

Fig. 201. Reconstruction of the same, showing the common fuse. Photo. Historical Military Museum, Peking.

conical in shape so as to ensure a wide area of dispersion of the points of impact (cf. Figs. 200, 201 and p. 483 above).[a] We translate but one passage, that concerning the portable bamboo rocket-arrow carrier with a sling (Fig. 202). The text says:[b]

The small bamboo rocket-arrow tube (*hsiao chu thung chien*[1]). Each tube holds ten short rocket-arrows, only 9 in. long, and poison is applied to the head of each. The total weight (of the tube and its contents) does not exceed 2 lb., and each soldier can carry four or five of them (on its sling) easily. The enemy would not know what exactly they were transporting. At a distance of some 100 paces (about 170 yards) away, the rocket-arrows are all fired as one. These arrows, though small, are fast, and the enemy cannot avoid them; so one soldier can do as much harm (with these arrows) as several dozen others (using more conventional arms). These rocket quivers can be carried by the personal guards of the commander, or by the detachment of soldiers surrounding the flag, or else by men scattered among ordinary fighting units. The rocket-arrows should be tested to ensure that they can penetrate thin wooden planks. If the bamboo tube is slightly raised at the time of firing, the arrows can reach over 200 paces (say 340 yards). This weapon should not be overlooked just because the arrows are so small.[c]

[a] This may have been suggested by the shape of the age-old quiver for carrying arrows about, as Mr Michael Rosen has intimated to us.
[b] *WPC*, ch. 126, pp. 14*b*, 15*a*; *PL*, ch. 12, pp. 49*b*, 50*a*, tr. auct.
[c] Rough translations of the entries for nos. 1, 9, 10, 12 and 13 have been given by Davis & Ware (1), pp. 532–3; Davis (10).

[1] 小竹筒箭

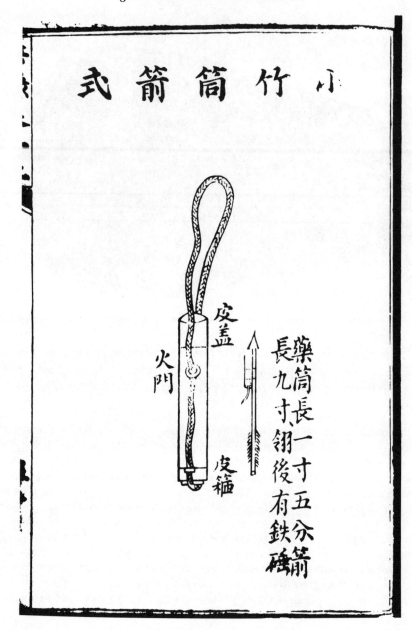

Fig. 202. Portable rocket-arrow carrier and launcher, with sling (*PL*, ch. 12, p. 50*a*).

Ho Ju-Pin gives much longer ranges for rockets, as much as 600 or 700 paces (up to 1150 yards) if made by expert technicians, but adds that they can also be let off at quite short ranges, 20 or 30 paces (c. 40 yards) when they will still do a lot of damage.[a]

Where the story becomes rather fascinating is the mounting of four flared rectangular wooden 'long-serpent' rocket-launchers (Fig. 203) in rows on wheel-barrows (Figs. 204, 205), together with two rectangular wooden 'hundred-tiger' rocket-launchers (Fig. 206), one on each side. Thus 320 rocket-arrows could be despatched almost at one time. Each wheelbarrow was further provided with three multiple-bullet proto-guns or fire-lances, two spears for repelling close attack, and curtains of leather for hiding the movements of the gunners. Two soldiers looked after the fighting and two others provided the motive power.[b] In this way veritable batteries of rocket-launchers (Fig. 207) could be wheeled into position, and (hopefully) away again, doubtless under cover of other troops.[c] Such manoeuvres, explicitly carried out in conjunction with true cannon, might form an interesting chapter, not yet written, so far as we know, in the history of artillery and rocketry.[d]

(iv) Winged rockets

Among the various stabilising devices which have been introduced in modern times for controlling rocket flight, fins and wings have been outstanding.[e] By +1741 fins were fitted to rocket-bombs by the pyrotechnist François Frézier,[f] and they have continued to be used in many recent types, such as the German 'V 2' of the Second World War.[g] But wings are also very often part of the design, as in the 'Styx',[h] 'Mace'[i] and 'Matador'[j] rocket-missiles, as well as the later

---

[a] *PL*, ch. 11, pp. 37a ff.

[b] A photograph of the scale model reconstruction of this combat-vehicle is given in Fig. 205.

[c] There is mention of a hundred such combat-vehicles working together as a battery. Cf. *Ming Shih*, ch. 92, p. 15a. On the history of the wheelbarrow (itself a Chinese invention) see Vol. 4, pt. 2, pp. 258 ff.; Vol. 1, p. 242.

[d] It may not be generally known that rocket frames or multiple launchers can still be seen at the present day if one goes to Yenshui[1] in Southern Taiwan at the time of the lantern festival (Yuan Hsiao[2]). The firework rockets are collected together in 'hives' (*fêng phao thai*[3] or *fêng tshai phao wo*[4]) and let off simultaneously. There is a graphic description by Jih Yüeh & Chung Yung-Ho (1, 1).

[e] Topologically the two are closely connected, the fin being a wing of reduced size, and generally placed towards the tail of the rocket rather than half-way along its length. Cf. Humphries (1), pp. 133 ff., 139, figs. 69, 71, 76, 77, 79.

[f] Frézier (1); Taylor (1), pp. 8–9. Frézier had many Chinese connections (perhaps without knowing it), for he made much use of iron filings in his fireworks, specialised in *tourbillons* (rotating rockets), and called his Roman candles *lances-à-feu*. Perhaps Reinhart de Solms was the first European to put wings on rockets, as he did in +1547 (Duhem (1), p. 288).

[g] Taylor (1), pp. 24–5; von Braun & Ordway (2), pp. 106–7; Baker (1), pp. 43 ff.

[h] Taylor (1), pp. 34–5.

[i] Taylor (1), pp. 32–3; Baker (1), p. 179.

[j] Taylor (1), *loc. cit.*; Baker (1), p. 178.

[1] 鹽水　　　[2] 元宵　　　[3] 蜂炮台　　　[4] 蜂探炮窩

Fig. 205. Scale model reconstruction of the assault-barrow in the previous picture (orig. photo. 1964, in the Nat. Historical Military Museum, Peking).

'Thunderbird'[a] and 'Nike–Zeus'[b] types; and here must also be numbered the Ohka Kamikaze winged (and manned) rocket-aircraft, also of World War II.[c] The 'Space Shuttle' of our own times is another case in point, launched as a rocket but capable of returning to earth as an airplane.[d]

Consequently it is very reasonable to ask, who first gave rockets wings? We find it in the oldest stratum of the *Huo Lung Ching*, which must mean the middle of the +14th century, and quite possibly soon after +1300. The passage (cf. Fig. 208) runs as follows:[e]

[a] Von Braun & Ordway (2), p. 147; Baker (1), p. 130.
[b] Von Braun & Ordway (2), p. 146; Baker (1), p. 178.
[c] Baker (1), p. 92; von Braun & Ordway (2), pp. 87–9.
[d] Baker (1), pp. 215, 248.
[e] *HLC*, pt. 1, ch. 3, p. 18a, b; *HKPY ibid.*; Hsiangyang ed., *HCT*, p. 34a, tr. auct. Passages in square brackets are from the slightly expanded *WPC* version, ch. 131, pp. 12b, 13a.

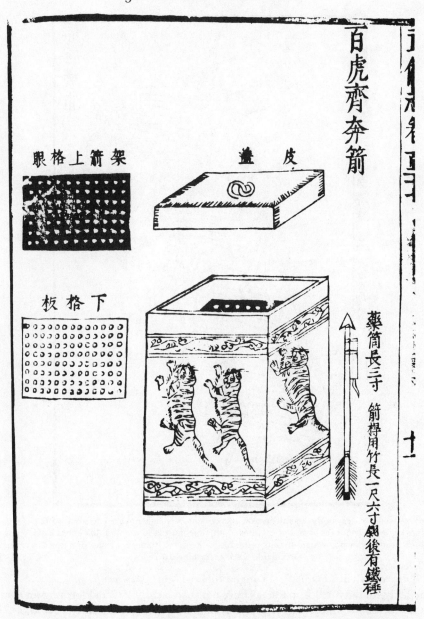

Fig. 206. The 'hundred-tigers running-side-by-side rocket-arrow launcher' (*pai hu chi pên chien*), from *WPC*, ch. 127, p. 11*b*. This is the type referred to in the caption for Fig. 204.

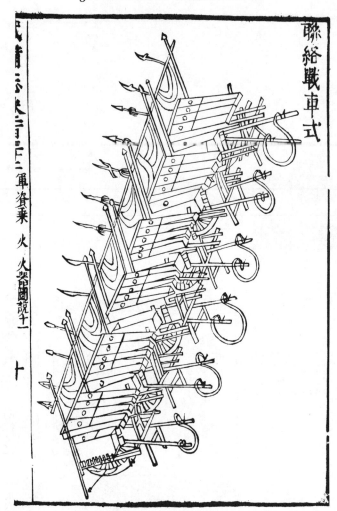

Fig. 207. A whole battery of the assault-barrow rocket-launchers facing to the left, from *WPC*, ch. 132, p. 10*a*. This drawing easily conduces to an optical illusion, but one must keep in mind that one is looking down upon the battery from a viewpoint high up and behind it to the right. Batteries of this kind must have been quite formidable when everything worked well.

The 'flying crow with magic fire' winged rocket-bomb (*shen huo fei ya*[1]).

The body (of the bird) is made of [fine] bamboo laths [or reeds] forming an elongated basketwork, in size and shape like a chicken, weighing over a catty (0·6 kilo.). It has paper glued over to strengthen it, and it is filled with explosive gunpowder (*ming huo cha yao*[2]). All is sealed up using more paper, with head and tail fixed on before and behind, and the two wings nailed firmly on both sides, so that it looks just like a flying crow.

Under each wing there are two [slanting] rockets (*ta chhi huo*[a] *erh chih*[3]). The fourfold (branching) fuse, connected with the rockets [and about a foot long], is put through a hole drilled on the back (of the bird). When in use, this [main fuse] is lit first.

---

[a] This refers to the 'rising' gunpowder rocket compositions; cf. p. 483 above.

[1] 神火飛鴉   [2] 明火炸藥   [3] 大起火二枝

Fig. 208. A winged rocket-bomb, the 'flying crow with magic fire' (*shen huo fei ya*) from *HLC*, pt. 1, ch. 3, p. 18a, and therefore at least as early as +1350, probably a century earlier. The idea was doubtless derived from the use of expendable birds (Fig. 38) carrying glowing tinder wherewith to set on fire the roofs of the enemy city. But the provision of wings or fins for increasing aerodynamic stability long preceded anything of the same kind elsewhere in the world. And the provision of an explosive payload was also a new development.

The bird flies away more than 1000 ft, and when it eventually falls to the ground, the explosive gunpowder in the cavity of the bird is (automatically) lit, and the flash can be seen miles away. [This weapon is used against enemy encampments to burn them, but also at sea to set ships on fire. It should never fail to bring victory].[a]

The illustration suggests that the shafts and feathering of rocket-arrows were retained, but the text does not say so. In any case this must surely be the oldest account of the invention of the winged rocket in any civilisation.

One must naturally suppose that the wings were fitted with the four rockets to the weak-casing bomb because it was found that they gave added stability and accuracy to the flight. But what suggested them in the first place? The answer is immediately at hand—namely the use of expendable birds as incendiary carriers. It must be significant that these always accompany and precede the winged rocket-bomb in the military compendia. There were, for example, the 'fire-bird' (*huo chhin*[1])[b] and the 'nut sparrow' (*chhiao hsing*[2])[c] both carrying nutfuls of burning moxa tinder attached to their necks or legs, so that when they perched on the housetops of the enemy city they would set the roofs on fire. Both these had come down with little or no change from the *Wu Ching Tsung Yao* of +1044,[d] but again significantly they were there followed by no rocket-propelled artificial bird. Going back further, we can find the former easily in the *Hu Chhien Ching*[e] of +1004, and even in the *Thai Pai Yin Ching*[f] of +759. The practice was probably age-old, and there is no point in pursuing it further. Thus the winged rocket had to await the latter part of the +13th century at earliest—but even so it long preceded the winged rockets of the West.

There is another winged rocket-bomb, or rather grenade, in the *Wu Pei Chih*, the 'free-flying enemy-pounding thunder-crash bomb' (*fei khung chi tsei chen-thien-lei phao*[3]).[g] A rocket-tube (*sung yao thung*[4]) is contrived within the body of it, and when the wind is favourable the fuse is lit, whereupon it flies over to the enemy.[h] When the rocket-composition is nearly burnt out, the charge is automatically ignited, releasing a poisonous and irritating smoke as well as water-calthrops the thorns of which are tipped with tiger-poison. The whole thing is no more than 3·5 in. in diameter, made of dozens of layers of oiled paper, but on each side it has artificial wind-borne wings (*hsia fêng chhih*[5]) which will take it, in suitable conditions, right over a city wall (Fig. 211).[i]

---

[a] We also give in Figs. 209, 210 reconstructions made by the National Historical Military Museum in Peking.

[b] *HLC.*, pt. 1, ch. 3, p. 16*a*, *b*, *HKPY*, ibid., Hsiangyang ed., *HCT*, p. 33*a*; *WPC*, ch. 131, pp. 10*b*, 11*a*.

[c] *HLC.*, pt. 1, ch. 3, p. 17*a*, *b*, *HKPY*, ibid., Hsiangyang ed., *HCT*, p. 33*b*; *WPC*, ch. 131, pp. 11*b*, 12*a*.

[d] *WCTY*, ch. 11, pp. 21*a*, *b*, 22*a*, *b*.

[e] Ch. 54 (ch. 6), p. 5*a*, *b*.   [f] Ch. 38 (ch. 4), p. 8*b*.

[g] *WPC*, ch. 123, pp. 22*b*, 23*a*. From p. 163 above we know that 'thunder-crash' was the key-word for a strong-casing bomb. Here perhaps it was loosely used.

[h] The stronger the wind the further it goes, says the text.

[i] We also give in Fig. 212 the reconstruction made by the National Historical Museum in Peking.

[1] 火禽   [2] 雀杏   [3] 飛空擊賊震天雷砲   [4] 送藥筒

[5] 轄風翅

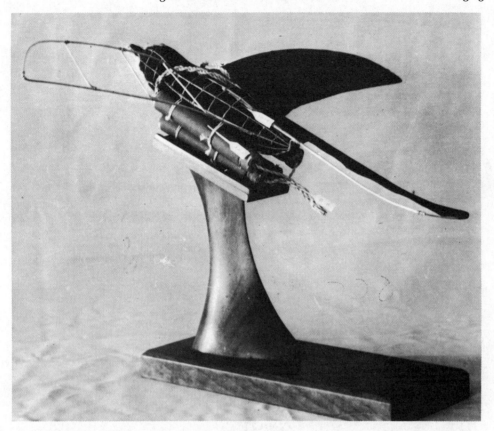

Fig. 209. Model of the winged rocket-bomb to show the structure and design (photo. Nat. Historical Museum, Peking, 1978).

Fig. 210. Model of the complete winged rocket-bomb (photo. Nat. Historical Military Museum, Peking).

Fig. 211. Another winged flying rocket-bomb, the 'free-flying enemy-pounding thundercrash bomb' (*fei khung chi tsei chen-thien-lei phao*), from *WPC*, ch. 123, p. 22*b*. From its name this was a strong-casing bomb, and the rocket tube was contrived within the body of it.

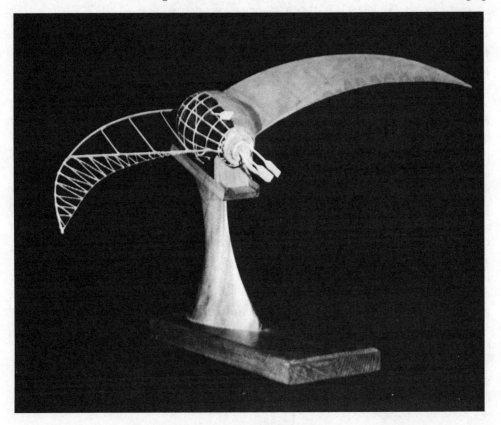

Fig. 212. Reconstruction of the rocket-containing flying bomb to show the structure of the wings. Photo. Nat. Historical Museum, Peking, 1978.

Such were the beginnings of the winged rockets of the present day that reach beyond the stratosphere.

(v) *Multi-stage rockets*

Today it is a commonplace, not only the pabulum of science fiction, that if we wish to leave the earth and travel into outer space, mankind can only do so by using rocket-craft with combustibles that fire in several successive stages, boosters to begin with, then smaller rockets, finally to take advantage of gravitational pulls within the emptiness, and cruise away among the stars and planets.[a] Artificial satellites launched by multi-stage rockets are now familiar to everyone,[b]

[a] Aided of course by small bursts from rocket-motors from time to time to change or adjust direction.
[b] They may circulate anywhere between 500 and 25,000 miles above the earth, and the higher they are the longer they will endure before descending and being burnt up by the friction of the earth's atmosphere. They must also avoid the van Allen radiation belt, which is most dangerous between 2000 and 12,000 miles. Cf. Taylor (1), pp. 82 ff.; von Braun & Ordway (2).
I shall always remember seeing the pin-point light of 'Sputnik I' crossing the sky, man's first artificial satellite, as we sat at dinner in the open air on the harbour mole at Valencia in Spain in 1957.

and space probes can be sent to remote inhospitable parts of the solar system where men are not yet ready to venture themselves.[a] In seeking for the origin of multi-stage rockets let us start from the present day and work backwards, tracing their development to its source.[b]

The 'Apollo' moon-landings of 1969 were accomplished by means of a three-stage rocket of enormous size, 'Saturn V'.[c] Such space projectiles have been developed along with those more menacing and dangerous missile carriers known as IRBM and ICBM.[d] Indeed it is an extraordinary fact that the very same rocket vehicles which can be, and have been, used for the exploration of extra-terrestrial space by human beings, can also be turned against themselves for fratricidal purposes of mass extermination—like fire itself, everything depends on what you do with it. The American 'Thor', 'Atlas', 'Titan' and 'Minuteman' have all been rockets of three or four stages, as also the Russian ones 'Scrag' and 'Sasin'.[e] It was the Russian pioneer Konstantin Tsiolkovsky about 1883 who first realised that space-flight would necessarily demand what he called 'rocket-trains' or multiple rockets firing in successive stages.[f] Only so could a sufficiently high speed be attained to overcome the pull of earth's gravity. Also in the nineteenth century came the application of Edward Boxer about 1855 of two-stage rockets for the purpose of life-saving at sea; they shot a cord over the endangered vessel so that a cable and a breeches-buoy could follow. By 1870 every British lifeboat station was equipped with these, and they are still in use at the present day, having saved many tens of thousands of lives.[g] Boxer based the design on the rockets of François Frézier, published in his book of +1741.

But the idea of two-stage rockets goes much further back, into the +17th and +16th centuries. It has long been known that the Lithuanian military engineer[h] Kazimierz Siemienowicz described them in his book *Ars Magna Artilleriae* pub-

---

[a] E.g. 'Mariner 4' and 'Venus 4'; Taylor (1), pp. 148–9.

[b] Dollfuss (1) suggests that the first payload-carrying rockets were those of French displays from +1772 onwards which shot live dogs and sheep high into the air, after which they descended safely by parachute (*parasol à feu*). Some of these seem to have been two- or three-stage rockets. Similar systems were used later on, for 'Verey lights' on battlefields from 1837, and for reconnaissance cameras from 1860. On the history of the parachute as such see Vol. 4, pt. 2, pp. 594 ff.
Duhem (1), pp. 292, 300, also describes the animal parachute experiments.

[c] This was eight years after the first human being had been put into space, Yuri Gagarin in 'Vostok I'. See Taylor (1), pp. 92 ff., 124 ff.; von Braun & Ordway (1), pp. 176 ff. By the time the last stage fell away, the 'Apollo' spacecraft was making 24,200 mph (Fig. 213).

[d] Intermediate-Range Ballistic Missiles and Inter-Continental Ballistic Missiles.

[e] Taylor (1), pp. 38, 72 ff.; Baker (1), pp. 109 ff.; von Braun & Ordway (2), pp. 135, 172–3 (1), pp. 175–6. China launched her first multi-stage carrier of modern type to a destined area in the South Pacific on 18 May 1980.

[f] See Baker (1), pp. 17 ff.; Olszewski (2); von Braun & Ordway (1), pp. 124–5; Ley (2), pp. 101 ff.; Taylor (1), pp. 14–5. Tsiolkovsky even envisaged liquid oxygen and hydrogen as fuels, the very solution adopted nearly a century later in 'Saturn V'.

[g] Taylor (1), pp. 11–2; Humphries (1), pp. 143 ff., 178. The second-stage rocket was lit by a small detonating charge of gunpowder.

[h] Siemienowicz spent his life in the Polish service.

Fig. 213. Multi-stage rockets; the blast-off of Saturn I in 1961 (from Baker (1), p. 157).

lished at Amsterdam in +1650.[a] But more recently a MS. conserved at Sibiu in
Rumania and written by Konrad Haas[b] about +1560 shows also a clear presentation of the same idea,[c] and it is thought to have reached Siemienowicz by
way of the book of Schmidlap (1), often printed in the second half of the +16th
century.[d] Less clear is the attribution to Biringuccio, which would take the matter back to +1540.[e]

But all these European devices were much posterior to the two-stage rocket
described in the *Huo Lung Ching*.[f] Since it occurs in the oldest portion of the book
it must be dated to the second half of the +14th century, and quite probably to
the first half also.[g] Describing the 'fire-dragon issuing from the water' (*huo lung
chhu shui*[1]), it says:[h]

A tube of bamboo (*mao chu*[2])[i] 5 ft long is taken, the septa removed, and the nodes scraped
smooth [with an iron knife]. A piece of wood is carved into the shape of a dragon's head
(and fitted on at the front) while a wooden dragon tail is made for the rear end. [The
mouth must be facing upwards, and] in the belly of the dragon there are several 'mysterious mechanism rocket-arrows' (*shen chi huo chien*[3]).[j] At the dragon head there is an
opening through which go all the fuses of the rockets (inside).

[Beneath the dragon head on both sides there are two (big) rocket-tubes weighing a
catty and a half each. Their fuses (and orifices) should face downwards (and backwards), and their front ends must face upwards (and forwards); and they are fixed tight
to the body by (bands of) hempen cloth secured with skin- and fish-glue. The fuses of the
rocket(-arrows) within the belly lead out from the head of the dragon, and they are
divided into two. Oiled paper is used to make them firm, and they are so arranged as to
be connected with the front (ends) of the (outside) rocket-tubes (*huo chien thung*[4]). And
under the tail of the dragon on each side there are also two (big) rocket-tubes, fastened
in the same style. The fuses of the four rockets are twisted into a single one. In a naval
battle] the apparatus can fly 3 or 4 ft above the water.

Upon lighting it will fly over the water as far as 2 or 3 *li*.[k] At a distance it really looks
like a flying dragon coming out of the water. When the gunpowder in the rocket-tubes is
nearly all finished (that in the rocket-arrows within the belly is ignited, so that) they fly
forth, destroying the enemy and his ships. [It can be used either on land or sea.]

[a] A French translation appeared in the following year. See Olszewski (1), p. 251; Barowa & Berbelicki (1),
p. 12 and opp. p. 9; Thor (2); Subotowicz (2); Berninger (1).
[b] +1529 to 69.
[c] Todericiu (1–5); Subotowicz (1); von Braun & Ordway (1), pp. 11 ff. Haas added delta-shaped fin-
stabilisers to his rocket tails.
[d] He designed three-stage rockets, on which see Subotowicz (1), as did Siemienowicz later.
[e] (1), Eng. ed. p. 442; see Thor (1).
[f] *HLC*, pt. 1, ch. 3, p. 23a, b, *HKPY*, *ibid.*; Hsiangyang ed., *HCT*, p. 36b; *WPC*, ch. 133, pp. 3a, b, 4a.
[g] It was only natural that it should have been earlier, in view of the antecedent development of all gunpowder devices and weapons in China.
[h] Tr. auct. Passages in square brackets come from the rather longer version in the *Wu Pei Chih*.
[i] *Phyllostachys*, probably *edulis*; cf. Chhen Jung (1), p. 78; Steward (2), p. 437—but in any case one of the
bamboos of large diameter.
[j] Cf. Table 6 above.
[k] This range is not so long as it sounds. The Yuan *li* was 0·344 mile, or 605 yards, so the maximum given
would only be 1816 yards.

¹ 火龍出水   ² 猫竹   ³ 神機火箭   ⁴ 火箭筒

Thus the automatic lighting of the second-stage rockets is clearly stated.[a] Although strangely prefiguring submarine-launched weapons of 'Polaris' type,[b] it was not in fact fired from under water, but rather from near the water-level on shipboard, and its trajectory was evidently kept very flat.[c] Fig. 214 shows the illustration from the *Huo Lung Ching*; those in later books simply re-draw it.[d] This invention has been noted by a few writers,[e] but its full significance has hardly ever been appreciated.

(vi) *The rise and fall, and rise again, of military rockets*

For reasons which have already been explained (p. 472), the origin and development of the rocket is an exceptionally difficult study in technological history. We must unravel it as best we can, but a definitive account will have to await further research.[f]

To begin with, we have two fixed points, +1264 when an empress was frightened by the 'ground-rats' or 'earth-rats' at a firework display (p. 135 above);[g] and the neighbourhood of +1280 when al-Rammāḥ in Syria described rocket-arrows as *sahm al-Khiṭāi*, 'arrows of China' (p. 41 above). Equally, in spite of arguments to the contrary, we do not believe that rockets were described in the *Wu Ching Tsung Yao* of +1044 (pp. 226 above); while on the other hand they were prominent among the fireworks mentioned by Fêng Ying-Ching and Shen Pang in +1592 (p. 134 above).[h] The details in the *Huo Lung Ching* affirm rockets clearly

[a] The same principle was even applied to fire-crackers in traditional China; cf. Ball (1), p. 282.

[b] Cf. Taylor (1), pp. 76–7. In Oct. 1982 the Chinese navy successfully tested a submarine-launched ballistic missile.

[c] It was thus the very model of a modern 'Exocet' missile (named from the flying fish *Exocetus*), so prominent in the Falklands campaign, as Dr Christopher Cullen remarked to us at Louvain.

[d] We also give in Fig. 215 the reconstruction made by Chiang Chêng-Lin for the National Historical Military Museum in Peking. Cf. Anon. (*209*).

[e] E.g. Hsi Tsê-Tsung (6); Hsü Hui-Lin (1); Chiang Chêng-Lin (1); Sandermann (1), p. 171.

[f] One meets from time to time in the Western literature with dubious stories about Chinese rocketry. For example, Hokeš (1) has written about 'Wan Hoo', a supposed official of the Ming period, who invented a kite-like monoplane powered by about 30 rockets, but perished in its first experimental flight. There is a whole series of uncritical references to this, as in Ley (2), pp. 84–5; Gibbs-Smith (10); Zim (1), etc. and it has even been entertained by Chinese writers such as Hsü Hui-Lin (1). But in spite of much correspondence, as with A. T. Philp in Australia, we have never been able to get any firm reference to Wan Hoo, and we suspect that he is a myth invented probably during or after the Chinoiserie period. The matter is reminiscent of a similar story about a dirigible airship ascribed to the Yuan (Vol. 4 pt. 2, p. 598) and probably equally without foundation.

The application of rocket-propulsion to land vehicles has never in fact been of much practical use (Taylor (1), pp. 18 ff.) except for test-track sleds (Humphries (1), p. 179, fig. 113), because although rocket thrust is so high per unit weight, and realisable with extreme rapidity, its fuel consumption is extremely great. But rocket-assisted take-off for aeroplanes has become commonplace (cf. Humphries (1), pp. 163 ff., fig. 100), and a glider like that ascribed to Wan Hoo was successfully flown by Fritz von Opel in 1928.

One can even find Wan Hoo in Norwegian; cf. Holmesland *et al.* (1), vol. 16, p. 508.

[g] Of course it does not follow that the ground-rats were a new invention of that year, nor that civil fireworks were their only employment. They may well have been a century or more old at the time. We have suggested (p. 474 above) that the incorporation of these mini-rockets in cavalry-confusing bombs was the most primitive form of the use of rockets in warfare.

[h] Their 'ascending fires' (*chhi huo*[1]) were undoubtedly rockets, and they also knew of the ground-rats (*ti lao shu*[2]) and the similar toys that whizzed about on water surfaces (*shui shu*[3]). Something like this last is in al-Rammāḥ (Partington (5), p. 203).

[1] 起火　　　[2] 地老鼠　　　[3] 水鼠

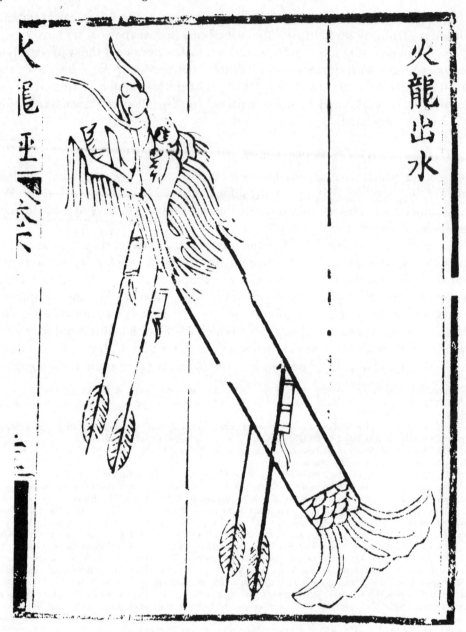

Fig. 214. The first of all multi-stage rockets, the 'fire-dragon issuing from the water' (*huo lung chhu shui*), a device from *HLC*, pt. 1, ch. 3, p. 23*a*. It therefore belongs to the middle, perhaps to the beginning, of the +14th century. It was a two-stage rocket, for when the carrier or booster rockets were about to burn out they automatically ignited a swarm of smaller rocket-arrows which issued through the dragon mouth and fell down upon the enemy. The design seems to have been for use mainly in naval warfare, and as the trajectory was very flat the weapon may be regarded as an ancestor of the modern 'Exocet'

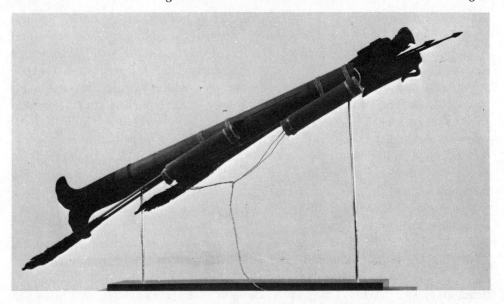

Fig. 215. Reconstruction of the two-stage rocket described in the previous illustration (photo. Nat. Historical Military Museum, Peking).

by about +1350 (p. 479 above),[a] so the period in which we mainly have to look lies between about +1050 and +1280.

Now it will be remembered (pp. 148ff. above) that between +969 and +1002 there was a crop of military inventions by Thang Fu, Yo I-Fang and others, in which new sorts of firearms figured, but we do not believe that these were rockets.[b] Fire-arrows were standard equipment on battleships in +1129, but again there is no justification for interpreting them as rockets.[c] By +1206 a term not previously used appears, 'gunpowder arrows' (huo yao chien[1]), fired off by Chao Shun's men during the defence of Hsiangyang against the Chin Tartars (p. 168 above), but though these may have been rockets the expression could easily have referred to low-nitrate gunpowder used on incendiary arrows, as it had been for at least a couple of centuries previously. On the other hand the 'fire-arrows' launched in +1245 during the military and naval exercises in the Chhien-thang estuary (p. 132 above) most probably were rockets. There is here a zone of probability which we can only assess in the light of the following circumstance.

This is the description of the fireworks used at festivals on the West Lake at

[a] And very complicated ones too, such as winged ones and two-stage ones.
[b] Wang Ling (1), pp. 165, 168; Goodrich & Fêng Chia-Shêng (1), p. 114, were uncertain about the nature of these. Průsek (4); Köhler (1), vol. 3, pt. 1, p. 169; and Hsü Hui-Lin (1), thought they were rockets.
[c] In spite of what we said in Vol. 4, pt. 3, pp. 575–6 above, which misled von Braun & Ordway (1), p. 41.

[1] 火藥箭

Hangchow around +1180 (p. 132 above).[a] Here lies what is probably our best starting-point. Though he was writing a hundred years later, Chou Mi would have been quite well informed of what went on, and among the pyrotechnic devices he named 'meteors' or 'comets' (*liu hsing*[1])[b] as well as 'water-crackers' (*shui pao*[2])[c] and others that flew in the air like pigeon-whistles on kites (*fêng chêng*[3]).[d] Like Fêng Chia-Shêng himself,[e] we are strongly inclined to take comets[f] here to mean rockets, because, though possible, it is surely less likely that the pyrotechnical masters took bows or crossbows and shot balls of fire into the air. At the other end of history we know from Chao Hsüeh-Min's monograph on fireworks, the *Huo Hsi Lüeh* of +1753, that *liu hsing*[1] was by then the common name for rocket;[g] but he also uses the extremely significant expression 'flying rats' (*fei shu*[5]).[h] This is reminiscent of another term, equally conjunctive though seemingly self-contradictory, the 'meteoric ground-rat' (*liu hsing ti lao shu*[6]) which we find in the *Wu Pien* of +1550.[i] Intermediate in date is a weapon described in the *Wu Pei Chih* and called the 'comet, or meteoric, bomb' (*liu hsing phao*[11]).[j] This was a rocket-arrow, with a shaft 4 ft 5 in. long, a poisoned arrowhead, and a small carton bomb about the same diameter as the rocket-tube fixed in front of it. As the rocket burnt out, it ignited the bomb (Fig. 216).

Unfortunately, it would be highly deceptive to take everything bearing the name *liu hsing* as a rocket. For example, we have already encountered (p. 180) the 'magic-fire meteoric bomb that goes against the wind' (*tsuan fêng shen huo liu hsing phao*[15]), certainly current by the mid +14th century;[k] it was probably thrown in antique style from a trebuchet, and perhaps got its name simply from

---

[a] In fact, the Shun-Hsi reign-period, +1174 to 89.

[b] *Wu Lin Chiu Shih*, ch. 3, p. 1 *b*.

[c] This probably means the water-rats or rocket-skimmers, perhaps igniting a small explosive charge as they burnt out.

[d] This sounds like Verey lights suspended in that way—or of course live birds could have carried them. On pigeon-whistles, see Vol. 4, pt. 2, p. 578.

[e] Letter to J. N. of 1 Jan. 1956.

[f] Strictly speaking, 'meteors' is the better word, for properly comets were *hui hsing*[4] (cf. Vol. 3, pp. 431, 433).

[g] Cf. Davis & Chao Yün-Tshung (9), p. 104.

[h] *Ibid.* p. 103. Earth-rats and water-rats are mentioned many times (pp. 101–2, 103–4).

[i] Ch. 5, pp. 63 *b* ff. Such names make one think of bats and other flying mammals. Indeed *fei shu* was an occasional synonymic name for the bat. 'Ground-rat' had always been a good term for the small rocket because it scuttled about at random. But there could never have been any confusion in the names of the airborne ones, partly because the flight was so different, and partly because they had long had their own special names. The commonest bat, *Vesperugo noctula*, was called *pien fu*[7] (or *thien shu*[8]); cf. *PTKM*, ch. 48, p. 43 *b*; R 288; Tu Ya-Chhüan *et al.* (1), p. 1956·2. Other species, such as the flying squirrel *Pteromys xanthipes* also had their special names, in this case *lei shu*[9] or *fu shu*[10]; cf. *PTKM*, ch. 48, p. 47 *b*; R 289.

[j] *WPC*, ch. 128, pp. 16 *b*, 17 *a*. The description says that the use of the weapon is a good way of causing commotion among enemy troops, especially cavalry, as well as doing some incidental damage, after which one should press the attack. But the artists forgot to put in the feathering (*ling*[12]), though it is mentioned in the text, and moreover the arm is called *chhiang*[13] rather than *chien*[14]—confusing features which led Davis & Ware (1), pp. 523–4 to regard it as a fire-lance or incendiary whip-arrow, i.e. javelin.

[k] Because of *HLC*, pt. 1, ch. 2, p. 7 *a*, *b*. Lit. 'wind-piercing'.

[1] 流星	[2] 水爆	[3] 風箏	[4] 彗星	[5] 飛鼠
[6] 流星地老鼠	[7] 蝙蝠	[8] 天鼠	[9] 鸓鼠	[10] 蝠鼠
[11] 流星砲	[12] 翎	[13] 鎗	[14] 箭	[15] 鑽風神火流星砲

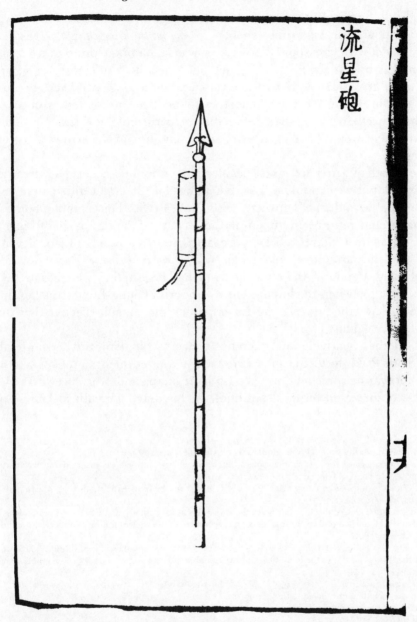

Fig. 216. That the earth-rat turned into the rocket is well illustrated by the expression 'meteoric ground-rat' (*liu hsing ti lao shu*) found in +1550, and another, the 'flying rat' (*fei shu*) of +1753. Here we have a confusing instance of similar nomenclature, the 'meteoric bomb' (*liu hsing phao*), from *WPC*, ch. 128, p. 16*b*. The bomb was simply a carton of gunpowder fixed forward of the rocket-tube head, which automatically set off the explosion as it was about to burn out.

the light of the burning fuse as it passed through the air. Equally there was the 'fire-crossbow meteoric arrow-(shooter)', (*huo nu liu-hsing chien*[1]).[a] This had nothing to do with crossbows either;[b] it was a bamboo proto-gun[c] firing ten poisoned arrows at a time, which came out 'like a flock of locusts' (Fig. 217). It happens that we can perhaps trace this weapon a long way back, because we read[d] that in +1049 a certain magistrate, Kuo Tzu[3], presented prototypes of a 'combat wheelbarrow'[e] and an 'invincible meteoric crossbow' (*wu ti liu hsing nu*[4]); at that time it would have been a fire-lance sending out the arrows as co-viative projectiles.[f]

The dearth of battle accounts specifically mentioning rockets has already been mentioned, but we can find a few, though not for the vital century that we have now been able to define, between +1180 and +1280. For example, bombs containing ground-rats are prominent in the account of the campaign of Liu Chi[5] in Chekiang against inland rebels and coastal pirates around +1340.[g] Launchers are in evidence around +1380, when 'wasps' nests' (*i wo fêng*[6]) are included in lists of army supplies.[h] And after the Ming had begun, they were much used in a battle of +1400 when the imperial army under Li Ching-Lung[7] was fighting the Prince of Yen[8] (the future Yung-Lo emperor), but though effective they did not save the day against him.[i]

Yet another relatively late reference concerns the Timurid Persian embassy from Shāh Rukh to China in +1419, when we find mention of rockets not so much for war as travelling on wires to light lamps and other fireworks at ceremonies to amaze glittering assemblies. In his diary Ghiyāth al-Dīn Naqqāsh wrote:[j]

---

[a] *HLC*, pt. 2, ch. 2, p. 20*a*, *b*; *WPC*, ch. 126, pp. 12*b*, 13*a*; *PL*, ch. 12, pp. 50*b*, 51*a*.

[b] The only similarity was that the handle was curved like a mark of interrogation.

[c] We say this because the caption mentions a plug (*tan ma*[2]), so that the bore was probably occluded in front of the propellant charge. The barrel was reinforced with iron straps.

[d] *YCLH*, ch. 226, p. 6*b*, quoting (via *Ping Lüeh Tshuan Wên*[9]) *Yü Hai* (+1267), ch. 150, p. 24*a*, *b*. Cf. Chou Chia-Hua (*1*), pp. 210–11.

[e] What relation this could have had with those just discussed (pp. 497 ff. above) we do not know. But we doubt that they launched rockets. One is, of course, reminded of the ancient military connections of the vehicle (Vol. 4, pt. 2, p. 260).

[f] True, it was not called a 'fire-crossbow', but at the same time another official, Sung Shou-Hsin[10], presented other fire-weapons, so the identification is reasonable. Of course it may have been a real crossbow with flaming incendiary bolts.

[g] *Hsi Hu Erh Chi*, ch. 17 (pp. 335–6). Cf. p. 183 above. The 'great wasps' nest' (*ta fêng kho*[11]) is here described as including ground-rats, though not in *HLC*, pt. 1, ch. 3, p. 11*a*, *b*, or *WPC*, ch. 130, p. 14*b*. The 'fire-brick' (*huo chuan*[12]) always has them.

[h] *HWHTK*, ch. 134 (p. 3994·3).

[i] *Ming Shih Lu* (Thai Tsung sect. 6), p. 5*b* (p. 64); cf. Goodrich & Fêng Chia-Shêng (*1*), p. 122, who give further references. See also Chang Hsüan's[13] *Hsi Yuan Wên Chien Lu*[14] (Things Heard and Seen in the Western Garden), ch. 73, pp. 3*b*, 4*a*, *b*, 5*b*.

[j] Tr. Quatremère (*3*), p. 387; Rehatsek (*1*), the latter reproduced in Yule (*2*), vol. 1, p. 282. The log of the expedition formed the appendix to the *Ruzat al-Safā* of Muḥammad Khāvend Shāh. An exactly similar passage occurs in the *Zubdatu't Tawārikh* of Hafīz-i Abrū, tr. Maitra (*1*), p. 90.

---

[1] 火弩流星箭	[2] 彈馬	[3] 郭諮	[4] 無敵流星弩	[5] 劉基
[6] 一窩蜂	[7] 李景隆	[8] 燕王	[9] 兵略纂聞	[10] 宋守信
[11] 大蜂窠	[12] 火磚	[13] 張萱	[14] 西園聞見錄	

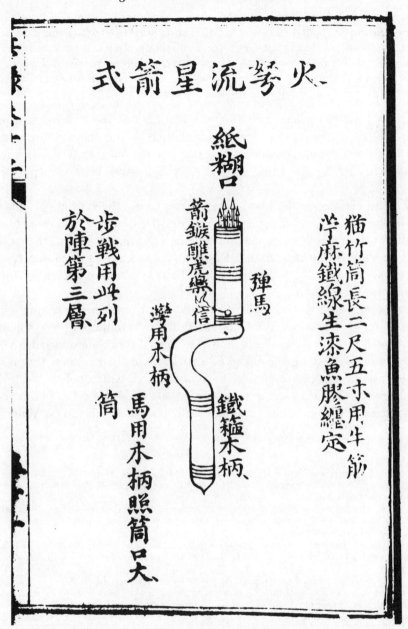

火砮流星箭式

紙糊口

貓竹筒長二尺五寸用牛筋
苧麻鐵線生漆魚膠纏定

箭鏃雕虎藥以信
彈馬

步戰用此列
於陣第三層

灣用木柄

鐵箍木柄

筒
馬用木柄照筒口大

Fig. 217. Another example of a weapon which though called 'meteoric' had nothing to do with rockets, the 'fire-crossbow meteoric arrow-shooter' (*huo nu liu hsing chien*), nor did it have anything to do with crossbows either. It was a bamboo fire-lance or proto-gun which shot forth arrows as co-viative projectiles. *PL*, ch. 12, p. 51a.

At that season the Feast of Lanterns takes place, when for seven days and nights, in the interior of the imperial palace, a wooden ball is suspended from which numberless chandeliers branch out, so that it appears to be a mountain of emeralds; and thousands of lamps are suspended from cords. Rats of naphtha are then prepared, and when lit they run along the cords and light every lamp they touch, so that in a single moment all the lamps from the top to the bottom of the ball are kindled.[a]

Actually this use of rockets travelling along cords has come down as a ploy in China to our own time, under various names such as 'the phoenix flitting among the peonies' (*fêng chhuan mou tan*[1]).[b] And it got to the West as well, since we find dragons propelled in the same way in +17th-century European pyrotechnic books.[c]

All in all therefore we shall be fairly safe in placing the Chinese origins of the rocket in the second half of the +12th century, no doubt when Hangchow had entered that period of great peace and prosperity which it had as the capital of Southern Sung.[d] By the time that al-Rammāḥ got to know of them they had been in use for something like a century and a half. When, one may ask, did their history in the West begin?

It is generally agreed that rockets[e] are first mentioned in connection with the Battle of Chioggia between the Genoese and the Venetians in +1380, though they may well have been used a little before that.[f] From then onwards there are many references. By +1405 Konrad Kyeser in his *Bellifortis* knew that a rocket must be a tubular gas-tight container open at one end, with a hollow 'Seele' bored in its charge, and a stick or arrow-shaft 'to steer it'.[g] In +1440 Giovanni da Fontana knew rocket-propelled missiles well,[h] as did Leonardo da Vinci in his

---

[a] There is something here reminiscent of the 'lamp-trees' which we discussed in the fireworks sub-section (p. 136 above).

[b] Sun Fang-To (1), p. 8 (pp. 302–3).

[c] E.g. in +1633, Leurechon, Henriot & Mydorge (1), p. 272. Cf. Brock (1), pp. 186–7. Later these rocket-propelled 'cable-cars' were called *courantins* (von Braun & Ordway (1), pp. 67–8). And in +1765 unmistakable 'water-rats' were described by Jones (1) as well.

[d] And in this case it does look as if the 'flying rats' were initially a civilian pyrotechnic device applied to warfare only rather later. Yet if the rocket stick derived from the rocket-arrow shaft (cf. p. 477 above) the two uses perhaps grew up together.
We are glad to be able to report that our estimate of dating is shared by our friends Mr Hu Tao-Ching, the eminent historian of science at Shanghai, and Mr Phan Chi-Hsing, of the Institute of the History of Science in Peking.

[e] This would be a suitable place to mention the origin of our word 'rocket'. In old Italian *rocca* was a distaff, or a quill or bobbin for silk-winding, hence a long thin tube (Skeat), and the same word was also used to denote a wooden sheath that covered the sharp points of lances during combat exercises (v. Braun & Ordway).

[f] *Danduli Chronicon*, in Muratori (1), vol. 12, p. 448 (*igne imissio cum rochetis*), vol. 15, p. 769 (*furono tirate molte rochette*); cf. Partington (5), pp. 174, 184; Hime (1), pp. 144 ff. The date is just about what one would expect for Europe.

[g] Partington (5), pp. 147–8.

[h] *Ibid.* pp. 161–2. Fontana also proposed a rocket-driven vehicle on four wheels; cf. von Braun & Ordway (2). opp. p. 68. This strange device reappeared in actuality during the Indian Mutiny of 1857 (*ibid.* p. 116). Giovanni da Fontana may well have drawn directly from Chinese sources, because in a work of +1454 he makes a reference to 'my true friend Constantine of Venice, who for many years travelled about in the realm of the Great Khan'. See Birkenmaier (2); Thorndike (12); Clagett (4) and Lynn White (20), p. 8. Other +15th-century references are given by Brock (2) pp. 158 ff.

[1] 鳳穿牡丹

Codex Atlanticus (+1514) and other MSS.[a] Rockets applied both for war and for peaceful pyrotechnics were now commonplace, and in the +17th century there grew up a large literature on them, from which one need only mention Ufano (1) in 1613, Appier-Hanzelet (1) in 1625, and Furtenbach (1, 2) in 1629 and 1650.[b]

But for some reason or other, probably the early and rapid development of gunnery in Europe, rockets played no great part in warfare after that, being mainly confined to firework displays.[c] India was the part of the world where the rocket-arrow achieved greatest prominence, and from the time of the Mogul emperor Akbar (r. +1556 to +1605) onwards.[d] No records which would fix the date at which India received the rocket-principle from China have been found, but it must have been some time in the +14th or +15th century, for the oldest literary reference which Gode[e] could find was of about +1500, the *Kautukacintāmaṇi* by Prataparudradeva of Orissa.[f] This agrees with the earliest historical references which Winter noted, namely in +1499, possibly +1452;[g] and it is certain that Duarte Barbosa saw pyrotechnic rockets when attending a Brahmin wedding in Gujerat in +1515.[h] The word for rocket in Sanskrit is *bāṇ*, *bāṇa*;[i] which explains the following passage written by François Bernier concerning an event of which he was an eye-witness in +1658. After describing the battle-array, cannon, swordsmen, etc. of the prodigious great Mogul armies in the combat of Aurungzeb against Dara at Samugarh, he goes on to say that 'they hardly made use of any more art than what hath now been related; only they placed here and there some men casting *bannes*, which is a kind of granado fastened to a stick, that may be cast very far through the cavalry, and which extremely terrifieth horse, and even hurts and kills sometimes'.[j]

But it was in the late +18th century that military rockets became really prominent, especially in the Second, Third and Fourth Mysore Wars,[k] during the last twenty years from +1780 onwards. Haidar Ali, the Rājā of Mysore, then invaded the Carnatic, but soon dying, his struggle against the British was carried on by Tipū Sahib his son. Before the fall of Seringapatam and Tipū's death in 1799, these princes had had 6000 rocketeers in their armies, and the East India Company's troops suffered severely from them.

[a] Partington (5), p. 175. Interestingly, he describes various kinds of ground-rat bombs (McCurdy (1), vol. 2, pp. 198, 203–4, 219).

[b] Cf. Kalmar (1); Partington (5), pp. 167–8, 177.

[c] Brock (1), pp. 181 ff.   [d] Cf. Elliott (1), vol. 6, p. 470.

[e] (7), pp. 12, 19.

[f] References continue in later works, such as the *Rukmiṇī Svayaṃvara* by Ekanātha (+1570) and the *Rāmadāsa Samagra Grantha* by Rāmadāsa (+1650).

[g] (1), pp. 9 ff. Winter lists fourteen other accounts, including the Battle of Gwalior in +1518, Akbar's expedition to Gujerat in 1572, Aurungzeb's campaigns of 1657 onwards, the fights against the French in 1750, the Maratha wars after 1792, and finally the last appearance of rocket-arrows in the attack on Jhānsi as late as 1858.

[h] (1), vol. 1, p. 117. Cf. Gode (7).

[i] Gode (7), p. 20, says that it may be connected with a similar earlier word meaning arrow, but suspects a borrowing from some other language for the meaning of rocket.

[j] (1), p. 40; 1671 ed., p. 109.

[k] See V. Smith (1), pp. 540 ff., 583 ff.

As Winter says,[a] the rocket became far more extensively employed in India than in any other nation during the +17th and +18th centuries, perhaps because of a certain lack of barrel firearms, especially light artillery. No one ever described it better than Quintin Craufurd, writing in +1790.[b]

It is certain, that even in those parts of Hindostan that never were frequented by Mahommedans or Europeans, we have met with rockets, a weapon which the natives almost universally employ in war. The rocket consists of a tube of iron, about 8 in. long, and one and a half inches in diameter, closed at one end. It is filled in the same manner as an ordinary sky-rocket, and fastened toward the end of a piece of bamboo, scarcely as thick as a walking-cane, and about 4 ft long, which is pointed with iron. At the opposite end of the tube from the iron point, or that towards the head of the shaft, is the match. The man who uses it, points the end that is shod with iron, to which the rocket is fixed, to the object to which he means to direct it; and setting fire to the match, it goes off with great velocity. By the irregularity of its motion, it is difficult to be avoided, and sometimes acts with considerable effect, especially among cavalry.

Craufurd even used a pile of Indian rockets for the cut on the title-page of his book (Fig. 218). Their average weight was about 9 lb., though it could go up to 30, and their usual range was 1000 yards or more, though they could in certain conditions carry two and a half times that distance.[c] The usual armament was an arrow-head, but the rockets sometimes bore automatically fused bombs, and were often provided with various kinds of launchers.

This Indian rocketry led directly, and perhaps unexpectedly, to a great development of military rockets in Europe.[d] William Congreve (+1772 to 1828) who rose to the rank of Major-General in the Hanoverian service, and shone in the dignity of F.R.S., was directly inspired by the Indian example,[e] and engaged in many experiments with (and much propaganda for) rockets from 1804 onwards, to such good effect that a Rocket Brigade or Regiment was formed in 1808.[f] It was urged that since no wheeled carriages were needed, rockets[g] gave 'to cavalry the power of artillery', and that when provided, every carriage, because of the lightness of the projectiles, was 'a volley-carriage, instead of being armed with a

---

[a] (1), p. 21.

[b] (1), pp. 294–5, 2nd ed., vol. 2, pp. 54 ff. Craufurd was a Scot who made a fortune in Asia and died in Paris; cf. Partington (5), p. 232.

[c] Winter (1); Baker (1), p. 12; Gode (6), p. 222 quoting Moor (1), p. 509. The sticks were in fact often 10 or 12 ft long. One of Moor's remarks suggests a connection with the Chinese ground-rats, for he says that 'others called ground-rockets have a serpentine motion and on striking the ground rise again and bound along till their force be spent'.

[d] The story has been told many times, as by Corréard (1); Gibbs-Smith (10); Brock (2); Hime (1); Winter (3); Baker (1), pp. 13 ff.; von Braun & Ordway (1), pp. 69 ff., pp. 93 ff., (2). pp. 30 ff.; Reid (1), pp. 184, 186; Katafiasz (1).

[e] As he himself tells us in his introduction; Congreve (3), p. 15.

[f] Brock (2), pp. 158 ff. Other armies soon followed suit, e.g. those of Austria, Russia, Switzerland, Mexico and Bengal.

[g] The Congreve rockets went up to 32 lb. with balancing poles 16 ft long, and carried incendiary, explosive or shrapnel war-heads; their range could exceed 3000 yards. They were fired from tripod launchers, or from specially graded ramps in fortifications, or from 'scuttles' within the hulls of ships. Cf. Congreve (1, 2).

SKETCHES

CHIEFLY RELATING TO THE

HISTORY, RELIGION, LEARNING,
AND MANNERS,

OF THE

HINDOOS.

WITH

A concife Account of the PRESENT STATE of the
NATIVE POWERS of HINDOSTAN.

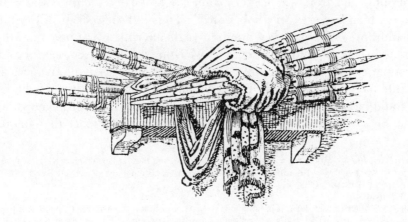

LONDON:

PRINTED FOR T. CADELL, IN THE STRAND.

MDCCXC.

Fig. 218. A pile of Indian rocket-arrows seen in a cut on the title-page of the book of Quintin Craufurd (1).

single *bouche-à-feu*'. Moreover, rockets carried their own recoil, as it were, with them, so that they were particularly suitable on shipboard for naval actions. In due course the Rocket Brigade saw a great deal of service,[a] including a considerable role at the Battle of Leipzig in 1813[b] and even a presence at Waterloo two years later. As time went on, further improvements were made, such as the invention of the spinning 'rotary' rockets (which needed no stick) by William Hale about 1840,[c] and these were used during the American–Mexican war of 1846–8. But the Achilles' heel of all the early nineteenth-century military rockets was their great inaccuracy of delivery, especially at long ranges,[d] as that after the thirties of the century the steadily increasing precision of conventional artillery and small-arms led to their virtual disappearance.[e] By 1850 the Rocket Brigades of most countries had been disbanded.

It was natural that rockets figured on both sides in China during the Opium Wars.[f] Stores of rocket-arrows were found when the Tinghai forts of Choushan were captured in 1840.[g] In the following year at Anson's Bay Congreve rockets were used, one of which set fire to the largest war-junk there, which blew up with all her crew on board.[h] A dozen years later, in the Canton River, in 1856, Admiral Kennedy wrote that 'as a rule the Chinese rockets did little harm, as often as not doubling back from whence they came', but 'one of our cutters was struck by a rocket, which burnt a large hole in her'.[i] Thus did the rockets of Europe contend with those of China seven hundred years after their first invention there.[j]

So now in our concluding discussion we come to the present century and the modern period, on which we must be very brief, even though advances almost incredible have been made. Neither the Chinese nor the British of the Opium Wars could have imagined it, but there is in fact only one vehicle known to man that can be navigated more easily in the vacuum of outer space than in our own

---

[a] Incendiary attacks, all too successful, occurred at Boulogne (1806), Copenhagen (1807), Callao (1809), Cadiz (1810), Washington and Baltimore (1814) hence the 'rocket's red glare' of F. Scott Key's poem; Danzig (1813), Algiers (1816) and Rangoon (1824).

[b] Cf. Whinyates (1).

[c] Hale (1); Winter (2); Taylor (1), p. 9; Baker (1), p. 14; von Braun & Ordway (1), p. 78, (2), p. 33.

[d] This was seen particularly clearly by Scoffern (1) in 1852. It accounts for their sparse use in the American Civil War (1861–5). There were also storage problems. Attempts were made to improve rockets, as by Boxer (1) in 1855, using two-stage ones, but these found permanent use only in life-saving equipment (p. 506 above), and for signalling, and whaling.

[e] They continued in use, however, in a sporadic fashion, in colonial African wars until the end of the nineteenth century (von Braun & Ordway (1), pp. 116 ff.). Here psychological effect was more important than actual destruction. Indeed, they might be said to live again in the anti-tank bazookas of contemporary times (Baker (1), p. 66; Reid (1), pp. 257–9; von Braun & Ordway (2), pp. 94 ff.)

[f] The first encounter of Europeans with Chinese war-rockets had occurred much earlier, in +1637, according to the journal of Peter Mundy, noted by Winter (5), p. 15. At 'Tayfoo' or Tiger Island, not far from Hongkong, a Chinese naval defence vessel assailed the English ship in that year. 'Balles of wyldefire, rocketts and fire-arrows flew thicke as they passed by us; butt God be praised, not one of us were toutched.'

[g] Jocelyn (1), p. 59.

[h] Ouchterlony (1), pp. 98–9.

[i] Kennedy (1), p. 51. His estimate of their inaccuracy may have been an exaggeration.

[j] Other descriptions will be found in Bingham (1), vol. 1, p. 345; Bernard (1), vol. 2, p. 20.

domestic atmosphere. This is the rocket, though far greater than those they knew. Jet-propulsion covers other engines, such as turbo-jets and ram-jets,[a] as well as rockets, but all the former need to take in air at the front so that it can burn the fuel and produce the exhaust that rushes out through the rear nozzle. The rocket alone needs no air to feed on, and carries within itself the oxidant and fuel necessary for combustion and the production of a powerful stream of exhaust gases. As a jet reaction motor it is thus absolutely independent of a surrounding atmosphere, and indeed in airless space it becomes much more efficient since it is free from the drag and resistance of a material medium. Moreover, its thrust is independent of its actual forward speed, and it gives full thrust at all altitudes, even in the near vacuum of space. With what amazement Chiao Yü or Mao Yuan-I would have learnt these things, could they have known of them. The rocket has been called the oldest of all practical heat-engines, yet the liquid-propellant type which is its modern form uses some of the most advanced engineering techniques and materials at present known.[b]

The words of this last sentence have taken us across a decisive step—beyond the classical solid charge of gunpowder. That notable mixture had its oxygen built in, as it were, but in the course of time it became clear that separately carried supplies of oxidant and fuel, held apart and combusted in an ignition chamber, would give far safer conditions and immeasurably more powerful thrusts. This was the gateway (it would not be too much to write) to the moon, the planets and the stars. The modern period of liquid propellants was ushered in by two great pioneers, a Russian and an American, and two engineer-propagandists, a German-Hungarian who worked in Rumania, and a Frenchman. The first we have had occasion to mention already (p. 506); he was Konstantin Eduardovitch Tsiolkovsky (1857 to 1936), a mathematician of deep insight, who was probably the first to work out the theory of rocket flight, and

[a] The history of jet-propulsion as such is a different question. As a principle it must have been obvious from the movements of coelenterates and cephalopods, but mankind seems to take many centuries to see the obvious. At an earlier point (Vol. 4, pt. 2, pp. 163–4, 575–6) we discussed possible explanations of the flying automata ascribed to many ancient thaumaturgical artisans, notably Archytas of Tarentum (*fl.* −380), the Alexandrian mechanicians, and Chang Hêng himself (*c.* +125); who might conceivably have used jet-streams of compressed air or steam, as Heron unquestionably did in his aeolipile (*ibid.* pp. 226, 407). Han Chih-Ho (+890) was almost too early for gunpowder, though Regiomontanus (*c.* +1450) could have used it. On Archytas and Regiomontanus see Duhem (1), pp. 125–8, 290 ff.

Duhem also tells us (pp. 295 ff.) of the Jesuit Honoratus Faber, who in +1669 proposed a flying-machine driven by a jet of air compressed by men working a pump inside. This idea was apparently continued in a notorious design by the Brazilian Jesuit Bartholomeu Lourenço de Gusmão, to which Duhem (1), pp. 297, 418 ff., (2), pp. 140 ff. has given minute attention. Then in +1715 Marc-Antoine Legrand turned to steam as the vapour to be employed in his jets (Duhem (1), p. 298). It is not quite clear how serious all these ideas were, but they certainly had a post-Renaissance character, and we have no Chinese parallels for them. In any case, the principle had no practical application until modern times, when large quantities of combustible fuel could be carried on board airplanes to provide the exhaust gases and their thrust.

It is interesting that we have a familiar example of the jet-principle, two thousand years after the aeolipile, in Segener's rotating garden-lawn water-sprinkler (Ley (2), p. 84).

[b] This paragraph is based on some formulations of Humphries (1) and Gibbs-Smith (10). Cf. Malina (1) and Anon. (161), vol. 1, pp. 578–9.

proposed as fuel liquid oxygen and kerosene or liquid hydrogen.[a] But if Tsiol-kovsky can now be called the father of rocket motor science, the father of rocket motor engineering was the American, Robert H. Goddard (1882 to 1945),[b] also a university professor, who worked for many years from 1907 onwards with dog-ged concentration and very limited support in search of the means of reaching 'infinite altitudes' beyond the earth's atmosphere.[c] The world's first liquid-fuel rocket[d] was successfully launched by him in March 1926, and four years later a height of 2000 ft was attained. The one who wrote in German was Hermann Oberth (1894 to 1982),[e] who was associated with the Verein f. Raumschifffahrt (Space-Flight Society)[f] founded in 1927 and taken over by the Nazis in 1934.[g] They changed the name of the Verein's A 4 to the now universally known V 2, and it was one of these vehicles which was the first to leave earth's atmosphere and reach airless outer space in October 1942, at an altitude of 52 miles.[h] Lastly the French contributor was Robert Esnault-Pelterie, who was active and widely read in the late twenties and early thirties of the present century.[i]

Long before this time of course the gunpowder rocket had become a common-place, universally familiar in pyrotechny. Congreve rockets had lingered on till almost the end of the nineteenth century, and they had acquired a tried and tested place for life-saving at sea, as also for averting hailstorms (cf. p. 528 below) from 1900 onwards. But rocket-borne aerial photography was being re-placed by airplane cameras, rocket signalling was superseded by radio, war-rockets were almost entirely out-matched by more accurate artillery, and there was only a limited scope in World War I for rockets carrying Verey lights or making smoke-screens. It seemed as though there was little future for the use of rockets in war. And indeed we are told that the main aim of the German Verein was originally the designing of meteorological rockets.[j]

Now it is a remarkable fact that the whole of the new movement, the study of liquid propellants, derived not from military rocketry, nor from traditional pyrotechnics, but rather from the idea of the 'plurality of worlds', and the con-

---

[a] See von Braun & Ordway (1), pp. 121 ff.; Baker (1), pp. 17 ff. His works have been translated into English (1–4). Cf. Petri (7).

[b] See Baker (1), pp. 22 ff.; von Braun & Ordway (2), pp. 43 ff.; Ley (2), pp. 106 ff.; Taylor (1), pp. 16–18.

[c] His classical papers came in 1919 and 1936; cf. Goddard (1, 2).

[d] Using liquid oxygen and petrol.

[e] See Ley (2), pp. 113 ff.; Baker (1), pp. 27 ff.; Taylor (1), pp. 16–18. For his influential books, Oberth (1, 2).

[f] Ley (2), pp. 121 ff. He was later at the Peenemünde base, where the German war-rockets were developed; cf. Ley (2), pp. 184 ff., 204 ff. He alone lived to see the Cape Canaveral operations.

[g] Cf. von Braun & Ordway (1), p. 139; Taylor (1), p. 21.

[h] This 46-ft rocket was driven by liquid oxygen and ethyl alcohol, led to the combustion chamber by turbo-pumps working on steam formed from hydrogen peroxide catalysed with sodium permanganate. See Ley (2), p. 226; von Braun & Ordway (1), p. 147; Taylor (1), p. 22.

[i] See Esnault-Pelterie (1, 2). There were other names of some honour in this roll-call too. Nikolai Ivano-vitch Kibalchich (d. 1882) developed the idea of vectored thrust, i.e. the swivelling of exhaust nozzles to change the direction of the rocket's flight-path. Hans Ganswindt, active about a decade later, designed (long ahead of its time) a reaction-powered space-ship. And Eugen Sänger continued the movement in the thirties. On these see Baker (1), p. 15; Ley (2), pp. 91 ff.

[j] Ley (2), pp. 169 ff. Cf. pp. 527 ff. below.

viction that reaction-motors were the only way that man could ever take to reach them. Goddard stands in a line of descent, not from Chiao Yü, Tipū Sahib and Congreve, but rather from Chang Hêng,[a] Lucian and de Fontenelle. At an earlier moment[b] we found a good deal to say about the role of Chinese thought in the dissolution of those so long dominant European notions, the Aristotelian crystalline celestial spheres, and the perfection and immutability of the heavens, after it became known in the West through the Jesuit mission in the +17th and +18th centuries.[c] Lucian's *True History* of the men in the moon, with Cicero's *Somnium Scipionis*, were written before these doctrines had become riveted on the world-view of Christendom, but in the +17th century Europe broke free, and a whole succession of writers described extra-terrestrial voyages.[d] Thus one could say that the Chinese invention of the rocket, coming to Europe in the +14th century, was complemented by Chinese ideas about infinite empty space which reached Europe by the end of the +16th.[e]

Indeed, as Schafer has put it, 'Tours of space were a commonplace in ancient China.'[f] Accounts of them[g] long preceded Chang Hêng; in the *Lun Hêng*[2], for instance (+83), we find one about a Taoist, Hsiang Man-Tu[3], who spent some years on the moon.[h] Recently, Cadorna (1) has translated one of the Tunhuang manuscripts in the Stein Collection[i] which tells how a famous Taoist astronautical master, Yeh Ching-Nêng[4], conducted the Thang emperor Hsüan Tsung, about +718, to view the palaces of the moon.[j] As Schafer says,[k] 'The great palace of the moon ... though not the abode of any deity of the first rank, was often rather fully portrayed in Chinese, both in poetry and prose, especially as a palace of ice crystals', an intensely cold, angular, crystalline, brittle habitation of extraordinary spirits, like, yet unlike, men. In spite of the appearance of

[a] The great +2nd-century astronomer himself wrote, in his *Ssu Hsüan Fu*[1], of an imaginary journey beyond the sun.
[b] Vol. 3, pp. 438 ff.
[c] There is an interesting recent book by Dick (1) on the notion of the plurality of worlds, though it ignores the role of Chinese thought in the liberation of European ideas.
[d] One need only name Francis Godwin, John Wilkins, F.R.S., Daniel Defoe and Miles Wilson. The genre of scientific romances has been brilliantly reviewed by Nicolson (1, 2). At the same time the ancient works, which had lain dormant during the millennium of dominance, were revived and broadened men's thinking once again.
[e] The case is reminiscent of some others previously encountered. For example, it has been said that 'just as Chinese gunpowder helped to shatter European feudalism [after the +15th century], so Chinese stirrups had originally helped to set it up' (Needham (47), pp. 286–9).
[f] (26), pp. 234 ff., cf. (27). Generally we are not told very much about the nature of the vehicles employed, and the extra-terrestrial travel is often magical, but the point is that for the ancient and medieval Chinese it was in no way unthinkable.
[g] Doubtless arising in the first place from the magic flights of shamans; cf. Vol. 2, pp. 132, 141; Vol. 4, pt. 2, pp. 568 ff.
[h] Tr. Forke (4), vol. 1, pp. 340–1. Wang Chhung[5] of course didn't believe it.
[i] S 6836. An earlier translation was that of Waley (31), pp. 139 ff.
[j] The same story is in the *Tao Tsang*'s *Thang Yeh Chen Jen Chuan*[6] (Biography of the Perfected Sage Yeh of the Thang), *TT* 771.
[k] (26), pp. 194–5.

[1] 思玄賦　　[2] 論衡　　[3] 項曼都　　[4] 葉淨能　　[5] 王充
[6] 唐葉眞人傳

some lunar beauties, the emperor could not stand the cold, and begged to be taken back home, which Master Yeh duly did.

In the nineteenth century all these traditions crystallised into what we now call science fiction, on which there is a large descriptive literature,[a] and it was works of this kind which, on their own explicit statements, had the greatest influence on the pioneers of modern rocketry. Reaction-motors, to be sure, were not the only means of inter-stellar flight envisaged;[b] there were also imaginary anti-gravity substances,[c] and of course great cannon pointing to the stars.[d] Tsiolkovsky was inspired by Eyraud, Jules Verne, Dumas and Greg; Goddard and Oberth in addition by Lasswitz and H. G. Wells. And not only was the cosmic navigational tradition primarily responsible; it would also be justifiable to say that the military rocket-missiles of World War II and subsequently were a spin-off or by-product of the peaceful urge for space research and exploration. May it be granted that the former do not overwhelm the latter.

In due course all the pioneers of liquid-fuel rocket flight were sucked into the maw of military preparations. Goddard was eventually aided by the American army and navy development establishment (1918), while the Verein's Berlin Raketenflugplatz was supported by the German military from 1932 onwards.[e] Four years later GALCIT[f] was formed, under the direction of Theodore von Kármán,[g] with Frank Malina and Chhien Hsüeh-Sên[1] among its staff;[h] significantly it became ORDCIT[i] in 1945, and applied itself almost entirely to war missiles. Among its achievements was the use of red fuming nitric acid and aniline or benzene as the self-igniting liquids;[j] as also the development of strange solid propellants such as mixtures of asphalt or polyurethane and potassium perchlorate, or sodium nitrate with ammonium picrate.[k] Other liquid propellants used today are fluorine, tetranitromethane, liquid ammonia, hydrazine hydrate, boron hydride, etc.[l] If the Russian 'Katyusha' and 'Stalin organ' war-rockets were so effective in World War II it was because they no longer used gunpowder charges, but rather guncotton[m] and nitroglycerine, still generally

[a] Cf. Flammarion (1); Ley (2), p. 41; Morgan (1); Anon. (162); Darko Suvin (1).
[b] But these occur in Achille Eyraud's *Voyage à Vénus* (1865) and Kurt Lasswitz' *Auf Zwei Planeten* (1908).
[c] As in Percy Greg's *Across the Zodiac* (1880) and H. G. Wells' *The First Men in the Moon* (1901). His *War of the Worlds* had appeared three years earlier.
[d] Here of course the type-specimen is Jules Verne's *De la Terre à la Lune* (1865). In the same year Alexandre Dumas wrote a novel with almost the same title.
[e] Taylor (1), p. 21; von Braun & Ordway (1), p. 138. Von Braun, Oberth and Ley, with many others, were all appropriated by the American rocketry organisation at the end of World War II.
[f] The Guggenheim Aeronautical Laboratory of the California Institute of Technology. On it see Malina (2, 5); Baker (1), pp. 2 ff.; Ley (3); von Braun & Ordway (2), pp. 84–5.
[g] Cf. Wattendorf & Malina (1).
[h] Alternatively, Tsien Hsue-Shen. Other Chinese scientists also worked there, notably W. Z. Chien and C. C. Lin.
[i] Ordnance Department Laboratory of the California Institute of Technology. See Malina (3, 4): Baker (1), pp. 73 ff.
[j] Ley (3).
[k] Humphries (1), p. 26; Anon. (161), vol. 1, pp. 580–1, vol. 2, pp. 363–4.
[l] Cf. Clark (1); Parker (1); Humphries (1), p. 40; Anon. (161), vol. 2, pp. 362–3.
[m] Nitrocellulose was discovered by Schönbein as long ago as 1845.

[1] 錢學森

with chemically built-in oxygen.[a] If the Russians were the first to launch a successful earth satellite (1957) and the first to put a man (Yuri Gagarin) into space (1961), it was perhaps because of their heavy atomic war-head payloads, which necessitated enormous rockets.[b] Yet nuclear energy may well be the ultimate answer to the demands of jet-propulsion for space-flight.[c] So here again we touch upon a paradox already mentioned (p. 506 above) that the very engines which would be capable of destroying civilisation itself are the same great rocket-motors as those which are opening the way to the planets and the stars.[d] It is common knowledge that space probes such as 'Mariner' have been sent out all over the solar system since 1962.[e] And finally the first Chinese artificial satellite went up in 1970, from the rocket's very homeland,[f] since when there have been at least eight more.

In the end the rocket motor could be the means of the preservation of the human race itself, removing it to other habitations as the sun of our solar system cools or overheats.[g] It might turn out that the rocket was the greatest single invention ever made by man. So in spite of all the perils of guided rocket missiles still impending, those Chinese who first experimented successfully with 'flying meteoric ground-rats', though we may never know their names, have been extraordinary benefactors of humanity, and citizens of no mean city.

(20) PEACEFUL USES OF GUNPOWDER

Since our mind has been running so much on rockets in the preceding pages, it will make an easy transition to begin with those same devices applied to religious observance and weather control, as also the exploration of the earth's upper atmosphere. Then we can go on to consider the even more universal role

[a] Taylor (1), pp. 23–5; Ley (2), pp. 190 ff.; von Braun & Ordway (1), p. 160; Popescu (1).

[b] Von Braun & Ordway (1), pp. 162, 176; Taylor (1), pp. 92, 144 ff.; Popescu (1). The American 'Apollo' moon landings followed from early 1963 onwards (Baker (1), pp. 165 ff.; von Braun & Ordway (2), pp. 172, 218–19). Fig. 219.

[c] Humphries (1), p. 194; Anon. (161), vol. 2, pp. 366–7.

[d] Sokolsky (1) takes the story to 1974, Buedeler (1) to 1979, and Cornelisse, Schöyer & Wakker (1) to 1981. Here we reach the truly professional level of current research. The reader equipped with mathematical, chemical or metallurgical expertise will find whole series of collective volumes which discuss the latest advances in our knowledge. For example, there is *Progress in Astronautics and Rocketry*, which began in 1960 and numbers some fifty volumes at the present time; all under the aegis of the American Rocket Society.

[e] Taylor (1), pp. 146 ff.; von Braun & Ordway (2), pp. 164 ff.; Baker (1), pp. 135 ff.; Ley & von Braun (1).

[f] The two-stage motors burnt dimethyl hydrazine as fuel and nitrogen tetroxide as oxidiser; they could reach at least 4000 miles in surface-to-surface flight, and with a third stage could put a satellite into geostationary orbit at an altitude of some 23,000 miles. Cf. Hewish (1); Anon. (163).

New launches of space rockets have taken place in 1982 (*Jen Min Jih Pao*, 14 Jan. reprinted in *CKKCSL*, 1982, no. 2, 90). And a submarine-based carrier rocket was successfully tested in October (*China Pictorial*, 1983, no. 1). On China's first communications satellite, lofted by a three-stage rocket on 16 April 1984, see Yang Wu-Min (1).

[g] Today the rocket vehicle looms very large in the imagination of all those who are conscious of the vastness of our universe, and inspires the engaging fantasies of eminent scientific men. For example, Francis Crick (1), the molecular biologist, finding difficulty in accounting for the origin of life on earth, imagines a rocket spacecraft which could have brought it (in the form of eukaryote bacteria) billions of years ago, from some other civilisation in our own, or some other, galaxy. Of course, this 'directed panspermia' only puts the problem back another remove.

Fig. 219. The Apollo 14 blast-off in 1971 (Baker (1), p. 219).

which gunpowder has played in rock-blasting by miners and civil engineers concerned with roads, railways and waterways.

(i) *Ceremonial and meteorological rockets*

The recreational use of gunpowder in fireworks, especially rockets, has been so widespread in all parts of the world for so long, and so many good histories of them exist,[a] that we need say no more of them here. But the meteorologists soon found rockets invaluable for exploring the nature of the upper air and the fringes of space. We have already had occasion to mention meteorological rockets (p. 522), and indeed they are in active use at the present day.[b] Sounding-rockets go up to the hundred-mile altitude level, launching-rockets that carry payloads such as satellites reach two or three hundred miles, and are effectively in outer space. A great many types have been used, such as the 'Viking' and the 'Datasonde'; and some, such as 'Aerobee' and 'Skua', still are.[c] The instruments with their readings are often recovered by parachute.[d] Their sensors have given meteorologists a great wealth of data, on winds, temperatures, the earth's magnetic field, the ionosphere, cosmic rays, infra-red and ultra-violet radiation, X-rays, etc.

But rockets have also played a part in that other, even more prestigious (if still in some sense equivocal), branch of meteorological endeavour known as 'weather modification'.[e] In November 1946 Vincent Schaefer made the fundamental discovery of glaciogenesis when he dropped dry ice[f] pellets from an airplane on to a cloud, which within five minutes gave a snow shower. It was quickly realised that the provision of nuclei for snowflake, hail or raindrop formation was the issue, and in the following year Bernand Vonnegut, searching the literature for the crystal forms most similar to ice, suggested the iodides of silver and lead.[g] So arose the technique of 'cloud-seeding', which since that time has become so world-wide a practice that laws have even been introduced to control it. The effect may also be produced just by the shock of an explosion, hence rockets carrying charges, as well as those with silver iodide, have come into use.

---

[a] E.g. Brock (1, 2).   [b] Cf. von Braun & Ordway (2), pp. 150 ff.

[c] Cf. Firiger (1); Almond, Walczewski *et al.* (1); Schmidlin, Ivanovsky *et al.* (1).

[d] Russian geophysical rockets have sometimes shot up, and safely recovered, living experimental animals, such as dogs, recalling the +18th-century experiments mentioned above (p. 506), but more usefully.

[e] See the book of Dennis (1).   [f] Solid CO_2.

[g] On the growth of snow-crystals see Mason (1). It was J. K. Wilcke in +1761 who first made snow-crystals artificially, and this bore fruit in the following century when iodoform and camphor were found to be nucleating agents; cf. Dogiel (1) and Spencer (1). In our own time Mason & Maybank (1) have shown that particles of clays and other minerals from the earth's surface are more probably the cause of precipitation than meteoric dust. One of these agents is gypsum (calcium sulphate), and this salt was mentioned as a six-pointed crystal, precisely in this context by Chu Hsi[1], the great Neo-Confucian philosopher and naturalist, in the +12th century. The hexagonal symmetry of ice-crystals was in fact known in China long before Europe; by Han Ying[2] about −135, and many other students of Nature, earlier than Johannes Kepler in +1610, as has been shown by Needham & Lu Gwei-Djen (5).

[1] 朱熹   [2] 韓嬰

Unfortunately the results are not always reliable and conclusive, nor statistically certain; controlled experiments are very difficult to carry out, so that the subject is still to a certain extent controversial. In a recent study Mason (2) has considered three programmes, one in Tasmania, one in Florida, and one in Israel, but only the last has shown consistently positive results over several years. Sometimes the technique works, sometimes not. Nevertheless it is very-widely employed, and when I used to holiday in the Sarthe, in France, years ago, I remember being shown rockets which the vineyard workers used to shoot off at impending hailstorm clouds in order to get them to discharge their hail before ruining the wine-grapes.[a] The technique is hard to use because of the height and speed of the clouds.

Horwitz (8), the only historian, so far as we know, who has tried to trace the origins of the empirical practice, thought that it became widespread only towards the end of the nineteenth century. There may have been more than one root of it, for example a general belief, hard to document, that big battles brought on unusual rain, and a folk-superstition that weather-witches causing whirlwinds could be shot if one fired guns into the air.[b] But Horwitz found in the autobiography of Benvenuto Cellini (+1500 to 71) an early reference to the letting off of cannon at storm-clouds, with good results;[c] and something similar occurs in a book on thunder by Abraham Hosemann in +1618.[d] No other relevant passage has been reported.

But in spite of its seeming modernity, the idea of weather modification does not appear to have originated in Europe. In South-east Asia we find a large group of folk customs which consist essentially in firing off large conventional rockets at the storm-clouds of the monsoon, with the double objective of honouring the tempest spirits and initiating the precipitation of rain. We may take as a typical example the ceremonies at villages near Chiengmai in northern Thailand.[e] First, the home-made rockets, mounted on bamboo biers or carriers decorated with branches, rest for some days in front of the *stupa* of the local Buddhist temple (*wat*) before being carried into the rice-fields[f] for firing; then to the accompaniment of gongs and drums they are taken to the launching-ramp

[a] See Foote & Knight (1); Dennis (1), pp. 75–6, 206–7, 208 ff., 232 ff. Hail-storms are among the greatest dangers for the *vignerons* (cf. Lichine, Fifield *et al.* (1), p. 35, 116, 433, 538; Schoonmaker (1), p. 162). They can wipe out a valuable crop completely. On the French anti-hail measures see Dessens (1); on the Russian anti-hail rockets see Bibilashvili *et al.* (1) and Dennis (1), fig. 5. 14. Fog dispersal is a related subject, on which see Dennis (1), pp. 163 ff.

[b] Wuttke (1), p. 283, §444.

[c] Bettoni ed., vol. 2, p. 56; Goethe tr. vol. 44 (vol. 2), p. 334.

[d] (1), p. 121.

[e] Studied and photographed in 1965 by Mr Hugh Gibb, to whom we are grateful for much information and many pictures. He also followed the ceremonies at Nongkhai in northern Thailand.

[f] There is no irrigation in these parts, so people are particularly dependent on the monsoon rain. In central Thailand, where extensive irrigation systems exist, the rocket custom does not. In Cambodia at Angkor the farmers also depend on rainfall because the Khmer irrigation systems of old have disappeared. Hugh Gibb witnessed rain-making ceremonies there in which water was splashed on kneeling women by Buddhist priests—there were no rockets. He suggests therefore that the rockets were Laotian or Thai (and ultimately Chinese) in origin, rather than Mon-Khmer.

tower (Fig. 220)[a] and at the right moment let off by a man who climbs up and lights the fuse. Very often this is a Buddhist priest or abbot,[b] and the *bhikkus* always supervise the firing.[c] This custom is attested from many places in Laos[d] as well as Thailand, and the Tai people of the Hsi-shuang-bana region in Yun-nan.[e]

Another first-hand description has been given of a similar custom at the village of Nong-song-hong near Nongkhai,[f] the *Bun-bang-fai* (Deed of Rocket-firing). The rockets were very large indeed[g] and sometimes decorated with wooden serpent-head (*naga*) carvings; the rationale was said to be the placating of the rain-gods, so ensuring a good harvest.[h] Winter (5) found further examples of large commemorative rockets in Burma, but there essentially at the funerals of Buddhist priests. This certainly goes back to the beginning of the last century, because William Carey[i] described it in 1816; the pyre was lit by rockets running along cords, and at the same time large rockets were set off.[j] This Burmese custom is attested from 1839 by Malcom[k], and certainly continued into this century;[l] indeed it is said to go on still. The roles of Chinese and Indian influence respectively in these intermediate countries have not yet been elucidated, but there is an interesting passage in the travels of Tavernier[m] which shows that in the +17th century China was still looked to as the home of pyrotechnic art.[n] About +1645, during his stay in Western Java, he was present on an occasion when at the court of the King of Bantam:

There were five or six captains seated round the room, examining some fire-works which the Chinese had brought with them, such as grenades, rockets, and other things of that kind, some running upon the surface of water[o]—for the Chinese surpass all other nations of the world in this respect.

---

[a] Many are over 40 ft high. The long bamboo stick or tail of the rocket is prominent.

[b] The rockets frequently have *nagas* painted on them, symbolising the serpent-king having dominion over all waters, parallel to the *lung*[1] or dragon, in China.

[c] See the description of the 'sky-rocket festival' (*Bun-bang-fai*) in NE Thailand by Klausner (1), pp. 89 ff.

[d] Cf. Winter (1), pp. 7–8.

[e] On these see Alley (9), p. 9. Here some of the rockets are, or were, three-stage ones.

[f] By Winter (5), pp. 12 ff.

[g] The bamboo guide-stick was measured as 45 ft in length.

[h] Local tradition took it back a thousand years, which would be an exaggeration, but not necessarily a very gross one.

[i] (1), pp. 188–90.

[j] He mentions the size as 7 to 8 ft long and 3 to 4 ft in circumference. But apparently these too were made to run along ropes more or less parallel with the ground, not pointed upwards at the sky and the clouds.

[k] (1), vol. 1, pp. 208–9. Here the rocket cylinders were 12 ft long. He cites another witness who said that a single one might contain ten thousand pounds of powder. Both Carey and Malcom described the rockets as made of hollowed-out logs bound with hoops of iron or rattan.

[l] See Hart (1), p. 124; Kelly (1), p. 174.

[m] (1), p. 360.

[n] The passage is quoted by Gode (7), p. 20; Winter (5), p. 14.

[o] These would have been the floating rocket-propelled *shui lao shu*[2] or 'water-rats', on which see p. 473 above.

[1] 龍   [2] 水老鼠

Fig. 220. The rocket-firing ceremony at a village a few miles east of Chiengmai in northern Thailand; an example of an ancient tradition of weather modification, since the large rockets are fired at the storm-clouds of the monsoon. The picture shows the rocket being carried up to the top of the tower or launching-pad, specially constructed for the occasion in the midst of the rice-fields which at this time of the year are parched and dry. The monsoon rains are necessary (in the absence of irrigation) before the rice-plants can be transplanted. The actual firing is supervised by the local Buddhist abbot or a *bhikku*. Photo. Hugh Gibb (courtesy of the BBC). For a Japanese (Shinto) parallel see Anon. (*262*).

The connection between the rocket-firing and the monsoon rains is not very close, for the rains may come before the rockets are let off, or be delayed for a good while afterwards—but this never dampened the belief that there was a connection. Some local people, however, deny it, saying that the custom is a form of reverence for the Hindu rain-god, Indra. Perhaps what was originally a practical act became converted with the passage of time into a religious observance.

It is not easy to find out how far back these weather-modification rockets go. But in view of the antiquity of the gunpowder rocket in China, it seems overwhelmingly probable that they came down from there, and it could have been at any time after about +1200.[a] Today anti-hailstorm rockets are widespread in China,[b] often locally made, and stationed at many thousands of posts among the communal farms up and down the country.[c] It looks as if a prediction could safely be made that somewhere in the vast literature of medieval China[d] a passage or two will be found which will point to the beginnings of the practice of using rockets for weather modification—unless indeed the Thais thought of it themselves, using the jet-propelled vehicle that came from further north. After all, why not?

Still, our belief is that there was a very ancient *ourano-bolic* tradition in China, as one might call it, a conviction that shooting missiles up into the heavens could give useful results. One of the earliest of Chinese legends concerns the ten suns[e] which appeared in the time of the Emperor Yao[7], and would have burnt up the face of the earth, if the Archer-Lord Hou Yi[8] had not skilfully shot nine of them down.[f] One of the commonplaces of Old China Hands was to make superior

[a] It would be good to find mention of them in Chou Ta-Kuan's[1] description of Cambodia in +1297, the *Chen-La Fêng Thu Chi*[2]. But though Pelliot (33), p. 21 translated *fusées* fired from high scaffolds (*ta phêng*[3]), the text itself (e.g. in *TSCC*, Pien i tien, ch. 101, p. 26b) says only *yen huo*[4] (smoke fireworks) and *pao chang*[5] (fire-crackers) as big as trebuchet bombs (*phao*[6]), let off at the New Year ceremonies. Chhen Chêng-Hsiang (2), p. 53, has no comment on this passage. Cf. Pelliot (59).

[b] And guns and anti-aircraft guns too.

[c] Cf. Anon. (164). Our friend the climatologist Thu Chhi-Phu confirms this information. Some can attain an altitude of a mile or more.

[d] The only clue we have is an intriguing note which occurs in Bastian's account (1) of his travels in China. In vol. 6, p. 410 he reports, without any reference, a remark said to have been made by the Khang-Hsi emperor about +1695: 'If the lamas seek vainly to drive away the rains on the Gobi desert by firing off cannon, it must be because such a practice is lacking in respect for the spirits of the lakes and rivers.' One would not expect the custom so far north, and one can only hope that further research will throw more light on the whole matter.

[e] There can in fact be nine supernumerary or 'mock' suns in a complete parhelic display, brought about by ice-crystals in the upper atmosphere; see Vol. 3, pp. 473 ff. The legend must surely have originated from this. Parhelic phenomena were first described in Europe by Christopher Scheiner in +1630, but (as Ho Ping-Yü & Needham (1) found) a full thousand years before that, all the components had been named and noted in the *Chin Shu*, ch. 12, pp. 8a–9b, tr. Ho Ping-Yü (1).

[f] The legend is very old in China; see Granet (1), pp. 377 ff. and passim. One of the earliest references is that in the poem *Thien Wên*[9] (Questions about the Heavens) perhaps of the −5th century (*Chhu Tzhu Ssu Chung*, p. 56; Hawkes (1) tr., p. 49). Another is in the *Chao Hun*[10] (Calling Back the Soul) probably by Ching Chhai,[11] c. −240 (*Chhu Tzhu Ssu Chung*, p. 120; Hawkes, *ibid.* p. 104). The *locus classicus* is in the *Huai Nan Tzu* book, ch. 8, pp. 5b, 6a, tr. Morgan (1), p. 88. See also Werner (1), pp. 181–2, (4), p. 470; Allan (1), pp. 301 ff.

[1] 周達觀   [2] 眞臘風土記   [3] 大棚   [4] 烟火   [5] 爆枚
[6] 砲   [7] 堯   [8] 后羿   [9] 天問   [10] 招魂
[11] 景差

comments on the noise of gongs, drums and fire-crackers[a] with which people continued to salute solar and lunar eclipses centuries after their true nature was well known among the literati, but this again had an ourano-bolic element because it was associated with an Immortal named Chang (Chang hsien[1]),[b] essentially an archer who used his bow to shoot at the Celestial Dog (Thien Kou[2]) supposed to be devouring the luminaries;[c] and guns were frequently let off at him.[d]

These practices go back very far. In the *Tso Chuan* for the 4th year of Duke Chao (−537) one can find an interesting reference to weather phenomena in connection with shooting.[e]

In the spring, in the first month of the year according to the imperial calendar, there fell much hail (*pao*[3]). Chi Wu Tzu[4] asked Shen Fêng[5], saying 'Can one stop hail from falling?' Shen Fêng replied 'When a great sage is in power hail does not fall, or if it does it causes no harm. Formerly, when the sun was in the northern part of its course, everyone gathered up the ice and piled it in the prince's ice-house.... When it was taken out for use, people took peach-wood bows, and arrows of jujube-wood, to shoot and chase away all evil influences[f].... If this was done, thunderstorms, frost and hail did no harm, and the people suffered no epidemic diseases.'[g]

Whether the bows were pointed upwards to the sky the text does not say, but very likely they were. Then at the other end of history, in +1695, we have a curious story describing the customs at a place in Kansu province, and now the link with gunpowder technology comes out clearly, though there is no word of rockets. Liu Hsien-Thing wrote as follows:[h]

Mr Tzu-Thêng[6] said that in the neighbourhood of Phing-liang, in the fifth and sixth months of summer, there were often violent gusty winds bringing yellow clouds down from the mountains along with icy hail, the biggest stones of which were the size of one's fist, and the smaller like chestnuts. It ruined people's crops like a diabolical disaster. As soon as the local people saw such a cloud coming they beat upon gongs and drums, and fired off cannon (*chhiang phao*[7]) to disperse it. Sometimes they hit it fair and square, and then yellowish (lit. blood-red) rain fell, and so the cloud gradually disappeared.

Sometimes it entered the mountain-caves, and people pursued it there and surrounded it, after which they smoked it out with gunpowder. After a long time what was in there died, and when they dug it out they found either a big snake or a big toad, and every one had a piece of ice in its belly.

---

[a] Cf. de Groot (2), vol. 6, pp. 941 ff.

[b] Patron saint of pregnancy and childbirth since at least the +10th century; see Doré (1), vol. 11, pp. 981 ff.; Werner (4), pp. 34 ff.

[c] Grosier (1), Eng. tr., vol. 2, p. 439; Werner (1), pp. 177 ff., (4), p. 469.

[d] Williams (1), vol. 1, p. 819.

[e] Our attention was drawn to both the following passages by Dr Wang Phêng-Fei of the National College of Meteorology, Nanking, at the request of our friend Dr Thu Chhi-Phu.

[f] Peach-wood was a classical demonifuge in China. Cf. Fig. 1362a in Vol. 5, pt. 3.

[g] Tr. Couvreur (1), vol. 3, pp. 70 ff., eng. auct.

[h] *Kuang-Yang Tsa Chi*, ch. 3 (p. 158), tr. auct.

[1] 張仙　　　[2] 天狗　　　[3] 雹　　　　[4] 季武子　　　[5] 申豐
[6] 子騰　　　[7] 鎗砲

Apart from the piece of mythological embroidery at the end, the passage, referring to about +1680, points rather clearly at ourano-bolic activities in which rockets could easily have been involved. The practice of using firearms against hail-storms is moreover attested by oral tradition from Hobei in the Ming period (+1400 onwards.) All this, in sum, may be regarded as supporting a Chinese origin for the weather-rockets of the south.

(ii) *Rock-blasting in mines and civil engineering*

But the greatest field for the use of explosives in a peaceful context has always been rock-blasting, whether in mining, quarrying or civil engineering. Just as the history of gunpowder blends into, and was indeed a continuation of, the history of incendiary weapons (p. 94 above), so also that of mining explosives was a continuation of that of the much earlier technique of 'fire-setting'. And we find ourselves here up against an exact parallel to some of the terminological difficulties already encountered (p. 130), for just as *huo chien*[1] originally meant an incendiary arrow but later on a rocket, the thing changing fundamentally though the name did not; so also now when *huo shou*[2] (firemen) or *huo chiang*[3] (fire-artisans) are mentioned, they were in early times assuredly engaged in breaking up rocks by fire-setting, while in later times they were using gunpowder. In order to trace, therefore, the development of the use of explosives in mining or civil engineering we must make a judgment on other grounds, weighing the probabilities in the light of what has already been established about the history of gunpowder, and the changes in its composition as time went on.

What exactly was fire-setting? It must have been noticed in many parts of the world in high antiquity that forest fires would crack, rend and split the hardest rocks, and it would soon have been found that pouring water over them while still hot would increase the effect because of the expansive force of steam formed in the cracks.[a] There may be a −7th-century mention in Jeremiah,[b] but the celebrated description of the Egyptian gold-mines by Agatharchidas of the −2nd century, reported in Diodorus Siculus,[c] is precise and unquestionable. Thenceforward fire-setting is found in mining all through the ages, in the +11th-century Pisan mines of argentiferous galena in Sardinia, or the Rammelsberg mines of Germany in +1359. Agricola's account in +1555 is very detailed. Indeed the art survived almost until our own time at Kongsberg in Norway, and in Burma, India and Korea, going on two or three centuries after the introduction

---

[a] Diodorus Siculus, v, 2, Booth tr., p. 320, says that the Phoenicians were led to silver mines in the Pyrenees by observing the cracked rocks and ore outcrops after such natural fires. Good accounts of the history of deliberate fire-setting in the Western world are to be found in Collins (1); Hoover & Hoover (1), pp. 118–19, translating Agricola, *De Re Metallica*, bk. 5; Sandström (1), pp. 28 ff., 271 ff.

[b] XXIII, 29: 'Is not my word like as a fire? saith the Lord, and like a hammer that breaketh the rock in pieces?' Perhaps also Job XXVIII, cf. Bromehead (9), pp. 565 ff.

[c] III, 1, Booth tr., pp. 89, 158, given in modified form by Hoover & Hoover (1), pp. 279–80.

[1] 火箭   [2] 火手   [3] 火匠

of gunpowder blasting. According to Collins, gunpowder did not kill it, but dynamite eventually did. 'Hot mining', as it was called, consisted of lighting veritable pyres of wood in the drifts (tunnels) and in stopes (halls) after which, when the fumes had cleared away sufficiently, the rock was found to have spalled off in large flat chunks.

Drenching with water was obviously easier in civil engineering than in mining operations, but for two millennia a persistent Western tradition held that vinegar was sometimes used instead of water. This seems to have started with the account in Livy (−59 to +17)[a] of the widening of gorges in the Alps by the Carthaginian general Hannibal during his descent into Italy in −218 during the Second Punic War. It was continued by Pliny (+23 to +79) in his description of gold-mining,[b] but most modern writers have considered it a myth. Hoover & Hoover suggested emending *infusa aceto* to *in fossa acuto*,[c] and Sandström pointed out that since the dry distillation of wood produces 'wood vinegar' or pyroligneous acid, dousing a large timber fire with water would give the characteristic smell, and hence a misunderstanding.[d] We might however be prepared to allow that if veins of calcium or other carbonate were present, the resulting gas might help the splitting; and this may gain some credence from the fact, generally unnoticed, that exactly the same tradition occurs in China, where it would seem to have originated independently. For the oldest references are Thang ones, for example, the cutting of Li Chhi-Wu's[1] by-pass canal around the San Mên Hsia[2] gorge channels on the Yellow River in +741, where vinegar (*hsi*[3, 4]) is distinctly mentioned.[e] Then in +839 Liu Yü-Hsi[5] (+772 to +842) was reconstructing a road in southern Shensi,[f] and had to remove large boulders which held up the project. In his *Liu Pin-Kho Wên Chi*[6] he wrote that 'blazing coal (or charcoal) was used to roast them, and strong vinegar (*yen hsi*[7]) poured on them,[g] whereupon the rocks were burst into fragments as fine as coal-dust, so that they could be swept away (and removed in barrows)'.[h] The same prescription occurs again a thousand years later in official instructions for building mountain roads in Szechuan.[i] Meanwhile, in +1189 Yang Wang-Hsiu[8] was on a tour of inspection of that province. He found that

a certain waterway was dangerous for merchant shipping, so he spent much money in paying men to remove the rocks, but they proved to be so hard that they could not be

[a] XXI, 37.

[b] *Nat. Hist.* XXXIII, xxi, 72, Loeb ed., vol. 9, p. 55; another mention is in XXIII, xxvii, 60, Loeb ed., vol. 6, p. 453.

[c] (1), p. 119.   [d] (1), p. 29.

[e] Cf. Vol. 4, pt. 3, p. 278 and especially Anon. (*33*), p. 69, with references.

[f] The 'Western post-station road south of the mountains', i.e. the Chhin-ling Shan.

[g] We may remember what was said about the concentration of vinegar by the freezing-out process, rather than distilling, in Vol. 5, pt. 4, pp. 152–3, 178–9.

[h] Ch. 8, pp. 9a–10a (p. 67), tr. auct. adjuv. Hartwell (3), p. 49. Unfortunately, Hartwell translated *hsi* in this and the next passage as a alcohol or 'strong spirits', thus missing the point about the use of vinegar.

[i] Dated +1739 and quoted in *Chao Hua Hsien Chih*[9], ch. 28, pp. 6a–8a; cf. Wiens (2), p. 147.

[1] 李齊物   [2] 三門峽   [3] 醯   [4] 醋   [5] 劉禹錫
[6] 劉賓客文集   [7] 嚴醯   [8] 楊王休   [9] 昭化縣志

broken up. When he saw this he started an iron-works to make sledge-hammers and chisels from scrap, after which he attacked the rocks with vinegar and blazing charcoal fires (*hsi chhih than i kung chih*[1]). To this they yielded, and broke into pieces, which could be carried away on trebuchet poles (*shih wei chih chieh, i phao kan i chhü*[2]).[a]

It is just possible, though not at all likely, that Liu Yü-Hsi was indebted to Pliny, whose book he could never have heard of; and it is possible, though not easily believable, that the vinegar idea could have persisted so long at both ends of the Old World if under all circumstances there was nothing in it.

Be this as it may, however, the introduction of true explosives for blasting is much more central to our interest. It did not begin in Europe before the +16th century,[b] but the use of gunpowder in mining by Kaspar Weindl in +1627 is well authenticated and widely accepted.[c] Thereafter there are many references to the practice, in +1635 at the Nasa silver mines of Lapland, where timber was

[a] *Kung Khuei Chi*[3] (+1210), ch. 91, p. 5*b*, tr. auct. The poles were used as the familiar shoulder-poles for two or more men to carry; one would like to think that Yang shot the pieces right out of the gorge, but that is not the meaning. The scrap-iron was obtained in the offending sandbank itself, probably from earlier wrecks. Hartwell (3), p. 50, who translated a little more of the passage, thought that the chisels were used to drill holes in the rocks for gunpowder charges, but unfortunately this is not warranted by the text. He too noticed the 'ballista' and supposed that it was used 'to lift the fragmented rocks', but we do not think that this is warranted either.

[b] Dates of +1548 and +1572 are given by Li Yen (*1*) for clearing silt from the River Niemen in Poland, but without references. Raffaello Vergani (priv. comm.) put it about +1575 in Italy. How right he was is shown by his subsequent publication, Vergani (1). Gunpowder blasting did in fact start in Italy, in the silver mines at Leogra below Schio north-west of Vicenza, where Giovanni-Batista Martinengo applied it from +1572 onwards. It must have been used a little earlier in Sicily, for Martinengo, in his application to the Most Serene Republic for a mining patent, refers to 'his new method' as following the example of the Sicilian miners who worked for the Spanish crown. Another instance comes from an account of the papal alum mines at Tolfa in Latium (inland from Civitavecchia) written in +1588, describing how charges were let off in holes drilled in exposed rock-faces. A third case concerns the mines of argentiferous lead ore in the Zoldo valley north-east of Belluno in the Veneto, but no exact date can be given. By +1586 the deviating of rivers and the moving of mountains is mentioned in the very title of the artillery manual of Luys Collado (1).

Seeking to explain the economic drive that led to the application, Vergani stresses (and documents) the timber shortage crisis which affected many parts of Europe at this time. He considers whether gunpowder rock-blasting spread north to influence Kaspar Weindl and others, as also whether news of the Central European successes diffused in time back south again. Spread was certainly slow—for example in his book of +1640 Barba (1) knows nothing of blasting, but it is regarded as a commonplace in that of delle Fratte e Montalbano (1) in +1678. Perhaps this was because the associated techniques were slow in developing; it seems that there were no adequate distance-fuses (*funiculi sulphurati*) until the end of the +16th century. There is no necessity to assume any direct influence from China for gunpowder rock-blasting. It was a natural extension of use once the explosive as such had become known in the West, from about +1300 onwards.

Two other discoveries are also due to Vergani. As early as +1481 the road between Bressanone and Bolzano in the Alto Adige had been opened or widened through the gorge of the Isarco by the Archduke Sigismund of Austria using gunpowder, as we know from the travel journal of the German Dominican Felix Faber (1), who passed that way two years later on his pilgrimage to the Holy Land. 'Dux fecit arte cum igne et bombardarum pulvere dividi petras, et scopulos abradi, et saxa grandia removeri.'

Secondly, in +1560 and +1563 there were two attempts to raise sunken ships by gunpowder in the port of Venice, presumably by shifting the mud in which they were stuck. Or perhaps the idea was to blow up the wrecks. The first was proposed by Bartolomeo Campi and his brothers, the second by Antonio Suriano, one of the Armenian colony at Venice. The fact that gunpowder would explode under water had been known in Europe since about +1440; cf. Partington (5), pp. 152, 157.

[c] Feldhaus (1), col. 1072; Darmstädter (1), p. 115; Zworykin *et al.* (1), pp. 103, 780; Watson (1), vol. 1, pp. 341 ff.; Sandström (1), p. 278; Hoover & Hoover (1), p. 119. Some say it was near Chemnitz in South Germany, others at Banská Stavnica in Czechoslovakia. But Schemnitz was the old German name for the Slovak mining town. Hollister-Short (4) has translated the relevant passage from the *Berg-Protocollbuch* preserved at the State Central Mining Archives there. He agrees with Vergani about the wood shortage, but calls for further documentation of the Italian +16th-century cases. See also Vozàr (1); Hollister-Short (7).

[1] 醯爇炭以攻之   [2] 石爲之解以砲竿移去   [3] 攻媿集

Fig. 221. The beginnings of gunpowder blasting in European mines. A medal struck to commemorate the visit of Prince William of Anhalt to the Elizabeth Albertine silver mine in +1694. The princely family is shown at the bottom or sole of the main shaft, having descended by the rope railway behind. In the upper cameo on the left, a figure is seen setting off a blasting-charge. Diameter 6·3 cm., wt. 84 gm. Photo. Graham Hollister-Short, by the kind cooperation of Werner Krober of the Bergbau-Museum, Bochum.

scarce, in +1643 at Freiberg, near Chemnitz, by Kaspar Morgenstern, and in +1644 at the Röros mines in Norway.[a] Later we hear of its use in +1682 by the Staffordshire copper miners,[b] and in +1690 for the boring of the great Languedoc Canal tunnel, planned by Leonardo da Vinci but realised two centuries later by P. P. Riquet.[c] We need not follow it further into the eighteenth century (Figs. 221, 222), but the dates of important accompanying developments are worth noting. The earliest and simplest method was 'plug-shooting', with holes drilled about 2 in. in diameter and 3 or 4 ft deep,[d] gunpowder put in loose, and

[a] Sandström (1), pp. 278 ff.

[b] Plot (1), p. 165; the tradition was that rock-blasting had been started there about +1640 by German miners brought over by Prince Rupert (Partington (5), p. 174).

[c] Sandström (1), p. 73. Li Yen (1) gives +1696 for road-making in Switzerland.

[d] Today the holes hardly exceed 1·5 in. in diameter and 1 ft 8 in. in depth. In the slate quarries of North Wales the chisel-ended boring tool was known as a 'jumper', and gunpowder was the only explosive available down to 1854. At one time they got through a ton of it a month at Blaenau. It is still preferred because it does least damage to the slate blocks (Morgan Rees (1), p. 6; Wynne-Jones (1), p. 7).

Fig. 222. Detail of the cameo showing the miner setting off the blasting-charge. Gunpowder was assuredly the explosive used.

the opening plugged with a wooden bung. In +1686 Henning Hutmann invented a machine for drilling the holes,[a] and in the following year Karl Zumbe introduced 'stemming', or tamping with clay.[b] It is often said that Hans Luft of Clausthal was the first to use carton cartridges fitting the hole in +1689, but that cannot be so because these were reported by Sir Robert Moray in the first volume of the *Philosophical Transactions* in +1665, translating a letter from Monsieur du Son.[c] Bickford safety fuse did not come until 1831 in Cornwall, where mining gunpowder had been used for a century at least, and then about 1863 came Alfred Nobel's dynamite.[d] This answered to the need for an explosive more violent, brisant and shattering than propellant gunpowder, and there were many later variants of it.[e] Such is the story, more or less, of gunpowder applied to mining and civil engineering.

[a] Darmstädter (1), p. 150. On modern developments see Lankton (1).
[b] *Ibid.* p. 152. The explosive could also be fired on a rock surface after being covered over with tamped clay.
[c] Here the hole was made self-closing by greased wedges at the moment of the explosion. An illustration of such a cartridge (+1774) is given in Sandström (1), fig. 117.
According to Vergani (2) the merit of introducing water-resistant brass cartridges belongs to Giacomo Conedera, in the Venetian mines, from +1694 onwards.
[d] A three-to-one mixture of nitroglycerine and *kieselguhr*. Cf. Anon. (161), vol. 1, pp. 450–1; Darmstädter (1), p. 629.
[e] Such as blasting gelatin, i.e. nitroglycerine and nitrocellulose. Detonator caps for the cartridges could be of picrate, mercuric fulminate or lead azide. Common salt, it was found, would slow the detonation down, and reduce the danger of igniting methane or coal dust in the air. At one time this was avoided by filling cartridges with quicklime instead of explosive, a device introduced by George Elliott in 1882 (Darmstädter (1), p. 800);

Robert Boyle's enthusiasm for the technique was well justified. In +1671 he wrote:[a]

It has long been, and still is in many places, a matter of much trouble and expence, as well of Time as Money, to cut out of Rocks of Alabaster and Marble, great pieces, to be afterwards squar'd or cut into other shapes; but what by the help of divers Tools and Instruments cannot in some Quarries be effected without much time and toyl, is in other places easily and readily perform'd, by making with a fit Instrument a small perforation into the Rock, which may reach a pretty way into the body of it, and have such a thicknesse of the Rock over it, as is thought convenient to be blown up at one time; for at the farther end of this Perforation (which tends upwards) there is plac'd a convenient quantity of Gunpowder, and then all the rest of the Cavity being fill'd with Stones and Rubbish strongly ramm'd in (except a little place that is left for a Train), the Powder by the help of that train being fir'd, and the impetuous flame being hindred from expanding itself downwards, by reason of the newly mention'd Obstacle, concurring with its own tending another way, displays its force against the upper parts of the Rock, which in making its self a passage, it cracks into several parts, most of them not too unweildy to be manageable by the Workmen.

And by this way of blowing up Rocks a little varied and improv'd, some ingenious Acquaintances of ours, imploy'd by the Publick to make vast Piles, have lately (as I receiv'd the account of themselves) blown up or scatter'd, with a few barrels of Powder, many hundred, not to say thousand, Tuns of common Rock.[b]

We are now in a position to follow the parallel evolution of the techniques in China, and we have to do them both together, because, as already pointed out, Chinese writers make no clear distinction between fire-setting and gunpowder-blasting, all, to the scholar-officials, being examples of attacking rocks by fire. We can only guess, therefore, what was happening, in the light of the knowledge already gained (pp. 358 ff. above) of the development of the explosive mixture.

The oldest account comes from before the time of Agatharchidas, and concerns the activities, about −270, of the great engineer-governor of Szechuan, Li Ping[1], he who built the marvellous irrigation-system of Kuanhsien.[c] In the *Hua Yang Kuo Chih*[2] (Records of the Country south of Mt Hua), a historical geography of Szechuan down to +138, Chhang Chhü[3] says:

In Chhing-I[4] the Mo[5] river took its rise in the Mêng[6] mountains and flowed through the country to Nan-an[7], where it joined the Min[8] river. There it burst against the mountain-side, its wild torrent foaming below the precipice and causing great damage to boat traffic. This long-standing evil Li Ping remedied by sending soldiers to chisel (*tsuo*[9]) away the rocks and so rectify the current. Legend relates that this angered the water-

---

upon slaking, great heat and a doubling of volume occurred (Smith (1), p. 595; Durrant (1), p. 351) with consequent rock-splitting. These 'lime-cartridges' are often mentioned in the *Transactions of the Federated Institution of Mining Engineers* around 1890–5, but they do not seem to have been a great success.

[a] *Of the Usefulnesses of Experimental Natural Philosophy*, Boyle (8), pt. 2, sect. 2, eassy 5, p. 14.
[b] Must this not have been a reference to the contemporary works of Christopher Wren?
[c] Already described in Vol. 4, pt. 3, pp. 288 ff.

[1] 李冰	[2] 華陽國志	[3] 常璩	[4] 青衣	[5] 沫
[6] 蒙	[7] 南安	[8] 岷江	[9] 鑿	

spirit (*hui shen*[1]) much, but Li Ping boldly entered the water brandishing a sword, and gave her battle. In any case, the work brought satisfaction to the people. Again, the ancient Princes of Shu had had a fortified barrier at a place on the river where nature had formed a great rapid under a huge overhanging cliff. Li Ping had his men start to remove the rocks by cutting and chiselling, but they did not succeed, so he collected great amounts of wood, and attacked them by fire (*chi hsin shao chih*[2]). That is why the rocks at this place still bear traces of red and white and the five colours.

During the Han period the technique of fire-setting continued, as witness the work of Yü Hsü[3] and Li Hsi[4] in *c.* +120 and +160 respectively. The former engineer-official was working on the upper reaches of the Chialing River, and made it navigable by 'causing the rocks to be fired, and the wood to be cut down, for several tens of *li*, so as to open the way for the boats to pass'.[b] Another source describes how he 'made his men set great fires to the rocks, and then lead water on to them, so that they split in pieces and could be removed with crowbars'.[c] Similarly, Li Hsi, when Governor of Wu-tu between +147 and +167,[d] 'brought the high places low, and made straight the way' in the widening of the mountain roads; he had his men use splay-drills (*sun*[7])[e] and split the rocky obstructions with fire.[f]

In the Thang there are many further examples (cf. Fig. 223). One finds them particularly in connection with the San Mên Hsia[8] gorges of the Yellow River,[g] where Li Chhi-Wu[9] used steam fire-setting for his by-pass canal in +741,[h] after Yang Wu-Lien[10] had done the same for trackers' galleries around +700.[i] Similar work was undertaken by the monk Tao-Yü[11] at the Lung Mên Hsia[12] gorge near Loyang, to make the rapids navigable in +844[j] Then of Kao Phien[13], about +862,[k] it is said that[l]

it was difficult to get supplies through, so he surveyed the waterway between Kuang (-chou, Canton) and Chiao (-chou), and found that there were many huge rocks which obstructed the channel. [Ma Yuan[14] in the Han[m] had not been able to do anything about

---

[a] Ch. 3, pp. 7*a*, 9*b*, 10*a*, tr. auct., adjuv Torrance (2), p. 68.

[b] *CHS*, ch. 88, p. 4*b*; a translation of the whole passage has been given in Vol. 4, pt. 3, p. 26.

[c] Hsiao Chhang's[5] *Hsü Hou Han Shu*[6], cf. Lo Jung-Pang (6), p. 60.

[d] His activity continued during the following reign, down to +188.

[e] A special tool normally used for drilling door-pintle gudgeons, i.e. round holes. Its proper technical name woud perhaps be some kind of fantail auger or reamer (cf. Mercer (1), pp. 190 ff.; Salaman (2), p. 390).

[f] Chu Chhi-Chhien & Liang Chhi-Hsiung (*5*), p. 64.

[g] See Vol. 4, pt. 3, pp. 274 ff. On all the works here see Anon. (*33*), esp. p. 69.

[h] As already mentioned; see *HTS*, ch. 53, p. 2*a*. Cf. Twitchett (4), pp. 89, 307; Pulleyblank (1), pp. 131, 206.

[i] *Chhao Yeh Chhien Tsai*, ch. 2, pp. 19*a* ff.; cf. Pulleyblank (1), p. 128; Twitchett (4); pp. 86, 302.

[j] Yang Lien-Shêng (11), pp. 15–16.

[k] Kao Phien, the general who campaigned successfully in Annam (+863 to 4), was subsequently legate and commissioner against the rebel forces of Huang Chhao[15] (+875).

[l] *CTS*, ch. 182, p. 5*b*, tr. auct.; phrases in square brackets from the parallel passage in *HTS*, ch. 224B, pp. 3*b*, 4*a*.

[m] The great general (*fl.* +20 to +49) who re-conquered Annam (+42 to +44).

---

[1] 水神	[2] 積薪燒之	[3] 虞詡	[4] 李翕	[5] 蕭常
[6] 續後漢書	[7] 鐉	[8] 三門陝	[9] 李齊物	[10] 楊務廉
[11] 道遇	[12] 龍門陝	[13] 高駢	[14] 馬援	[15] 黃巢

campaign against Huang Chhao[1].[a] Wang Shen-Chih had Taoism in the family, for an ancestor of his was a Taoist of I-Shan[2] near Fuchow, and prophesied the rise of his descendants to imperial, or at least kingly, rank.[b] Later, about +887, other Taoists such as Hsü Hsüan-Ching[3] and Hsü Ching-Li[4] assisted Wang Shen-Chih with advice, perhaps participating in the rock clearance of +904; and indeed the whole dynasty of Min was much given to Taoism.[c]

These connections are quite important, because alchemy and early empirical experimental chemistry were so strongly Taoist in character, and since gunpowder was so great a discovery it would assuredly have been kept secret for many decades afterwards. Our chief hesitation in accepting the events associated with Kao Phien and Wang Shen-Chih as true rock-blasting concerns the proportion of nitrate probable in the mixture, but since such an action can be brought about at 50% saltpetre or even rather less, given adequate confinement, the possibility is just admissible. If it should be true, it would have one rather important corollary, namely that gunpowder was after all first used in China for peaceful ends, not indeed for recreational fireworks (as convention had it)[d], but for the assurance of transport and communications, enhancing human physical strength. At any rate, from this point onwards the case is altered, and gunpowder explosions for civil engineering purposes can reasonably be looked for.

Quite early in the new period we come upon something which constitutes a conundrum of considerable interest. In +1066 Thang Chi[6] wrote a book on the inkstones[e] of Hsichow, and the mines or quarries whence they were procured,[f] entitled *Hsi-Chou Yen Phu*[7]. The last chapter in this[g] lists the tools used for 'attacking' the rocks (*kung chhi*[8]), and besides many obvious instruments such as large and small iron hammers (*thieh ta hsiao chhui*[9]), long and short chisels (*chhang tuan tsuo*[10]), crow-beak picks (*ya tsui chhü*[11]), shovels and baskets, we find also *chhung*[12]. This is of course the word later used for the metal-barrelled hand-gun, arquebus or musket, first thus appearing (p. 294) in +1288. What could it be doing here? Now *chhung*[12] anciently meant the hole of an axe in which the handle is fitted, but that would not make any sense in the present context. One then remembers how cartridges of paper or carton to hold the gunpowder in the drilled rock-holes were current in Europe by +1665 (p. 537), and that suggests

---

[a] On this group see the study of Miyakawa Hisayuki (1). They could indeed have been the very circle in which gunpowder originated.

[b] *Shih Kuo Chhun Chhiu*, ch. 90, p. 5b. According to Schafer (25), p. 107, this was Wang Pa[5], who was an alchemist as well.

[c] Schafer (25), pp. 93, 96 ff., 104 ff.

[d] Cf. p. 128 above.　　[e] Cf. Vol. 3, pp. 645–6.

[f] We learn from ch. 1 that the deposit was first found by a hunter named Yeh[13] in the first half of the +8th century, and that the stones, presented by Li Shao-Wei[14], were much prized at court during the Southern Thang (+943 to 58). Access was dangerous (ch. 3) because of poisonous insects and wild beasts.

[g] Ch. 10, p. 6a. Unfortunately the text, if there was one, is lost, and only the catalogue remains.

[1] 黃巢	[2] 怡山	[3] 徐玄景	[4] 徐景立	[5] 王霸
[6] 唐積	[7] 歙州硯譜	[8] 攻器	[9] 鐵大小鎚	[10] 長短鑿
[11] 鴉觜鋤	[12] 銃	[13] 葉氏	[14] 李少微	

that just possibly such tubes of gunpowder were used here. Against this there is the consideration that gunpowder sufficiently high in nitrate might not have been available by +1066, only twenty years after the *Wu Ching Tsung Yao* (pp. 149 ff.). Yet two of the formulae there (pp. 117 ff.) had a saltpetre content of 60% or above, and we know that modern slow blasting powders normally contain between 60 and 70%.[a] We incline, therefore, to the belief that at this date gunpowder was used in the Hsichow mines to break up the rocks[b] and liberate the mineral suitable for polishing into inkstones.[c] Perhaps the word *chhung*[1] was re-introduced for the metal barrel when it came in two centuries later.

During the following centuries gunpowder blasting could have been used quite often, but we have not come across many examples of it. In +1310 Wang Chhêng-Tê[4] organised a great deal of rock-cutting, as for the Hung Khou Chhü[5] in Shensi, part of the Wei Pei project,[d] using hundreds of masons with metal tools (*chin chiang*[6]) and 'fire-artisans' (*huo chiang*[7]). These latter could have been blasting experts, but there is no specific mention of explosives, and they 'used fire to burn and water to splash on' the rocks (*yung huo fên shui tshui*[8]), so the work, which accounted for 500 ft of progress each day, was very likely just fire-setting.[e] On the other hand, by +1541, gunpowder blasting, followed by dredging of the detritus, was clearly employed by Chhen Mu when improving the waterway at the point where the Grand Canal crossed the Yellow River.[f] Thus one might at least say that the use of gunpowder in civil engineering began as early in China as it did in Europe, and perhaps a good while earlier.[g]

Where mining was concerned, however, there may have been many hesitations about the use of gunpowder, as Golas (1) has pointed out. According to the general view (p. 534 above) it was dynamite that killed fire-setting, not gunpowder. In modern times, for which we have eye-witness accounts, gunpowder was only very sparingly used by Chinese miners, neither for gold[h] nor silver,[i] nor yet for coal[j] or iron[k]; only in quicksilver mining is it attested.[l] Golas gives four reasons for these hesitations; (1) in general Chinese miners had access only to inferior gunpowders, and these rarely or never granulated (corned); (2) the cost

---

[a] Partington (5), pp. 326–7 gives figures averaging at 63·96%.

[b] It might have been used also to facilitate the approaches to the mines, for ch. 8 speaks of the 'narrow paths winding through the mountains' (*phan chhü niao tao*[2]).

[c] Steel dust or filings (*kang hsiao*[3]) is mentioned among the materials used.

[d] Cf. Vol. 4, pt. 3, pp. 285 ff.   [e] *Yuan Shih*, ch. 65, p. 13b.

[f] H. Li (1), p. 72; but unfortunately he gave no exact reference.

[g] Mention is made of rock-blasting in Japan by Collins (1) and Perrin (1), but without reference, nor does it appear in the latter's source around those pages, namely Tuge Hideomi (1). Of course a special research would clear the matter up, but it would be particularly interesting to know if such an application was made between +1540 and +1640, i.e. before the closure (cf. pp. 465 ff. above).

[h] Anon. (170); Louis (1, 2).   [i] Wu Yang-Tsang (1); Dawes (1).

[j] Read (14).   [k] Jameson (1).   [l] Anon. (169); Moller (1).

¹ 銃   ² 盤屈鳥道   ³ 鋼屑   ⁴ 王承德   ⁵ 洪口渠
⁶ 金匠   ⁷ 火匠   ⁸ 用火焚水淬

and labour involved in shothole-drilling was rather high, especially in hard rock;[a] (3) there was great danger of igniting fire-damp (methane) in coal-mines (the most numerous of all Chinese workings), and of delayed charges in small mines where each miner sold what he himself got;[b] (4) finally, the fire-setting technique continued to be available and was much employed. For the safe use of gunpowder, specialist miners, as well as a disciplined work-force, were necessary, yet very often the labour in small Chinese mines was essentially that of peasant-farmers turning an honest penny during seasonal periods slack in agricultural work. Thus there were several reasons why we could expect only a very slow development of rock-blasting in China in the specifically mining context.

At an earlier point (p. 506) we spoke of the ubiquitous life-saving rockets introduced by Edward Boxer in 1855 for throwing life-lines to shipwrecked sailors. But there was another beneficent action of gunpowder which has perhaps saved even more lives, namely its embodiment in the humble railway fog-signal,[c] which alerts the locomotive crew to danger ahead under conditions of minimal visibility. As White (1) has found, this valuable little detonation was introduced by Edward A. Cowper (1819 to 93) in England in 1837, and since then it has come into use wherever there are railways all over the world.

(iii) *Gunpowder as the Fourth Force; its role in the history of heat engines*

All these peaceful applications of gunpowder, however, tend to pale into insignificance by comparison with the part it played in the genesis of the steam-engine; and by the same token its predecessor, Greek Fire, the oil that we call petrol or gasoline, was destined to fuel, when the time was ripe, the internal-combustion engine. The title of this sub-section is taken from Varagnac (1), who distinguished seven successive energy-sources discovered and applied by mankind, the first three being fire, agriculture and the working of metals, then gunpowder followed by steam, electricity and sub-atomic power.

For half a dozen decades past the idea has been hovering among the minds of historians that the cylinder and the cannon-barrel are essentially analogous, and that the piston and piston-rod may be considered a tethered cannon-ball.[d] The piston and cylinder of course long preceded the metal-barrel gun and bombard, going back to the Alexandrian mechanicians and the Roman force-pumps, as also to the Chinese piston-bellows;[e] but the military engineers of China can have had little idea of what they were starting when they first used metal to make

---

[a] Successful mechanical rock-drilling was not introduced until the middle of the nineteenth century, and then in North America, cf. Lankton (1).

[b] Cf. Anon (169).

[c] In American parlance, railroad torpedo.

[d] No one seems to have said this in so many words, but the idea was current coin in Cambridge in the thirties, when Desmond Bernal was writing his *Social Function of Science*. Cf. Needham (66) p. 99; Needham & Needham (1), p. 250; also Needham (48), p. 7, (64), p. 143.

[e] I.e. to the −3rd or −4th century. Cf. Vol. 4 pt. 2, pp. 135 ff., 141–2; Needham (48), p. 12.

their fire-lance tubes (cf. p. 234 above), and then later their metal-barrel hand-guns and bombards (p. 289 above). One can now see the significance of our definition of true guns (p. 488 above) as opposed to co-viative projectiles and proto-guns, for only when the projectile exactly fitted the bore did the analogy with cylinder and piston make itself manifest. It all began with incendiary and hurtful fire as such, but when it ended with propellant explosion, then the door was open for all piston engines. The djinn was now well and truly in the bottle, and it was the Chinese military inventors who put it there in the first place.

There was really a convergence here of two strains of cylindrical structures; in the pumps and bellows the force was applied from the exterior to the contents,[a] but in the cannon the force, and a very great one, was applied from the inside outwards, doing work. We already long ago came across this antithesis when we found that the morphology of the rotary steam-engine of the early nineteenth century, with its classical solution of the problem of inter-conversion of longitu-dinal and rotary motion (piston-rod, connecting-rod and eccentric crank) had been anticipated by that of the water-powered reciprocating blowing-engines of China.[b] But the physiology was exactly the inverse, for the water-power bellows applied the force to the piston from outside, while the steam-engine applied it from inside. As for the dating, we used to say that the reciprocating furnace-bellows and flour-sifters were in general use by about +1300, but we now know that the whole assembly developed much earlier. First it was pushed back to the +10th century by Chêng Wei (1) who studied a painting of +965; and then Jenner[c], translating ch. 3 of the *Loyang Chhieh-Lan Chi* (Description of the Bud-dhist Temples and Monasteries of Loyang), found unmistakable terminological evidence of it—a bolting- or sifting-machine about +530. The book says that at Ching Ming Ssu[1], south of the city:

there were roller-mills and mills (for grinding), trip-hammers (for pounding), and bolting-machines (for sifting and shaking), all driven by water-power (*yu nien wei chhung pho, chieh yung shui kung*[2]). Of all the marvels of the monasteries these were considered the most remarkable (*chhieh lan chih miao, tsui wei chhêng shou*[3]).

This water-powered shaker or sifter assuredly worked by the mechanism which we find depicted later on.[d] The reciprocating conversion design thus preceded the rotary steam-engine by no less than thirteen centuries. And the steam-engine, and later the internal-combustion engine were in the truest sense chil-dren of the cannon.[e] But now the work they did was beneficent work.

---

[a] This was of course true also of all the late +17th-century air-pumps. It was precisely the exploration of the properties of the vacuum which led to the atmospheric steam-engine and ultimately to the fully developed steam-engine.

[b] Vol. 4, pt. 2, pp. 369 ff., 373, 759, Figs. 602, 603, 627b; Needham (48), fig. 8.

[c] (1), pp. 109, 207, 281 ff.   [d] See Vol. 4, pt. 2, Fig. 461.

[e] How direct the genesis was will appear a few pages below.

[1] 景明寺   [2] 有䃺磑舂簸, 皆用水功   [3] 伽藍之妙, 最爲稱首

Perhaps Bernal was the first to formulate the analogy when he remarked that

the steam-engine has a very mixed origin; its material parents might be said to be the cannon and the pump. Awareness of the latent energy of gunpowder persistently suggested that uses other than warfare might be found for it, and when it proved intractable there was a natural tendency to use the less violent agents of fire and steam.[a]

And he also wrote:

A new and important connection between science and war appeared at the breakdown of the Middle Ages with the introduction ... of gunpowder, itself a product of the half-technical half-scientific study of salt mixtures ... In their physical aspect the phenomena of explosion led to the study of the expansion of gases, and thus to the steam-engine; and this was suggested even more directly by the idea of harnessing the terrific force that was seen to drive the ball out of the cannon, to the less violent function of doing useful civil work.[b]

Seven years later, Vacca was speaking at an Italian symposium on the origins of specifically modern science in Europe rather than in China.[c]

A further advance was made [he said] by the invention of firearms, and of machines to use the expansive force of steam. Gradual familiarisation with the mechanics of explosions as they occur in firearms led to an almost ceaseless series of attempts to harness their power, from Papin's rudimentary efforts down to the modern internal-combustion engine.[d]

In 1948 Bernal discussed the connection again.

Ultimately [he wrote] it was the effects of gunpowder on science rather than on warfare which were to have the greatest influence in bringing about the Machine Age. Gunpowder and the cannon not only blew up the mediaeval world economically and politically; they were major forces in destroying its system of ideas. As John Mayow put it: 'Nitre, that admirable salt, hath made as much noise in philosophy as it hath in war, all the world being filled with its thunder.'[e] The force of the explosion itself, and the expulsion of the ball from the barrel of the cannon, was a powerful indication of the possibility of making practical use of natural forces, particularly of fire, and was the inspiration behind the development of the steam-engine.[f]

---

[a] (3), p. 24. It may have been natural, but it occupied many men's minds for many years in Europe all the same.

[b] (3), p. 166. But as we shall see, the way to steam power lay not through gaseous expansion, rather through the partial vacuum created after an explosion.

[c] Giovanni Vacca (1872 to 1933) was an Italian sinologist whose work we have often quoted in previous volumes.

[d] (9), p. 11. We shall explain the reference to Papin shortly, but the same remark about expansive force applies. Expansive pressure on both sides of the piston alternately only came in after the atmospheric engine had succeeded.

[e] 'Quasi nimirum in Fatis esset, ut Sal hoc admirabile non minus in Philosophia quam Bello Strepitus ederet, omniaque sonitu suo impleret.' Mayow (1), Tractatus 1, *De Sal Nitro et spiritu nitro-aerea*, p. 2. Cf. p. 102 above.

[f] (1), pp. 238–9, 2nd ed., pp. 322–3. On pp. 414 ff. (577 ff.) he mentions Papin's steam cylinder, but pushes the analogy with the cannon no further. So also in (2), pp. 256 ff., 262 ff. he tells the story of the steam-engine without emphasising this connection.

But it was left for Lynn White in 1962 to express the matter even more clearly.[a]

The cannon [he wrote] was not only important in itself as a power-machine applied to warfare; it is a one-cylinder internal-combustion engine, and all of our more modern motors of this type are descended from it. The first effort to substitute a piston for a cannon-ball, that of Leonardo da Vinci, used gunpowder for fuel, as did Samuel Morland's patent of +1661, Huygens' experimental piston-engine of +1673, and a Parisian air-pump of +1674.[b] Indeed, the conscious derivation of such devices from the cannon continued to handicap the development [of internal-combustion engines] until the nineteenth century, when liquid fuels were substituted for powdered.[c]

And in 1977 he returned to the same theme, saying that

Francis Bacon had more reason to be excited about a cannon than perhaps he himself realised. The cannon constitutes a one-cylinder internal-combustion engine, the first of its genus.... Lamentably, inventors along this line of technological growth fell into the very trap against which Bacon had warned them; focus on tradition rather than on the qualities of Nature itself. They were so conscious of the cannon and gunpowder as precedents for their efforts that it was not until the mid-nineteenth century that they finally realised that the Chinese chemical mixture ... was inherently too awkward to give power to continuously operating engines. Only then did they turn—and with immense technical success—to the lighter distillates of petroleum which during the Middle Ages had been developed by the alchemists of Byzantium, Islam [and China] primarily for use as 'Greek Fire'. Two of the more conspicuous results were the automobile and the piston-engined aeroplane.[d]

But in the meantime the gunpowder-engine's failure had led directly, as we shall see, to the steam-engine's success.

So far we have been dealing in generalities. By +1500 it was becoming clear that the force of gunpowder ought to be made to do something useful, instead of just propelling projectiles. It was the merit of Hollister-Short[e] that he recognised the first appearance of such a use in the 'gunpowder triers', devices introduced by gunners to test the quality of their powder by making it perform some effect, some measurable work;[f] and that he then saw the close relation of these to the gunpowder-engines which appeared rather later. Broadly speaking, one may say that the heyday of the triers or testers[g] occupied the century +1550 to +1650,

[a] (7), p. 100.

[b] We shall follow these developments more closely in what follows.

[c] Perhaps Lynn White was thinking here of the *pyréolophore* of the Niepce brothers invented in 1806, a very complicated machine which used lycopodium powder as the fuel, combusted within the cylinder; cf. Daumas (3). This carbonaceous material, long used by pharmacists as a dusting-powder and for pill-coatings, consists of the spores of the club-moss, *Lycopodium* spp. (Lawrence (1), pp. 337–8; Stuart (1), p. 251; R 795–6).

[d] (20), p. 3.

[e] (4, 6); much of the information in the following paragraphs comes from his researches. The only earlier paper on the subject known to us is that of Fischler (1), and that we have also used.

[f] Here one can see a good example of the definition of specifically modern science as the mathematisation of hypotheses about Nature, combined with relentless experiment, for this at once led to the quantisation of phenomena, describing effects in terms of measure and number.

[g] *Pulverprober* or *éprouvettes*.

while that of the engines followed in the half-century ending about +1700.[a] The latter were thus directly proemial to the development of the steam-engine, and intimately connected with it.

In +1540 Biringuccio was still taking a piece of paper and burning a small amount of gunpowder on it to see whether it would go off in a puff without burning the paper or no.[b] But soon afterwards designs for more sophisticated mechanical triers were beginning to be pondered.[c] The oldest which has some down to us is that of William Bourne in his book of +1578, *Inventions or Devises....*[d] He had a cylindrical metal box within which the powder was set off, and according to its strength the explosion pushed up the hinged metal lid so that it caught on one or other tooth of a quadrant ratchet, giving thereby a crude quantitative measurement. This device was again described by John Bate in his *Mysteries of Nature and Art* of +1634; his cut is reproduced in Fig. 224.[e] 'So, by firing the same quantity of divers kindes of powders at severall times, you may know which is the strongest.' The hinged cover appears again in the fine plate of John Babington's *Pyrotechnia* (+1635) which we give in Fig. 225; it is (A) below on the right, but now the lid when blown off upwards, rotates a graduated discoidal plate which was braked by a spring (Fleming, 1) so that the strength of the gunpowder could be empirically measured.[f] Finally, the hinged cover reached its apogee in the trier which Robert Hooke demonstrated to the Royal Society, a much more workmanlike machine than any that had gone before.[g] On 9 September 1663 'Mr Hooke brought in a scheme of the instrument for determining the force of gunpowder by weight, together with an explication thereof; which was ordered to be registered as follows ...'. The design is shown in Fig. 226. The explosion cylinder had a hinged lid with a touch-hole closed by a strong spring, and at the other end the lid narrowed to a tooth which engaged with a cam or wheel-ratchet;[h] this was on the same axle as a beam or arm which could be loaded with a variable weight.[i] During the following months several tests were

---

[a] There was one engine which preceded all the triers, as we shall see (p. 553 below), but Leonardo was a law unto himself.

[b] (1), p. 415. We discussed at an earlier point (p. 359) the Chinese statements about this test. It was still recommended by Collado (1) in +1586, but in +1627 Furtenberg (2) said that one might tell good from bad powder in this way but not much more. Still, it lingered on till the end of the century, as in Mieth (1) and de St Remy (1), vol. 2, p. 112, the latter even perpetuating the test that used the palm of the hand (cf. p. 105 above); but he recommended it only when you knew that your powder was good and dry. Of course there was always the inspection of the residues of explosion, if any, as recommended by Fronsperger (1) in +1555 and later.

[c] It will be convenient to follow them through according to their principles irrespective of strict chronological order.

[d] (3), pp. 39–40, device no. 54.

[e] (1), bk. 2, pp. 55–6, 2nd ed. pp. 95–6. The quadrant ratchet is now held against the lid by a spring.

[f] (1), pp. 69 ff., ch. 64. Babington remarks: 'hee that will make a good rocket must be certain of the strength of his powder, which if it bee to strong, will break; if too weake, it will not rise to that heighth it should ...'. If too weak add 'peter', if too strong add 'coale'.

[g] Birch (1), vol. 1, p. 302.

[h] Hooke called it the 'nick of the nut'.

[i] As Hollister-Short remarks, this could have been the origin of Papin's safety-valve (+1681), used on his digester or steam pressure-cooker. Hooke's trier was loaded with different weights so as to see what a given amount of a gunpowder mixture would do.

56   *The second Booke*

lid ioynted unto it. The box ought to be made of iron,
braſſe, or copper, and to bee faſtned unto a good thick
plank, and to haue a touch-hole at the bottom, as O, and
that end of the box where the hinge of the lid is, there
muſt ſtand up from the box a peece of iron or braſſe, in
length anſwerable unto the lid of the box : this peece of

iron muſt haue a hole quite
through it, towards the top,
and a ſpring, as, A, G, muſt bee
ſcrewed or riueted, ſo that the
one end may couer the ſayd
hole. On the top of all this i-
ron, or braſſe that ſtandeth up
from the box, there muſt bee
ioynted a peece of iron (made as
you ſee in the figure) the hinder
part of which is bent down-
ward, and entreth the hole that the ſpring couereth ; the
other part reſteth upon the lid of the box. Open this
box lid, and put in a quantity of powder, and then ſhut
the lid down, and put fire to the touch-hole at the bottom, and the powder in the box being fired, will blow
the box lid up the notches more or leſſe, according as the
ſtrength of the powder is . ſo by firing the ſame quantity
of diuers kindes of powders at ſeuerall times, you may
know which is the ſtrongeſt. Now perhaps it will bee

Fig. 224. Gunpowder triers and testers as the first explosion machines to do some useful work. They were
designed to give gunners an indication of the strength of their powder. The second to be described was that of
John Bate in +1634; the explosion pushed up the hinged metal lid of the box so that it caught on one or other
tooth of a quadrant ratchet.

made but all failed, yet when notices of gunpowder experiments resume in
+1667 the deficiencies of construction had been overcome, and the emphasis
had shifted to making the engine do some other kinds of useful work. In January
of that year an experiment was ordered for the applying of the strength of gunpowder to the bending of springs, thus storing energy, and this was successfully
accomplished. Hooke was also asked to see if weights could not be raised by
gunpowder. Robert Boyle suggested that the force of gunpowder might be tried
by making it raise a weight of water (which it would expel out of a vessel). How
exactly the springs were wound up, or the weights raised, the Journal Books of

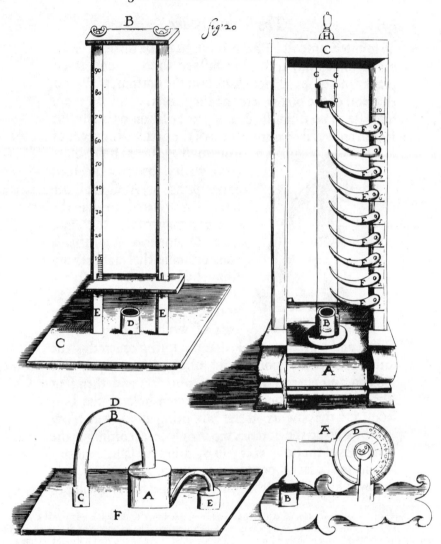

Fig. 225. The plate of John Babington's triers (+1635). A variant of the hinged lid appears at the bottom on the right, but now the explosion is empirically more quantified by having it rotate a graduated disc against spring resistance. The ratchet teeth appear again at the top on the right, where the lid of the box is blown upwards by the explosion and, guided by wires, comes to rest upon one or other of the movable ratchet hooks. At the top on the left is a device where the lid itself is blown upwards until it comes to rest at one of the graduations marked on the left-hand column. Finally at the bottom on the left is a device where the gases of the explosion displace different amounts of water in the reservoir A, thereby indicating the strangth of the powder.

the Royal Society do not say, and on subsequent experiments they are silent too, but the whole sequence is of the greatest interest for it shows a gunpowder trier in the very act of turning into a gunpowder engine.

The blowing off of lids continued to the end of the century and beyond, as can be seen in the book of Surirey de St Remy (1) published in +1697. Though often mounted like pistols, they were quite similar to Babington's device, for

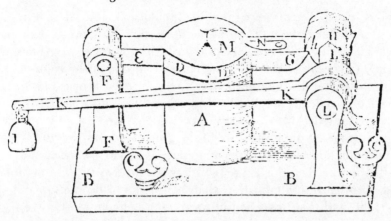

Fig. 226. The gunpowder trier of Robert Hooke (+1663). It was in fact almost a gunpowder-engine, because it was capable of raising a variable weight, and was later applied to the bending of springs. Design taken from the Royal Society Minute-Books, after Hollister-Short (4), p. 12. Description in text.

upon firing, the cap of the explosion chamber rotated a graduated wheel,[a] which came to rest upon a ratchet tooth and so assessed the force of the charge.[b]

Boyle's suggestion reminds us that another of Babington's triers involved precisely the expulsion of water from one vessel to another.[c] The set-up is seen in Fig. 225 (D, below on the left); a given weight of a gunpowder mixture exploded in (C) sent its gases into the vessel (A) and expelled a measurable quantity of water into (E). For comparing powders, this, said Babington, was 'the certainest way, although the most troublesome'. But now he was measuring the volume of gases formed rather than the mechanical force of the explosion, and the result was directly proportional to the percentage of nitrate in the composition,[d] since that was the oxygen-provider giving mostly CO_2 with smaller amounts of the oxides of sulphur and nitrogen.[e] The displacement of water by air or steam was an ancient principle, going back to the Alexandrians,[f] and therefore very familiar, but in +1635 the properties of the vacuum had still not been explored, so that Babington's device was only obliquely a predecessor of the water-raising systems of de Hautefeuille and Savory.[g]

[a] This probably explains the reference to trier 'wheels' in Mieth (1), +1683, ch. 55, not further elucidated.

[b] Fischler (1) gives two pictures from de St Remy's book and figures two extant examples from the am Rhyn Collection. De St Remy's preferred method, however, was to cast brass balls of known weight from a mortar at a known elevation (30° to 45°) and measure their range. Fischler figures a late standard mortar of this kind from the Luzerner Waffensammlung in fig. 4.

[c] (1), ch. 68.

[d] See our arguments relating to this on pp. 110 ff., 342 ff. above.

[e] Johannes Bernoulli (1), in a Basel dissertation of +1690, described how he ignited with a burning-glass a small amount of gunpowder in a glass bulb which connected with a tube dipping into water. After the boiling which ensued the water was found to be acidified, so he concluded, not knowing what the gaseous combustion products were, that fire was itself an acid (*haud incongrue dici potest, quod ignis sit acidum*). But this early chemical experiment was not related to the triers.

[f] Cf. Woodcroft (1), pp. 26, 57.

[g] We shall look at these on pp. 562 ff. below.

The other two pieces of apparatus in Babington's plate derive from experiments of Joseph Furtenberg (2) published in +1627. Both are at the top (B, C in Fig. 225). The former was not very practical, driving up a cover-plate with two holes along a vertical graduated scale marked on one of the columns,[a] but the latter was a useful and workable device.[b] Here the cap of the explosion chamber was blown up vertically guided by two wires, tripping as it went a series of twenty hinged ratchet-arms or 'keys', upon one or other of which it eventually came to rest, thus giving a measure of the gunpowder's strength.[c] This system has a descendant among the pieces of apparatus used for determining explosive force at the present day; this is the 'whirling height *éprouvette*'.[d] The upper conical rifled opening of a combustion chamber is closed by the tapered end of a 10 kg. weight, and this is whirled upwards by the explosion between a cage of slide-bars, clicking in at the culmination point by means of a catch.

It should be mentioned here, however, that the most widely used contemporary device for measuring explosive force is the 'ballistic pendulum' developed and used at Fort Halstead.[e] This employs a principle quite different from those used in any of the old triers, namely retro-active rocket propulsion.[f] A mass of steel 150 kg. in weight is suspended from a rigid framework by wire, just over two metres long, and within this mass there is a steel tube of 25 mm. bore taking a charge of about 10 gm. and with an unconfined orifice.[g] When detonated electrically the heavy weight is propelled forwards in an arc by the energy release and its swing recorded by a stylus; then the excursion is expressed in percentage terms of a standard charge of picric acid. This has quite superseded the rather qualitative Trauzl test, which assessed the deformation produced by explosions set off within a block of lead.

Hollister-Short remarks that Furtenberg's flying-cap trier could have been a precursor of, or at least a stimulus for, Huygens' gunpowder engine of +1673, since the guide-wires directed the cap just as the cylinder-walls directed the piston.[h] We should not lightly dismiss this idea, which after all is no more farfetched than the comparison of the piston and cylinder with the cannon-ball and cannon, an analogy accepted on all hands as justified.

But Huygens was not the first to make a gunpowder-engine; he had been anticipated by Leonardo da Vinci (as so often happened with that great

---

[a] Babington (1), ch. 66.

[b] (1), ch. 67. The Waffensammlung at Vienna has a fine contemporary example from Ambras Castle. It is 57 cm. in height.

[c] De St Remy (1) described in +1697 a variant in which the cap was weighted and carried a vertical shaft the teeth of which were also caught on ratchets. A late example of this device is in the Luzerner Waffensammlung (Fischler (1), fig. 2).

[d] Described by Hahn, Hintze & Treumann (1).

[e] For our information on these matters we are much indebted to Dr Nigel Davies.

[f] Cf. Connor (1). The test does not distinguish between deflagration and true detonation, but there is a cartridge case deformation test which will do so; Connor (2). Cf. Hughes (1), p. 46.

[g] The degree of tamping may be varied at will.

[h] The arresting catches might also provide a clue to the origin of the retaining devices talked of by Huygens and in +1690 described by Papin.

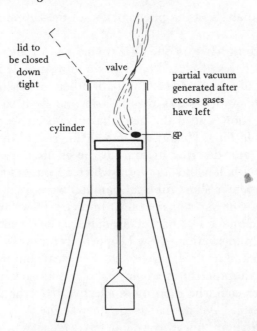

Fig. 227. The weight-lifting gunpowder-engine of Leonardo da Vinci described and figured by him in +1508. In this suggested reconstruction diagram a gunpowder explosion above the piston generated a partial vacuum, which then pulled up the weight suspended underneath.

Renaissance inventive genius) and Leonardo had a cylinder and piston.[a] We must compare what he wrote with the diagram which he drew (Fig. 227). His words of +1508 were these:

To lift a weight with fire, like a horn or a cupping-glass. The vessel [i.e. the cylinder] should be 1 braccio wide and 10 in length, and it should be strong. It should be lit from below like a bombard, and the touch-hole rapidly closed, and then [all] immediately closed at the top. [You will see] the bottom [i.e. the piston], which has a very strong leather [ring] like a [pump-] bellows, rise; and this is the way to lift any heavy weight.[b]

As the diagram shows, a weight was suspended from the downward-pointing piston-rod, and gunpowder ignited above the leather-packed piston, then as soon as the gases had rushed out (expelling most of the air) all openings were closed, and as the remaining gases cooled and contracted, a partial vacuum was generated, thus sucking up the piston and raising the weight.[c] Here Leonardo came nearer to the ultimate gateway of success, the vacuum, than any before him, anticipating in a sense the +17th-century physicists by 150 years or so; but

---

[a] See Reti (2), p. 29, fig. 20; Reti & Dibner (1), pp. 94 ff.
[b] Tr. Hart (4), p. 299, mod. auct. seq. Reti. Codex F (Institut de France), 16 v; cf. Codex Atlanticus, 5 r.a., 7 r.a.
[c] Doubtless this was then to be secured in its new position with blocks and wedges.

without a deeper analysis of the phenomenon of the cupping-glass he could go no further.[a]

What was this thing? It was one of mankind's most ancient medical instruments, a cup-shaped vessel placed on the skin at a suitable site and emptied of air by the burning of a small piece of wool or other combustible material inside it. The vacuum so formed sucks up the skin and flesh so as to bring about cutaneous vaso-dilation, and if the place has first been scarified it encourages transudation and bleeding. Historians of medicine regard the procedure as prehistoric in origin,[b] and describe it from all the civilisations.[c] The oldest vessel used was probably a hollow buffalo horn, which accounts for one of its Chinese names, chio fa[1], but later short tubes of bamboo were used, and these are still called huo kuan[2],[d] tow or paper being burnt in them. There are many early literary mentions, and cupping was part of one of the seven departments (kho[3]) of the Thang Medical Administration.[e] Now Leonardo certainly did not have the concept of the vacuum and its uses, subsequently so clear, but it was a fine thing to recognise that certain procedures would make vessels suck other things in, and burning gunpowder could be even more effective than the small combustibles used by the physicians. All he had to go on was the Aristotelian truism that 'nature abhors a vacuum', but it was enough.[f]

Perhaps the most extraordinary aspect of the situation was that Leonardo also conceived what one might call the standard experimental set-up afterwards used by Huygens and others, namely a cylinder and piston, the piston-rod of which was attached to a cord passing over two pulleys and then suspending a counterbalance weight.[g] But he did not use this for weight-raising by gunpowder, he set it up about +1505 in order to see how much steam coming off from heated water would expand.[h] All this was connected, no doubt, with his steam cannon, the Architronito, in which a jet of high-pressure steam was suddenly admitted behind a ball to shoot it forth through a long barrel.[i] Here again was a striking link in

[a] Needham (48), p. 10, (64), p. 147.

[b] Garrison (3), p. 28; Sigerist (1), vol. 1, p. 116. The latter, pp. 193, 202 gives anthropological evidence for its use among many primitive peoples.

[c] For Babylonia Mettler (1), p. 320; for India p. 333 and Castiglioni (1), p. 89; for Hippocrates and later Greeks p. 174 and Mettler (1), pp. 529; for Arabic medicine Garrison (3), p. 134; Mettler (1), p. 537. A Roman reference is Celsus, II, 11 and IV, 7, 2 (Mettler, pp. 338, 509). Perhaps the best account of it in the Middle Ages and later Europe is that of Brockbank (1), pp. 67 ff.

[d] It is significant, in view of pp. 221 ff. above, that this name of 'fire-tube' was always medical, and never used for any gunpowder weapon. On cupping in China see Pálos (1), pp. 94, 158–9, 182.

[e] See Lu Gwei-Djen & Needham (2).

[f] There is an interesting book by Grant (2) on ancient and medieval arguments about empty space.

[g] He also used a lever suspended from a central fulcrum.

[h] Reti (2), p. 17; Hart (4), pp. 249 ff. Leicester Codex, 10 r, 15 r. The cylinders here were of square cross-section. Leonardo got a figure of 1 : 1500 for the comparison of the volume of water and steam; the true relation is about 1 : 1700.

[i] Codex B, 33 r; cf. Reti (2), pp. 21 ff.; Hart (4), pp. 295–6, pl. 100.

[1] 角法   [2] 火管   [3] 科

the connections we are unravelling between the cannon barrel and the steam-engine cylinder.[a]

Yet still the expansive force of steam was not the clue or key which would open the gate into the future; that key was nothing at all, the absence even of air, just emptiness. Ctesibius had been responsible, about −230, for a simple and fundamental machine, the piston air-pump, known from the descriptions of later mechanicians.[b] This simplest of pumps entered upon a new incarnation in the +17th century, when the virtuosi began to explore with excitement the properties of vacuous spaces, for what had been invented originally as a bellows for pumping air into something now found fresh employment as *the* 'air-pump' for getting as much air as possible out of it. The closer scrutiny of the alleged *horror vacui* began with Galileo himself in +1638, and was continued in +1643 by his disciple Evangelista Torricelli (+1608 to 47), whose mercury barometer was the start of many experiments showing that the air weighs down on everything with a pressure of some 14 lb. per square inch.[c] This opened men's minds to the recognition of the fact that air has weight, and that the vacuum was a physical reality. Then came the long-continued work of Otto von Guericke (+1602 to 86) who invented the evacuating air-pump by about +1650, and then four years later performed the sensational experiment of the 'Magdeburg hemispheres'.[d] He also demonstrated the weight required to tear them apart, the crumpling of evacuated copper globes, and the pistons which raised men or weights into the air when sucked down by the vacuum.[e] In +1659 there followed the improved air-pump of Robert Boyle (+1627 to 91),[f] and with assistance from Robert Hooke it had attained its final form by +1667.[g]

Thus was established that all these effects were due to an omnipresent force—'the spring and weight of the air', from which man might draw infinite

[a] As is well known, Leonardo ascribed the invention to Archimedes. The background to this strange story has been examined by Simms (3) and Clagett (5).

[b] See Heron, *Pneumatica*, ch. 1, no. 42; Philon, *Pneumatica*, App. 1; Vitruvius, x, 8. Discussion in Beck (1), pp. 24 ff; Drachmann (2) pp. 7 ff., 100, (9), p. 206. Cf. Woodcroft (1), chs. 76, 77, pp. 105, 108. Ctesibius' pump powered what is confusingly known as a 'water-organ' because the air was pumped into a reservoir where it was held at approximately constant pressure by means of a water-seal. Though the Vitruvian text describes two cylinders, the Philonic and the Heronic speak only of one, and the apparatus has nearly always been so reconstructed—in any case the pump or pumps were invariably single-acting, 'inhaling' and 'exhaling' on alternate strokes. The double-acting piston-pump 'air-bellows' of China probably goes back well before Ctesibius (Vol. 4, pt. 2, p. 139) but it was not destined to stimulate the vacuum pumps of the +17th century because Europe did not have it until the +18th (*ibid.* pp. 151, 380).

[c] There is a monograph on the role of the barometer in the history of physics by de Waard (1). Cf. Wolf (1), pp. 92 ff.

[d] *Ibid.* pp. 99 ff.

[e] Cf. Dickinson (4), pp. 7 ff., Gerland & Traumüller (1), pp. 129 ff. Von Guericke demonstrated his experiments at the Imperial Diet of Ratisbon in +1654, and a preliminary account of them was published by the Jesuit Caspar Schott (2) three years later. Von Guericke's own account did not appear till +1672 under the title *Experimenta Nova (ut vocantur) Magdeburgica de Vacuo Spatio*. There is a German translation of this in Ostwald's *Klassiker d. exakten Naturwissenschaften*, no. 59.

[f] Hence Boyle (6); cf. Wolf (1), pp. 102 ff.

[g] Hence Boyle (7), *A Continuation of New Experiments Physico-Mechanicall touching the Spring and Weight of the Air* (+1669). Cf. Wolf (1), p. 107.

'I think [he wrote] that a flexible wooden tail at the stern of a boat, as I have visualised it, and as I believe they make use of in China, would be a good application for this, if moved by the force generated in this cylinder.' Here was unquestionably a reference to the *yuloh* scull (*yao lu*[1]) or self-feathering propulsion-oar propeller,[a] characteristic of small Chinese craft from Han times onwards. Huygens also sketched a form of ballista operated by linkwork as the piston descended.

So matters remained until Denis Papin, now occupying a chair at Marburg, returned to the problem of improving the gunpowder-engine. In +1688 he published a new version (1), the chief difference in which was that the piston was now furnished with a spring valve closed by atmospheric pressure when the gases had left, after which it was allowed to make a powerful down-stroke (Fig. 229).[b] But the fifth or sixth part of the air and gases always remained. Ruminating on this, Papin made a pregnant statement in his paper (2) of +1690:

Since it is a property of water that a small quantity of it, turned into vapour by heat, has an elastic force like that of air, but upon cold supervening is again resolved into water, so that no trace of the said elastic force remains, I readily concluded that machines could be constructed wherein water, by the help of no very intense heat, and at little cost, could produce that perfect vacuum which could by no means be obtained by the aid of gunpowder....

Thus was born the first of all steam-engines.[c] It looked just like the earlier gunpowder-engines, but a spring catch fitted into a notch on the piston-rod so that the down-stroke could be delayed until it was as powerful as possible,[d] and then it went down right to the bottom of the cylinder (Fig. 230). Here the boiler, engine-cylinder and condenser were all in one; it was given to Thomas Newcomen to separate the boiler from the cylinder, and to James Watt to introduce a separate condenser—otherwise all the essential parts were present (Fig. 231).[e] Here at last was an effective cycle, the removal of air and the condensation of steam, so that the way was open to the 'atmospheric, or vacuum, steam-engine' (+1712). Though Denis Papin never harnessed his piston-rod to anything,[f] his historical position in the transition from gunpowder to steam is a central one,

---

[a] Vol. 4, pt. 3, pp. 622 ff. The *yuloh* has a significant place in the history of screw-propulsion, and in +1790 there was an unsuccessful attempt to apply steam-power to it.

[b] Cf. Galloway (1), pp. 41 ff., 44 ff.; Gerland & Traumüller (1), pp. 227 ff.

[c] Dickinson (4), pp. 10 ff., (6), p. 171; Galloway (1), pp. 47 ff.; Thurston (1), pp. 50 ff.; Matschoss (2), pp. 32 ff., 354 ff., 359 ff.; Reti (2), p. 28; Garland & Traumüller (1), pp. 228 ff.

[d] This way of getting a violent stroke probably derived from Huygens' ballista design.

[e] The double-acting steam-engine, using fully the expansive force of steam, and the high-pressure steam-engine all of course lay in the future. Cf. Needham (48), p. 11; (64), p. 149.

[f] At any rate, not successfully. There are accounts that towards the end of his life, in +1707, Papin experimented with a small paddle-driven steamboat on the Fulda R. near Cassel, but it is not easy to see how it could have worked even with a Newcomen beam-engine, as the stroke frequency would have been so slow. Our information remains obscure and somewhat contradictory on this last phase. See Galloway (1), pp. 76 ff.; Thurston (1), pp. 224 ff., where more detailed references will be found.

[1] 搖櫓

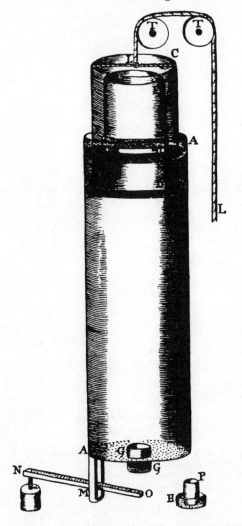

C	attachment of piston to cord
D	ventilator for the gases
F	edge of ventilator
HP	gunpowder holder fitting in easily to
GG	tube in the base of the cylinder so that successive charges could be introduced
MNO	safety valve lever weighted at N
LL	cord bearing weight
TT	pulleys

Fig. 229. Denis Papin's gunpowder-engine of + 1688, after Gerland & Traumüller (1), fig. 219. It was similar to that of Huygens except that it had a spring valve closed by atmospheric pressure after the gases had left, so that the piston could then make a powerful down-stroke and raise the weight.

and Thomas Newcomen himself would surely never grudge him his statue among the flower-sellers and vegetable-stalls that overlook the Loire on the great flight of steps at Blois.[a]

Steam had been on Papin's mind for quite a long time. Nine or ten years before, he had produced his steam pressure-cooker or 'digester'. In China steam had traditionally been used for many things, especially in cooking,[b] and bread

[a] I find it almost impossible to believe that Newcomen never knew of Papin's steam cylinder, at least by hearsay. Papin published several other papers (3, 4, 5) and maybe there was someone at Dartmouth, one of the gentry perhaps, who could read French, if not Latin, and gave Newcomen access to Papin's work.

[b] Cf. Vol. 1, pp. 81–2.

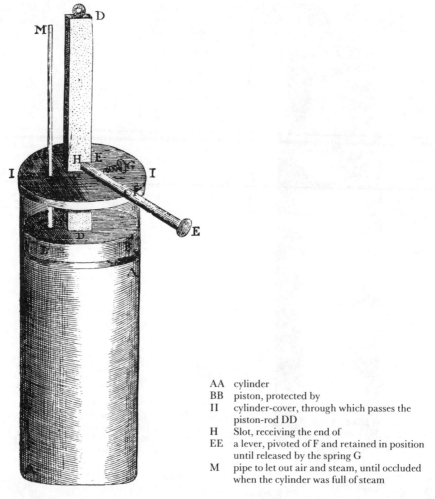

AA cylinder
BB piston, protected by
II cylinder-cover, through which passes the
     piston-rod DD
H   Slot, receiving the end of
EE a lever, pivoted of F and retained in position
     until released by the spring G
M   pipe to let out air and steam, until occluded
     when the cylinder was full of steam

Fig. 230. Denis Papin's steam-engine of +1690, after Gerland & Traumüller (1), fig. 220. It looked just like the earlier gunpowder-engines, but the spring catch on the piston-rod delayed the down-stroke until the cooling condensed the steam and created an approach to a perfect vacuum, after which the suspended weight would be drawn up to the maximum extent. This was the invention which led to Thomas Newcomen's first successful 'atmospheric', or vacuum, steam-engine of +1712.

was (and is) generally steamed there rather than baked. In his *Travels in China* (1804) John Barrow wrote:[a]

In like manner they (the Chinese) are well acquainted with the effect of steam upon certain bodies that are immersed in it; that its heat is much greater than that of boiling water. Yet although for ages they have been in the habit of confining it in close vessels, something like Papin's digester, for the purpose of softening horn, from which their thin, transparent and capacious lanterns are made, they seem not to have discovered its

---

[a] (1), p. 298.

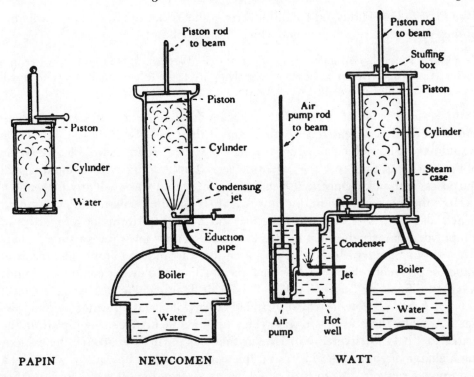

PAPIN   NEWCOMEN   WATT

Fig. 231. Diagram from Dickinson (4), p. 67, illustrating the progressive differentiation of function in the steam-engine. On the left, Papin's cylinder filled with steam, in the centre, Thomas Newcomen's separate boiler. On the right, James Watt's separate condenser, which periodically drained the cylinder of steam, thus permitting the down-stroke, while at the same time its walls were kept hot by the steam case, and any air removed from the system by the air-pump automatically operated.

extraordinary force when thus pent up; at least, they have never thought of applying that power to purposes which animal strength has not been adequate to effect.[a]

What Barrow perhaps failed to appreciate was that the way to the steam-engine historically lay not directly through high-pressure cookers, but indirectly through evacuated vessels, and the understanding of the vacuum was a characteristic result of the methods of modern science born in the Scientific Revolution. In other words the way to high-pressure steam, and all that it could do, lay dialectically through its precise opposite; and the whole historical process was an extraordinary justification of the classical idea of Taoist philosophy that emptiness would be the gateway to all power.

Papin would never have guessed that seventeen centuries earlier in China some experimentalist stumbled upon the creation of a vacuum by condensing steam, and then proceeded no further. We have told this story before[b] but it deserves to be touched upon again here. Among the procedures in the *Huai Nan*

[a] On horn-working in China see Section 42 in Vol. 6. There is an account of the craft in Grosier (1), vol. 7, pp. 237 ff.
[b] Vol. 4, pt. 1, pp. 69 ff.

Wan Pi Shu (Ten Thousand Infallible Arts of the Prince of Huai-Nan), probably of the −2nd century, there is one which runs as follows:[a]

To make a sound like thunder in a copper vessel (*thung wêng*[1]). Put boiling water into such a vessel [which must be closed extremely tightly],[b] and then sink it in a well. It will make a noise which can be heard several dozen *li* away.

If the vessel was full of steam when it was let down into the cold water, condensation would have created a vacuum, and if the copper was thin an implosion would have followed, echoing far beyond the well. Perhaps it was characteristic of the place and time that the invention served only military or thaumaturgical purposes, with no attempt to use the strong force that was evidently present.[c]

Here there was no piston, but nor was there any in a collateral development which also preceded the steam-engine, and also arose from the properties of gunpowder, namely the vacuum displacement systems for water-raising. As early as +1661 Samuel Morland got a patent or warrant for pumping water from mines or pits more effectively 'by the force of Aire and Powder conjointly', but it was never finalised.[d] Then in +1678 Jean de Hautefeuille published his tract entitled *Pendule Perpetuelle* which included 'a way of elevating water by gunpowder'.[e] Actually there were two ways (Fig. 232). In the first, a rising pipe from the water 30 ft below delivered into a vessel that was partially evacuated by exploding a charge of gunpowder in it,[f] and the water so sucked up was drawn off by a tap into a reservoir; this in turn could act as a second-stage sump for a further 30 ft lift arranged in the same way. Such cisterns in pairs, with gunpowder successively let off, would give a continuous discharge. But this was doing no better than a set of suction-pumps, so a second system was described, for use where force-pumps were necessary. Here a horizontal pipe was set under the water-surface down below, with an inlet-valve at its central point. At one end of this pipe there rose above the water-level a short vertical tube leading to a gunpowder combustion chamber. At the other end a much longer tube rose up having a succession of non-return valves. As one charge after another was ignited, the water was driven up the rising main as high as the materials would stand. In this second system there was no dependence on the partial vacuum,[g] and the gunpowder could be supplied in culasses like breech-loading cannon.[h] All in all,

---

[a] *TPYL*, ch. 736, p. 8*b*, tr. auct.　　[b] This clause is from ch. 758, p. 3*b*.

[c] It reminds one of Shen Kua valuing petroleum only for the black soot it would make, so suitable for ink, and for no other purpose (Vol. 3, p. 669).　　[d] Rhys Jenkins (3), p. 44; Dickinson (4), p. 16.

[e] Cf. Rolt (1), p. 33; Rolt & Allen (1), p. 24; Galloway (1), pp. 18 ff.; Thurston (1), pp. 24 ff. The second tract (1682) adds various improvements.

[f] The gases and air were exhausted through four non-return valves.

[g] But de Hautefeuille could well have known of Babington's fourth trier method, where water was displaced by the combustion gases.

[h] Cf. pp. 365 ff. above. How close the connection is between the cannon and the steam-engine appears when one realises that Newcomen had to get his cylinders bored smooth by the gun-founders; Rolt (1), p. 80.

[1] 銅甕

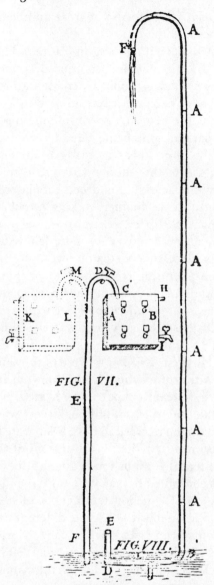

Fig. 232. Jean de Hautefeuille's methods of raising water by means of gunpowder explosions, after Hollister-Short (4), p. 16. The drawing is from the rare tract of +1678. In Fig. VII a rising pipe delivered into a vessel that was partially evacuated by exploding a gunpowder charge in it, and the water so raised was drawn off by a tap into a reservoir. In Fig. VIII, however, the force of the gunpowder explosions brought about the raising of the water through non-return valves as high as the materials would stand. Again, these systems were precursory to Thomas Savery's 'water-commanding engine' of +1698, which made use of the more convenient expansion and condensation of steam.

EF	rising pipe
AB, KL	tanks alternately evacuated by gunpowder charges
A	rising pipe with non-return valves
E	pipe from gunpowder combustion chamber

these devices would obviate the expense of great numbers of men and horses in mines, drainage-schemes and the like.

But the vacuum came back with a bang in Captain Thomas Savery's 'water-commanding engine'.[a] In this machine, so often described,[b] water was sucked up some 30 ft into a vessel made vacuous by condensation of steam, then forced higher still by a second admission of steam, suitable cocks being turned by hand at the several phases of the system.[c] A continuous discharge was gained by having two vessels in parallel, one being filled by suction while the other was emptied upwards by pressure. This was ready by +1698, but ran into many difficulties largely because of the inferior strength of the materials available. William Blakey improved it nearly a hundred years later,[d] but by then Thomas Newcomen's atmospheric steam-engine, working a rocking beam, the ancestor of all later steam-engines, had been set up in many places since +1712, and the need for displacement systems was no more felt.[e] Still, they are justifiably numbered among the predecessors of steam power.

From all that has how been said it will be evident that the explosive force of gunpowder played a fundamentally important part in the development of the steam-engine. But there is a second chapter yet to relate, that of the internal-combustion engines. Of these the very first was the hand-gun and cannon or bombard itself, which we have traced back to China about +1285; and the gunpowder-engines of Huygens and Papin were of the same category since they exploded the mixture within the cylinder itself and not in any separate vessel. The water of Papin's first steam cylinder was heated directly in it, so his experimental engine was an 'internal' one although there was no 'combustion', but as soon as Thomas Newcomen decided to have a separate boiler, as he did in the early years of the +18th century, the line of descent of the steam-engine separated off from all true internal-combustion engines. Still, for a century and more thereafter men's minds continued to be haunted by the idea of having an explosion right in the cylinder, and somehow taming its violence to give useful power. But the purpose of the explosion was now quite different from that of Huygens; it was no longer to drive out air and gases with a view to forming a vacuum so that the piston would get sucked in (at least some way), it was rather to effect a working stroke more closely similar to that of the cannon itself—though the piston was not free to depart from the machine.

[a] Savery was born about +1650 and died in +1715. His title may have come from mining; ultimately he became F.R.S. in +1705.

[b] E.g. Dickinson (4), pp. 18 ff.; (6), pp. 171 f.; Matschoss (1), pp. 39 ff.; Wolf (1), pp. 551 ff.; Rolt (1), pp. 35 ff.; Rolt & Allen (1), pp. 24 ff. Savery may have been preceded by other inventors such as David Ramsay in +1631, Edward Somerset (Marquis of Worcester) in +1663, and Samuel Morland in +1685; but the descriptions of their pumps are not clear enough to be sure.

[c] Note that this combined both the principles which de Hautefeuille's gunpowder designs had kept separate.

[d] In +1776; Dickinson (4), p. 28. Denis Papin, in +1707, a few years before his death, also had a go, and put in to the Royal Society for research funds, but Savery refereed it and the grant was withheld.

[e] Savery's principle persists, however, in Hall's pulsometer pump (1876), still made.

It is interesting that the evolution of the steam-engine was just about complete by the time that engineering inventors began producing designs for internal-combustion engines.[a] The separate condenser had been evolved by James Watt between +1765 and +1776,[b] the double-acting principle[c] came in about the same time as reciprocating rotary motion[d], c. +1783, and high-pressure steam was introduced by Richard Trevithick from 1811 onwards.[e] In the light of this it is quite interesting that gas-engines date from about 1826, and all the oil-engines (among which one must include those running on Diesel oil and petrol) from about 1841.

The first way of getting an explosion in the cylinder was to make a mixture of air and coal gas,[f] and then to ignite it on each stroke.[g] This was accomplished more or less by Samuel Brown from 1823 onwards, but his engine was not a success.[h] Ignition at reliable intervals was always the problem, and William Barnett used coupled gas flames in 1838. Others turned to different gases, such as hydrogen and air, or pure methane and air,[i] as in the work of Eugenio Barsanti and Felice Mateucci between 1843 and 1854, and it was these inventors who were the first to introduce that electrical ignition to which the future belonged.[j] But the Newcomen of gas engines was J. J. E. Lenoir (1822–1900)[k] who in 1859 made the first practical types, resembling horizontal double-acting steam-engines, with flywheels, slide-valves and water-cooling.[l] The next greatest step forward came, however, when Alphonse Beau de Rochas (1815 to 91) described in 1862 the four-stroke cycle basic to the successful operation of all internal-combustion engines. The first outward stroke of the piston draws the explosive mixture into the cylinder and the first inward stroke compresses it; ignition then takes place at or about the dead-centre position and the explosion

[a] Part of their impetus undoubtedly came from the fact that the power/weight ratio of steam-engines was so low, so that they were not adaptable for small factories and workshops, nor for road transport, let alone for air.

[b] Dickinson (4), pp. 60 ff.; Galloway (1), pp. 142 ff.; Wolf (2), pp. 618 ff.; Needham (48), p. 11.

[c] In which steam is admitted on both sides of the piston alternately so that each stroke does useful work (Dickinson (4), pp. 79 ff., (7), pp. 124 ff., 134 ff.; Galloway (1), pp. 162 ff.; Wolf (2), pp. 621 ff.). This principle was much more ancient in China, as the history of the double-acting piston-bellows goes to show (Vol. 4, pt. 2, pp. 135 ff.; Needham (48), pp. 15 ff.).

[d] This again was ancient in China; p. 545 above. Cf. Needham (48), pp. 29 ff.

[e] Dickinson & Titley (1), pp. 127 ff., 144; Pole (1), p. 51; Galloway (1), p. 192.

[f] This is mostly methane, with small amounts of CO, CO_2, ethylene and acetylene. Here we cannot go into the history of gases, which forms so large a part of that of modern chemistry itself, nor do more than recall John Baptist van Helmont's coining of the word, but a quick reference to coal gas would be the book of Clow & Clow (1), pp. 389 ff.

[g] In what follows we use the expositions of Field (1); le Gallec (1); Uccelli (1), pp. 373 ff., 377 ff., 381 ff.; Usher (1), pp. 370 ff., 2nd ed. 406 ff.; Burstall (1), pp. 333 ff.; Gille & Burty (1); and Day (1).

[h] There had been a string of similar projects and patents almost from the time when gas-lighting was introduced (Clow & Clow (1), p. 429). Robert Street's ideas (+1794) were vague, but Philippe le Bon d'Humbersin in +1799 got rather further.

[i] James Johnston in 1841 dared to try hydrogen and oxygen, but in those days the liquid forms were (perhaps fortunately) not available.

[j] Something of the sort had already been suggested by Alessandro Volta in +1776.

[k] See Leprince-Ringuet et al. (1), p. 148.

[l] This was always a problem. In 1862 M. Hugon made an engine in which a fine spray of cold water was injected into the cylinder after each explosion, but it was unsuccessful.

drives the piston on its second outward stroke, after which its second inward stroke expels the burnt gases from the cylinder.[a] Now at last the engineers had got their explosions under control, so to that extent the cannon was by 1860 firmly mastered. The real dénouement from our present point of view was, however, yet to come, as we shall see. A few gas-engines are still running, though most of them exist today only in museums;[b] naturally they could never go far from gas supplies,[c] though of course there is a sense in which all internal-combustion engines are gas-engines since the combustible material enters the cylinder as a fine spray mixed with air.

There followed an entr'acte or deviation somewhat analogous to the steam-vacuum displacement water-raising systems in the history of the steam-engine—namely that of the hot-air engine. John Stirling in 1826 and Eric Ericsson in 1849 had the thought of substituting for steam some new motor fluid more economical and easy to deal with. They therefore fell back on air itself, noting that its volume increases by a third between 0° and 100°, doubles by 272° and triples by 544°. Most of the older generation have memories of seeing part of an engine heated by a blow-torch, after which a swing of the flywheel would set the machine going; but although a number of engineers sought to perfect it, there were many disadvantages, such as fire danger and deformations of the working parts, with the result that like the gas-engine it now survives only in museums, and on a small scale for toys and working models.[d] The hot-air engine lies on a siding because no explosion, no internal combustion, was involved, only the expansion of heated air; but some source of heat remained imperative, so a heat-engine it certainly was. But the motive power of the future it was not.[e]

The dénouement of the whole story came in 1836, when Luigi de Cristoforis (+1798 to 1862) began to think of making an internal-combustion engine run on naphtha, a project which he perfected by 1841. Now at last those light fractions of distilled petroleum, originally as Greek Fire so hurtful an incendiary weapon, burning men as well as things, were to become a beneficent power-source for daily use. What the Byzantine +7th century had begun and the Chinese +10th century had continued, now, after a thousand years, found its ideal place within the cylinders of internal-combustion engines. Perhaps we should pause here an instant to consider all the oily substances of this kind which can be used as combustibles; for oil-engines,[f] Diesel engines and petrol engines form a single family. We can tabulate the boiling points of these hydrocarbon fuels as follows:[g]

[a] This is generally known as the Otto Cycle, after A. N. Otto (1832–91) who re-invented it in 1877.

[b] We have one in the Cambridge Museum of Technology at Cheddar's Lane.

[c] This was no doubt the greatest limiting factor for their ubiquitous mobility. Only in difficult conditions when liquid fuels were in short supply, did vans and buses carry balloons of gas on their roofs, or, as in war-time China, water-gas generators alongside the driver's seat.

[d] Mr John Shaw, to whom we are indebted for much information about present-day practice, remembers seeing a 0·25 h.p. hot-air engine driving laundry apparatus in an Irish country house for many years.

[e] Nevertheless it had obliquely a magnificent descendant, as we shall shortly see.

[f] Many oil-engines are still in service, and John Shaw tells us that an earlier generation of torpedoes ran on shale oil, and their engines were actually started with an explosive charge.

[g] We take them from the classical Perkin & Kipping (1), pp. 71 ff., 336 ff.

	b.p. °C.	
petroleum ether or petrol	40–70	
gasoline	70–90	
ligroin or light petroleum	80–120	
benzene and toluene (from coal-tar)	82–110	
cleaning oil (turpentine-substitute)	120–150	
naphtha (from coal-tar)	140–170	(mostly xylene, pseudocumene, mesitylene)
kerosene (paraffin oil)	150–250	
Diesel fuel oil	250–300	
carbolic oil (from coal-tar)	170–230	(mostly naphthalene and carbolic acid)
creosote oil (from coal-tar)	230–270	
anthracene oil (from coal-tar)	270–	
lubricating oil	300–	

Many of the lighter fractions of these oils have been used in internal-combustion engines at one time or another, but eventually engineers settled for the lower b.p. oils in what we universally know nowadays as the automobile and aero engine. Higher hydrocarbons are commonly 'cracked' to give the lower lighter ones.

The history of these power-sources can be briefly told. In 1873 J. Hock made an engine work with kerosene, and two years later Siegfried Marcus introduced petrol much like that of today; both worked in Austria.[a] At the same time another petrol engine was improved by Enrico Bernardi of Verona, and in the following decade Gottlieb Daimler and Karl Benz (1883–5) brought it almost to its present form, attaining 800 r.p.m.[b] In a parallel development many types of oil-engine appeared,[c] but the greatest advance was made by Rudolph Diesel (1858 to 1913)[d] who in a certain sense married the hot-air engine to the oil or pertrol engine by compressing air violently to a temperature of 800°, sufficient to ignite spontaneously a quite heavy oil injected into the cylinder. As everyone knows, there has been a vast expansion in the use of Diesel engines, especially for railway locomotives. Meanwhile, by 1895 the internal-combustion petrol-burning high-speed automobile engine had reached essentially modern design in the hands of the Count de Dion and M. Bouton.[e]

So now, reflecting on what we have found, we can see that the inventions of Greek Fire in the +7th century and of gunpowder in the +9th were not the unmitigated disasters that many people, even Shakespeare, speaking through

[a] Larsen (1), pp. 149 ff.   [b] Field (1), pp. 164 ff.
[c] Dent & Priestman (1886), Capitaine (1893), Hornsby, Crossley, etc.
[d] See Leprince-Ringuet (1), p. 152.
[e] I remember well the De Dion-Bouton limousine which my father, who was then in general practice in South London, bought to visit his patients in the first decade of the present century.

his characters, have thought. Without them we might have had neither the steam-engine nor the internal-combustion engine. And the moral is the same as that which we saw in the case of the rocket—all depends on what you do with it. Like fire itself, which can be used either for cooking food and warming people, or alternatively for torturing and killing people, the uses of every invention depend upon human ethical judgments; a problem for mankind as a whole, and common to all the civilisations. But the tragic aspect of history is that it should take so many centuries to find out the good use of inventions, and to refrain from the evil.

(21) INTER-CULTURAL TRANSMISSIONS

Looking back over the long countryside through which we have come, the outstanding impression one has is that what took 400 years to develop in China was then conveyed to the Arabic countries and Europe within 40 years or rather less.[a] Two fuses in particular led into this gunpowder train, a previous 600 years of the isolation and purification of saltpetre in China alone,[b] and a previous 200 years of the distillation of petroleum,[c] first in Byzantium, then in Middle and South-east Asian lands and China. All the long preparations and tentative experiments were made in China,[d] and everything came to Islam and the West fully fledged,[e] whether it was the fire-lance or the explosive bomb, the rocket[f] or the metal-barrel hand-gun and bombard. It reminds one of the old rhyme:

> The bible and Puritans, hops and beer,
> Came into England all in one year.

A multitude of traits there are which betray the derivativeness—the use of the term for a vegetable drug,[g] the persistence of mineral, plant and animal poisons in the powder,[h] the trumps as the fire-lances of Europe,[i] and the vase-shape of the early bombards.[j] Striking parallelisms there are too, notably the warnings of the fate which might befall the early powder-makers.[k] All in all, the gunpowder

---

[a] Cf. pp. 274, 294, 304 above.   [b] Cf. p. 107 above.

[c] The reason why Greek Fire was important was twofold: first its prefiguration of the incendiary properties of low-nitrate gunpowder, and secondly the fact that the first appearance of gunpowder in war was in the form of slow-match for petrol flame-throwers. See p. 92 above.

[d] This can be seen well in the study of the gunpowder compositions, pp. 346 ff. above.

[e] Cf. p. 348 above. There was a difference, however, in that the Arabs, receiving gunpowder weapon technology first, used the mixture (as had been the case in China much earlier) primarily as a more effective sort of incendiary (cf. Ayalon (1), pp. 9, 14, 24–6). Above (p. 45) we noted how the terms *naft*, and then *bārūd*, were used for centuries in Arabic to denote gunpowder, ignoring the sulphur (*kibrīt*) and the charcoal (*faḥm*). Partington (5), p. 197, acutely remarked that this was explicable if gunpowder reached the Arabs from China, but not if it came to them from Europe, where gunpowder was not used as an incendiary to begin with.

[f] Cf. p. 472 above. Transmissions were going on here as late as +1450.

[g] P. 108 above.   [h] See p. 353 above.   [i] See pp. 261 ff. above.

[j] See pp. 325 ff. above. It is also striking that these were used to shoot arrows, in the West as in China previously (cf. pp. 287–8, 307 ff. above). Partington (5), p. 101, found a dozen examples down to +1588.

[k] Pp. 111–2 above. And we could add here the use of live expendable birds (Cf. pp. 211 ff.), recommended by John Arderne about +1350 (Partington (5), p. 324).

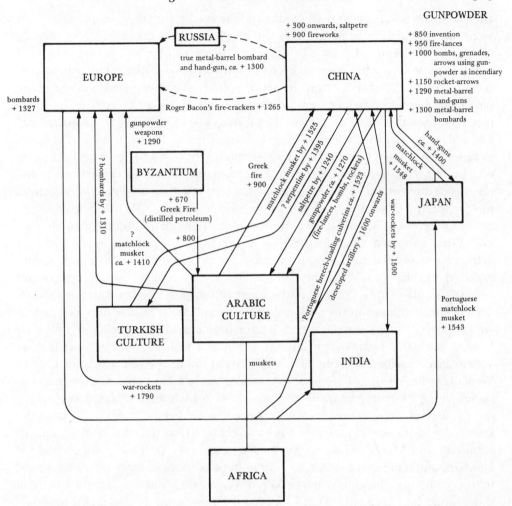

Fig. 233. Chart to illustrate the inter-cultural transmission of gunpowder technology in the Old World. For discussion, see text.

formula was China's equivocal gift to the rest of the world, passing through channels which we have attempted to depict in the chart of Fig. 233.[a]

It may also be appropriate to consider the environment or accompanying circumstances in which the basic transmission occurred. From all our work we have been able to distinguish particular 'transmission clusters', times when several important inventions and discoveries came westwards together.[b] For example, there were several which accompanied the transmission of the magnetic

[a] Transmissions in the reverse direction when Europe was developing capitalism are also shown in this; cf. pp. 365 ff. above for the breech-loading principle and improved artillery, pp. 429 ff. for the matchlock musket, and pp. 465–6 for flint-and-steel ignition.

[b] Cf. Vol. 4, pt. 3, pp. 695 ff.; Vol. 5, pt. 2, pp. 123–4, pt. 4, pp. 157, 492 ff. See also Needham (64), pp. 22, 24, 32, 33–4, 61–2, 133, 210, 300.

compass, the windmill and the axial rudder in the +12th century; and there were others which went along with the mechanical clock, the blast-furnace for cast iron,[a] the segmental arch bridge, and the helicopter top, in the +14th. It remains to be seen what transmissions exactly we should place with gunpowder in the +13th. Probably certain forms of textile machinery were among them, paper-making and printing were on the way, but above all there was that deep conviction emanating from China that if men knew more about chemistry untold longevity could be achieved. Roger Bacon (+1214 to 92), the first European to talk like a Taoist, represented this outstandingly—and yet by a strange paradox he himself was one of the first Europeans to record the constituents and effects of gunpowder.[b] This was neither the first nor the last occasion in human history when men would touch and know, and come to handle, the double-edged powers inherent in Nature, pregnant with almost unlimited might for good or for evil. What gunpowder was for Brother Roger, nuclear energy is for us.

So now it is time to draw all the threads of this sub-section together, and tackle the problems of how the Chinese discoveries and inventions spread out over the whole world. The great advantage of pin-pointing the dates of appearance of specific things in the regions of Europe and West Asia is that we know in what decades to look for the means of transmission; and by the same token we have to be clearly aware of the dates at which specific inventions first made their appearance in China. People have often talked about the passage from East to West of the knowledge of gunpowder as such, but in fact it looks as if we ought to be searching for three separate transmissions:[c] (a) whoever it was that brought the present of fire-crackers to Roger Bacon, soon before +1265; (b) how the knowledge of fire-lances, bombs and rockets got into the hands of Ḥasan al-Rammāḥ and Marcus Graecus by +1280; and lastly (c) how the metal-barrel bombard and hand-gun found their way to the European military by about +1300 in time to get into the picture in the MS. of Walter de Milamete. The first three of these things had been current in China, as we know from the abundant evidence already given, from the +10th century onwards, the fourth (rockets) since the second half of the +12th, and the last only from about +1290. Increasing complexity and effectiveness were thus mirrored in increased speed of transmission.

Perhaps the first of these passages is the easiest to understand. As we noted already (p. 49), when Roger Bacon was writing about his fire-crackers, it had been just thirty years since the first friars had visited the Mongol court at Kara-koron. In the intervening time several eminent ecclesiastical travellers had followed them, notably John of Plano Carpini, who went as envoy from Innocent IV to the Mongol khan in +1245, returning two years later;[d] and André de Longjumeau, who went again on the same mission in +1249.[e] Above all there

---

[a] We know now that this belongs rather to the former cluster, and first in Scandinavia; see Tylecote (1); Wagner (1).
[b] Cf. pp. 47–8 above.
[c] This conviction was already forming in the course of a conversation which we had with Fêng Chia-Shêng in Peking in 1952.
[d] Rockhill (5); Beazley (3); Komroff (1).   [e] Pelliot (10); Sinor (7).

was William Ruysbroeck, another Franciscan, sent by King Louis in +1252 and returning by +1256; particularly important not only because he wrote up all his travels but because he knew Roger Bacon personally in Paris.[a] Nor were all the voyagers friars, for there was a layman, a French knight, Baldwin of Hainault, sent out to treat with the Mongols by the Latin emperor of Byzantium Baldwin II about +1250. Much better known was Guillaume Boucher, goldsmith, metalworker and engineer, who had employment at the court of the Great Khan at Karakoron under Küyük (r. +1246–9) and his successor Mangu (r. +1250–9),[b] during which time he was personally known to William Ruysbroeck. He worked alongside many Chinese artisans, and he would have been particularly interested in any piece of technology emanating from Cathay. All in all, there were many channels, some quite direct, through which the fire-crackers (and the knowledge of what was in them) could have reached Brother Roger.

But West European friars and knights were not the only people in the picture. Herbert Franke (20, 26) discovered a very interesting account of the appearance of Scandinavian traders in +1261 at the Mongol court, which by then had moved a long way east, from Karakoron (Ho-lin[1])[c] to Shangtu[2], north of Peking.[d] One of the Chinese court secretaries, Wang Yün[3], kept a diary of those years, the *Chung Thang Shih Chi*[4], and in it he recorded the following event:[e]

In June there came merchant-envoys from the Fa-Lang[5] country (Frankistan) and (Khubilai Khan[f]) received them in audience. They presented garments made of vegetable fibres[g] and other gifts. Their home, they said, which they had left three years before,[h] was in the Far West, beyond the lands of the Uighurs. In that country there is always daylight (*chhang hua pu yeh*[6]), and you can only tell the evening time by seeing when the field-mice (*yeh shu*[7]) come out of their holes....[i]

The women are very beautiful, and the men generally have blue (*pi*[8]) eyes and blond (*huang*[9]) hair....

Their ships are large, carrying between 50 and 100 men. These people presented a wine-beaker made from the egg-shell of a sea-bird, and wine poured into it became warm at once. It was called a 'warm-cool cup' (*wên liang chan*[10]).[j]

[a] Cordier (1), vol. 2, p. 398; Komroff (1); Dawson (3); Beazley (3). Chambers (1), pp. 166–7, believes that the intermediary was most probably Ruysbroeck.

[b] Olschki (4) devoted a monograph to him. There was a small colony of Latins resident there, including a woman, Paquette of Metz, and the interpreter Basil the Hungarian, son of an Englishman. Cf. Komroff (1), pp. 134–5, 157, 160; Dawson (3), pp. 157, 176–7.

[c] A place N.E. of the Altai Mountains and S. of Lake Baikal, on the Orkhon River.

[d] This place, Dolon Nor, was the summer capital from +1260. The Jurchen Chin State had already been conquered in +1234.

[e] In *Chhiu Chien hsien-sêng Ta Chhüan Wên Chi*, ch. 81, pp. 9b, 10a; tr. auct. adjuv. Franke (20, 26). The visit is also mentioned in *Hsin Yuan Shih*, ch. 7, p. 10b.

[f] He had been enthroned just the year before.

[g] Perhaps cotton, more likely linen from flax.

[h] I.e. in +1258.

[i] An early observation of circadian rhythms?

[j] The suggestion that the effect was due to quicklime in the shell is not scientifically plausible.

[1] 和林   [2] 上都   [3] 王惲   [4] 中堂事記   [5] 發郎
[6] 常晝不夜   [7] 野鼠   [8] 碧   [9] 黃   [10] 溫涼盞

The emperor was very pleased that this group had come so far, and gave them liberal gifts of gold and textile materials.

Could he perhaps have given them some fire-crackers also? In any case it is obvious that these must have been yellow-haired Norsemen, from the 'white nights' of Scandinavia, probably coming by way of Novgorod,[a] at that time the centre of an independent State. And this was several years before the elder Polo brothers reached the court of the Great Khan. It would have been just in time for Roger Bacon's description.

Next comes the second problem, that of the transmissions of the more complicated devices which reached Ḥasan al-Rammāḥ and Marcus Graecus by +1280 or so. Paradoxically, these do not seem to have taken place during the European campaigns of the Mongol armies, which lasted for about a decade from +1236 onwards.[b] In that year Bulgaria was overrun, and in the next all Russia was devastated.[c] Kiev was taken in +1240, and the greatest fight, at Liegnitz, was in the following year, when an army of 10,000 Germans, Teutonic Knights, Poles and Silesians, under Henry the Pious, Duke of Silesia, was overwhelmed. After this the Mongols faltered, failing to take Olmutz, and sheering off from Austria, instead going down to the Adriatic coast, avoiding Dubrovnik (Ragusa) but sacking Kotor and many other places. By +1246, when Küyük was elected Great Khan, the westward push was over, though Poland was invaded again, Kraków burnt in +1259, and Budapest destroyed as late as +1285.

Now the Mongols were essentially mounted archers with strong tactical discipline,[d] and on the whole made little use even of trebuchet artillery, though some such engines appear from time to time.[e] Incendiary arrows occur in the accounts, however.[f] There is no mention of gunpowder in the narrative of John

[a] S. of Leningrad and W. of Moscow. Scandinavian–Russian commerical connections were intimate all through the Middle Ages.
[b] See Cordier (1), vol. 2, pp. 246 ff.   [c] With the notable exception of Novgorod.
[d] Liddell Hart (2) remarked that the role of tanks and planes in modern warfare, exemplifying the theory of 'fire *and* mobility', was a natural development of the tactics of the Mongolian mounted archers. Both Rommel and Patton, he says, were students and admirers of generals like Bātū, Bayan and Mangu, Ögötäi and Subotai.
[e] As in Howorth (1), pt. 1, p. 149; Martin (2), p. 67; d'Ohsson (1). Yule (1), vol. 2, p. 168 quotes a fugitive Russian archbishop as saying of the Mongols in +1244: 'Machinas habent multiplices, recte et fortiter jacientes'. Trebuchet artillerists are much more in evidence during Hulagu's campaign against the Muslims of Persia and Iraq from +1253 onwards. Indeed he mobilised whole regiments of Chinese engineers, with *arcubalistae* as well; Yule (1), vol. 2, p. 168; Reinaud & Favé (2), pp. 294–5; Huuri (1), pp. 123, 181; Howorth (1), pt. 3, p. 97; Boyle (1), vol. 2, p. 608.
We have already given a translation of the interesting passage Howorth quoted (and cf. p. 89 above). Presumably by this time they fired gunpowder bombs with strong cast-iron casings (*chen thien lei*[1]), cf. p. 171 above. Later still, in the unsuccessful campaigns of the Ilkhān Ghāzān against the Mamluk Caliphate in +1299 and +1303 for the control of Syria, there must have been many opportunities for the transfer of gunpowder technology to the Arabic armies, but the dates are by then rather too late for our present purpose. Cf. Ayalon (1).
[f] And also hot-air balloons or wind-socks like fire-breathing dragons used for signalling or as standards. This is a curious subject demanding further research; we collected and discussed a number of references in Vol. 4, pt. 2, pp. 597–8.

[1] 震天雷

of Plano Carpini (+1247) though he describes the European campaign in some detail; but he does talk of Greek Fire (or naphtha) in pots thrown over the walls of besieged forts or cities.[a] Prawdin alone asserts[b] that the Mongol forces under Bātū used gunpowder at the Battle of the Sajo River against the Hungarian King Bela two days after Liegnitz; Goodrich & Fêng Chia-Shêng took leave to question this,[c] and asked for evidence, but we do not know of any subsequently provided.[d] On the whole therefore it seems fairly safe to say that the wars of the Mongols against Europe were not the means of transmission of gunpowder technology to the West.[e]

But after +1260 or so the case is altered. Many men from Persia, Syria and the Arab lands entered the Mongolian service in China, and some were military technicians. As is well known, the Yuan dynasty under Khubilai preferred to employ foreigners as far as possible to run the Chinese State, not trusting the scholars with their difficult written language, nationalist sentiments, and age-old administrative customs.[f] This was why Sáīd Ajall Shams al-Dīn (Sai-Tien-Chhih Shan-Ssu-Ting[1]) found himself from +1274 onwards Governor of Yunnan, where he accomplished many valuable works especially in hydraulic engineering besides establishing a Confucian temple, schools and libraries.[g] Later, another Shams al-Dīn (Shan Ssu[2]) achieved great fame as a geographer and engineer. In +1263 a Nestorian Arab physician, 'Īsa Tarjaman (Ai Hsüeh[3]) had been appointed Director of the Astronomical Bureau, and all his five sons were in the Mongol service. From +1266 onwards the Arab architect Ikhtiyar al-Dīn (Yeh-Hei-Tieh-Erh[4]), a great master of the Chinese style, was laying out lakes, palaces and city walls and buildings at Khubilai's capital, Peking.

Any of these men could have had a hand in conveying to the people of Islam and Christendom the knowledge which is at issue here. How much more so, then, could the professional military men have done it—and they were not few. One of their periods of greatest prominence was at the siege of Hsiangyang[6] (Saian-fu) between +1268 and +1273, when they made great use of those counterweighted trebuchets which seem to have been an Arab invention

---

[a] Beazley (3); Komroff (1); Dawson (3), p. 37. He also remarks (p. 46) that the Mongols feared the crossbow. Cf. Prawdin (1), pp. 263–4.

[b] (1), p. 259.

[c] (1), p. 118. Professor Owen Lattimore (priv. comm.) recalls reading of the use of gunpowder when the Mongols stormed Merv and Samarqand, but perhaps only in mines set off below the walls.

[d] Saunders (1), pp. 176–9 made a special study of the question, and decided negatively.

[e] Lot (1), vol. 2, p. 393 remarked that although firearms gave the Russians superiority over the Tartars in the +15th century, there was no evidence of the transmission of gunpowder technology from the Mongols in the +13th, though presumably they would have known of it. The more recent researches of Chambers (1), pp. 57, 63–4, 166–7, support the negative conclusion.

[f] Cf. p. 209 above. The classical work on all these foreigners and their gradual, even sometimes rapid, sinisation, is that of Chhen Yuan (3). Goodrich (26) on their acculturation is well worth reading too. See also Chhen Yuan (1).

[g] His son Ḥuṣain (Hu-Hsin[5]) continued his benevolent activities. Cf. Vol. 4, pt. 3, p. 297.

[1] 賽典赤瞻思丁　　　　[2] 瞻思　　　[3] 愛薛　　　[4] 也黑迭兒
[5] 忽辛　　　　[6] 襄陽

and which afterwards took their Chinese name (*hui-hui phao*[1] or *hsiang-yang phao*[2]) from their operators and the Sung city that was being attacked.[a] The oldest of these engineers was ‘Alī Yaḥyā (A-Li-Hai-Ya[3], d. *c.* +1280) the Uighur artillery general of Khubilai, and it was he who suggested the summoning of the two experts from Persia and Syria ‘Alā al-Dīn of Mosul (A-Lao-Wa-Ting[4], d. *c.* +1295) and ’Ismā‘īl of Herat or Shiraz (I-Ssu-Ma-Yin[5], d. +1274). The former had one son, Abū’l Mojid (Fu-Mou-Chê[6], d. +1312) who succeeded his father; and the latter two, Abū Bakr (Pu-Pai[7], d. *c.* +1295) and Ibrāhīm (I-Pu-La-Chin[8], d. +1329)—all became artillery generals in the Yuan service. Whether or not their counterweighted trebuchets hurled explosive bombs, we do know that gunpowder weapons, such as fire-lances, were abundantly used in the operations connected with the siege;[b] and since it is unlikely that the Muslim commanders would have been entirely cut off from their original home-lands, it would seem extremely probable that they were among the means of conveyance of the technology, at least to the Islamic peoples.[c]

In connection with all this it is interesting to reflect that the first soldiers anywhere in the world to use metal-barrel hand-guns were the Chinese detachments in the Mongol service a couple of decades after the fall of Hsiangyang. Gunpowder weapons had certainly helped to put Khubilai on the Chinese throne in the fifties. But although the appreciation of this new technology rose to a certain height among the Mongols of the end of the +13th century, later on, towards the end of the Yuan dynasty in the fifties of the +14th, they undervalued it again, and the great success of Chu Yuan-Chang in driving them out, and establishing the Ming, was partly because he supported all efforts to improve artillery and gunpowder weapons in general (cf. p. 26).

So much for the soldiers, but we have still two other tribes of men to consider—the ecclesiastics and the merchants. Both could have had some part to play in the transmission of knowledge about gunpowder bombs, mines, fire-lances and rocket-arrows before about +1280. Let us review once more the course of events in this turbulent century. The Mongols were on the up and up. First the Hsi-Hsia Tangut State was conquered, next the Western Liao kingdom of Qarā-khitāi, and then the Turkic lands of Khwarizm. When Chinghiz died in +1227, four descendants took over, Ögötäi to rule East Asia, Chagatai to govern

---

[a] Cf. p. 175 above. In spite of what Rusticianus put in Marco Polo's book, it is as good as certain that the Polos were not present at the siege (cf. p. 277 above) nor their European trebuchet artillerist friends either. See Moule (13).

[b] See p. 174 above.

[c] We need not think here only of the land route. One recalls that there had been colonies of Arabic merchants at several of the great southern ports, notably Canton and Chhüanchow, since the +7th or +8th century (cf. Vol. 1, p. 180). From the remarkable work of Kuwabara (1) we know a good deal about the Commissioner of Merchant Shipping at the latter place between +1250 and +1275, Phu Shou-Kêng[9], himself a man of Arab or Persian descent. Here were obvious opportunities for contact with Syria, and the second transmission to al-Rammāḥ and Marcus Graecus—to say nothing of the fireworks which the Muslim merchants might have thought fit to take along with them.

[1] 回回砲   [2] 襄陽砲   [3] 阿里海牙   [4] 阿老瓦丁   [5] 亦思馬因
[6] 富謀尺   [7] 布百   [8] 亦不剌金   [9] 蒲壽庚

Turkestan, Hulagu in charge of Persia, and Bātū leading the 'Golden Horde' on the Volga in South Russia. The Jurchen Tartar Chin dynasty in North China was overthrown in +1234, and far away to the West, Mangu invaded Armenia in +1236. The following year saw the fall of Russian Ryazān, and the Mongols invaded Poland. In +1241, along with the victory of Liegnitz, there was the siege and capture of Budapest, but also the death of Ögötäi, to be succeeded by Küyük and then Mangu ten years later. In +1253 Mongol dignitaries went to register the population of South Russia for fiscal purposes,[a] and in the reverse direction (as we have seen) came the journeys of William Ruysbroeck and other Franciscan friars to the Mongolian court at Karakoron.[b] They were diplomatic envoys quite as much as missionaries, sent to seek the help of the Mongols against the Muslims, traditional foes of the Frankish Christians. It was a classic case of that encircling strategy by which one seeks to mobilise the forces of allies whose lands lie beyond those of one's immediate enemy.[c] One would give a good deal to know what exactly the friars saw of gunpowder and fire-weapons during their wanderings in Mongolia and China. Although such interests would have consorted ill with their habit, they might have felt it their duty to bring back knowledge and skills which could conserve the safety and power of Christendom against the 'infidel'. With this transmission in mind, the activities of the friars need looking at more closely than hitherto. One of them might even have been accompanied on the way home by some Chinese gunner who knew the multifarious devices of the previous half-dozen centuries as well as the latest inventions, and was not averse to seeking his fortune in strange foreign lands—but so far history has not had word of his name or his activities.[d]

The overall strategy of the friars, directed against Islam, succeeded beyond all expectation, apart from the fact that the Mongols did the job for themselves, and made no firm alliances with Christian powers.[e] Having subdued Persia, they invaded Iraq at the head of the Persian Gulf, and Baghdad fell in +1258. Two years previously the Mongolian Ilkhānate, centred on Iran, had been established, and the great astronomical observatory of Marāghah founded.[f] Then came a second possible medium of transmission, also ecclesiastical, the travels of Rabban Bar Sauma and his friend, the fascinating account of which was translated from the Syriac long ago by Wallis Budge (2). These two young men were

---

[a] Cf. Franke (20); Lot (1), vol. 2, p. 386.

[b] Ascelin the Dominican and Guiscard of Cremona were there with Simon de St Quentin in +1247, André de Longjumeau in +1249, and Bartholomew of Cremona accompanied William Ruysbroeck.

[c] Cf. Vol. 1, pp. 223 ff.

[d] Only in one place have we found a possible name, that of Chhi-Wu-Wên[1], a Mongol, who was said to have taken the knowledge of gunpowder technology to the Western world in the early Yuan time, the second half of the +13th century. It was Yü Wei[2], an otherwise little-known scholar, who let this name drop, and Miu Yu-Sun recorded it in his *O Yu Hui Pien*, whence it got into the *Ko Chih Ku Wei* (ch. 2, p. 28a). How firm this tradition would be is anyone's guess, but it does seem worth mentioning.

[e] *En revanche*, of course, from +1282 onwards they increasingly embraced Islam.

[f] Cf. Vol. 3, pp. 372 ff.; Howorth (1), pt. 3, pp. 137 ff. It will be remembered that the astronomical delegation to China headed by Jamal al-Dīn (Cha-Ma-Lu-Ting[3]) took place in +1267.

[1] 奇渥溫   [2] 余威   [3] 札馬魯丁

Chinese Christian (Nestorian) priests of Uighur stock, born and educated in Peking, who pined to go on a pilgrimage to Jerusalem.[a] Neither of them ever got there, but they did travel the whole length of the Old World between +1278 and +1289 before returning to settle down in Persia and Iraq. The friend, Marqos Bayniel, was unexpectedly elected to a bishopric at Baghdad, as Metropolitan of Cathay, and then a year later Catholicos (Patriarch) of all the Nestorian Churches, as Mar Yabhallaha III, so his duties detained him there indefinitely.[b] But Bar Sauma travelled on to the West as an envoy from the Ilkhān Arghun,[c] visited Italy, and in +1287 was warmly received at Rome (where no unduly tactless doctrinal questions were asked), finally reaching Bordeaux (where he celebrated the liturgy in the presence of the King of England). Eventually he returned to Persia by way of Italy, and built a church at Marāghah, where he died in +1294. The purpose of this pilgrimage was again partly political, to get Western assistance for ousting the Muslims from Jerusalem, but it never had the slightest chance of success. The dates are rather late for the transmission we are looking for, but our shadowy Chinese gunner might conceivably have come along with the two priests, and handed on his knowledge to discreet persons in the Mediterranean region capable of receiving it.

Lastly, we have to think not only of soldiers, or West Asian scholars, or ecclesiastics whether Latin or Nestorian, but of the European merchants. The name which springs to mind of course is that of Marco Polo, 'Il Milione' (the man who averred that there were millions of ships on China's rivers, and millions of bridges in Hangchow—and fundamentally he was not wrong).[d] But he did not leave China till +1292,[e] which makes him too late for the second transmission, though he might just have accomplished the third. His father Niccolò and his uncle Maffeo, who were in China first between +1261 and +1269, could on the other hand have been responsible for the second, the bringing of news of fire-lances, bombs and rockets. Marco was with them on their second visit (+1271 to 95), during which he served Khubilai Khan, sometimes on secret service missions, more often in the salt administration; and when he left it was by sea, accompanying a Mongolian princess proceeding with a great fleet to become the Ilkhān Arghun's second wife.[f] This might have been an even more appropriate scenario for the Chinese gunner we have in mind, and now he could have been a gunner in the fullest sense, acquainted with metal-barrel bombards and hand-guns.

Much less well known is the colony of Italian merchants established at Tabriz in the Ilkhānate.[g] Though the silk trade had been active since +1257,[h] the first

---

[a] Cf. Vol. 1, pp. 221, 225.
[b] Chabot (1).   [c] Chabot (2).
[d] Yule (1); Moule & Pelliot (1); Olschki (10).   [e] A better date is +1291.
[f] Doubt has sometimes been expressed as to whether Marco Polo was ever in China at all, and certainly no one has found a reference to him in Chinese historical writings, but perhaps that is because he was too unimportant a person. Ho Yung-Chi (1), however, has brought forward a number of Chinese references to the sea-voyage of the princess and her entourage, which began in +1291, so that Marco's account of the circumstances of his departure is thus far independently substantiated.
[g] See Petech (5).   [h] Cf. Lopez (3, 5).

name we know is that of the Venetian Pietro Vilioni, who died there in +1264. In +1269 Mongol ambassadors from the Ilkhān arrived at Genoa, and a Genoese merchant, Luchetto de Recco, was stationed in Tabriz in +1280. From +1274 onwards Buscarello Ghisolfi played an important diplomatic role between the Ilkhāns, the Italian city-States and the Pope; he was even twice in London (+1289 and +1300) on the usual ploy of constructing Mongol–Christian alliances against the Muslims, and accompanied an Englishman, Sir Geoffrey Langley, on a visit to the Ilkhān in +1292. Many other names of Italian merchants trafficking about this time in the Ilkhānate are known, both Venetian and Genoese.[a] The colony continued to prosper until about +1336. Its members could certainly have played a part in the second and third transmissions of knowledge which we are considering.[b]

Perhaps there is room for speculation that the third, i.e. that of the true metal-barrel bombard and hand-gun, reached Europe directly overland and not through the Arabs at all. Lattimore acutely noted[c] that the Russian word for cannon is *pushka*, and that since the Slavs, unlike the Germans, do not confuse *p* with *b* in borrowed words,[d] the usual derivation from German *Büchse*, cannot hold water.[e] But *phao*[1] would go some way to meet the case, so perhaps the transition was *phao —pushka—Büchse*, and the usually assumed origin from Gr. *pyxis* (πυξίς), a box, is wrong.[f] It is only fair to mention here a persistent Chinese tradition[g] that the Russians were the intermediaries in the travel of gunnery to Europe.[h] The trouble with Arabic intermediation is that it is so hard

---

[a] As also indeed Florentine, Pisan and Sienese.

[b] But the Tartar (Mongolian) slaves, or domestic servants, who reached the Florentine markets for a century after about +1325 could not, for the dating is just too late (cf. Vol. 1, p. 189). True, Dr Alice Kehoe (priv. comm.) tells us that sugar-cane plantations in Cyprus, owned by Crusaders, were worked by such slave labour in the late +13th century, and if this can be substantiated, it would have constituted a possible channel.

[c] (10), p. 10. Prof. Owen Lattimore had already discussed the point in correspondence with us in the autumn of 1954.

[d] Preobrazhensky (1), *s.v.* He notes similar forms in Bulgarian, Serbo-Croat and Albanian.

[e] The usual view, expressed by Lot (1), vol. 2, p. 392, was that cannon were unknown in Russia before +1389, when they were acquired from Germany.

[f] Mavrodin (1) argued long ago that some Turkish word might have been the origin. The paper of Vilinbakhov (1) deals only with naphtha pots thrown from trebuchets and *arcuballistae* in Russia, while that of Vilinbakhov & Kholmovskaia (1) discusses Chinese sources only. Both are rather confused.

[g] E.g. in *Ko Chih Ku Wei*, ch. 2, p. 28a.

[h] Dr Michael Hendy suggests to us that the derivation of the Russian silver rouble (O.R. *rublǐ*, a block or lump) from the Chinese silver ingots used in currency might perhaps be a parallel. In the Thang and Sung, these silver *ting*[2] or *ping*[3] weighed 50 oz. each, but in the Ming the weight of the ingots (*kho tzu*[4]) fell to 5 oz., no doubt for greater convenience in transactions, and in the Chhing it was but a single ounce (*liang*[5]). This was the *tael*, so prominent in the writings of Old China Hands, a word derived, it seems, from Hindi *tola*, a weight, via the Portuguese. Although in universal use for centuries, the only issue of these by a government as official currency, took place in +1197 in the J/Chin State, when silver pieces were cast in five weights varying from 1 to 10 oz. On all this see Yang Lien-Shêng (3), pp. 43 ff.

What is not so generally known is that before the +15th century silver ingots, in the form of elongated rods the size and length of a table-knife handle, circulated in Russia, as also in Rumania; they were actually made in Byzantium, exported north, and much used in the trade between the Slavs and the Mongolians. Here then would be another example of the influence of Chinese ways on the Slavonic peoples, and it would be compatible with the date for the cannon transmission about +1300.

[1] 砲　　　[2] 錠　　　[3] 餅　　　[4] 錁子　　　[5] 兩

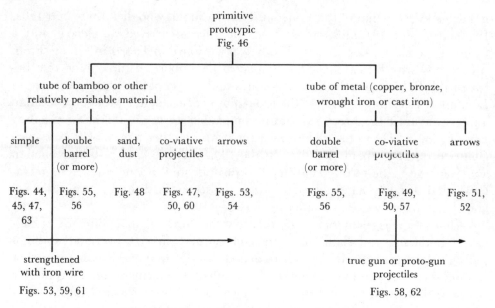

Fig. 234. Stages in the development of the Fire-lance.

to tell when *midfaʻ* as the name for the fire-lance emitting co-viative projectiles (or proto-gun) turned into *midfaʻ* as the name for the true metal-barrel bombard or hand-gun. This transition had certainly not happened by the time of al-Rammāḥ (*c.* +1280), but it probably did happen during the following few decades. Possibly, therefore, the Arabs received the bombard from Russia, Eastern Europe, including the Balkans, or Germany, rather than directly from China. The earliest date for the bombard in Spain has been held to be +1359,[a] but Lavin (1) makes it +1343, when during the siege of Algeciras the Moors within used iron cannon (*tiros de hierro*), and *truenos* (bombards).[b] This was well after Walter de Milamete's picture.[c]

So here we are back again at the deadline of +1327, or better, a dozen or more years earlier.[d] We can ignore all the events that happened after that, however exciting they are in themselves, such as the colony of Italian merchants at Yang-

---

[a] Partington (5), p. 123.

[b] Though this word was also used for the projectiles themselves. Another trait betraying knowledge of Chinese usage? Cf. Partington (5), pp. 193–4. Even the word 'gunne' could be used for a projectile, as Burtt (1) noticed in *The Avowynge of King Arthur* (+14th cent.), st. 65:

> '... there came fliand a gunne
> And lemet as the levyn....'

[c] Cf. Partington (5), pp. 200 ff., 204 ff.

[d] We say this not because of the Ghent reference of +1313 to *bussen met kruyt*, which, though accepted by Hime (1), p. 119, was rejected by Partington (5), p. 97 as a forgery, but because there must have been bombards and hand-guns in Europe some little time before Walter de Milamete's picture. A Florentine reference of +1326 is more acceptable, though Arima (1), p. 339, expressed scepticism about it. On the general development of artillery in Europe after this time, see Partington (5), pp. 98 ff.

chow, and the tombstone of Catherine de Vilioni there dated +1342,[a] or the activities of the Loredans in China around +1339,[b] or the embassy of the Genoese merchant Andalò de Savignone from the Yuan emperor Shun Ti (Toghan Timur) to the Pope in +1336.[c] Nor need we be concerned with the book of Messer Francisco Balducci Pegolotti (1) about travel and merchandising through the length and breadth of Asia (though he never actually went to Cathay himself), since he did not write it till about +1340.[d] Similarly, the journeys of the Latin bishops John of Monte Corvino,[e] John de Marignolli,[f] and Guillaume du Prat,[g] are all too late to be included in our story. By the beginning of that century the bell had rung, the curtain had come down, and the Western world was set upon the fateful road to all the techniques of managing explosions. Hence all later small-arms and artillery, but not only that, all heat-engines too, and all space travel.

[a] Cf. Rouleau (1); Foster (1); Rudolph (12).
[b] Petech (5), p. 556.
[c] *Ibid.* pp. 554–5. See also Vol. 4, pt. 2, pp. 507 ff.
[d] Cf. Vol. 1, pp. 188–9.
[e] See v. d. Wyngaert (1); Cordier (1), vol. 2, p. 411.
[f] Moule (1), pp. 257–8; Fuchs (7).
[g] Petech (5), p. 558.

APPENDIX A   THE OLDEST
REPRESENTATION OF A HAND-GUN?

An outstanding discovery was made by Robin Yates in June 1985 when visiting the Buddhist cave-temples at Ta-tsu[1] in Szechuan.[a] In the Pei-Shan[2] (Lung-Kang[3]) complex (one of seven) he found (cave no. 146) a relief of a hang-gun held by a small demon with two horns (Fig. 235).[b] The hand-gun is being let off, as appears from the blast issuing to the right from its muzzle, and a projectile is also represented in the flames.

As will be seen from the illustration, the figure is at the bottom on the right of a group of seventeen, with a many-armed bodhisattva or Buddha at the top and the back.[c] Twelve of the figures seem to be robed saints, but five have skull- or demon-faces and carry weapons, among which one can make out a spear, a mace, a hammer and a sword, as well as the hand-gun. Perhaps they are demons converted to sainthood. Probably all are attendants of either Kuan Yin or the Buddha of Medicine, whose seated image is the central figure of the shrine.

The object of interest to us here seems at first sight to be some sort of musical instrument, with the right hand of the figure plucking the strings, but a second look makes out the flames coming from the muzzle, and even the spherical ball or bullet among them. Of course, the sculpture cannot have been done by any-one who knew anything about hand-guns, because the explosion-chamber would have been much too hot to hold, and usually there was a socket cast on behind it,[d] into which a wooden 'tiller' was fitted, for grasping.[e] All the same, we may well have here the oldest representation in the world of a hand-gun, using the propellant power of high-nitrate gunpowder, similar to the larger bombards and primitive cannon which followed so quickly afterwards (cf. Table 1 on p. 290 above). The bulbous shape of the thickened metal wall around the explosion-chamber is too characteristic of these early gunpowder-weapons to be mistaken.[f]

The dating of the carving is somewhat obscure, but from pp. 293–4 above we know that the oldest hand-gun excavated so far is datable at ca. +1288, so that one would expect the date of the sculptured group to be any time between

---

[a] It parallels the discovery made at the Musée Guimet in Paris by Clayton Bredt, who found a clear representation of a fire-lance on a Buddhist temple banner of about +950 from Tunhuang (see pp. 222–3).

[b] It may be significant that the wielder of the fire-lance on the banner (Fig. 45 above) also has horns.

[c] On many-armed images cf. Vol. 4, pt. 1, p. 123 and Fig. 296.

[d] But Walter de Milamete's guns have none.

[e] The same is true of the fire-lance on the banner, because the demon is using one hand to hold the hot tube itself.

[f] On the other hand, historians of Buddhist art, such as Li Ssu-Shêng and Wang Kung-I, prefer to interpret the figure as the Szechuanese god of the winds, with his bag. If the date of +1128 is substantiated, they might well be right. But could that representation perhaps have influenced the designers of the earliest hand-guns and bombards?

¹ 大足   ² 北山   ³ 龍岡

Fig. 235. A group of figures at the Ta-tsu cave-temples (Pei-Shan Section) which contains what may be the oldest depiction of a hand-gun in any civilisation. The typical bulbous thickening of the metal wall around the explosion-chamber is seen, together with a stream of flame and gases issuing from the muzzle to the right, and a bore-occluding projectile. Date uncertain, but perhaps +1250 to +1280. Photo. Robin Yates.

+1250 and +1280. It is generally agreed that the figures are of the Sung period, though Yang Chia-Lo (3) and other experts such as Anon. (263) tend to place them between +1130 and +1170, while yet others, such as Angela Howard, put them even earlier, in the Northern Sung, from the late +10th to the early +11th centuries. Such dates would be too early for a hand-gun, though not for a fire-lance; nevertheless the relief has the form so typical of the earliest hand-guns and bombards, bulbous or pear-shaped (cf. Figs. 82–4, 90, 92, 94–5, 97, 100, 107–110, 116), while an approximately bore-fitting projectile is visible in the flames of the blast. A neighbouring inscription records the name of Wang Tzu-I[1], whose *floruit*, as we know from another inscription, was +1186. But even this date would be rather too early to expect a relief of a hand-gun. Possibly the content of the sculpture may help to date the ensemble.

At all events, we may well have here the earliest representation of a hand-gun in any civilisation, and the relief is therefore worthy of close attention.

[1] 王子惷

APPENDIX B THE DEVELOPMENT OF THE *MIDFA'*

On p. 43 above we discuss this Arabic word, which seems to have the general sense of a tube or cylinder. The illuminating work of Donald Hill (2), which translates and analyses the 'Book of Ingenious Mechanical Devices' (*Kitāb fī Ma'rifat al-Ḥiyal al-Handasiya*), written by Ibn al-Razzaz al-Jazarī in +1206, needs to be taken into special account here. Al-Jazarī, speaking of his slot-rod water-raising pump, says that 'this machine resembles the ejectors (or projectors, i.e. pumps) of naphtha (*zaraqāt al-naft*), except that it is larger', (Hall (2), p. 188). To understand this, one must remember the Chinese petrol flame-thrower described and illustrated on pp. 82 ff. and Fig. 7 above (and also Vol. 4, pt. 2, pp. 144 ff. and Figs. 433, 434) together with our account of the slot-rod water-raising pump (Vol. 4, pt. 2, p. 381 and Fig. 609). A critique of the reconstruction of this by Aubrey Burstall (depicted in our Fig. 610) is given by Hall (2), p. 273.

Again, when in his 'Key of the Sciences' (*Mafātīḥ al-'Ulūm*) Abū 'Abdallāh al-Khwārizmī al-Kātib (+976) speaks of *bāb al-midfa'* and *bāb al-mustaq*, both parts of the naphtha-projectors (*al-naffatāt wa'l-zarāqāt*), the word *bāb* (gate) means technically a valve, rather than just a mouth or opening (Hall (2), p. 274). From this we can conclude that the word *midfa'* originally meant the tube or cylinder of the naphtha-projector; then after the invention of gunpowder in China and its passage to the Arabs it meant the tube of the fire-lance; finally it was applied to the cylinder of the hand-gun and cannon.[a] It still retains this meaning in Arabic today. The fact that already in +1206 al-Jazarī recognised the affinity between the cylinder of his water-raising engine and the tube of the flame-thrower casts an interesting light on the connection between guns and engine-cylinders which we explore on pp. 544 ff. above.

---

[a] As we originally suspected when we first discussed the term.

BIBLIOGRAPHIES

A CHINESE AND JAPANESE BOOKS BEFORE +1800

B CHINESE AND JAPANESE BOOKS AND JOURNAL ARTICLES SINCE +1800

C BOOKS AND JOURNAL ARTICLES IN WESTERN LANGUAGES

In Bibliographies A and B there are two modifications of the Roman alphabetical sequence: transliterated *Chh-* comes after all other entries under *Ch-*, and transliterated *Hs-* comes after all other entries under *H-*. Thus *Chhen* comes after *Chung* and *Hsi* comes after *Huai*. This system applies only to the first words of the titles. Moreover, where *Chh-* and *Hs-* occur in words used in Bibliography C, i.e. in a Western language context, the normal sequence of the Roman alphabet is observed.

When obsolete or unusual romanisations of Chinese words occur in entries in Bibliography C, they are followed, wherever possible, by the romanisations adopted as standard in the present work. If inserted in the title, these are enclosed in square brackets; if they follow it, in round brackets. When Chinese words or phrases occur romanised according to the Wade–Giles system or related systems, they are assimilated to the system here adopted (cf. Vol. 1, p. 26) without indication of any change. Additional notes are added in round brackets. The reference numbers do not necessarily begin with (1), nor are they necessarily consecutive, because only those references required for this volume of the series are given.

Korean and Vietnamese books and papers are included in Bibliographies A and B. As explained in Vol. 1, pp. 21 ff., reference numbers in italics imply that the work is in one or other of the East Asian languages.

ABBREVIATIONS

See also p. xxiv

A	Archeion	ARUSNM	Annual Reports of the U.S. National Museum
AA	Artibus Asiae		
AAA	Archaeologia	AQ	Antiquity
AAAG	Annals of the Assoc. of American Geographers	AQR	Asiatic Quarterly Review
		AS/BIHP	Bulletin of the Institute of History and Philology, Academia Sinica
AAN	American Anthropologist		
AAS	Arts Asiatiques (continuation of Revue des Arts Asiatiques)	ASKR	Asiatick Researches (Calcutta, 1788 to 1839)
ACANT	Archaeologia Cantiana	ASTRA	Astronautica Acta
ACASA	Archives of the Chinese Art Soc. of America	AX	Ambix
ACP	Annales de Chimie et Physique	B	Byzantion
ACSS	Annual of the China Society of Singapore	BAU	Belleten Ankara Univ.
		BGP	Bulletin Catholique de Pékin
ACTAS	Acta Asiatica (Bull. of Eastern Culture, Tōhō Gakkai, Tokyo)	BE/AMG	Bibliographie d'Études (Annales du Musée Guimet)
ADVS	Advancement of Science (British Assoc. London)	BEC	Bulletin de l'École des Chartes (Paris)
		BEDM	Boletim Ecclesiástico da Diocese de Macao
AEHW	Archiv. f. d. Eisenhüttenwesen		
AER	Acta Eruditorum (Leipzig, 1682 to 1731)	BEFEO	Bulletin de l'École Française de l'Extrême Orient (Hanoi)
AGNT	Archiv. f. d. Gesch. d. Naturwiss. u. d. Technik (cont. as AGMNT)	BEO/IFD	Bull. Études Orientales (Institut Français de Damas)
AGWG/PH	Abhdl. d. Gesell. d. Wiss. z. Göttingen (Phil.-Hist. Kl.)	BGTI	Beiträge z. Gesch. d. Technik u. Industrie (cont. as Technik Geschichte; see BGTI/TG)
AHES/AESC	Annales; Economies, sociétés, civilisations	BGTI/TG	Technik Geschichte (see above)
AHSNM	Acta Historica Scientiarum Naturalium et Medicinalium	BLM	Blackwood's Magazine
		BLSOAS	Bulletin of the London School of Oriental and African Studies
AIMSS	Annali dell'Istituto e Museo di Storia della Scienza (Florence)	BMFEA	Bulletin of the Museum of Far Eastern Antiquities (Stockholm)
AJOP	Amer. Journ. Physiol.	BMQ	British Museum Quarterly
AJP	American Journ. Philology	BSRCA	Bull. Soc. Research in Chinese Architecture
AJSC	American Journ. Science and Arts (Silliman's)		
AM	Asia Major	BV	Bharatiya Vidya (Bombay)
ANA	All-Nippon Airways In-Flight Magazine	BYZ	see B
		BZJ	Bonner Zeitschrift f. Japanologie
ANTIQ	The Antiquary		
ANTJ	Antiquaries Journal	CA	Chemical Abstracts
APAW/PH	Abhandlungen d. preuss. Akad. Wiss. Berlin (Phil.-Hist. Klasse)	CAMR	Cambridge Review
		CCL	Chê Chiang Lu (Biographies of Chinese Engineers, Architects, Technologists and Master-Craftsmen, by Chu Chhi-Chhien and collaborators, q.v. [a series, not a journal].)
AP/HJ	Historical Journal, National Peiping Academy		
ARAB	Arabica		
ARIL	Atti (Annale) delli reale Istituto Lombardo		
ARJ	Archaeological Journal	CHEM	Chemistry (Easton, Pa.)
ARLC/DO	Annual Reports of the Librarian of Congress (Division of Orientalia)	CHI	Cambridge History of India
		CHJ	Chhing-Hua Hsüeh Pao (Chhing-Hua (Ts'ing-Hua) University Journal of Chinese Studies)
ARMA	Armi Antiche (Bull. dell'Accad. di San Marciano), Turin		
ARO	Archiv Orientalní (Prague)	CHYM	Chymia
ARSI	Annual Reports of the Smithsonian Institution	CHZ	Chemiker Zeitung
		CIB	China Institute Bulletin (New York)

584

CJ	China Journal of Science and Arts		JA	Journal asiatique
CKHW	Chung-Kuo Hsin Wên (= NCNA Bulletin)		JAAS	Journal of the Arms and Armour Soc.
			JAAR	Journ. Amer. Acad. Religion
CKKCSL	Chung-Kuo Kho Chi Shih Liao		JAEROS	Journ. Aeronautical Sciences
CMS	Chartered Mechanical Engineer		JAHIST	Journ. Asian History (International)
CR	China Review (HongKong and Shanghai)		JANS	Journ. Astronautical Sciences
			JAOS	Journal of the American Oriental Society
CRAS	Comptes Rendus de l'Académie des Sciences (Paris)		JATMOS	Journ. Atmospheric Science
CREC	China Reconstructs		JCE	Journal of Chemical Education
			JCR(M)	Journ. Chem. Research (Microfiches)
DCRI	Bulletin of the Deccan College Research Institute (Poona)		JCR(S)	Journ. Chem. Research (Synopses)
			JEPH	Journ. Ethnopharmacology
DHT	Documents pour l'histoire des Techniques (Paris)		JGLGA	Jahrbuch d. Gesellschaft. f. löthringen Geschichte u. Altertumskunde
DI	Die Islam		JHAS	Journ. Hist. Arabic Science
			JHPHARM	Journ. Hist. Pharmacol.
EAST	The East		JMATS	Journ. Materials Science
EG	Economic Geology		JOP	Journal of Physiology
EHR	Economic History Review		JOS/HK	Journal of Oriental Studies (Hongkong)
EMJ	Engineering and Mining Journal			
ESA	Eurasia Septentrionalis Antiqua		JOSA	Journ. Oriental Soc. Australia
ESCI	Engineering and Science		JPOS	Journal of the Peking Oriental Society
ETH	Ethnos		JRA	Journal of the Royal Artillery
			JRAES	Journal of the Royal Aeronautical Society (formerly Aeronautical Journal)
FCLT	Fu-Chien Lan Than (Fukien Forum)			
FEQ	Far Eastern Quarterly (continued as Journal of Asian Studies)		JRAI	Journal of the Royal Anthropological Institute
FSH	Fuji Chikurni Shokobutsu-en Hōkoku (Bull. Fuji Bamboo Bot. Gdn.)		JRAS	Journal of the Royal Asiatic Society
			JRAS/B	Journal of the (Royal) Asiatic Society of Bengal
GLAD	Gladius (Études sur les Armes Anciennes, etc.)		JRAS/HKB	Journal of the Hong Kong Branch of the Royal Asiatic Society
GR	Geographical Review		JRAS/KB	Journal (or Transactions) of the Korea Branch of the Royal Asiatic Society
GTIG	Geschichtsblätter f. Technik, Industrie u. Gewerbe			
GUNC	The Gun Collector (U.S.A.)		JRAS/M	Journal of the Malayan Branch of the Royal Asiatic Society
GUND	The Gun Digest		JRAS/NCB	Journal of the North China Branch of the Royal Asiatic Society
HBAS	Hauszeitschrift d. Badischen Anilin & Soda Fabrik AG		JRI	Journ. Royal Institution (London)
HEM	Hemisphere		JRUSI	Journ. Royal United Services Institution (London)
HHSTP	Hua Hsüeh Thung Pao (Chemical Intelligencer)		JS	Journal des Savants
HJAS	Harvard Journal of Asiatic Studies		JSCI	Journ. Soc. Chem. Industry
HKH	Hanguk Kwahaksa Hakhoechi (Journ. Korean Hist. of Sci. Soc.)		JSHS	Japanese Studies in the History of Science (Tokyo)
HMM	Harper's Monthly Magazine (New York)		JWCBRS	Journal of the West China Border Research Society
HORIZ	Horizon (New York)		JWH	Journal of World History (UNESCO)
HOSC	History of Science (annual)		JWM	Journ. Weather Modification
HOT	History of Technology (annual)			
			KGZ	Kahei Gakkai Zasshi (Journ. Soc. Technol. Arms and Ammunition Manufacture)
IAE	Internationales Archiv f. Ethnographie			
IAQ	Indian Antiquary			
IDSR	Interdisciplinary Science Reviews		KHCK	Kuo Hsüeh Chi Khan (Chinese Classical Quarterly)
IHQ	Indian Historical Quarterly			
ILN	Illustrated London News		KHNT	Kwartalnik Historii Nauki i Techniki (Warsaw)
ISIS	Isis			
ISL	Islam		KKPT	Kertas-Kertas Pengajian Tionghua (Papers on Chinese Studies, University of Malaya)
ISP/WSFK	I Shih Pao (Wên Shih Fu Khan); Literary Supplement of "Benefitting the Age" Periodical.			
			KKJL	Khao-Ku Jen Lei Hsüeh Chi-Khan

	(Bull. Dept. of Archaeol. and Anthropol. Univ. Thaiwan)	*NJKA*	*Neue Jahrbücher f. d. klass. Altertum, Geschichte, deutsch. Literatur u. f. Pädagogik*
KKTH	*Khao Ku Thung Hsün* (Archaeological Correspondent)	*NKKZ*	*Nihon Kagaku Koten Zensho*
KKWW	*Khao-Ku yü Wên-Wu Chi Khan* (Journ. Cultural Archaeology)	*NR*	*Numismatic Review*
KJ	*Korea Journal*	*NS*	*New Scientist*
KMJP	*Kuang Ming Jih Pao*	*NTM*	*Schriftenreihe f. Gesch. d. Naturwiss. Technik, u. Med.* (East Germ.)
KS	*Keleti Szemle* (Budapest)	*NYR*	*New Yorker*
KYHY	*Kung Yeh Huo Yao Hsieh Hui Chih* (Journ. of the Japanese Gunpowder Industry Association)	*NYTHP*	*Nan-Yang Ta-Hsüeh Hsüeh Pao* (Nanyang Univ. Journal, Singapore)
		OAZ	*Ostasiatische Zeitung*
LHHP	*Li Hsüeh Hsüeh Pao* (Journal of Physics)	*OLZ*	*Orientalische Literatur-Zeitung*
LI	*Listener* (B.B.C.)	*OPO*	*Oriente Poliano*
LIFE	*Life* (New York)	*OR*	*Oriens*
LN	*La Nature*	*ORA*	*Oriental Art*
LSCY	*Li Shih Yen Chiu* (Pkg.) J. Historical Research	*ORD*	*Ordnance*
		ORE	*Oriens Extremus*
MA	*Man*	*ORG*	*Organon* (Warsaw)
MAF	*Mémorial de l'Artillerie de France*	*OV*	*Orientalia Venetiana*
MAI/NEM	*Mémorial de l'Académie des Inscriptions et Belles-Lettres*, Paris (Notices et Extraits des MSS.)	*PAA*	*Progress in Astronautics and Aeronautics*
MART	*Memorial de Artilleria* (Madrid)	*PAAAS*	*Proceedings of the British Academy*
MAS/MPDS	*Mémoires de Mathématique et de Physique presentés à l'Académie Royale des Sciences* (Paris) *par Divers Sçavans et lus dans les Assemblées*	*PAE*	*Propellants and Explosives*
		PAR	*Parabola* (*Myth and the Quest for Meaning*)
		PFEH	*Papers on Far Eastern History* (Canberra)
MBLB	*May & Baker Laboratory Bulletin*	*PKCS*	*Pai Kho Chih Shih* (Peking)
MCHSAMUC	*Mémoires concernant l'Histoire, les Sciences, les Arts, les Moeurs et les Usages, des Chinois, par les Missionaires de Pékin* (Paris 1776–)	*PKR*	*Peking Review*
		PP	*Past and Present*
		PRAI	*Proc. Royal Artillery Institution* (contd. as JRA)
MC/TC	*Techniques et Civilisations* (originally *Métaux et Civilisations*)	*PRS*	*Proceedings of the Royal Society*
MDGNVO	*Mitteilungen d. deutsch. Gesellschaft f. Natur. u. Volkskunde Ostasiens*	*PTRS*	*Philosophical Transactions of the Royal Society*
		PVS	*Preuves* (Paris)
MEM	*Meteorological Magazine*	*QJRMS*	*Quarterly Journal of the Royal Meteorological Society*
MGK	*Manshū Gakuhō* (Dairen)		
MIE	*Mémoires de l'Institut d'Egypte* (Cairo)	*QJSLA*	*Quart. Journ. Science, Literature and the Arts* (cont. as *JRI, Journ. Roy. Inst.*)
MIMG	*Mining Magazine*		
MINGS	*Ming Studies*	*QSGNM*	*Quellen u. Studien z. Gesch. d. Naturwiss. u. d. Medizin*
MJ/UP	see *MUJ*		
MMI	*Mariner's Mirror*		
MMO	*Mammō* (Dairen)	*RBS*	*Revue Bibliographique de Sinologie*
MPCASP	*Mélanges de Phys. et Chim. de l'Acad. de St. Petersbourg*	*RC*	*Revista de Universidade de Coimbra* (Portugal)
MRAS/P	*Mémoires de l'Académie des Sciences* (Paris)	*RDI*	*Rivista d'Ingegneria*
		RDM	*Revue des Mines* (later *Revue Universelle des Mines*)
MS	*Monumenta Serica*		
MSOS	*Mitteilungen d. Seminar f. orientalischen Sprachen* (Berlin)	*REA*	*Revue des Études Anciennes*
		REG	*Revue des Études Grecques*
MUJ	*Museum Journal* (Philadelphia)	*RHSID*	*Revue d'Histoire de la Sidérurgie* (Nancy)
N	*Nature*	*ROC*	*Revue de l'Orient Chrétien*
NCR	*New China Review*	*ROL*	*Revue de l'Orient Latin*
NFR	*Nat. Fireworks Review*	*RQS*	*Revue des Questions Scientifiques* (Brussels)
NGM	*National Geographic Magazine*		

RRH	*Revue Roumaine d'Histoire* (Bucarest)	*TFIME*	*Trans. Federated Institution of Mining Engineers* (cont. as *TIME*)
RROWC	*Research Reports of the Okasaki Women's Junior College*, near Nagoya	*TFTC*	*Tung Fang Tsa Chih* (*Eastern Miscellany*)
RTPT	*Revista Transporturilor* (Rumania)	*TG/K*	*Tōhō Gakuhō, Kyōto* (*Kyoto Journal of Oriental Studies*)
SA	*Sinica* (originally *Chinesische Blätter f. Wissenschaft u. Kunst*)	*TGUOS*	*Transactions of the Glasgow University Oriental Society*
SAM	*Scientific American*	*TH*	*Thien Hsia Monthly* (Shanghai)
SARCH	*Sovietskaya Archaeologia*	*THSH*	*Ta Hsüeh Shêng Huo*
SBAW/PP & H	*Sitzungsberichte d. Bayerischen Akad. d. Wiss./Philos.-Philol. u. Hist. Kl.*	*TIME*	*Transactions of the Institution of Mining Engineers*
SCIS	*Sciences* (Paris)	*TJKHSYC*	*Tzu-Jan Khao-Hsüeh Shih Yen-Chiu*
SCSML	*Smith College Studies in Modern Languages*	*TJPCF*	*Tzu-Jan Pien Chêng Fa Thung Hsün* (Dialectics of Nature)
SE	*Stahl und Eisen*	*TK*	*Tōyōshi Kenkyū* (Researches in Oriental History)
SHHH	*Shih Hsüeh Hsiao Hsi*		
SHKS	*Shê Hui Kho-Hsüeh* (Chhinghua Journ. Soc. Sci.)	*TNS*	*Transactions of the Newcomen Society*
SHS	*Studia Historica Slovaca*	*TP*	*T'oung Pao* (*Archives concernant l'Histoire, les Langues, la Géographie, l'Ethnographie et les Arts de l'Asie Orientale*, Leiden)
SINRA	*Sinorama* (= *Kuang Hua*)		
SINT	*Sbornik Istorii Nauki i Techniki* (Moscow)	*TR*	*Technology Review*
SKSL	*Skrifter som udi det Kjøbenhavnske Selskab af Laerdoms...*	*TSHU*	*Tu Shu*
SMC	*Smithsonian* (*Institution*) *Miscellaneous Collections* (Quarterly Issue)	*UC/PAAA*	*Univ. of Calif./Publications in Amer. Arch. and Anth.*
SMITH	*The Smithsonian* (Magazine)	*UM*	*Universal Magazine of Knowledge and Pleasure*
SOF	*Studia Orientalia* (Fennica)		
SP	*Speculum*	*USNIP*	*United States Naval Institute Proceedings*
SPAW/PH	*Sitzungsber. d. preuss. Akad. d. Wissenschaften* (Phil.-Hist. Kl.)	*UZWKL*	*Universitas; Zeitschr. f. Wissenschaft, Kunst und Literatur*
SPCK	*Society for the Promotion of Christian Knowledge*	*VBGA*	*Verhandlungen d. Berliner Gesellschaft f. Anth., Eth. und Vorgeschichte* (see *ZFE*)
SPFL	*Spaceflight*		
SPMSE	*Sitzungsberichte d. physik. med. Soc. Erlangen*	*VH*	*Voprosy Historii* (Moscow)
		VIAT	*Viator*
SRFAOU	*Science Reports of the Faculty of Agriculture of Okayama University*	*VK*	*Vijnan Karmee*
		VS	*Variétés Sinologiques*
SUJCAH	*Suchow University Journ. Chinese Art History*		
SV	*Studi Veneziani*	*W*	*Weather*
STC	*Studi Colombiani*	*WW*	*Wên Wu*
SWAW/PH	*Sitzungsberichte d. k. Akad. d. Wissenschaften Wien* (Phil.-Hist. Klasse), Vienna	*WWTK*	*Wên Wu Tshan Khao Tzu Liao* (Reference Materials for History and Archaeology)
		WWTLTK	*Wên Wu Tzu Liao Tshung Khan*
TAIME	*Trans. Amer. Inst. Mining Engineers* (cont. as *TAIMME*)	*YCHP*	*Yenching Hsüeh Pao* (Yenching University Journal of Chinese Studies)
TAIMME	*Trans. Amer. Inst. Mining and Metallurgical Engineers*	*YJBM*	*Yale Journal of Biology and Medicine*
TBG	*Tijdschrift van het Bataavsche Genootschap van Kunsten en Wetenschappen* (later incorporated in *Tijdschrift voor Indische Taal-Land-, en Volkskunde*)	*ZAC*	*Zeitschr. f. angewandte chemie*
		ZDMG	*Zeitschrift d. deutsch. Morgenländischen Gesellschaft*
		ZFE	*Zeitschr. f. Ethnol.* (see *VBGA*)
TBGZ	*Tōkyō Butsuri Gakko Zasshi* (Journ. Tokyo College of Physics)	*ZGSS*	*Zeitschr. f. d. gesamte Schiess- und Sprengstoffwesen; Nitrocellulose*
TCC	*Tzu Chin Chhêng* (Forbidden City) Hongkong	*ZHWK*	*Zeitschrift. f. historische Wappenkunde* (cont. as *Zeitschr. f. hist. Wappen- und Kostumkunde*)
TCULT	*Technology and Culture*		

A. CHINESE AND JAPANESE BOOKS BEFORE +1800

Each entry gives particulars in the following order:
  (a) title, alphabetically arranged, with characters;
  (b) alternative title, if any;
  (c) translation of title;
  (d) cross-reference to closely related book, if any;
  (e) dynasty;
  (f) date as accurate as possible;
  (g) name of author or editor, with characters;
  (h) title of other book, if the text of the work now exists only incorporated therein; or, in special cases, references to sinological studies of it;
  (i) references to translations, if any, given by the name of the translator in Bibliography C;
  (j) notice of any index or concordance to the book if such a work exists;
  (k) reference to the number of the book in the *Tao Tsang* catalogue of Wieger (6), if applicable;
  (l) reference to the number of the book in the *San Tsang* (Tripitaka) catalogues of Nanjio (1) and Takakusu & Watanabe, if applicable.

Words which assist in the translation of titles are added in round brackets.

Alternative titles or explanatory additions to the titles are added in square brackets.

It will be remembered (p. 305 above) that in Chinese indexes words beginning *Chh-* are all listed together after *Ch-*, and *Hs-* after *H-*, but that this applies to initial words of titles only.

Where there are any differences between the entries in these bibliographies and those in Vols. 1–4, the information here given is to be taken as more correct.

An interim list of references to the editions used in the present work, and to the *tshung-shu* collections in which books are available, has been given in Vol. 4, pt. 3, pp. 913 ff., and is available as a separate brochure.

Chang Tzu-Yeh Tzhu Pu I 張子野詞補遺.
  Remaining Additional Poetical Works of Chang Tzu-Yeh.
  Sung, c. +1080.
  Cheng Tzu-Yeh 張子野.

Chao Chung Lu 昭忠錄.
  Book of Examples of Illustrious Loyalty.
  Yuan, c. +1290.
  Writer unknown.
  Cf. Balazs & Hervouet (1), p. 124.

Chao-Hua Hsien Chih 昭化縣志.
  Gazetteer of Chao-hua (in Szechuan).
  Chhing.
  Chang Shao-Ling (ed.) 張紹齡.
  Revised 1845, 1864.

Chao Hun 招魂.
  The Calling Back of the Soul [perhaps a ritual ode].
  Chou, c. −240.
  Attrib. Sung Yü 宋玉.
  Prob. by Ching Chhai (or Tsho) 景差.
  Tr. Hawkes (1).

Chen Chi 陣紀.
  Record of Army Drill and Tactics
  Ming, c. +1546.
  Ho Liang-Chhen 何艮臣.

Chen-La Fêng Thu Chi 眞臘風土記.
  Description of Cambodia.
  Yuan, +1297.
  Chou Ta-Kuan 周達觀.

Chen Yuan Miao Tao Yao Lüeh 眞元妙道要略.
  Classified Essentials of the Mysterious Tao of the True Origin (of Things) [alchemy and chemistry].

  Ascr. Chin, +3rd, but probably mostly Thang, +8th and +9th, at any rate after +7th as it quotes Li Chi.
  Attrib. Chêng Ssu-Yuan 鄭思遠.
  TT/917.

Chi Hsiao Hsin Shu 紀效新書.
  A New Treatise on Military and Naval Efficiency.
  Ming, +1560, pr. +1562, often repr.
  Chhi Chi-Kuang 戚繼光.

Chi Jan
  See *Chi Ni Tzu*.

Chia-Thai Kuei-Chi Chih 嘉泰會稽志.
  Records of Kuei-Chi (Shao-hsing in Chekiang) during the Chia-Thai reign-period (+1201 to +1205).
  Sung, not long after +1205.
  Shih Hsiu 施宿.

Chiang-Nan Ching Lüeh 江南經略.
  Military Strategies in Chiang-nan.
  Ming, +1566.
  Chêng Jo-Tsêng 鄭若曾.

Chien-Yen Tê-An Shou Yü Lu 建炎德安守禦錄.
  An Account of the Defence and Resistance of Tê-an (City) in the Chien-Yen reign-period [+1127 to +1132], (by the Sung against the J/Chin).
  Sung, +1172.
  Liu Hsün 劉荀.
  This book, now lost as such, was probably absorbed in that of the same name by Thang Tao (q.v.).
  Cf. Balazs & Hervouet (1), p. 237.

Chien-Yen Tê-An Shou Yü Lu 建炎德安守禦錄.

Chien-Yen Tê-An Shou Yü Lu (*cont.*)
An Account of the Defence and Resistance of Tê-an (City) in the Chien-Yen reign-period [+1127 to +1132], (by the Sung against the J/Chin).
Sung, +1193.
Original name of the book by Thang Tao which was combined with the *Shou Chhêng Lu* as chs. 3 and 4 in +1225 (q.v.).
Cf. Balazs & Hervouet (1), p. 237.

Chih Shêng Lu 制勝錄.
Records of the Rules for Victory.
Ming, *c.* +1430.
Writer unknown.
Now extant only in quotations.

Chin Phing Mei 金瓶梅.
Golden Lotus [novel].
(Cf. *Hsü Chin Phing Mei*) Ming.
Writer unknown.
Tr. Egerton (1), Kuhn (2) (Miall). See Hightower (1), p. 95.

Chin Shih 金史.
History of the Chin (Jurchen) Dynasty [+1115 to +1234].
Yuan, *c.* +1345.
Tho-Tho (Toktaga) 脫脫 & Ouyang Hsüan 歐陽玄.
Yin-Tê Index, no. 35.

Chin Shih Pu Wu Chiu Shu Chüeh 金石簿五九數訣.
Explanation of the Inventory of Metals and Minerals according to the Numbers Five (Earth) and Nine (Metal) [catalogue of substances with provenances, including some from foreign countries].
Thang, perhaps *c.* +670 (contains a story relating to +664).
Writer unknown.
TT/900.

Chin Thang Chieh Chu Shih-erh Chhou 金湯借箸十二籌.
Twelve Suggestions for Impregnable Defence.
Ming, *c.* +1630.
Li Phan 李盤.
The first two words of the title recall the phrase *chin chhêng thang chhih*, adamantine walls and scalding moats, hence impregnable.

Ching Chhu Sui Shih Chi 荊楚歲時記.
Annual Folk Customs of the States of Ching and Chhu [i.e. of the districts corresponding to those ancient States; Hupei, Hunan and Chiangsi].
Prob. Liang, *c.* +550, but perhaps partly Sui, *c.* +610.
Tsung Lin 宗懍.
See des Rotours (1), p. cii.

Ching-Khang Chhuan Hsin Lu 靖康傳信錄.
Record of Events in the Ching-Khang reign-period [+1126, year of the fall of Khaifêng to the Chin Tartars].
Sung, *c.* +1130.
Li Kang 李綱.

Chiu Kuo Chih 九國志.
Historical Memoir on the Nine States (Wu, Nan Thang, Wu-Yüeh Chhien Shu, Hou Shu, Tung Han, Nan Han, Min, Chhu and Pei Chhu, in the Wu Tai Period).
Sung, *c.* +1064.
Lu Chen 路振.

Chiu Ming Shu 救命書.
See *Hsiang Ping Chiu Ming Shu* and *Shou Chhêng Chiu Ming Shu*.

Chu Chia Shen Phin Tan Fa 諸家神品丹法.
Methods of the Various Schools for Magical Elixir Preparations (an alchemical anthology).
Sung.
Mêng Yao-Fu 孟要甫 (Hsüan Chen Tzu) 玄眞子 and others.
TT/911

Chu Shih 麈史.
Conversations on Historical Subjects (lit. while yak's-tail fly-whisks are waving).
Sung, pref. +1115.
Wang Tê-Chhen 王得臣.

Chuang Lou Chi 妝樓記.
Records of the Ornamental Pavilion.
Wu Tai or Sung, *c.* +960.
Chang Pi 張泌.

Chung Hsi Pien Yung Ping 中西邊用兵.
Military Practice on the Central and Western (Fronts)
Sung, *c.* +1150.
Fang Pao-Yuan 方寶元.
Now extant only in quotations.

Chung Thang Shih Chi 中堂事記.
Personal Recollections of Affairs at the Court [of Khubilai Khan, +1260 and +1261].
Yuan, *c.* +1280.
Wang Yün 王惲.
Cf. H. Franke (20, 26)

Chhao Yeh Chhien Yen 朝野僉言.
Narratives of the Court and the Country.
Sung, +1126.
Hsia Shao-Tsêng 夏少曾.
Now extant only in quotations.

Chhê Chhung Thu 車銃圖.
Illustrated Account of Muskets, Field Artillery and Mobile Shields, etc. (Appendix to *Wo Chhing Thun Thien Chhê Chhung I* and *Pei Pien Thun Thien Chhê Chhung I*, q.v.)
Ming, *c.* +1585.
Chao Shih-Chên 趙士禎.
(In *I Hai Chu Chhen, i chi*, pt. 1 藝海珠塵, 乙集).

Chhêng Chai Chi 誠齋集.
Collected Writings of (Yang) Chhêng-Chai (Yang Wan-Li).
Sung, *c.* +1200.
Yang Wan-Li 楊萬里.

Chhi Hsiu Lei Kao 七修類稿.
Seven Compilations of Classified Manuscripts.
Ming, +1555 to +1567.
Lang Ying 郎瑛.

Chhi Hsiu Lei Kao (*cont.*)
Cf. W. Franke (4), p. 106.

Chhi-tan Kuo Chih 契丹國志.
Memoir of the Liao (Chhi-tan Tartar
Kingdom).
Sung & Yuan, mid. +13th century.
Yeh Lung-Li 葉隆禮.

Chhi Tung Yeh Yü 齊東野語.
Rustic Talks in Eastern Chhi.
Sung, *c.* +1290.
Chou Mi 周密.

Chhien Hung Chia Kêng Chih Pao Chi Chhêng 鉛汞
甲庚至寶集成.
Complete Compendium on the Perfected Treasure of Lead, Mercury, Wood and Metal
[with illustrations of alchemical apparatus].
On the translation of this title, cf. p. 116. Has
been considered Thang +808; but perhaps
more probably Wu Tai or Sung. Cf. p. 116.
Chao Nai-An 趙耐菴.
TT/912.

Chhien-Thang I Shih 錢塘遺事.
Memorabilia of Hangchow and the Chhien-
thang River.
Yuan.
Liu I-Chhing 劉一清.

Chhing Hsiang Tsa Chi 靑箱雜記.
Miscellaneous Record on Green Bamboo
Tablets.
Sung, *c.* +1070.
Wu Chhu-Hou 吳處厚.

Chhing Shih Kao 清史稿.
Draft History of the Chhing Dynasty.
See Chao Erh-Hsün & Kho Shao-Min.

Chhing-Tai Chhou-Pan I-Wu Shih-Mo 清代籌辦
夷務始末.
See Anon. (*212*).

Chhiu Chien Hsien-sêng Ta Chhüan Wên Chi 秋澗先
生大全文集.
Complete Literary Works of Mr Autumn-
Torrents [Wang Yün].
Yuan, *c.* +1304.
Wang Yün 王惲.
Cf. H. Franke (20, 26).

Chhiu Shêng Khu Hai 求生苦海.
Saving Souls from Hell.
Chhing, +18th.
Writer unknown.

Chhou Hai Thu Pien 籌海圖編.
Illustrated Seaboard Strategy and Tactics.
Ming, +1562. Repr. +1572, +1592, +1624, etc.
Chêng Jo-Tsêng 鄭若曾.
Cf. W. Franke (4), p. 223; Goodrich & Fang
Chao-Ying (1), p. 204.

Chhu Tzhu 楚辭.
Elegies of Chhu (State) [or, Songs of the South].
Chou, *c.* −300 (with Han additions).
Chhü Yuan 屈原 (& Chia I 賈誼 Yen
Chi 嚴忌 Sung Yü 宋玉 Huainan
Hsiao-Shan 淮南小山 *et al.*).
Partial tr. Waley (23); tr. Hawkes (1).

Fan Tzu Chi Jan 范子計然.
See *Chi Ni Tzu*.

Fêng Shen Pang 封神榜.
Pass-Lists of the Deified Heroes.
Popular form of the title *Fêng Shen Yen I*, q.v.

Fêng Shen Yen I 封神演義.
Stories of the Promotions of the Martial Genii
[novel].
Ming.
Hsü Chung-Lin 許仲琳.
Tr. Grabe (1).

Fêng Su Thung I 風俗通義.
The Meaning of Popular Traditions and
Customs.
H/Han, +175.
Ying Shao 應劭.
Chung-Fa Index, no. 3.

Fu Hung Thu 伏汞圖.
Illustrated Manual on the Subduing of Mercury.
Sui, Thang, Wu Tai, J/Chin (or possibly, in
some parts, Ming).
Shêng Hsüan Tzu 昇玄子.
Survives now only in quotations.

Fu kien Thung Chih 福建通志.
Gazetteer of Fukien Province.
Chhing, completed 1833, pr. 1867.
See Chhen Shou-Chhi (1), (ed.).

Hachiman Gudō-Kun (or -*Ki*) 八幡愚童訓(記).
Tales of the God of War told to the Simple [a
military history, including details of the Mongol invasions of +1274 and +1281].
Japan, late +14th or somewhat earlier ed. used
dates from between +1469 and +1486.
Writer unknown.
In *Gunsho Ruiji* collection (ch. 13, p. 328)
羣書類從.

Hai-Chhiu Fu Hou Hsü 海鰌賦後序.
Postface to the Rhapsodic Ode on the 'Sea-Eel'
(Warships) [and their role at the Battle of
Tshai-Shih, +1161].
Sung, *c.* +1170
Yang Wan-Li 楊萬里.
In *Chhêng-Chai Chi*, ch. 44, pp. 6 *b* ff.

Hai Fang Tsung Lun 海防總論.
A General Discourse on Coastal Defence.
Ming, before +1621.
Chou Hung-Tsu 周宏祖.

Hai Kuo Thu Chih 海國志.
See Wei Yuan & Lin Tsê-Hsü (1).

Ho-Hsien Thuan Lien Thiao Kuei 賀縣團練條規.
Rules for Training the Militia Bands at Ho-
hsien.
Ming, *c.* +1615.
Author uncertain.

Hōjō Godai-Ki 北條五代記.
Chronicles of the Hōjō Family through Five
Generations.
Japan, *c.* +1600.
Writer unknown.
In *Shiseki Shūran* 史籍集覽.

Hōjō Godai-Ki (*cont.*)
(Collection of Historical Materials).
Ed. Kondō Heijō 近藤瓶城.
3rd. ed. Kondō Shuppan-bu, Tokyo, 1907.

Honchō Gunkikō 本朝軍器考.
Investigation of the Military Weapons and
Machines of the Present Dynasty.
Japan preface +1709, postface +1722, printed
+1737.
Arai Hakuseki 新井白石.
Autobiography tr. J. Ackroyd (1).

Hou Han Shu 後漢書.
History of the later Han Dynasty [+25 to
+220].
L/Sung, +450.
Fan Yeh 范曄.
The monograph chapters by Ssuma Piao
司馬彪 (d. +305), with commentary by
Liu Chao 劉昭 (*c.*+510), who first incor-
porated them in the work.
A few chs. tr. Chavannes (6, 16); Pfizmaier (52,
53).
Yin-Tê Index, no. 41.

Hu Chhien Ching 虎鈐經.
Tiger Seal Manual [military encyclopaedia]
Sung, begun +962, finished +1004.
Hsü Tung 許洞.
Cf. Balazs & Hervouet (1), p. 236.

Hu Khou Yü Shêng Chi 虎口餘生記.
Record of Life Regained out of the Tiger's
Mouth.
Chhing, +1645.
Pien Ta-Shou 邊大綬.
Cf. Hummel (2), p. 741.

Hua I Hua Mu Niao Shou Chen Wan Khao 華夷花
木鳥獸珍玩考.
A Useful Examination of the Flowers, Trees,
Birds and Beasts found among the Chinese
and neighbouring Peoples (lit. Barbarians).
Ming, +1581.
Shen Mou-Kuan 慎懋官.
WY/135.

Huang Chhao Ma Chêng Chi 皇朝馬政記.
Record of Army Remount Organisation in the
Ming Dynasty.
Ming, +1596.
Yang Shih-Chhiao 楊時喬.

Huang Hsiao Tzu Wan Li Chi Chhêng 黃孝子萬里
紀程.
Memories of the Thousand-Mile Peregrinations
of a Filial Son named Huang.
Ming and Chhing, pref. of +1643, but not
finished till *c.* +1652.
Huang Hsiang-Chien 黃向堅.

Huang Ming Ching Shih Shih Yung Pien 皇明經世
實用編.
Political Encyclopaedia of Ming Dynasty
Materials (down to the Wan-Li Reign-Period,
including border defence and maritime
defence); or, Imperial Ming Handbook of
Practical Statesmanship.

Ming, +1603.
Fêng Ying-Ching (ed.) 馮應京.
Cf. W. Franke (4), p. 195; *GF*/1141.

Huang Ming Shih Fa Lu 皇明世法錄.
Political Encyclopaedia of the Ming Dynasty
(containing imperial edicts, military history,
and treatises on astronomy and calendar,
music and ceremonies, financial administration, economics, agriculture, communications,
etc.).
Ming, +1630, pr. after +1632.
Chhen Jen-Hsi (ed.) 陳仁錫.
Cf. W Franke (4), p. 196; *WY*/420; *GF*/162.

Huang Ti Chiu Ting Shen Tan Ching Chüeh 黃帝九
鼎神丹經訣.
The Yellow Emperor's Canon of the Nine-
Vessel spiritual Elixir, with Explanations.
Early Thang or early Sung, but incorporating as
ch. 1 a canonical work probably of the +2nd
century
Writer unknown
TT/878. Also, abridged, in *YCCC*, ch. 67, pp.
1 a ff.

Hui-An Hsien-sêng Chu Wê Kung Chi 晦菴先生朱
文公集.
Collected Writings of Chu Hsi (lit. Mr. (Chu)
Hui-An's Records of the Ven. Chu Wên
Kung).
Sung, *c.* +1200.
Chu Hsi 朱熹.

Hui-chán Ryōsa 彙纂麗史.
Collected, Compiled and Edited History of
Korea, especially the Koryō Kingdom.
Korea, +18th.
Hong Yeha 洪汝河.
Courant (1), no. 1863.

Huo Chhê Chen Thu Shuo 火車陣圖說.
Illustrated Accounts of the Formations in which
Mobile Shields should be used with Guns and
Cannon.
Ming, prob. +16th.
Chhen Phei 陳裴.
Cf. Lu Ta-Chieh (1), p. 138.

Huo Chhi Chen Chüeh Chieh Chêng 火器眞訣解證.
Analytical Explanations of Firearms and In-
structions for using them.
Chhing.
Shen Shan-Chêng 沈善蒸.
Now extant only in quotations.
Cf. Lu Ta-Chieh (1), p. 164, (2), p. 19.

Huo Chhi Lüeh Shuo 火器略說.
(= *Tshao Shêng Yao Lan*)
Classified Explanations of Firearms.
Chhing.
Wang Ta-Chhüan 王達權 & Wang
Thao 王韜.
Cf. Lu Ta-Chieh (1), p. 161, (2), p. 18.
Now extant only in quotations.

Huo Chhi Ta Chhüan 火器大全.
Everything one needs to know about Gunpowder Weapons.

Huo Chhi Ta Chhüan (*cont.*)
  Date unknown.
  Writer unknown.
  Title known only from *Tu Shu Min Chhiu Chi*, q.v.
  Cf. Lu Ta-Chieh (*1*), p. 169.
Huo Chhi Thu 火器圖.
  Illustrated Account of Gunpowder Weapons
    and Firearms.
  Running-head title of the Hsiang-yang edition of
    Huo Lung Ching (q.v.).
Huo Chhi Thu 火器圖.
  Illustrated Account of Gunpowder Weapons
    and Firearms.
  Ming, *c.* +1620.
  Ku Pin 顧斌.
  Cf. Lu Ta-Chieh (*1*), p. 128.
Huo Chhi Thu Shuo 火器圖說.
  Illustrated Account of Fire- (and Gunpowder-)
    Weapons.
  Ming, prob. +16th.
  Huang Ying-Chia 黃應甲.
  Lu Ta-chieh (*1*), p. 122.
Huo Kung Chen Fa 火攻陣法.
  Troop Formations for Combat with Firearms.
  Title of the book which was given to Chiao Yü,
    the writer of the *Huo Lung Ching* (pt. 1) by the
    old Taoist of Thien-thai Shan.
  Cf. *CCL* (7), p. 86.
Huo Kung Chen Fa 火攻陣法.
  Tactical Formations for Attack by Fire- (includ-
    ing Gunpowder-) Weapons.
  Ming.
  Writer unknown.
  Cf. Lu Ta-Chieh (*1*), p. 149.
Huo Kung Chhieh Yao 火攻挈要.
  [or, Tsê Kho Lu 則克錄].
  Essentials of Gunnery.
  [or, Book of Instantaneous Victory].
  Ming, +1643.
  Chiao Hsü 焦勗.
  With the collaboration of Thang Jo-Wang (J.A.
    Schall von Bell) 湯若望.
  Bernard-Maître (18), no. 334; Pelliot (55).
Huo Kung Pei Yao 火攻備要.
  Essential Knowledge for the Making of Gun-
    powder Weapons.
  Alt. title of Pt. 1 of the *Huo Lung Ching*, q.v.
Huo Kung [*Wên*] *Ta* 火攻問答.
  Answers (to Questions) on Fire-Weapons and
    Firearms.
  Ming, *c.* +1598.
  Wang Ming-Hao 王鳴鶴.
  In *Huang Ming Ching Shih Shih Yung Pien*,
    ch. 16 (p. 1287).
Huo Lung Ching 火龍經.
  The Fire-Drake (Artillery) Manual (of Gunpow-
    der Weapons).
  Ming, +1412.
  Chiao Yü 焦玉.
  The first part of this book, in three sections,
    is attributed fancifully to Chuko Wu-ou
    (i.e. Chuko Liang), and Liu Chi 劉基

(+1311/+1375) appears as co-editor, really
  perhaps co-author.
The second part, also in three sections, is
  attributed to Liu Chi alone, but edited,
  probably written, by Mao Hsi-Ping 毛希秉
  in +1632.
The third part, in two sections, is by Mao Yuan-
  I 茅元儀 (fl. +1628) and edited by Chuko
  Kuang-Jung 諸葛光榮. whose preface is of
  +1644, Fang Yuan-Chuang 方元壯 &
  Chung Fu-Wu 鍾伏武.
This work should be considered a main nucleus
  with two supplements, summarising the development of successive gunpowder weapons
  between about +1280 and +1644. The first
  part, i.e. the book itself, is the work of Chiao
  Yü, who had been a leading artillery officer in
  the army of Chu Yuan-Chang which finally
  conquered China for the Ming dynasty in
  +1367.
Huo Lung Ching Chhüan Chi 火龍經全集.
  Complete Materials of the 'Fire Drake Manual'
    (Nanyang edition).
  = *Huo Kung Pei Yao*, q.v.
Huo Lung Shen Chhi Chen Fa 火龍神器陣法.
  Fire-Drake Manual of Military Formations
    using Magically (Efficacious) Weapons
    (i.e. Muskets).
  Date uncertain; a +16th century MS.
  Perhaps an early version of *Huo Lung Ching* (q.v.)
    copied and re-copied.
Huo Lung Shen Chhi Thu Fa 火龍神器圖法.
  Fire-Drake Illustrated Technology of Magically
    (Efficacious) Weapons.
  Yuan, perhaps *c.* +1330.
  Writer unknown.
  Listed in the *Liao, Chin, Yuan, I Wên Chih* of Lu
    Wên-Chhao, *c.* +1770.
  Possibly the earliest form of the *Huo Lung Ching*,
    q.v.
  Now extant only in quotations.
  Cf. Lu Ta-chieh (*1*), p. 108.
Huo Lung Shen Chhi Yao Fa Pien 火龍神器藥
    法編.
  Fire-Drake Book of Magically (Efficacious)
    Weapons, with the Method of Making Gun-
    powder.
  Date uncertain, perhaps Yuan.
  Writer unknown.
  MS in the Library of the History of Science In-
    stitute, Academia Sinica, Peking, with illus-
    trations more delicate and precise than those
    in any printed edition of the *Huo Lung Ching*, of
    which it may represent an early version.
Huo Lung Wan Shêng Shen Yao Thu 火龍萬勝神藥圖.
  Illustrated Fire-Drake Technology for a Myriad
    Victories using the Magically (Efficacious)
    Gunpowder.
  Date unknown.
  Writer unknown.
  Title known only from *Tu Shu Min Chhiu Chi*, q.v.
  Cf. Lu Ta-Chieh (*1*), p. 169.

Huo Yao Fu 火藥賦.
Rhapsodic Ode (or, Poetical Essay) on Gunpowder.
Ming, *c.* +1620.
Mao Yuan-I 茅元儀.
In *TSCC*, Jung chêng tien, ch. 96, i wên i, p. 2*a*, *b*, 3*a*.

Huo Yao Miao Phin 火藥妙品.
The Wonderful Uses of Gunpowder.
Ming.
Writer unknown.
Cf. Lu Ta-Chieh (*1*), p. 149.

Hsi Chhi Tshung Hua 西溪叢話(語).
(*SKCS* has *Yü.*)
Western Pool Collected Remarks.
Sung, *c.* +1150.
Yao Khuan 姚寬.

Hsi Chou Yen Phu 歙州硯譜.
Hsichow Inkstone Record.
Sung, +1066.
Thang Chi 唐積.

Hsi Hu Chih Yü 西湖志餘.
Additional Records of the Traditions of West Lake (at Hangchow).
Ming, *c.* +1570.
Thien I-Hêng 田藝蘅.

Hsi Hu Erh Chi 西湖二集.
Second Collection of Materials about West Lake [at Hangchow, and the neighbourhood].
Ming, *c.* +1620.
Chou Chhing-Yuan 周清源.

Hsi-Yang Huo Kung Thu Shuo 西洋火攻圖說.
Illustrated Treatise on European Gunnery.
Ming, before +1625.
Chang Tao 張燾 & Sun Hsüeh-Shih 孫學詩.
Cf. Pelliot (55).
Now extant only in quotations.

Hsi Yuan Wên Chien Lu 西園聞見錄.
Things Seen and Heard in the Western Garden (the Imperial Library), [a work of notes for the history of the Ming, +1368 to +1620].
Ming, +1627; first printed 1940.
Chang Hsüan 張萱.
Cf. Goodrich & Fang Chao-Ying (*1*), p. 79.

Hsiang Ping Chiu Ming Shu 鄉兵救命書.
On Saving the Situation by (the Raising of) Militia.
Ming, +1607.
Lü Khun 呂坤.
Cf. Goodrich & Fang Chao-Ying (*1*), p. 1006.

Hsiang-Yang Shou Chhêng Lu 襄陽守城錄.
An Account of the Defence of Hsiang-yang (City) [+1206 to +1207], (by the Sung against the J/Chin).
Sung, *c.* +1210.
Chao Wan-Nien 趙萬年.
This siege was not by the Mongols, as in the more famous one of +1268/+1273.
Cf. Balazs & Hervouet (*1*), p. 95.

Hsin-Ssu Chhi Chhi Lu 辛巳泣蘄錄.
The Sorrowful Record of (the Siege of) Chhi

(-chou) in the Hsin-Ssu Year (+1221), (by the Chin Tartars).
Sung, *c.* +1230.
Chao Yü-Jung 趙與裒.

Hsin Wu Tai Shih 新五代史.
New History of the Five Dynasties [+907 to +959].
Sung, *c.* +1070.
Ouyang Hsiu 歐陽修.
For translations of passages see the index of Frankel (*1*).

Hsin Yuan Shih 新元史.
See Kho Shao-Min (*1*).

Hsing Chün Hsü Chih 行軍須知.
What an Army Commander in the Field should Know.
Sung, *c.* +1230; repr. +1410, +1439.
Writer unknown.
Preface by Li Chin (Ming ed.) 李進.
Appended to the Ming ed. of *Wu Ching Tsung Yao, Hou Chi.*
Cf. Fêng Chia-Shêng (*1*), p. 61.

Hsü Chin Phing Mei 續金瓶梅.
Golden Lotus, Continued [novel] (cf. *Chin Phing Mei*).
Chhing, +17th century.
Tzu Yang Tao-Jen 紫陽道人.
Tr. Kulm (*1*).

Hsü Hou Han Shu 續後漢書.
Supplement to the History of the Later Han.
Sung.
Hsiao Chhang 蕭常.

Hsü I Chien Chih 續夷堅志.
More Strange Stories from I-Chien.
J/Chin, *c.* +1240.
Yuan Hao-Wên 元好問.

Hsü Sung Chung Hsing Pien Nien Tzu Chih Thung Chien 續宋中興編年資治通鑑.
Continuation of the 'Mirror of History for Aid in Government' for the Sung Dynasty from its Restoration onwards [i.e. Southern Sung from +1126].
Sung, *c.* +1250.
Liu Shih-Chü 劉時舉.
Cf. Balazs & Hervouet (*1*), p. 77.

Hsü Sung Pien Nien Tzu Chih Thung Chien 續宋編年資治通鑑.
Alt. title of *Hsü Sung Chung Hsing Pien Nien Tzu Chih Thung Chien*, q.v.

Hsü Tzu Chih Thung Chien Chhang Pien 續資治通鑑長編.
Continuation of the *Comprehensive Mirror (of History) for Aid in Government* [+960 to +1126].
Sung, +1183.
Li Tao 李燾.

Hsü Wên Hsien Thung Khao 續文獻通考.
Continuation of the *Comprehensive Study of (the History of Civilisation)* (cf. *Wên Hsien Thung Khao* and *Chhin Ting Hsü Wên Hsien Thung Khao*).
Ming, +1586; pr. +1603.
Ed. Wang Chhi 王圻.

Hsüan Kuai Hsü Lu 玄怪續錄.
The *Record of Things Dark and Strange*, continued.
Thang.
Li Fu-Yen 李復言.

I Hai Chu Chhen 藝海珠塵.
Pearls from the Dust; a Collection (of Tractates)
from the Ocean of Artistry [*a tshung-shu*].
Chhing, *c.* +1760.
Ed. Wu Shêng-Lan 吳省蘭.

Inatomi-ryu Teppō Densho 稻富流鐵砲傳書.
Record of Matchlock Muskets current in the
Inatomi Family.
Japan, +1595; never printed.
Nagasawa Shigetsune 長澤七右衛 for
Kawakami Mosuke 河上茂介殿.
An MS. of +1607 is in the New York Public Lib-
rary (Spencer Colln. no. 53).

Kai Wên Lu 該聞錄.
Things Heard Worthy of Record.
Sung, *c.* +990.
Li Thien 李畋.

Kai Yü Tshung Khao 陔餘叢考.
Miscellaneous Notes made while attending his
aged Mother.
Chhing, +1790.
Chao I 趙翼.

Kaisan-ki 改算記.
Book of Improved Mathematics.
Japan, +1659.
Yamada Shigemasa 山田重正.

Kaisan-ki Kōmoku 改算記綱目.
Comprehensive Summary of Integration
[early calculus].
Japan, +1687.
Mochinaga, Toyotsugu 持永豐次 &
Ōhashi, Takusei 大橋宅清.

Keisei Hisaku 經世秘策.
A Secret Plan for Managing the Country.
Japan (Yedo) +1798, pr. after 1821.
Honda Toshiski 本多利明.
Cf. Keene (1).

Khai-Hsi Tê-An Shou Chhêng Lu 開禧德安守城錄.
An Account of the Defence of Tê-an (City) in the
Khai-Hsi reign-period [+1206 to +1207], (by
the Sung against the J/Chin).
Sung, +1224.
Wang Chih-Yuan 王致遠.
Tr. K. Hana (1).

Kho Chai Tsa Kao, Hsü Kao Hou 可齋雜槀,
續稿後.
Miscellaneous Matters recorded in the Ability
Studio, Second Addendum.
Sung, *c.* +1265.
Li Tsêng-Po 李曾伯.

Khua Ao Chi 跨鼇集.
Collected Memorabilia of Mr Khua-Ao.
Sung, *c.* +1100.
Li Hsin 李新 (Khua Ao chü
shih 跨鼇居士).
He called himself the Recluse of the Atlas-

bestriding stone steles, because such monuments are generally placed upon sculptures of
tortoises, and one of these was in mythology
the supporter of the world, hence a symbol of
longevity.

Kikai Kanran.
See Aoji Rinsō (1).

Ko Chih Ku Wei.
See Wang Jen-Chün (1).

Ko Wu Hsü Chih 格物須知.
What One should Know about Natural
Phenomena.
Chhing, +18th.
Chu Pên-Chung 朱本中.

Kōrai Sensenki 高麗舩戰記.
A Record of the Sea-Fights against Korea.
Japan, +1592.
Soto-oka Jinjaimon 外岡甚左衛門.
MS preserved in the Nabeshima family, and
now in the Library of Kyushu University.
Cf. Pak Hae-ill (2).

Koryŏ-sa 高麗史.
History of the Koryō Kingdom [+918 to
+1392].
Korea, first compiled in +1395; oldest extant
version commissioned +1445, completed
+1451.
Ed. Chŏng Inji 鄭麟趾.
Courant (1), no. 1846.

Ku Chin Shuo Hai 古今說海.
Sea of Sayings Old and New [florilegium].
Ming, +1544.
Lu Chi (ed.) 陸楫.

Kuang Po Wu Chih 廣博物志.
Enlargement of the *Records of the Investigation of
Things* (by Chang Hua, *c.* +290).
Ming, +1607.
Tung Ssu-Chang 董斯張.

Kuang-Yang Tsa Chi 廣陽雜記.
Collected Miscellanea of Master Kuang-Yang
(Liu Hsien-Thing).
Chhing, *c.* +1695.
Liu Hsien-Thing 劉獻廷.

Kuei Chhien Chih 歸潛志.
On Returning to a Life of Obscurity.
J/Chin, +1235.
Liu Chhi 劉祁.

Kuei Hsin Tsa Chih 癸辛雜識.
Miscellaneous Information from Kuei-Hsin
Street (in Hangchow).
Sung, late +13th-century, perhaps not finished
before +1308.
Chou Mi 周密.
See des Rotours (1), p. cxii; H. Franke (14).

Kuei Thien Shih Hua 歸田詩話.
Poems of Return to Farm and Tillage.
Ming, +1425.
Chhü Yu 瞿佑.

Kuei Tung 鬼董.
The Control of Spirits.
Sung, prob. *c.* +1185; pub. +1218 or later.
Mr Shen 沈氏.

Kukcho Orye-ŭi 國朝五禮儀.
Instruments for the Five Ceremonies of the (Ko-
rean) Court.
Korea (Chosŏn), +1474.
Sin Sukju 申叔舟 & Chŏng Ch'ăk 鄭陟.
Cf. Trollope (1), p. 21; Courant (1), no. 1047.

Kukcho Pogam 國朝寶鑑.
The Precious Dynastic Mirror [official history of
the Yi Dynasty, +1392 to 1910].
Korea (Chosŏn), begun *c.* +1460, commissioned
by King Sejo.
Kwŏn Nam 權擥 and many subsequent wri-
ters.
Courant (1), no. 1894, 1897.

Kukcho Sok Orye-ŭi 國朝續五禮儀.
A Continuation of the *Instruments for the Five Cere-
monies of the* (Korean) *Court*.
Korea (Chosŏn) +1744.
Ed. Courant (1), no. 1047.

Kukcho Sok Orye-ŭi Po 國朝續五禮儀補.
An Extension of the *Continuation of the Instruments
for the Five Ceremonies of the* (Korean) *Court*.
Korea (Chosŏn), +1751.
Ed. Courant (1); no. 1047.

Kung Khuei Chi 攻媿集.
Bashfulness Overcome; Recollections of My Life
and Times.
Sung, *c.* +1210.
Lou Yo 樓鑰.

Kung Pu Chhang Khu Hsü Chih 工部廠庫須知.
What should be known (to officials) about the
Factories, Workshops and Storehouses of the
Ministry of Works.
Ming, +1615.
Ho-Shih-Chin 何士晉.

Kung-Sha Hsiao Chung Chi 公沙効忠紀.
Eulogy of the Loyal and Gallant Gonçalvo
[Teixeira-Correa, Captain of Artillery in the
Chinese Service].
Ming, +1633.
Lu Jo-Han (João Rodrigues, S. J.) 陸若漢.
Pfister (1), p. 25* (add.)

Kuo Chhao Ming Chhen Shih Lüeh 國朝名臣事略.
Biographies of (47) Famous Statesmen and
Generals of the Present Dynasty (Yuan)
Yuan, *c.* +1360.
Su Thien-Chio 蘇天爵.
Cf. H. Franke (14), p. 119

Kuo Chhao Wên Lei 國朝文類.
Classified Prose of the Present Dynasty
(Yuan).
Yuan, *c.* +1340.
Ed. Satula (Thien Hsi) 薩都拉(天錫)
& Su Thien-Chio 蘇天爵.
Cf. H. Franke (14), p. 119.

Kuo Chhao Wu Li I.
See *Kukcho Orye-ŭi.*

Lang Chi Tshung Than 浪跡叢談.
See Liang Chang-Chü (1).

Lao Hsüeh An Pi Chi 老學庵筆記.
Notes from the Hall of Learned Old Age.

Sung, *c.* +1190.
Lu Yu 陸游.

Li Shao Phien 蠡勺編.
Measuring the Ocean with a Calabash-Ladle
[title taken from a diatribe against narrow-
minded views in the biography of Tungfang
Shuo in *CHS*].
Chhing, *c.* +1799.
Ling Yang Tsao 淩揚藻.

Li Wei Kung Wên Tui 李衛公問對.
The Answers of Li Wei Kung to Questions (of
the emperor Thang Thai Tsung) (on the Art
of War).
Supposedly Thang, but more probably pro-
duced in the Sung, +11th.
Writer unknown.
Perhaps composed by Juan I 阮逸.

Liao, Chin, Yuan I Wên Chih 遼金元藝文志.
Bibliography of the Liao, J/Chin and Yuan
Dynasties [the official histories of which lack *i
wên chih*].
Chhing.
Huang Yü-Chi (+1629 to +1691) 黃虞稷.
Ni Tshan (+1704 to 1841) & 倪燦.
Chhien Ta-Hsin (+1728 to 1804) 錢大昕.
and others.

Liao Shih 遼史.
History of the Liao (Chhi-tan) Dynasty [+916
to +1125].
Yuan, +1343 to +1345.
Tho-Tho (Toktaga) 脫脫 & Ouyang Hsüan
歐陽玄.
Partial tr. Wittfogel, Fêng Chia-Shêng *et al.*
Yin-Tê Index, no. 35.

Lieh Hsien Chuan 列仙傳.
Lives of Famous Immortals (cf. *Shen Hsien
Chuan*).
Chin, +3rd or +4th century, though certain
parts date from about −35 and shortly after
+167.
Attrib. Liu Hsiang 劉向.
Tr. Kaltenmark (2).

Lien Ping Shih Chi 練兵實紀.
Treatise on Military Training.
Ming, +1568; pr. +1571, often repr.
Chhi Chi-Kuang 戚繼光.

Lien Ping Shih Chi Tsa Chi 練兵實紀雜集.
Miscellaneous Records concerning Military
Training (and Equipment) [the addendum to
Lien Ping Shih Chi, q.v., in 6 chs. following the
9 chs. of the main work].
Ming, +1568; pr. +1571.
Chhi Chi-Kuang 戚繼光.

Lien Yüeh Huo Chhi Chen Chi 練閱火器陣紀.
An Examination of Training in the Use of Gun-
powder Weapons, Cannon and Catapults.
Chhing, +1696.
Hsüeh Hsi 薛熙.

Liu Pin-Kho Wên Chi 劉賓客文集.
Literary Records of the Imperial Tutor Liu.
Thang, after +842.
Liu Yü-Hsi 劉禹錫.

Liu Po-Wên Chien Hsien Phing Chê Chung 劉伯溫薦
    賢平浙中.
    The Pacification of central Chekiang by the
    Able Officers recommended by (Commander)
    Liu Po-Wên [Liu Chi, in +1340 to + 1350,
    acting as a Yuan officer against the rebels and
    pirates of the region.]
    Ch. 17 of Chou Chhing-Yuan's *Hsi Hu Erh Chi*,
    q.v.
Liu Thao 六韜.
    The Six Quivers [treatise on the art of war].
    H/Han, +2nd century, incorporating material
    as early as the −3rd.
    Writer unknown.
    See Haloun (5); L. Giles (11).
Lo-Yang Chhieh Lan Chi 洛陽伽藍記.
    (or '*Loyang Ka-Lan Chi*'; *sêng ka-lan* transliterat-
    ing *sanghārāma*).
    Description of the Buddhist Temples and
    Monasteries at Loyang.
    N/Wei, *c.* +547.
    Yang Hsüan-Chih 楊衒之.
Lü Li Yuan Yuan 律曆淵源.
    Calendrical and Acoustic, Ocean of Calculations
    (compiled by Imperial Order) [includes *Li
    Hsiang Khao Chhêng, Shu Li Ching Yün*, Lü Lü
    Chêng I, q.v.].
    Chhing, +1723; printing probably not finished
    before +1730.
    Ed. Mei Ku-Chhêng 梅穀成 & Ho Kuo-
    Tsung 何國宗.
    Cf. Hummel (2), p. 285; Wylie (1), pp. 96 ff.
Lü Lü Chêng I 律呂正義.
    Collected Basic Principles of Music (compiled
    by Imperial Order) [part of *Lü Li Yuan Yuan*,
    q.v.].
    Chhing, +1713 (+1723).
    Ed. Mei Ku-Chhêng 梅穀成 & Ho Kuo-
    Tsung 何國宗.
    Cf. Hummel (2), p. 285.
Lun Hêng 論衡.
    Discourses Weighed in the Balance.
    H/Han, +82 or +83.
    Wang Chhung 王充.
    Tr. Forke (4); cf. Leslie (3).
    Chung-Fa Index, no. 1.
Lung Hu Huan Tan Chüeh 龍虎還丹訣.
    Explanation of the Dragon-and-Tiger Cyclically
    Transformed Elixir.
    Wu Tai, Sung, or later.
    Chin Ling Tzu 金陵子.
    TT/902.

Man-Chou Shih Lu Thu 滿州實錄圖.
    Veritable Records of the Manchus, with Illus-
    trations [depicting the martial exploits of
    Nurhachi, Thai Tsu of the Chhing, d. +1626].
    Alt. title of *Thai Tsu Shih Lu Thu*, q.v.
Man Shu 蠻書.
    Book of the Barbarians [itineraries].
    Thang, *c.* +862.
    Fan Chho 樊綽.

Mêng Hua Lu
    See *Tung Ching Mêng Hua Lu*.
Mêng Liang Lu 夢粱錄.
    Dreaming of the Capital while the Rice is Cook-
    ing [description of Hangchow towards the end
    of the Sung].
    Sung, +1275.
    Wu Tzu-Mu 吳自牧.
Mōko Shūrai Ekotoba 蒙古襲來繪詞.
    Illustrated Narrative of the Mongol Invasions
    (of Japan) [+1274 and +1281].
    Japan, +1293; facsim. ed. ed. Kubota Beisan
    (Kubota Yonenari), Tokyo, 1916.
    Painted by some unknown master to illustrate
    the experiences of Takezaki Sueriaga
    竹崎季長.
Mu An Chi 牧菴集.
    Literary Collections of (Yao) Mu-An.
    Yuan, *c.* +1310.
    Yao Sui 姚燧.
Muye Tobo T'ongji Ōnhae 武藝圖譜通志諺解.
    Illustrated Encyclopaedia of Military Arts (the
    Korean translation of the *Wu I Thu Phu Thung
    Chih*).
    Korea, after +1790.
    Editor not known.
    Courant (1), no. 2467.

Nan Thang Shu 南唐書.
    History of the Southern Thang Dynasty [+923
    to +936].
    Sung, +11th.
    Ma Ling 馬令.
Nan Thang Shu 南唐書.
    History of the Southern Thang Dynasty [+923
    to +936].
    Sung, +12th.
    Lu Yu 陸游.
Nihon Kokujokushi 日本國辱史.
    History of Japan's Humiliation [the Mongol in-
    vasions of +1274 and +1281].
    Japan, *c.* +1300.
    Writer unknown.
No Kao Chi 諾皋記.
    Records of No-Kao [collected popular beliefs
    concerning spirits, genii and Taoist gods].
    Thang, *c.* +850.
    Tuan Chhêng-Shih 段成式.
    No-Kao was a Taoist military archangel analo-
    gous to St Michael, mentioned in *Pao Phu Tzu*,
    ch. 17, p. 4*b* (Ware tr. (5), p. 285).
Nung Chi 農紀.
    Agricultural Record.
    Sung, Yuan or Ming.
    Writer unknown.
    Not in Wang Yü-Hu (1).

O Yu Hui Phien.
    See Miu Yu-Sun (1).

Pa Pien Lei Tsuan 八編類纂.
    Classified Florilegium of Eight Literary

Pa Pien Lei Tsuan (*cont.*)
　Collections.
　Ming, *c.* +1620.
　Chhen Jen-Hsi　陳仁錫.
　Now extant only in quotations.
Pa Shih Ching Chi Chih　八史經藉志.
　Bibliography of the Eight Histories (includes the
　lists in six dynastic histories and four sup-
　plementary bibliographies compiled during
　the Chhing period).
　Chhing, pr. 1825 and 1883.
　See Têng & Biggerstaff (1), 1st ed. p. 15, 2nd ed.
　p. 10.
Pai Chan Chi Fa　百戰奇法.
　Wonderful Methods for (Victory in) a Hundred
　Combats.
　Sung, *c.* +1260.
　Writer unknown.
Pai Chan Ching.
　See *Ping Fa Pai Chan Ching.*
Pai Pien　稗編.
　Leaves of Grass [encyclopaedia].
　Ming, +1581.
　Ed. Thang Shun-Chih　唐順之.
Pao Yüeh Lu　保越錄.
　The Defence of the City of Yüeh (Shao-Hsing)
　[+1358].
　Yuan, +1359.
　Hsü Mien-Chih　徐勉之.
Pei Mêng So Yen　北夢瑣言.
　Fragmentary Notes Indited North of (Lake)
　Mêng.
　Wu Tai (S/Phing), *c.* +950.
　Sun Kuang-Hsien　孫光憲.
　See des Rotours (4), p. 38.
Pei Pien Thun Thien Chhê Chhung I　備邊屯田車
　銃議.
　Discussions on the Use of Military-Agricultural
　Settlements, Muskets, Field Artillery and
　Mobile Shields in the Defence of the Frontiers.
　Ming, *c.* +1585.
　Chao Shih-Chên　趙士禎.
Phing Han Lu　平漢錄.
　Records of the Pacification of Han [the cam-
　paign of Chu Yuan-Chang and his generals in
　+1363 which overthrew the Han State of
　Chhen Yu-Liang in the Yangtse Valley
　and established the power of the Ming
　dynasty].
　Ming, *c.* +1521.
　Thung Chhêng-Hsü　童承敍.
Phing Hsia Lu　平夏錄.
　Records of the Pacification of Hsia [the cam-
　paign of Chu Yuan-Chang and his generals in
　+1371 which overthrew the Hsia State of
　Ming Shêng in Szechuan and established the
　power of the Ming Dynasty].
　Ming, *c.* +1544.
　Huang Piao　黃標.
　Cf. W. Franke (4), p. 56.
Phing Phi Pai Chin Fang　洴澼百金方.
　The Washerman's Precious Salve; (Appropri-

ate) Techniques (of Successful Warfare)
　[military encyclopaedia].
　Ming, after +1626.
　Ed. Hui Lu　惠麓.
　The title is taken from a story in *Chuang Tzu*,
　ch. 1, tr. Legge (5), Vol. 1, p. 173; Fêng Yu-
　Lan (5), p. 39. A man of Sung State invented
　a salve for chapped hands, and it was used in
　his family, professional washers of silk, for
　several generations. A stranger bought the
　formula for 100 pieces of gold, went down to
　Wu State, and being made Admiral there, em-
　ployed it for the sailors so that they gained a
　great victory over Yüeh. One application
　brought little gain; the other won great reward
　and a noble title.
　The work seems to be rare (not in *SKCS/
　TMTY*).
Phing Wu Lu　平吳錄.
　Records of the Pacification of Wu [the campaign
　by Chu Yuan-Chang and his generals in
　+1366 which overthrew the Chou State of
　Chang Shih-Chhêng and established the
　power of the Ming Dynasty].
　Ming, *c.* +1472.
　Wu Khuan　吳寬.
　Cf. W. Franke (4), p. 57.
Pi Chou Kao Lüeh　敝帚稿略.
　Classified Reminiscences swept up by an Old
　Broom.
　Sung, *c.* +1250.
　Pao Hui　包恢.
Ping Chhien　兵鈐.
　Key to Military Affaris; or, Key of Martial Art.
　Chhing, +1675.
　Lü Phan　呂磻 & Lu Chhêng-Ên　盧承恩.
Ping Fa Pai Chan Ching　兵法百戰經.
　Manual of Military Strategy for a Hundred
　Battles.
　Ming, *c.* +1590.
　Wang Ming-Hao　王鳴鶴.
　Ed. Ho Chung-Shu　何仲叔.
Ping Lu　兵錄.
　Records of Military Art.
　Ming, +1606; pr. +1628. Later eds. have pre-
　faces of +1630 and +1632.
　Ho Ju-Pin　何汝賓.
　Cf. Wang Chung-Min & Yuan Thung-Li (1), i,
　pp. 472, 475.
Ping Lüeh Tshuan Wên　兵略纂聞.
　Classified Compendium of Things Seen and
　Heard on Military Matters.
　Ming, late +16th.
　Chhü Ju-Chi　瞿汝稷.
　Cf. Lu Ta-Chieh (1), p. 127.
Pu Liao Chin Yuan I Wên Chih　補遼金元藝文志.
　Additional Bibliography of the Liao, Chin and
　Yuan Dynasties.
　A continuation of *Liao Chin Yuan I Wên Chih*, q.v.
　by many Chhing scholars especially Lu Wên-
　Chao　盧文炤 (or Chhao　弨) *c.* +1770.
　In *Pa Shih Ching Chi Chih*, q.v.

San Chhao Pei Mêng Hui Pien 三朝北盟會編.
　　Collected Records of the Northern Alliance
　　　during Three Reigns.
　　Sung, +1196.
　　Hsü Mêng-Hsin 徐夢莘.
San Kyūkai 算九回.
　　Mathematics in Nine Chapters [in each of three
　　　volumes or parts].
　　Japan, +1677.
　　Nozawa Sodanaga 野沢定長.
　　Cf. Itakura (1).
San Tshai Thu Hui 三才圖會.
　　Universal Encyclopaedia.
　　Ming, +1609.
　　Wang Chhi 王圻.
Shan Tso Chin Shih Chih 山左金石志.
　　Record of Inscriptions on Metal and Store from
　　　the Left-hand Side of the Mountain.
　　Chhing, +1796.
　　Pi Yuan 畢沅 & Juan Yuan 阮元.
Shen Chhi Phu 神器譜.
　　Treatise on Extraordinary (lit. Magical)
　　　Weapons [musketry].
　　Ming, +1598.
　　Chao Shih-Chên 趙士禎.
　　Cf. W. Franke (3) no. 255, (4), p. 208; Goodrich
　　　(15).
Shen Chhi Phu Huo Wên 神器譜或問.
　　Miscellaneous Questions (and Answers arising
　　　out of) the Treatise on Guns.
　　Ming, +1599.
　　Chao Shih-Chên 趙士禎.
　　Cf. W. Franke (3) no. 255, (4), p. 208; Goodrich
　　　(15).
Shen Chi Chih Ti Thai Pai Ying Ching 神機制敵太
　　白陰經.
　　Secret Contrivances for the Defeat of Enemies;
　　　the Manual of the White Planet.
　　Full title of *Thai Pai Yin Ching*, q.v.
Shen I Chi 神異記.
　　(Probably an alternative title of *Shen I Ching*,
　　　q.v.)
　　Records of the Spiritual and the Strange.
　　Chin, c. +290.
　　Wang Fou 王浮.
Shen I Ching 神異經.
　　Book of the Spiritual and the Strange.
　　Ascr. Han, but prob. +3rd, +4th or +5th
　　　century.
　　Attrib. Tungfang Shuo (−2nd.) 東方朔.
　　Probable author, Wang Fou 王浮.
Shen Wei Thu Shuo 神威圖說.
　　Illustrated Account of the Magically Over-
　　　awing (Weapon, i.e. the Cannon).
　　Chhing, +1681.
　　Nan Huai-Jen (Ferdinand Verbiest, S. J.)
　　　南懷仁.
　　This book, if it still exists at all, must be ex-
　　　ceedingly rare; we know of no copy either in
　　　China or elsewhere.
Shih Chin Shih 試金石.
　　On the Testing of (what is meant by) 'Metal'

and 'Mineral'.
　　See Fu Chin-Chüan (5).
Shih Hu Shih Chi 石湖詩集.
　　Collected Works of the Lakeside Poet.
　　Sung, c. +1190.
　　Fan Chhêng-Ta 范成大.
Shih Kuo Chhun Chhiu 十國春秋.
　　Spring and Autumm Annals of the Ten King-
　　　doms (the States of the Five Dynasties Period,
　　　+10th cent.).
　　Chhing, +1678.
　　Wu Jen-Chhen 吳仕臣.
Shiseki Shūran 史籍集覽.
　　Collection of Historical Materials.
　　+15th to +18th centuries.
　　Ed. Kondō Heijō 近藤瓶城.
　　Kondō Shuppan-bu, Tokyo, 1907.
Shou Chhêng Chiu Ming Shu 守城救命書.
　　On Saving the Situation by the (Successful) De-
　　　fence of Cities.
　　Ming, +1607.
　　Lü Khun 呂坤.
　　Cf. Goodrich & Fang Chao Ying (1), p. 1006.
Shou Chhêng Lu 守城錄.
　　Guide to the Defence of Cities [lessons of the
　　　sieges of Tê-an in Hupei, +1127 to +1132].
　　Sung, c. +1140 and +1193 (combined in
　　　+1225).
　　Chhen Kuei 陳規 & Thang Tao 湯璹.
　　Cf. Balazs & Hervouet (1), p. 237.
Shu Nan Hsü Lüeh 蜀難敘略.
　　Collected Records of the Difficulties of
　　　Szechuan.
　　Chhing, c. +1663, but dealing with events in
　　　+1642 and after.
　　Shen Hsün-Wei 沈荀蔚.
　　Cf. Struve (1), pp. 346, 362.
Shu Yü Chou Tzu Lu 殊域周咨錄.
　　Record of Despatches concerning the Different
　　　Countries.
　　Ming, +1574.
　　Yen Tshung-Chien 嚴從簡.
Shui Hu Chuan 水滸傳.
　　Stories of the River-Banks [novel 'All Men are
　　　Brothers' and 'Water Margin'].
　　Ming, first collected c. +1380, but derived from
　　　older plays and stories. Oldest extant 100-ch.
　　　version, +1589, a reprint of an original earlier
　　　than +1550. Oldest extant 120-ch. version,
　　　+1614.
　　Ascr. Shih Nai-An 施耐庵.
　　Tr. Buck (1); Jackson (1).
Shui Lei Thu Shuo 水雷圖說.
　　Illustrated Account of Sea Mines.
　　See Phan Shih-Chhêng (1).
Ssu Hsüan Fu 思玄賦.
　　Thought the Transcender [ode on an imaginary
　　　journey beyond the sun].
　　H/Han, +135.
　　Chang Hêng 張衡.
Ssuma Fa 司馬法.
　　The Marshal's Art (of War).

Ssuma Fa (*cont.*)
Chou (late), prob. −4th or −3rd.
Writer unknown.

Su-a Munjip 西崖文集
Essays from the Western Cliff (one of Yu's names).
Korea, *c.* +1605, preface of +1633.
Yu Sŏngnyong 柳成龍
Courant (1), no. 624.

Suan Fa Thung Tsung 算法統宗
Systematic Treatise on Arithmetic.
Ming, +1592.
Chhêng Ta-Wei 程大位

Sun Tzu Ping Fa 孫子兵法
Master Sun's Art of War.
Chou (Chhi), *c.* −345.
Attrib. Sun Wu 孫武, more probably by Sun Pin 孫臏

Sung Chi Chao Chung Lu 宋季昭忠錄
Records of Distinguished Patriots of the [Second Half of the Southern] Sung Dynasty.
Alt. title of *Chao Chung Lu*, q.v.

Sung Chi San Chhao Chêng Yao 宋季三朝政要
The Most Important Aspects of Government as seen under the Last Three Courts of the (Southern) Sung Dynasty.
Yuan, *c.* +1285.
Writer unknown.
Cf. Balazs & Hervouet (1), p. 83.

Sung Hsüeh Shih Chhüan Chi 宋學士全集
Complete Record of Sung Scholars.
Ming, *c.* +1371.
Sung Lien 宋濂

Sung Hsüeh Shih Chhüan Chi Pu I 宋學士全集補遺
Additions to the Complete Record of Sung Scholars.
Ming, *c.* +1375.
Sung Lien 宋濂

Sung Hui Yao Kao 宋會要稿
Drafts for the *History of the Administrative Statutes of the Sung Dynasty*.
Sung.
Collected by Hsü Sung (1809) 徐松
From the *Yung-Lo Ta Tien*.

Sung Shih 宋史
History of the Sung Dynasty [+960 to +1279].
Yuan, *c.* +1345.
Tho-Tho (Toktaga) 脫脫 & Ouyang Hsüan 歐陽玄
Yin-Tê Index, no. 34.

Sung Shu 宋書
History of the (Liu) Sung Dynasty [+420 to +478].
S/Chhi, +500.
Shen Yo 沈約
A few chs. tr. Pfizmaier (58).
For translations of passages see the index of Frankel (1).

Sung Thung Chien Chhang Phien Chi Shih Pên Mo 宋通鑑長編紀事本末
Comprehensive Mirror Chronological History of the Sung Dynasty from Beginning to End.
Sung, +1253.
Yang Chung-Liang 楊仲良

Ta Chhing Shêng Tsu Jen Huang Ti Shih Lu 大清聖祖仁皇帝實錄
Veritable Records of the Benevolent Emperor of the Great Chhing Dynasty Shêng Tsu [Sage Ancestor = Khang-Hsi, r. +1661 to +1722].
Chhing, *c.* +1729.
Ed. Chiang Thing-Hsi *et al.* 蔣廷錫
Hu/143, 327.

Ta Hsüeh Yen I 大學衍義
Extension of the Ideas of the Great Learning [Neo-Confucian ethics].
Sung, +1229.
Chen Tê-Hsiu 眞德秀

Ta Hsüeh Yen I Pu 大學衍義補
Restoration and Extension of the Ideas of the *Great Learning* [contains many chapters of interest for the history of technology].
Ming, *c.* +1480.
Chhiu Chün 丘濬

Taiheiki 太平記
Records of the Reign of Great Peace [a romance history of one of the most troubled periods of Japanese history, +1318 to +1368].
Japan, *c.* +1370.
Attrib. Kojuma (monk) 小嶋

Tê-An Shou Chhêng Lu 德安守城錄
See *Shou Chhêng Lu*.

Têng Than Pi Chiu 登壇必究
Knowledge Necessary for (Army) Commanders.
Ming, +1599.
Wang Ming-Hao 王鳴鶴
Cf. W. Franke (4), p. 208.

Têng Wu Shê Phien 登吳社編
Records of a Journey up to the Cities of Wu (Chiangsu).
Sung.
Wang Chih 王穉

Teppō-ki 鐵砲記
Record of Iron Guns.
Japan, +1606; pr. +1649.
Nampo Bunshi 南浦文之
Cf. Arima (*1*), pp. 617 ff.

Thai-Chhing Ching Thien-Shih Khou Chüeh 太清經天師口訣
Oral Instructions from the Heavenly Masters [Taoist Patriarchs] on the Thai-Chhing Scriptures.
Date unknown, but must be after the mid +5th cent. and before Yuan.
Writer unknown.
TT/876.

Thai-Chhing Tan Ching Yao Chüeh 太清丹經要訣 (= *Thai-Chhing Chen Jen Ta Tan*)
Essentials of the Elixir Manuals, for Oral Transmission; a Thai-Chhing Scripture.
Thang, mid +7th (*c.* +640).
Prob. Sun Ssu-Mo 孫思邈
In *YCCC*, ch. 71.
Tr. Sivin (1), pp. 145 ff.

Thai Pai Yin Ching 太白陰經.
　　Manual of the White (and Gloomy) Planet (of
　　　War; Venus) [military encyclopaedia]
　　Thang, +759.
　　Li Chhüan 李筌.
Thai Tsu Shih Lu Thu 太祖實錄圖.
　　Veritable Records of the Great Ancestor
　　　(Nurhachi, d. +1626, retrospectively emperor
　　　of the Chhing), with Illustrations.
　　Ming +1635; revised Chhing, +1781.
　　Writer unknown.
　　MS of +1740 reproduced by NE University,
　　　Mukden, 1930; in Chinese with captions in
　　　Chinese and Manchu for the illustrations.
Thang Yeh Chen Jen Chuan 唐葉真人傳.
　　Biography of the Perfected Sage Yeh (Ching-
　　　Nêng) of the Thang.
　　Prob. Sung.
　　Chang Tao-Thung 張道統.
　　TT/771
Thien Kung Khai Wu 天工開物.
　　The Exploitation of the Works of Nature.
　　Ming, +1637.
　　Sung Ying-Hsing 宋應星.
　　Tr. Sun Jen I-Tu & Sun Hsüeh-Chuan (1)
Thien Wên 天問.
　　Questions about Heaven ['ode', perhaps a ritual
　　　catechism].
　　Chou, generally ascr. late +4th, but perhaps
　　　−5th century.
　　Attrib. Chhü Yuan, but probably earlier
　　　屈原.
　　Tr. Erkes (8); Hawkes (1).
Thing Hsün Ko Yen 庭訓格言.
　　Talks on Experiences in the Hall of Edicts.
　　Chhing, c. +1722.
　　Aihsin-chüeh lo Hsüan-Yeh (Khang-Hsi emper-
　　　or of the Chhing) 愛新覺羅玄燁.
Thung Tien 通典.
　　Comprehensive Institutes [a reservoir of source
　　　material on political and social history]
　　c. +812 (events down to +801).
　　Embodied the earlier *Chêng Tien* of Liu Chih.
　　Tu Yu 杜佑.
　　Têng & Biggerstaff (1) p. 148.
Thung Ya 通雅.
　　Helps to the Understanding of the *Literary Ex-
　　positor* [general encyclopaedia with much of
　　　scientific and technological interest].
　　Ming and Chhing, finished +1636, pr. +1666.
　　Fang I-Chih 方以智.
Tiao Chi Li Than 釣磯立談.
　　Talks at Fisherman's Rock.
　　Wu Tai (S/Thang) & Sung, begun c. +935,
　　　finished after +975.
　　Shih Hsü-Pai 史虛白.
Tsao Chia Fa 造甲法.
　　Treatise on Armour-Making.
　　Sung, c. +1150.
　　Writer unknown.
　　Now extant only in quotations.
Tsao Shen Pei Kung Fa 造神臂弓法.

　　Treatise on the Making of the Strong Bow.
　　Sung, c. +1150.
　　Writer unknown.
　　Now extant only in quotations.
Tsê Kho Lu 則克錄.
　　Book of Instantaneous Victory.
　　Alt. title of *Huo Kung Chhieh Yao*, (q.v.) given
　　　only to the reprint of 1841 (Pelliot (55),
　　　p. 192), which introduced many mistakes and
　　　suppressed some of the illustrations.
Tshao Shêng Yao Lan 操勝要覽.
　　Important Perspectives for the Attainment of
　　　Victory.
　　= *Huo Chhi Lüeh Shuo*, q.v.
Tso Mêng Lu 昨夢錄.
　　Dreaming of the Good Old Days [written in the
　　　South after the victory of the Chin Tartars,
　　　recalling life in the former capital city of
　　　Khaifêng (Pien-ching) under the Northern
　　　Sung].
　　Sung, c. +1137.
　　Khang Yü-Chih 康譽之.
Tu Chhêng Chi Shêng 都城紀勝.
　　The Wonder of the Capital (Hangchow).
　　Sung, +1235.
　　Mr Chao 趙氏 [Kuan Pu Nai
　　　Tê Ong 灌圃耐得翁; The Old Gentleman
　　　of the Water-Garden who achieved Success
　　　through Forbearance]
Tu Shih Ping Lüeh 讀史兵略.
　　Accounts of Battles in the Official Histories.
　　See Hu Lin-I (1).
Tu Shu Min Chhiu Chi, Chiao Chêng 讀書敏求記
　　校證.
　　Record of Diligently Sought for and Carefully
　　　Collated Books.
　　Chhing, +1684; first pr. +1726.
　　Chhien Tsêng 錢曾.
　　Cf. Têng & Biggerstaff (1), 1st ed. p. 42.
Tung Ching Chi 東京記.
　　Records of the Eastern Capital.
　　Sung, c. +1065.
　　Sung Min-Chhiu 宋敏求.
　　Now extant only in quotations.
Tung Ching Mêng Hua Lu 東京夢華錄.
　　Dreams of the Glories of the Eastern Capital
　　　(Khaifêng).
　　S/Sung, +1148 (referring to the two decades
　　　which ended with the fall of the capital of
　　　N/Sung in +1126 and the completion of
　　　the move to Hangchow in +1135).
　　Mêng Yuan-Lao 孟元老.

Wan Shu Chi 宛署記.
　　See *Yuan Shu Tsa Chi*.
Wang Wên Chhêng Kung Chhüan Shu 王文成公
　　全書.
　　Collected Writings of Wang Shou-Jen (Wang
　　　Yang-Ming).
　　Ming, +1574.
　　Ed. Hsieh Thing-Chieh 謝廷傑.
　　Cf. Franke (4), p. 138.

Wei Kung Ping Fa Chi Pên 衛公兵法輯本.
  Military Treatise of (Li) Wei-Kung.
  Thang, +7th.
  Li Ching 李靖.
  Fragments collected by Wang Tsung-I
    (Chhing) 汪宗沂.
Wei Lüeh 魏略.
  Memorable Things of the Wei Kingdom (San
    Kuo).
  San Kuo (Wei) or Chin, +3rd or +4th century.
  Yü Huan 魚豢.
Wên Li Su 問禮俗.
  Questions on Popular Ceremonies and Beliefs.
  San Kuo/Wei, c. +225.
  Tung Hsün 董勛.
  In *YHSF* ch. 28, pp. 72 a ff.
Wo Chhing Thun Thien Chhê Chhung I 倭情屯田
  車銃議.
  Discussions on the Use of Military-Agricultural
    Settlements, Muskets, Field Artillery and
    Mobile Shields against the Japanese (Pirates).
  Ming, c. +1585.
  Chao Shih-Chên 趙士禎.
  Chhê Chhung Thu is a supplement to this.
Wu Ching Shêng Lüeh 五經聖略.
  The (Essence of the) Five (Military) Classics, for
    Imperial Consultation.
  Sung, c. +1150.
  Wang Shu 王洙.
  Now extant only in quotations.
Wu Ching Tsung Yao 武經總要.
  Collection of the Most Important Military Tech-
    niques [compiled by Imperial Order].
  Sung, +1040 (+1044). Repr. +1231 and c.
    +1510. This Ming edition is the oldest extant
    now.
  Ed. Tsêng Kung-Liang 曾公亮 assisted by
    Yang Wei-Tê 楊惟德 and Ting
    Tu 丁度.
Wu Ching Yao Lan 武經要覽.
  Essential Readings in the Most Important Mili-
    tary Techniques (lit. Classics).
  Title of one of the Wan-Li editions of *Wu Ching
    Tsung Yao*.
Wu Hsien Chih 吳縣志.
  Local History and Geography of Wu-hsien
    (Suchow in Chiangsu).
  Chhing, +1691 (2nd ed.).
  Ed. Sun Phei 孫佩.
Wu I Thu Phu Thung Chih 武藝圖譜通志.
  Illustrated Encyclopaedia of Military Arts.
  Korea (Chosŏn), +1790.
  Ed. Pak Chega 朴齊家 & Yi Tŏngmu
    李德懋.
  Based on an earlier draft by Han Kyo 韓嶠
    done in the +1590s in consultation with
    Chinese military technologists then in Korea
    fighting the Japanese under Hideyoshi (*Chong-
    jo Sillok*, 30/31 a).
  Cf. *Muye Tobo T'ongji Ŏnhae*, the Korean version
    of the text.
Wu Li Hsiao Shih 物理小識.

Small Encyclopaedia of the Principles of Things.
  Chhing, +1664.
  Fang I-Chih 方以智.
  Cf. Hirth (17).
Wu Lin Chiu Shih 武林舊事.
  Institutions and Customs of the Old Capital
    (Hangchow).
  Sung, c. +1270 (but referring to events from
    about +1165 onwards).
  Chou Mi 周密.
Wu Lüeh Huo Chhi Thu Shuo 武略火器圖說.
  Illustrated Account of Gunpowder Weapons
    and their Use in Various Tactical Situations.
  Ming, c. +1560.
  Incorporated in *Wu Pei Chhüan Shu*, q.v.
  Hu Tsung-Hsien 胡宗憲.
Wu Lüeh Shen Chi 武略神機.
  The Magically(Effective) Arm in Various
    Tactical Situations [musketry].
  Ming, c. +1550.
  Hu Hsien-Chung 胡獻忠 (perhaps Hu
    Hsien-Chung 胡憲仲).
  Cf. Lu Ta-Chieh (1), p. 139.
Wu Lüeh Shen Chi Huo Yao 武略神機火藥.
  On Gunpowder for Muskets and their Use in
    Various Tactical Situations.
  Ming, c. +1560.
  Incorporated in *Wu Pei Chhüan Shu*, q.v.
  Hu Tsung-Hsien 胡宗憲.
Wu Pei Chhüan Shu 武備全書.
  Complete Collection of Works on Armament
    Technology (including Gunpowder
    Weapons).
  Ming, +1621.
  Ed. Phan Khang 潘康.
Wu Pei Chih 武備志.
  Treatise on Armament Technology.
  Ming, prefaces of +1621, pr. +1628.
  Mao Yuan-I 茅元儀.
  Cf. Franke (4), p. 209.
Wu Pei Chih Lüeh 武備志略.
  Classified Material from the Treatise on Arma-
    ment Technology.
  Chhing, c. +1660.
  Fu Yü 傅禹.
Wu Pei Chih Shêng Chih 武備制勝志.
  The Best Designs in Armament Technology.
  Ming, c. +1628.
  Mao Yuan-I 茅元儀.
  MS of 1843 in the Cambridge University
    Library.
  Cf. Franke (4), p. 209.
Wu Pei Hsin Shu 武備新書.
  New Book on Armament Technology [very simi-
    lar to *Chi Hsiao Hsin Shu*, q.v.].
  Ming, +1630.
  Attrib. Chhi Chi-Kuang 戚繼光.
  True compiler unknown.
Wu Pei Huo Lung Ching 武備火龍經.
  The Fire-Drake Manual and Armament Tech-
    nology [gunpowder weapons and firearms].
  Ming, completed after +1628, but containing

Wu Pei Huo Lung Ching (*cont.*)
　　much material from earlier versions of the *Huo Lung Ching*.
　　Attrib. Chiao Yü　焦玉.
　　Cf. Ho Ping-Yü & Wang Ling (1).
Wu Pei Pi Shu　武備秘書.
　　Confidential Treatise on Armament Technology [a compilation of selections from earlier works on the same subject].
　　Chhing, late +17th; (repr. 1800).
　　Shih Yung-Thu　施永圖.
Wu Plen　武編.
　　Military Compendium [technology and equipment, including Western-influenced firearms].
　　Ming, *c.* +1550.
　　Thang Shun-Chih　唐順之.
Wu Shih Pên Tshao　吳氏本草.
　　Mr. Wu's Pharmaceutical Natural History.
　　San Kuo (Wei), *c.* +235.
　　Wu Phu　吳普.
　　Extant only in quotations in later literature.
Wu Shih Thao Lüeh　武試韜略.
　　A Classified Quiverful of Military Texts.
　　Ming, before +1621.
　　Wang Wan-Chhing　汪萬頃.
Wu Shu Ta Chhüan　武書大全.
　　Complete Collection of the Military Books.
　　Ming, +1636.
　　Ed. Yin Shang　尹商.
　　Cf. Lu Ta-Chieh (2), p. 12.
Wu Tai Shih Chi　五代史記.
　　See *Hsin Wu Tai Shih*.
Wu Ti Chen Chhüan　無敵眞詮.
　　Reliable Explanations of Invincibility.
　　Ming, *c.* +1430.
　　Writer unknown.
　　Now extant only in quotations.
Wu Tu Fu　吳都賦.
　　Ode on the Capital of Wu (Kingdom).
　　Chin, *c.* +270.
　　Tso Ssu　左思.
　　Tr. von Zach (6).
Wu Yüeh Pei Shih　吳越備史.
　　Materials for the History of the Wu-Yüeh State (in the Five Dynasties Period).
　　Sung, *c.* +995.
　　Lin Yü　林禹.
Yawata Gudōki.
　　See *Hachiman Gudō-Kun*.
Yin Fu Ching　陰符經.
　　The Harmony of the Seen and the Unseen.
　　Thang, *c*, +735 (unless in essence a preserved late Warring States document).
　　Li Chhüan　李筌.
　　TT/30. Cf. *TT*/105–24. Also in *TTCY* (Tou Chi, 6).
　　Tr. Legge (5). Cf. Maspero (7), p. 222.
Yü Chhien Chün Chhi Chi Mu　御前軍器集模.
　　Imperial Specifications for Army Equipment.

　　Sung, *c.* +1150.
　　Writer unknown.
　　Now extant only in quotations.
Yü Chih Thang Than Wei　玉芝堂談薈.
　　Thickets of Talk from the Jade-Mushroom Hall.
　　Ming, *c.* +1620.
　　Hsü Ying-Chhiu (ed.)　徐應秋.
Yu Chu Shih I　友助事宜.
　　The Organisation of Friends for Mutual Protection (in the Ming militia).
　　Ming, *c.* +1600.
　　Chin Shêng　金聲.
Yü Ssu Chi　玉笥集.
　　Jade Box Collection [poetry].
　　Yuan, *c.* +1341.
　　Chang Hsien　張憲.
Yü Tung Hsü Lu　餘冬序錄.
　　Late Winter Talks.
　　Ming, +1528.
　　Ho Mêng-Chhun　何孟春.
　　Cf. Franke (4), p. 105.
Yü Tung Hsü Lu Tsê Chhao Wai Phien　餘冬序 (or 緒) 錄摘抄外扁.
　　Further Collection of Selected Excerpts from the 'Late Winter Talks'.
　　Orig. Ming, +1528.
　　Ho Mêng-Chhun　何孟春.
Yuan Shih　元史.
　　History of the Yuan (Mongol) Dynasty [+1206 to +1367].
　　Ming, *c.* +1370.
　　Sung Lien　宋濂 *et al.*
　　Yin-Tê Index, no. 35.
Yuan Shu Tsa Chi　宛署雜記.
　　Records of the Seat of Government at Yuan (-phing), (Peking). [or, Miscellaneous Records of a Minor Office].
　　Ming, +1593.
　　Shen Pang　沈榜.
Yüeh Ling Kuang I　月令廣義.
　　Amplifications of the 'Monthly Ordinances'.
　　Ming, soon after +1592.
　　Fêng Ying-Ching　馮應京.
Yüeh Shan Tshung Than　月山叢談.
　　Collected Discourses of Mr Moon-Mountain (i.e. Li Wên-Fêng; Yüeh Shan Tzu).
　　Ming, *c.* +1545.
　　Li Wên-Fêng　李文鳳.
Yün Lu Man Chhao　雲麓漫抄.
　　Random Jottings at Yün-Lu.
　　Sung, +1206 (referring to events of about +1170 onwards).
　　Chao Yen-Wei　趙彥衛.
Yünnan Chi Wu Chhao Huang　雲南機務鈔黃.
　　A Rough Statement of the Course of Affairs in Yunnan.
　　Ming, +1388.
　　Chang Tan　張紞.
　　Cf. W. Franke (4), p. 56.

B. CHINESE AND JAPANESE BOOKS AND JOURNAL
ARTICLES SINCE +1800

Anon. (*33*).
San Mên Hsia Tshao Yün I Chi 三門峽漕運
遺跡.
The Remains of the Canal [and the Trackers'
Galleries] in the San Mên Gorge (of the
Yellow River).
Kho-Hsüeh, Peking, 1959.
(Academia Sinica, Archaeological Field Studies,
no. 8.)

Anon. (*209*).
Chung-Kuo Li-Shih Po-wu-kuan Chhen Lieh ti i phi
Ming-Tai Huo-Chhi Fu-Yuan Mo-Hsing 中國
歷史博物館陳列的一批明代火器復
原模型.
Comments on the Reconstructions and Models
of the Fire-Weapons of the Ming period displayed in the National Historical Museum.
WWTK, 1959, no. 10 (no. 110), 53.

Anon. (*210*).
Yin Chhiao Shan Han Mu Chu Chien 'Sun Tzu Ping
Fa' 銀雀山漢墓竹簡「孫子兵法」.
The Versions of 'Master Sun's Art of War' found
on Bamboo strips in a Han Tomb at Silver-
sparrows Mountain.
Wên-wu, Peking, 1975, 1976.

Anon. (*211*) (ed.).
Nei Mêng-Ku Wên-Wu Tzu-Liao Hsüan Chi 內蒙
古文物資料選集.
Choice Collection of Cultural Objects of Inner
Mongolia [album].
Huhihot, 1964.

Anon. (*212*).
Chhing-Tai Chhou-Pan I-Wu Shih-Mo 清代籌
辦夷務始末.
Complete Record of the Management of Barba-
rian Affairs during the Chhing Dynasty.
Peking, 1930.
Cf. Hummel (2), p. 383.

Anon. (*213*).
Mao Sê Chhiang Yung Fa Thu Shuo 毛瑟槍用
法圖說.
Illustrated Manual of the Use of the Mauser
Rifle.
c. +1890.

Anon. (*214*).
Hua Hsüeh Fa-Chan Chien Shih 化學發展
簡史.
A Simple Introduction to the History of
Chemistry.
Kho-Hsüeh Chhu-pan-shê, Peking, 1980.

Anon. (*262*).
Chichibu Muku Jinja no Ryūsei 秩父椋神社
の龍勢.
The Dragon-Power Festival at the Muku (Shin-
to) Shrine in Chichibu District (formerly
Saitama)—rocket-launching displays.
ANA 1983, no. 170, 9.

Anon. (*263*).
Ta-Tsu Shih Kho 大足石刻.
The Stone-Carvings (in the cave Temples) of
Tu-Tsu (Szechuan).
Szechuan Fine Arts Academy, Chhêngtu, 1962.

Aoji Rinsō (*1*) 青地林宗.
Kikai Kanran 氣海觀瀾.
A Survey of the Ocean of Pneuma [astronomy
and meteorological physics].
Japan, 1825; enlarged in 1851 by Kawamoto
Kōmin.
In. NKKZ, vol. 6.
Cf. Tuge Hideomi (1), p. 81.

Arakawa Hideyoshi (*1*)
Bunkyu no Eki ni Mohogun wa Roketto o Rioy shita
ka?
Were Rockets used in the Mongol Invasions
(+1274 and +1281)?
NR, 1960, no. 148, 86.

Arima Seihō (*1*) 有馬成甫.
Kahō no Kigen to Sono Denryū 火砲の起原と
その傳流.
On the Origin and Diffusion of Cannon and
Firearms.
Yoshikawa Kōbunka, Tokyo, 1962.

Arima Seihō 有馬成甫 & Kuroda Genji (*1*)
黒田源次.
Kōbu Zaimeihō ni tsuite 洪武在銘砲について.
On Self-dated Inscribed Cannon of the Hung-
Wu reign-period (+1368 to +1398).
MMO, 1934, **16**, 1.

Arizaka Shōzō (*1*) 有坂鉊藏.
Heiki Enkaku Zusetsu 兵器沿革圖說.
Illustrated Account of the Development of Mili-
tary Weapons.
Tokyo.

Arisaka Shozō (*2*) 有坂鉊藏.
Heikikō 兵器考.
A Study of Military Armaments, 4 vols.
Yuzankaku, Tokyo, 1935–7.

Chang Chou-Hsün (*1*) 張焯焄.
Chhi-Shih Nien Lai Chung-Kuo Ping Chhi Chih Chih-
Tsao 七十年來中國兵器之制造.
The Manufacture of Weapons in China during
the past Seventy Years.
TFTC, 1936, **33** (no. 2), 21 (104053).

Chang Hui (*1*) 章回.
Huo-Yao ti Fa-Ming 火藥的發明
On the Invention of Gunpowder.
Thung-Su Tu-Wu, Peking, 1956.

Chang Wei-Hua (*1*) 張維華.
Ming Shih Fo-Lang-Chi Lü-Sung, Ho-Lan, I-Ta-Li-

BIBLIOGRAPHY B

Chang Wei-Hua (1) (cont.)
Ya, Ssu Chuan Chu Shih 明史佛郎機呂
宋和蘭意大里亞四傳注釋.
A Commentary on the Four Chapters on Portug-
al, Spain, Holland and Italy in the *History of
the Ming Dynasty*.
YCHP, Monograph Series, no. 7.
Peiping, 1934.
Chang Wên-Hu (1) 張文虎.
Shu I Shih Shih Tshun 舒藝室詩存.
Selected Poems from the Pavilion of Relaxed
Aesthetic Contemplation.
Nanking, c. +1860.
Chang Yün-Ming (1) 張運明.
Hei Huo-Yao Shih Yung Thien Jam Liu-Huang Phei
Chih ti Ma? 黑火藥是用天然硫磺配制
的嗎.
Was Black (Gun-)Powder made from native
Sulphur?
CKKC, 1982 (no. 1), 32.
Chao Erh-Hsün 趙爾巽 & Kho Shao-Min
(1) 柯劭忞.
Chhing Shih Kao 清史稿.
Draft History of the Chhing Dynasty.
Peking, 1914–27; pr. 1927–8.
Chao Hua-Shan (1) 晁華山.
Sian Chhu-Thu-ti Yuan-Tai Thung Shou Chhung yü
Hei Huo-Yao 西安出土的元代銅手銃與
黑火藥.
A Bronze Hand-Gun excavated at Sian and still
containing Traces of Gunpowder.
KKWW, 1981, no. 3 (no. 7), 73.
Chao Thieh-Han (1) 趙鐵寒.
Huo-Yao ti Fa-Ming 火藥的發明.
On the Invention of Gunpowder.
Nat. Hist. Museum, Thaipei, 1960.
With Engl. summary, in many ways misleading
(*Coll. Papers on Hist. and Art of China*, 1st series,
no. 4)
Abstr. J. Needham, RBS, 1967, **6**, no. 686.
Chêng Chen-To (1) (ed.) 鄭振鐸.
Chhüan Kuo Chi-Pên Chien-Shê Kung-Chhêng Chung
Chhu-Thu Wên-Wu Chan-Lan Thu Lu 全國基
本建設工程中出土文物展覽圖錄.
Illustrated Catalogue of an Exhibition of
Archaeological Objects discovered during the
Course of Engineering Operations in the
National Basic Reconstruction Programme.
Exhibition Committee, Peking, 1954.
Chêng Shao-Tsung (1) 鄭紹宗.
Jehol Hsing-lung (Hsien) Fa-Hsien-ti Chan-Kuo
Sêng-Chhan Kung-Chü Chu Fan 熱河興隆發
現的戰國生產工具鑄範.
(Cast-Iron) Casting Moulds for Tool Production
of the Warring States Period found at Hsing-
lung (Hsien) in Jehol (Province).
KKTH, 1956, **2** (no. 1), 29.
Chêng Tso-Hsin (2) 鄭作新.
Chung-Kuo Niao Lei Fên Pu Ming Lu 中國鳥
類分布名錄.
A Dictionary of Chinese Birds.
2nd ed. Kho-Hsüeh, Peking, 1976.

Chêng Wei (1) 鄭爲.
'Chia Khou Phan Chhê' Thu Chüan 閘口盤車
圖卷.
The Scroll-Painting entitled 'The Horizontal
Water-Wheels beside the Sluice-Gate' [by
Wei Hsien, c. +970].
WWTK, 1966 (no. 2), 17.
Chhêng Tung (1) 成東.
Chiao Yü ti Chen Shih Shen Fên Chi Chhi Huo Kung
Shu ti Shih Liao Chieh Chih 焦玉的真實身
份及其火攻書史料價值.
On Chiao Yü's Identity and Biography, with a
Discussion of the Value of his Writings in the
History of Firearms and Artillery.
Mimeographed article issued by the Chinese
Arsenals Association (Chung Kuo Ping
Kung Hsüeh Hui), Peking, 1983.
Chiang Chhen-Ying (1) 姜宸英.
Chan Yuan Cha Chi 湛園札記.
Notes (on the Classics) from the Still Garden.
1829.
Chou Chia-Hua (1) 周嘉華.
Huo Yao ho Huo Yao Wu Chhi 火藥和火藥
武器.
On the History of Gunpowder and Firearms.
Art. in Anon (202), *Chung-Kuo Ku-Tai Kho Chi
Chhêng-Chiu*, 1978, p. 203.
Chou Wei (1) 周緯.
Chung-Kuo Ping Chhi Shih Kao 中國兵器
史稿.
A Draft History of Chinese Weapons.
Posthumous pub., ed. Kuo Pao-Chün
郭寶鈞.
San-Lien, Peking, 1957.
Chu Chhi-Chhien 朱啟鈐, Liang Chhi-
Hsiung 梁啟雄 and Liu Ju-Lin (1)
劉儒林.
Chê Chiang Lu [part 7] 哲匠錄.
Biographies of [Chinese] Engineers, Architects,
Technologists and Master-Craftsmen (con-
tinued).
BSRCA, 1934, 5 (no. 2), 74.
Chu Shêng (1) 朱晟.
Huo Yao, Shen, Yang ti Fa-Hsien yü Lien Tan Shu ti
'Fu Huo' yu Kuan 火藥砷氧的發現與煉
丹術的「伏火」有關.
The Relationship of the Concept of 'Subduing
by Fire' in Alchemical Elixir-making to the
Discovery of Gunpowder, Arsenic and
Oxygen.
Paper given at the Second National Conference
on the History of Chemisty (Chinese Society
for the History of Science and Technology).
Chu Shêng & Ho Tuan-Shêng (1) 朱晟 &
何端生.
Huo Yao, Yang, ti Fa-Hsien yü Lian Tan Shu ti
'Fu Huo' yu kuan 火藥, 氧的發現與煉
丹術的「伏火」有關.
How the Concept of 'Subduing by Fire' in
Alchemical Elixir-making relates to the Dis-
coveries of Gunpowder and Oxygen.
Typed paper without indication of origin, in

Chu Shêng & Ho Tuan-Shêng (*1*) (*cont.*)
   East Asian History of Science Library, Cambridge.
Chung Thai (*1*)　鍾泰.
   Chung-Kuo Chê-Hsüeh Shih　中國哲學史
   A History of Chinese Philosophy.
   Com. Press, Shanghai, *c.* 1930.
Chhen Chêng-Hsiang (*2*)　陳正祥.
   '*Chen-La Feng Thu Chi' ti Yen-Chiu*　眞臘風土記
   的研究.
   Researches on [Chou Tai-Kuan's] 'Description of Cambodia' [+1297].
   Chinese University Press, Shatin, Hongkong, 1975.
Chhen Shou-Chhi (*1*) (ed.)　陳壽祺.
   Fukien Thung Chih　福建通志
   Gazetteer of Fukien Province.
   Completed 1833, pr. 1867.
Chhen Thing-Yuan　陳廷元 & Li Chen (*1*)
   (ed.)　李震.
   Chung-Kuo Li-Tai Chan-Chêng Shih　中國歷代戰
   爭史.
   A History of Wars and Military Campaigns in China, 16 vols. (with abundant maps).
   Armed Forces University and Li-Ming Wên-Hua Shih-Yeh, Thaipei, 1963; repr. 1972; finalised 1976.
Chhen Wên-Shih (*1*)　陳文石.
   Chhing-Jen Ju Kuan Chhien ti Shou Kung Yeh
   清人入關前的手工業.
   The Technical Handicrafts and Industries of the Manchus before their Invasion of China.
   AS/BIHP, 1962, **34** (Hu Shih Festschrift), 291.
   Abstr. *RBS*, 1969, **8**, no. 246.
Chhen Yuan (*3*)　陳垣.
   Yuan Hsi-Yü Jên Hua Hua Khao　元西域人華
   化考.
   On the Sinisation of 'Western People' During the Yuan Dynasty.
   Pt I. *KHCK*, 1923, **1**, 573.
   Pt II. *YCHP*, 1927, **2**, 171.
Fang Hao (*3*)　方豪.
   Fu Miu Yen-Wei hsien-sêng Lun Pei Chhao Hu Su
   Shu　復繆彥威先生論北朝俗書.
   A Letter in answer to a Letter of Mr Miu Yen-Wei discussing the Foreign Customs Prevalent in the Northern Dynasties.
   ISP/WSFK, no. 26.
Fêng Chia-Shêng (*1*)　馮家昇.
   Huo-Yao ti Fa-Hsien chi chhi Chhuan Pu　火藥的
   發現及其傳佈.
   The Discovery of Gunpowder and its Diffusion.
   AP/HJ, 1947, 5, 29.
   Abstr. in *KMJP* 1952, 7 June (Li Shih Chiao Hsüeh, no. 34).
Fêng Chia-Shêng (*2*)　馮家昇.
   Hui Chiao Kuo Wei Huo-Yao Yu Chung-Kuo Chhuan
   Ju Ou-Chou ti Chhiao Liang　回教國爲火藥
   由中國傳入歐洲的橋樑.
   The Muslims as the Transmitters of Gunpowder from China to Europe.
   AP/HJ, 1949, 1.

Fêng Chia-Shêng (*3*)　馮家昇.
   Tu Hsi-Yang ti Chi Chung Huo-Chhi Shih Hou
   讀西洋的幾種火器史後.
   Notes on reading some of the Western Histories of Firearms.
   AP/HJ, 1947, 5, 279.
Fêng Chia-Shêng (*4*)　馮家昇.
   Huo-Yao ti Yu Lai chi chhi Chhuan Ju Ou-chou ti
   Ching Kuo　火藥的由來及其傳入歐洲的
   經過.
   On the Origin of Gunpowder and its Transmission to Europe.
   Essay in Li Kuang-Pi & Chhien Chün-Yeh, (q.v.), p. 33.
   Peking 1955.
Fêng Chia-Shêng (*6*)　馮家昇.
   Huo-Yao ti Fa-Ming ho Hsi Chhuan　火藥的發
   明和西傳.
   The Discovery of Gunpowder and its Transmission to the West.
   Hua-tung, Shanghai, 1954.
   Revised eds. Jen-min, Shanghai, 1962, 1978.
Fêng Chia-Shêng (*8*)　馮家昇.
   Tu Hsi-Yang ti Chi Phien Huo Yao Huo Chhi Wên
   Hou　讀西洋的幾篇火藥火器文後.
   Notes on Reading some of the Western Histories of Gunpowder and Firearms.
   AP/HJ, 1949, **7**, 241.
Fu Chin-Chhüan (*5*)　傅金銓.
   Shih Chin Shih　試金石.
   On the Testing of (what is meant by) 'Metal' and 'Mineral'.
   c. 1820.
   In *Wu Chen Phien Ssu Chu* ed.
Han Kuo-Chün (*1*)　韓國鈞.
   Wu Wang Chang Shih-Chhêng Tsai Chi　吳王張
   士誠載紀.
   A Memoir on Chang Shih-Chhêng, Prince of Wu (rebel against the Yuan dynasty, and founder of the short ill-fated dynasty of Chou, +1354 to +1357).
   Shanghai, 1932.
Hara Tomio (*1*)　洞富雄.
   Teppō Denrai to son Eikyo　鐵砲傳來.
   The Social Effects of the Coming of Muskets to Japan.
   Azekura Shobō, Tokyo, 1959.
Hara Tomio (*2*)　洞富雄.
   Tanegashima Jū　種子島銃.
   The Muskets of Tanegashima Island.
   Awaji Shobō Shinsha, Tokyo, 1958.
Harada Yoshito　原田淑人 & Komai Kazuchika
   (*1*) = (*1*)　駒井和愛.
   Shina Koki Zukō　支那古器圖考.
   Chinese Antiquities (Pt. 1, Arms and Armour; Pt. 2 Vessels [Ships] and Vehicles).
   Tōhō Bunka Gakuin, Tokyo, 1937.
Harada Yoshito　原田淑人 & Komai Kazuchika
   (*2*) = (*2*)　駒井和愛.
   Jōto, Mōko Ronnōru ni okeru Gendai miyako-ato no
   Chōsa　上都濛古ロニノールに於ける
   元代都址の調查.

Harada Yoshito (*cont.*)
Shangtu; the Summer Capital of the Yuan Dynasty at Dolon Nor, Mongolia.
Toa-Koko Gaku Kwai, Tokyo, 1941 (Archaeologia Orientalis, B, no. 2). With English résumé.

Ho Ping-Yü (*1*) 何丙郁.
Sung Ming Ping Shu so Chien ti 'Tu Yen', 'Tu Wu' ho 'Yen Mu' 宋明兵書所見的「毒煙」, 「毒霧」和「煙幕」.
On the 'Poison-smoke', 'Poison-fog' and 'Smoke-screen', described in the Military Texts of the Sung and Ming Periods.
KKPT, 1983, **2**, 1 (Twentieth Anniversary Commemoration Volume).

Hu Chien-Chung (*1*) 胡建中.
Cho-Lung ti Shêng Tan Hsia 卓隆的盛彈匣.
The Best Type of Magazine for Artillery Pieces (discusses the use of bronze casting, surrounded by cast-iron for cannons in Chhing tewes, "composite castings", called *thieh hsin thung* 鐵心銅).
TCC, 1984, no. 2, 24.

Hu Lin-I (*1*) 胡林翼.
Tu Shih Ping Lüeh 讀史兵略.
Accounts of Battles in the Official Histories.
Peking, 1861.
Cf. Lu Ta-Chieh (*1*), p. 159.

Huang Shang (*1*) 黃裳.
Shu ti Ku Shih 書的故事.
A Bookman's Reminiscences.
TSHU, 1979, **4**, 126.

Hung Huan-Chhun (*1*) 洪煥椿.
Shih chih Shih-san Shih-Chi Chung-Kuo Kho-Hsüeh-ti Chu-Yao Chhêng-Chiu 十至十三世紀中國科學的主要成就.
The Principal Scientific (and Technological) Achievements in China from the +10th to the +13th centuries (inclusive), [the Sung Period].
LSYC, 1959, **5** (no. 3), 27.
Abstr. J. Needham, *RBS*, 1965, **5**, no. 809.

Huang Thien-Chu, Tshai Chhang-Chhi & Liao Yuan-Chhüan (*1*) 黃天柱, 蔡長溪 & 廖淵泉.
Kho Hsüeh Chia Ting Kung-Chhen, Chung-Kuo Chin Tai Chün Huo 科學家丁拱辰中國近代軍火.
The Scientist (and Engineer) Ting Kung-Chhen and Modern Chinese Fire-Weapons.
FCLT 1982, no. 1, 81.

Hsi Tsê-Tsung (*6*) 席澤宗.
Huo Chien ti Chia Shih 火箭的家世.
Our Rocket Heritage.
CKHW, 7 June 1962.

Hsiang Ta (*5*) 向達.
Liang Chung Hai Tao Chen Ching 兩種海道針經.
An Edition of Two Rutters [the *Shun Fêng Hsiang Sung* (Fair Winds for Escort), perhaps *c.* +1430, from a MS of *c.* +1575; and the *Chih Nan Chêng Fa* (General Compass-Bearing Sailing Directions) of *c.* +1660].
Chung-Hua, Peking, 1961.

Hsiung Fang-Shu (*1*) 熊方樞.
Huo Chhi Ming Chung 火器命中.
Artillery Exercises.
c. 1850.
Cf. Lu Ta-Chieh (*1*), p. 163, (*2*), p. 19.

Hsü Chi-Yü (*1*) 徐繼畬.
Ying Huan Chih Lüeh 瀛環志略.
Classified Records of the Encircling Oceans (world geography).
1850; repr. 1866.

Hsü Hui-Lin (*1*) 許會林.
Chung-Kuo Huo-Yao Huo-Chhi Shih Hua 中國火藥火器史話.
Historical Talks on (the Development of) Gunpowders and Fire-Weapons in China.
Kho Hsüeh, Peking, 1986.

Inosaki Takaoki (*1*) 井崎隆興.
Gendai no Take no Sembai to Sono Shikō-igi 元代の竹の專とその施行意義.
On the Bamboo Monopoly [for bows, crossbows, arrows and even charcoal for gunpowder] in the Yuan period [between +1267 and +1292].
TK, 1957, **16**, 135.
Abstr. *RBS*, 1962, **3**, no. 263.

Iwasaki Tetsushi (*1*) 岩崎鐵志.
Takeshima-ryu Hojutsu Denpan no Kenkyu; Mikawa Tawara Hanshi Murakami Sadahe-o Chūshin-ne 高島流砲術伝播の研究；三河田原藩士村上定平を中心に.
A Study of the Diffusion of the Knowledge of Gunnery from Takeshima, with special reference to the Master Murakami Sadahe of the Tawara Clan (1808 to 1872).
Art. in *Higashi-asia no Kagaku*, ed. Yoshida Tadashi (*1*), p. 109.

Jih Yüeh 日月 & Chung Yung-Ho 鍾永和 (*1*) = (*1*).
Yen-shui Kuan Fêng Phao 鹽水觀蜂炮.
'Rocket Hives' in Yen-shui; a Peculiar Tradition.
SINRA, 1984, **9** (no. 3), 106.

Kawamoto Kōmin (*1*) 川本幸民.
Kikai Kanran Kōgi 氣海觀瀾廣義.
Enlargement of the 'Survey of the Ocean of Pneuma'.
Japan, 1851.
Cf. Tuge Hideomi (*1*), p. 81.

Kao Chih-Hsi 高至喜, Liu Lien-Yin 劉廉銀 *et al.* (*1*).
Chhangsha Shih Tung-pei Chiao Ku Mu Tsang Fa-Chüeh Chien-Pao 長沙市東北郊古墓葬發掘簡報.
Short Report on the Excavations of Tombs (of Warring States and Later Periods) in the North-eastern Suburbs of Chhangsha.
KKTH, 1959 (no. 12), 649.

Kho Shao-Min (1) 柯紹忞.
 Hsin Yuan Shih 新元史.
 New History of the Yuan Dynasty (issued as an
 official dynastic history by Presidential Order).
 Peiping, 1922.
Ku Yün-Chhüan (1) 古連泉.
 *Kuangtung Kao-yao Hsien Fa-Hsien Ming Chhu
 Thung Thieh Chhung* 廣東高要縣發現明
 初銅鐵銃.
 Hand-guns of Bronze and Iron, Early Ming in
 Date, found at Kao-yao Hsien in Kuangtung.
 WW, 1981, no. 4 (no. 299), 94.
Kuan Chhêng-Hsüeh & Wang Yü (1) 管成學 &
 王禹.
 *Tsung 'Khua Fu Chu Jih' Than Chhi; Wo Kuo
 Ku-Tai Chieh-Chhü-ti Kho-Hsüeh Fa-
 Ming* 從「夸父逐日」談起；我國古代
 傑出的科學發明.
 'Khua Fu pursuing the sun'; Talks on the Great
 Scientific Discoveries made in our Country in
 Older Times. (Pp. 110–110 has a brief
 account of the Invention of Gunpowder and
 Firearms).
 Chhilin Jen Jen Min, Chhangchhuen, 1978.
Kung Chen-Lin (1) 龔振麟.
 Thieh Mu Thu Shuo 鐵模圖說.
 Illustrated Account of a Method of (Casting
 Cannon in Cast-) Iron Moulds.
 Shanghai, 1846.
 Reprinted in *Hai Kuo Thu Chih*, ch. 86, pp. 1 a ff.
Kuo Chêng-I (1) 郭正誼.
 Kuan-Yu Huo-Yao Fa-Ming ti I-Hsieh Shih-Liao
 關於火藥發明的一些史料.
 Some New Materials on the Discovery of Gun-
 powder.
 HHSTP, 1981, no. 6.
Kuo Chêng-I (2) 郭正誼.
 Sun Ssu-Mo pu shih Huo Yuo Fa-Ming Jen 孫思
 邈不是火藥發明人.
 That Sun Ssu-Mo was not the Discoverer of
 Gunpowder.
 TJPCF, 1981 (no. 5), 62.
Kuo Chêng-I (3) 郭正誼.
 Huo-Yao Fa-Ming ti Hsin Than Thao 火藥發明
 的新探討.
 New Investigations on the Discovery of Gun-
 powder.
 Offprint without indication of origin (in
 EAHoSL).
Kuo Hua-Jo (1) 郭化若.
 Chin I Hsin Pien 'Sun Tzu Ping Fa' 今譯新編
 「孫子兵法」.
 A New Transcription of the 'Art of War of Mas-
 ter Sun' into Modern Chinese.
 Jen-Min, Peking, 1957; repr. Chung-hua, Shang-
 hai, 1964; Jen-Min, Shanghai, 1977.
Kuroda Genji (1) 黑田源次.
 Shinki Kahōron 神機火砲論.
 On the Magically Efficacious Weapons (Early
 Chinese Hand-guns and Bombards).
 MGK, 1936, **4**, 43.
 Abstr. in *SHHH*, 1937, **1** (no. 4), 24.

Lei Hai-Tsung (1) 雷海宗.
 Chung-Kuo ti Ping 中國的兵.
 The Historical Development of the Chinese
 Soldier.
 SHKS, 1935, **1**, 1.
 Eng. abstr. in *CIB*, 1936, **1**, 5 (abstr. no. 12).
Li Chhung-Chou (3) 李崇州.
 *Chung-Kuo Ming-Tai-ti Shiu Lei-Shih-Chieh
 Shui Lei-ti Pi Tsu* 中國明代的水雷；
 世界水雷的鼻祖.
 The Chinese Naval Mine of the Ming Period; a
 Pioneer Device in the World History of Naval
 Mines.
 CKKCSL 1985, **6** (no. 2), 32.
Li I-Yu (1) 李逸友.
 *Nei Mêng-Ku Tho-Kho-Tho Chhêng ti Khao-Ku Fa-
 Hsien* 內蒙古托克托城的考古發現.
 Archaeological Discoveries at Thokhto City in
 Inner Mongolia.
 WWTLTK, 1981, **4**, 210.
Li Shan-Lan (1) 李善蘭.
 Huo Chhi Chen Chüeh 火器眞訣.
 Instructions on Artillery.
 c. 1845.
 Cf. Lu Ta-Chieh (1), p. 162, (2), p. 19.
Li Shao-I (1) 李少一.
 Shuo Phao 說礮.
 Brief Discourse on Trebuchets.
 PKCS, 1981, no. 5 (no. 22), 32 (1312).
Li Ti (1) 李迪.
 *Chung-Kuo Jen Min tsai Huo-Chien Fang-Mien-ti
 Fa-Ming chhuang-Tsao* 中國人民在火箭方
 面的發明創造.
 On the Discovery and Development of Rocket
 Propulsion among the Chinese People.
 LHHP, 1978 (no. 1), 81.
Li Yen (1) 李岩.
 *Tshung Hei Huo Yao tao Hsien-Tai Kung Yeh Cha
 Yao* 從黑火藥到現代工業炸藥.
 From Black Powder to the Contemporary Explo-
 sives used in Industry.
 PKCS, 1980, no. 2, 63 (143).
Liang Chang-Chü (1) 梁章鉅.
 Lang Chi Tshung Than 浪跡叢談.
 Impressions Collected during Official Travels.
 c. 1845.
Lin Tsê-Hsü (2) 林則徐.
 Chao Phao Fa 炸礮法.
 On the Manufacture of Artillery Shells.
 In *HKTC*, ch. 87, pp. 6 a, b.
Lin Wên-Chao & Kuo Yung-Fang (1) 林文照 &
 郭永芳.
 *Kuan-yü Fo-lang-chi Huo-Chhung tsui-tsao
 Chhuan Ju Chung-Kuo-ti Shih Chien Khao*
 關于佛郎機火銃最早傳入中國的時
 間考.
 Textual Researches on the Date of the spread of
 the Fo-lang-chi (Portuguese breech-loading
 culverin) into China.
 3rd. International Conference on the History
 of Science in China, Peking, 1984, Paper
 no. 69.

Liu Hsien-Chou (*12*) 劉仙洲.
  Wo Kuo Ku-Tai Man Phao, Ti Lei ho Shui Lei Tzu-Tung Fa Huo Chuang Chih ti Fa Ming 我國古代慢炮, 地雷和水雷, 自動發火裝置的發明.
  Chinese Inventions in the Field of the Construction and Timing of Bombs, Land Mines, Sea Moves and Limpet Mines.
  WWTK, 1973, no. 11 (no. 210), 46.
Liu Yao-Han (*1*) 劉堯漢.
  Yi Tsu ti Huo Chhi; 'Hu-Lu Fei Lei' 彝族的火器; 胡盧飛雷.
  A Fire-weapon of the Yi (Minority) People; the 'Gourd-shaped Flying Thunder' [an explosive bomb containing lead pellets].
  Art. in Anon (*202*), *Chung-Kuo Ku-Tai Kho Chi Chhêng-Chiu*, 1978, p. 699.
Lo Hsiang-Lin (*6*) 羅香林.
  Researches on a Cannon made in the 4th year of the Yang-Li reign-period of the Southern Ming (+1650) in Hongkong.
  THSH, 1957, **2** (no. 10).
Lu Mou-Tê (*1*) 陸懋德.
  Chung-Kuo Jen Fa-Ming Huo-Yao Huo-Phao Khao 中國人發明火藥火礮考.
  A Study of the Invention of Gunpowder and Gunpowder Weapons by the Chinese.
  CHJ, 1928, **5** (no. 1), 1489.
Lu Ta-Chieh (*1*) 陸達節.
  Li-Tai Ping Shu Mu-Lu 歷代兵書目錄.
  A Bibliography of (Chinese) Books on Military Science [all periods].
  Nanking, 1933; repr. Thaipei, 1970.
Lu Ta-Chieh (*2*) 陸達節.
  Chung-Kuo Ping-Hsüeh Hsien-Tsun Shu Mu; fu Li-Tai Ping Shu Kai Lun 中國兵學現存書目; 附歷代兵書概論.
  A Bibliography of Extant Books on Military Science; with an Appended Essay on this genre of literature.
  Kuangchow, 1944; repr. 1949.
Lung Wên-Pin (*1*) (ed.) 龍文彬.
  Ming Hui Yao 明會要.
  History of the Administrative Statutes of the Ming Dynasty.
  c. 1870; pr. 1887.
  Cf. Têng & Biggerstaff (1), p. 163.

Mêng Sên (*1*) 孟森.
  Ming Tai Shih 明代史.
  A History of the Ming Period.
  Chunghua Tshung-Shu, Thaipei, 1957.
Mikami, Yoshio (*21*) 三上義夫.
  Sō no Chinki no 'Shujōroku' notosekin ki no Kansetsu-shageki 宋の陳規の守城錄と投石機の間接射擊.
  The *Shou Chhêng Lu* (Guide to the Defence of Cities) of Chhen Kuei of the Sung Dynasty, and the Use of Trebuchets.
  TBGZ, 1941, no. 600.
Mikami Yoshio (*22*) 三上義夫.
  'Kaisanki' no Dandō Mondai 改算記の彈道
  問題.
  Problems of Ballistics in the *Kaisanki* (+1659).
  KGZ, 1913, **8** (no. 4), 251.
Mikami Yoshio (*23*) 三上義夫.
  Shizuki Tadao yaku 'Kaki Happō-den' no Dandō Mondai 志筑忠雄譯「火器發法傳」の彈道問題.
  Ballistic Problems in the *Kaki Happō-den* translated by Shizuki Tadao (from the Dutch).
  KGZ, 1915, **10** (no. 1), 1.
Mikami Yoshio (*24*) 三上義夫.
  Koide Shūki no Dandō ni Kansuru Kenkyū 小出修喜の彈道に關する研究.
  Studies on the Ballistics of Koide Shūki (1847).
  KGZ, 1913, **8** (no. 1), 13.
Mikami Yoshio (*25*) 三上義夫.
  'Kikai Kanran Kōgi' no Dandō-ron 「氣海顜瀾廣義」の彈道論.
  Ballistics in the *Kikai Kanran Kōgi* (by Kawamoto Kōmin, following Aoji Rinsō).
  KGZ, 1914, **9** (no. 4), 117.
Mikami Yoshio (*26*) 三上義夫.
  Ikebe Harutsune no Dandō ni Kansuru Kōshiki 池部春常の彈に關する公式.
  The Ballistic Formulae of Ikebe Harutsune.
  KGZ, 1913, **8** (no. 3).
Miu Yu-Sun (*1*) 繆祐孫.
  O Yu Hui Pien 俄遊彙編.
  Narrative of a Journey into Russia (with Diary).
  Peking, 1889.

Naganuma Kenkai (*1*) 長沼賢海.
  Teppō no Denrai 鐵砲の傳來.
  On the Introduction of Firearms to Japan.
  RC, 1914, **23** (no. 6), 623.
Naganuma Kenkai (*2*) 長紹賢海.
  Teppō no Denrai Hosetsu 鐵砲の傳來補說.
  Further Evidence on the Introduction of Firearms to Japan.
  RC, 1914, **24** (no. 2), 131.
Naganuma Kenkai (*3*) 長紹賢海.
  Teppō no Deurai Otō 鐵砲の傳來應答.
  A Reply on the Introduction of Firearms to Japan.
  RC, 1915, **25** (no. 1), 52.
Naganuma Kenkai (*4*) 長紹賢海.
  Dai Nihonshi Kōza 大日本史講座.
  Lectures on the History of Japan.
  Tokyo, 1929.
Nambō Heizō (*1*) 南坊平造.
  Kayaku-wa Darego Hatsumei Shitaka? 火藥は誰が發明したか.
  Who Really Invented Explosives?
  KYHY, 1969, **28** (no. 4), 322; (no. 5), 403.

Okada Noboru (*1*) 岡田登.
  Chūgoku Sōdai ni okeru Kaki to Kayaku Heiki 中國宋代における火器と火藥兵器.
  Chinese Firearms and Guns in the Sung Dynasty.
  JHPHARM, 1981, **16** (no. 2), 50.
Okada Noboru (*2*) 岡田登.

Okada Noboru (2) (cont.)

Chūgoku ni okeru Bakuchiku, Bakujō, Enka no Kigen to sono shoki no Haten　中國における爆竹，爆仗，煙火の起源とその初期の發展.

The Origin and Development of Crackers, Fire-Crackers and Fireworks in China.

RROWC, 1982, **15**, 63.

Okada Noboru (3)　岡田登.

Hekireki Kakyū no Kigen to Haten　霹靂火毬の起源と發展.

The Origin and Development of the Thunderbolt Fire-Ball.

FSH, 1980, **7**, 66.

Okada Noboru (4)　岡田登.

Gendai Shoki ni okeru, Gengun no Chūgoku Kokunai ni okeru Kaki to Kayaku Heiki　元代初期における, 元軍の中國國內における火器と火藥兵器.

Firearms and Gunpowder Weapons of the Yuan Army used in China at the beginning of the Yuan Period.

RROWC, 1983, **16**, 47.

Okamura Shōji (1)　奧村正二.

Hinawajū Kara Karofune made: Edojidai Gijutsu-Shi　火繩銃から黑船まで；江戶時代技術史.

The Matchlook Musket and the "Black (European) Ships'; as aspect of the History of Technology of the Yedo Period. (Includes chapters on Shipbuilding, Metals Mining and Hydraulic Machinery together with Clockwork).

Tokyo, 1970, repr. 1973.

岩波新書 no. 750.

Ong Thung-Wên (1)　翁同文.

'Chen Yuan Miao Tao Yao Lüeh' ti Chhêng Shu Shih-Tai chi hsiang-kuan-ti Huo Yao Shih Wên-Thi　「眞元妙道要略」的成書時代及相關的火藥史問題.

The Dating of the Chen Yuan Miao Tao Yao Lüeh in relation to the History of Gunpowder.

NYTHP, 1975, **5**, 73 (Engl. abstr. 80).

Pak Hae-ill (1)　朴惠一.

李舜臣龜船의鐵裝甲과李朝鐵甲의現存原型의對比.

Parallelisms between the Iron Cladding of Admiral Yi Sunsin's Combat Turtle-Ships and Extant Iron Armouring of Yi Dynasty [City Gates].

HKH, 1979, **1** (no. 1), 27.

Pak Hae-ill (2)　朴惠一.

李舜臣龜船의鐵裝甲에對한補遺의註釋.

Supplementary Remarks on the Armoured Iron-clad Combat Turtle-Ships of Admiral Yi Sunsin (+1592).

HKH, 1982, **4** (no. 1), 26.

Phan Chi-Hsing (13)　潘吉星.

Lun Huo-Chien-ti Chhi Yuan　論火箭的起源.

On the Origin of the Rocket.

TJKHSYC 1985, **4** (no. 1), 64.

Engl. abstract in Proc. 3rd. International Conference on the History of Chinese Science, Peking, 1984, paper no. 68.

Phan Shih-Chhêng (1)　潘仕成.

Shui Lei Thu Shuo　水雷圖說.

Illustrated Account of Sea Mines.

Canton, 1843.

Later incorporated as chs. 92 and 93 of Hai Kuo Thu Chih.

Sakamoto Shunjō (1)　坂本俊奘.

Taihō Chūzōhō　大砲鑄造法.

On the Casting of Great Cannon.

Japan, begun 1804, finished c. 1835 (repr. NKKZ, vol.10, ch. 10, no. 4, p. 463; 2nd ed. vol. 5, ch. 10, p. 463).

Shen Li-Shêng (1) (ed.)　申力生.

Chung-Kuo Shih Yu Kung Yeh Fa Chan Shih　中國石油工業發展史.

Pt. 1 Ku-Tai-ti Shih Yu yü Thien-Jan Chhi　古代的石油與天然氣.

A History of the Development of the Oil Industry in China Pt. 1 Petroleum and Natural Gas in Ancient and Mediaeval Times.

Shih Yu Kung Yeh Chhu-Pan-Shih, Peking, 1984.

Tanaka Katsumi (1)　田中克己.

Tai Kokusenya sen ni okeru Kangun no yakuwari　對國姓爺戰における漢軍の役割.

On the Warfare of the Chinese Army at the beginning of the Chhing Dynasty.

Wada hakase koki kinen tōyōshi ronsō (Wada Festschrift), p. 589.

Abstr. RBS, 1968 (for 1961), **7**, no. 243.

Than Tan-Chhiung (2)　譚旦冏.

Chhêngtu Kung Chien Chih Tso Thiao-Chha Pao-Kao　成都弓箭製作調查報告.

Report of an Investigation of the Bow and Arrow Making Industry in Chhêngtu (Szechuan).

AS/BIHP, 1951, **23** (no. 1), 199 (Fu Ssu-Nien Festschrift).

Tr. C. Swinburne (unpub.).

Abstr. H. Franke (22), p. 238.

Thang Mei-Chün (1)　唐美君.

Thai-wan Thu Chu Min Tsu chih Nu chi Nu chih Fên Pu Yü Chhi-Yuan　臺灣土著民族之弩及弩之分佈與起源.

The Crossbows of the Aboriginal Peoples of Formosa, and the Origin and Diffusion of the Crossbow.

KKJL, 1958, no. 11, 5.

Thu Chi (1)　屠寄.

Mêng Wu Erh Shih Chi　蒙兀兒史記.

A History of the Mongols.

Peking, 1912.

Ting Fu-Pao (3), (d.)　丁福保.

Chhüan Han San-Kuo Chin Nan-Pei-Chhao Shih　全漢三國晉南北朝詩.

Complete Collection of Poetry from the Han, Three Kingdoms, Chin and Northern and Southern Kingdoms periods. (i.e. from the be-

Ting Fu-Pao (*3*), (d.) (*cont.*)
    ginning of Han to the beginning of Sui).
    Peking, *c.* 1935.
Tshao Yuan-Yü (*4*) 曹元宇.
    Chung-Kuo Hua-Hsüeh Shih Hua 中國化學
    史話.
    Talks on the History of Chemistry in China.
    Chiangsu Kho-Hsüeh Chi-Shu, Nanking, 1979.
Tshên Chia-Wu (*1*) 岑家梧.
    *Liao Tai Chhi-Tan ho Han Tsu chi Chhi-Tha Min-
    Tsu-ti Ching-Chi Wên-Hua Lien-Hsi* 遼代契
    丹和漢族及其他民族的經濟文化
    聯繫.
    The Relations in Economics and Culture be-
    tween the Chinese, the Liao (Chhi-Tan Tar-
    tars) and other minorities.
    LSYC, 1981, no. 1, 114.
Tshui Hsüan (*1*) 崔璿.
    *Nei Mêng-Ku Fa-Hsien-ti Ming Chhu Thung Huo
    Chhung* 內蒙古發現的明初銅火銃.
    Metal-Barrel Bombards of Bronze Discovered in
    Inner Mongolia.
    WWTK, 1973, no. 11 (no. 210), 55.
Tu Ya-Chhüan, 杜亞泉, Tu Chiu-Thien 杜就田 *et
    al.* (*1*).
    Tung Wu Hsüeh Ta Tzhu Tien 動物學大辭典.
    A Zoological Dictionary.
    Com. Press, Shanghai, 1932; repr. 1933.

Udagawa Yōan (*2*) *et al.* 宇田川榕庵.
    Kaijo Hojitsu Zenshō 海上砲術全書.
    Complete Treatise on Naval Artillery (contains
    a translation of J. N. Caltin's book (1) on this
    subject).
    Tokyo, *c.* 1847.

Wan Pai-Wu (*1*) 萬百五.
    *Wo Kuo Ku-Tai Tzu-Tung Chuang-Chih-ti Yuan-Li
    Fên-Hsi chi chhi Chhêng-Chiu-ti Than-Thao*
    我國古代自動裝置的原理分析及其
    成就的探討.
    On certain Automatic Devices and Machines in
    Ancient (and Medieval) China—A Discus-
    sion of their Principles and Achievements.
    AAS, 1965, **3** (no. 2), 57.
Wang Hsien-Chhen & Hsü Pao-Lin (*1*) 王顯臣
    & 許保林.
    Chung-Kuo Ku-Tai Ping Shu Tsa Than
    中國古代兵書雜談.
    A Discussion of the Ancient and Mediaeval
    Chinese Military Books.
    Chan Shih Chhu-Fan-Shih, Peking, 1983.
Wang Jen-Chün (*1*) 王仁俊.
    Ko Chih Ku Wei 格致古微.
    Scientific Traces in Olden Times.
    1896.
Wang Jung (*1*) 王榮.
    Yuan Ming Huo Chhung-ti Chuang-chih Fu-Yuan
    元明火銃的裝置復原.
    On the Restoration of the Carriage Mountings of
    the Yuan and Ming Bombards.
    WWTK, 1962, no. 3 (no. 137), 41.

Wang Khuei-Kho 王奎克 & Chu Shêng
    (*1*) 朱晟.
    *Chin-Tai Ta Lien Tan Chia Ko Hung yu Kuan Chih
    Chhü Tan Chih Shen ho Huo Yao Chhi Yuan Wên-
    Thi ti Chi Tsai* 晉代大煉丹家葛洪有關
    制取單質砷和火藥起源問題的記載.
    On the Great Alchemist Ko Hung of the Chin
    Period, and the Problem of Getting Pure
    Metallic Arsenic in connection with the
    Origin of Gunpowder.
    HHSTP, 1982, no. 1.
Wang Thao (*1*) 王韜.
    Tshao Shêng Yao Lan 操勝要覽.
    A Review of the Important Factors for Gaining
    Victories [cannon, cannon-founding, gun-
    powder manufacture, etc.]
    Chhing, *c.* 1870.
    In *THTYW*, 2, 3.
Wang Yü (*1*) 王愚.
    *Huo-Yao ti Fa-Ming Jen, Ma Chün; San Kuo Shih
    Fa-Ming Ta Chia chih I* 火藥的發明人馬鈞；
    三國時發明大家之一.
    The Inventor of Gunpowder, Ma Chün; one of
    the Greatest Inventive Geniuses of the Three
    Kingdoms Period.
    Art. in a journal about 1944, p. 39.
Wei Kuo-Chung (*1*) 魏國忠.
    *Heilungchiang A-chhêng Hsien Pan-la-chhêng-tzu
    Chhu-Thu-ti Thung Huo Chhung* 黑龍江阿城
    縣半拉城子出土的銅火銃.
    A Bronze Bombard excavated at Pan-la-chhêng-
    tzu in A-chhêng Hsien in Heilungchiang pro-
    vince (datable before +1290 because its
    accompanying objects were J/Chin in char-
    acter).
    WWTK, 1973, no. 11 (no. 210), 52.
Wei Yuan 魏源 & Lin Tsê-Hsü (*1*) 林則徐.
    Hai Kuo Thu Chih 海國圖志.
    Illustrated Record of the Maritime [Occidental]
    Nations
    1844, enlarged 1847, further enlarged 1852,
    abridged edition 1855.
    For the problem of authorship see Chhen Chhi-
    Thien (1) and Hummel (2), p. 851.
Wei Yûn-Kung (*1*) (ed.)
    Chiang-nan Chih Tsao Chü Chi 江南製造局記.
    A Record of the Kiangnan Arsenal.
    Wên-Pao, Shanghai, 1905.
Wu Chü-Hsien (*7*) 衛聚賢.
    Huo Yü Huo Yao 火與火藥.
    Fire and Fire-Weapons (lit. Gunpowder).
    Shuo Wên Press, Hsin Cha, Taiwan, 1979.

Yang Chia-Lo (*3*) 楊家駱.
    Ta-Tsu Thang Sung Shih Kho 大足唐宋石刻.
    The Thang and Sung Rock-carved (Temples) at
    Ta-tsu (Szechuen).
    Taipei, 1968.
Yang Hung (*1*) 楊泓.
    Chung-Kuo Ku Ping Chhi Lun Tshung 中國古
    兵器論叢.
    Ancient Chinese Weapons and War-Gear.

Yang Hung (*1*) (*cont.*)
　Wên Wu, Peking, 1980.
Yang Khuan (*1*) 楊寬.
　Chung-Kuo Ku-Tai Yeh-Thieh Ku-Fêng Lu ho Shui-
　Li Yeh Thieh Ku Fêng Lu ti Fa-Ming 中國古
　代冶鐵鼓風爐和水力冶鐵鼓風爐的
　發明.
　On the Blast Furnaces used for making Cast
　Iron in Ancient China, and the Invention of
　Hydraulic Blowing Engines for them.
　Essay in Li Kuang-Pi & Chhien Chün Yeh,
　　(q.v.), p. 71.
　Peking, 1955.
Yang Khuan (*6*) 楊寬.
　Chung-Kuo Ku-Tai Yeh Thieh Chi-Shu ti Fa-Ming ho
　Fa-Chan 中國古代冶鐵技術的發明
　和發展.
　The Origins, Inventions, and Development of
　Iron [and steel] Technology in Ancient and
　Medieval China.
　Jen-Min, Shanghai, 1956.

Yano Jinichi (*1*) 矢野仁一.
　Kindai Shina Seiji Oyobi Bunka 近代支那政
　治及文化.
　Political and Cultural Trends in China during
　the Modern Age.
　Tokyo, 1926.
Yoshida Mitsukuni (*7*) 吉田光邦.
　Chūgoku Kagaku-gijitsu-shi Ronshū 中國科學
　技術史論集.
　Collected Essays on the History of Science and
　Technology in China.
　Tokyo, 1972.
Yoshida Mitsukuni (*8*) 吉田光邦.
　Sōgen no Gunji Gijutsu 宋元の軍事技術.
　Military Technology in the Sung and Yuan
　Periods.
　Art. in Yabuuehi (26), p. 211.
Yoshida Tadashi (*1*) (ed.) 吉田忠.
　Higashi-asia no Kagaku 東アジアの科学.
　Science and its History in Eastern Asia.
　Tokyo, 1982.

C. BOOKS AND JOURNAL ARTICLES IN
WESTERN LANGUAGES

ACKER, WILLIAM R. B. (1). 'The Fundamentals of Japanese Archery.' Pr. pr. Kyoto, 1937 (with introduction in Japanese by Toshisuke Nasu).

ACKROYD, JOYCE (1) (tr.). *Told round a Brushwood Fire; the Autobiography of Arai Hakuseki* (+1657 to +1725). (Writer of the *Honchō Gunkikō*.) Princeton Univ. Press, Princeton, N.J., 1980; Tokyo Univ. Press, Tokyo, 1980. Rev. I. J. McMullen, *JRAS*, 1982 (no. 1), 95.

ADLER, B, (1). 'Das nordasiatische Pfeil; ein Beitrag zur Kenntnis d. Anthropo-Geographie des asiatischen Nordens.' *IAE*, 1901, **14** (Suppl.) 1.

ADLER, B. (2). 'Die Bogen Nordasiens' [including the Chinese bow]. *IAE*, 1902, **15**, 1 (with notes by G. Schlegel, pp. 31 ff. and footnotes in the paper itself by Ratzel and Conrady).

ALEXANDER, A. E. & JOHNSON, P. (1). *Colloid Science*. 2 vols. Oxford, 1949.

ALLAN, SARAH (1). 'Sons of Suns; Myth and Totemism in Early China.' *BLSOAS*, 1981, *44*, 290.

ALLEN, W. G. B. (1). *Pistols, Rifles and Machine-Guns; a Straightforward Explanation of their Mechanism, Construction and Role in Battle*. Eng. Univ. Press, London, 1953.

ALMOND, R., WALCZEWSKI, J., GOSSETT, R. W., MATTHEWS, A. J. & ROFE, B. (1). 'Meteorological Rocket Systems.' *PAA*, 1969, **22**, 29.

AMIOT, J. J. M. *See* de Rochemonteux (1).

AMIOT, J. J. M. (2). 'Sur l'Art Militaire des Chinois.' *MCHSAMUC*, 1782, **7**, 1–397+xx; Supplément, 1782, **8**, 327–75. (The translations of *Sun Tzu Ping Fa* and *Wu Tzu* were first sent to Europe in 1766.) The material of the main part first appeared in a separate book: *Art Militaire des Chinois, ou Recueil d'anciens Traités sur la Guerre, composés avant l'Ère Chrétienne, par différents Généraux Chinois*. Didot and Nyon, Paris, 1772. This prompted the work of de St Maurice & de Puy-Ségur (q.v.). The Supplement was stimulated by remarks in the *Recherches Philosophiques sur les Égyptiens et les Chinois* of de Pauw (q.v.).

ANON. (29). *The Arms and Armour of Old Japan*. Japan Society, London, 1905 (Catalogue of an exhibition).

ANON. (157). 'Feuerwerkbuch (in diesem buch das da heisset das fürwerckbuch).' A collection of gunpowder techniques of many decades from about +1430 onwards, contained in at least five MSS (*see* Partington (5), p. 152). First printed as an appendix to *Flavii Vegetii Renati Vier Büchern von der Ritterschaft*. Augsburg, 1529. Ed. Hassenstein (1).

ANON. (158). *Livre de Canonnerie et Artifice de Feu*, with appendix: 'Petit Traicté contenant plusieurs Artifices du Feu, très-utile pour l'Estat de Canonnerie, recueilly d'un vieil Livre éscrit à la main et nouvellement mis en lumière.' Paris, 1561. The MS in question (Paris BN 4653) is largely a translation of the German *Feuerwerkbuch*, and dates from about +1430. It was entitled: *Le Livre de Secret de l'Art de l'Artillerie et Canonnerie. See* Partington (5), pp. 101, 154.

ANON. (159). 'The Manner of Making Flowers in the Chinese Fire-Workes, illustrated with an elegantly engraved Copper-Plate—from the 4th Volume [just published] of the Memoirs Presented to the Academy of Sciences [at Paris].' *UM*, 1764, **34**, 21.

ANON. (162) *Twentieth-Century Science Fiction Writers*. Macmillan, London, 1982.

ANON. (163). 'China Successfully Launches another Man-made Earth Satellite.' *PKR*, 1976 (no. 50, Dec. 10), 5.

ANON. (164). 'Man Beats Hailstorms.' *PKR*, 1975 (no. 33, Aug. 15), 30.

ANON. (167). 'Apuntes Historicos sobre la Invencion de la Polvora.' *MART*, 1847, **3**, 19.

ANON. (169). 'Quicksilver in China.' *EMJ*, 1907, **84**, 152.

ANON. (170). 'Silver and Gold Mining in China.' *EMJ*, 1888, **46**, 194.

ANON. (196). 'The History of Making Gunpowder.' Art. in T. Sprat (1), 1667, p. 277.

ANON. (197). *How Things Work; the Universal Encyclopaedia of Machines*. 2 vols. Allen & Unwin, London, 1967; Granada–Paladin, London, 1972; reprinted many times. Tr. from the German of 1963 by C. van Amerongen.

APPIER (dit HANZELET), JEAN (1). *La Pyrotechnie de Hanzelet, ou sont representez les plus rare et plus appreuvez Secrets des Machines et des Feux Artificiels propres pour assieger, battre, surprendre et deffendre toutes places*. Bernard, Pont-à-Mousson, 1630.

APPIER (dit HANZELET), JEAN & THYBOUREL, FRANÇOIS, (Maître Chyrurgien), (1). *Receuil de plusieurs Machines Militaires, et Feux Artificiels pour la Guerre, et Récréation. Avec l'Alphabet de Tritemius, par laquelle chacun qui sçait écrire, peut promptement composer congrument en Latin. Aussi le moyen d'escrire la nuict à son amy absent*. Marchant, Pont-à-Mousson, 1620. See Partington (5), p. 176.

ARENDT, W. W. (1). 'The Scythian Stirrup [on the Chertomlyk Vase].' *ESA*, 1934, **9**, 206.

ARPAD, HORVÁTH (1). *Az Agyu Históriája* (A History of Gunpowder) (in Magyar). Zrínyi Katansi Kiadó, Budapest, 1966.

ASHTON, T. S. (1). 'Iron and Steel in the Industrial Revolution.' *MUP*, Manchester, 1924.

ATKINSON, W. C. (1) (tr.). *Camoens' 'The Lusiads'*. Penguin, London, 1952.

ATTWATER, R. *See* Duhr, J. (1), whose book he adapted for English readers.

AUBIN, FRANÇOISE (1) (ed.). *Études Song in Memoriam Étienne Balazs*. Sér. 1: 'Histoire et Institutions' 3 fascicles. Mouton, Paris, 1970–76. Sér. 2: 'Civilisation'. École des Hautes Études en Sciences Sociales, Paris, 1980– .

[AUDOT, L. E.] (1). *L'Art de Faire, à peu de Frais, les Feux d'Artifice*. Paris, 1818.

AYALON, DAVID (1). *Gunpowder and Firearms in the Mamluk Kingdom*. Vallentine Mitchell, London, 1956.

AYALON, DAVID (2). *The Mamlūk Military Society; Collected Studies*. Variorum, London, 1979. Contains a reprint of Ayalon (3).

AYALON, DAVID (3). 'A Reply to Professor J. R. Partington.' *ARAB*, 1963, **10**, 64. Critique of Partington (5).

BADDELEY, F. (1). *The Manufacture of Gunpowder*. HMSO, Waltham Abbey, 1830, 1857.

BAK HAE-ILL. *See* Pak Hae-ill.

BAKER, DAVID (1). *The Rocket; a History and Development of Rocket and Missile Technology*. New Cavendish, London, 1978.

BALAZS, E. & HERVOUET, Y. (1). *A Sung Bibliography*. Chinese University Press, Hongkong, 1978.

BALBI DA CORREGGIO, FRANCISCO. *See* di Correggio, F. Balbi (1).

BALFOUR, H. (3). 'On the Structure and Affinities of the Composite Bow.' *JRAI*, 1889, **19**, 220.

BALFOUR, H. (4). 'On a Remarkable Ancient Bow and Arrows believed to be of Assyrian Origin.' *JRAI*, 1897, **26**, 210.

BALFOUR, H. (5). 'The Archer's Bow in the Homeric Poems.' *JRAI*, 1921, **51**, 291.

BALL, J. DYER (1). *Things Chinese; being Notes on Various Subjects connected with China*. Hongkong, 1892; Murray, London, 1904; 5th ed. revised by E. T. C. Werner, Kelly & Walsh, Shanghai, 1925; repr. London, 1926.

BANKS, G. (1). 'Chinese Guns.' *ILN*, 1861, 325 (5 Apr.).

BARBA, ALVARO ALONZO (1). *El Arte de los Metales*. Madrid, 1640. Fr. tr. *Métallurgie, ou l'Art de tirer et de purifier les Métaux*. Paris, 1751.

BARBOSA, DUARTE (1). *A Description of the Coasts of East Africa and Malabar in the Beginning of the +16th Century, by D. B., a Portuguese*. Eng. tr. from a Spanish MS, by H. E. J. Stanley (Lord Stanley of Alderley). Hakluyt Society, London, 1866 (Hakluyt Soc. Pubs., 1st ser., no. 35). Eng. tr. from the Portuguese, by Hakluyt Society, London, 1918 (Hakluyt Soc. Pubs., 2nd. ser., no. 44).

BARBOTIN, A. (1). *L'Industrie des Pétards au Tonkin*. Paris, 1913.

BARNETT, R. D. & FALKNER, M. (1). *The Sculptures of Aššur-naṣir-apli II* (Ashurnasirpal; r. −883 to −859), *Tiglath-pileser III and Esarhaddon from the Central and Southwest Palaces at Nimrud*. London, 1962.

BAROWA, I. & BERBELICKI, WL. (1). *The Polish Scientific Book of the +15th to +18th Centuries; Exhibition Catalogue of the Jagellonian Library* [on the occasion of the International Congress of the History of Science]. Kraków, 1965.

BARROW, SIR JOHN (1). *Travels in China, containing Descriptions, Observations and Comparisons, made and collected in the Course of a Short Residence at the Imperial Palace of Yuen-Min-Yuen, and on a subsequent Journey through the Country from Pekin to Canton; in which it is attempted to appreciate the Rank that this Extraordinary Empire may be considered to Hold in the Scale of Civilised Nations*. Cadell & Davies, London, 1804. German tr. 1804; French tr. 1805; Dutch tr. 1809.

BASTIAN, A. (1). 'Die Völker des ostlichen Asiens; Studien und Reisen', 6 vols. Vol. 6. *Reisen in China; von Peking zur mongolischen Grenze, und Rückkehr nach Europa*. Hermann Costenoble, Jena, 1871.

BATE, JOHN (1). *The Mysteryes of Nature and Art: conteined in foure severall Tretises, the first of Water Workes, the second of Fyer Workes, the third of Drawing, Colouring, Painting and Engraving, the fourth of Divers Experiments, as wel serviceable, as delightful; partly collected, and partly of the authors peculiar practice and invention*. Harper, Mab, Jackson & Church, London, 1634, 1635, 1654. Photolitho reproduction, Theatrum Orbis Terrarum, Amsterdam, 1977 (The English Experience, no. 845). Bibliography in John Ferguson (2).

BAUERMEISTER, H. (1). 'Geschichte d. See-mine.' *BGTI/TG*, 1938, **27**, 98.

BEAL, S. (4). 'Account of the *Shui Lui* [*Shui Lei*] or Infernal Machine, described in the 58th volume [chapter] of the *Hoi Kwak To Chi* [*Haikuo Thu Chih*].' *JRAS* (trans.)/*NCB* 1859, **2** (no. 6), 53.

BEAZLEY, C. R. (3). *The Texts and Versions of John de Plano Carpini and William de Rubruquis*. Hakluyt Society, London, 1903.

BECK, T. (1). *Beiträge z. Geschichte d. Maschinenbaues*. Springer, Berlin, 1900.

BELL OF ANTERMONY, JOHN (1). *Travels from St Petersburg in Russia to Diverse Parts of Asia*. Vol. 1, *A Journey to Ispahan in Persia*, 1715 to 1718; *Part of a Journey to Pekin in China, through Siberia*, 1719 to 1721. Vol. 2, *Continuation of the Journey between Mosco and Pekin; to which is added, a translation of the Journal of Mr de Lange, Resident of Russia at the Court of Pekin*, 1721 and 1722, etc., etc. Foulis, Glasgow, 1763; repr. 1806. Repr. as *A Journey from St Petersburg to Pekin*, ed. J.L. Stevenson. University Press, Edinburgh, 1965.

BERGMAN, FOLKE (3). 'A Note on Ancient Laminar Armour.' *ETH*, 1936, **1** (no. 5).

BERNAL, J. D. (1). *Science in History*. Watts, London, 1954 (Beard Lectures at Ruskin College, Oxford, 1948). Repr. 4 vols. Penguin, London, 1969.

BERNAL, J. D. (2). *The Extension of Man; a History of Physics before 1900*. Weidenfeld & Nicolson, London, 1972. (Lectures at Birkbeck College, London, posthumously published.)

BERNAL, J. D. (3). *The Social Function of Science*. Routledge, London, 1939.

BERNARD, W. D. (1). *Narrative of the Voyages and Services of the 'Nemesis' from 1840 to 1843, and of the Combined Naval and Military Operations in China; comprising a complete account of the Colony of Hongkong, and Remarks on the Character of the Chinese, from the Notes of Cdr. W. H. Hall R.N., with personal observations*. 2 vols. Colburn, London, 1844.

BERNARD-MAITRE, H. (7). 'L'Encyclopédie Astronomique du Père Schall, *Chhung-Chêng Li Shu* (+1629) et *Hsi Yang Hsin Fa Li Shu* (+1645). La Réforme du Calendrier Chinois sous l'Influence de Clavius, Galilée et Kepler.' *MS*, 1937, **3**, 35, 441.

BERNARD-MAITRE, H. (18). 'Les Adaptations Chinoises d'Ouvrages Européens; Bibliographie chronologique depuis la venue des Portugais à Canton jusqu'à la Mission française de Pékin (+1514 à +1688).' *MS*, 1945, **10**, 1–57, 309–88.

BERNIER, FRANÇOIS (1). *Bernier's Voyage to the East Indies; containing The History of the Late Revolution of the Empire of the Great Mogul; together with the most considerable passages for five years following in that Empire; to which is added A Letter to the Lord Colbert, touching the extent of Hindostan, the Circulation of the Gold and Silver of the world, to discharge itself there, as also the Riches Forces and Justice of the Same, and the principal Cause of the Decay of the States of Asia—with an Exact Description of Delhi and Agra; together with (1) Some Particulars making known the Court and Genius of the Moguls and Indians; as also the Doctrine and Extravagant Superstitions and Customs of the Heathens of Hindustan, (2) The Emperor of Mogul's Voyage to the Kingdom of Kashmere, in 1664, called the Paradise of the Indies* ... Dass (for SPCK), Calcutta, 1909. [Substantially the same title-page as the editions of 1671 and 1672.]

BERNINGER, E. H. (1). Introduction to the *Vollkommene Geschütz-, Feuerwerck-, und Büchsenmeisterey-Kunst* (+1676) *of Casimir Siemienowicz* (d. +1650). Akad. Druck u. Verlagsanstalt, Graz, 1976. (Veröfftl. d. Forschungsinst. d. Deutschen Museums, A, 186.)

BERNOULLI, JOHANNES, the elder (1). *Dissertatio de Effervescentia et Fermentatione nova hypothesi fundata*. Bertsch, Basel, 1690. Often reprinted with *De Motu Musculorum*, e.g. Venice, 1721. Also in *Opera Omnia*, Lausanne & Geneva, 1742.

BERTHELOT, M. (4). 'Pour l'Histoire des Arts Mécaniques et de l'Artillerie Vers la Fin du Moyen Age (I).' *ACP*, 1891 (6⁰ sér.), **24**, 433. (Descr. of Latin MS Munich, no. 197, the Anonymous Hussite engineer (German), *c.* +1430; of Ital. MS Munich, no. 197: Marianus Jacobus Taccola of Siena, *c.* +1440; of *De Machinis*, Marcianus, no. XIX, 5, *c.* +1449; and of *De Re Militari*, Paris, no. 7239: Paulus Sanctinus, *c.* +1450, the MS from Istanbul.)

BERTHELOT, M. (5). 'Histoire des Machines de Guerre et des Arts Mécaniques au Moyen Age; (II) Le Livre d'un Ingénieur Militaire à la Fin du 14ème Siècle.' *ACP*, 1900 (7⁰ sér.), **19**, 289. (Descr. of MS *Bellifortis*, Göttingen, no. 63, Phil. K. Kyeser +1395 to +1405 and of Paris, MS no. 11015, Latin, Guido da Vigevano, *c.* +1335.)

BERTHELOT, M. (6). 'Le Livre d'un Ingénieur Militaire à la Fin du 14ème Siècle.' *JS*, 1900, 1 & 85. (Konrad Kyeser and his *Bellifortis*.)

BERTHELOT, M. (7). 'Sur le Traité *De Rebus Bellicis* qui accompagne le *Notitia Dignitatum* dans les Manuscrits.' *JS*, 1900, 171.

BERTHELOT, M. (8). 'Les Manuscrits de Léonard da Vinci et les Machines de Guerre.' *JS*, 1902, 116. (Argument that L. da Vinci knew the drawings in the +4th-century anonymous *De Rebus Bellicis* and also many inventions and drawings of them by the +14th- and early +15th-century military engineers.)

BERTHELOT, M. (9). 'Les Compositions Incendiaires dans l'Antiquité et Moyen Ages.' *RDM*, 1891, **106**, 786. Crit. Fêng Chia-Shêng (**8**).

BERTHELOT, M. (10). *La Chimie au Moyen Age*; vol. 1, *Essai sur la Transmission de la Science Antique au Moyen Age* (Latin texts). Impr. Nat. Paris, 1893. Photo-repr. Zeller, Osnabrück; Philo, Amsterdam, 1967. Rev. W. P[agel], *AX*, 1967, **14**, 203.

BERTHELOT, M. (12). 'Archéologie et Histoire des Sciences; avec Publication nouvelle du Papyrus Grec chimique de Leyde, et Impression originale du *Liber de Septuaginta* de Geber.' *MRAS/P*, 1906, **49**, 1–377. Sep. pub. 197.

BERTHELOT, M. (13). *Sur la Force des Matières Explosives d'après la Thermochimie*. 3rd ed. Paris, 1883. Vol. 1, pp. 352 ff. has section: 'Des Origines de la Poudre et des Matières Explosives'. Crit. Fêng Chia-Shêng (**8**).

BERTHELOT, M. (14). 'Histoire des Corps Explosifs' (review of S. J. M. von Romocki's book). *JS*, 1895, 684.

BERTHELOT, M. & DUVAL, R. (1). *La Chimie au Moyen Age*; vol. 2. *l'Alchimie Syriaque*. Impr. Nat. Paris, 1893. Photo-repr. Zeller, Osnabrück; Philo, Amsterdam, 1967. Rev. W. P[agel], *AX*, 1967, **14**, 203.

BERTHELOT, M. & HOUDAS, M. O. (1). *La Chimie au Moyen Age*: vol. 3, *l'Alchimie Arabe*. Impr. Nat. Paris, 1893. Photo-repr. Zeller, Osnabrück; Philo, Amsterdam, 1967. Rev. W. P[agel], *AX*, 1967, **14**, 203.

BEVERIDGE, H. (1). 'Oriental Crossbows.' *AQR*, 1911, **32** (no. 3), 344.

BIBILASHVILI, N. S. *et al.* (1). *Anti-hail Rockets and Shells.* Proc. WMO/IAMAP Scientific Conference on Weather Modification, Tashkent, Oct. 1973. (WMO Paper no. 399, 1974, pp. 333–41.)

BINGHAM, Cdr. J. ELLIOTT (1). *Narrative of the Expedition to China, from the Commencement of the War to the present Period.* 2 vols. Colburn, London, 1842.

BIOT, E. (18). 'Mémoire sur les Colonies Militaires et Agricoles des Chinois.' *JA*, 1850 (4ᵉ Ser.), **15**, 338 & 529.

BIRCH, THOMAS (1). *The History of the Royal Society of London, for Improving of Natural Knowledge, from its first rise; in which The most considerable of those Papers communicated to the Society, which have hitherto not been published, are inserted in their proper order, As a Supplement to the Philosophical Transactions.* 4 vols. Millar, London, 1756–7.

BIRINGUCCIO, VANNOCCIO (1). *Pirotechnia.* Venice, 1540, 1559. Eng. tr. C. S. Smith & M. T. Gnudi, Amer. Inst. Mining Engineers, New York, 1942. Account in Beck (1), ch. 7. *See* Sarton (1), vol. 3, p. 1554, 1555. Bibliography in John Ferguson (2).

BIRKENMAIER, A. (2). 'Zur Lebensgeschichte und wissenschaftlichen Tätigkeit von Giovanni Fontana (c. +1395 bis c. +1455).' *ISIS*, 1932, **17**, 34.

BISHOP, C. W. (10). 'Notes on the Tomb of Ho Chhü-Ping' [Han general, d. –117]. *AA*, 1928, **4**, 34.

BISHOP, C. W. (11). 'The Horses of Thang Thai Tsung; on the Antecedents of the Chinese Horse.' *MUJ* 1918, **9**, 244.

BISHOP, C. W. (14). 'An Ancient Chinese Cannon.' *ARSI*, 1940, 431 (and pl. 10).

BISSET, N. G. (1). 'Arrow Poisons in China, Pt. I.' *JEPH*, 1979, **1**, 325.

BISSET, N. G. (2). 'Arrow Poisons in China, Pt. II; *Aconitum*, its Botany, Chemistry and Pharmacology.' *JEPH*, 1981, **4**, 247.

BISWAS, ASIT K. (3). *A History of Hydrology.* North-Holland, Amsterdam and London, 1970.

BLACKMORE, HOWARD L. (1). *Guns and Rifles of the World.* Batsford, London, 1965.

BLACKMORE, HOWARD L. (2). *The Armouries of the Tower of London; I. Ordnance.* HMSO, for the Dept. of the Environment, London, 1976.

BLACKMORE, HOWARD L. (3). 'The Seven-Barrel Guns.' *JAAS*, 1954, **1** (no. 10), 165.

BLACKMORE, HOWARD L. (4). *Firearms.* Dutton Vista, New York, 1964; Studio Vista, London, 1964.

BLACKMORE, HOWARD L. (5). *Hunting Weapons.* Walker, London, 1971.

BLAIR, C. (1). 'A Note on the Early History of the Wheel-Lock.' *JAAS*, 1961, **3**, 221.

BLOCH, M. R. (5). 'Two Ancient Saltpetre Plants by the Dead Sea.' Introduction to *Report on the Excavations at Um Baraque* [on the Western Coast of the Dead Sea], by M. Gichon. In the press.

BLOCHET, E. (2). *Introduction à l'Histoire des Mongols* (tr. from the *Jami 'al-Tawārīkh* of Fadl Allah Raschid ed-Din al-Hamadānī), E. J. W. Gibb Memorial series, vol. XII. Leyden, London, 1910.

BÖCKMANN, F. (1). *Die Explosiven Stoffe...* Hartleben, Leipzig, 1880.

BODDE, DERK (25). *Festivals in Classical China: New Year and Other Annual Observances During the Han Dynasty, 206 B.C.–A.D. 220.* Princeton Univ. Press, 1975.

BOEHEIM, WENDELIN (1). *Handbuch d. Waffenkunde; das Waffenwesen in seiner historischen Entwickelung vom Beginn des Mittelalters bis zum Ende des 18. Jahrhunderts.* Seemann, Leipzig, 1890. Photolitho reprint, Zentralantiquariat d. Deutschen Demokratischen Republik, Leipzig, 1982.

BOERHAAVE, H. (1). *Elementa Chemiai, quae anniversario labore docuit, in publicis, privatisque, scholis.* 2 vols. Severinus and Imhoff, Leiden, 1732. Eng. tr. by P. Shaw: *A New Method of Chemistry; including the History, Theory and Practice of the Art.* 2 vols. Longman, London, 1741, 1753.

BOGUE, R. H. (1). *The Chemistry and Technology of Gelatin and Glue.* McGraw-Hill, New York, 1922.

BONAPARTE, PRINCE NAPOLÉON-LOUIS & FAVÉ, Col. I. (1). *Études sur le Passé et l'Avenir de l'Artillerie.* 4 vols., the first two written by Bonaparte, the second two by Favé, on the basis of the emperor's notes. Dumaine, Paris, 1846–51 and 1862–63. Vol. 1. First Part, chs. 1–4, Artillery on the Battlefield 1328 to 1643. Vol. 2. Second Part, chs. 1–4, Artillery in Siege Warfare 1328 to 1643. Vol. 3. Third Part, chs. 1–9, History of Gunpowder and Artillery to 1650. Vol. 4. Third Part, chs. 10–14, History of Artillery 1650–1793. Schneider (1) p. 10 gives date and place of first publication Liège, 1847 and Sarton (1), vol. 3, p. 726, describes 6 vols to 1871.

BONGARS, JACQUES (1) (ed.). *Gesta Dei per Francos, sive Orientalium Expeditionum, et Regni Francorum Hiero-solimitani Historia.* 2 vols. Hannover, 1611.

BOODBERG, P. A. (5). *The Art of War in Ancient China; a Study based upon the 'Dialogues of Li, Duke of Wei'* [*Li Weikung Wên Tui*]. Inaug. Diss., Berkeley, 1930.

BOOTS, J. L. (1). 'Korean Weapons and Armour [including Firearms].' *JRAS/KB*, 1934, **23** (no. 2), 1–37.

BORNET, P. (1). 'La Préface des *Novissima Sinica*.' *MS*, 1956, **15**, 328.

BORNET, P. (2). 'Au Service de la Chine; Schall et Verbiest, maîtres-fondeurs, I. Les Canons.' *BCP*, 1946 (no. 389), 160.

BORNET, P. (3) (tr.). '"Relation Historique" [de Johann Adam Schall von Bell S.J.]; Texte Latin avec Traduction française.' Hautes Études, Tientsin, 1942 (part of *Lettres et Mémoires d'Adam Schall S.J.*, ed. H. Bernard-[Maître]).

BOSIO, GIACOMO (1). *Dell'Istoria della Sacra Religione e Illustra Militia di San Giovanni Gierosolimitano*. 2 vols. Rome, 1594.

BOSMANS, H. (2). 'Ferdinand Verbiest, Directeur de l'Observatoire de Pékin.' *RQS*, 1912, **71**, 196 & 375. Sep. pub. Louvain, 1912.

BOSMANS, H. (4). *Documents relatifs à Verbiest*. Bruges, 1912.

BOURNE, WILLIAM (2). *The Art of Shooting in Great Ordnaunce*. London, 1587.

BOURNE, WILLIAM (3). *Inventions or Devises, Very Necessary for all Generalles and Captaines, or Leaders of Men, as wel by Sea as by Land, Written by W.B.* Woodcock, London, 1578.

BOVILL, E. W. (1). 'Queen Elizabeth's Gunpowder.' *MMI*, 1947, **33** (no. 3), 179.

BOWERS, F. (1). *The Dramatic Works in the Beaumont & Fletcher Canon*. 2 vols. Cambridge, 1966.

BOXER, C. R. (7). *The Christian Century in Japan*. Univ. Calif. Press, Berkeley, Calif., 1951.

BOXER, C. R. (8). 'Notes on Chinese abroad in the Late Ming and Early Manchu Periods, compiled from contemporary European Sources (+1500 to +1700).' *TH*, 1939, 454.

BOXER, C. R. (11). 'Asian Potentates and European Artillery in the +16th to +18th Centuries; a Footnote to Gibson-Hill.' *JRAS/M*, 1965, **38**, 156.

BOXER, C. R. (12). 'Portuguese Military Expeditions in aid of the Ming against the Manchus, +1621 to +1647.' *TH*, 1938, 24.

BOXER, CAPT. [EDWARD] (1). *The Congreve Rocket*. 1853. Pr. in *Treatise on Artillery* as Sect. 1, pt. 2. Eyre & Spottiswoode, London, 1860. Repr. Museum Restoration Service, Ottawa, 1970.

BOYLE, J. A. (1) (tr.). *The 'Ta'rīkh-i Jahān-Gushā' (History of the World Conqueror, Chingiz Khan), by 'Alā'al-Dīn 'Aṭā-Malik [al-]Jurvaynī [+1233 to +1283]*. 2 vols. Harvard Univ. Press, Cambridge, Mass., 1958.

BOYLE, ROBERT (6). *New Experiments Physico-Mechanicall touching the Spring of the Air*. Oxford, 1660.

BOYLE, ROBERT (7). *A Continuation of New Experiments Physico–Mechanicall touching the Spring and Weight of the Air*. Oxford, 1669.

BOYLE, ROBERT (8). *Some Considerations touching the Usefulnesse of Experimental Natural Philosophy, propos'd in a Familiar Discourse to a Friend, by way of Invitation to the Study of it*. Hall & Davis, Oxford, 1663; 2nd ed. 1664.
I, Pt. 1, 126 pp. followed by an index. Pt. 2, Sect. 1, 416 pp. followed by 'Citations Englisht'. Pt. 2, Sect. 2 (1671), 'Generall Considerations', followed by separate title-pages for:
II, 'Of the Usefulnesse of Mathematicks to Naturall Philosophy'.
III, 'Of the Usefulnesse of Mechanicall Disciplines to Naturall Philosophy'.
IV, 'That the Goods of Mankind may be much encreased by the Naturalists Insight into Trades' (with an Appendix).
V, 'Of doing by Physicall Knowledge what is wont to require Manuall Skill'.
VI, 'Of Mens great Ignorance of the Uses of Naturall Things'.

BRACKENBURG, SIR HENRY (1). 'Ancient Cannon in Europe; Pt. 1, from their First Employment to +1350.' *PRAI*, 1865–6, **4**, 287. Pt. 2 'From +1351 to +1400'. *PRAI*, 1867, **5**, 1.

BRADFORD, ERNLE (1). *The Great Siege; Malta, +1565*. Hodder & Stoughton, London, 1961; Penguin, London, 1964; repr. 1966, 1968, 1970, 1971.

BRADFORD, ERNLE (2). *The Shield and the Sword; the Knights of Malta*. Hodder & Stoughton, London, 1972; Penguin, London, 1974.

BRAUDEL, F. (2). *Civilisation Matérielle et Capitalisme (+15e au +18e Siècle)*. 2 vols. Colin, Paris, 1967. (Deals with the early history of artillery in vol. 1, pp. 294 ff.)

VON BRAUN, WERNHER & ORDWAY, FREDERICK I. (1). *The Rockets' Red Glare; an Illustrated History of Rocketry through the Ages*. Anchor Doubleday, New York, 1976.

VON BRAUN, WERNHER & ORDWAY, FREDERICK I. (2). *A History of Rocketry and Space Travel*. Nelson, London, 1966. (With illustrations by H. H. K. Lange.)

BREDON, J. & MITROPHANOV, I. (1). *The Moon Year; a Record of Chinese Customs and Festivals*. Kelly & Walsh, Shanghai, 1927.

BREDT, CLAYTON (1). 'Fighting for Fun, and in Earnest' (Notes on the History of Martial Arts, Gunpowder and Firearms in China). *HEM*, 1977, **21** (no. 10), 9.

BREWER, J. S. (1) (ed.). *Fr. Rogeri Bacon: 'Opus Tertium, Opus Minus, Compendium Philosophiae'*. Longman Green Longman & Roberts, London, 1859 (Rolls Series no. 15).

BREWER, W. H. (1). *Up and down California in 1860–64* [a journal], ed. F. P. Farquhar. New York, 1900.

BRINK, C. O. (2). *Horace on Poetry*, with text of *De Arte Poetica*. 3 vols. Cambridge, 1963, 1971, 1982.

BROCK, A. St H. (1). *A History of Fireworks*. Harrap, London, 1949.

BROCK, A. St H. (2). *Pyrotechnics; the History and Art of Firework Making*. O'Connor, London, 1922.

BROCKBANK, W. (1). *Ancient Therapeutic Arts*. Heinemann, London, 1954. (Fitzpatrick Lectures, Royal College of Physicians, 1950–1.)

Bromehead, C. E. N. (9). 'Mining and Quarrying [from Early Times to the Fall of the Ancient Empires].' Art. in *A History of Technology*, ed. C. Singer *et al.*, Oxford, 1954, vol. 1, p. 558.

Brown, Delmer M. (1). 'The Impact of Firearms on Japanese Warfare, +1543 to +1598.' *FEQ*, 1947, 7, 236.

Brown, M. L. (1). *Firearms in Colonial America; the Impact on History and Technology, 1492 to 1792*. Smithsonian Inst. Press, Washington, D.C., 1980. Rev. V. Foley, *TCULT*, 1982, 23, 118.

Buck, P. (1) (tr.). *All Men are Brothers (Shui Hu Chuan)*. London, 1937. A translation based on the 70-chapter version but without the verses.

Buckle, H. T. (1). *A History of Civilisation in England*. 2 vols. Parker, London, 1857.

Budge, Sir E. A. Wallis (2). *The Monks of Kûblâi Khân, Emperor of China; or, The History of the Life and Travels of Rabban Ṣawmâ, Envoy and Plenipotentiary of the Mongol Khâns to the Kings of Europe, and Markôs who as Mâr Yahbh-Allâhâ III became Patriarch of the Nestorian Church in Asia; translated from the Syriac...* Religious Tract Society, London, 1928. Reprod. AMS Press, New York, 1973. (Tr. from Yish'iata demâr Yahbalâdha vderaban Ṣauma.)

Buedelor, W. (1). *Geschichte der Raumfahrt*. Sigloch, Kuenzelsau, 1979. Rev. E. F. M. Rees, *TCULT* 1981, 22 (no. 4), 816.

Bull, H. B. (1). *Physical Biochemistry*. Wiley, New York, 1943.

Burton, Sir Richard (1). *The Book of the Sword*. Chatto & Windus, London, 1884.

Burtt, Joseph (1). 'Extracts from the Pipe Roll of the Exchequer for 27 Edw. III (+1353) relating to the Early Use of Guns and Gunpowder in the English Army.' *ARJ*, 1862, 19, 68.

Butler, A. R., Glidewell, C. & Needham, Joseph (1). 'The Solubilisation of Cinnabar; Explanation of a Sixth-Century Chinese Alchemical Recipe.' *JCR(S)*, 1980, 47; (M), 1980, 0817–0832.

Butler, A. R. & Needham, Joseph (1). 'An Experimental Comparison of the East Asian, Hellenistic and Indian (Gandhāran) stills in relation to the Distillation of Ethanol and Acetic Acid.' *AX*, 1980.

Byron, Robert (1). *The Byzantine Achievement; an historical perspective A.D. 330–1454*. London, 1929.

Cable, M. & French, F. (1). *The Gobi Desert*. Hodder & Stoughton, London, 1942; Macmillan, New York, 1944.

Cadonna, A. (1). 'Astronauti' Taoisti da Chhang-an alla Luna; Note sul Manoscritto di Dunhuang S 6836 alla Luce di alcuni Lavori di Edward H. Schafer.' *OV*, 1983.

'Cafari et Continuatorum Annales Januae.' *See* Pertz, G. H. (1).

Cahen, C. (1) 'Un Traité d'Armurerie composé pour Saladin' (Salaḥ al-Dīn, +1138/+1193) (the *al-Tabṣira...fī'l Ḥurûb....* (Explanations of Defence and Descriptions of Military Equipment) written by Murdā ibn 'Alī ibn Murdā al-Tarsūsī, an Armenian of Alexandria, about +1185). *BEO/IFD*, 1947, 12, 103.

Caillois, R. (1) 'Lois de la Guerre en Chine' (a study of *Sun Tzu, Wu Tzu, Ssuma Ping Fa*, etc. after Amiot, 2). *PVS*, 1956, 16

Caillot, A. (1). *Curiosités Naturelles, Historiques et Morales de l'Empire de la Chine; ou, Choix des Traits les plus Intéressans de l'Histoire de ce Pays, et des Relations des Voyageurs qui l'ont-Visité, à l'Usage de la Jeunesse*. 2 vols. Ledentu, Paris, 1818.

Calten, J. N. (1). *Leiddraad bij het Onderrigt in de Zee-artillerie...* Zweesaardt, Amsterdam, 1832. 2nd ed. 1847.

Camden, William (1). *Remaines concerning Britaine, reviewed, corrected and encreased*. Legatt & Waterson, London, 1614.

de Camoens, Luis (1). *Os Lusiados*. Lisbon, 1572 (facsimile Lisbon, 1943). Eng. trns. *see* Fanshawe, R. (1); Aubertin, J. J. (1) (with Portuguese); Burton, R. F. (2); Atkinson, W. C. (1) (prose); Mickle, W. J. (1) (Popian couplets, not recommended, and contains insertions and deletions, but with interesting notes).

Cardwell, D. S. L. (1). *Steam Power in the +18th Century; a Case Study in the Application of Science*. Sheed & Ward, London, 1963. (Newman History & Philosophy of Science Series, no. 12.)

Cardwell, Robert (1). 'Pirate-Fighters of the South China Sea.' *NGM*, 1946, 89 (no. 6), 787.

Carey, W. (1). 'An Account of the Funeral Ceremonies of a Burman Priest.' *ASKR*, 1816, 12, 186.

Carnat, G. (1). *Le Fer à Cheval à travers l'Histoire et l'Archéologie; contributions à l'histoire de la Civilisation*. Paris, 1951.

Caron, François & Schouten, Joost (1). *A True Description of the Mighty Kingdoms of Japan and Siam. Written...in Dutch... and now rendred into English by Capt. Roger Manley*. London, 1663, 1671. Repr. ed. C. R. Boxer, Argonaut, London, 1935.

Carpeaux, C. (1). *Le Bayon d'Angkor Thom; Bas-Reliefs publiés... d'après les documents recueillis par le Mission Henri Dufour*. Leroux, Paris, 1910.

Casiri, M. (1). *Bibliotheca Arabico-Hispana Escurialensis sive Librorum Omnium MSS quos Arabicè ab Auctoribus magnam partem Arabo-Hispanis compositos Bibliotheca Coenobii Escurialensis complectitur. Recensio et Explanatio Operâ et Studio Michaelis Casiri*. 2 vols. Madrid, 1770.

Cellini, Benevenuto (1). Autobiography, *Vita di Benvenuto Cellini, Orefice e Scultore Fiorentino, da lui medesimo Scritta*. Ed. N. Bettoni. Milan, 1821. Eng. tr. by R. H. H. Cust. 2 vols. Bell, London, 1910.

CHABOT, J. B. (1). *Histoire de Mar Jabalaha III*. Paris, 1895.

CHABOT, J. B. (2). 'Note sur les Relations du Roi Arghun [the Mongol Ilkhan of Persia] avec l'Occident.' *ROL*, 1894, **2**, 570.

CHAKRAVARTI, P. C. (1). *The Art of War in Ancient India*. Dacca, 1941.

CHAMBERS, JAMES (1). *The Devil's Horsemen; the Mongol Invasion of Europe*. Weidenfeld & Nicolson, London, 1979.

CHAN HOK-LAM. *See* Chhen Ho-Lin.

CHANG THIEN-TSÊ (1). *Sino-Portuguese Trade from +1514 to +1644; a Synthesis of Portuguese and Chinese Sources*. Brill, Leyden, 1933; repr. 1969.

CHAVANNES, E. (20). 'Une Stèle de l'Année +554' in *Six Monuments de la Sculpture Chinoise*. van Oest, Paris & Brussels, 1914 (Ars Asiatica, no. 2).

CHAVANNES, E. (22). 'Leou Ki [Lou Chi] et Sa Famille.' *TP*, 1914, **15**, 193. (Includes a discussion of Lou Chi's descendant, Lou Chhien-Hsia, who held Nan-ning against the Mongols for the Sung, and who blew up his men and himself with gunpowder bombs in +1277, seeing that further resistance was useless.)

CHÊNG CHEN-TO (1). 'Building the New, Uncovering the Old.' *CREC*, 1954, **3** (no. 6), 18.

CHÊNG LIN (2) (tr.). *The Art of War, a Military Manual [Sun Tzu Ping Fa] written c. −510 (with original Chinese Text)*. World Encyclopaedia Institute, Chungking, 1945. Repr. World Book Co., Shanghai, 1946.

CHÊNG TÊ-KHUN (18). 'Cannons of the Opium War on the Campus of Amoy University, Fukien.' *CJ*, 1937. Adapted from *Hsia-Mên Ta-Hsüeh Hsiao-Chih Khao* (in Chinese). *BAU*, 1936, **22** (no. 15), 3. Repr. in Chêng Tê-Khun (19), p. 119.

CHÊNG TÊ-KHUN (19). *Studies in Chinese Archaeology*. Chinese Univ. Press, Shatin, Hongkong, 1982 (Centre for Chinese Archaeology and Art Studies Series, no. 3).

CHERONIS, N. D. (1). 'Chemical Warfare in the Middle Ages; Kallinikos' "Prepared Fire".' *JCE*, 1937, **14**, 360.

CHERTIER, F. M. (1). *Nouvelles Recherches sur les Feux d'Artifice*. Paris, 1854.

CHHEN CHHI-THIEN (1). *Liu Tsê-Hsü; Pioneer Promoter of the Adoption of Western Means of Maritime Defence in China*. Dept. of Economics, Yenching Univ., Vetch (French Bookstore), Peiping, 1934. ([Studies in] Modern Industrial Technique in China, no. 1.)

CHHEN CHHI-THIEN (2). *Tsêng Kuo-Fan; Pioneer Promoter of the Steamship in China*. Dept. of Economics, Yenching Univ., Vetch (French Bookstore), Peiping, 1935. ([Studies in] Modern Industrial Technique in China, no. 2.)

CHHEN CHHI-THIEN (3). *Tso Tsung-Thang; Pioneer Promoter of the Modern Dockyard and the Woollen Mill in China*. Dept. of Economics, Yenching Univ., Vetch (French Bookstore), Peiping, 1938. ([Studies in] Modern Industrial Technique in China, no. 3.)

CHHEN HO-LIN (1). 'Liu Chi (+1311 to +1375) in the *Ying Lieh Chuan*; the Fictionalisation of a Scholar-Hero.' *JOSA*, 1967, **5**, 25.

CHHEN YUAN (1). *Western and Central Asians in China under the Mongols; their Transformation into Chinese*. Tr. and annot. Chhien Hsing-Hai & L. C. Goodrich. Monumenta Serica, Univ. Calif., Los Angeles, 1966. (Monumenta Serica Monographs, no. 15.)

CHHIEN HSÜEH-SÊN (Tsien Hsue-Shen) & MALINA, F. J. (1). 'Flight Analysis of a Sounding Rocket with Special Reference to Propulsion by Successive Impulses.' *JAEROS*, 1938, **6**, 50.

CHIANG CHÊNG-LIN (1). *Ancient Chinese Rockets*. Unpub. typescript issued by China Features Office (Chhen Lung), 1963.

CIOLKOVSKIJ, KONSTANTIN E. *See* Tsiolkovsky.

CIPOLLA, C. M. (1). *Guns and Sails in the Early Phase of European Expansion, +1400 to +1700*. Collins, London, 1965. Ital. tr. 1969. Subsequently published together with (2) as 'European Culture and Overseas Expansion'. Penguin, London, 1970.

CLAGETT, MARSHALL (4). 'The Life and Works of Giovanni Fontana.' *AIMSS*, 1976, 1, 5.

CLAGETT, MARSHALL (5). *Archimedes in the Middle Ages*. 3 vols. Philadelphia, 1978.

CLARK, BRACY (1). *Essay on the Knowledge of the Ancients respecting the Art of Shoeing the Horse, and of the probable period of the Commencement of the Art*. London, 1831.

CLARK, GRAHAME (2). 'Horses and Battle-Axes.' *AQ*, 1941, **15**, 50.

CLARK, JOHN D. (1). *Ignition; an Informal History of Liquid Rocket Propellants*. Rutgers Univ. Press, New Brunswick, N.J., 1972.

CLEPHAN, R. COLTMAN (1). 'An Outline of the History of Gunpowder and that of the Hand-Gun, from the Epoch of the Earliest Records to the end of the 15th century.' *ARJ*, 1909, **66** (2nd ser.), **16**, 145.

CLEPHAN, R. COLTMAN (2). 'The Ordnance of the +14th and +15th Centuries.' *ARJ*, 1911, **68** (2nd ser.), **18**, 49–138.

CLEPHAN, R. COLTMAN (3). 'The Military Handgun of the +16th Century.' *ARJ*, 1910, **67** (2nd ser.), **17**, 109.

CLEPHAN, R. COLTMAN (4). 'Some Very Early Types of Hand-Guns.' *ANTIQ*, 1909, **45**, 93. Crit. Fêng Chia-Shêng (8).

CLEPHAN, R. COLTMAN (5). *An Outline of the History and Development of Hand Firearms*. London and Felling-on-Tyne, 1906.

CLOW, A. & CLOW, NAN L. (1). *The Chemical Revolution; a Contribution to Social Technology*. Batchworth, London, 1952.

COLIN, G. S., AYALON, DAVID, PARRY, V. J., SAVORY, R. M. & YAR MUHAMMAD KHAN (1). Art. 'Bārūd' (saltpetre, later gunpowder itself) in *Enc. of Islam*. Brill and Luzac, Leiden and London, 1958, vol. 1, pt. 2, pp. 1055 ff.

COLLADO, LUYS (1). *Pratica Manuale di Artiglieria, nella quale si tratta della Inventione di essa, dell'ordine di condurla & piantarla sotto à qualcunque Fortezza, fabricar Mine da far volar im alto le Fortezze, spiaviar le Montagne, divertir l'acque offensive à i Regni & Provincie, tirar co i pezzi in molti & diversi modi, far fuochi artificieli*, etc. Venice, 1586.

COLLINS, A. L. (1). 'Fire-setting; the Art of Mining by Fire.' *TFIME*, 1892, **5**, 82.

CONGREVE, WILLIAM (1). *A Concise Account of the Origin and Progress of the Rocket System; with a View of the apparent advantages, both as to the Effect produced, and the ... Saving of Expence, arising from the peculiar facilities of application which it possesses, as well for Naval as Military purposes*. Whiting, London, 1807.

CONGREVE, COL. [WILLIAM] (2). *The Details of the Rocket System: Shewing the Various Applications of this Weapon, both for Land and Sea Service, and its Different Uses in the Field and in Sieges; illustrated by Plates of the Principal Equipments, Exercises, and Cases of Actual Service, with General Instructions for its Application, and Demonstration of the Comparative Economy of the System ...* Whiting, London, 1814. Repr. Museum Restoration Service, Ottawa, 1970.

CONGREVE, MAJOR-GENERAL SIR W[ILLIAM] (3). *A Treatise on the General Principles, Powers, and Facility of Application of the Congreve Rocket System, as compared with Artillery: showing the Various Applications of this Weapon, both for Land and Sea Service ...* Enlarged 2nd. ed. of Congreve (2) but more elegantly printed, with folding plates of engravings. Longman, Rees, Orme, Brown and Green, London, 1827.

CONNOLLY, P. (1). *Greece and Rome at War*. Macdonald Phoebus, London, 1981.

CONNOR, J. (1). 'Tests of Response to Detonative Stimuli.' Paper in *Proc. 1st International Loss Prevention Symposium*. The Hague, Netherlands, 1974, p. 104.

CONNOR, J. (2). 'Explosion Risk of Unstable Substances.' Paper in *Proc. 1st International Loss Prevention Symposium*. The Hague, Netherlands, 1974, p. 265.

COOK, M. A. (1) (ed.). *Studies in the Economic History of the Middle East, from the Rise of Islam to the Present Day*. Oxford, 1970.

COOMARASWAMY, A. K. (7). 'The Blow-pipe in Persia and India.' *AAN*, 1943, **45**, 311.

COOPER, MICHAEL (1). *Rodrigues the Interpreter; an Early Jesuit in Japan and China*. Weatherhill, Tokyo and New York, 1974.

CORDIER, H. (1). *Histoire Générale de la Chine*. 4 vols. Geuthner, Paris, 1920.

CORDIER, H. (8). *Essai d'une Bibliographie des Ouvrages publiés en Chine par les Européens au 17e et au 18e siècle*. Leroux, Paris, 1883.

CORNELISSE, J. W., SCHÖYER, H. F. R. & WAKKER, K. F. (1). *Rocket Propulsion and Spaceflight Dynamics*. Pitman, London, 1981 (1st ed. 1979).

CORRÉARD, J. (1). *Histoire des Fusées de Guerre; ou, Recueil de tout ce qui a été publié ou écrit sur ce Projectile*. Paris, 1841.

DI CORREGGIO, FRANCISCO BALBI (1). *La Verdadera Relacion de todo lo que esto Año de MDLXV ha sucedido en la Isla de Malta*. Alcala de Henares, 1567; Barcelona, 1568. Eng. tr. by H. A. Balbi, *The Siege of Malta, 1565*. Copenhagen, 1961.

CORTESAO, A. (2) (tr. & ed.). *The Suma Oriental of Tomé Pires, an Account of the East from the Red Sea to Japan ... written in ... 1512 to 1515 ...* London, 1944 (Hakluyt Society Pubs., 2nd ser. nos. 89, 90).

COURANT, M. (1). *Bibliographie Coréenne*. Paris, 1894–6, 3 vols. with one suppl. vol. (Pub. École. Langues Or. Viv. (3° sér.), nos. 18, 19, 20.) Photolitho reproduction, Burt Franklin, New York, 1975.

COWPER, H. S. (1). *The Art of Attack, being a Study in the Development of Weapons and Appliances of Offence, from the Earliest Times to the Age of Gunpowder*. Holmes, Ulverston, 1906.

CRANSTONE, B. A. L. (1). 'The Blow-gun in Europe.' *MA*, 1949, **49**, 119.

CRAUFURD, QUINTIN (1). *Sketches chiefly relating to the History, Religion, Learning and Manners of the Hindoos*. London, 1790; 2nd ed. 1792.

CRICK, FRANCIS (1). *Life Itself; its Origin and Nature*. McDonald, London and Sydney, 1981.

DE CRISTOFORIS, LUIGI (1). 'Di Una Macchina Igneo-pneumatica.' *ARIL*, 1841, **2**, 22.

CRUCQ, K. C. (1). 'Cannon surviving in Indonesia.' *TBG*, 1930, **70**, 195; 1937, **77**, 105; 1938, **78**, 93, 359; 1940, **80**, 34; 1941, **81**, 74.

DA CRUZ, GASPAR (1). *Tractado em que se cotam muito por esteco as cousas da China*. Evora, 1569, 1570 (the first book on China printed in Europe), tr. Boxer (1), and originally in *Purchas his Pilgrimes*, vol. 3, p. 81. London, 1625.

CURWEN, C. A. (1). *Taiping Rebel; the Deposition of Li Hsiu-Chhêng.* Cambridge, 1977.
CURWEN, M. D. (1), (ed.). *Chemistry and Commerce,* 4 vols. Newnes, London, 1935.
CUSHING, F. H. (1). 'The Arrow.' *AAN,* 1895, **8**, 344.
CUTBUSH, JAMES (1). *A System of Pyrotechny, Comprehending the Theory and Practice, with the Application of Chemistry; Designed for Exhibition and for War.* Philadelphia, 1825.
CUTBUSH, JAMES (2). 'Remarks on the Composition and Properties of the Chinese Fire, and on the so-called Brilliant Fires.' *AJSC,* 1823, **7**, 118.

DARDESS, J. W. (1). *Conquerors and Confucians; Aspects of Political Change in Late Yuan China.* Columbia Univ. Press, New York and London, 1973.
DARMSTÄDTER, L. (1) (with the collaboration of R. duBois-Reymond & C. Schäfer). *Handbuch zur Geschichte d. Naturwissenschaften u. d. Technik.* Springer, Berlin, 1908.
DATE, Q. T. (1). *The Art of War in Ancient India.* Mysore and Oxford, 1929.
DAUMAS, M. (3). 'Le Brevet du Pyréolophore des Frères Niepce (1806).' *DHT,* 1961, **1**, 23.
DAVIS, J. F. (1). *The Chinese; a General Description of China and its Inhabitants.* 1st ed. 1836. 2 vols. Knight, London, 1844, 3 vols., 1847, 2 vols. French tr. by A. Pichard, Paris, 1837, 2 vols. Germ. trs. by M. Wesenfeld, Magdeburg, 1843, 2 vols. and M. Drugulin, Stuttgart, 1847, 4 vols.
DAVIS, TENNEY L. (10). 'Early Chinese Rockets.' *TR,* 1948, **51**, 101, 120, 122.
DAVIS, TENNEY L. (11). 'Early Pyrotechnics; I, Fire for the Wars of China, II, Evolution of the Gun, III, Chemical Warfare in Ancient China.' *ORD,* 1948, **33**, 52, 180, 396.
DAVIS, TENNEY L. (15). 'The Cultural Relationships of Explosives.' *NFR,* 1944, **1**, 11.
DAVIS, TENNEY L. (17). *The Chemistry of Powder and Explosives.* 2 vols. Wiley, New York, Chapman & Hall, London, 1941, 1943. 4th repr. 1956 (bound in one vol.).
DAVIS, TENNEY L. (18). 'The Early Use of Potassium Chlorate in Pyrotechny; Dr Moritz Meyer's Coloured Flame Compositions.' *CHYM,* 1948, **1**, 75.
DAVIS, TENNEY L. & CHAO YÜN-TSHUNG (9) (tr.). 'Chao Hsüeh-Min's Outline of Pyrotechnics [*Huo Hsi Lüeh*]; a Contribution to the History of Fireworks.' *PAAAS,* 1943, **75**, 95.
DAVIS, TENNEY L. & WARE, J. R. (1). 'Early Chinese Military Pyrotechnics' (analysis of *Wu Pei Chih,* chs. 119 to 134). *JCE,* 1947, **24**, 522.
DAWES, H. F. (1). 'Chinese Silver Mining in Mongolia.' *EMJ,* 1891, **54**, 335.
DAWSON, H. CHRISTOPHER (3) (ed.). *The Mongol Mission; Narratives and Letters of the Franciscan Missionaries in Mongolia and China in the +13th and +14th Centuries, translated by a Nun of Stanbrook Abbey.* Sheed & Ward, London, 1955.
DAY, D. J. H. (1). 'Early Non-Electric Ignition Systems for Internal Combustion Engines.' *TNS,* 1981, **53**, 171.
DEBUS, A. G. (9). 'The Aerial Nitre in the +16th and early +17th Centuries.' Communication to the Xth International Congress of the History of Science, Ithaca, N.Y., 1962. In *Communications,* p. 835.
DEBUS, A. G. (10). The Paracelsian Aerial Nitre.' *ISIS,* 1964, **55**, 43.
DEBUS, A. G. (13). 'Solution Analyses Prior to Robert Boyle.' *CHYM,* 1962, **8**, 41.
DEBUS, A. G. (18). *The English Paracelsians.* Oldbourne, London, 1965; Watts, New York, 1966. Rev. W. Pagel, *HOSC,* 1966, **5**, 100.
DEBUS, H. (1). 'The Chemical Theory of Gunpowder' (Bakerian Lecture). *PRS,* 1882, no. 218, 361.
DELLRÜCK, H. (1). *Geschichte d. Kriegskunst im Rahmen der politischen Geschichte.* Stilke, Berlin, 1907–36. 7 vols.
DEMMIN, A. (1). *Die Kriegswaffen in ihrer historischen Entwicklung.* Leipzig, 1886, then 1893. Eng. tr. C. C. Black, *Weapons of War, being a history of arms and armour from the earliest period to the present time.* Bell & Daldy, London, 1870. Re-issued, with changed title: *Illustrated History of Arms and Armour from the earliest period to the present time.* Bell, London, 1877.
DENING, W. (1). *The Life of Toyotomi Hideyoshi.* 5 parts, Tokyo, 1888–90. 3rd ed. ed. M. E. Dening, Thompson, Kobe, and Kegan Paul, London, 1930.
DENIS, E. (1). *Hus et la Guerre des Hussites.* Paris, 1930.
DENNIS, A. S. (1). *Weather Modification by Cloud Seeding.* Academic Press, New York, 1980 (International Geophysics Series, no. 24).
DENWOOD, P. (1) (ed.). *Arts of the Eurasian Steppelands.* David Foundation, London, 1978 (Colloquies on Art and Archaeology in Asia, No. 7).
DESSENS, J. (1). 'Cloud seeding for hail suppression (programme of the Association Nationale de Lutte contre les Fléaux Atmosphériques).' *JWM,* 1979, **11**, 4.
DIBNER, B. (1). 'Leonardo da Vinci; Military Engineer' in *Studies and Essays in the History of Science and Learning* (Sarton Presentation Volume), ed. M. F. Ashley-Montagu. Schuman, New York, 1944, p. 85. Sep. pub. Burndy Library, New York, 1946.
DICK, STEVEN J. (1). *Plurality of Worlds; the Extra-terrestrial Life Debate, from Democritus to Kant.* Cambridge, 1982.

DICKINSON, H. W. (4). *A Short History of the Steam-Engine*. Cambridge, 1939. Re-issued, with introduction by A. E. Musson, Cass, London, 1963. Rev. L. T. C. Rolt, *TCULT*, 1965, **6**, 115.

DICKINSON, H. W. (6). 'The Steam-Engine to 1830.' Art. in *A History of Technology*, ed. C. Singer *et al.*, vol. 4, p. 168. Oxford, 1958.

DICKINSON, H. W. (7). 'James Watt, Craftsman and Engineer.' Cambridge, 1936.

DICKINSON, H. W. & TITLEY, A. (1). *Richard Trevithick, the Engineer and the Man*. Cambridge, 1934.

DIELS, H. (1). *Antike Technik*. Teubner, Leipzig & Berlin, 1914. 2nd ed. 1920. Rev. B. Laufer, *AAN*, 1917, **19**, 71.

DIELS, H. & SCHRAMM, E. (3) (ed. & tr.). 'Excerpte aus Philons Mechanik [Bks. VII, *Paraskeuastika*, and VIII, *Poliorketika*] vulgo Bk IV' (mostly on tactical use of artillery). *APAW/PH*, 1919, no. 12.

DIKSHITAR, V. R. R. (1). *War in Ancient India*. Macmillan, Madras, 1944 (2nd edition 1948).

DOGIEL, I. (1). 'Ein Mittel, die Gestalten der Schneeflocken künstlich zu erzeugen.' *MPCASP*, 1879, **9**, 266.

DOLLFUS, C. (1). *The First Burden- or Payload-carrying Rockets*. Allocution à la Réunion de la Commission d'Histoire de l'Académie Internationale Astronautique, Sept. 1963.

DORÉ, H. (1). *Recherches sur les Superstitions en Chine*. 15 vols. T'u-Se-Wei Press, Shanghai, 1914-29.
  Pt. I, vol. 1, pp. 1-146: 'Superstitious' practices, birth, marriage and death customs (*VS*, no. 32).
  Pt. I, vol. 2, pp. 147-216: talismans, exorcisms and charms (*VS*, no. 33).
  Pt. I, vol. 3, pp. 217-322: divination methods (*VS*, no. 34).
  Pt. I, vol. 4, pp. 323-488: seasonal festivals and miscellaneous magic (*VS*, no. 35).
  Pt. I, vol. 5, sep. pagination: analysis of Taoist talismans (*VS*, no. 36).
  Pt. II, vol. 6, pp. 1-196: Pantheon (*VS*, no. 39).
  Pt. II, vol. 7, pp. 197-298: Pantheon (*VS*, no. 41).
  Pt. II, vol. 8, pp. 299-462: Pantheon (*VS*, no. 42).
  Pt. II, vol. 9, pp. 463-680: Pantheon, Taoist (*VS*, no. 44).
  Pt. II, vol. 10, pp. 681-859: Taoist celestial bureaucracy (*VS*, no. 45).
  Pt. II, vol. 11, pp. 860-1052: city-gods, field-gods, trade-gods (*VS*, no. 46).
  Pt. II, vol. 12, pp. 1053-1286: miscellaneous spirits, stellar deities (*VS*, no. 48).
  Pt. III, vol. 13, pp. 1-263: popular Confucianism, sages of the Wên miao (*VS*, no. 49).
  Pt. III, vol. 14, pp. 264-606: popular Confucianism, historical figures (*VS*, no. 51).
  Pt. III, vol. 15, sep. pagination: popular Buddhism, life of Gautama (*VS*, no. 57).

DRACHMANN, A. G. (2). 'Ktesibios, Philon and Heron; a Study in Ancient Pneumatics.' *AHSNM*, 1948, **4**, 1-197.

DRACHMANN, A. G. (4). 'Remarks on the Ancient Catapults [Calibration Formulae].' *Actes du VIIe Congrès International d'Histoire des Sciences, Jerusalem, 1953*, p. 279.

DRACHMANN, A. G. (9). 'The Mechanical Technology of Greek and Roman Antiquity; a Study of the Literary Sources.' Munksgaard, Copenhagen, 1963.

DREW, R. B. (1). 'The Hide and Bone Glue Industries' in *Chemistry and Commerce*, ed. M. D. Curwen (1), vol. 4.

DREYER, E. L. (1). *The Emergence of Chu Yuan-Chang, +1360 to +1365*. Inaug. Diss., Harvard, 1970.

DREYER, E. L. (2). 'The Po-yang [Lake] Campaign, +1363; Inland Naval Warfare in the Founding of the Ming Dynasty.' Art. in Kierman & Fairbank (1) (ed.), *Chinese Ways in Warfare*, p. 202.

DUBOIS-REYMOND, C. (2). 'Notes on Chinese Archery.' *JRAS/NCB*, 1912, **43**, 32.

DUHEM, P. (4). *Un Fragment Inédit de 'l'Opus Tertium' de Roger Bacon*. Quaracchi, Florence, 1909.

DUHR, J. (1). *Un Jésuite en Chine, Adam Schall*. Desclée de Brouwer, Paris, 1936. Eng. adaptation by R. Attwater, *Adam Schall, a Jesuit at the Court of China, 1592 to 1666*. Geoffrey Chapman, London, 1963. Not very reliable sinologically.

DUNLOP, D. M. (10). 'Hāfiz-i Abru's Version of the Timurid Embassy to China in +1420.' *TGUOS*, 1948, **11**, 15.

DURRANT, P. J. (1). *General and Inorganic Chemistry*. 2nd ed. repr. Longmans Green, London, 1956.

DUTHEIL, G. *See* de la Porte du Theil, G. (1).

DUYVENDAK, J. J. L. (19). 'Desultory Notes on the *Hsi Yang Chi* [Lo Mou-Têng's novel of +1597 based on the Voyages of Chêng Ho]' (concerns spectacles and bombards). *TP*, 1953, **42**, 1.

EBBUTT, M. I. (1). *Hero Myths and Legends of the British Race*. Harrap, New York. (*Beowulf* repr. in *PAR*, 1982, **7** (no. 4), 20.)

EBERHARD, W. (2). 'Lokalkulturen im alten China.' *TP* (Suppl.), 1943, **37**; *MS* Monograph no. 3, 1942. (Crit. H. Wilhelm, *MS*, 1944, **9**, 209.)

EBERHARD, W. (3). 'Early Chinese Cultures and their Development, a Working Hypothesis.' *ARSI*, 1937, 513 (Pub. no. 3476).

EBERHARD, W. (27). *The Local Cultures of South and East China*. Brill, Leiden, 1968.

EBERHARD, W. (31). *Chinese Festivals*. Orient Cultural Service, Thaipei, 1972 (Asian Folklore and Social Life Monographs, no. 38).

EGERTON OF TATTON, LORD (1). *A Description of Indian and Oriental Armour.* Allen, London, 1896.

ELLERN, H. & LANCASTER, R. (1). *Military and Civilian Pyrotechnics.* Chem. Pub. Co., New York, 1968. Rev. *SAM*, 1969, **220** (no. 4), 140.

ELLIOTT, SIR H. M. (1). *The History of India as told by its own Historians; the Muhammedan Period.* (Posthumous papers edited and continued by J. Dowson.) 6 vols. London, 1867–75.

EMME, E. M. (1) (ed.). *The History of Rocket Technology; Essays on Research, Development and Utility.* Wayne State Univ. Press, Detroit, 1970.

ERBEN, W. (1). 'Beiträge z. Geschichte des Geschützwesens im Mittelalter.' *ZHWK*, 1916, **7**, 85, 117.

ERCKER, L. (1). *Beschreibung Alle-fürnemsten Mineralischen Ertzt und Berckwercks Arten....* Prague, 1574. 2nd ed. Frankfurt, 1580. Eng. tr. by Sir John Pettus, as *Fleta Minor, or, the Laws of Art and Nature, in Knowing, Judging, Assaying, Fining, Refining and Inlarging the Bodies of confin'd Metals...* Dawks, London, 1683. *See* Sisco & Smith (2); Partington (7), vol. 2, pp. 104 ff.

VON ERDBERG-CONSTEN E. (1). 'A *Hu* with Pictorial Decoration' (late Chou battle scenes). *ACASA*, 1952, **6**, 18.

ESNAULT-PELTERIE, R. (1). *L'Exploration par Fusée de la très haute Atmosphère et la Possibilité des Voyages Inter-planétaires.* Soc. Astron. de France, Paris, 1928.

ESNAULT-PELTERIE, R. (2). *L'Astronautique.* Lahure, Paris, 1930.

ESPÉRANDIEU, E. (2). *Recueil Général des Bas-Reliefs, Statues et Bustes de la Gaule Romaine.* Imp. Nat. Paris, 1908, 1913.

VON ESSENWEIN, A. (2). *Quellen zur Geschichte der Feuerwaffen.* Leipzig, 1877.

EVANS, J. (1). *The History of St. Louis* [by Jean de Joinville, 1309]. London, 1938.

ÉVRARD, R. & DESCY, A. (1). *Histoire de l'Usine des Vennes, suivie de Considérations sur les Fontes Anciennes.* Éditions Solédi, Liège, 1948.

FABER, FELIX (1). *F. Fabri Evagatorium in Terrae Sanctae, Arabiae et Egyptae Peregrinationem* [+1483]. Ed. C. D. Hassler. Stuttgart, 1843.

FABER, H. B. (1). *Military Pyrotechnics; the Manufacture of Military Pyrotechnics, an Exposition of the Present Methods and Materials Used.* 3 vols. Govt. Printing Office, Washington D.C., 1919.

FAIRBANK, J. K. (3). 'Varieties of the Chinese Military Experience.' Art. in Kierman & Fairbank (1) (ed.), *Chinese Ways in Warfare*, p. 1.

FAIRBANK, J. K. (4). 'The Creation of the Treaty System.' Art. in *Cambridge History of China*, ed. D. Twitchett & J. K. Fairbank, vol. 10, p. 213. Cambridge, 1978.

FALK, H. S. & TORP, ALF (1). *Norwegisch-Dänisches Etymologisches Wörterbuch.* Universitetsforlaget, Oslo, 1960.

FANSHAWE, RICHARD (1) (tr.). *The Lusiad, or Portugalls Historicall Poem...* Moseley, London, 1655. Ed. and repr. J. D. M. Ford. Harvard Univ. Press, Cambridge, Mass., 1940.

DE FARIA Y SOUSA, MANUEL (1). *Imperio de la China y Cultura Evangelica en el por los Religiosos de la Compañia de Jesus....* Officinà Herreriana, Lisbon, 1731.

FARIS, NABIH AMIN & ELMER, R. P. (1). *Arab Archery; an Arabic MS of about +1500: 'A Book on the Excellence of the Bow and Arrow' and the Description thereof.* Princeton Univ. Press, Princeton, N.J., 1945.

FELDHAUS, F. M. (1). *Die Technik der Vorzeit, der Geschichtlichen Zeit, und der Naturvölker* (encyclopaedia). Engelmann, Leipzig and Berlin, 1914. Photographic reprint; Liebing, Würzburg, 1965.

FELDHAUS, F. M. (2). *Die Technik d. Antike u. d. Mittelalter.* Athenaion, Potsdam, 1931. Crit. H. T. Horwitz, *ZHWK*, 1933, **13** (NF **4**), 170.

FELDHAUS, F. M. (28). 'Eine Chinesische Stangenbüchse von +1421.' *ZHWK*, 1907, **4**, 256.

FELDHAUS, F. M. (29). 'Geschützkonstruktionen von Leonardo da Vinci.' *ZHWK*, 1913, **6**, 128.

FELDHAUS, F. M. (30). 'Zur ältesten Geschichte des Schiesspulvers in Europa.' *ZAC*, 1906, **19**, 465.

FELDHAUS, F. M. (31). 'Die ältesten Nachrichten über Berthold den Schwarzen, den angeblichen Erfinder des Schiesspulvers.' *CHZ*, 1907, **31**, 831.

FELDHAUS, F. M. (32). 'Der Pulvermönch Berthold, +1313 oder +1393?' *ZAC*, 1908, **21**, 639, *CHZ*, 1908, **32**, 316.

FÊNG CHIA-SHÊNG (1). 'The Origin of Gunpowder and its Diffusion Westwards.' Unpub. MS.

FÊNG TA-JAN & KILBORN, L. G. (1). 'Nosu and Miao Arrow Poisons.' *JWCBRS*, 1937, **9**, 130.

FENTON, H. J. H. (1). *Notes on Qualitative Analysis, Concise and Explanatory.* Cambridge, 1916.

FERGUSON, J. C. (8) [& CHANG, S. K.]. 'The Six Horses of Thang Thai Tsung.' *JRAS/NCB*, 1936, **67**, 1.

FERGUSON, J. C. (9). 'The Tomb of Ho Chhü-Ping' [Han general, d. −117]. *AA*, 1928, **4**, 228.

FERNALD, H. E. (2). 'The Horses of Thang Thai Tsung and the Stele of Yü [Shih-Hsiung].' *JAOS*, 1935, **55**, 420.

FFOULKES, C. (1). *Armour and Weapons.* Oxford, 1909.

FFOULKES, C. (2). *Arms and Armament, an Historical Survey of the Weapons of the British Army.* Harrap, London, 1945.

FIELD, D. C. (1). 'Internal-Combustion Engines' [in the Late Nineteenth Century]. Art. in *A History of Technology*, ed. C. Singer *et al.*, Oxford, 1958, vol. 4, p. 157.

FINDLAY, A. (1). *Practical Physical Chemistry*. Longmans Green, London, 1923.

FIRIGER, F. G. *et al.* (1). 'The Compatibility of Meteorological Rocketsonde Data as indicated by International Comparison Tests.' *JATMOS*, 1975, **32**, 1705.

FISCHER-WIERUSZOWSKI, F. (1). 'Kriegerischer Einfall d. Mongolen in Japan; eine japanische Bildrolle.' *OAZ*, 1935 (NF), **11**, 121.

FISCHLER, G. (1). 'Über Pulverproben früherer Zeiten.' *ZHWK*, 1927, **11**, 49.

FISHER, I. (1). 'A New Method for Indicating Food Values.' *AJOP*, 1906, **15**, 417.

FLAMMARION, CAMILLE (1). *Mondes Imaginaires et Mondes Réels*. Paris, 1865.

DE FLURANCE, RIVAULT (1). *Les Elemens de l'Artillerie*. 1607.

FOLEY, V. & PERRY, K. (1). 'In Defence of *Liber Ignium*; Arab Alchemy, Roger Bacon, and the Introduction of Gunpowder into the West.' *JHAS*, 1979, **3** (no. 2), 200.

DE FONTENELLE, B. (1). *Entretiens sur la Pluralité des Mondes*. Brunet, Paris, 1698. (1st ed. 1686; Eng. trs. 1688, 1702.)

FOOTE, G. B. & KNIGHT, C. A. (1). *Hail; a Review of Hail Science and Hail Suppression*. Amer. Meteorol. Soc., Boston, 1977 (Meteorological Monographs, no. 38).

[FORBES, R. J.] (4*a*). *Histoire des Bitumes, des Époques les plus Reculées jusqu'à l'an 1800*. Shell, Leiden, n.d.

FORBES, R. J. (4*b*). *Bitumen and Petroleum in Antiquity*. Brill, Leiden, 1936.

FORBES, R. J. (8). 'Metallurgy [in the Mediterranean Civilisations and the Middle Ages].' In *A History of Technology*, ed. C. Singer *et al.*, vol. 2, p. 41. Oxford, 1956.

FORBES, R. J. (20). *Studies in Early Petroleum History*. Brill, Leiden, 1958.

FORBES, R. J. (21). *More Studies in Early Petroleum History*. Brill, Leiden, 1959.

FORDHAM, S. (1). *High Explosives and Propellants*. Pergamon, Oxford, 1981.

FORKE, A. (17). 'Der Festungskrieg im alten China.' *OAZ*, 1919, **8**, 103. (Repr. from Forke (3), pp. 99 ff.)

FORKE, A. (18). 'Über d. chinesischen Armbrust.' *ZFE*, 1896, **28**, 272.

FOSTER, J. (1). 'Crosses from the Walls of Zaitun [Chhüanchow].' *JRAS*, 1954, 1–25.

FRANKE, H. (14). 'Some Aspects of Chinese Private Historiography in the + 13th and + 14th Centuries.' Art. in *Historians of China and Japan*, ed. W. G. Beasley & E. G. Pulleyblank, p. 115. London, 1961.

FRANKE, H. (20). *Sino-Western Contacts under the Mongol Empire*. Hume Memorial Lecture, Yale University, New Haven, 1965.

FRANKE, H. (22). 'Besprechungen ostasiatische Neuerscheinungen.' *ORE*, 1956, **3** (no. 2), 234.

FRANKE, H. (23) (tr.). 'Die Verteidigung von Shao-Hsing; *Pao Yüeh Lu* von Hsü Mien-Chih.' Unpub. typescript, 1957.

FRANKE, H. (24). 'Siege and Defence of Towns in Mediaeval China.' Art. in Kierman & Fairbank (1) (ed.), *Chinese Ways in Warfare*, p. 151.

FRANKE, H. (25). 'Die Belagerung von Hsiang-yang; eine Episode aus dem Krieg zwischen Sung und Chin, + 1206 bis + 1207.' Art. in *Society and History* (Wittfögel Festschrift), ed. G. L. Ulmen, 1978.

FRANKE, H. (26). Review of L. Olschki's *Marco Polo's Asia* (10). *ZDMG*, 1962, **112** (NF **37**), 228.

FRANKE, O. (1). *Geschichte d. chinesischen Reiches*. 5 vols. de Gruyter, Berlin, 1930–53. Crit. O. B. van der Sprenkel, *BLSOAS*, 1956, **18**, 312.

FRANKE, W. (3). *Preliminary Notes on Important Literary Sources for the History of the Ming Dynasty Chhêngtu, 1948*. (SSE Monographs Ser. A, no. 2.)

FRANKE, W. (4). *An Introduction to the Sources of Ming History*. Univ. Malaya Press, Kuala Lumpur and Singapore, 1968.

FRANKEL, SIGMUND (1). *Die Aramäische Fremdwörter in Arabische*. Brill, Leiden, 1886.

DELLA FRATTE E MONTALBANO, MARIO ANTONIO (1). *Pratica Minerale*. Bologna, 1678.

FRÉDÉRIC, LOUIS. *See* Louis-Frédéric.

FRÉZIER, A. FRANÇOIS (1). *Traité des Feux d'Artifice, pour le Spectacle*. Paris, 1706; repr. 1747.

FROISSART, JEAN (1). *Chronicles*, 1400, ed. J. A. Buchon. Paris, 1824. K. de Lettenhove, Brussels, 1867. Cf. Sarton (1), vol. 5, p. 1751.

FRONSPERGER, LEONHARD (1). *Grossem Kriegsbuch; dass ist Fünff Bücher, vonn Kriegs-Regiment und Ordnung, wie sich ein jeder Kriegssmann in seinem Ampt unnd Beuelch halten soll*. Frankfurt a/Main, 1555, 1558, 1564, 1596, 1598. The second edition has woodcuts by Jost Amman. The title varies greatly according to the edition, some include mention of Wagenburg tactics.

FRONSPERGER, LEONHARD (2). *Vonn Geschütz und Fewerwerck, wie dasselb zuwerffen und schiessen; Auch von gründtlicher Zubereytung allerley Gezeugs, unnd rechtem Gebrauch der Fewerwerck... Das ander Buch. Von erbawung, erhaltung, besatzung und profantierung der wehrlichen Beuestigungen...* Lechler, Frankfurt a/Main, 1557, 1564.

FUCHS, W. (7). 'Ein Gesandschaftsbericht ü. Fu-Lin in chinesischer Wiedergabe aus den Jahren + 1314 bis + 1320.' *ORE*, 1959, **6**, 123.

FUJIKAWA YU (1). *Geschichte der Medizin in Japan; Kurzgefasste Darstellung der Entwicklung der japanischen*

Medizin, mit besonderer Berücksichtigung der Einführung der europäischen Heilkunde in Japan. Imper. Jap, Ministry of Education, Tokyo, 1911.

FULLER, J. F. C. (1). *Dragon's Teeth; a Study of War and Peace.* Constable, London, 1932.

FULLMER, J. Z. (1). 'Technology, Chemistry and the Law in Early 19th-Century England.' *TCULT*, 1980. **21** (no. 1), 1.

FURTENBACH, J. (1). *Architectura Navalis. Das ist: Von dem Schiff Bebäw.* Ulm, 1629.

FURTENBACH, J. (2). *Halinitro Pyrbolia: Schreibung eine newen Büchsenmeisterey, nemlichen: Gründlicher Bericht wie die Salpeter, Schwefel, Kohlen unnd das Pulfer zu praepariren, zu probieren, auch langwirzig gut zu behalten: Das Fewerwerck zu Kurtzweil und Ernst zu laboriren. . . Alles aufz eygener Experientza.* Ulm, 1627.

FURTTENBACH, J. *See* Furtenbach, J.

LE GALLEC, Y. (1). 'Les Origines du Moteur à Combustion Interne.' *MC/TC*, 1951, **2**, 28.

GALLOWAY, R. L. (1). *The Steam-Engine and its Inventors; a Historical Sketch.* Macmillan, London, 1881.

GARDNER, G. B. (1). *Keris (Kris) and other Malay Weapons.* Progressive Pub. Co., Singapore, 1936.

GARLAN, Y. (1). *Recherches de Poliorcétique Grecque.* Boccard, Limoges, 1974. (Bibl. des Écoles Françaises d'Athènes et de Rome, no. 223.)

GARNIER, JOSEPH (1). *L'Artillerie des Ducs de Bourgogne, d'après les Documents conservés aux Archives de la Côte-d'Or.* Paris, 1895.

GARNIER, JOSEPH (2). *L'Artillerie de la Commune de Dijon, d'après les Documents conservés dans les Archives.* Paris, 1863.

GARRISON, F. H. (3). *An Introduction to the History of Medicine.* Saunders, Philadelphia, 1913; 4th ed. 1929.

GASSER, ACHILLES (1). *Chronica der Weitberempten Keyserlichen Freyen. . . Statt Augspurg.* Frankfurt, 1595.

GAUBIL, A. (12). *Histoire de Gentchiscan [Chingiz Khan] et de toute la Dinastie des Mongous ses Successeurs, Conquérans de la Chine.* Briasson & Piget, Paris, 1739.

DE GAYA, LOUIS (1). *Traité des Armes.* Paris, 1678. Repr. and ed. C. Ffoulkes, Oxford, 1911. Eng. tr. by R. Harford: *A Treatise of Arms, of Engines, Artificial Fires, Ensignes, and of all Military Instruments.* Harford, London, 1680.

GEIL, W. E. (3). *The Great Wall of China.* Murray, London, 1909.

GENOESE ANNALS. *See* Pertz, G. H. (1).

GERLAND, E. & TRAUMÜLLER, F. (1). *Geschichte d. physikalischen Experimentierkunst.* Engelmann, Leipzig, 1899.

GIBB, HUGH (1). *The River Mekong.* Script for a BBC Television programme, 1965.

GIBBS-SMITH, C. H. (10). 'The Rockets' Red Glare.' *LI*, 1956, **55** (no. 1409), 320.

GIBSON, C. E. (2). *The Story of the Ship.* Schuman, New York, 1948; Abelard–Schuman, New York, 1958.

GIBSON-HILL, C. A. (1). 'Notes on the Old Cannon found in Malaya and known to be of Dutch Origin.' *JRAS/M*, 1953, **26**, 145.

GILES, L. (11) (tr.). *Sun Tzu on the Art of War* ['*Sun Tzu Ping Fa*']; *the Oldest Military Treatise in the World.* Luzac, London, 1910 (with original Chinese text). Repr. without notes, Nanfang, Chungking, 1945; also repr. in *Roots of Strategy*, ed. Phillips, T. R. (q.v.).

GILLE, B. (14). 'Machines [in the Mediterranean Civilisations and the Middle Ages].' Art. in *A History of Technology*, ed. C. Singer *et al.*, vol. 2, p. 629, Oxford, 1956.

GILLE, B. & BURTY, P. (1). *Les Moteurs à Combustion Interne.* Paris, 1948.

GLUCKMAN, COL. ARCADI (1). *United States Rifles, Muskets and Carbines.* Ulbrich, Buffalo, N.Y., 1948.

GODDARD, R. H. (1). *Rockets.* American Rocket Society, New York, 1946 (reprints of Goddard's two classical papers originally published in 1919 and 1936; both in *SMC*: 'A Method of Reaching Extreme Altitudes', **71** (no. 2) and 'Liquid-Propellant Rocket Development').

GODDARD, R. H. (2). *Rocket Development; Liquid-Fuel Rocket Research, 1929–41.* Prentice-Hall, New York, 1948. (Extracts from Goddard's notebooks, 1929 to 1941.)

GODDARD SPACE FLIGHT CENTER. *See* Rosenthal, Alfred (1).

GODE, P. K. (1). 'The Mounted Bowman on Indian Battlefields from the Invasion of Alexander (−326) to the Battle of Panipat (+1761).' *DCRI*, 1947, **8** (no. 1), 1 (K. N. Dikshit Memorial Volume). Repr. in *Studies in Indian Cultural History* vol. 2 (Gode Studies, vol. 5), p. 57.

GODE, P. K. (6). 'The Manufacture and Use of Firearms in India between +1450 and 1850.' Art. in *Munshi Diamond Jubilee Indological Volume*, Pt. 1. *BV*, 1948, **9**, 202.

GODE, P. K. (7). *The History of Fireworks in India between +1400 and 1900.* Rasavangudi, for Indian Institute of Culture, Bangalore, 1953 (Transaction, no. 17).

GOETZ, HERMANN (1). 'Das Aufkommen der Feuerwaffen in Indien.' *OAZ*, 1923, NF **2** (no. 2/3), 226.

GOHLKE, W. (1). *Geschichte d. gesamten Feuerwaffen bis 1850; die Entwicklung der Feuerwaffen von ihren ersten Auftreten bis zur Einführung der gezogenen Hinterlader, unter besonderer Berücksichtigung der Heeresbewaffnung.* Göschen, Leipzig, 1911.

GOHLKE, W. (2). 'Das älteste Latierte Gewehr.' *ZWHK*, 1916, **7**, 205.

GOHLKE, W. (3). 'Handbrandgeschosse aus Ton.' *ZWHK*, 1913, **6**, 377.

GOLAS, P. (1). 'Chinese Mining: where was the Gunpowder?' Art. in *Explorations in the History of Science and Technology in China*, ed. Li Kuo-Hao, Chang Mêng-Wên, Tshao Thien-Chhin & Hu Tao-Ching, Shanghai, 1982, p. 453.

GOLONBEV, V. (2). 'Quelques Sculptures Chinoises.' *OAZ*, 1913, **2**, 326.

GOODALL, A. M. (1). 'Gunnery and Firearms.' *CME*, 1961, **8** (no. 8), 489.

GOODRICH, L. CARRINGTON (15). 'Firearms among the Chinese; a supplementary note.' [Dated bombards preserved in China.] *ISIS*, 1948, **39**, 63.

GOODRICH, L. CARRINGTON (23). 'A Cannon from the end of the Ming Period' [+1650]. *JRAS/HKB*, 1967, **7**, 152.

GOODRICH, L. CARRINGTON (24). 'Note on a few Early Chinese Bombards.' *ISIS*, 1944, **35**, 211. Reply to Sarton (14), a query.

GOODRICH, L. CARRINGTON (25). 'Early Cannon in China' [Report on Wang Jung (*1*)]. *ISIS*, 1964, **55**, 193.

GOODRICH, L. CARRINGTON (26). 'Westerners and Central Asians in Yuan China.' *OPO*, 1957, 1–21.

GOODRICH, L. CARRINGTON & FÉNG CHIA-SHÊNG (1). 'The Early Development of Firearms in China.' *ISIS*, 1946, **36**, 114. With important addendum giving a missing page, *ISIS*, 1946, **36**, 250.

GORDON, D. H. (1). 'Swords, Rapiers and Horse-Riders.' *AQ*, 1953, **27**, 67.

GRAM, H. (1). 'Om Bysse-Krud [gunpowder], naar det er opfundet i Europa, hvorlaenge det har voeret i Brug i Danmark.' *SKSL*, 1745, pt. 1, 213. For Latin and German translations, *see* Partington (5), p. 131.

GRANT, EDWARD (2). *Much Ado about Nothing; Theories of Space and Vacuum from the Middle Ages to the Scientific Revolution*. Cambridge, 1981.

GRAY, EILEEN (1). *Charcoals for Gunpowder*. Inaug. Diss., Newcastle-upon-Tyne, 1982.

GRAY, EILEEN, MARSH, H. & McLAREN, M. (1). 'A Short History of Gunpowder, and the Role of Charcoal in its Manufacture.' *JMATS*, 1982, **17**, 3385.

GREENER, W. W. (1). *The Gun and its Development*. New York, 1881. 9th ed. New York, 1910.

GRIFFITH, S. B. (1) (tr.). *Sun Tzu; the Art of War*. Oxford, 1963. With foreword by R. H. Liddell Hart.

DE GROOT, J. J. M. (2). *The Religious System of China*. Brill, Leiden, 1892.

    Vol. 1, Funeral rites and ideas of resurrection.

    Vols. 2, 3, Graves, tombs and *fêng-shui*.

    Vol. 4, The soul, and nature-spirits.

    Vol. 5, Demonology and sorcery.

    Vol. 6, The animistic priesthood (*wu*).

GROSIER, J. B. G. A. ABBÉ (1). *De la Chine; ou, Description Générale de Cet Empire, redigée d'après les Mémoires de la Mission de Pé-kin—Ouvrage qui contient la Description Topographique des quinze Provinces de la Chine... les trois Règnes de son Histoire Naturelle...; et l'Exposé de toutes les Connoissances acquises et parvenues jusqu'içi en Europe sur le Gouvernement, la Religion, les Lois, les Moeurs, les Usages, les Sciences et les Arts des Chinois.* 3rd ed. 7 vols. Pillet, Paris 1818–20. 1st ed. 1785; 2nd ed. 1787. Eng. tr.: *General Description of China, containing the Topography of the Fifteen Provinces which compose this vast Empire, that of Tartary... the Natural History of its Animals, Vegetables and Minerals, together with the latest Accounts which have reached Europe, of the Government, Religion, Manners, Customs, Arts and Sciences of the Chinese—illustrated by a New and Correct Map of China, and other Copper-Plates*. 2 vols. Robinson, London, 1788. Partial Eng. tr. by Lana Castellano & Christina Campbell-Thomson: *The World of Ancient China*. Gifford, London, 1972 (well illustrated but contains many mistakes, and does not make clear which portions of Grosier's original text were drawn upon).

GROSLIER, G. (1). *Recherches sur les Cambodgiens*. Challamel, Paris, 1921.

GROUSSET, R. (1) *Histoire de l'Extrême-Orient*. 2 vols. Geuthner, Paris, 1929. (Also appeared in *BE/AMG*, nos. 39, 40.)

GRUBE, W. (1) (tr.). *Die Metamorphosen der Götter (Fêng Shen Yen I)*, [Stories of the Promotions of the Martial Genii], chs. 1–46, with summary of chs. 47–100. 2 vols. Brill, Leiden, 1912.

GUERLAC, H. (1). 'The Poets' Nitre; Studies in the Chemistry of John Mayow, II.' *ISIS*, 1954, **45**, 243.

GUERLAC, H. (2). 'John Mayow and the Aerial Nitre; Studies in the Chemistry of John Mayow, I.' *Actes du VIIe Congrès International d'Histoire des Sciences*, Jerusalem. 1953, p. 332.

GUILMARTIN, J. F. (1). *Gunpowder and Galleys; Changing Technology and Mediterranean Warfare at Sea in the +16th Century*. Cambridge, 1974.

GUTTMANN, OSCAR (1). '*Monumenta Pulveris Pyrii*'; *Reproductions of Ancient Pictures concerning the History of Gunpowder, with Explanatory Notes*. Pr. pr., Artists Press, London, 1906. ('The text is useless': Partington (5), p. 129.)

GUTTMANN, OSCAR (2). 'The Oldest Document in the History of Gunpowder.' *JSCI*, 1904, **23**, 591. Crit. Fêng Chia-Shêng (8).

GUTTMANN, OSCAR (3). *The Manufacture of Explosives*. 2 vols. Macmillan, New York, 1895.

HAGERMAN, CAPT. G. M. (1). 'Lord of the Turtle-Boats.' *USNIP*, 1967, **93**, 69.

HAHN, H., HINTZE, W. & TREUMANN, H. (1). 'Safety and Technological Aspects of Black Powder.' *PAE*, 1980, **5**, 129.

DU HALDE, J. B. (1). *Description Géographique, Historique, Chronologique, Politique et Physique de l'Empire de la Chine et de la Tartare Chinoise.* 4 vols. Paris, 1735, 1739; The Hague, 1736. Eng. tr. R. Brookes, London, 1736, 1741. Germ. tr. Rostock, 1748.

HALE, WILLIAM (1). *A Treatise on the Comparative Merits of a Rifle Gun and a Rotary Rocket.* Mitchell, London, 1863.

HALEY, A. G. (1). *Rocketry.* Van Nostrand, New York, 1970. Rev. *LI*, 1970 (21 May), 687.

HALL, A. R. (1). *Ballistics in the Seventeenth Century; a Study in the Relations of Science and War, with reference principally to England.* Cambridge, 1951. Rev. T. S. Kuhn, *ISIS*, 1953, **44**, 284.

HALL, A. R. (2). 'A Note on Military Pyrotechnics [in the Middle Ages].' Art. in *A History of Technology*, ed. C. Singer *et al.*, vol. 2, p. 374. Oxford, 1956.

HALL, A. R. (3). *The Military Inventions of Guido da Vigevano.* Proc. VIIIth International Congress of the History of Science, p. 966. Florence, 1956.

HALL, A. R. (5). 'Military Technology [from the Renaissance to the Industrial Revolution].' Art. in *A History of Technology*, ed. C. Singer *et al.*, vol. 3, p. 347. Oxford, 1957.

HALL, A. R. (6). 'Military Technology [in the Mediterranean Civilisations and the Middle Ages].' Art. in *A History of Technology*, ed. C. Singer *et al.*, vol. 2, p. 695. Oxford, 1956.

HALL, BERT, S. & WEST, DELMO C. (1) (ed.). *On Pre-Modern Technology and Science; Studies in Honour of Lynn White, Jr.*; being vol. 1 of *Humana Civilitas*; Sources and Studies relating to the Middle Ages and the Renaissance. Center for Mediaeval & Renaissance Studies, Univ. of California, Los Angeles, 1976.

HALL, F. CARGILL (1) (ed.). *Essays on the History of Rocketry and Astronautics.* Proc. 3rd to 6th. History Symposia of the International Academy of Astronautics, NASA Conference Pub. no. 2014. NASA, Washington, D.C., 1977.

HALLAM, HENRY (1). *A View of the State of Europe during The Middle Ages—History of Ecclesiastical Power—Constitutional History of England—The State of Society in Europe.* London 1875; repr. 1877. First pub. 3 vols., 1819 and at least eight other editions.

HALOUN, G. (5). 'Legalist Fragments, I; *Kuan Tzu* ch. 55, and related texts.' *AM*, 1951 (n.s.), **2**, 85.

VON HAMMER-PURGSTALL, J. (2). 'Ü. d. Verfertigung und den Gebrauch von Bogen und Pfeil bei den Arabern und Türken.' *SWAW/PH*, 1851, **6**, 239, 278.

HANA, KORINNA (1) (tr.). *Bericht über die Verteidigung der Stadt Tê-an während der Periode Khai-Hsi (+1205 bis +1208); der 'Khai-Hsi Tê-An Shou Chhêng Lu' von Wang Chih-Yuan—ein Beitrag zur privaten Historiographie des 13.-Jahrhunderts in China.* Steiner, Wiesbaden, 1970. (Münchener Ostasiatische Studien, no. 1.)

HANČAR, M. (1). 'Das Pferd in Mittelasien.' *KS/WBKL*, 1952.

HANSJAKOB, HEINRICH (1). *Der Schwarze Berthold, der Erfinder des Schiesspulvers und der Feuerwaffen.* Freiburg i/Breisgau, 1891. Summaries in von Romocki (1), vol. 1, pp. 106 ff. von Lippmann (22) and Hime (1), p. 124.

HANZELET, JEAN. *See* Appier, Jean (his original name).

HARADA, YOSHITO & KOMAI, KAZUCHIKA (1) = (1). *Chinese Antiquities.* Pt. 1, *Arms and Armour*; Pt. 2, *Vessels [Ships] and Vehicles.* Academy of Oriental Culture, Tokyo Institute, Tokyo, 1937.

HARADA, YOSHITO & KOMAI, KAZUCHIKA (2) = (2). *Shangtu, the Summer Capital of the Yuan Dynasty at Dolon Nor, Mongolia.* Toa-Koko Gakukwai, Tokyo, 1941. (Archaeologia Orientalis B, no. 2.)

HARDING, D. (1) (ed.). *Weapons; an International Encyclopaedia from −5000 to +2000.* Macmillan, London, 1980.

HARDY, SIR WILLIAM BATE (1). 'On the Mechanism of Gelation in Reversible Colloidal Systems.' *PRS*, 1899, **66**, 95. Repr. in Hardy (3), p. 322.

HARDY, SIR WILLIAM BATE (2). 'On the Coagulation of Proteid by Electrolytes.' *JOP*, 1899, **24**, 288. Repr. in Hardy (3), p. 294.

HARDY, SIR WILLIAM BATE (3). *Collected Scientific Papers of Sir W. B. Hardy.* Cambridge, 1936.

HARPER, DONALD J. (2). 'Chinese Divination and Portent Interpretation'. Paper given to the ACLS Conference, 1983.

HARRISON, H. S. (4). 'Fire-making, Fuel and Lighting [from Early Times to the Fall of the Ancient Empires].' Art. in *A History of Technology*, ed. C. Singer *et al.*, Oxford, 1954., vol. 1, p. 216.

HART, MRS ERNEST (1) (ALICE MARION). *Picturesque Burma, Past and Present.* Dent, London, 1897.

HART, B. H. LIDDELL (1). *The Decisive Wars of History; a Study in Strategy.* Bell, London, 1929. New ed. *The Strategy of Indirect Approach.* London, 1941.

HART, B. H. LIDDELL (2). *Great Captains Unveiled.* London, 1927.

HART, C. (1). 'Mediaeval Kites and Windsocks.' *JRAES*, 1969, **73**, 1019.

HART, I. B. (2). *The Mechanical Investigations of Leonardo da Vinci.* Chapman & Hall, London; Open Court, Chicago, 1925. 2nd ed. Univ. Calif. Press, Berkeley & Los Angeles, 1963.

HART, I. B. (4). *The World of Leonardo da Vinci, Man of Science, Engineer, and Dreamer of Flight* (with a note on Leonardo's Helicopter Model, by C. H. Gibbs-Smith). McDonald, London, 1961. Rev. K. T. Steinitz, *TCULT*, 1963, **4**, 84.

HASSAN, AHMAD YUSUF (1). 'A Note on Gunpowder and Cannon in Arabic Culture.' *Proc. XVIth Internat. Congress Hist. of Science, Bucharest*, 1981, vol. 1, p. 51.

HASSENSTEIN, W. (1) (ed.). *Das Feuerwerkbuch von 1420; Sechshundert Jahre Deutsche Pulverwaffen und Büchsenmeisterei. Neudruck des Erstdruckes aus dem Jahre 1529 mit Übertragung ins hochdeutsche und Erläuterungen....* Verlag d. Deutschen Technik, München, 1941.

HASSENSTEIN, W. (2). 'Die Chinesen und die Erfindung des Pulvers.' *ZGSS*, 1944, **39** (no. 1), 1; (no. 2), 22.

HASSENSTEIN, W. (3). 'Zur Entwicklungsgeschichte der Rakete.' *ZGSS*, 1939, **34**, 172.

HATTAWAY, M. (1) (ed.). *The Knight of the Burning Pestle* (Beaumont & Fletcher). Benn, London, 1969.

DE HAUTEFEUILLE, JEAN (1). *Pendule Perpétuelle, avec un nouveau Balancier, et la Manière d'élever l'eau par le Moyen de la Poudre à Canon...* Paris, 1678.

DE HAUTEFEUILLE, JEAN (2). *Réflexions sur quelques Machines à élever les Eaux, avec la Description d'une nouvelle Pompe.* Paris, 1682.

HAWKINS, W. M. (1). 'Japanese Swords' (chart).

HAYWARD, J. F. (1). *Die Kunst d. alten Büchsenmeister.* Germ. tr., by G. Espig, of *The Art of the Gun-Maker.* 2 vols. Vol. 1: *From +1500 to +1660.* Vol. 2: *From +1660 to 1830.* London, 1962–3; 2nd ed. London, 1965.

VAN HÉE, L. (17). *Ferdinand Verbiest, Mandarin Chinois.* Bruges, 1913.

HEIM, J. (1). [Archery among the Osmanli.] *DI*, 1925.

HEMMERLIN, FELIX (1). *Felicis Malleoli...De Nobilitate et Rusticate Dialogus.* Basel, 1490 (?). Later editions 1495 (?) and 1497.

HENTZE, C. (7). '*Ko' und 'Chhi' Waffen in China und Amerika; Studien z. frühchinesischen Kulturgeschichte.* 1943.

D'HERBELOT, BARTHÉLEMY (1). *Bibliothèque Orientale, ou Dictionnaire Universel, contenant généralement tout ce qui regarde la Connoissance des Peuples de l'Orient, leurs Histoires et Traditions véritables ou fabuleuses...leurs Sciences et leurs Arts, leur Théologie, Mythologie, Magie, Physique, Morale, Médecine, Mathématiques, Histoire Naturelle, Chronologie, Géographie, Observations Astronomiques, Grammaire et Rhétorique, les Vies et Actions remarquables de tous leurs Saints, Docteurs, Philosophes, etc...* With Supplements by C. de Visdelou and A. Galand. 4 vols. Dufour & Roux, Maestricht, 1776–80. Suppl. vol. p. 117 has article: 'De l'Invention des Canons en Chine'.

HEWISH, M. (1). 'China makes Ground in the Space Race.' *NS*, 1980, **86** (no. 1207), 378.

HEYMANN, R. E. (2). 'Prehistoric spear-throwers.' Paper to the XIth International Congress of the History of Science, Warsaw, 1965.

HILL, DONALD R. (1). 'Trebuchets.' *VIAT*, 1973, **4**, 99.

HIME, H. W. L., COL. (1). *The Origin of Artillery.* Longmans Green, London, 1915. Pts. 1 and 3 are the second and revised edition of Hime (2), pt. 2 was taken from Hime (3). Crit. Fêng Chia-Shêng (*3*).

HIME, H. W. L., COL. (2). *Gunpowder and Ammunition; their Origin and Progress.* Longmans Green, London, 1904.

HIME, H. W. L., COL. (3). 'Our Earliest Cannon, +1314 to +1346.' *JRA*, 1904–5, **31**, 489.

HIME, H. W. L., COL. (4). 'Roger Bacon and Gunpowder.' Art. in *Roger Bacon: Essays*, ed. A. G. Little (2), 1914, p. 321. Crit. Fêng Chia-Shêng (*8*).

HISAMATSU SEIICHI (1). *Biographical Dictionary of Japanese Literature.* Kodansha International, Tokyo and New York, 1976.

HO PING-YÜ (1) (tr.). *Astronomy in the 'Chin Shu' and the 'Sui Shu'.* (Inaug. Diss. Singapore, 1955.) Paris, 1966.

HO PING YÜ & NEEDHAM, JOSEPH (1). 'Ancient Chinese Observations of Solar Haloes and Parhelia.' *W*, 1959, **14**, 124.

HO PING-YÜ & WANG LING (1). 'On the *Karyūkyō* [*Huo Lung Ching*], the "Fire-Dragon Manual".' *PFEH*, 1977, **16**, 147.

HO YUNG-CHI (1). 'Marco Polo—was he ever in China?' *ACSS*, 1953, 45.

HOEFER, F. (1). *Histoire de la Chimie.* 2 vols., Paris, 1842–3; 2nd ed. 2 vols., Paris, 1866–9.

HOKES, E. S. (1). 'Chinese Rocket Aircraft of the +16th Century.' Unpub. MS deposited in the East Asian History of Science Library, Sept. 1980.

HOLLISTER-SHORT, G. (1). 'The Sector and Chain; a Historical Enquiry.' *HOT*, 1979, **4**, 149.

HOLLISTER-SHORT, G. (2). 'The Vocabulary of Technology.' *HOT*, 1977, **2**, 125.

HOLLISTER-SHORT, G. (3). 'Leads and Lags in Late +17th-Century English Technology.' *HOT*, 1976, **1**, 159.

HOLLISTER-SHORT, G. (4). 'The Civil Uses of Gunpowder.' MS, June 1982. 'The Use of Gunpowder in Mining; a Document of +1627.' *HOT*, 1983, **8**, 111.

HOLLISTER-SHORT, G. (5). 'Antecedents and Anticipations of the Newcomen Engine.' *TNS*, 1980, **52**, 103.

HOLMESLAND, A., STØRMER, L., TVETERÅS, E. & VOGT, H. (1). *Aschehoug's Konversasjonslexikon.* 5th ed. Aschehoug, Oslo, 1974.

HOMMEL, R. P. (2). 'Notes on Chinese Sword Furniture.' *CJ*, 1928, **8**, 3.

HOOVER, H. C. & HOOVER, L. H. (1) (tr.). *Georgius Agricola 'De Re Metallica' translated from the 1st Latin edition of 1556, with biographical introduction, annotations and appendices upon the development of mining methods, metallurgical processes, geology, mineralogy and mining law from the earliest times to the 16th century.* 1st ed. *Mining Magazine*, London, 1912; 2nd ed. Dover, New York, 1950.

HOPKINS, E. W. (1). 'The Social and Military Position of the Ruling Class in India, as represented by the Sanskrit Epic.' *JAOS*, 1889, **13**, 57–372 (with index). Military techniques, pp. 181–329.

HOPKINS, E. W. (2). 'On Firearms in Ancient India.' *JAOS*, 1889, **13**, cxciv (the thesis of Oppert exploded). Cf. 'The Princes and Peoples of the Epic Poems.' *CHI*, ch. 11, p. 271. Cf. Hopkins, E. W. (1), pp. 296 ff.

HORN, J. (1). *Ü. die ältesten Hufschutz d. Pferdes; ein Beitrag z. Gesch. d. Hufbeschlages.* Dresden, 1912.

HORNELL, J. (25). 'South Indian Blow-guns, Boomerangs and Crossbows.' *JRAI*, 1924, **54**, 326.

HORVÁTH, ÁRPÁD (9). *Az Ágyú Históriája; Kepek a Tüzértechnika Történetéböl* (A History of Firearms and Artillery). Zrínyi Katonai Kiadó, Budapest, 1966 (in Magyar).

HORWITZ, H. T. (6). 'Beiträge z. aussereuropäischen u. vorgeschichtlichen Technik.' *BGTI*, 1916, **7**, 169.

HORWITZ, H. T. (8) (with a note by F. M. FELDHAUS). 'Zur Geschichte d. Wetterschiessens.' *GTIG*, 1915, **2**, 122.

HORWITZ, H. T. (13) [with the assistance of Hsiao Yü-Mei & Chu Chia-Hua]. 'Die Armbrust in Ostasien.' *ZHWK*, 1916, **7**, 155.

HORWITZ, H. T. (14). 'Zur Entwicklungsgeschichte d. Armbrust.' *ZHWK*, 1919, **8**, 311. With two supplementary notes under the same title, *ZHWK*, 1921, **9**, 73 & 114.

HORWITZ, H. T. (15). 'Über die Konstruktion von Fallen und Selbstschüssen.' *BGTI*, 1924, **14**, 85. Rev. K. Himmelsbach, *ZHWK*, 1927, **11**, 291.

HORWITZ, H. T. (16). 'Ein chinesisches Armbrustschloss in amerikanischem Besitz.' *ZHWK*, 1927, **11**, 286.

HORWITZ, H. T. & SCHRAMM, E. (1). 'Schieber an antiken Geschützen.' *ZHWK*, 1921, **9**, 139.

HOSEMANN, ABRAHAM (1). *De Tonitru.* Magdeburg, 1618.

HOSSMANN, ABRAHAM. *See* Hosemann, Abraham.

HOU, K. C. (perhaps HOU KUANG-CHAO or HOU KUANG-CHHIUNG) (1). 'Profile Descriptions of Soils of Hsü-chang Hsien, Honan.' Ref. in KOVDA (1), p. 122 but no exact details, nor whether the paper is in Chinese or English. Thorp (1), p. 515, lists two papers with this title as MSS in 1936.

HOUGH, W. (3). 'Primitive American Armour.' *ARUSNM*, 1893, 627.

HOWORTH, SIR HENRY H. (1). *History of the Mongols; from the 9th to the 19th Century.* 3 vols. Longmans Green, London, 1876–1927. Repr. in 5 vols. Chhêng Wên, Thaipei, 1970.

HSIAO CHHI-CHHING (1). *The Military Establishment of the Yuan Dynasty.* Harvard Univ. Press, Cambridge, Mass., 1978.

HSÜ HUI-LIN (1). 'Gunpowder and Ancient Rockets.' *CREC*, 1980, **29** (no. 10), 58.

HUANG JEN-YÜ (5). *1587, a Year of No Importance; the Ming Dynasty in Decline.* Yale Univ. Press, New Haven and London, 1981. Chinese tr.: *Wan-Li Shih-wu Nien*, Chung-Hua, Peking, 1981.

HUANG JEN-YÜ (6). 'The Liaotung Campaign of +1619.' *ORE*, 1981, **28**, 30.

HUC, R. E., ABBÉ (2). *The Chinese Empire; forming a Sequel to 'Recollections of a Journey through Tartary and Thibet'.* 2 vols. Longmans, London, 1855, 1859.

HUCKER, C. O. (5). 'Hu Tsung-Hsien's Campaign against Hsü Hai, +1556.' Art. in Kierman & Fairbank (1) (ed.), *Chinese Ways in Warfare*, p. 273.

HUCKER, C. O. (6). 'Governmental Organisation of the Ming Dynasty.' *HJAS*, 1958, **21**, 1–66.

HUCKER, C. O. (7). 'An Index of Terms and Titles in "Governmental Organisation of the Ming Dynasty".' *HJAS*, 1961, **23**, 127.

HULBERT, H. B. (2). *History of Korea.* Seoul, 1905. (Revised edition ed. C. N. Weems, 2 vols. Hilary House, New York, 1962.)

HULBERT, H. B. (3). 'Korean Inventions.' *HMM*, 1899, 102.

HUMPHRIES, J. (1). *Rockets and Guided Missiles.* Benn, London, 1956.

HUNTER, JOSEPH (1). 'Proofs of the Early Use of Gunpowder in the English Army.' *AAA*, 1802, **32**, 379.

HÜTTEROTT, G. (1). 'Das japanische Schwert.' *MDGNVO*, 1885, **4**, 33.

HUURI, K. (1). 'Zur Geschichte des mittelalterlichen Geschützwesens aus orientalischen Quellen.' *SOF*, 1941, **9**, no. 3.

HUYGENS, CHRISTIAAN (2). *Oeuvres Complètes.* 22 vols. Nijhoff, The Hague, 1897–1950.

IMBAULT-HUART, C. (4). 'L'Introduction des Torpilles en Chine.' *LN*, 1884, **12** (pt. 1), 114.

IMBERT, H. (1). *Les Négritos de la Chine.* Impr. d'Extr. Orient, Hanoi, 1923.

INALCIK, HALIL (1). 'The Socio-Political Effects of the Diffusion of Fire-arms in the Middle East.' Art. in Parry & Yapp (1), *War, Technology and Society in the Middle East*, Oxford, 1975, p. 195. Repr. in Inalcik (2), no. XIV.

INALCIK, HALIL (2). *The Ottoman Empire; Conquest, Organisation and Economy—Collected Studies.* Variorum, London, 1978.

D'INCARVILLE, P. (1). 'Manière de faire les Fleurs dans les Feux d'Artifice Chinois.' *MAS/MPDS*, 1763, **4**, 66. Anon. Eng. abstr. with plate, *UM*, 1764, **34**, 21.

ITAKURA KIYONOBU (1). 'The First Ballistic Laws developed by a Japanese Mathematician, and their Origin.' *JSHS*, 1963, **2**, 136.

ITAKURA KIYONOBU & ITAKURA REIKO (1). 'Studies of Trajectory [Ballistics] in Japan before the Days of the Dutch Learning.' *JSHS*, 1962, **1**, 83.

IWAO SEIICHI (1). *Biographical Dictionary of Japanese History*. Tr. B. Watson. Kodansha International, Tokyo, 1978.

JACKSON, W. (1). 'Some Inquiries concerning the Salt-Springs and the Way of Salt-Making at Nantwich in Cheshire; Answer'd by the Learned and Observing William Jackson, Dr of Physick.' *PTRS*, 1669, **4** (no. 53), 1060; (no. 54), 1077.

JACOB, G. (2). *Der Einfluss d. Morgenlandes auf das Abendland, Vornehmlich während des Mittelalters*. Hanover, 1924.

JACOB, G. (3). *Östliche Kulturelemente im Abendland*. Berlin, 1902.

JACOB, G. (4). 'Ostasiens Kultureinfluss auf das Abendland.' *SA*, 1931, **6**, 146.

JAGNAUX, R. (1). *Histoire de la Chimie*. 2 vols. Paris, 1891.

JÄHNS, M. (1). *Geschichte d. Kriegswissenschaften, vornehmlich in Deutschland*. Oldenbourg, München & Leipzig 1889.

Vol. 1. *Altertum, Mittelalter, 15. & 16. Jahrhundert*;
2. *17. & 18. Jahrh. bis zum Auftreten Friedrichs d. Grossen* (+1740);
3. *Das 18. Jahrh. seit dem Auftr. Fr. d. Gross.* (1740/1800).

JÄHNS, M. (2). *Handbuch eine Geschichte d. Kriegswesens von der Urzeit bis zu der Renaissance; technische Teil, Bewaffnung, Kampfweise, Befestigung, Belagerung*. 2 vols., Leipzig, 1880. Repr. Berlin, 1897.

JÄHNS, M. (3). *Entwicklungsgeschichte d. alten Trutzwaffen mit einem Anhang u. d. Feuerwaffen*. Berlin, 1899.

JAMES, MONTAGUE R. (2) (ed.). *The Treatise of Walter de Milamete, 'De Nobilitatibus, Sapientis et Prudentiis Regum', reproduced in facsimile from the unique MS.* [1326–7] *preserved at Christ Church, Oxford; together with a Selection of Pages from the Companion MS. of the Treatise 'De Secretum Secretorum Aristotelis', preserved in the Library of the Earl of Leicester at Holkham Hall* [Norfolk], *with an Introduction...* Roxburghe Club, London, 1913.

JAMESON, C. D. (1). 'Coal and Iron in Eastern China.' *EMJ*, 1898, **66**, 367.

JANSE, O. R. T. (1). 'Notes sur quelques Epées Anciennes trouvées en Chine.' *BMFEA*, 1930, **2**, 67.

JANSE, O. R. T. (4). 'Quelques Antiquités Chinoises d'un caractère Hallstattien.' *BMFEA*, 1930, **2**, 177.

JENKINS, RHYS (3). *Links in the History of Engineering and Technology from Tudor Times: Collected Papers of R.J.— comprising articles in the professional and technical press mainly prior to 1920, and a Catalogue of other published work*. (Newcomen Society), Cambridge, 1936.

JENKINS, RHYS (4). 'A Contribution to the History of the Steam Engine: (1) The Notebook of Roger North, (2) The Work of Sir Samuel Morland.' Repr. in (3), p. 40.

JENNER, W. J. F. (1) (tr.). *Memories of Loyang; Yang Hsüan-Chih and the Lost Capital* (+493 to +534). [Translation of the *Loyang Chhieh-Lan Chi* with annotations.] Oxford, 1981.

JETT, S. C. (2). 'The Development and Distribution of the Blow-Gun.' *AAAG*, 1970, **60**, 662.

JOCELYN, R. LORD (1). *Six Months with the Chinese Expedition; Leaves from a Soldier's Notebook*. Murray, London, 1841.

JOHANNSEN, O. (3). 'Die Erfindung der Eisengusstechnik.' *SE*, 1919, **39**, 1457 & 1625.

JOHANNSEN, O. (4). 'Die Quellen zur Geschichte des Eisengusses im Mittelalter und in d. neueren Zeit bis zum Jahre 1530.' *AGNT*, 1911, **3**, 365; 1915, **5**, 127; 1918, **8**, 66.

JOHNSTON, R. F. (3). 'The Cult of Military Heroes in China.' *NCR*, 1921, **3**, 41 & 79.

JOINVILLE, JEAN (1). *Historie de Saint Loys*, 1309. Orig. old French version in Petitot's *Collection Complète des Mémoires relatifs à l'Histoire de France*. Paris, 1824, vol. 2. Mod. French version W. de Wailly (1). Eng. tr. by J. Evans (1), vol. 4, p. 928. Cf. Sarton (1), vol. 4, p. 928.

JONES, ROBERT (1). *A New Treatise on Artificial Fire-Works*. London, 1765; 2nd ed. London, 1766.

JULIEN, STANISLAS (4). 'Notes sur l'Emploi Militaire des Cerfs-Volants, et sur les Bateaux et Vaisseaux en Fer et en Cuivre, tirées des Livres Chinois.' *CRAS*, 1847, **24**, 1070.

JULIEN, STANISLAS (8). Translations from *TCKM* relative to +13th-century sieges in China (in Reinaud & Favé, 2). *JA*, 1849 (4e sér.), **14**, 284 ff.

KAHANE & TIETZE (1). *The Lingua Franca in the Levant*. Urbana, Ill., 1958.

KALMAR, J. (1). 'Die Raketentechnik im 17. Jahrhundert, auf Grund einschlägiger Materials im Grazer Zeughaus.' *ZHWK*, 1933, **13** (NF **4**), 102.

KAO LEI-SSU (1) (Aloysius Ko, S.J.). 'Remarques sur un Écrit de M. P[auw] intitulé "Recherches sur les Égyptiens et les Chinois" (1775).' *MCHSAMUC*, 1777, **2**, 365–574 (in some editions, 2nd pagination, 1–174).

KARLGREN, B. (13). 'Weapons and Tools of the Yin [Shang] Dynasty.' *BMFEA*, 1945, **17**, 101.

KATAFIASZ, T. (1). 'Recherches sur les Acquisitions Polonais dans le Domaine de la Technique des Fusées de Guerre au 19e Siècle' (in Polish, with French summary). *KHNT*, 1982, **27** (no. 2), 379.

KEDESDY, E. (1). *Die Sprengstoffe; Darstellung und Untersuchung der Sprengstoffe und Schiesspulver.* Jänecke, Hannover, 1909.

KEEGAN, J. (1). *The Face of Battle.* Cape, London, 1976; Penguin, London, 1978.

KELLY, R. TARBOT (1). *Burma, the Land and the People.* Millet, Boston and Tokyo, 1910.

KEMP, PETER (1) (ed.). *The Oxford Companion to Ships and the Sea.* Oxford, 1976.

KENNEDY, SIR WM., ADMIRAL (1). *Hurrah for the Life of a Sailor; Fifty years in the Royal Navy.* Nash, London, 1910.

KIERMAN, F. A. (1) (tr.). *Ssuma Chhien's Historiographical Attitude as reflected in Four Late Warring States Biographies [in Shih Chi].* Harrassowitz, Wiesbaden, 1962. (Studies on Asia, Far Eastern and Russian Institute, Univ. of Washington, Seattle, no. 1.)

KIERMAN, F. A. (2). 'Phases and Modes of Combat in Early China' (Chou and Warring States periods). Art. in Kierman & Fairbank (1) (ed.), *Chinese Ways in Warfare*, p. 27.

KIERMAN, F. A. & FAIRBANK, J. K. (1) (ed.). *Chinese Ways in Warfare.* Harvard Univ. Press, Cambridge, Mass., 1974.

KIKUOKA TADASHI (1) (tr.). *Teppō-ki;* the "Chronicle of the Arquebus".' *EAST*, 1981, 47.

KIMBROUGH, R. E. (1). 'Japanese Firearms.' *GUNC*, 1950, no. 33, 445.

KIRCHER, ATHANASIUS (1). *China Monumentis qua Sacris qua Profanis Illustrata.* Amsterdam, 1667. (French tr. Amsterdam, 1670.)

KIRCHER, ATHANASIUS (5). *Mundus Subterraneus, in XII Libros Digestus.* Jansson & Weyerstraten, Amsterdam, 1665. Cf. Thorndike (1), vol. 7, pp. 567 ff.

KLAEBER, F. (1). *Beowulf, and the Fight at Finnsburg, edited, with Introduction, Bibliography, Notes, Glossary and Appendices.* 3rd ed., with 1st and 2nd supplements. Heath, Boston, 1950.

KLAUSNER, W. J. (1). 'Popular Buddhism in Northeast Thailand.' Art. in *Cross-Cultural Understanding...,* ed. Northrop & Livingston (1), p. 69.

KLEMM, G. (1). *Werkzeuge und Waffen; ihre Entstefung und Ausbildung.* Sondershausen, 1858. Repr. Zentral-antiquariat, Leipzig, 1978.

KLOPSTEG, P. E. (1). *Turkish Archery and the Composite Bow.* Pr. pr. Evanston, Ill., 1947.

KÖCHLY, H. & RÜSTOW, W. (1) (tr.). *Griechische Kriegsschriftsteller.* 3 vols. Engelmann, Leipzig, 1853–5.

KÖHLER, G. (1). *Die Entwickelung des Kriegswesens und der Kriegführung in der Ritterzeit von Mitte des 11. Jahrh. bis zu den Huesitenkriegen.* Koebner, Berlin, 1886–90. 5 vols. (Vol. 3 pt. 1 contains his argument in favour of the view that the torsion catapult of antiquity remained in use till the end of the middle ages.) Crit. Fêng Chia-Shêng (3).

KÖHLER, G. (2). *Geschichte der Explosivstoffe.* 2 vols. 1895.

KOMROFF, M. (1) (ed.). *Contemporaries of Marco Polo; consisting of the 'Travel Records in the Eastern Parts of the World', of William of Rubruck (+1253 to +1255); the 'Journey' of John of Pian de Carpini (+1245 to +1247); the 'Journal' of Friar Odoric [of Pordenone] (+1318 to +1330); and the 'Oriental Travels' of Rabbi Benjamin of Tudela (+1160 to +1173).* Cape, London, 1928. Boni & Liveright, New York, 1928.

KOSAMBI, D. D. (2). *An Introduction to the Study of Indian History.* Popular, Bombay, 1956.

KRÄTZ, O. (1). 'Elektrische Pistolen; eine Kuriosität der Gas-chemie des ausgehenden 18. Jahrhunderts.' *HBAS*, 1973, **23**, 3. (Veröfftl d. Forschungsinst. d. Deutschen Museums, A, 130.)

KRAUS, P. (2). 'Jābir ibn Ḥayyān; Contributions à l'Histoire des Idées Scientifiques dans l'Islam; I, Le Corpus des Écrits Jābiriens.' *MIE*, 1943, **44**, 1–214. Rev. M. Meyerhof, *ISIS*, 1944, **35**, 213.

KRAUS, P. (3). 'Jābir ibn Ḥayyān; Contributions à l'Histoire des Idées Scientifiques dans l'Islam; II, Jābir et la Science Grecque.' *MIE*, 1942, **45**, 1–406. Rev. M. Meyerhof, *ISIS*, 1944, **35**, 213.

KRAUSE, F. (1). 'Fluss- und Seegefechte nach chinesischen Quellen aus der Zeit der Chou- und Han-Dynastie und der Drei Reiche.' *MSOS*, 1915, **18**, 61.

KROEBER, A. L. (1). *Anthropology.* Harcourt Brace, New York, 1948.

KROEBER, A. L. (7). 'Arrow Release Distributions.' *UC/PAAA*, 1927, **23**, 283.

KROMAYER, J. & VEITH, G. (1) (ed.). 'Heerwesen und Kriegsführung d. Griechen u. Römer.' (*Handbuch d. Altertumswissenschaft*, ed. I. v. Müller & W. Otto, Section IV, Pt. 3, vol. 2.). Beck, München, 1928.

KUHN, P. A. (1). 'The Taiping Rebellion.' Art. in *Cambridge History of China*, ed. D. Twitchett & J. K. Fairbank, Cambridge, 1978, vol. 10, p. 264.

KUO TING-YI & LIU KUANG-CHING (1). 'Self-strengthening; the Pursuit of Western Technology.' Art. in *Cambridge History of China*, ed. D. Twitchett & J. K. Fairbank, Cambridge, 1978, vol. 10, p. 491.

KYESER, KONRAD (1). *Bellifortis* (the earliest of the +15th-century illustrated handbooks of military engineering, begun +1396, completed +1410). MS Göttingen Cod. Phil. 63 and others. See Sarton (1) vol. 3, p. 1550; Berthelot (5), (6).

LACABANE, L. (1). 'De la Poudre à Canon et de son Introduction en France au 14ème Siècle.' *BEC*, 1844, **1** (2ᵉ sér.), 28. Crit. Fêng Chia-Shêng (**8**).

LACOSTE, E. (1) (tr.). 'La Poliorcétique d'Apollodore de Damas.' *REG*, 1890, **3**, 268.

LAFFIN, J. (1). *The Face of War.* London, 1963.

LAKING, SIR GUY, F. (1). *A Record of European Arms and Armour through Seven Centuries.* 3 vols. Bell, London, 1920.

LALANNE, L. (1). 'Essai sur le Feu Grégeois et sur l'Introduction de la Poudre à Canon en Europe, et principalement en France.' *MAI/NEM* 1843 (2ᵉ sér.), **1**, 294–363. Sep. pub., 1841. 2nd ed. sep. pub. Paris, 1845, under title: *Recherches sur....*

LALANNE, L. (2). 'Controverse à propos du Feu Grégeois.' *BEC*, 1846, **3** (2ᵉ sér.), 338. Reply by J. T. Reinaud, pp. 427, 534 and rejoinder by Lalanne, p. 440.

[LALANNE, L.] (3). 'Greek Fire and Gunpowder' (a review of Reinaud & Favé's book). *BLM*, 1846, **59**, 749.

LAMB, HAROLD (1). *Tamerlane the Earth-Shaker.* Butterworth, London, 1929.

DE LANA, FRANCESCO TERTÜ (1). *Magisterium Naturae et Artis; Opus Physico-Mathematicum...* 3 vols. Ricciardus, Brescia, 1684–92.

LANCASTER, O. E. (1) (ed.). *Jet Propulsion Engines.* Princeton Univ. Press, Princeton, N.J., 1959. (High Speed Aerodynamics and Jet Propulsion, vol. 12.)

LANKTON, L. D. (1). 'The Machine under the Garden; Rock Drills arrive at the Lake Superior Copper Mines (1868 to 1883).' *TCULT*, 1983, **24**, 1.

LARCHEY, LOREDAN (1). *Les Origines de l'Artillerie française (+1324 à +1394).* Paris, 1882.

LARSEN, E. (1). *A History of Invention.* Phoenix, London, 1961.

LASSEN, TAGE (1). 'From Hand-Cannon to Flint-Lock.' *GUND*, 1956, **10**, 33.

LATTIMORE, O. (10). *Nationalism and Revolution in Mongolia; with a Translation from the Mongol of S. Nachukdorji's Life of Sukebatur, by O. L. & U. Onon.* Brill, Leiden, 1955.

LAUFER, B. (47). Review of Diels (1), *Antike Technik*, 1914. *AAN*, 1917, **19**, 71.

LAVIN, J. (1). 'An Examination of some Early Documents regarding the Use of Gunpowder in Spain.' *JAAS*, 1964, **4**, 163.

LEBEAU, CHARLES (1). *Histoire du Bas-Empire.* 27 vols. Paris, 1757–1811. 2nd ed. (St. Martin & Brosset): 21 vols. Paris, 1824–36.

LECLERC, L. (1) (tr.). 'Le Traité des Simples par Ibn al-Beithar.' *MAI/NEM*, 1877, **23, 25**; 1883, **26**.

LENZ, E. (1). 'Handgranaten oder Quecksilbergefässe?' *ZHWK*, 1913, **6**, 367.

LEPRINCE-RINGUET, LOUIS *et al.* (1). *Les Inventeurs Célèbres; Sciences Physiques et Applications.* Mazenod, Paris, 1950.

DE LETTENHOVE, KERVYN (1) (ed.). *Oeuvres de Froissart.* Brussels, 1867, 1868, 1873.

[LEURECHON, J., HENRIOT, F. & MYDORGE, C.] (1). *Récréations Mathématiques: composées de plusieurs Problèmes plaisans et facétieux d'Arithmétique, Géometrie, Astrologie, Optique, Perspective, Mechanique, Chymie et d'autres rares et curieux Secrets; plusieurs desquels n'ont jamais esté Imprimez.* Rouen, 1630. The third part is entitled 'Recueil de plusieurs plaisantes et récréatives inventions de Feux d'Artifice; Plus, la manière de faire toutes sortes de Fuzées...' Eng. tr. *Mathematicall Recreations; or, a Collection of sundrie Problemes, extracted out of the Ancient and Moderne Philosophers, as secrets in nature, and experiments in Arithmetique, Geometrie, Cosmographie, Horologographie, Astronomie, Navigation, Musicke, Opticks, Architecture, Staticke, Mechanicks, Chimestrie, Waterworkes, Fireworks, etc.* Cotes & Hawkins, London, 1633. The last section (p. 265) is entitled: 'Artificiall Fire-Workes: or the manner of making Rockets and Balls of Fire, as well for the Water, as for the Ayre: with the Composition of Stars, Golden-raine, Serpents, Lances, Wheeles of fire, and such like, pleasant and Recreative'. Repr. Leake, London, 1653, (with William Oughtred's 'Double Horizontall Dyall'.) 1674 (-do-), etc.

LEY, W. (1). *Rockets.* Viking, New York, 1944. Enlarged ed. *Rockets, Missiles, and Men in Space.* Viking, New York, 1968.

LEY, W. (2). *Rockets and Space Travel; the Future of Flight beyond the Stratosphere.* Chapman & Hall, London, 1948. Rev. R. A. Rankin, *N*, 1949, **163**, 820.

LEY, W. (3). 'Rockets.' *SAM*, 1949, **181**, 31.

LEY, W. (4). *Die Fahrt ins Weltall.* Hachmeister & Thal, Leipzig, 1926. Enlarged ed. *Die Möglichkeit der Weltraumfahrt.* Hachmeister & Thal, Leipzig, 1928.

LEY, W. & VON BRAUN, W. (1). *The Exploration of Mars.* Viking, New York, 1966.

LI CHHIAO-PHING (2) (ed. & tr., with 14 collaborators). '*Thien Kung Khai Wu*' (*The Exploitation of the Works of Nature*); *Chinese Agriculture and Technology in the Seventeenth Century, by Sung Ying-Hsing.* China Academy, Thaipei, 1980. (Chinese Culture Series II, no. 3.)

LI HSIEH (LI I-CHIH) (1). 'Die Geschichte des Wasserbaues in China.' *BGTI*, 1932, **21**, 59.

LI KUO-HAO, CHANG MÊNG-WÊN, TSHAO THIEN-CHHIN & HU TAO-CHING (1) (ed.). *Explorations in the History of Science and Technology in China; a Special Number of the 'Collections of Essays on Chinese Literature and History'* (compiled in honour of the eightieth birthday of Joseph Needham). Chinese Classics Publishing House, Shanghai, 1982.

LICHINE, ALEXIS, FIFIELD, W. *et al.* (1). *Encyclopaedia of Wines and Spirits.* Cassell, London, 1974.

LIDDELL-HART, B. H. *See* Hart, B. H. Liddell.

LINDSAY, M. (1). *One Hundred Great Guns; an Illustrated History of Firearms.* London, 1968.

LINDSAY, M. (2). 'Pistols Shed Light on Famed Duel.' *SMITH*, 1976, **7** (no. 8), 94. (The duel of 1804 between Aaron Burr, Vice-President of the United States under Jefferson, who survived, and Alexander Hamilton, former Secretary of the Treasury, who was killed. The pistols had hair-triggers, and Hamilton probably fired too soon.)

VON LIPPMANN, E. O. (9). *Beiträge z. Geschichte d. Naturwissenschaften u. d. Technik.* 2 vols. Vol. 1, Springer, Berlin, 1925. Vol. 2, Verlag Chemie, Weinheim, 1953 (posthumous, ed. R. von Lippmann). Both vols. photographically reproduced, Sändig, Niederwalluf, 1971.

VON LIPPMANN, E. O. (21). 'Zur Geschichte des Schiesspulvers und des Salpeters.' *CHZ*, 1928, **52**, 2. Abstr. *CA*, 1928, **22**, 894. Repr. v. Lippmann (9), p. 83.

VON LIPPMANN, E. O. (22). 'Zur Geschichte des Schiesspulvers und der älteren Feuerwaffen.' Lecture at Halle, 1898. Repr. in von Lippmann (3), vol. 1, p. 125.

LITTLE, A. G. (2) (ed.). *Roger Bacon, Essays.* Oxford, 1914.

LIU KUANG-CHING (1). 'The Chhing Restoration.' Art. in *Cambridge History of China*, ed. D. Twitchett & J. K. Fairbank, Cambridge, 1978, vol. 10, p. 409.

LO JUNG-PANG (10). *The Art of War in the Chhin and Han Periods; −221 to +220* (in the press).

LO JUNG-PANG (12). 'Missile Weapons in pre-modern China.' Contribution to the Meeting of the Association for Asian Studies, Chicago, 1967.

LOEHR, M. (1). 'The Earliest Chinese Swords and the [Scythian] *Akinakes*.' *ORA*, 1948, **1**, 132.

LOEHR, M. (2). *Chinese Bronze Age Weapons; the Werner Jannings Collection in the Chinese National Palace Museum, Peking.* Univ. Michigan Press, Ann Arbor, 1956.

LOEWE, M. (11). 'The Campaigns of Han Wu Ti.' Art. in Kierman & Fairbank (1) (ed.), *Chinese Ways in Warfare*, p. 67.

LONGMAN, C. J. (1). 'The Bows of the Ancient Assyrians and Egyptians.' *JRAI*, 1894, **24**, 49.

LONGMAN, C. J., WALROND, H. *et al.* (1). *Archery.* Longmans Green, London, 1894.

LOPEZ, R. S. (3). 'Venezia e le grandi Linee dell'espanzione commerciale nel Secolo 13.' Art. in *La Civiltà Veneziana del Secolo di Marco Polo.* Sansoni, Florence, 1955, pp. 37–82.

LOPEZ, R. S. (5). 'Nuove Luci sugli Italiani in Estremo Oriente prima di Colombo.' *STC*, 1951, **3**, 350.

LORRAIN, HANZELET. *See* Appier, Jean (Lorraine was his place of origin).

LOT, F. (1). *L'Art Militaire et les Armées au Moyen-Age en Europe et dans le Proche Orient.* 2 vols. Payot, Paris, 1946.

LOTZ, A. (1). *Das Feuerwerk, seine Geschichte und Bibliographie in sieben Jahrhunderten.* Leipzig, 1940.

LOUIS, H. (1). 'A Chinese System of Gold Milling [in Malaysia].' *EMJ*, 1891, **54**, 640.

LOUIS, H. (2). 'A Chinese System of Gold Mining [in Malaysia].' *EMJ*, 1892, **55**, 629.

LOUIS-FRÉDÉRIC (ps.) (1) (FRÉDÉRIC, LOUIS). *Daily Life in Japan at the Time of the Samurai, +1185 to +1603.* Tr. from the French ed. of 1968 by E. M. Lowe, Allen & Unwin, London, 1972. (Daily life series, no. 17.)

LU MAU-DÊ. *See* Lu Mou-Tê.

LU MOU-TÊ (1). 'Untersuchung ü. d. Erfindung der Geschütze u. d. Schiesspulvers in China.' *SA*, 1938, **13**, 25 and 99b. A translation of Lu Mou-Tê (1) by Liao Pao-Shêng.

LUCIAN OF SAMOSATA (1). *True History.* Bullen, London, 1902. *Certaine Select Dialogues of Lucian, Together with his True Historie, translated from the Greeke into English by Mr Francis Hickes, Whereunto is added the Life of Lucian gathered out of his own Writings, with briefe Notes and Illustrations upon each Dialogue and Booke.* Oxford, 1634.

LULHAM, R. (1). *An Introduction to Zoology.* Macmillan, London, 1913.

McCRINDLE, J. W. (2). *Ancient India as described by Ktesias the Knidian; being a translation of the abridgement of his Indica by Photios, and of the fragments of that work preserved in other Writers, with notes, etc.* Thacker & Spink, Calcutta, 1882.

McCULLOCH, J. (1). 'Conjectures respecting the Greek Fire of the Middle Ages.' *Q JSLA*, 1823, **14**, 29.

McCURDY, E. (1). *The Notebooks of Leonardo da Vinci, Arranged, Rendered into English, and Introduced by . . .* 2 vols. Cape, London, 1938.

McGOWAN, D. J. (7). 'Blood-Sweating Horses in Ancient Turkestan.' *JPOS*, 1887, **1**, 196.

McGRATH, J. (1). 'Explosives [in the Late Nineteenth Century].' Art. in *A History of Technology*, ed. C. Singer *et al.*, Oxford, 1958, vol. 4, p. 284.

MACHELL-COX, E. (1) (tr.). *The Principles of War, by Sun Tzu; a classic of the Military Art [Sun Tzu Ping Fa].* RAF Welfare Publications, Colombo, Ceylon, 1943.

McLAGAN, GEN. R. (1). 'Early Asiatic Fire Weapons.' *JRAS/B*, 1876, **45**, 30.

DE MAILLA, J. A. M. DE MOYRIAC (1) (tr.). *Histoire Générale de la Chine, ou Annales de cet Empire, traduites du 'Tong Kien Kang Mou' [Thung Chien Kang Mu].* 13 vols. Pierres & Clousier, Paris, 1777. (This translation was made from the edition of +1708; Hummel (2), p. 689.)

MAINWARING, SIR HENRY (1). *The Seaman's Dictionary.* London, 1644.

MAITRA, K. M. (1) (tr.). *A Persian Embassy to China; being an Extract from the 'Zubdatu't Tawārikh' of Hafīz-i Abrū*... Lahore. 1934 (with introduction by L. C. Goodrich). Paragon Reprint, New York, 1970.

MALCOM, HOWARD (1). *Travels in South-eastern Asia*. 2 vols. Gould, Kendall & Lincoln, Boston, 1839; London, 1839.

MALINA, F. J. (1). 'A Short History of the Development of Rockets and Jet Propulsion Engines down to 1945.' Art. in *High Speed Aerodynamics and Jet Propulsion*, ed. O. E. Lancaster (1). Princeton Univ. Press, Princeton N.J., 1953–9, vol. 12.

MALINA, F. J. (2). 'Memoir on the GALCIT Rocket Research Project, 1936–38.' *Proceedings of the 1st Internat. Symp. on History of Astronautics*, 1967 (Belgrade). Abridged version *ESCI*, 1968, **31**, 9. Russian tr. 1970.

MALINA, F. J. (3). 'Memoir on the U.S. Army Air Corps Jet Propulsion Research Project, GALCIT No. 1, 1939–46.' *Proceedings of the 3rd Internat. Symp. on History of Astronautics*, 1969.

MALINA, F. J. (4). 'America's First Long-Range Missile and Space Exploration Programme; the ORDCIT project of the Jet Propulsion Laboratory, 1943–46.' *SPFL*, 1973, **15**, 442.

MALINA, F. J. (5). 'The Jet Propulsion Laboratory; its Origins and First Decade of Work.' Art. in Emme (1). Also in *SPFL*, 1964, **6**, 160, 193.

MALTHUS, F. *See* de Malthe, François.

MANN, SIR JAMES G. (1). *European Arms and Armour* (Catalogue of the Wallace Collection). 3 vols. Clowes, London, 1945. (Contains a useful glossary of terms.)

MANSI, J. D. *et al.* (1). *Sacrorum Conciliorum Nova et Amplissima Collectio*. Florence, 1759–98.

MARSDEN, E. W. (1). *Greek and Roman Artillery; Historical Development*. Oxford, 1969.

MARSDEN, E. W. (2). *Greek and Roman Artillery; Technical Treatises*. Oxford, 1971.

MARSHALL, ARTHUR (1). *Explosives*. 2 vols. Churchill, London, 1917.
  Vol. 1. *History and Manufacture*.
  Vol. 2. *Properties and Tests*.

MARSHALL, A. M. & HURST, C. H. (1). *A Junior Course of Practical Zoology*. Smith Elder, London, 1916.

MARTIN, H. D. (1). 'The Mongol Wars with [the] Hsi-Hsia (+1205/+1227).' *JRAS*, 1942, 195.

MARTIN, H. D. (2). 'The Mongol Army.' *JRAS*, 1943, 46.

MARTIN, H. D. (3). 'Chingiz Khan's First Invasion of the Chin Empire (+1211/+1213).' *JRAS*, 1943, 182.

MARTIN, H. D. (4). *The Rise of Chinghiz Khan and his Conquest of North China*. Johns Hopkins Univ. Press, Baltimore, 1950. Repr. Octagon, New York, 1971.

MASON, B. J. (SIR JOHN) (1). 'The Growth of Snow Crystals.' *SAM*, 1961, **204** (no. 1), 120.

MASON, B. J. (SIR JOHN) (2). 'A Review of Three Long-Term Cloud-Seeding Experiments.' *MEM*, 1980, **109**, 335.

MASON, B. J. (SIR JOHN) & MAYBANK, J. (1). 'Ice-Nucleating Properties of Some Natural Mineral Dusts.' *QJRMS*, 1958, **84**, 235.

MATHENHEIMER, A. (1). *Die Rückladungs-Gewehr*. Darmstadt and Leipzig, 1876.

MATSCHOSS, C. (2) *Geschichte der Dampfmaschine, ihre Kulturelle Bedeutung, technische Entwicklung und ihre grossen Männer*. Springer, Berlin, 1901. Photolitho reproduction, Gesstenberg, Hildesheim, 1978.

MAVRODIN, V. (1). 'O Poyarlenii Ognestrel'nogo Oruzhiya na Rusi' (on the Origin of Firearms in Russia). *VH*, 1946 (no. 8/9), 98. Crit. Fêng Chia-Shêng (8).

MAYERS, W. F. (6). 'On the Introduction and Use of Gunpowder and Firearms among the Chinese, with Notes on some Ancient Engines of Warfare, and Illustrations.' *JRAS/NCB*, 1870 (NS), **6**, 73. Comment by E. H. Parker, *CR*, 1887, **15**, 183.

MAYOW, JOHN (1). *Tractatus Quinque Medico-Physici....* Sheldonian, Oxford, 1674. Repr. *Opera Omnia Medico-Physica Tractatibus Quinque comprehensa...* Leers, The Hague, 1681.

MÉAUTIS, G. (1). 'Les Romains connaissent-ils le Fer à Cheval?' *REA*, 1934, **36**, 88.

MELLOR, J. W. (1). *Modern Inorganic Chemistry*. Longmans Green, London, 1916; often reprinted.

MELLOR, J. W. (2). *Comprehensive Treatise on Inorganic and Theoretical Chemistry*. 15 vols. Longmans Green, London, 1923.

DE MENDOZA, JUAN GONZALES (1). *Historia de las Cosas mas notables, Ritos y Costumbres del Gran Reyno de la China, sabidas assi por los libros de los mesmos Chinas, como por relacion de religiosos y oltras personas que an estado en el dicho Reyno*. Rome, 1585 (in Spanish). Eng. tr. Robert Parke, *The Historie of the Great & Mightie Kingdome of China and the Situation theoreof; Togither with the Great Riches, Huge Citties, Politike Gowernement and Rare Inventions in the same* [undertaken 'at the earnest request and encouragement of my worshipfull friend Master Richard Hakluyt, late of Oxforde']. London, 1588 (1589). Reprinted in Spanish, Medina del Campo, 1595; Antwerp, 1596 and 1655; Ital. tr. Venice (3 editions), 1586; Fr. tr. Paris, 1588, 1589 and 1600; Germ. and Latin tr. Frankfurt, 1589. New ed. G. T. Staunton, London, 1853 (Hakluyt Soc. Pubs. 1st ser. nos 14, 15). Spanish text again Ed. P. F. García, Madrid, 1944 (España Misionera, no. 2.).

MERCER, H. C. (1). *Ancient Carpenter's Tools illustrated and explained, together with the Implements of the Lumberman, Joiner, and Cabinet-Maker, in use in the Eighteenth Century*. Bucks County Historical Society, Doylestown, Pennsylvania, 1929.

MERCIER, M. (1). *Le Feu Grégeois; Les Feux de Guerre depuis l'Antiquité; La Poudre à Canon.* Geuthner, Paris, 1952; Aubarel, Avignon, 1952.

METTLER, CECILIA C. (1). *History of Medicine; a Correlative Text arranged according to Subjects.* Blakiston, Philadelphia, 1947.

MICKLE, W. J. (1) (tr.). *The 'Lusiad', or the Discovery of India; an Epic Poem* [by Luis de Camoëns], *translated from the original Portuguese by W. J. M.* Jackson & Lister, Oxford, 1776; repr. 1778; 5th ed. London, 1877.

MIETH, MICHAEL (1). '*Artilleriae Recentior Praxis*', *oder neuere Geschütz-Beschreibung worinnen von allen vornehmsten Haupt-Puncten der Artillerie gründlich...gehandelt...mit vielen Kupffer-Stücken erkläret wird...* Pr. pr. Frankfurt a/Main & Leipzig, 1683. Re-issued 1684. 2nd ed. with new title, Dresden & Leipzig, 1736.

MILLS, J. V. (6). MS translation of part of ch. 13 of the *Chhou Hai Thu Pien* (on shipbuilding, etc.). Unpub. MS.

MILSKY, M. (1). 'Les Souscripteurs de "l'Histoire Générale de la Chine" du P. de Mailla; Aperçus du Milieu Sinophile Français.' Art. in *Les Rapports entre la Chine et l'Europe au Temps des Lumières* (Actes du 2ᵉ Colloque International de Sinologie, Chantilly), ed. J. Sainsaulieu. Cathasia, Paris, 1980, p. 101.

MINORSKY, V. F. (4) (ed. & tr.). *Sharaf al-Zamān Ṭāhir al-Marwazī on China, the Turks and India* (c. +1120). Royal Asiatic Soc., London, 1942. (Forlong Fund series, no. 22.)

MITRA, HARIDAS (1). *The Fire-works and Fire Festivals in Ancient India.* Abhedananda Academy, Calcutta, 1963.

MIYAKAWA HISAYUKI (1). 'The Legate Kao Phien [d. +887] and a Taoist Magician, Lü Yung-Chih, in the Time of Huang Chhao's Rebellion [+875 to +884].' *ACTAS*, 1974, no. 27, 75. (The Taoist entourage of the general who suppressed it, including alchemists, Chuko Yin, Tshai Thien and Shenthu Shêng = Pieh-Chia.)

MOLINARI, E. & QUARTIERI, F. (1). *Notizie sugli Esplodenti in Italia.* Milano, 1913.

MOLLER, W. A. (1). 'Mining in Manchuria.' *TIME*, 1903, **25**, 144.

MONTANDON, G. (1). *l'Ologénèse Culturelle; Traité d'Ethnologie Cyclo-Culturelle et d'Ergologie Systématique.* Payot, Paris, 1934.

MONTANUS, ARNOLDUS (1). '*Atlas Japannensis*'; *being remarkable Addresses by way of Embassy from the East-India Company...to the Emperor of Japan...* English'd by John Ogilby, Esq. Johnson, London, 1670.

MONTEIL, V. (1) (tr.). *Ibn Khaldūn: Discours sur l'Histoire Universelle—'al-Muqaddimah'.* 4 vols. Unesco, Beirut, 1967.

MOOR, EDWARD (1). *Narrative of the Operations of Capt. Little's Detachment...during the late Confederacy in India, aginst the Nawab Tippas Sultan Bahadur.* London, 1794.

MORAY, SIR ROBERT (1). 'A Way to break easily and speedily the hardest Rocks, communicated by the same Person (Sir R. M.) as he received it from Monsieur du Son the Inventor.' *PTRS*, 1665, **1** (no. 5), 82.

MORGAN, C. (1). *The Shape of Futures Past; the Story of Prediction* (a history of science fiction). Webb & Bower, Exeter, 1980.

MORGAN, E. (1) (tr.). *Tao the Great Luminant; Essays from Huai Nan Tzu, with introductory articles, notes and analyses.* Kelly and Walsh, Shanghai, n.d. (1933?).

MORLAND, SIR SAMUEL (1). *Élevation des Eaux par toute sorte de Machines...* Paris, 1685.

MORRIS, WILLIAM & WYATT, A. J. (1). *The Tale of Beowulf, sometime King of the Folk of the Weder Geats.* Kelmscott, Hammersmith, 1895. Repr. London, 1898.

MOTE, F. W. (3). 'The Thu-mu Incident of +1449.' Art. in Kierman & Fairbank (1) (ed.), *Chinese Ways in Warfare*, p. 243.

MOULE, A. C. (1). *Christians in China before the year 1550.* SPCK, London, 1930.

MOULE, A. C. (13). 'The Siege of Saianfu [Hsiang-yang] and the Murder of Achmach Bailo; two Chapters of Marco Polo.' *JRAS/NCB*, 1927, **58**, 1, 1928, **59**, 256. (Deals in detail with the Muslim trebuchet engineers and the alleged presence of Marco Polo at the siege of Hsiang-yang.)

MOULE, A. C. & PELLIOT, P. (1) (tr. & annot.). *Marco Polo* (+1254 to +1325); *The Description of the World.* 2 vols. Routledge, London, 1938. Repr. AMS Press. New York, 1976. Further notes by P. Pelliot (posthumously pub.). 2 vols. Impr. Nat. Paris, 1960.

MUIRHEAD, J. P. (1). *The Origin and Progress of the Mechanical Inventions of James Watt, illustrated by his Correspondence with his Friends and the Specifications of his Patents.* 3 vols. London, 1854.

MULLER, JOHN (1). *Treatise on Artillery; to which is prefixed, a Theory of Powder applied to Firearms...* Millan, London, 1757; repr. 1768.

MULTHAUF, R. P. (5). *The Origins of Chemistry.* Oldbourne, London, 1967.

MULTHAUF, R. P. (9). 'An Enquiry into Saltpetre Supply and the Early Use of Firearms.' *Abstracts of Scientific Section Papers 15th Internat. Congress Hist. of Sci.* Edinburgh, 1977, p. 37.

MUNDY, PETER (1). *Travels in Europe and Asia* (+1608 to +1667). 5 vols. in 6, Hakluyt Soc., Cambridge, 1907; London, 1914–36 (Hakluyt series, nos. 17, 35, 45, 46, 55, 78). Ed. Lt.-Col. Sir Richard Carnac Temple & L. M. Anstey. Repr. 3 vols. Kraus, Liechtenstein, 1957.

MUNSTER, SEBASTIAN (1). *Cosmographiae Universalis, Libri VI.* Petri, Basel, 1550, 1552, 1554, 1556, 1572.

MURATORI, L. A. (1) (ed.). *Rerum Italicarum Scriptores, ex Codicibus L.A.M. Collegit, Ordinavit et Praefationibus Auxit.* 25 vols. Milan, 1728–51. 2nd ed. Città di Castello, 1900– , ed. G. Carducci & V. Fiorini.

MURDOCH, JAMES (1) (with the collaboration of I. Yamagata). *A History of Japan*. First pub. Yokohama, 1910. 3 vols. ed. J. H. Longford, Kegan Paul, London, 1925.

MUS, P. (2). 'Les Ballistes du Bayon.' *BEFEO*, 1929, **29**, 331 (Études Indiennes et Indochinoises, pt. 3).

MUTHESIUS, V. (1). *Zur Geschichte der Sprengstoffe und des Pulvers*. Pr. pr. Berlin, 1941.

NAMBO HEIZO (1) = (1). 'Who Invented Explosives?' *JSHS*, 1970, **9**, 49–98. (This paper contains many errors, both sinological and historical; it should be used with caution.)

NAPOLEON III (EMPEROR OF FRANCE) & FAVÉ, I. CAPT. *See* Bonaparte & Favé (1).

NAUMANN, K. (1). 'Untersuchung eines Luristanischen Kurzschwertes.' *AEHW*, 1957, **28**, 575.

NEEDHAM, JOSEPH (2). *A History of Embryology*. Cambridge Univ. Press, 1934. 2nd ed., revised with the assistance of A. Hughes. Cambridge, 1959; Abelard–Schuman, New York, 1959.

NEEDHAM, JOSEPH (27). 'Limiting Factors in the History of Science, as observed in the History of Embryology' (Carmalt Lecture at Yale University, 1935). *YJBM*, 1935, **8**, 1. Reprinted in Needham (3).

NEEDHAM, JOSEPH (32). *The Development of Iron and Steel Technology in China*. Newcomen Soc., London, 1958. (Second Biennial Dickinson Memorial Lecture, Newcomen Society.) Précis in *TNS*, 1960, **30**, 141; rev. L. C. Goodrich, *ISIS*, 1960, **51**, 108. Repr. Heffer, Cambridge, 1964, French tr. (unrevised, with some illustrations omitted and others added by the editors) *RHSID*, 1961, **2**, 187, 235; 1962, **3**, 1, 62.

NEEDHAM, JOSEPH (47). 'Science and China's Influence on the West.' Art. in *The Legacy of China*, ed. R. N. Dawson. Oxford, 1964, p. 234.

NEEDHAM, JOSEPH (48). 'The Prenatal History of the Steam-Engine.' (Newcomen Centenary Lecture.) *TNS*, 1963, **35**, 3–58.

NEEDHAM, JOSEPH (59). 'The Roles of Europe and China in the Evolution of Oecumenical Science.' *JAHIST*, 1966, **1**, 1. As Presidential Address to Section X, British Association, Leeds, 1967, *ADVS*, 1967, **24**, 83.

NEEDHAM, JOSEPH (60). 'Chinese Priorities in Cast Iron Metallurgy.' *TCULT*, 1964, **5**, 398.

NEEDHAM, JOSEPH (65). *The Grand Titration; Science and Society in China and the West* (Collected Addresses). Allen & Unwin, London, 1969.

NEEDHAM, JOSEPH (80). Notes on the Shansi Provincial Museum at Thaiyuan. Unpub.

NEEDHAM, JOSEPH (81). 'China's Trebuchets, Manned and Counterweighted.' Art. in Lynn White Festschrift *Humana Civilitas*, ed. Hall & West, 1976, p. 107.

NEEDHAM, JOSEPH (82). Notes of an Archaeological Study-Tour in China, 1958. Unpub.

NEEDHAM, JOSEPH (84). 'L'Alchimie en Chine; Pratique et Théorie.' *AHES/AESC*, 1975, no. 5, 1045.

NEEDHAM, JOSEPH (85). *China and the Origins of Immunology*. Centre of Asian Studies, Univ. of Hongkong, Hongkong, 1980. (First S.T. Huang-Chan Memorial Lecture.)

NEEDHAM, JOSEPH (86). 'Science and Civilisation in China; State of the Project.' *IDSR*, 1980, **5** (no. 4), 263. Chinese tr. *Chung-Kuo Kho Chi Shih Liao*, 1981, no. 3, 5.

NEF, JOHN U. (1). *La Route de la Guerre Totale; Essai sur les Relations entre la Guerre et le Progrès Humain*. Colin, Paris, 1949. (Cahiers de la Fondation Nationale des Sciences Politiques, no. 11.) Eng. tr., enlarged and revised: *Western Civilisation since the Renaissance; Peace, War, Industry and the Arts* (small print edition), Harper, New York, 1963. Eng. tr. again enlarged and revised. *War and Human Progress; an Essay on the Rise of Industrial Civilisation*. Russell & Russell, New York, 1968.

NUBURGER, A. (1). *The Technical Arts and Sciences of the Ancients*. Methuen, London, 1930. Tr. by H. L. Brose from *Die Technik d. Altertums*. Voigtländer, Leipzig, 1919 (with a drastically abbreviated index and the total omission of the bibliographies appended to each chapter, the general bibliography, and the table of sources of the illustrations).

NICOLAS, SIR HARRIS (1). *History of the Royal Navy, from the Earliest Times to the Wars of the French Revolution*. 2 vols. London, 1847.

NICOLSON, M. H. (1). *Voyages to the Moon*. Macmillan, New York, 1948.

NICOLSON, M. H. (2). 'A World in the Moon; a Study of the Changing Attitude towards the Moon in the +17th and +18th Centuries.' *SCSML*, 1936, **17** (no. 2), 1–72.

NIELSEN, NIELS AGE (1). *Dansk Etymologisk Ordbog*. 2nd. ed. Copenhagen, 1966.

NORTHROP, F. S. C. & LIVINGSTON, H. H. (1) (ed.). *Cross-Cultural Understanding; Epistemology in Anthropology* (a Wenner-Gren Foundation Symposium). Harper & Row, New York, 1964.

NORTON, ROBERT (1). *The Gunner*. London, 1628.

NYE, NATHANIEL, MASTER-GUNNER (1). *The Art of Gunnery*. London, 1647.

O'NEILL, B. H. St J. (1). *Castles and Cannon; a Study of Early Artillery Fortifications in England*. Oxford, 1960. Rev. J. Beeler, *SP*, 1962, **37**, 146.

O'NEILL, B. H. St J. (2). *Castles*. HMSO, London, 1953.

OBERTH, H. (1). *Die Rakete an den Planetenraümen*. Oldenbourg, München, 1923. Eng. tr. *The Rocket into Interplanetary Space*.

OBERTH, H. (2). *Wege zur Raumschifffahrt.* Oldenbourg, München, 1929. Eng. tr. *Ways to Spaceflight.* Agence Tunisienne des Relations Publiques, Tunis, 1972. NASA document TT/F 622. Photolitho reproduction, Edwards, Ann Arbor, 1945.

D'OHSSON, MOURADJA (1). *Histoire des Mongols depuis Tchinguiz Khan jusqu'à Timour Bey ou Tamerlan.* 4 vols., van Cleef, The Hague and Amsterdam, 1834–52.

OLSCHKI, L. (4). *Guillaume Boucher; a French Artist at the Court of the Khans.* Johns Hopkins Univ. Press, Baltimore, 1946. Rev. H. Franke, *OR*, 1950, **3**, 135.

OLSCHKI, L. (10). *L'Asia di Marco Polo.* Sansoni, Florence, 1957. Eng. tr. by J. A. Scott; '*Marco Polo's Asia; an Introduction to his "Description of the World", called "Il Milione".*' Univ. Calif. Press, Berkeley & Los Angeles, 1960.

OLSHAUSEN, O. & HIRTH, F. (1). '(1) Ü. einen Grabfund von Hedehusum auf Föhr; (2) Zur Kenntnis d. Schnallen; (3) Beitrag z. Geschichte d. Reitersporns; (4) Bemerkungen ü. Steigbügel.' With comments by Hirth on Stirrups in China. *ZFE/VBGA*, 1890, **22**, (170) & (209).

OLSZEWSKI, EUGENIUSZ (1). 'An Outline of the Development of Polish Science.' *ORG*, 1965, no. 2, 249.

OLSZEWSKI, EUGENIUSZ (2). 'To Commemorate the Centenary of the Birth of Konstantin Tsiolkovsky.' *KHNT*, 1957–8, Special issue, 25 (on the occasion of the First Polish National Science Congress).

OMAN, C. W. C. (1). *A History of the Art of War in the Middle Ages.* 1st ed. 1 vol. 1898; 2nd ed. 2 vols. 1924 (much enlarged); vol. 1, +378 to +1278; vol. 2, +1278 to +1485. Methuen, London (the original publication had been a prize essay printed at Oxford in 1885; this was reprinted in 1953 by the Cornell University Press, Ithaca N.Y., with editorial notes and additions by J. H. Beeler). Crit. Fêng Chia-Shêng (3).

OPPERT, G. (1). *On the Weapons, Army Organisation, and Political Maxims of the Ancient Hindus, with special reference to Gunpowder and Firearms.* Madras, 1880.

OUCHTERLONY, J. (1). *The Chinese War; an Account of all the Operations of the British Forces from its Commencement to the Treaty of Nanking.* Saunders & Otley, London, 1844.

PAK HAE-ILL (1) = (1). 'Parallelisms between the Iron Cladding of Admiral Yi Sunsin's Combat Turtle-Ships and Extant Iron Armouring of Yi Dynasty [City Gates].' *HKH*, 1979, **1** (no. 1), 27.

PAK HAE-ILL (2). 'A Short Note on the Iron-clad Turtle-Ships of Admiral Yi Sunsin.' *KJ*, 1977, **17** (no. 1), 34.

PÁLOS, S. (1). *Chinesische Heilkunst; Rückbesinnung auf eine grosse Tradition,* tr. from the Hungarian by W. Kronfuss. Delp, München, 1963; 2nd ed. 1966. Eng. tr. *The Chinese Art of Healing.* Bantam, New York, 1972.

PANCIROLI, GUIDO (1). *Rerum Memorabilium sive Deperditarum pars prior (et secundus) Commentariis illustrata et locis prope innumeris postremum aucta ab Henrico Salmuth.*' Amberg, 1599 and 1607; Schonvetter Vid. et Haered. Frankfurt, 1617, 1646, 1660. Eng. tr. *The History of many Memorable Things lost, which were in Use among the Ancients; and an Account of many Excellent Things found, now in Use among the Moderns, both Natural and Artificial...now done into English....To this English edition is added, first, a Supplement to the Chapter of Printing, shewing the Time of its Beginning, and the first Book printed in each City before the Year 1500. Secondly, what the Moderns have found, the Ancients never knew; extracted from Dr Sprat's History of the Royal Society, the Writings of the Honourable Mr Boyle, The Royal-Academy at Paris, etc...* London, 1715, 1727. French tr. Lyon, 1608. Bibliography in John Ferguson (2).

PAPIN, DENIS (1). '*De Novo Pulveris Pyri Usu* (on a New Application of Gunpowder).' *AER*, 1688, 497.

PAPIN, DENIS (2). '*Nova Methodus ad vires Motrices validissimas Levi Pretio Comparendas* (Papin's New Method of obtaining very great Moving Powers at small cost).' *AER*, 1690, 410. Latin text reprinted in Muirhead (1), vol. 3, p. 139 with an English translation. French text, probably the original, in *Recueil de Diverses Pièces* (1695). Cf. Galloway (1), pp. 14 ff.

PAPIN, DENIS (3). In *Fasciculus Dissertationum de Novis Quibusdam Machinis...* Marburg, 1695.

PAPIN, DENIS (4). In *Recueil de Diverses Pièces touchant quelques Nouvelles Machines.* Cassel, 1695.

PAPIN, DENIS (5). In *Traité de plusieurs Nouvelles Machines et Inventions Extraordinaires sur différens Sujets.* Paris, 1698.

PAPINOT, E. (1). *Historical and Geographical Dictionary of Japan.* Overbeck, Ann Arbor, Mich., 1948. Lithoprinted from original edn. Kelly & Walsh, Yokohama, 1910. Eng. tr. of *Dictionnaire d'Histoire et de Géographie du Japon.* Sanseido, Tokyo, 1906; Kelly & Walsh, Yokohama, 1906.

PARKER, E. H. (6). 'Military Engines.' *CR*, 1887, **15**, 253.

PARKER, E. H. (7). 'The Invention of Firearms.' *CR*, 1890, **18** (no. 6), 379.

PARKER, E. H. (8). 'The Military Organisation of China prior to 1842 as described by Wei Yuan.' *JRAS/NCB*, 1887, **22**, 1.

PARKER, E. H. (9). 'Military Engineering.' *CR*, 1885, **14**, 217.

PARKER, E. H. (10). 'Greek Fire and Firearms.' *CR*, 1887, **15**, 183.

PARKER, W. G. S. (1). 'Fuels for Research Rockets and Space Vehicles.' *MBLB*, 1965, **6**, 41.

PARRY, V. J. (1). 'Materials of War in the Ottoman Empire.' Art. in Cook (1) (ed.), *Studies in the Economic History of the Middle East.* Oxford, 1970, p. 219.

PARRY, V. J. & YAPP, M. E. (1) (ed.). *War, Technology and Society in the Middle East.* Oxford, 1975.

PARTINGTON, J. R. (5). *A History of Greek Fire and Gunpowder.* Heffer, Cambridge, 1960.

PARTINGTON, J. R. (10). *General and Inorganic Chemistry*... 2nd ed. Macmillan, London, 1951.

PARTINGTON, J. R. (20). 'The Life and Work of John Mayow (+1641 to +1679).' *ISIS*, 1956, **47**, 217, 405.

DE PAUW, C. (1). *Recherches Philosophiques sur les Égyptiens et les Chinois*... (Vols. IV and V of *Oeuvres Philosophiques*) Cailler, Geneva, 1774. 2nd ed., Bastien, Paris, Rep. An. III, (1795). Crit. Kao Lei-Ssu [Aloysius Ko, S.J.], *MCHSAMUC*, 1777, **2**, 365 (2nd pagination), 1–174.

VON PAWLIKOWSKI-CHOLEWA, A. (1). *Die Heere des Morgenlandes.* de Gruyter, Berlin, 1940.

PAYNE-GALLWEY, SIR RALPH (1). *The Crossbow, Mediaeval and Modern, Military and Sporting; its Construction, History and Management, with a Treatise on the Balista and Catapult of the Ancients.* Longmans Green, London, 1903: repr. Holland, London, 1958.

PAYNE-GALLWEY, SIR RALPH (2). *A Summary of the History, Construction and Effects in Warfare of the Projectile-Throwing Engines of the Ancients; with a Treatise on the Structure, Power and Management of Turkish and other Oriental Bows of Mediaeval and Later Times.* Longmans Green, London, 1907 (separately paged, no index). Practically identical with: *Appendix to the Book of the Crossbow and Ancient Projectile Engines*, Longmans Green, London, 1907. The *Summary* is more richly illustrated and has a fuller text than the *Appendix* yet its preface is dated Dec. 1906 while that of the latter is dated Jan. 1907.

PEGGE, S. (1). 'On Shoeing of Horses among the Ancients.' *A*, 1775, **3**, 39.

PEGOLOTTÍ, FRANCESCO BALDUCCI (1). *La Pratica della Mercatura, c.* +1340. Ed. A. EVANS, Cambridge, Mass., 1936.

PELLIOT, P. (10). 'Les Mongols et la Papauté.' *ROC*, 1922 (3e sér.), **3**, 3; **4**, 225, 1923 (3e sér.), **8**, 3.

PELLIOT, P. (33) (tr.). *Mémoire sur les Coutumes de Cambodge de Tcheou Ta-Kouan [Chou Ta-Kuan]; Version Nouvelle, suivie d'un Commentaire inachevé.* Maisonneuve, Paris, 1951. (Oeuvres Posthumes, no. 3.)

PELLIOT, P. (49). Note on gunpowder and firearms in a review of C. A. S. Williams (1) q.v. *TP*, 1922, **21**, 432.

PELLIOT, P. (53). 'Le Hōja et le Sayyid Husain de l'Histoire des Ming.' *TP*, 1948, **38**, 81–292.

PELLIOT, P. (55). 'Henri Bosmans, S.J.' *TP*, 1928, **26**, 190. (Includes material on the *Huo Kung Chhieh Yao*, and similar books.)

PELLIOT, P. (56). Review of Cordier (12), *l'Imprimerie Sino-Européenne en Chine. BEFEO*, 1903, **3**, 108.

PELLIOT, P. (59). Review of G. Schlegel (12), *On the Invention and Use of Firearms and Gunpowder in China*... *BEFEO*, 1902, **2**, 407.

PEPYS, SAMUEL (1). *The Diary of Samuel Pepys.* Everyman ed. Ed. J. Warrington, 2 vols. Dent, London, 1953.

PERCY, THOMAS (Bishop of Dromore) (1). *Reliques of Ancient English Poetry.* First pub. 1765. 3 vols. Washbourne, London, 1847.

PERRIN, NOEL (1) (with the assistance of Kuroda Eishoku & Sato Kiyondo). *Giving up the Gun; Japan's Reversion to the Sword, 1543 to 1879.* Godine, Boston, 1979. Pre-pub. abstr. in *NYR*, 1965, 20 Nov., 211. Rev. J. R. Bartholomew, *SCIS*, 1979, **19** (no. 7), 25.

PERTUSI, A. (1). *La Caduta di Constantinopoli; le Testimonianze dei Contemporanei.* 2 vols. Verona, 1976.

PERTZ, G. H. (1) (ed.). 'Cafari et Continuatorum Annales Januae' in *Monumenta Germaniae Historica*, vol. 18. Hannover, 1863 [MSS Paris, nos. 773 and 10136].

PETECH, L. (5). 'Les Marchands Italiens dans l'Empire Mongol.' *JA*, 1962, 549.

PETERSON, C. A. (1). 'Regional Defence against the Central Power; the Huai-hei Campaign, +815 to +817.' Art. in Kierman & Fairbank (1) (ed.), *Chinese Ways in Warfare*, p. 123.

PETERSON, H. L. (1). *The Book of the Gun.* London, 1962.

PETERSON, MENDEL L. (1). 'Richest Treasure Trove; a Bermuda Skin-diver discovers Sunken Bonanza Three Hundred Years Old. The Significance of Edward Tucker's Undersea Finds.' *LIFE*, 1956, **20** (no. 5), 43.

PETERSON, W. J. (2). *Bitter Gourd; Fang I-Chih [+1611 to +1671] and the Impetus for Intellectual Change.* Yale Univ. Press, New Haven, Conn., 1979.

PETRI, W. (7). 'Die Zukunft des Raumfahrtzeitalters in Sowjetischer Sicht; Prognosen und wissenschaftlich-kosmischer Utopien.' *UZWKL*, 1972, **27**, 1173. (Veröfftl. d. Forschungsinst. d. Deutschen Museums, A, 128.)

PETROVIC, DJURDJICA (1). 'Fire-arms in the Balkans on the Eve of and after the Ottoman Conquests of the +14th and +15th Centuries.' Art. in Parry & Yapp (1), *War, Technology and Society in the Middle East*, Oxford, 1975, p. 164.

PFIZMAIER, A. (34) (tr.). 'Die Feldherren Han Sin, Pêng Yue, und King Pu' (Han Hsin, Phêng Yüeh & Ching Pu). *SWAW/PH*, 1860, **34**, 371, 411, 418. (Tr. chs. 90 (in part), 91, 92, *Shih Chi*, ch. 34; *Chhien Han Shu*; not in Chavannes (1).)

PFIZMAIER, A. (37) (tr). 'Die Gewaltherrschaft Hiang Yü's' (Hsiang Yü). *SWAW/PH*, 1860, **32**, 7. (Tr. ch. 31, *Chhien Han Shu.*)

PFIZMAIER, A. (42) (tr.). 'Die Heerführer Li Kuang und Li Ling.' *SWAW/PH*, 1863, **44**, 511. (Tr. ch. 54, *Chhien Han Shu.*)

PFIZMAIER, A. (44) (tr.). 'Die Heerführer Wei Tsing und Ho Khiu-Ping' (Wei Chhing and Ho Chhü-Ping). *SWAW/PH*, 1864, **45**, 139. (Tr. ch. 55, *Chhien Han Shu.*)

PFIZMAIER, A. (98) (tr.). 'Die Anwendung und d. Zufälligkeiten des Feuers in d. alten China.' *SWAW/PH*, 1870, **65**, 767, 777, 786, 799. (Tr. chs. 868, 869 (fire and fire-wells), 870 (lamps, candles and torches), 871 (coal), of *Thai-Phing Yü Lan.*)

PFIZMAIER, A. (107) (tr.). 'Der Feldzug der Japaner gegen Corea im Jahre 1597' [translation of the *Chōsen Monogatari*]. Vienna, 1875.

PHILLIPS, T. R. (1) (ed.). *Roots of Strategy.* Lane, London, 1943. (A collection of classical Tactica, including Sun Tzu, Vegetius, de Saxe, Frederick the Great, and Napoleon.)

PITT-RIVERS, A. H. LANE-FOX (3). 'Primitive Warfare.' *JRUSI*, 1867, **11**; 1868, **12**, 399; 1869, **13**, 509. Reprinted in Pitt-Rivers (4).

PITT-RIVERS, A. H. LANE-FOX (4). *The Evolution of Culture, and other Essays.* OUP, Oxford, 1906. Ed. J. L. Myres with introdn. by H. Balfour. The title essay was first printed in *PRI*, 1875, **7**, 496.

PLATH, L. (2). 'Das Kriegswesen d. alten Chinesen.' *SBAW/PH*, 1873, **3**, 275.

PLOT, ROBERT (1). *The Natural History of Staffordshire.* Oxford, 1686.

POLE, W. (1). *A Treatise on the Cornish Pumping Engine.* London, 1844.

POLLARD, H. B. C. (1). *A History of Fire-arms.* Bles, London, 1926. Houghton Mifflin, Boston, 1936. Repr. Country Life, London, 1983, ed. Claud Blair, with three chapters by Howard Blackmore.

POPESCU, JULIAN (1). *Russian Space Exploration; the First Twenty-one Years.* Gothard, Henley-on-Thames, 1979.

DE LA PORTE DU THEIL, GABRIEL (1). '*Liber Ignium ad comburendos Hostes*', auctore Marco Graeco; ou, Traité des Feux propre à détruire les Ennemies, composé par Marcus le Grec; publié d'après deux manuscrits de la Bibliothèque Nationale. Delance & Lesueur, Paris, 1804.

PORTER, WHITWORTH, MAJ. (1). *A History of the Knights of Malta.* 2 vols. London, 1858.

POST, P. (1). 'Die früheste Geschützdarstellung von etwa +1330.' *ZHWK*, 1938, **15** (NF 6), 137.

POWER, d'ARCY (1). 'The Lesser Writings of John Arderne (+1307 to *c.* +1380).' *Proc. XVIIth Internat. Congr. Med.* Sect. 23, p. 107. London, 1913.

PRATT, PETER (1). *History of Japan, compiled from Records of the English East India Company at the instance of the Court of Directors.* London, 1822. Ed. M. B. T. Paske-Smith, 2 vols. in 1, Thompson, Kobe, 1931. 2nd ed. Curzon, London, 1972; Barnes & Noble, New York, 1972.

PRAWDIN, M. (1). *The Mongol Empire, its Rise and Legacy.* Tr. from the German of 1938 by E. & C. Paul. Allen & Unwin, London, 1940 (twice repr. 1952).

PREOBRAZHENSKY, A. G. (1). *Etymological Dictionary of the Russian Language.* Repr. Columbia Univ. Press, New York, 1951.

PREVITÉ-ORTON, C. W. (1) (ed.). *The Shorter Cambridge Medieval History.* Vol. 1. *The Later Roman Empire to the +12th Century*; vol. 2. *The +12th Century to the Renaissance.* Cambridge, 1953.

PRŮŠEK, J. (4). 'Quelques Remarques sur l'Emploi de la Poudre à Canon en Chine.' *ARO*, 1952, **20**, 250.

PULLEYBLANK, E. G. (5). 'A Geographical Text of the Eighth Century' [in ch. 3 (ch. 34) of the *Thai Pai Yin Ching*, +759]. Art. in Silver Jubilee Volume of the Zinbun Kagaku Kenkyusho, Kyoto University, 1954, p. 301 (*TG/K*, 1954, **25**, pt. 1).

QUATREMÈRE, E. M. (1) (tr.). *Histoire des Mongols de la Perse; écrite en Persan par Raschid-el-din* (part of the *Jami'al-Tawārīkh* of Rashid al-Dīn). Imp. Roy., Paris, 1836. (Vol. 1; only one vol. published.)

QUATREMÈRE, E. M. (2). 'Observations sur le Feu Grégeois.' *JA*, 1850 (4ᵉ sér.), **15**, 214–74. (A polemic against Reinaud & Favé (1), whom he considered had attacked him.) Short reply by J. T. Reinaud, p. 371. Crit. Fêng Chia-Shêng (**8**).

QUATREMÈRE, E. M. (3) (tr.). 'Notice de l'Ouvrage Persan qui a pour Titre *Matla Assaadeïn ou-madjina-albahreïn* et qui contient l'Histoire des deux Sultans Schah-rokh et Abou-Saïd' (The account by Ghiyāth al-Dīn-i Naqqāsh of the embassy from Shāh Rukh to the Ming emperor). *MAI/NEM*, 1843, **14**, pt. 1, 1–514 (387).

RAFEQ, ABDUL KARIM (1). 'The Local Forces in Syria in the Seventeenth and Eighteenth Centuries.' Art. in *War, Technology and Society in the Middle East*, ed. Parry & Yapp (1), p. 277.

RANDALL, J. T. & JACKSON, S. F. (ed.). *The Nature and Structure of Collagen.* Butterworth, London, 1953.

RATHGEN, B. (1). *Das Geschütz im Mittelalter; Quellenkritische Untersuchungen...* VDI Verlag, Berlin, 1928. Crit. Fêng Chia-Shêng (**3**).

RATHGEN, B. (2). 'Der deutsche Büchsenmeister Merckln Gast, der erste urkundlich erwähnte Eisengiesser.' *SE*, 1920, **40**, 148.

RATHGEN, B. (3). 'Das Drehkraftgeschütz in Deutschland.' *ZHWK*, 1919, **8**, 54.

RATHGEN, B. (4). 'Eisenguss und Urkundenbuch der Waffengeschichte.' *ZHWK*, 1919, **8**, 343.

RATHGEN, B. (5). 'Die Pulverwaffen in Indien.' *OAZ*, 1925, **2**, 9, 196.

RAVERTY, H. G. (1) (tr.). *Ṭabaḳāt-i Nāṣirī; a general History of the Muhammedan Dynasties of Asia, including Hindustan from +810 to +1260, and the Irruption of the Infidel Mughals [Mongols] into Islam, by the Maulānā, Minhāj ud-Dīn, Abū 'Umar-i 'Uṣmān [al- Juzjānī].* Gilbert & Rivington, London, 1881. (Bibliotheca Indica, for the Asiatic Society of Bengal.)

RAY, J. C. (1). 'Firearms in Ancient India.' *IHQ*, 1932, **8**, 268.

RAY, P. C. (1). *A History of Hindu Chemistry, from the Earliest Times to the middle of the 16th cent. A.D., with Sanskrit Texts, Variants, Translation and Illustrations.* 2 vols. Chuckerverty & Chatterjee, Calcutta, 1902, 1904, repr. 1925. New enlarged and revised edition in one volume, ed. P. Ray, retitled *History of Chemistry in Ancient and Medieval India*, Indian Chemical Society, Calcutta, 1956. Revs. J. Filliozat, *ISIS*, 1958, **49**, 362; A. Rahman, *VK*, 1957, 18.

READ, BERNARD E. (1) (with LIU JU-CHHIANG). *Chinese Medicinal Plants from the 'Pên Tshao Kang Mu'* (+1596)...*a Botanical, Chemical and Pharmacological Reference List.* (Publication of the Peking Nat. Hist. Bull.) French Bookstore, Peiping, 1936 (chs. 12–37 of *PTKM*). Rev. W. T. Swingle, *ARLC/DO*, 1937, 191. Originally published as *Flora Sinensis*, Ser. A, vol. 1, *Plantae Medicinalis Sinensis*, 2nd ed., *Bibliography of Chinese Medicinal Plants from the Pên Tshao Kang Mu*, +1596, by B. E. Read & Liu Ju-Chhiang. Dept. of Pharmacol. Peking Union Med. Coll. & Peking Lab. of Nat. Hist. Peking, 1927. First ed. Peking Union Med. Coll. 1923.

READ, J. (3). *Explosives.* Penguin, London, 1942.

READ, T. T. (4). 'The Early Casting of Iron; a Stage in Iron Age Civilisation.' *GR*, 1934, **24**, 544.

READ, T. T. (14). 'Coal-Mining in Manchuria.' *MIMG*, 1909, **1**, 217.

Reconstructions of Chinese Gunpowder Weapons. *See* Chiang Chêng-Lin (1) and Anon. (209).

REES, D. MORGAN (1). *The North Wales Quarrying Museum*, [Llanberis] *Gwynneth.* H.M.S.O. Cardiff, 1975, several times repr. (Welsh Office Official Handbook.)

REHATSEK, E. (1) (tr.). 'An Embassy to Khatā or China, A.D. 1419; from the Appendix to the *Ruzat al-Safā* of Muḥammed Khāvend Shāh, or Mirkhond, translated from the Persian....' *IAQ*, 1873, 75. (The embassy from Shāh Rukh, son of Tīmūr, to the Ming emperor; narrative written by Ghiyāth al-Dīn-i Naqqāsh.)

REID, W. (1). *The Lore of Arms.* Beazley, London, 1976.

REID, W. (2). 'Samuel Johannes Pauly, Gun Designer.' *JAAS*, 1957, **2**, 181.

REIFFERSCHEID, M. (1) (ed.). *Annae Comnenae Porphyrogenitae 'Alexias'.* 2 vols. Leipzig, 1884.

REILLY, JOSEPH (1). *Explosives, Matches and Fireworks.* Gurney & Jackson, London, 1938.

REINAUD, J. T. (3). 'De l'Art Militaire chez les Arabes au Moyen Age.' *JA*, 1848, (4ᵉ sér.), **12**, 193.

REINAUD, J. T. (4). *Extraits des Histoires Arabes relatifs aux Guerres des Croisades.* Paris, 1829. In J. F. Michaud's *Bibliothèque des Croisades*, 4 vols.

REINAUD, J. T. & FAVÉ, I. (1). 'Histoire de l'Artillerie, pt. 1; *Du Feu Grégeois, des Feux de Guerre, et des Origines de la Poudre à Canon, d'après des Textes Nouveaux.*' Dumaine, Paris, 1845. Crit. rev. by D[efrémer]y, *JA*, 1846 (4ᵉ sér.), **7**, 572; E. Chevreul, *JS*, 1847, 87, 140, 209.

REINAUD, J. T. & FAVÉ, I. (2). 'Du Feu Grégeois, des Feux de Guerre, et des Origines de la Poudre à Canon chez les Arabes, les Persans et les Chinois.' *JA*, 1849 (4ᵉ sér.), **14**, 257–327. Crit. Fêng Chia-Shêng (**3**) and (**8**).

REINAUD, J. T. & FAVÉ, I. (3). 'Controverse à propos du Feu Grégeois; Réponse aux Objections de M. Ludovic Lalanne.' *BEC*, 1847 (2ᵉ sér.), **3**, 427.

REINAUD, J. T., QUATREMÈRE, E. M., BEUGNOT, DE SACY, S. *et al.* (1) (ed.). *Recueil des Historiens des Croisades.* 17 vols. (5 vols. Occidentaux, 5 vols. Orientaux, 2 vols. Grecs, 2 vols. Arméniens). Acad. des Inscriptions, Paris, 1841–1906.

RÉMUSAT, J. P. A. (12). *Nouveaux Mélanges Asiatiques; ou, Recueil de Morceaux de Critique et de Mémoires relatifs aux Religions, aux Sciences, aux Coutumes, à l'Histoire et à la Géographie des Nations Orientales.* 2 vols. Schubart & Heideloff and Dondey-Dupré, Paris, 1829.

RENN, L. (1). *Warfare and the Relation of War to Society.* Faber, London, 1939.

RETI, LADISLAO (2). 'Leonardo da Vinci nella Storia della Macchina a Vapore.' *RDI*, 1957, 21.

RETI, LADISLAO & DIBNER, BERN (1). *Leonardo da Vinci, Technologist; Three Essays on some Designs and Projects of the Florentine Master in adapting Machinery and Technology to Problems in Art, Industry and War.* Burndy Library, Norwalk, Conn., 1969.

REYNIERS, COL. (1). 'Vues Anciennes et Nouvelles sur les Origines de l'Artillerie et de la Balistique.' *MAF*, 1956, **30** (no. 2), 511.

RICHARDSON, J. C. (1). 'On the ignition of petroleum by the heat of quicklime in contact with water; one of the proposed explanations of Greek fire—an experimental demonstration.' *N*, 1927, **120**, 165.

RIDGEWAY, SIR WILLIAM (2). *Origin and Influence of the Thoroughbred Horse.* CUP, Cambridge, 1905.

RITTER, H. (4). '"La Parure des Cavaliers" und die Literatur über die ritterliche Künste [the Arabic *furūsīya* literature].' *DI*, 1929, **19**, 116.

ROBINS, BENJAMIN (1). *New Principles of Gunnery.* London, 1742.

ROBINSON, H. R. (1). *Oriental Armour.* Jenkins, London, 1967.

DE ROCHEMONTEUX, C. (1). *Joseph Amiot et les Derniers Survivants de la Mission Française à Pékin (1750 à 1795); Nombreux Documents inédits, avec Carte.* Picard, Paris, 1915.

ROCK, JOSEPH F. (2). 'Konka Risumgongba, Holy Mountain of the Outlaws.' *NGM*, 1931, **60** (no. 1), 1.

ROCKHILL, W. W. (5) (tr. & ed.). *The Journey of William of Rubruck to the Eastern Parts of the World* (+1253 to +1255) *as narrated by himself; with Two Accounts of the earlier Journey of John of Pian de Carpine.* Hakluyt Soc., London, 1900 (second series, no. 4).

RODRIGUES, JOAO (1) (RODRIGUES TÇUZZU). *This Island of Japon; Joao Rodrigues' Account of 16th-century Japan* (+1577 to +1610). Tr. & ed. M. Cooper. Kodansha, Tokyo, 1973.

ROGERS, S. (1). 'On the Antiquity of Horse-shoes; a Letter to the Rev. J. Milles.' *A*, 1775, **3**, 35.

ROGERS, S. L. (1). 'The Aboriginal Bow and Arrow in North America and East Asia.' *AAN*, 1940, **42**, 255.

ROHDE, F, (2), 'Die Abzugsvorrichtung der frühen Armbrust und ihre Entwicklung.' *ZHWK*, 1933, **13** (NF **4**), 100.

ROLT, L. T. C. (1). *Thomas Newcomen; the Prehistory of the Steam Engine.* David & Charles, Dawlish, 1963; Macdonald, London, 1963.

ROLT, L. T. C. & ALLEN, J. S. (1). *The Steam Engine of Thomas Newcomen.* Moorland, Hartington, 1977; Neale Watson, New York, 1977.

VON ROMOCKI, S. J. (1). *Geschichte d. Explosivstoffe.* 2 vols. (usually bound in one). Oppenheim (Schmidt), Berlin, 1895, repr. Jannecke, Hannover, 1896. Vol. 1. *Geschichte der Sprengstoffchemie, der Sprengtechnik und des Torpedowesens bis zum Beginn der neuesten Zeit* (with introduction by M. Jähns). Vol. 2. *Die rauchschwachen Pulver in ihrer Entwickelung bis zur Gegenwart.* Two vols. Photolitho repr. Gerstenberg, Hildesheim, 1976. Crit. Fêng Chia-Shêng (**3**).

RONDOT, NATALIS (2). 'Lettre de M. Natalis Rondot à M. Reinaud sur le Feu Grégeois, etc.' *JA*, 1850, (4e sér.), **16**, 100. Also sep. pub. *Lettre à M. Reinaud; la Fabrication de la Poudre à Canon et de l'Acide Azotique en Chine.* Paris, 1850.

ROSE, W. (1). 'Anna Comnena über die Bewaffnung der Kreuzfahrer.' *ZHWK*, 1921, **9**, 1.

ROSENTHAL, ALFRED (1) *et al. The Early Years; the Goddard Space Flight Center—Historical Origins and Activities through December 1962.* Nat. Aeronautics & Space Admin., Washington, D.C., 1964. 2nd, enlarged, edition, retitled: *Venture into Space; Early Years of the Goddard Space Flight Center.* Nat. Aeronautics & Space Admin., Washington, D.C., 1968.

ROSENTHAL, F. (1) (tr.). '*The " Muqaddimah" [of Ibn Khaldūn]; an Introduction to History* [+1377].' Bollingen, New York, 1958. Abridgement by N. J. Dawood, Routledge & Kegan Paul, London, 1967.

ROSZAK, T. (1). *The Making of a Counter-Culture; Reflections on the Technocratic Society and its Youthful Opposition.* New York, 1968, repr. 1969; Faber & Faber, London, 1970, repr. 1971.

ROSZAK, T. (2). *Where the Wasteland Ends; Politics and Transcendence in Post-Industrial Society.* New York and London, 1972–3.

ROULEAU, F. A. (1). 'The Yangchow Latin Tombstone as a Landmark of Mediaeval Christianity in China.' *HJAS*, 1954, **17**, 346.

ROUSE, H. & INCE, S. (1). *A History of Hydraulics.* Iowa Univ. Press, Iowa City, 1957.

RUDOLPH, R. C. (12). 'A Second Fourteenth-Century Italian Tombstone in Yangchow' [+1344]. *JOS*, 1975, **13**, 133.

RUGGIERI, CLAUDE FORTUNÉ (1). *Élémens de Pyrotechnie...* Paris 1801; repr. 1821.

RUGGIERI, CLAUDE-FORTUNÉ (2). *Pyrotechnie Militaire...* Paris, 1812.

RUNCIMAN, STEVEN (3). *The Fall of Constantinople, +1453.* Cambridge, 1965; paperback ed. Cambridge, 1969.

RÜSTOW, W. & KÖCHLY, H. (1). *Geschichte d. griechischen Kriegswesens von der ältesten Zeit bis auf Pyrrhos.* Aarau, 1852.

RUSKA, J. (14). 'Übersetzung und Bearbeitungen von al-Rāzī's Buch "Geheimnis der Geheimnisse" [*Kitāb Sirr al-Asrār*].' *QSGNM*, 1935, **4**, 153–238; 1937, **6**, 1–246.

RUSKA, J. (24). *Das Steinbuch aus der 'Kosmographie' des Zakariya ibn Mahmūd al-Qazwīnī [c. +1250] übersetzt und mit Anmerkungen versehen...* Schmersow (Zahn & Baendel), Kirchhain N-L, 1897. (Beilage zum Jahresbericht 1895-6 der prov. Oberrealschule Heidelberg.)

DE ST. MAURICE, DE ST. LEU, COL. & DE PUY-SÉGUR, MARQUIS, LT. GEN. (1). *État Actuel de la Science Militaire à la Chine, dans lequel se trouve une analyse critique de 'l'Art Militaire des Chinois'.* Nyon, Paris, 1773. Cf. Milsky (1), pp. 104–5.

DE ST. REMY, SURIREY (1). *Mémoires d'Artillerie.* 2 vols. Paris, 1697; repr. Amsterdam, 1702. 2nd. ed. The Hague, 1741. 3rd ed., 3 vols., Paris, 1745.

SADLER, A. L. (1). *The Maker of Modern Japan; the Life of Tokugawa Ieyasu.* Allen & Unwin, London, 1937.

SALAMAN, R. A. (2). *A Dictionary of Tools used in the Wood-working and Allied Trades, +1790 to 1970.* Allen & Unwin, London, 1975.

SANDERMANN, W. (1). *Das erste Eisen fiel vom Himmel; die grossen Erfindungen der frühen Kulturen.* Bertelsmann, München, 1978.

SÄNGER, E. (1). *Raketen-Flugtechnik*. Oldenbourg, München, 1933.

SARTON, GEORGE (1). *Introduction to the History of Science*. Vol. 1, 1927; vol. 2, 1931 (2 parts); vol. 3, 1947 (2 parts). Williams and Wilkins, Baltimore (Carnegie Institution Publ. no. 376).

SARTON, G. (14). 'A Chinese Gun of +1378?' *ISIS*, 1944, **35**, 177.

SAUNDERS, J. J. (1). *The History of the Mongol Conquests*. Routledge & Kegan Paul, London, 1971.

SAVERY, THOMAS (1). *The Miner's Friend; or, an Engine to raise Water by Fire describ'd, and the Manner of fixing it in Mines, with an Account of the severall other Uses it is applicable unto; and an Answer to the Objections made against it*. London, 1702. Also *PTRS*, 1699, **21** (no. 253), 189, 228.

SCHAFER, E. H. (13). *The Golden Peaches of Samarkand; a Study of Thang Exotics*. Univ. of Calif. Press, Berkeley and Los Angeles, 1963. Rev. J. Chmielewski, *OLZ*, 1966, **61**, 497.

SCHAFER, E. H. (25). *The Empire of Min*. Tuttle, Rutland, Vt. and Tokyo, 1954 (Harvard-Yenching Institute).

SCHAFER, E. H. (26). *Pacing the Void; Thang Approaches to the Stars*. Univ. California Press, Berkeley, etc., 1977.

SCHAFER, E. H. (27). 'A Trip to the Moon.' *JAOS*, 1976, **96** (n. 1), 27. Repr. *PAR*, 1983, *8* (no. 4), 68.

SCHALL VON BELL, JOHN ADAM (1). *Historica Relatio de Initio et Progressu Missionis Societatis Jesu apud Sinenses, ac praesertim in Regia Pekinensi, ex Litteris R. P. Adami Schall, ex eadem Societate, supremi ac regii Mathematum Tribunalis ibidem Praesidiis*. Vienna, 1665; Hauckwitz, Ratisbon, 1672.

SCHLEGEL, G. (12). 'On the Invention and Use of Firearms and Gunpowder in China, prior to the arrival of Europeans.' *TP*, 1902, **3**, 1. Rev. P. Pelliot, *BEFEO*, 1902, **2**, 407.

SCHMIDLAP, J. (1). *Künstliche und rechtschaffene Feuerwerk...* Nürnberg, 1561. Repr. 1590, 1591, 1608.

SCHMIDLIN, F. J., DUKE, J. R., IVANOVSKY, A. I. & CHERNYSHENKO, Y. M. (1). *Results of the August 1977 Soviet and American Meteorological Rocketsonde Intercomparison held at Wallops Island, Virginia* [Feb. 1980]. NASA Reference Pubs. no. 1053.

SCHMIDT, I. J. (1). *Geschichte der Ostmongolen*. St Petersburg, 1829. Eng. tr. by J. R. Krueger, *The Story of the Eastern Mongols* in Occasional Papers no. 2, Pubs. of the Mongolia Society, Univ. Indiana Press, Bloomington, Ind., 1964.

SCHNEIDER, RUDOLF (1). *Die Artillerie des Mittelalters, nach den Angaben der Zeitgenossen dargestellt*. Weidmann, Berlin, 1910.

SCHNEIDER, RUDOLF (2). *Geschütze nach handschriftlichen Bildern*. Metz, 1907.

SCHNEIDER, R. (3). *Anonymi 'De Rebus Bellicis' Liber; Text und Erläuterungen*. Weidmann, Berlin, 1908.

SCHNEIDER, RUDOLF (4). 'Griechischer Poliorketiker.' *AGWG/PH*, 1908 (NF), **10**, no. 1; 1908, **11**, no. 1; 1912, **12**, no. 5. *JGLG*, 1905, **17**, 284.

SCHNEIDER, RUDOLF (5). 'Geschütze.' Art. in Pauly-Wissowa, *Realenzyklopädie d. Klass. Altertumswissenschaft*. Vol. 7 (1), pp. 1298 ff.

SCHNEIDER, RUDOLF (6). 'Anfang und Ende der Torsionsgeschütze.' *NJKA*, 1909, **23**, 133.

SCHOONMAKER, FRANK (1). *Encyclopaedia of Wine*. Black, London, 1975; 2nd. ed. 1977.

SCHOTT, CASPAR (2). *Mechanica Hydraulico-Pneumatica*. 1657. The first published account of Otto von Guericke's experiments, and the work which stimulated Robert Boyle to construct his new and improved air-pump.

SCHRAMM, E. (2). 'Poliorketik [d. Griechen u. Römer].' Art. in Kromayer & Veith, *Heerwesen und Kriegsführung d. G. u. R.* (q.v.), pp. 209–47.

SCHULTZ, ALWIN (1). *Das höfische Leben zur Zeit der Minnesinger [12th & 13th cents.]*. 2 vols. 2nd ed. Hirzel, Leipzig, 1889.

SCHUMPETER, J. A. (1). *Theory of Economic Development*. 1912.

SCHUMPETER, J. A. (2). *Business Cycles*. 1939.

SCOFFERN, J. (1). *Projectile Weapons of War and Explosive Compounds*. Cook & Whitley, London, 1852.

VON SENFFTENBERG, WULFF (1). 'Von allerlei Kriegsgewehr und Geschütz.' MS in the Dépôt Général de la Guerre *c.* +1580. Cf. Partington (5), pp. 170, 183; Bonaparte & Favé (1), vol. 1, p. 166. vol. 3. pp. 265 ff.; v. Romocki (1), vol. 1, pp. 263 ff.

SERRUYS, H. (2). 'Towers in the Northern Frontier Defences of the Ming.' *MINGS*, 1982, **14**, 8.

SETTON, K. M. (1) (ed.). *A History of the Crusades*. 3 vols. Madison, Wisconsin, 1975.

SHAMASASTRY, R. (1) (tr.). *Kautilya's 'Arthasāstra*. With introdn. by J. F. Fleet. Wesleyan Mission Press, Mysore, 1929.

SHARPE, MITCHELL R. (1). 'Non-Military Applications of the Rocket between the +17th and 20th Centuries [in Europe].' Paper presented at the 4th History Symposium of the International Academy of Astronautics, Constance, Germany, 1970. Abbreviated version in F. Cargill Hall (1), vol. 1, p. 51.

SHAW, PETER (1) (tr.). 'A New Method of Chemistry; including the History, Theory and Practice of the Art.' From Hermann Boerhaave's *Elementa Chemiae...* (1732), 2 vols. Longman, London, 1741, 1753.

SHERLOCK, T. P. (1). 'The Chemical Work of Paracelsus.' *AX*, 1948, **3**, 33.

SHIPLEY, A. E. & McBRIDE, E. W. (1). *Zoology, an Elementary Textbook*. 1st ed. Cambridge, 1901; 4th ed. Cambridge, 1920.

SHKOLYAR, S. A. *See* Školjar, S. A.

SIEMIENOWICZ, KAZIMIERZ (1). *Ars Magna Artilleriae, Pars Prima*. Amsterdam, 1650. French tr. by P. Noizet: 'Grand Art d'Artillerie, par le Sieur Casimir Siemienowicz, Chevalier litvanien; jadis Lieutenant-General de l'Artillerie dans le Royaume de Pologne.' Jansson, Amsterdam, 1651.

SIGERIST, HENRY E. (1). *A History of Medicine*. 2 vols. Oxford (New York), 1951, 1961. Vol. 1 'Primitive and Archaic Medicine'; vol. 2 'Early Greek, Hindu and Persian Medicine'. (Yale Medical Library Pubs. no. 27 and no. 38.)

SIMMS, D. L. (3). 'Archimedes and the Invention of Artillery and Gunpowder.' *TCULT*, in the press.

SINGER, C., HOLMYARD, E. J., HALL, A. R. & WILLIAMS, T. I. (1) (ed.). *A History of Technology*. 5 vols. Oxford, 1954-8.

SINGER, D. W. (5). 'On a +16th-Century Cartoon concerning the Devilish Weapon of Gunpowder; some Mediaeval Reactions to Guns and Gunpowder.' *AX*, 1959, **7** (no. 1), 25.

SINGER, E. (1). *Raketenflugtechnik*. Oldenbourg, München, 1933.

SINHA, B. P. (1). 'The Art of War in Ancient India, −600 to +300.' *JWH*, 1957, **4**, 123.

SINOR, DENIS (3). 'Les Relations entre les Mongols et l'Europe jusqu'à la Mort d'Arghoun et de Bela IV.' *JWH*, 1956, **3** (no. 1), 39. Repr. in Sinor (9), no. x.

SINOR, DENIS (7). 'The Mongols and Western Europe.' Art. in *A History of the Crusades*, ed. K. M. Setton, vol. 3, p. 513. Repr. in Sinor (9), no. ix.

SINOR, DENIS (8). 'Un Voyageur du Treizième Siècle; le Dominicain Julien de Hongrie.' *BLSOS*, 1952, **14** (no. 3), 589. Repr. in Sinor (9), no. xi.

SINOR, DENIS, (9). *Inner Asia and its Contacts with Mediaeval Europe*. Variorum, London, 1977.

SISCO, A. G. & SMITH, C. S. (2) (tr.). *Lazarus Ercker's Treatise on Ores and Assaying (Prague, 1574), translated from the German edition of 1580*. Univ. Chicago Press, Chicago, 1951.

ŠKOLJAR, S. A. (1). 'L'Artillerie de Jet à l'Époque Sung.' Art. in Balazs Festschrift, ed. Aubin, F. *Études Song in Memoriam Étiènne Balazs*. Sér. 1, Histoire et Institutions, no. 2, p. 119.

ŠKOLJAR, S. A. (2). *Kitaiskaia Doogniestrelvnaia Artillerii* [Chinese Pre-Gunpowder Artillery], in Russian. Isdatelstvo Nauka (Glavnaia Redakshnia Vostochnoi Literatury), Moscow, 1980.

DE SLANE, BARON McGUCKIN (3) (tr.). *Ibn Khaldūn: 'Histoire des Berbères' et les Dynasties Musulmanes de l'Afrique Septentrionale* [c. +1382]. Govt. Printing House, Algiers, 1852-3. 2nd ed., with P. Casanova and indexes by H. Pérès. Geuthner, Paris, 1956.

SMITH, ALEXANDER (1). *Introduction to Inorganic Chemistry*. Bell, London, 1912.

SMITH, C. S. & GNUDI, M. T. (1) (tr. & ed.). *Biringuccio's 'De La Pirotechnia' of +1540, translated with an introduction and notes*. Amer. Inst. of Mining and Metallurgical Engineers, New York, 1942, repr. 1943. Reissued, with new introductory material. Basic Books, New York, 1959.

SMITH, J. E. (1). *Small Arms of the World*. London, 1960. 10th ed. London, 1973.

SMITH, V. A. (1). *Oxford History of India, from the earliest times to 1911*. 2nd ed. Ed. S. M. Edwardes. Oxford, 1923.

SNODGRASS, A. M. (1). *Arms and Armour of the Greeks*. London, 1967.

SOKOLSKY, V. N. (1) (ed.). *Research Work on the History of Rocketry and Astronautics, 1972-3*. Moscow, 1974. (International Academy of Astronautics; Committee on the History of the Development of Rockets and Astronautics, Information Bulletin, no. 1.)

VON SOMOGYI, JOSEPH (1). 'Ein arabischer Bericht über die Tataren im *Ta'rīkh al-Islām* von al-Dhahabī.' *ISL*, 1937, **24**, 105.

SOWERBY, A. DE C. (2). 'The Horse and other Beasts of Burden in China.' *CJ*, 1937, **26**, 282.

SPAK, F. A. (1). *Öfversigt öfver Artilleriets Uppkomst*. Stockholm, 1878.

SPENCE, J. D. (1). *Emperor of China; the Self-Portrait of Khang-Hsi* [r. +1661 to +1722]. Cape, London, 1974. Penguin (Peregrine), London, 1974.

SPENCE, J. D. & WILLS, J. F. (1) (ed.). *From Ming to Chhing; Conquest, Region and Continuity in Seventeenth-century China*. Yale Univ. Press, New Haven and London, 1979.

SPENCER, J. (1). *On the Similarity of Form observed in Snow Crystals as compared with Campher*. London, 1856.

SPRAT, THOMAS (1). *The History of the Royal Society of London, for the Improving of Natural Knowledge*. London, 1667. 3rd ed. Knapton *et al*. London, 1722.

STEELE, R. (4). 'Luru Vopo Vir Can Utriet' (the cipher attributed to Roger Bacon on gunpowder, in late versions of the *De Secretis Operibus*...). *N*, 1928, **121**, 208.

STERNE, LAURENCE (1). *The Life and Opinions of Tristram Shandy, Gentleman*. (First pub. vols. i and ii, 1760, vols. iii to vi, 1762, vols. vii and viii, 1765, vol. ix, 1767.) Oxford, 1903. Often reprinted.

STONE, G. C. (1). *A Glossary of the Construction, Decoration and Use of Arms and Armour in all Countries and all Times, together with some closely related Subjects*. New York, 1931; Southworth, Portland, Maine, 1934. Repr. Brussel, New York, 1961.

STRUBELL, W. (1). 'Die Geschichte der Rakete im alten China.' *NTM*, 1965, **2**, 84.

STRUVE, L. A. (1). 'Ambivalence and Action; some Frustrated Scholars of the Khang-Hsi Period.' Art. in Spence & Wills (1), *From Ming to Chhing* ... 1979, p. 321.

STUART, G. A. (1). *Chinese Materia Medica; Vegetable Kingdom, extensively revised from Dr F. Porter Smith's work.* Amer. Presbyt. Mission Press, Shanghai, 1911. An expansion of Smith, F. P. (1)

SUBOTOWICZ, M. (1). 'K. Haas (1529–1569), V. Buringuccio (1540), J. Schmidlap (1561), K. Siemienowicz (1650); Rakiety Wielostopniowe, Baterie Rakietowe, Stabilizatory Lotu Typu Delta (Multi-Stage Rockets, Rocket Batteries and Delta-shaped Flight Stabilisers).' *KHNT*, 1968, **13** (no. 4), 805. In Polish, with English summary.

SUBOTOWICZ, M. (2). 'Kazimierz Siemienowicz (+1650) and his Contributions to Rocket Science.' *KHNT*, 1957–8, Special issue, 5. (On the occasion of the First Polish National Science Congress.)

SUBOTOWICZ, M. (3). 'Remarks on Some Important Polish Contributions to the Development of the Rocket and Space Research.' In booklet circulated at the 13th International Congress of History of Science, Moscow, 1971.

SUBOTOWICZ, M. (4). 'The Development of the Technology of Rocketry and Space Research in Poland.' Paper presented at the 23rd International Astronautical Federation Congress, Vienna, 1972.

SUGIMOTO MASAYOSHI & SWAIN, D. L. (1). *Science and Culture in Traditional Japan, A.D. 600–1854.* M.I.T. Press, Cambridge, Mass. 1978. (M.I.T. East Asian Science Series, no. 6.)

SUN FANG-TO (1). 'Rockets and Rocket Propulsion Devices in Ancient China.' *Proc. XXXIst Congress of the International Astronautical Federation, Tokyo,* Sept. 1980. Revised version, *JANS*, 1981. 29 (no. 3), 289.

SUN FANG TO (2). 'On Gunpowder, Rockets, and Related Firearms and Peaceful Devices in Ancient China.' Typed abstract 12 Mar. 82.

SUN JEN I-TU & SUN HSÜEH-CHUAN (1) (tr.). '*Thien Kung Khai Wu*', *Chinese Technology in the Seventeenth Century, by Sung Ying-Hsing.* Pennsylvania State Univ. Press; University Park and London, Penn. 1966.

SUVIN, DARKO (1). *Metamorphoses of Science Fiction; On the Poetics and History of a Literary Genre.* Yale Univ, Press, New Haven, Conn., etc. 1979.

SWINFORD, C. B. (1) (tr.). 'Than Tan-Chhiung's "An Investigation of the Bow and Arrow Industry in Chhêngtu, Szechuan".' Unpub. MS.

SWORYKIN, A. A., OSMOWA, N. I., TSCHERNYSCHEV, W. I. & SUCHARDIN, S. W. *See* Zworykin, Osmova, Chernychev & Suchardin.

TANG, M. *See* Thang Mei-Chün (1).

TAVERNIER, J. B. (1). *Les Six Voyages de J-B. T....* Paris, 1676. Eng. tr. (current) *Collection of Travels through Turkey into Persia and the East Indies...being the Travels of Monsieur Tavernier, Monsieur Bornier, and other Great Men.* London, 1884. But the first English translation appeared in 1678 with the following title: *The Six Voyages of John Baptist Tavernier, a Noble Man of France now living, through Turky into Persia and the East-Indies, finished in the year 1670, giving an Account of the State of those Countries, illustrated with Sculptures; together with a New Relation of the Present Grand Seignor's Seraglio, by the same Author; made English by J. P. – to which is added A Description of all the Kingdoms which Encompass the Euxine and Caspian Seas, by an English Traveller, never before printed.* R. L. & M. P., Starkey and Pitt, London, 1678.

TAYLOR F. SHERWOOD (4). *A History of Industrial Chemistry.* Heinemann, London, 1957.

TAYLOR, J. W. R. (1). *Rockets and Missiles.* Hamlyn, London, 1970; Sun, Melbourne, Australia, 1970.

TÊNG SSU-YÜ (3). *The Nien Army and their Guerilla Warfare, 1851 to 1868.* Mouton, The Hague, 1961. (Le Monde d'Outre-Mer Passé et Présent, 1st series, Études, no. 13.)

TÊNG SSU-YÜ & FAIRBANK, J. K. (1), with Sun Chhen I-Tu (E-tu Zen Sun), Fang Chao-Ying et al. *China's Response to the West; a Documentary Survey, 1839 to 1923.* (Medium 8vo) Harvard Univ. Press, Cambridge, Mass., 1954. Sep. pub. *Research Guide for 'China's Response to the West; a Documentary Survey...* (Small Demy 4to). Harvard University Press, Cambridge, Mass., 1954.

TERWIEL, B. J. (1). *Monks and Magic; an Analysis of Religious Ceremonies in Central Thailand.* Copenhagen, 1978. (Scandinavian Institute of Asian Studies Monographs, no. 24.)

TESSIER, M. (1). *Chimie Pyrotechnique, ou Traité Pratique des Feux Colorés.* Paris, 1859.

THAN TAN-CHHIUNG (1) = (2). 'Investigative Report on Bow and Arrow Manufacture in Chhêngtu, Szechuan.' *SUJCAH*, 1981, **11**, 143–216. Tr. by C. B. Swinford.

THOMPSON, A. H. (1). *Military Architecture in England during the Middle Ages.* OUP, Oxford, 1912.

T[HOR], I. (1). 'Bibliographical Notes.' *KHNT*, 1964, **9** (no. 2), 322–3.

THOR, I. (2). 'Tłumaczenia "Artis Magnae Artilleriae" K. Siemienowicza.' *KHNT*, 1968, **13** (no. 1), 91. English summary in Thor (3).

THOR, J. (3). 'Casimir Siemienowicz's Contribution to the Development of +17th- and +18th-century Rockets.' *Proc. XIth International Congress of the History of Science, Warsaw,* 1965, p. 507.

THORNDIKE, LYNN (12). 'An Unidentified Work by Giovanni da' Fontana, *Liber De Omnibus Rebus Naturalibus* [+1454].' *ISIS*, 1931, **15**, 31. (No MS known, but pr. Venice, 1544, and ascribed wrongly to one Pompilius Azalus.)

THUDICHUM, J. L. W. (1). *The Spirit of Cookery; a Popular Treatise on the History, Science, Practice and Ethical and Medical Import of Culinary Art—with a Dictionary of Culinary Terms.* Baillière, Tindall & Cox, London, 1895.

THURSTON, R. H. (1). *A History of the Growth of the Steam-Engine* (1878). Centennial edition, with a supplementary chapter by W. N. Barnard. Cornell Univ. Press, Ithaca, N.Y., 1939.

TISSANDIER, G. (7). 'La Chimie dans l'Extrême-Orient; Feux d'Artifices [Chinois et] Japonais.' *LN*, 1884, **12** (pt. 1), 267.

TODERICIU, DORU (1). *Preistoria Rachetei Moderne; Manuscrisul de la Sibiu* (+*1400 à* +*1569*]. Ed. Acad. Rep. Soc. Rumania, Bucarest, 1969.

TODERICIU, DORU (2). 'Raketentechnik im 16. Jahrhundert; Bemerkungen zu einer in Sibiu (Hermannstadt) vorhandenen Handschrift des Conrad Haas.' *BGTI/TG*, 1967, **34**, 97.

TODERICIU, DORU (3). 'Niezany Mechanik z 16 Wieku Prekursorem Novoczesnej Rakiety.' *KHNT*, 1969, **14** (no. 3), 475.

TODERICIU, DORU (4). 'The Sibiu Manuscript.' *RRH*, 1967 (no. 3), 333.

TODERICIU, DORU (5). 'Racheta in Trepte, creata in Tara Nostra in Secolui al 16-lea.' *RTPT*, 1964, **8**, 376.

TOMKINSON, L. (1). *Studies in the Theory and Practice of Peace and War in Chinese History and Literature*. Friends' Centre, Shanghai, 1940.

TOPPING, A. & NEEDHAM, JOSEPH (1). 'Clay Soldiers; the Army of Emperor Chhin.' *HORIZ*, 1977, **19** (no. 1), 4.

TORGASHEV, B. P. (1). *The Mineral Industry of the Far East*. Chali, Shanghai, 1930.

TORRANCE, T. (2). 'The Origin and History of the Irrigation Work of the Chêngtu Plain.' *JRAS/NCB*, 1924, **55**, 60. With addendum: 'The History of [the State of] Shu; a free translation of [part of] the *Shu Chih* [ch. 3 of *Hua Yang Kuo Chih*].'

TOUT, T. F. (1). 'Firearms in England in the Fourteenth Century.' *EHR*, 1911, **26**, 666. Repr. in *Collected Papers*, vol. 2, p. 233 (Manchester, 1934). Crit. Fêng Chia-Shêng (**8**).

TOY, S. (1). *Castles; a Short History of Fortifications*, −*1600 to* +*1600*. Heinemann, London, 1939.

TRENCH, C. CHEVENIX (1). *A History of Marksmanship*. London, 1972.

TROLLOPE, M. N. Bp. of Seoul (1). 'Korean Books and their Authors; with 'A Catalogue of Some Korean Books in the Chosen Christian College Library.' *JRAS/KB*, 1932, **21**, 1 & 59–104.

TSIEN HSUE-SHEN. *See* Chhien Hsüeh-Sên.

TSIOLKOVSKY, KONSTANTIN E. (1). *Sobranie Sochinenie* (*Collected Works*). Izd. Akad. Nauk USSR, Moscow, 1951, 1954, 1959. Eng. tr. by NASA (Washington, D.C.) 1965: Technical Translations F 236, 237, 238. Eng. tr. of articles of 1903, 1911 and 1926 (2, 3, 4): Technical Translations F 243.

TSIOLKOVSKY, KONSTANTIN E. (2). *A Rocket into Cosmic Space*. 1903.

TSIOLKOVSKY, KONSTANTIN E. (3). *The Investigation of Universal Space by means of Reactive Devices*. 1911.

TSIOLKOVSKY, KONSTANTIN E. (4). *The Investigation of Universal Space by Reactive Devices*. 1926.

TSIOLKOVSKY, KONSTANTIN E. (5). 'Vne Zemli.' MS of 1896. 1st ed. (incomplete) 1916; 2nd ed. Moscow, 1920; repr. 1958. German tr. by W. Petri: *Ausserhalb der Erde*. Heyne, München, 1977.

TSUNODA RYUSAKU (1). *Japan in the Chinese Dynastic Histories* (Later Han to and including Ming). Ed. L. C. Goodrich. Perkins, South Pasadena, 1951. (Perkins Asiatic Monographs, no. 2.)

TSUNODA RYUSAKU (2) (ed.) (with the collaboration of W. T. de Bary & D. Keene). *Sources of the Japanese Tradition*, Columbia Univ. Press. New York, 1958. (Columbia Univ. History Dept. Records of Civilisation Sources and Studies, no. 54.)

TUGE, HIDESMI (1). *The Historical Development of Science and Technology in Japan*. Kokusai Bunka Shinkokai (Society for International Cultural Relations). Tokyo, 1961. (Series on Japanese Life and Culture, no. 5.) Rev. Watanabe Masas, *ISIS*, 1964, **55**, 233.

TURNBULL, S. R. (1). *The Samurai; a Military History*. Macmillan, New York, 1977.

TURNER, SIR J. (1). *Pallas Armata*. London, 1670.

TWITCHETT, D. & FAIRBANK, J. K. (1) (ed.). *The Cambridge History of China*. 10 or more vols. Cambridge, 1978–

UBBELOHDE, A. R. J. P. (1). 'The Beginnings of the Change from Craft Mystery to Science as a Basis for Technology.' Art. in *A History of Technology*, ed. C. Singer *et al.*, Oxford, 1958, vol. 4, p. 663.

UCCELLI, A. (1) (ed.) (with the collaboration of G. Somigli, G. Strobino, E. Clausetti, G. Albenga, I. Gismondi, G. Canestrini, E. Gianni & R. Giacomell). *Storia della Tecnica dal Medio Evo ai nostri Giorni*. Hoeppli, Milan, 1945.

UFANO, DIEGO (1). *Tratado de la Artilleria y uso del Practicado*. Antwerp, 1613. French tr. *Artillerie; c'est à dire: Vraye Instruction de l'Artillerie de toutes ses Appartenances···le tout recueilly de l'Experience es Guerres du Pays bas et publié en langue Espagnolle...* Emmel, Frankfurt, 1614.

ULMEN, G. L. (1) (ed.). *Society and History* (Wittfögel Festschrift). Mouton, The Hague, 1978.

UNDERWOOD, LEON (1). '"Le Bâton de Commandement" [on throwing-sticks, *atlatl*].' *MA*, 1965, no. 142/143, 140.

UNKOVSKY, J. (1). 'Summary of an ambassador's diary of 1723 concerning finds of golden stirrups and other objects in grave-mounds of the Irtysh steppe.' *ZFE/VBGA*, 1895, **27**, (267).

URBANSKI, TADEUSZ (1). *The Chemistry and Technology of Explosives.* Tr. from the Polish by I. Jeczalikove, W. Ornaf & S. Laverton. 2 vols. Pergamon, Oxford, 1964, 1965.

URE, A. (1). *A Dictionary of Arts, Manufactures and Mines.* 1st American ed. Philadelphia, 1821; 1st ed., 2 vols. London, 1839. 5th ed., 3 vols., ed. R. Hunt. Longmans Green, Longman & Roberts, London, 1860.

USHER, A. P. (1). *A History of Mechanical Inventions.* McGraw-Hill, New York, 1929; 2nd ed. revised Harvard Univ. Press, Cambridge, Mass., 1954. Rev. Lynn White, *ISIS*, 1955, **46**, 290.

VACCA, G. (9). *Origini della Scienza, I. Perchè non si é Sviluppata la Scienza in Cina...* Quaderni di Sintesi (ed. A. C. Blanc), no. 1. Partenia, Rome, 1946. (With contributions to a discussion by A. C. Blanc, G. Bonarelli, P. Mingazzini & G. Rabbeno.)

VARAGNAC, A. (1). *La Conquête des Énergies; les Sept Révolutions Énergétiques.* Hachette, Paris, 1972. Rev. A. Herlea, *TCULT*, 1975, **16**, 79.

VÄTH, A. (1) (with the collaboration of L. van Hée). *Johann Adam Schall von Bell, S. J., Missionar in China, Kaiserlicher Astronom und Ratgeber am Hofe von Peking; ein Lebens- und Zeitbild.* Bachem, Köln, 1933. (Veröffentlichungen des Rheinischen Museums in Köln, no. 2.) Crit. P. Pelliot, *TP*, 1934, 178.

VENTURI, G. B. (1). *Recherches expérimentales sur le Principe de Communication latérale du Mouvement dans les Fluides, appliqué à l'Explication de différens Phénomènes Hydrauliques.* Paris, 1797.

VERGANI, RAFFAELLO (1). 'Gli Inizi dell'Uso della Polvere da Sparo nell'Attività Mineraria; il Caso Veneziano.' *SV*, 1979, **3**, 97.

VERGANI, RAFFAELLO (2). 'Lavoro e Creatività Operaia; una "Invenzione" Mineraria di Fine Seicento.' Art. in *Studi in Memoria di Luigi dal Pane.* Bologna, 1982, p. 487.

VERGIL, POLYDORE. *De Rerum Inventoribus.* Chr. de Pensis, Venice, 1499; and many later editions. Eng. tr. by T. Langley, *An Abridgement of the notable Worke of Polidore Vergile, conteygnyng the Devisers and first finders out as well of Arts and Mysteries as of kites and ceremonies, commonly used in the Churche.* Grafton, London, 1546. Repr. 1551.

VERGNAUD, A. D. & VERGNAUD, P. (1). *Nouveau Manuel Complet de l'Artificier; Pyrotechnie Civile.* Roret, Paris, 1906.

VIDEIRA-PIRES, B. (1). 'Os Trés Heróis do IV Centenario.' *BEDM*, 1964, **62**, 687.

VIEILLEFOND, J. R. (1). *Jules l'Africain; Fragments des 'Cestes' provenant de la Collection des Tacticiens Grecs.* Paris, 1932.

VILINBAKHOV, V. B. (1). 'A Contribution to the History of Fire-Weapons in Ancient Russia' (in Russian). *SARCH*, 1960, **25** (no. 1).

VILINBAKHOV, V. B. & KHOLMOVSKAIA, T. N. (1). 'The Fire-Weapons of Mediaeval China.' *SINT*, 1960, pt. 1.

VIOLLET-LE-DUC, E. E. (1). *Dictionnaire Raisonné de l'Architecture française du 11ème au 16ème Siècles.* 10 vols. Bance, Paris, 1861.

DE VISDELOU, C. (1). 'Histoire Abrégée de la Grande Tartarie.' Supplement to d'Herbelot's *Bibliothèque Orientale*, 1779, vol. 4, 42–296 (q.v.).

VOZÁR, J. (1). 'Der erste Gebrauch von Schiesspulver im Bergbau; Die Legende von Freiberg, die Wirklichkeit von Banská Štiavnica.' *SHS*, 1978, **10**, 257.

DE WAARD, C. (1). 'L'Expérience Barométrique; ses Antécédents et ses Explications.' Imp. Nouv. Thouars, 1936. Rev. G. Sarton *ISIS*, 1939, **26**, 212.

DE WAILLY, W. (1). *Jean, Sieur de Joinville: 'Histoire de St. Louis'* [1309]. Paris, 1874.

WAKEMAN, F. (2). 'The Canton Trade and the Opium War.' Art. in *Cambridge History of China*, ed. D. Twitchett & J. K. Fairbank, vol. 10, p. 163. Cambridge, 1978.

WALES, H. G. QUARITCH (3). *Ancient South-East Asian Warfare.* Quaritch, London, 1952.

WALEY, A. (28). *The Secret History of the Mongols, and Other Pieces.* Allen & Unwin, London, 1963; Barnes & Noble, New York, 1964.

WALEY, A. (31). *Ballads and Stories from Tunhuang.* Allen & Unwin, London, 1960.

WANG CHUNG-SHU (1). *Han Civilisation.* Tr. by Chang Kuang-Chih *et al.* Yale Univ. Press, New Haven, 1982.

WANG LING (1). 'On the Invention and Use of Gunpowder and Firearms in China.' *ISIS*, 1947, **37**, 160.

WANG ZHONGSHU. *See* Wang Chung-Shu.

WARD, G. (1). *ANTJ*, 1941, **21**, 9.

WARD, ROBERT (1). *Animadversions of Warre; or, a Militarie Magazine of...Rules and...Instructions for the Managing of Warre...* London, 1639.

WATERHOUSE, D. B. (1). 'Fire-arms in Japanese History; with Notes on a Japanese Wall-Gun.' *BMQ*, 1963, **27**, 94.

WATSON, R., Bishop of Llandaff (1). *Chemical Essays.* 2 vols. Cambridge, 1781; vol. 3, 1782; vol. 4, 1786; vol. 5, 1787. 2nd ed. 3 vols. Dublin, 1783. 3rd ed. Evans, London, 1788. 5th ed. 5 vols. Evans, London, 1789. 6th ed. London, 1793–6.

WATTENDORF, F. L. & MALINA, F. J. (1). 'Theodore von Kármán, 1881 to 1963.' *ASTRA*, 1964, **10** (no. 2), 81.

WEBB, H. J. (1). 'The Science of Gunnery in Elizabethan England.' *ISIS*, 1954, **45**, 10.

WEI CHOU-YUAN (VEI CHOW JUAN) (1). 'The Mineral Resources of China.' *EG*, 1946, **41**, 399–474.

WEIG, J. (1). *The Chinese Calendar of Festivals.* Catholic Mission Press, Tsingtao, 1929.

WEINGART, G. W. (1). *Dictionary and Manual of Pyrotechny...* Pr. pr., New Orleans [1937]. Re-issued as *Pyrotechnics, Civil and Military,* Chem. Pub. Co. Brooklyn, 1943. 2nd ed. rev. and enlarged: *Pyrotechnics,* Chem. Pub. Co., Brooklyn, 1947.

WELBORN, M. C. (1). '"The Errors of the Physicians" [De Erroribus Medicorum], according to Friar Roger Bacon of the Minor Order.' *ISIS*, 1932, **18**, 26.

WERHAHN-MEES, K. (1). *Chhi Chi-Kuang; Praxis der Chinesischen Kriegsführung.* Bernard & Graefe, München, 1980.

WERNER, E. T. C. (3). *Chinese Weapons.* Royal Asiatic Society (North China Branch), Shanghai, 1932.

WERNER, E. T. C. (4). *A Dictionary of Chinese Mythology.* Kelly & Walsh, Shanghai, 1932.

WERTIME, T. A. & MUHLY, J. D. (1) (ed.). *The Coming of the Age of Iron.* Yale Univ. Press, New Haven, 1980. (Cyril Stanley Smith Presentation Volume.)

WESCHER, C. (1) (ed.). *Poliorcétique des Grecs.* (Texts only.) Imp. Imp. Paris, 1867.

WHINYATES, F. A. (1). 'Captain Bogue and the Rocket Brigade.' *PRAI*, 1897, **24**, 131.

WHITE, JOHN H. (1). 'Safety with a Bang; the Railway Torpedo [Fog Signal].' *TCULT*, 1982, **23**, 195.

WHITE, LYNN (7). *Mediaeval Technology and Social Change.* Oxford, 1962. Revs. A. R. Bridbury, *EHR*, 1962, **15**, 371; R. H. Hilton & P. H. Sawyer, *PP*, 1963 (no. 24), 90; J. Needham, *ISIS*, 1963, **54** (no. 4).

WHITE, LYNN (20). 'The Eurasian Context of Mediaeval Europe.' *Proc. XIIth Congress of the International Musicological Society, Berkeley, Calif.* 1977. Ed. D. Heartey & B. Wade. Kossel, Bärenreiter, 1982, p. 1.

WHITEHORNE, PETER (1). *Certain Waies for the Orderyng of Souldiers in Battelray...and moreover, howe to make Saltpeter, Gunpowder and divers sortes of Fireworkes or wilde Fyre.* Kingston & Englande, London, 1562. The first edition of 1560 does not include the powder compositions.

WIEDEMANN, E. (7). 'Beiträge z. Gesch. d. Naturwiss.; VI, Zur Mechanik und Technik bei d. Arabern.' *SPMSE*, 1906, **38**, 1. Repr. in (23), vol. 1, p. 173.

WIEDEMANN, E. (23). *Aufsätze zur arabischen Wissenschaftsgeschichte* (a reprint of his 79 contributions in the series 'Beiträge z. Gesch. d. Naturwissenschaften' in *SPMSE*), ed. W. Fischer, with full indexes, 2 vols. Olm, Hildesheim and New York, 1970.

WIEDEMANN, E. (28). 'Beiträge z. Gesch. d. Naturwiss. XL; Über Verfälschungen von Drogen usw. nach Ibn Bassām und al-Nabarāwī.' *SPMSE*, 1914, **46**, 172. Repr. in (23), vol. 2, p. 102.

WIEDEMANN, E. & GROHMANN, A. (1). 'Beiträge z. Gesch. d. Naturwiss. XLIX; Über von den Arabern benutzte Drogen.' *SPMSE*, 1917, **48/49**, 16. Repr. in Wiedemann (23), vol. 2, p. 230.

WILBUR, C. M. (2). 'The History of the Crossbow.' *ARSI*, 1936, 427. (Smithsonian Institution Pub. no. 3438).

WILLIAMS, A. (1). 'Some Firing Tests with Simulated Fifteenth-century Hand-Guns.' *JAAS*, 1974, **8**, 114.

WILLIAMS, A. R. (1). 'The Production of Saltpetre in the Middle Ages.' *AX*, 1975, **22**, 125.

WILLIAMS, S. WELLS (1). *The Middle Kingdom; a Survey of the Geography, Government, Literature [or Education], Social Life, Arts, [Religion] and History, [etc.] of the Chinese Empire and its Inhabitants.* 2 vols. Wiley, New York, 1848; later eds. 1861, 1900; London, 1883.

WINTER, FRANK H. (1). 'On the Origins and Development of the Rocket in India.' *TCULT* (1977). 'Rocketry in India from the Earliest Times to the Nineteenth Century.' *Proc. XIVth Internat. Congr. Hist. of Science, Tokyo,* 1975. Vol. 3, p. 360.

WINTER, FRANK H. (2). 'William Hale; a Forgotten British Rocket Pioneer.' *SPFL*, 1973, **15** (no. 1), 31. Abstract in *The History of the Science and Technology of Aeronautics, Rockets and Space-flight.* Booklet circulated at the 13th International Congress of History of Science, Moscow, 1971, p. 63.

WINTER, FRANK H. (3). 'Sir William Congreve; a Bicentennial Memorial.' *SPFL*, 1972, **14** (no. 9), 333.

WINTER, FRANK H. (4). 'On the Origin of Rockets.' *CHEM*, 1976, **49** (no. 2), 8.

WINTER, FRANK H. (5). 'The Genesis of the Rocket in China, and its Spread to the East and West.' *Proc. XXXth Congress of the International Astronautical Federation, München,* 1979.

WINTER, FRANK H. (6). 'A History of Italian Rocketry during the Nineteenth Century.' *ARMA*, 1965, 181.

WINTER, FRANK H. & SHARPE, MITCHELL R. (1). 'William Moore; Pioneer in Rocket Ballistics.' *SPFL*, 1976, **18**, 180.

WINTRINGHAM, T. (1). *Weapons and Tactics.* Faber & Faber, London, 1943.

WOLF, A. (1) (with the co-operation of F. Dannemann & A. Armitage). *A History of Science, Technology and Philosophy in the 16th and 17th Centuries.* Allen & Unwin, London, 1935; 2nd ed., revised by D. McKie, London, 1950. Rev. G. Sarton, *ISIS*, 1935, **24**, 164.

WOLF, A. (2). *A History of Science, Technology and Philosophy in the 18th Century.* Allen & Unwin, London, 1938; 2nd ed., revised by D. McKie, London, 1952.

WOLFF, ELDON, G. (1). *Air Guns.* Milwaukee Public Museum, Milwaukee, 1958. (Museum Pubs. in History, no. 1.)

[Wong, K. C.] (1). 'Ancient Jade Sword and Scabbard Parts.' *CJ*, 1927, **6**, 295.

Woo, Y. T. *See* Wu Yang-Tsang.

Woodcroft, B. (1) (tr.). *The 'Pneumatics' of Heron of Alexandria*. Whittingham, London, 1851.

Wu Yang-Tsang (1). 'Silver Mining and Smelting in Mongolia' [Jehol]. *TAIME*, 1903, **33**, 755. With discussion by B. S. Lyman on p. 1038. *EMJ*, 1903, **75**, 147 (abridged version).

Wüstenfeld, F. (1) (tr.). 'Calcaschandi's (al-Qalquashandi) Geographie und Verwaltung von Ägypten.' *AGWG/PH*, 1879, **25**, 1.

Wuttke, A. (1). *Der Deutsche Volksaberglaube der Gegenwart*. Wiegand & Grüben, Berlin, 1869, repr. 1900.

van der Wyngaert, Anastasius (1). *Jean de Monte Corvin, O.F.M., premier Évêque de Khanbalig (+1247 à +1328)*. Lille, 1924.

Wynne-Jones, I. (1). *The Llechwedd Slate Caverns*. Llechwedd, Blaenau Ffestiniog, Gwynedd, 1980.

Yadin, Yigael (1). *The Art of Warfare in Biblical Lands*. London, 1963.

Yamoda, Nakaba (1). *Ghenkō; the Mongol Invasion of Japan*. Smith Elder, London, 1916; Dutton, New York, 1916.

Yang Lien-Shêng (14). 'The Form of the Paper Note [Money] Hui-tzu of the Southern Sung Dynasty.' *HJAS*, 1957, **16**, 365. Repr. in (9), p. 216.

Yates, Robin D. S. (2). 'Towards a Reconstruction of the Tactical Chapters of *Mo Tzu* (ch. 14).' Inaug. Diss. (M.A.), Univ. of California, Berkeley, 1975.

Yates, Robin D. S. (3). 'Siege Engines and Late Chou Military Technology.' Art. in *Explorations in the History of Science and Technology in China*, ed. Li Kuo-Hao, Chang Mêng-Wên, Tshao Thiën-Chhin & Hu Tao-Ching, p. 409. Shanghai, 1982.

Yates, Robin D. S. (4). 'The Mohists on Warfare; Technology, Technique, and Justification.' *JAAR*, 1979, **47** (no. 35, Thematic Issue), 549.

Yetts, W. P. (13). 'The Horse; a Factor in Early Chinese History.' *ESA*, 1934, **9**, 231.

Yoneda, S. (1). 'A Study of Saltpetre and Soda produced in the Saline and Alkali Soils of Northern Honan.' *SRFAOU*, 1953, **3**, 52.

Yoshiska, Shin-ichi (1). *Collection of Ancient Guns*. Tokyo, 1965.

Yule, Sir Henry (1) (ed.). *The Book of Ser Marco Polo the Venetian, concerning the Kingdoms and Marvels of the East, translated and edited, with Notes*, by H. Y....，1st ed. 1871, repr. 1875. 3rd ed., 2 vols. ed. H. Cordier; Murray, London, 1903 (reprinted 1921), 3rd ed. also issued Scribner, New York, 1929. With a third Volume, *Notes and Addenda to Sir Henry Yule's Edition of Sir Marco Polo*, by H. Cordier. Murray, London, 1920. Photolitho offset reprint in 2 vols., Armorica, St Helier and Philo, Amsterdam, 1975.

Yule, Sir Henry (2). *Cathay and the Way Thither; being a Collection of Mediaeval Notices of China*. 2 vols. Hakluyt Society Pubs. (2nd ser.) London, 1913–15 (1st ed. 1866). Revised by H. Cordier, 4 vols. Vol. 1, (no. 38), *Introduction; Preliminary Essay on the Intercourse between China and the Western Nations previous to the Discovery of the Cape Route*. Vol. 2, (no. 33), *Odoric of Pordenone*. Vol. 3, (no. 37), *John of Monte Corvino and others*. Vol. 4, (no. 41), *Ibn Baṭṭuṭah and Benedict of Goes*. Photolitho reprint, Peiping, 1942.

Zahn, Johann (1). *Oculus Artificialis Teledioptricus...* 1702.

Zenghelis, C. (1). 'Le Feu Grégeois et les Armes à Feu des Byzantins' (on the 'Strepta' or Byzantine Gun). *BYZ*, 1932, **7**, 265.

Zim, H. S. (1). *Rockets and Jets*. Harcourt, New York, 1945.

Zimmermann, Samuel (1). 'Dialogus oder Gespräch zweier Personen, nämlich einer Büchsenmeisters mit einem Feuerwerkskünstler, von der wahren Kunst und rechten Gebrauch des Büchsengeschosses und Feuerwerks.' MSS of 1574, 1575 and 1577.

Ziolkovsky. *See* Tsiolkovsky.

Zworykin, A. A., Osmova, N. I., Chernychev, W. I. & Suchardin, S. V. (1). *Geschichte der Technik*. Fachbuchverlag, Leipzig, 1964. Tr. from the Russian *Istoria Techniki*, Acad. Sci. Moscow, 1962 by P. Hüter, G. Hoppe & M. Brandt under the editorship of R. Ludloff and the advisory collaboration of A. Kraus & W. Lohse. Rev. J. Payen, *DHT*, 1965 (no. **5**), 319.

SUPPLEMENT TO BIBLIOGRAPHY C

ALLEY, REWI (9). 'A Visit to Hsishuangbana and the Thai Folk of Yunnan.' *EHOR*, 1966, **5** (no. 5), 6.

ALLOM, T. & WRIGHT, G. N. (1). *China, in a Series of Views, displaying the Scenery, Architecture and Social Habits of that Ancient Empire, drawn from original and authentic Sketches by T. A— Esq., with historical and descriptive Notices by Rev. G. N. W—*. Fisher, 4 vols. London & Paris, 1843.

ANON. (15). *Mittelälterliches Hausbuch*. Album of an arquebus-maker, containing various engineering drawings. MS Wolfsee Castle. Ed. A. von Essenwein, Frankfurt 1887; H. T. Bossent & W. F. Storck, Leipzig, 1912. See Sarton (1), vol. 3, p. 1553.

ANON (198). 'Early Inventions of the Chinese.' *HMM*, 1869, **39** (no. 234), 909.

BABINGTON, JOHN (1). *Pyrotechnia; or a Discourse of Artificiall Fire-works...; whereunto is annexed a Short Treatise of Geometrie*. Harper and Mab, London, 1635. Facsimile reproduction, 2 vols. Theatrum Orbis Terrarum Ltd., Amsterdam, 1971. (English Experience, nos. 295, 296.)

BELLEW, H. W. (1). *A Narrative of the Journey of the Embassy to Kashgar in 1873-4*. London, 1875.

BOSSERT, H. T. & STORCK, W. F. (1). *Das mittelälterliches Hausbuch...* (Anon. 15). Seemann, Leipzig, 1912.

BOWDEN, F. P. & YOFFE, A. D. (1). *Initiation and Growth of Explosion in Liquids and Solids*. Cambridge, 1952, repr. with new preface, 1985.

BOWDEN, F. P. & YOFFE, A. D. (2). *Fast Reactions in Solids*. Butterworth, London, 1958.

BRUNT, P. A. (1) (tr.). *Arrian; 'Anabasis Alexandri' and 'Indica'*. 2 vols. Harvard Univ. Press, Cambridge, Mass., 1983; Heinemann, London, 1983. (Loeb Classics, no. 269.)

BURKILL, I. H. (1). *A Dictionary of the Economic Products of the Malay Peninsula* (with contributions by W. Birtwhistle, F. W. Foxworthy, J. B. Scrivenor & J. G. Watson). 2 vols. Crown Agents for the Colonies, London, 1935.

BURSTALL, A. F. (1). *A History of Mechanical Engineering*. Faber & Faber, London, 1963.

CASTIGLIONI, ARTURO (1). *A History of Medicine*, tr. and ed. E. B. Krumbhaar. 2nd ed. revised and enlarged. Knopf, New York, 1947.

CHANG YÜN-MING (1). 'Ancient Chinese Sulphur Manufacturing Processes.' *Proc. XVIIth International Congress of the History of Sciences*, Berkeley, 1985. Abstracts, vol. 2, p. Tb 3a.

CHAVANNES, E. (2). 'Le Royaume de Wou et de Yue.' *TP*, 1916, **17**, 129.

COUVREUR, F. S. (1) (tr.). *Tch'ouen Ts'iou* [Chhun Chhiu] et *Tso Tchouan; Texte Chinois avec Traduction Française*. 3 vols. Mission Press, Hochienfu, 1914.

VON ESSENWEIN, A. (1). *Mittelälterliches Hausbuch; Bilderhandschrift des 15 Jahrh...* (Anon. 15). Keller, Frankfurt-am-Main, 1887.

FINÓ, J. F. (1). 'Le Feu et ses Usages Militaires.' *GLAD*, 1970, **9**, 15.

FLEMING, H. F. (1). *Der Volkommene Teutsche Jäger*. Leipzig, 1724.

GAIT, E. A. (1). *A History of Assam*. Govt. Printing Superintendency, Calcutta, 1906.

GRANET, M. (1). *Danses et Légendes de la Chine Ancienne*. 2 vols. Alcan, Paris, 1926.

HALL, BERT S. (2). *The Technological Illustrations of the So-called 'Anonymus of the Hussite Wars.'* Reichert, Wiesbaden, 1979. Rev. A. Keller, *TCULT*, 1984, **25**, 109.

HAWKES, D. (1) (tr.). *Chhu Tzhu; the Songs of the South – an Ancient Chinese Anthology*. Oxford, 1959. rev. J. Needham, *NSN*, 18 July 1959.

HILL, DONALD R. (2) (tr.). "'The Book of Knowledge of Ingenious Mechanical Devices' [*Kitāb fī Ma'rifat al-Hiyal al-Handasiya*, by Ibn al-Razzār al-Jazarī, + 1206], with a Foreword by Lynn White." Reidel, Dordrecht, 1974.

HO PING-YU & NEEDHAM, JOSEPH (2). *Theories of Categories in Early Mediaeval Chinese Alchemy* (with transl. of the *Tshan Thung Chhi Wu Hsiang Lei Pi Yao, c.* +6th to +8th century). *JWCI*, 1959, **22**, 173.

HOFF, A. (1). *Air-guns and Other Pneumatic Arms*. London, 1972.

DE HOFFMEYER, ADA BRUHN (4). 'Military Equipment in the Byzantine Manuscript of Scylitzes in the Biblioteca Nacional of Madrid.' *GLAD*, 1966, **5**, 1–160, with 51 pll.

HOLLISTER-SHORT, G. (6). 'Early Gunpowder Testers.' *ANTC*, 1984, **55** (no. 11), 104.

HOLLISTER-SHORT, G. (7). 'The Use of Gunpowder in Mining; a Document of +1627.' *HOT*, 1983, **8**, 111.

HUANG HSING-TSUNG (1). 'Peregrinations with Joseph Needham in China, 1943-4', in *Explorations in the History of Science and Technology in China*, ed. Li Kuottao *et al.* (1), p. 39.

HUGHES, B. P. (1). *British Smooth-bore Artillery*. London, 1969.

JIH YÜEH & CHUNG YUNG-HO (1) = (1). '"Rocket Hives" in Yen-shui [Thaiwan]; a Peculiar Tradition.' *SINRA*, 1984, **9** (no. 3), 106.

KEENE, DONALD (1). *The Japanese Discovery of Europe; Honda Tosheki and other Discoverers, +1720 to +1798*. Grove, New York, 1954.

LAWRENCE, G. H. M. (1). *Taxonomy of Vascular Plants*. Macmillan, New York, 1951.

Li Kuo-Hao, Chang Mêng-Wên, Tshao Thien-Chhin & Hu Tao-Ching (1) (ed.). *Explorations in the History of Science and Technology in China*; a special number of the *Collections of Essays on Chinese Literature and History* (compiled in honour of the eightieth birthday of Joseph Needham). Chinese Classics Publishing House, Shanghai, 1982.

van Linschoten, Jan Huyghen (1). *Itinerario, Voyage ofte Schipvaert van J. H. van L. naer oost ofte Portugaels Indien* (+1579 to +1592), Amsterdam, 1596, 1598; ed. H. Kern, 's Gravenhage, 1910. Repr. C. E. Warnsinck-Delprat, 5 vols., 1955. Eng. tr. by W. Phillips, *John Huighen Van Linschoten his discours of voyages into ye Easte and West Indies*, Wolfe, London, 1598. (Hakluyt Soc. Pub., 1st ser., nos. 70, 71. London, 1885. Ed. A. C. Burnell & P. A. Tiele. (Information about the Chinese coast dating from c. +1550 to +1588 collected at Goa c. +1583 to +1589).)

von Lippmann, E. O. (3). *Abhandlungen und Vorträge zur Geschichte d. Naturwissenschaften*, 2 vols. Vol. 1, Veit, Leipzig, 1906. Vol. 2, Veit, Leipzig, 1913.

Lo Chê-Wên (1). 'The Great Wall; a New Survey.' *CREC*, 1985, **34** (no. 3), 8.

Lu Gwei-Djen (4). 'The First Half-life of Joseph Needham.' In *Explorations in the History of Science and Technology in China*, ed. Li Kuo-Hao et al. (1), p. 1.

Luo Zhewen (1). *See* Lo Chê-Wên.

Malleolus, Felix, *See* Hemmerlin, Felix.

Mayer, K. P. (1). 'On Variations in the Shapes of the Components of the Chinese *me chi*' (crossbar 'latch' [trigger mechanism]).' *TP*, 1965, **52**, 7, with correction in *TP*, 1967, **53**, 293.

Mehren, A. F. M. (1). Tr. *Manuel de la Cosmographie du Moyen Age*, traduit de l'Arabe 'Nokhbet ed dahr fi 'adjaib-il-birr, wal-bah'r de Shems ed-din abou- 'Abdallah Mohammed de Damas. Copenhagen, 1874.

Melchett, Lord; Mann, J. G. et al. (1).
Correspondence in the *Times*, 1947, on the origin and historical significance of stirrups.
24 Feb. Melchett on the role of the stirrup in the transformation of the Roman legionary into the mounted knight.
26 Feb. J. G. Mann (Master of the Armouries, Tower of London) on the couching of the lance.
14 Mar. Lord Mersey on the battle of Adrianople.
20 Mar. R. W. Moore on the date and place of the invention.
27 Mar. Publication of a photograph of the Wei dynasty statuette presented by Sir William Ridgeway to the Cambridge University Museum of Archaeology and Ethnology.
31 Mar. Sir John Marshall on the thong stirrups of the Sanchi carvings in India.
14 Apr. Leading article.

Miller, Martin (1). *The Collector's Illustrated Guide to Firearms*. Barrie & Jenkins, London, 1978.

Needham, Joseph (32). *The Development of Iron and Steel Technology in China*. Newcomen Soc. London, 1958. (Second Biennial Dickinson Memorial Lecture, Newcomen Society.) Précis in *TNS*, 1960, **30**, 141. Rev. L. C. Goodrich, *ISIS*, 1960, **51**, 108. Repr. Heffer, Cambridge, 1964. French tr. (unrevised, with some illustrations omitted and others added by the editors) *RHSID*, 1961, **2**, 187, 235; 1962, **3**, 1, 62.

Needham, Joseph (66). *Within the Four Seas; the Dialogue of East and West* (Collected Addresses). Allen and Unwin, London, 1969.

Needham, Joseph (12). *Biochemistry and Morphogenesis*. Cambridge, 1942; Repr. 1950; repr. 1966, with historical survey as Foreword. Cf. Porkert (1); Needham & Lu (9).

Needham, Joseph & Needham, Dorothy M. (1) (ed.). *Science Outpost*. Pilot Press, London, 1948.

Needham, Joseph & Lu Gwei-Djen (5). 'The Earliest Snow Crystal Observations.' *W*, 1961, **16**, 319.

Needham, Joseph & Needham, Dorothy M. (1) (ed.). *Science Outpost*, Pilot Press, London, 1948.

Nowak, T. M. (1). 'O Wplywie Walk z Turcja i Tatarami na Rozwój Polskiej Techniki Wojskowej 16–17 W. (The Influence of the Wars with the Turks and Tartars on Polish Military Technology of the 16th and 17th Centuries).' *KHNT*, 1983, **28** (no. 3–4), 589.

Pauer, E. (1). 'Japans Industrielle Lehrzeit; Die Bedeutung des Flammofeus in der wirtschaftlichen und technischen Entwicklung Japans für den Beginn der industriellen Revolution.' *BZJ*, 1983, **4** (no. 1), 1–254; (no. 2), 257–576. Eng. Summary, pp. 494–506.

Perkin, W. H. & Kipping, F. S. (1). *Organic Chemistry*, rev. ed. Chambers, London and Edinburgh, 1917.

Prigogine, Ilya & Stengers, I. (1). *Order out of Chaos; Man's New Dialogue with Nature*. Heinemann, London, 1984.

Schukking, W. H. (1) (ed.). *The Principal Works of Simon Stevin*. Vol. 4. 'The Art of War.' Swets & Zeitlinger, Amsterdam, 1964.

So Kwan-Wai (1). *Japanese Piracy in Ming China during the +16th Century*. Michigan State Univ. Press, East Lansing, Mich., 1975.

Starikov, V. S. (1). 'Larion Rossochin and the Beginning of the Study of Chinese Pyrotechnics in Russia.' [With reproduction of a text of +1756 mainly on rockets] (in Russian). Art. in *Iz Istorii Nauki i Techniki v Strana Vostoka*. Vol. 2, p. 100. Academy of Sciences, Moscow, 1961.

Swift, Jonathan (2). *Gulliver's Travels* [+1726]. Encyclopaedia Britannica, Chicago, 1952.

TARN, W. W. (4). *Hellenistic Military and Naval Developments.* Cambridge, 1930.

TEMPLE, ROBERT K. G. (1). 'Discovery of a MS. eye-witness Account of the Battle of Maidstone [+ 1648].' *ACANT*, 1981, **97**, 209.

TSOU I-JEN (1). 'Historical Examples of China's Quality Control of Weapons, starting from the *Khao Kung Chi* chapter of the *Chou Li*.' MS. in English and Chinese, held by the East Asian History of Science Library, Cambridge.

TYLECOTE, R. F. (1). 'The Early History of the Iron Blast-Furnace in Europe [Scandinavia]; a Case of East–West Contact?'. Paper for the symposium 'Mediaeval Iron in Society', Norberg, 1985.

WAGNER, D. B. (1). 'The Transmission of the Blast-Furnace from China to Europe; Some Notes to complement Prof. Tylecote's Paper.' Paper for the symposium 'Mediaeval Iron in Society'. Norberg, 1985.

WERNER, E. T. C. (1). *Myths and Legends of China.* Harrap, London, 1922.

WILLIAMSON, H. R. (1). *Wang An-Shih, A Chinese Statesman and Educationialist of the Sung Dynasty.* 2 vols. Arthur Probsthain, London, 1935.

YANG WU-MIN (1). 'China's First Communications Satellite.' *CREC*, 1984, **33** (no. 7), 24.

ZAKY, ABDEL RAHMAN (4). 'Gunpowder and Arab Firearms in the Middle Ages.' *GLAD*, 1967, **6**, 45.

GENERAL INDEX

by Christine Outhwaite

Spain (*cont.*)
earliest date for the bombard in, 578
Spak (1), 347 (b)
Spencer (1), 527 (g)
Sprat (1), 347 (a), 349 (b)
'spurting tube'. See *phên thung*
Ssu Hsüan Fu (Thought the Transcender) by Chang Hêng, 523 (a)
Ssu Khu Chhüan Shu, 20, 21, 23, 33, 34, 35 (b) (g), 90 (c), 158 (c)
Ssu Khu Chhüan Shu Thi Tao, 21
Ssu-Lun-Fa (Shan Burmese prince, + 14th century), 307
Ssuma Fa (The Marshal's Art), 90 (e)
Stadlin, François Louis (Lin Chi-Ko) (Jesuit lay brother), 127
Starikov (1), 141 (d)
steam
use in China, 2, 559–62
gunpowder and the development of steam engines, 3, 5 (a), 544–8, 558–9, 565; vacuum displacement system pumps, 562–4
for jet-propulsion, 521 (a)
steam fire-setting, 539
for a pressure-cooker, 555 (d), 559
steel, 141, 339, 466 (g)
See also flint and steel
Steele (4), 50 (a)
sterilisation, medical, 2
Steward (2), 508 (i)
'stick'/cudgel (*kun*) in relation to fire-arms, 247
stills, 77, 92
still-heads, 126 (i)
stink, 125
'stink-pots', 189–92
Stirling, John, and the hot-air engine, 566
stirrups, foot- (*têng*), 17
Stone (1), 401 (a)
Street, Robert, 565 (h)
Streydtbuch von Pixen, Kriegsrüstung, Sturmzeuch und Feuerwerckh, 33
strontium, in coloured smokes, 138, 145
strophanthine, as a blow-gun poison, 272 (e)
Strubel (1), 473 (c)
Struve (1), 407 (b)
strychnine, as a blow-gun poison, 272 (e)
Stuart (1), 120 (g), 123 (c), 547 (c)
Su-a Munjip (Essays from the Western Cliff) by Yu Sŏnguyong, 289 (c)
Su Wei-Tao, poem by, 137
Suan Fa Thung Tsung (Systematic Treatise on Arithmetic) by Chhêng Ta-Wei, 391
'subduing by fire'. See *fu huo*
Subotai (Mongol general at the siege of Khaifêng, + 1232), 171, 572 (d)
Subotowicz (1), 508 (c) (d); (2), 508 (a)
Suchow, 306, 429
Sugimoto & Swain (1), 430 (i)
Sui dynasty, 61 (c)
and fireworks, 136, 146 (f)
sulphur, 1, 34, 48, 56, 57, 166, 261 (e), 262
in 'automatic fire', 39 (h), 67; for '*ignis vola-*

tilis', 40; in Greek and Roman incendiaries, 66, 74; to thicken petroleum in incendiaries, 77
as a constituent of gunpowder, 108–11
in early experiments using gunpowder ingredients, 111–17 *passim*; in the Sung formulae for gunpowder, 117–26 *passim*
restrictions on the sale of, 126
produced from iron pyrites, 126
for colouring smokes, 145
proportions in gunpowder compositions, 342–58
traditional theories concerning the nature of, 360–4 *passim*
exported to South-east Asia, 365 (e)
Sumatra, 88, 443
sumpitan, diffusion of the Malay term, 273
Sun Chhüan, 28
Sun Fang-To (1), 187 (d), 225 (c), 516 (b)
Sun Ssu-Mo (alchemist and physician, +581–+682), 115–16
Sun & Sun (1), 104 (f), 187 (a), 205 (c), 339 (f), 362 (f), 378 (f), 437 (g) (h) (i)
Sun Tzu Ping Fa (Master Sun's Art of War), 61, 69
Sun Yuan-Hua (governor of Têngchow in Shangtung), 393
Sun Yün-Chhiu and the telescope, 412–13
Ching Shih, 413
Sung Chi San Chhao Chêng Yao (The Most Important Aspects of Government as seen under the Last Three Courts of the (Southern) Sung Dynasty), 156 (e), 175 (a)
Sung dynasty
toxic smokes, 2
fire-arrows persist, 7
fire-lances, 8
'true' gun, 10
wide military use of gunpowder, 16
extant military writing, 19
use of the term *shen chhi*, 24 (a)
petrol flame-throwers, 89
saltpetre, 98
gunpowder formulae, 117–25
restrictions on sale of explosives, 126
gunpowder fire-crackers, 146
new type of fire-arrow, 148
gunpowder weapons, trebuchets, 154–7
explosive gunpowder, 163
thunderclap bomb, 165, 168
lime bombs, 165–7
bombs, fire-lances and trebuchets, 173–6
fire-lances at Tê-an, 222; Hsiang-yang and Yangchow, 227
fire-tube, 230
appearance of the term *huo phao*, 276
water mill, 359 (e)
See also Northern Sung, Southern Sung
Sung Hsüeh Shih Chhüan Chi (Complete Record of Sung Scholars) by Sung Lien, 210 (b), 306 (k)
Sung Hui Yao, 92 (f)
Sung Hui Yao Chih Kao, 92
Sung Hui Yao Kao (Drafts for the *History of the Administrative Statutes of the Sung Dynasty*) collected by Hsü Sung, 149 (a), 156 (k)

夏　Hsɪᴀ kingdom (legendary?)　　　　　　　　　　　　　_c._ −2000 to _c._ −1520
商　Sʜᴀɴɢ (Yɪɴ) kingdom　　　　　　　　　　　　　　　_c._ −1520 to _c._ −1030

周　Cʜᴏᴜ dynasty (Feudal　{ Early Chou period　　　　　　_c._ −1030 to −722
　　　　　　　　　　　　　　Chhun Chhiu period　春秋　−722 to −480
　　Age)　　　　　　　　　　Warring States (Chan　　　−480 to −221
　　　　　　　　　　　　　　　Kuo) period　戰國

First Unification　秦　Cʜʜɪɴ dynasty　　　　　　　　　　−221 to −207
　　　　　　　　　　　{ Chhien Han (Earlier or Western)　−202 to +9
　漢　Hᴀɴ dynasty{ Hsin interregnum　　　　　　　　　+9 to +23
　　　　　　　　　　　{ Hou Han (Later or Eastern)　　　+25 to +220

　　　三 國　Sᴀɴ Kᴜᴏ (Three Kingdoms period)　　　+221 to +265
First　　　　　　蜀　Sʜᴜ (Hᴀɴ)　　　　　+221 to +264
　Partition　　　魏　Wᴇɪ　　　　　　　　+220 to +265
　　　　　　　　吳　Wᴜ　　　　　　　　　+222 to +280

Second　　晉　Cʜɪɴ dynasty: Western　　　　　　　+265 to +317
　Unification　　　　　　Eastern　　　　　　　+317 to +420
　　　劉宋　(Liu) Sᴜɴɢ dynasty　　　　　　　　+420 to +479
Second　　　Northern and Southern Dynasties (Nan Pei chhao)
　Partition　　　　　齊　Cʜʜɪ dynasty　　　　+479 to +502
　　　　　　　　　　梁　Lɪᴀɴɢ dynasty　　　+502 to +557
　　　　　　　　　　陳　Cʜʜᴇɴ dynasty　　　+557 to +589
　　　　　　　　　　{ Northern (Thopa) Wᴇɪ dynasty　　+386 to +535
　　　　　　魏　{ Western (Thopa) Wᴇɪ dynasty　　+535 to +556
　　　　　　　　　　{ Eastern (Thopa) Wᴇɪ dynasty　　+534 to +550
　　　　　北齊　Northern Cʜʜɪ dynasty　　　+550 to +577
　　　　　北周　Northern Cʜᴏᴜ (Hsienpi) dynasty　+557 to +581
Third　　隋　Sᴜɪ dynasty　　　　　　　　　　+581 to +618
　Unification 唐　Tʜᴀɴɢ dynasty　　　　　　　　+618 to +906
Third　　五代　Wᴜ Tᴀɪ (Five Dynasty period) (Later Liang,　+907 to +960
　Partition　　　　Later Thang (Turkic), Later Chin (Turkic),
　　　　　　　　　Later Han (Turkic) and Later Chou)
　　　　　　　遼　Lɪᴀᴏ (Chhitan Tartar) dynasty　　+907 to +1124
　　　　　West Lɪᴀᴏ dynasty (Qarā-Khiṭāi)　　+1124 to +1211
　　　　　西夏　Hsi Hsia (Tangut Tibetan) state　　+986 to +1227
Fourth　　宋　Northern Sᴜɴɢ dynasty　　　　　+960 to +1126
　Unification 宋　Southern Sᴜɴɢ dynasty　　　　+1127 to +1279
　　　　　　金　Cʜɪɴ (Jurchen Tartar) dynasty　　+1115 to +1234
　　　　　元　Yᴜᴀɴ (Mongol) dynasty　　　　+1260 to +1368
　　　　　明　Mɪɴɢ dynasty　　　　　　　　+1368 to +1644
　　　　　清　Cʜʜɪɴɢ (Manchu) dynasty　　　+1644 to +1911
　　　　　民國　Republic　　　　　　　　　+1912

N.B. When no modifying term in brackets is given, the dynasty was purely Chinese. Where the overlapping of dynasties and independent states becomes particularly confused, the tables of Wieger (1) will be found useful. For such periods, especially the Second and Third Partitions, the best guide is Eberhard (9). During the Eastern Chin period there were no less than eighteen independent States (Hunnish, Tibetan, Hsienpi, Turkic, etc.) in the north. The term 'Liu chhao' (Six Dynasties) is often used by historians of literature. It refers to the south and covers the period from the beginning of the +3rd to the end of the +6th centuries, including (San Kuo) Wu, Chin, (Liu) Sung, Chhi, Liang and Chhen. For all details of reigns and rulers see Moule & Yetts (1).

ROMANISATION CONVERSION TABLES

by Robin Brilliant

PINYIN/MODIFIED WADE–GILES

Pinyin	Modified Wade–Giles	Pinyin	Modified Wade–Giles
a	a	chou	chhou
ai	ai	chu	chhu
an	an	chuai	chhuai
ang	ang	chuan	chhuan
ao	aɔ	chuang	chhuang
ba	pa	chui	chhui
bai	pai	chun	chhun
ban	pan	chuo	chho
bang	pang	ci	tzhu
bao	pao	cong	tshung
bei	pei	cou	tshou
ben	pên	cu	tshu
beng	pêng	cuan	tshuan
bi	pi	cui	tshui
bian	pien	cun	tshun
biao	piao	cuo	tsho
bie	pieh	da	ta
bin	pin	dai	tai
bing	ping	dan	tan
bo	po	dang	tang
bu	pu	dao	tao
ca	tsha	de	tê
cai	tshai	dei	tei
can	tshan	den	tên
cang	tshang	deng	têng
cao	tshao	di	ti
ce	tshê	dian	tien
cen	tshên	diao	tiao
ceng	tshêng	die	dieh
cha	chha	ding	ting
chai	chhai	diu	tiu
chan	chhan	dong	tung
chang	chhang	dou	tou
chao	chhao	du	tu
che	chhê	duan	tuan
chen	chhên	dui	tui
cheng	chhêng	dun	tun
chi	chhih	duo	to
chong	chhung	e	ê, o

Pinyin	Modified Wade–Giles	Pinyin	Modified Wade–Giles
en	ên	jia	chia
eng	êng	jian	chien
er	êrh	jiang	chiang
fa	fa	jiao	chiao
fan	fan	jie	chieh
fang	fang	jin	chin
fei	fei	jing	ching
fen	fên	jiong	chiung
feng	fêng	jiu	chiu
fo	fo	ju	chü
fou	fou	juan	chüan
fu	fu	jue	chüeh, chio
ga	ka	jun	chün
gai	kai	ka	kha
gan	kan	kai	khai
gang	kang	kan	khan
gao	kao	kang	khang
ge	ko	kao	khao
gei	kei	ke	kho
gen	kên	kei	khei
geng	kêng	ken	khên
gong	kung	keng	khêng
gou	kou	kong	khung
gu	ku	kou	khou
gua	kua	ku	khu
guai	kuai	kua	khua
guan	kuan	kuai	khuai
guang	kuang	kuan	khuan
gui	kuei	kuang	khuang
gun	kun	kui	khuei
guo	kuo	kun	khun
ha	ha	kuo	khuo
hai	hai	la	la
han	han	lai	lai
hang	hang	lan	lan
hao	hao	lang	lang
he	ho	lao	lao
hei	hei	le	lê
hen	hên	lei	lei
heng	hêng	leng	lêng
hong	hung	li	li
hou	hou	lia	lia
hu	hu	lian	lien
hua	hua	liang	liang
huai	huai	liao	liao
huan	huan	lie	lieh
huang	huang	lin	lin
hui	hui	ling	ling
hun	hun	liu	liu
huo	huo	lo	lo
ji	chi	long	lung

Pinyin	Modified Wade–Giles	Pinyin	Modified Wade–Giles
lou	lou	pa	pha
lu	lu	pai	phai
lü	lü	pan	phan
luan	luan	pang	phang
lüe	lüeh	pao	phao
lun	lun	pei	phei
luo	lo	pen	phên
ma	ma	peng	phêng
mai	mai	pi	phi
man	man	pian	phien
mang	mang	piao	phiao
mao	mao	pie	phieh
mei	mei	pin	phin
men	mên	ping	phing
meng	mêng	po	pho
mi	mi	pou	phou
mian	mien	pu	phu
miao	miao	qi	chhi
mie	mieh	qia	chhia
min	min	qian	chhien
ming	ming	qiang	chhiang
miu	miu	qiao	chhiao
mo	mo	qie	chhieh
mou	mou	qin	chhin
mu	mu	qing	chhing
na	na	qiong	chhiung
nai	nai	qiu	chhiu
nan	nan	qu	chhü
nang	nang	quan	chhüan
nao	nao	que	chhüeh, chhio
nei	nei	qun	chhün
nen	nên	ran	jan
neng	nêng	rang	jang
ng	ng	rao	jao
ni	ni	re	jê
nian	nien	ren	jên
niang	niang	reng	jêng
niao	niao	ri	jih
nie	nieh	rong	jung
nin	nin	rou	jou
ning	ning	ru	ju
niu	niu	rua	jua
nong	nung	ruan	juan
nou	nou	rui	jui
nu	nu	run	jun
nü	nü	ruo	jo
nuan	nuan	sa	sa
nüe	nio	sai	sai
nuo	no	san	san
o	o, ê	sang	sang
ou	ou	sao	sao

Pinyin	Modified Wade–Giles	Pinyin	Modified Wade–Giles
se	sê	wan	wan
sen	sên	wang	wang
seng	sêng	wei	wei
sha	sha	wen	wên
shai	shai	weng	ong
shan	shan	wo	wo
shang	shang	wu	wu
shao	shao	xi	hsi
she	shê	xia	hsia
shei	shei	xian	hsien
shen	shen	xiang	hsiang
sheng	shêng, sêng	xiao	hsiao
shi	shih	xie	hsieh
shou	shou	xin	hsin
shu	shu	xing	hsing
shua	shua	xiong	hsiung
shuai	shuai	xiu	hsiu
shuan	shuan	xu	hsü
shuang	shuang	xuan	hsüan
shui	shui	xue	hsüeh, hsio
shun	shun	xun	hsün
shuo	shuo	ya	ya
si	ssu	yan	yen
song	sung	yang	yang
sou	sou	yao	yao
su	su	ye	yeh
suan	suan	yi	i
sui	sui	yin	yin
sun	sun	ying	ying
suo	so	yo	yo
ta	tha	yong	yung
tai	thai	you	yu
tan	than	yu	yü
tang	thang	yuan	yüan
tao	thao	yue	yüeh, yo
te	thê	yun	yün
teng	thêng	za	tsa
ti	thi	zai	tsai
tian	thien	zan	tsan
tiao	thiao	zang	tsang
tie	thieh	zao	tsao
ting	thing	ze	tsê
tong	thung	zei	tsei
tou	thou	zen	tsên
tu	thu	zeng	tsêng
tuan	thuan	zha	cha
tui	thui	zhai	chai
tun	thun	zhan	chan
tuo	tho	zhang	chang
wa	wa	zhao	chao
wai	wai	zhe	chê

Pinyin	Modified Wade–Giles	Pinyin	Modified Wade–Giles
zhei	chei	zhui	chui
zhen	chên	zhun	chun
zheng	chêng	zhuo	cho
zhi	chih	zi	tzu
zhong	chung	zong	tsung
zhou	chou	zou	tsou
zhu	chu	zu	tsu
zhua	chua	zuan	tsuan
zhuai	chuai	zui	tsui
zhuan	chuan	zun	tsun
zhuang	chuang	zuo	tso

MODIFIED WADE–GILES/PINYIN

Modified Wade–Giles	Pinyin	Modified Wade–Giles	Pinyin
a	a	chhio	que
ai	ai	chhiu	qiu
an	an	chhiung	qiong
ang	ang	chho	chuo
ao	ao	chhou	chou
cha	zha	chhu	chu
chai	chai	chhuai	chuai
chan	zhan	chhuan	chuan
chang	zhang	chhuang	chuang
chao	zhao	chhui	chui
chê	zhe	chhun	chun
chei	zhei	chhung	chong
chên	zhen	chhü	qu
chêng	zheng	chhüan	quan
chha	cha	chhüeh	que
chhai	chai	chhün	qun
chhan	chan	chi	ji
chhang	chang	chia	jia
chhao	chao	chiang	jiang
chhê	che	chiao	jiao
chhên	chen	chieh	jie
chhêng	cheng	chien	jian
chhi	qi	chih	zhi
chhia	qia	chin	jin
chhiang	qiang	ching	jing
chhiao	qiao	chio	jue
chhieh	qie	chiu	jiu
chhien	qian	chiung	jiong
chhih	chi	cho	zhuo
chhin	qin	chou	zhou
chhing	qing	chu	zhu

Modified Wade–Giles	Pinyin	Modified Wade–Giles	Pinyin
chua	zhua	huan	huan
chuai	zhuai	huang	huang
chuan	zhuan	hui	hui
chuang	zhuang	hun	hun
chui	zhui	hung	hong
chun	zhun	huo	huo
chung	zhong	i	yi
chü	ju	jan	ran
chüan	juan	jang	rang
chüeh	jue	jao	rao
chün	jun	jê	re
ê	e, o	jên	ren
ên	en	jêng	reng
êng	eng	jih	ri
êrh	er	jo	ruo
fa	fa	jou	rou
fan	fan	ju	ru
fang	fang	jua	rua
fei	fei	juan	ruan
fên	fen	jui	rui
fêng	feng	jun	run
fo	fo	jung	rong
fou	fou	ka	ga
fu	fu	kai	gai
ha	ha	kan	gan
hai	hai	kang	gang
han	han	kao	gao
hang	hang	kei	gei
hao	hao	kên	gen
hên	hen	kêng	geng
hêng	heng	kha	ka
ho	he	khai	kai
hou	hou	khan	kan
hsi	xi	khang	kang
hsia	xia	khao	kao
hsiang	xiang	khei	kei
hsiao	xiao	khên	ken
hsieh	xie	khêng	keng
hsien	xian	kho	ke
hsin	xin	khou	kou
hsing	xing	khu	ku
hsio	xue	khua	kua
hsiu	xiu	khuai	kuai
hsiung	xiong	khuan	kuan
hsü	xu	khuang	kuang
hsüan	xuan	khuei	kui
hsüeh	xue	khun	kun
hsün	xun	khung	kong
hu	hu	khuo	kuo
hua	hua	ko	ge
huai	huai	kou	gou

Modified Wade–Giles	Pinyin	Modified Wade–Giles	Pinyin
ku	gu	mu	mu
kua	gua	na	na
kuai	guai	nai	nai
kuan	guan	nan	nan
kuang	guang	nang	nang
kuei	gui	nao	nao
kun	gun	nei	nei
kung	gong	nên	nen
kuo	guo	nêng	neng
la	la	ni	ni
lai	lai	niang	niang
lan	lan	niao	niao
lang	lang	nieh	nie
lao	lao	nien	nian
lê	le	nin	nin
lei	lei	ning	ning
lêng	leng	niu	nüe
li	li	niu	niu
lia	lia	no	nuo
liang	liang	nou	nou
liao	liao	nu	nu
lieh	lie	nuan	nuan
lien	lian	nung	nong
lin	lin	nü	nü
ling	ling	o	e, o
liu	liu	ong	weng
lo	luo, lo	ou	ou
lou	lou	pa	ba
lu	lu	pai	bai
luan	luan	pan	ban
lun	lun	pang	bang
lung	long	pao	bao
lü	lü	pei	bei
lüeh	lüe	pên	ben
ma	ma	pêng	beng
mai	mai	pha	pa
man	man	phai	pai
mang	mang	phan	pan
mao	mao	phang	pang
mei	mei	phao	pao
mên	men	phei	pei
mêng	meng	phên	pen
mi	mi	phêng	peng
miao	miao	phi	pi
mieh	mie	phiao	piao
mien	mian	phieh	pie
min	min	phien	pian
ming	ming	phin	pin
miu	miu	phing	ping
mo	mo	pho	po
mou	mou	phou	pou

Modified Wade–Giles	Pinyin	Modified Wade–Giles	Pinyin
phu	pu	tên	den
pi	bi	têng	deng
piao	biao	tha	ta
pieh	bie	thai	tai
pien	bian	than	tan
pin	bin	thang	tang
ping	bing	thao	tao
po	bo	thê	te
pu	bu	thêng	teng
sa	sa	thi	ti
sai	sai	thiao	tiao
san	san	thieh	tie
sang	sang	thien	tian
sao	sao	thing	ting
sê	se	tho	tuo
sên	sen	thou	tou
sêng	seng, sheng	thu	tu
sha	sha	thuan	tuan
shai	shai	thui	tui
shan	shan	thun	tun
shang	shang	thung	tong
shao	shao	ti	di
shê	she	tiao	diao
shei	shei	tieh	die
shên	shen	tien	dian
shêng	sheng	ting	ding
shih	shi	tiu	diu
shou	shou	to	duo
shu	shu	tou	dou
shua	shua	tsa	za
shuai	shuai	tsai	zai
shuan	shuan	tsan	zan
shuang	shuang	tsang	zang
shui	shui	tsao	zao
shun	shun	tsê	ze
shuo	shuo	tsei	zei
so	suo	tsên	zen
sou	sou	tsêng	zeng
ssu	si	tsha	ca
su	su	tshai	cai
suan	suan	tshan	can
sui	sui	tshang	cang
sun	sun	tshao	cao
sung	song	tshê	ce
ta	da	tshên	cen
tai	dai	tshêng	ceng
tan	dan	tsho	cuo
tang	dang	tshou	cou
tao	dao	tshu	cu
tê	de	tshuan	cuan
tei	dei	tshui	cui

Modified Wade–Giles	Pinyin	Modified Wade–Giles	Pinyin
tshun	cun	wang	wang
tshung	cong	wei	wei
tso	zuo	wên	wen
tsou	zou	wo	wo
tsu	zu	wu	wu
tsuan	zuan	ya	ya
tsui	zui	yang	yang
tsun	zun	yao	yao
tsung	zong	yeh	ye
tu	du	yen	yan
tuan	duan	yin	yin
tui	dui	ying	ying
tun	dun	yo	yue, yo
tung	dong	yu	you
tzhu	ci	yung	yong
tzu	zi	yü	yu
wa	wa	yüan	yuan
wai	wai	yüeh	yue
wan	wan	yün	yun